W9-BHU-263
St. Olaf College
JAN 0 8 1988
Science Library

Practical Protein Chemistry—A Handbook

Practical Protein Chemistry—A Handbook

Edited by
A. DARBRE
Department of Biochemistry
King's College London

A Wiley–Interscience Publication

JOHN WILEY & SONS
Chichester . New York . Brisbane . Toronto . Singapore

Copyright © 1986 by John Wiley & Sons Ltd.
Reprinted April 1987

All rights reserved.

No part of this book may be reproduced by any means, nor transmitted, nor translated into a machine language without the written permission of the publisher.

***Library of Congress Cataloging in Publication Data*:**
Main entry under title:

Practical protein chemistry.

'A Wiley–Interscience publication.'
Includes index.
1. Proteins—Analysis—Handbooks, manuals, etc.
2. Amino acid sequence—Handbooks, manuals, etc.
3. Biological chemistry—Technique—Handbooks, manuals etc. I. Darbre, A. [DNLM: 1. Biochemistry—handbooks.
2. Proteins—analysis—handbooks. QU 39 P895]
QP551.P65 1986 574.19′245 84-26942
ISBN 0 471 90673 5

***British Library Cataloguing in Publication Data*:**

Practical protein chemistry: a handbook.—
(A Wiley–Interscience publication)
1. Proteins
Darbre, A.
574.19′245 QP551

ISBN 0 471 90673 5

Printed and bound in Great Britain

To Hazel and Philippa

Contents

List of Contributors

E. APPELLA, Laboratory of Cell Biology, National Cancer Institute, National Institutes of Health, Bethesda, Maryland 20205, U.S.A.

J. C. BENNETT, Department of Microbiology, University of Alabama in Birmingham, Schools of Medicine and Dentistry, Birmingham, Alabama, U.S.A.

J. D. CAPRA, Department of Microbiology, The University of Texas Health Science Center at Dallas, 5323 Harry Hines Boulevard, Dallas, Texas 75235, U.S.A.

J.-Y. CHANG, Pharmaceuticals Research Laboratories, CIBA-GEIGY Limited, 4002 Basle, Switzerland.

J. R. CLAMP, Department of Medicine, Medical School, University of Bristol, University Walk, Bistol BS8 1TD, U.K.

A. DARBRE, Department of Biochemistry, King's College London, Strand, London WC2R 2LS, U.K.

A. DELL, Department of Biochemistry, Imperial College of Science and Technology, London SW7 2AZ, U.K.

D. DOVER, Department of Biophysics, King's College, London, Strand, London WC2R 2LS, U.K.

M. H. ENGEL, Geophysical Laboratory, Carnegie Institution of Washington, Washington D.C., U.S.A.

A. FONTANA, Institute of Organic Chemistry (Biopolymer Research Centre, C.N.R.), University of Padova, Padova, Italy.

*E. GROSS, National Institute of Child Health and Human Development, Endocrinology and Reproduction Research Branch, Section on Molecular Structure, Bethesda, Maryland 20205, U.S.A.

W. J. GULLICK, Protein Chemistry Laboratory, Imperial Cancer Research Fund, Lincoln's Inn Fields, London WC2A 3PX, U.K.

P. E. HARE, Geophysical Laboratory, Carnegie Institution of Washington, Washington D.C., U.S.A.

R. M. HEWICK, Genetics Institute, 225 Longwood Avenue, Boston, Massachusetts 02115, U.S.A.

* Deceased

M. W. HUNKAPILLER, Applied Biosystems Inc., 850 Lincoln Center Drive, Foster City, California 94404, U.S.A.

A. S. INGLIS, Division of Protein Chemistry, CSIRO, 343 Royal Parade, Parkville, 3052 Victoria, Australia.

J. K. INMAN, Laboratory of Immunology, National Institute of Allergy and Infectious Diseases, National Institutes of Health, Bethesda, Maryland 20205, U.S.A.

R. C. MARSHALL, Division of Protein Chemistry, CSIRO, 343 Royal Parade, Parkville, 3052 Victoria, Australia.

B. ROBSON, Department of Biochemistry, The Medical School, University of Manchester, Oxford Road, Manchester M13 9PL, U.K.

G. SCRACE, Protein Chemistry Laboratory, Imperial Cancer Research Fund, Lincoln's Inn Fields, London WC23A 3PX, U.K.

R. SCHROHENLOHER, Division of Clinical Immunology and Rheumatology, Department of Medicine, University of Alabama in Birmingham, Schools of Medicine and Dentistry, Birmingham, Alabama, U.S.A.

M. SIEGELMAN, Department of Microbiology, The University of Texas Health Science Center at Dallas, 5323 Harry Hines Boulevard, Dallas, Texas 75235, U.S.A.

A. D. STROSBERG, Molecular Immunology Laboratory, IRBM–CNRS and University of Paris VII, Place Jussieu 2, Paris 75251, France.

N. TOTTY, Protein Chemistry Laboratory, Imperial Cancer Research Fund, Lincoln's Inn Fields, London WC23A 3PX, U.K.

M. D. WATERFIELD, Protein Chemistry Laboratory, Imperial Cancer Research Fund, Lincoln's Inn Fields, London WC23A 3PX, U.K.

C. W. WARD, Division of Protein Chemistry, CSIRO, Parkville, 3052 Victoria, Australia.

S. WEINSTEIN, Department of Organic Chemistry, The Weizmann Institute of Science, Rehovet, Israel.

J. M. WILKINSON, Department of Biochemistry, Royal College of Surgeons of England, 35/43, Lincoln's Inn Fields, London WC2A 3PN, U.K.

G. WINTER, MRC Laboratory of Molecular Biology, Hills Road, Cambridge, CB2 2QH, U.K.

B. WITTMANN-LIEBOLD, Max-Planck-Institut für Molekulare Genetik, Abteilung Wittmann, Ihnestrasse 63–73, D-1000 Berlin 33 (Dahlem), Germany.

Preface

The idea for this book suggested itself to the editor as a result of his own time-consuming and sometimes fruitless search of the literature for methods. There are of course series of volumes such as *Methods in Enzymology* and *Methods of Biochemical Analysis* which are brim-full with practical details. However, discussion with colleagues indicated that there is a need for a compilation in one volume of concise and up-to-date information. This volume does not attempt to deal with all chemical methods for studying proteins. For instance, the synthesis of peptides in free solution and by the Merrifield process and the chemical modification of proteins have been adequately considered in other publications. The reader may find his or her favourite method to be missing from this book. The editor regrets such an omission, but the contributors have restricted themselves to methods with which they are familiar and which are widely used. The last three chapters in this book have been included because of their general interest to researchers in protein study.

For a brief period, the immense success of nucleotide sequencing seemed to indicate that protein sequencing might become a little passé, if not redundant. In fact, there was an article in *Nature* posing the question 'Decline and fall of protein chemistry?' (Malcolm, 1978). The mosaic organization of eucaryotic DNA with its 'exon' and 'intron' sequences (Gilbert, 1978) has led to a consideration of the corresponding protein structures. Where the protein structures have been determined the role of exon recombination in their evolution has been clarified (Blake, 1983).

Information on post-translational modification of proteins, with signal sequences, pre- and pro-peptide sequences has shown that there is no short cut to determination of protein structure, and this requires many different approaches and methods. Biotechnology is concerned with recombinant DNA techniques, but the ultimate objective of genetic manipultion is the production of specific proteins and peptides and these will require full characterization by the protein chemist.

Studies with pharmacologically-active peptides, antibiotics, hormone–protein and protein–protein receptor interactions, enzymes, structural proteins, evolutionary relationships and the recently discovered primary sequence homology between some oncogenes and normal proteins, all offer exciting research possibilities. Thus, many research workers in diverse biological disciplines have become involved in work with proteins. So much has been written on protein methodology that the novice may well wonder where to begin.

The contributors to this volume are all active researchers in various aspects of protein study and they have attempted to guide the reader by introducing briefly their subject and giving practical details for experimental work, which in their own experience give good results. Undoubtedly, many methods have stood the test of time, such as total protein hydrolysis with 6 M-hydrochloric acid or the colorimetric determination of proteins introduced by Lowry *et al.* (1951), but other newer methods have been introduced which offer big advantages in certain circumstances.

During the preparation of this volume the tragic death was announced of one of our contributors, Dr. Erhard Gross (1928–1981). We are much indebted to him for his contributions in protein chemistry. In particular, Angelo de Fontana (a coauthor in this volume) has expressed his great sorrow at the loss of both a great scientist and a friend. A tribute to Dr. Gross was written by Christian Birr (1983).

Also, the demise of Stanford Moore (who gave his personal encouragement for the preparation of this book) was another shock and some personal recollections (Hirs, 1984) and articles dedicated to him were published in an issue of *Analytical Biochemistry* (Kaplan & McElroy, 1984). A volume of *Methods in Enzymology* (Hirs & Timasheff, 1983) was dedicated to the memory of Stanford Moore and William Howard Stein.

I wish to thank all the contributors to this volume for their dedication and hard work. In particular I am much indebted to Michael Waterfield for his very valuable advice and the time he gave unstintingly in the early stages of planning this book, as well as Peter Williams of Heyden Press for his enthusiasm and help in reaching the publication stage.

King's College London, André Darbre
1985

References

Birr, C. (1983). *Int. J. Peptide Prot. Res.*, **21,** 1–2.
Blake, C. (1983). *Trends Biochem. Sci.*, **8,** 11–13.
Gilbert, W. (1978). *Nature* (*Lond.*), **271,** 501.
Hirs, C. H. W. (1984). *Anal. Biochem.*, **136,** 3–6.
Hirs, C. H. W. & Timasheff, S. N. (Eds.) (1983). *Methods Enzymol.*, **91,** 1–693.
Kaplan, N. O. & McElroy, W. D. (1984). *Anal. Biochem.*, **136,** 1.
Lowry, O. H., Rosebrough, N. J., Farr, A. L. & Randall, R. J. (1951). *J. Biol. Chem.*, **193,** 265–275.
Malcolm, A. D. B. (1978). *Nature* (*Lond.*), **275,** 90–91.

Foreword

I am delighted that Dr. Darbre has had the initiative to put together the contributions in this book so as to make a coherent whole of the techniques now available to the protein chemist. This is particularly timely because the impact of genetic engineering in general and protein engineering in particular has paradoxically increased the importance of the roles that the protein chemist can play.

Of course it is now obvious that the fastest route to a complete amino acid sequence is via the DNA sequence of the cloned gene or the cloned cDNA. But the fastest and safest route to cloning is often by probing a gene or cDNA bank with oligonucleotide probes of minimum redundancy designed from carefully chosen peptide sequences. Moreover few genes are expressed naturally without some degree of post-translational processing by proteolysis and/or side-chain modification, so the protein chemist must elucidate these in the natural product. And even when this proves to be entirely unnecessary, it is remarkably comforting to see peptide sequences that match the inferred coding sequence.

To many of us from the early days of molecular biology, protein engineering appears the culmination of the collaboration between molecular geneticists, protein chemists and X-ray crystallographers that we strove to create. Fig. 1 adumbrates some of the ways in which the protein chemist can assist the genetic

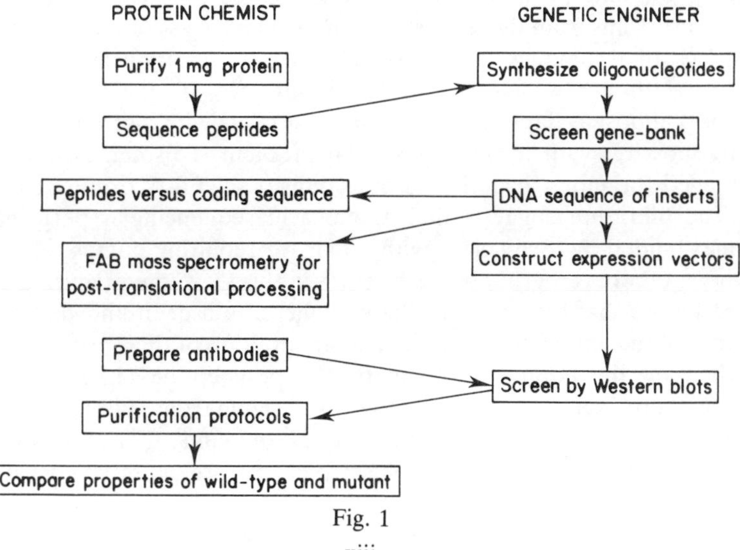

Fig. 1

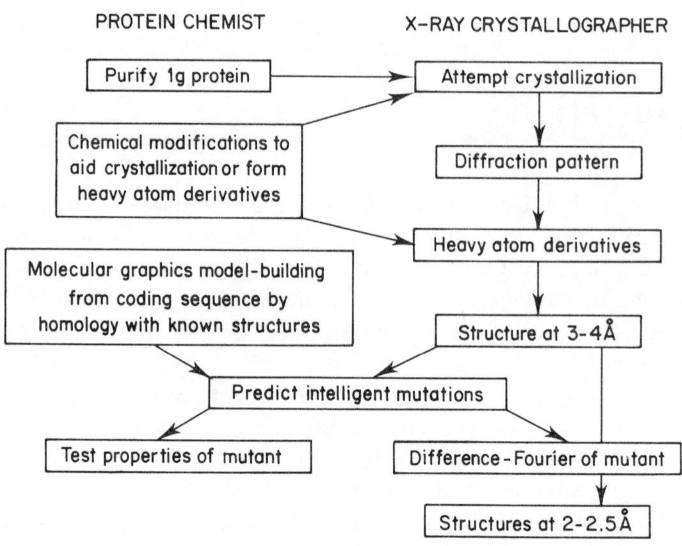

Fig. 2

engineer and Fig. 2 shows ways in which he is the ally of the X-ray crystallographer. Together such a team can begin to tackle detailed principles of protein folding and enzyme catalysis so that one day the DNA sequence of a gene might be theoretically translated into a primary structure, then to a tertiary structure and even to predictions of the catalytic properties of the product.

But the budding protein chemist who is excited by these prospects faces different problems from those presented in the Dark Ages of our subject. Then it was lack of techniques for peptide cleavage, purification and sequencing, so the practitioner was forced to invent his own. Today the plethora of methods is almost as daunting, since one must select with discrimination those that are most relevant to one's particular problem. To some extent this has discouraged the more inventive of our younger scientists from entering this field. The literature suggests that amino acid sequencing is performed by machines tended by technicians with a flair for repairing pumps.

Although there is regrettably some truth in this view, the practising protein chemist knows that his true skills lie elsewhere, in a deep knowledge of the chemistry of the amino acid residues and an intuition of their properties when perturbed by the complex folding of the polypeptide chain. He knows also that there is aesthetic satisfaction in designing the simple but elegant experiment to answer a particular question and that his automatic analysers, sequenators and mass spectrometers are merely aids and not absolute requirements for his operations.

To this end, therefore, this book will be valued by the experts, since beneath the inevitable quantity of detailed cookery lie the discriminating hints for which he is searching. But its main value will be for the average biochemist who feels the need to practise some protein chemistry but has been hitherto discouraged by the need for a single *vade mecum*. This is it.

Director, Centre for Biotechnology,
Imperial College of Science and Technology,
London SW7 2AZ

B. S. HARTLEY F.R.S.

Abbreviations

AAPTS	β-aminoethyl-(3-aminopropyl)-trimethoxysilane
APG	aminopropyl glass
β-APG	β-*N*-aminoethyl-(3-aminopropyl) glass
APS	aminopolystyrene
APTS	3-aminopropyl-triethoxysilane
A.u.f.s.	absorbance units full scale
ATZ	anilinothiazolinone
BAWP	butanol-acid-water-phenol
Boc	*t*-butyloxycarbonyl
BNPS-skatole	2-(2-nitrophenylsulphenyl)-3 methyl-3-bromoindolenine
BTI	bis (1,1-trifluoroacetoxy) iodobenzene
BTU	*S*-(n-butyl) thiuronium iodide
CBB	Coomassie Brilliant Blue
CNBr	cyanogen bromide
CPG	controlled pore glass
CTC	copper tartrate carbonate
DABITC	4-dimethylaminoazobenzene-4′-isothiocyanate
DABTH	dimethylaminoazobenzenethiohydantoin
Dansyl	5-dimethylaminonaphthalene-1-sulphonyl
DCC	dicyclohexyl carbodiimide
DFP	diisopropyl fluorophosphate
DITC	*p*-phenylene diisothiocyanate
DMAA	dimethylallylamine
DMBA	dimethylbenzylamine
DMF	dimethylformamide
DMSO	dimethylsulphoxide
DOC	deoxycholate
DTNB	5,5-dithiobis-(2-nitrobenzoic acid); Ellman's reagent
DTT	dithiothreitol
EDC	l-ethyl-3-(3-dimethylaminopropyl) carbodiimide
EDTA	ethylenediaminetetraacetic acid
EHT	extended Huckle Theory
EI	electron impact
ET	ethanethiol
FPLC	fast protein liquid chromatography
GLC	gas liquid chromatography

HFBA	heptafluorobutyric acid
HNB-Br	2-hydroxy-5-nitrobenzyl bromide
HPLC	high pressure liquid chromatography
HOBt	1-hydroxybenzotriazole
HOSu	*N*-hydroxysuccinimide
HVE	high voltage electrophoresis
IEHPLC	ion-exchange high pressure liquid chromatography
i.r.	infrared
MITC	methylisothiocyanate
MS	mass spectrometry
NBD	4-chloro-7-nitrobenzofuran
NBD-Cl	7-chloro-4-nitrobenzo-2-oxa-l,3-diazole
NBS	*N*-bromosuccinimide
NCS	*N*-chlorosuccinimide
NEM	*N*-ethylmorpholine
NMM	*N*-methylmorpholine
NMR	nuclear magnetic resonance
NTB	2-nitro-5-thiobenzoate
NTCB	2-nitro-5-thiocyanobenzoate
NTSB	2-nitro-5-thiosulphobenzoate
OD	optical density
OPA	*o*-phthalaldehyde
OPHA	*o*-pivaloylhydroxylamine
PAGE	polyacrylamide gel electrophoresis
PFB-Br	α-bromo-2,3,4,5,6-pentafluorotoluene
PITC	phenylisothiocyanate
PMA	pyromellitic acid
PMSF	phenylmethylsulphonyl fluoride
Polybrene	1,5-dimethyl-1,5-diazaundecamethylenepolymethobromide
p.s.i.	pounds per square inch
PTH	phenylthiohydantoin
PTI	pancreatic trypsin inhibitor
Quadrol	*N*,*N*,*N*′,*N*′-tetrakis (2-hydroxypropyl) ethylenediamine
RMS	root mean square
RPHPLC	reverse phase high pressure liquid chromatography
S	sensitivity
SCOT	support-coated open tubular
SDA	sodium dihydro-bis-(2-methoxyethoxy) aluminate
SDS	sodium dodecyl sulphate
TCA	trichloroacetic acid
TEMED	tetramethylenediamine
TETA	triethylenetetramine
TFA	trifluoroacetic acid
THEED	*N*,*N*,*N*′,*N*′-tetrakis (2-hydroxyethyl) ethylenediamine

THF	tetrahydrofuran
TLC	thin layer chromatography
TMA	trimethylamine
TNBS	2,4,6-trinitrobenzene sulphonic acid
Tris	tris (hydroxymethyl) aminomethane
u.v.	ultraviolet
WCOT	wall-coated open tubular

ABBREVIATIONS OF AMINO ACIDS, WITH SINGLE-LETTER SYMBOLS WHERE APPLICABLE

See Cohn, W. E. (1984), *Methods Enzymol.*, **106**, 1–17.

Alanine	Ala	A	Allo-isoleucine	aIle
Arginine	Arg	R	Allo-isothreonine	aThr
Asparagine	Asn	N	4-Carboxy glutamic acid	Gla
Aspartic acid	Asp	D	Cysteine sulphinic acid	Csa
Cysteic acid	Cya		Homocysteine	Hcy
Cysteine	Cys	C	Homoserine	Hse
Cystine	Cys Cys		Norleucine	Nle
Glutamine	Gln	Q	Norvaline	Nva
Glutamic acid	Glu	E	5-OH lysine	5Hyl
Glycine	Gly	G	4-OH proline	4Hyp
Histidine	His	H	3-OH proline	3Hyp
Isoleucine	Ile	I	Ornithine	Orn
Leucine	Leu	L	Phosphoserine	Ser(P)
Lysine	Lys	K	Phosphothreonine	Thr(P)
Methionine	Met	M	Phosphotyrosine	Tyr(P)
Methionine *S*-oxide	MetO			
Methionine *S,S*-dioxide	$MetO_2$			
Phenylalanine	Phe	F		
Proline	Pro	P		
Serine	Ser	S		
Threonine	Thr	T		
Tryptophan	Trp	W		
Tyrosine	Tyr	Y		
Valine	Val	V		

Practical Protein Chemistry—A Handbook
Edited by A. Darbre
© 1986, John Wiley & Sons Ltd.

ROBERT C. MARSHALL and ADAM S. INGLIS
Division of Protein Chemistry, CSIRO,
343 Royal Parade, Parkville, 3052,
Victoria, Australia.

1

Protein Oligomer Composition, Preparation of Monomers and Constituent Chains

CONTENTS

1.1. INTRODUCTION

In this chapter we are concerned largely with practical methods for determining the composition of a protein oligomer and the characterization and separation of the monomers and constituent chains. An oligomeric protein consists of non-covalently linked *monomers* (or subunits). These polypeptide chains may contain *constituent* chains covalently linked by disulfide bonds (or other covalent cross-links). So schematically, a tetramer composed of two pairs of different monomers, one containing two disulfide-linked polypeptide chains, can be illustrated as follows:

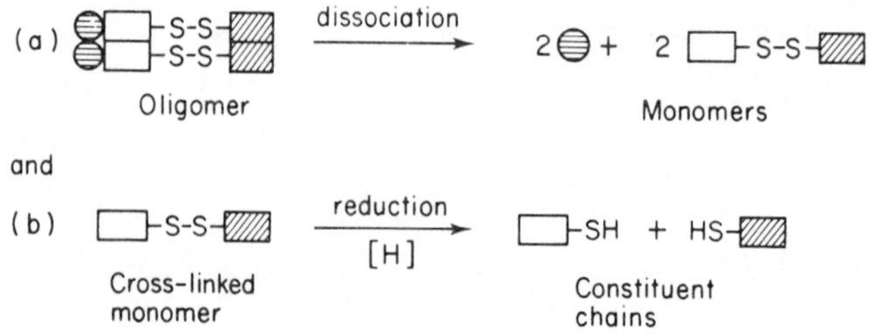

The methods described herein enable such stoichiometry to be established, while the amount, size and nature of the components of the oligomer should provide valuable primary information for the elucidation of the tertiary and questionary structures of the oligomer. However, consideration of symmetry arguments for accounting for the number of subunits involved (Perham, 1975) are outside the scope of the chapter. For additional information readers are referred to reviews on the quaternary structure of proteins by Klotz *et al.* (1975), protein–protein interactions (Frieden & Nicol, 1981) and the determination of the subunit structure of proteins (Thomas, 1978).

Since the methods used to characterize the monomers and constituent chains are an integral part of the solution of the structure of the oligomer, some degree of overlap with other chapters is inevitable. Methods described here were selected as a result of our own experience or that of our colleagues and are relatively inexpensive and easy to perform. Techniques involving mass spectrometry, ultracentrifugation or X-ray diffraction analysis are not described here and readers are referred to Chapters 19 and 20 in this book and other reviews (Klotz *et al.*, 1975, Van Holde, 1975; Amos, 1978; Jeffrey; 1981).

Procedures for the resolution and characterization of the monomers of a protein oligomer will give information concerning the mol. wt. of the oligomer and its subunits and the nature and numbers of the latter and lead to pure preparations of monomers suitable for primary structure determinations.

The advantages of beginning a primary structure determination with a crystalline protein and so expediting the amino acid sequence analysis by combining information obtained from the chemical procedures and X-ray crystallography, have been demonstrated by Schirmer *et al.* (1977). The opportunities for such strategic planning are rare. Frequently the preparation of a sufficiently pure sample of the protein or subunit for sequence analysis is a major difficulty, either because the protein occurs in trace amounts with a complex mixture of other proteins or their degradation products or because closely related proteins are present. Heterogeneous preparations may be obtained at the outset if the protein is insoluble and requires relatively harsh treatments to solubilize it. During such treatments charge heterogeneity is most likely to arise from the deamidation of the amide groups of asparagine and glutamine which occurs readily at extremes of pH. Urea solutions containing cyanate can cause carbamylation of some of the ε-amino groups on the side chains of lysyl residues in the protein with consequent production of differently charged molecules (Cole & Mecham, 1966). Chemical modification reactions can also generate charge heterogeneity if they are not specific for the target amino acid, e.g. in the alkylation of cysteinyl residues or the modification of methionine and histidine.

Size heterogeneity can be caused by cleavage of labile bonds in the peptide chain. Asp-Pro bonds are particularly sensitive to acid conditions and can be broken when using solvents such as 80% formic acid (Landon, 1977) and so the presence of an *N*-terminal proline residue should alert one to this possibility.

Complete exclusion or removal of proteolytic enzyme contaminants is essential since traces may cause limited proteolysis of particularly susceptible peptide bonds leading to low apparent mol. wts. of subunits (Kirschner & Bisswanger, 1976).

It is also relatively easy to block the α-amino group of the *N*-terminal amino acid. This can severely impede the complete primary structure determination using the Edman-type degradations. The most common blocking reactions that occur are those associated with the cyclization of *N*-terminal glutamine residues in acid solution and the carbamylation of the *N*-terminal amino acid in urea solution. Modification of the *N*-terminal α-amino group may also result from aldehyde impurities in reagents (Edman & Begg, 1967).

All of these difficulties can be overcome by careful selection of reaction conditions and methods for handling the proteins and their subunits.

1.2. CHARACTERIZATION OF OLIGOMER

The mol. wt. of the intact oligomer can be determined using non-denaturing conditions in the ultracentrifuge, gel filtration chromatography or polyacrylamide gradient gel electrophoresis. The stability and solubility of many oligomeric protein systems depend on ionic strength and pH. Although there are limitations to the buffers that can be used in polyacrylamide gel electrophoresis, e.g. ionic strength must not exceed 0.1, the elution buffer for gel filtration can usually be chosen to preserve the oligomer, e.g. by addition of metal ions or cofactors. Classical gel filtration chromatography on cross-linked dextran, polyacrylamide or agarose has been widely used to determine the mol. wts. of proteins, but recently new gel media capable of withstanding high pressure have been developed and these allow a substantial reduction in analysis time, e.g. 30 min instead of 1–2 days (see Chapter 6).

1.2.1. Polyacrylamide gradient gel electrophoresis

The mol. wt. of an oligomeric protein can be estimated after electrophoresis in a gel of increasing acrylamide concentration (gradient gel electrophoresis) (Margolis & Kenrick, 1968; Andersson *et al.*, 1972; Manwell, 1977; Lambin & Fine, 1979). The protein migrates through progressively smaller pores determined by the acrylamide concentration until it is virtually halted by the pore size of the gel. Migration distance is thus related to size and the mol. wt. can be determined by comparing migration distances of the test and standard proteins on the same slab gel.

Although proteins with relatively low charge move more slowly than those with high charge the final position of the protein is not significantly altered by the magnitude of the charge providing that the pH is not too close to the isoelectric points of the proteins and the charge on all proteins has the same sign.

1.2.1.1. Methods

Vertical slab electrophoretic apparatus of varying degrees of versatility and cost are readily available. Designs and instructions for constructing slab apparatus are available in the literature (e.g. Reid & Bieleski, 1968; Studier, 1973; Broadmeadow & Wilce, 1979; Hames & Rickwood, 1981). Temperature controlled equipment is usually not needed. A power supply with constant voltage and constant current output should be purchased.

Polyacrylamide gels have largely superseded other support media such as starch for gel electrophoresis. The polymer is formed with a range of pore sizes by varying the amounts of the monomer, acrylamide ($CH_2 = CHCONH_2$) and *N,N'*-methylenebisacrylamide ($CH_2 = CHCONHCH_2NHCOCH = CH_2$), the bifunctional (cross-linking) monomer (see section 1.3.1.1). Some authors suggest that the best results are obtained if these chemicals are recrystallized (see Chrambach *et al.*, 1976). However, we routinely use unrecrystallized products (Eastman) with satisfactory results. *Take care*: acrylamide is a neurotoxin. All other chemicals are analytical reagent grade.

Linear concentration gradients can be prepared by using two connected cylindrical vessels (Margolis & Kenrick, 1968), a peristaltic pump or commercially available equipment. Laboratories not routinely preparing gradient gels may encounter difficulties in forming reliable and reproducible gradients but preformed gradient gels of high quality are available (Gradipore, LKB, Pharmacia).

Wide range gels (4–30%) are suitable for examining proteins in the mol. wt. range 13 000 to 1 000 000. Proteins in the mol. wt. range from 200 000 to 1 000 000 can be examined on the narrow range gel (2–16%). Preformed gradient gels should be equilibrated by electrophoresis at a constant 150 V for about 1 h before application of the samples.

Electrophoresis

Tris/EDTA/borate stock electrode buffer is prepared by dissolving 121 g of Tris, 15.6 g of EDTA (disodium salt) and 9.2 g of boric acid in a total volume of 1 l. Before electrophoresis the stock solution is diluted 1:10 with water. A sample is prepared from the diluted electrode buffer by adding bromophenol blue (0.1% w/v) as a tracking dye and glycerol (10% w/v) (to make the sample solution denser than the electrode buffer).

Before loading, a protein solution should be dialyzed against the sample solution while a dry protein is dissolved in the sample solution. Protein concentrations should be about 0.5 μg/μl. Samples of 2–5 μl are loaded into slots in the gel with a microsyringe or a disposable micropipette and a constant potential of 150 V is applied for 7–16 h. Use standard proteins of known mol. wt. to calibrate the gel (see Klotz *et al.*, 1975). Kits are available containing ready-to-use lyophilized mixtures of proteins, e.g. thyroglobulin (669 000), ferritin

(440 000), catalase (232 000), lactate dehydrogenase (140 000) and albumin (67 000) (Pharmacia).

Coomassie Blue staining

After electrophoresis, the proteins are usually located by staining with Coomassie Brilliant Blue (CBB). The staining solution contains 0.15% (w/v) CBB G250 in methanol/glacial acetic acid/water (50:10:40, by vol.). The CBB is dissolved in methanol before adding acetic acid and water and the solution is filtered. Destaining is conveniently carried out overnight in a solution containing methanol/glacial acetic acid/water (5:7.5:87.5, by vol.)

The gel is placed in a photographic tray and is covered with staining solution. The tray, covered with perspex to minimize methanol evaporation, is rocked gently for about 1 h. After pouring off the staining solution (which can be reused about 20 times) the gel is covered with destaining solution. If a stain absorbing material (e.g. a 100 × 100 mm piece of woollen blanket, foam rubber or ion-exchange resin beads) is not added to the destaining solution, the solution should be changed 2–3 times in 24 h. After destaining overnight, the gels are stored in acetic acid solution (10% v/v). Although gels may also be destained electrophoretically, this procedure sometimes strips stain from the protein.

Silver staining

A more sensitive detection method involves equilibration of the gel with silver nitrate solution, then treatment with a reducing solution followed by equilibration in a solution containing sodium carbonate which enhances the appearance of the polypeptide-silver complexes. It may be used after the Coomassie Blue procedure with a large increase in sensitivity if the dye is carefully washed out with methanol/acetic acid (Poehling & Neuhoff, 1981).

Commercial kits for silver staining are available from Upjohn (Gelcode) and Biorad. These will stain with a sensitivity of about 250 and 50 times, respectively, greater than Coomassie Blue. Gelcode, developed from the work of Sammons *et al.* (1981) enables samples of 5–20 ng to be visualized. Both the colour and the intensity of staining are dependent on the protein analyzed. This is particularly useful when following the purification of components from complex mixtures. (See Chapter 8, section 19 for staining methods.)

Because the method is sensitive, scrupulous cleanliness is required in the preparation of the glass plates (see Chapter 8, section 2). Tracking of the samples between the plate and the gel can be a serious problem if the gel does not adhere to the glass.

Photography

After CBB staining the gels are photographed over a light box with Polaroid film or with a high contrast black and white film (e.g. Agfa Copex or Kodak Recordak, but not line film) and a range of exposure times, e.g. 1–6 s. Normally a Wratten Number 16 yellow filter is used, but if the bands are faint use Wratten Number 2 orange filter or if the bands are dark, use no filter. The film is developed with Agfa Rodinal (1:25 dilution, 3.5 min, 20°C), for example. The grade of glossy paper used in the printing depends on the nature of the negative and original gel. The prints are convenient for measuring migration distances as well as providing a record of data.

1.2.1.2. Analysis of results

A calibration curve for the gel is established by plotting log mol. wt. against migration distance (mm) or relative mobility (R_f) for each of the proteins of known mol. wt. (Fig. 1.1). The R_f may be relative to a dye or a standard protein. Determine the mol. wt. of the unknown protein from its R_f and the calibration curve. To avoid errors due to variability between gels it is advisable to run the sample and standards on the same slab gel.

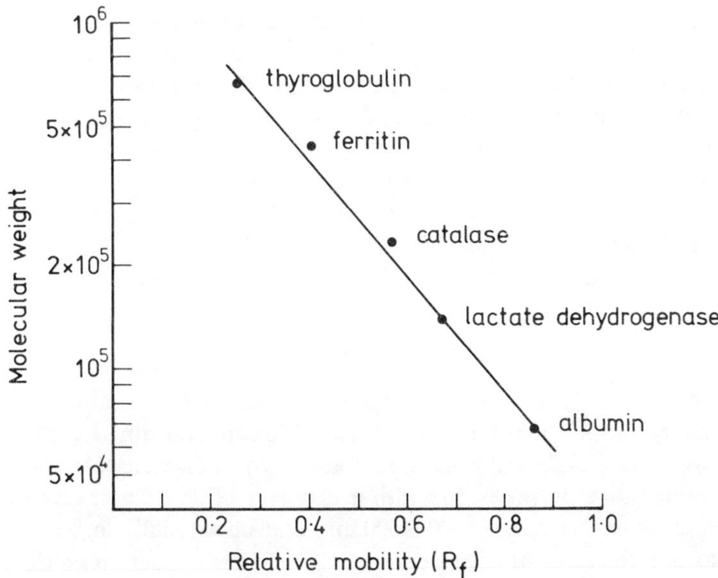

Fig. 1.1. Calibration curve for a Pharmacia 4–30% gradient gel established using thyroglobulin, ferritin, catalase, lactate dehydrogenase and bovine serum albumin (data from Lasky, 1978).

Small deviations from the calibration curve are frequently observed and this may lead to an error of up to 25% in the mol. wt. (Lambin & Fine, 1979). Large deviations will occur if the oligomer separates into its monomers during the run or if the protein is non-globular, e.g. fibrinogen with its extended structure has a larger hydrodynamic volume (and hence an anomalous mol. wt.) than would be expected for a globular protein with the same mol. wt.

1.2.2. Gel filtration

Gel filtration chromatography separates molecules primarily according to differences in molecular size (Andrews, 1964; Fischer, 1980). In this chapter open column methods are discussed, but the use of HPLC has become more important with the development of new columns, packings and instrumentation (see Chapter 6). The principles involved in open and closed columns for gel filtration are essentially identical.

The separating medium, or gel matrix, contains pores of controlled dimensions and the volume of the column occupied by the packed gel matrix is termed the total bed volume (V_t). V_t can be calculated by simple geometry or by filling with water and measuring the volume. As elution proceeds, sample molecules too large to enter the pores of the gel matrix pass rapidly through the bed in the space surrounding the gel particles and are eluted in a narrow zone at the beginning of the elution profile. The volume of elution buffer corresponding to this is the void volume (V_o). Smaller molecules move more slowly down the column because they penetrate the gel particles to an increasing extent. Molecules are therefore eluted in order of decreasing molecular size. The mol. wt. of the protein of interest is determined by comparing some elution volume parameter such as its elution volume (V_e) with the values obtained for several proteins of known mol. wt. Excellent reviews have been published (Ackers, 1975; Fischer, 1980). Booklets on experimental techniques are also available from manufacturers (e.g. Pharmacia, 1979).

1.2.2.1. Methods

Chromatography columns must be properly designed and of suitable dimensions for optimal results. Columns are readily available commercially (e.g. Pharmacia, LKB, Wright). For analytical work use a column with a diameter of about 15 mm and for preparative purposes one with a diameter of 25–50 mm or more. The length of a column is usually 500–1000 mm and occasionally longer with gels which do not collapse. Ancillary equipment includes a fraction collector, u.v. monitor and peristaltic pump.

Use a gel matrix such that the expected mol. wt. of the sample falls in the middle of the mol. wt. range. Gel media produced from dextran, acrylamide and/or agarose and containing different degrees of cross-linking fractionate in different mol. wt. ranges. Agarose gels (Sepharose, Bio-Gel A) are used for large proteins up to several million daltons and by cross-linking the individual agarose chains as in Sepharose CL increased stability and improved flow rates can be obtained. For proteins with a mol. wt. less than one million it is preferable to use Sephadex, Bio-

Gel P or Ultragel (LKB). Decreasing the density of cross-linking reduces the rigidity and flow rate of the gels but increases the exclusion limit. However, a more rigid gel with a higher flow rate (5–10 times) can be prepared by linking allyl dextran with *N*,*N*′-methylenebisacrylamide (known as Sephacryl) (Morgan & Ramsden, 1978; Johansson & Lundgren, 1979).

Best results are obtained when the gel matrix is prepared and packed according to the manufacturer's instructions. Dry gels, e.g. Sephadex, should be swollen in the column buffer for up to three days or autoclaved according to the manufacturer's instructions depending on the gel. Gels supplied in pre-swollen form, e.g. Sephacryl, are resuspended in the column buffer just before use. The space under the bed support in the column is filled with buffer and the degassed gel suspension is poured carefully down the side of the vertically mounted column. A thin layer of pre-swollen Sephadex G25 can be used to help prevent blocking of the filter membranes. If the volume of the prepared gel suspension is greater than the total column volume an extension tube should be mounted on the column so that all the gel suspension can be added in one pouring. The column is allowed to pack at a flow rate slightly higher than that to be used in subsequent fractionations. Care must be taken that the flow rate and hydrostatic pressure head do not exceed the manufacturer's recommendations. At least two bed volumes of eluent should be allowed to pass through the column to equilibrate the gel bed. Upward flow is often advisable for longer column life. Suitable adaptors are available (Pharmacia, LKB, Wright).

Normally a buffer with pH 6 to 8 and an ionic strength $\geqslant 0.1$ (e.g. 0.1 M Tris-HCl/0.1 M-NaCl, pH 8.0) is used, but it may be necessary to change the buffer to suit the requirements for the protein under study, e.g. a higher ionic strength may stop aggregation of the oligomers. However, since salt concentration and pH have been shown to affect the resolution on Sepharose (Lin & Castell, 1978), it may be necessary to compromise between protein requirements and resolution. An ionic strength > 0.5 may be advantageous when using Sephacryl since there is a report (Belew *et al.*, 1978b) that this gel exhibits ionic absorption properties. Another way to circumvent the adsorption problem on Sephacryl is to carry out the chromatography at pH 5.5.

To maintain a constant flow rate, it is desirable to use a peristaltic pump. The column eluate, which is directed to a fraction collector, can be monitored continuously or individual fractions assayed for protein. Monitoring is usually at 280 nm. A different method is used when tryptophan and tyrosine are absent, or only small quantities of protein are available, or absorption of the eluate at 280 nm is high, e.g. due to the presence of a cofactor.

Monitor at 220 nm if the eluent does not absorb strongly, by fluorescence if the protein contains a fluorophor or by radioactive measurement if the protein is radiolabelled either intrinsically (Ballou *et al.*, 1976; Schreiber *et al.*, 1979) or by chemical modification (see section 1.4.5.1). Alternatively, protein in the eluate can be assayed by alkaline hydrolysis of an aliquot and the ninhydrin reaction

(p. 260), the Lowry assay for total protein (p. 287) or dye-binding methods (p. 294). The dye-binding method is highly recommended due to its simplicity, sensitivity and relative freedom from interference by chemical reagents. Eluates can also be monitored by polyacrylamide gel analysis.

The void volume (or the dead volume) (V_o) is normally measured by determining the elution volume of Blue Dextran 2000 (Pharmacia), tobacco mosaic virus, ferritin or other material of very high mol. wt. which is excluded from the pores of the gel matrix. The theoretical value for V_o is about 35% of the total bed volume. A well-formed band of Blue Dextran as chromatography proceeds is a useful check on the homogeneity of packing of the column. A fresh solution of Blue Dextran (1 mg/ml) or protein sample (5–20 mg/ml) is prepared in the eluent buffer containing 10% w/v sucrose (to make the solution denser than the eluent buffer). The sample is carefully applied with a syringe or by using a sample loading loop and valve, without disturbing the top surface of the gel bed. To decrease adsorption of Blue Dextran to the gel matrix, the pH of the buffer should not be less than 5 when using Sephadex or Sepharose and not less than 7 for Sephacryl. The sample volume should not be more than about 1–2% of the total bed volume (V_t). To characterize the behavior of the protein in gel filtration, its elution volume V_e is measured. The position of the maximum of the peak in the u.v. chromatogram is taken as corresponding to the elution volume. Large sample volumes should be avoided because they may give a plateau region leading to difficulties in determining V_e. To determine V_e the flow rate is obtained either from a direct measurement or indirectly by measuring or weighing the eluent collected in a given time. Alternatively, since V_e is proportional to the time taken for a protein to elute, the time read from the u.v. chromatogram may be used as a basis for V_e.

1.2.2.2. Analysis of results

In determining mol. wts. by gel filtration chromatography a calibration curve for the particular gel column is established by plotting log mol. wt. against the elution volume or a derived parameter for a series of known proteins. Commercial kits contain ribonunclease A (mol. wt. 13 700), chymotrypsinogen A (25 000), ovalbumin (43 000) and albumin (67 000), or aldolase (158 000), catalase (232 000), ferritin (440 000) and thyroglobulin (669 000) (Pharmacia). Frequently the elution volume, V_e, itself is used in the calculation but the partition coefficient between liquid and gel phases, K_{AV}, which is independent of the volume of the gel bed has some advantages since

$$K_{AV} = \frac{V_e - V_o}{V_t - V_o}.$$

The calibration curve or log mol. wt. versus K_{AV} for a mixture of proteins separated on Sephadex G-200 is shown in Fig. 1.2. It is also possible to obtain a linear relationship by plotting log 100 K_{AV} against $N^{2/3}$ (where N represents the

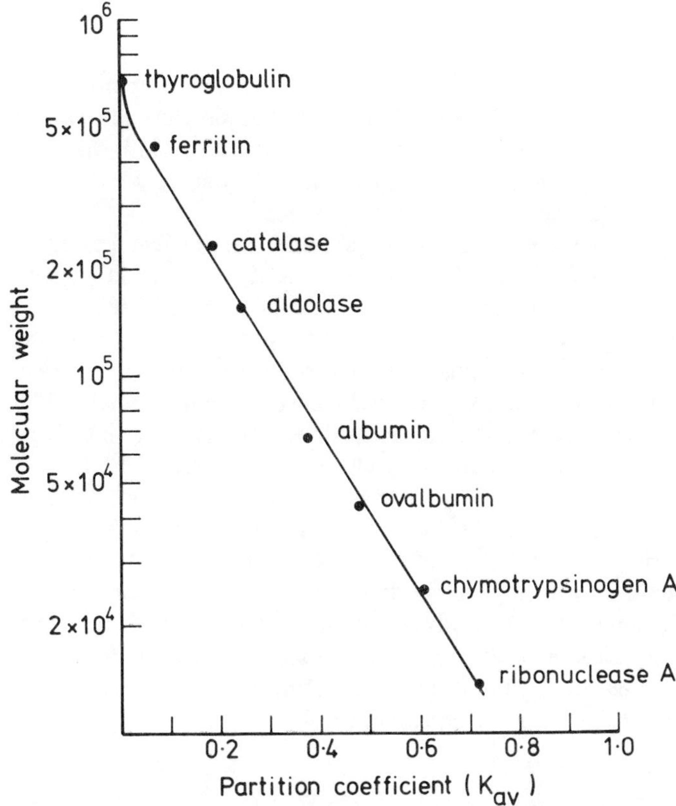

Fig. 1.2. Sephadex G-200 calibration curve established using Pharmacia low and high molecular weight gel filtration calibration kits (redrawn from Gel Filtration Calibration Kit Instruction Manual, Pharmacia).

number of amino acid residues in the calibrating protein) (Belew *et al.*, 1978a). It is recommended that when calibrating the column, the proteins chosen for a particular run should be well resolved to ensure accurate measurement of their elution volumes.

Most proteins fall on the calibration curve but deviations will occur if the oligomer aggregates, dissociates into its monomers or is non-globular (Axelsson, 1978). Care must be taken that the protein is not adsorbing to the gel matrix thus giving a spuriously low mol. wt. Proteins with a high content of either aromatic amino acids or carbohydrate are known to adsorb on Sephadex. Loss of material on the column, e.g. due to precipitation or binding, can be ascertained by comparing the total absorbance at 280 nm in the eluate with that of the sample loaded on the column.

1.3. CHARACTERIZATION OF MONOMER

To characterise the monomers the first necessary step in most procedures is effective dissociation of the oligomer. A mild method is preferable and this can be sometimes achieved simply with a pH change but more usually with denaturants such as urea, guanidine HCl or SDS. Electrostatic effects introduced by chemical modification, for example reaction with dicarboxylic anhydrides (sections 1.4.3 and 1.4.4), can also cause effective dissociation.

During the characterization of the monomer, the possible presence of disulfide bonds should be taken into account. If a thiol group is present disulfide interchange may occur under favourable conditions between and within monomers leading to anomalous electrophoretic and chromatographic separations. If the disulfide bonds are reduced, similar problems may occur if the constituent polypeptide chains are not kept in a reducing environment, e.g. by the addition of 0.05 M 2-mercaptoethanol. After reduction, it is advisable to convert cysteinyl residues to more stable residues by alkylation (see section 1.5.1) with iodoacetamide which leaves the protein charge unaffected or iodoacetic acid which introduces negative charges (see also Chapter 2, section 2.2.1). This usually increases the solubility of the protein under neutral conditions and may be an advantage in introducing charge differences between monomers. Alkylation of cysteinyl residues cannot usually be reversed and this modified residue may inhibit attempts to reassociate the oligomer (see section 1.4.4). As a general rule, it is advisable to use freshly prepared protein solutions to eliminate reactions such as aggregation that can occur on standing.

1.3.1. Molecular weight

Methods for determining the mol. wt. of monomers include polyacrylamide gel electrophoresis in the presence of SDS and gel filtration (classical open column and HPLC). SDS electrophoresis or HPLC is preferred, but at this stage of development HPLC is probably inferior for resolving closely similar mol. wt. species (see Chapter 6). Gel filtration on agarose gels is particularly useful if the protein contains more than 10% carbohydrate since SDS gel electrophoresis may not give a reliable estimate due to a difference in binding of SDS to carbohydrate and protein. It must be remembered that as a monomer may contain disulfide-bonded polypeptide chains, it should be examined both before and after reduction.

Mol. wts. may also be determined by quantitative determination of the *N*-terminal amino acid of the protein(s) (section 1.3.1.4). This normally requires about the same amount of protein as that used for gel filtration. However, smaller quantities can be used, particularly if automatic equipment is used. (See Chapters 12, 15 and 17.)

The mol. wts. of the monomers are most likely to be in the range 10 000–60 000 based on a statistical study of 500 proteins in living systems (Gianazza & Righetti, 1980). The subunits of more than two-thirds of the proteins so far analysed have mol. wts. in this range. Approximately 50% of these proteins are dimeric, 30% tetrameric and 8% hexameric.

1.3.1.1. SDS polyacrylamide gel electrophoresis

Most oligomeric proteins dissociate into their subunits in solutions of SDS. Subunits comprising two or more chains linked by disulfide bonds can be further dissociated by the incorporation of a reagent, such as 2-mercaptoethanol, capable of reducing disulfide bonds (section 1.5.1). Polypeptide chains in general bind similar amounts of SDS (~ 1.4 g SDS/1 g polypeptide (Reynolds & Tanford, 1970) and become highly negatively charged by the addition of strongly acidic sulfonic acid groups. Thus the charge of the native protein is masked and proteins show a relatively constant charge to mass ratio. SDS induces conformational changes in polypeptides although the absolute configuration in SDS is not known with certainty (rod-like, necklace, prolate ellipsoid and random coil models have been proposed). Polypeptide chains saturated with SDS have been shown to migrate upon electrophoresis in polyacrylamide gel according to their mol. wt. (Weber & Osborn, 1969). The sizes of the polypeptide chains are determined by comparing their electrophoretic mobilities to the mobilities of well-characterized proteins of known mol. wt. The following three different SDS polyacrylamide gel electrophoretic methods are currently in use:

1. A continuous buffer system (normally phosphate, according to Weber & Osborn, 1969) with a uniform gel.
2. A discontinuous buffer system (according to Laemmli, 1970) with a uniform gel.
3. Continuous or discontinuous buffer systems with gradient gels.

There are advantages and disadvantages for each method. There are comprehensive articles on general principles and experimental techniques (Allen & Maurer, 1974; Chrambach *et al.*, 1976; Stegemann, 1979; Weber & Osborn, 1975; Hames & Rickwood, 1981). In the discontinuous system, samples are concentrated in a stacking gel and enter the separation gel as very narrow zones, whereas in the continuous system this does not occur and large sample volumes normally produce broad bands. If small sample volumes are loaded, the results from continuous and discontinuous systems should be identical. Differences are sometimes observed possibly due to the removal of the SDS from the polypeptide during electrophoresis in the discontinuous system.

Many proteins do not bind the normal amount of SDS for a number of reasons, including unusual composition and the presence of carbohydrate (Segrest & Jackson, 1972). Proteins with a high negative or positive charge do not bind the same amount of SDS as normal proteins of comparable size; single amino acid substitutions and conformational changes in proteins also may have large effects on SDS-electrophoretic mobilities (De Jong *et al.*, 1978; Noel *et al.*, 1979; Crewther *et al.*, 1980). Such anomalous behavior can be detected and possibly overcome by examining proteins in gels containing different acrylamide concentrations, and then constructing a Ferguson plot (see later). Thus for reliable estimates of apparent mol. wt. in either continuous or discontinuous systems, it is necessary to make measurements using at least four different acrylamide concentrations.

In gradient gel electrophoresis the polypeptides move through the acrylamide gradient until a particular pore size is reached which restricts any further migration. If all SDS-polypeptide complexes have similar shapes, this acrylamide concentration will be characteristic of the mol. wt. (Lambin *et al.*, 1976; Lambin, 1978; Lasky, 1978, Poduslo & Rodbard, 1980). SDS may be stripped off especially as the SDS-polypeptide complex nears its equilibrium position but if this occurs it does not appear to affect the result. Sample volume should not affect the width of the protein bands, although in practice it is not advisable to load large volumes. Gradient electrophoresis has the disadvantage of requiring special equipment for producing the gel gradient and there may be problems in forming reliable and reproducible gradients. However, preformed gels of high quality are commercially available.

Apparatus and chemicals

Polyacrylamide gels may be prepared in either glass tubes or slab equipment, the latter being more suitable for comparative studies on molecular weight. Observed mobilities in tubes may vary as a result of differences in degree of polymerization, length of gel and conditions of electrophoresis. The procedure for removing gel rods from tubes is discussed by Chrambach *et al.* (1976) and gels with an acrylamide concentration greater than 10 % may be difficult to remove. The choice of equipment and precautions to be taken with chemicals are discussed in section 1.2.1.1.

A 30 % stock solution containing 29.2 % (w/v) acrylamide and 0.8 % (w/v) *N*, *N'*-methylenebisacrylamide is prepared, filtered and stored at 4°C (stable for several months). The solution is allowed to warm to room temperature and degassed before using.

N,*N*,*N'*,*N'*-tetramethylenediamine (TEMED) is distilled *in vacuo* and stored at 4°C in a dark bottle. High quality reagents are available from Biorad and other suppliers. We routinely use SDS from BDH (specially pure grade). The resolution and results obtained may be dependent on impurities, e.g. sodium tetradecyl sulphate in the SDS (Birdi, 1976; Matheka *et al.*, 1977; Dohnal & Garvin, 1979). All other chemicals are analytical reagent grade. SDS can be recrystallized after charcoal treatment to obtain highest purity (600 g in 3 l of 80 % EtOH).

Polyacrylamide gels

Gels range from soft (low polyacrylamide) to hard (high polyacrylamide) with the properties of a particular gel depending on the total weight concentration (%*T*) of the monomers (acrylamide + *N*,*N*′-methylenebisacrylamide) and the proportion by weight (%*C*) of the cross-linking reagent (*N*,*N*′-methylenebisacrylamide) (Chrambach & Rodbard, 1971). These are prepared by polymerizing an aqueous solution of acrylamide and *N*,*N*′-methylenebisacrylamide (prepared from the stock solution) using a free radical-generating catalyst system, e.g. persulfate–TEMED (see below).

We use the persulfate–TEMED catalyst system to polymerize the acrylamide monomers. After the persulfate is added, the solution is stirred to dissolve the persulfate and quickly transferred to the slab apparatus with a syringe. Care must be taken to avoid bubbles in the gel solution by using an intravenous cannula on a syringe and having the end of the cannula below the surface of the gel solution. The comb to form the sample wells is carefully inserted to avoid bubbles on the comb teeth and removed after the gel has polymerized. The rate of polymerization of the gel can be varied by the amounts of persulfate and TEMED used, but the gel should set within 10–25 min after adding the persulfate. Longer or shorter gelling times may lead to irregular and irreproducible pore sizes.

Ingredients to make 10 ml of acrylamide gel are given below.

	5%	7.5%	10%	12.5%
Stock 30% acrylamide solution	1.67 ml	2.50 ml	3.33 ml	4.17 ml
Stock phosphate-SDS buffer	2.0 ml	2.0 ml	2.0 ml	2.0 ml
Water	6.33 ml	5.50 ml	4.67 ml	3.83 ml
TEMED	5 μl	5 μl	5 μl	5 μl
Ammonium persulfate	10 mg	10 mg	10 mg	10 mg

Sample preparation

A solid protein sample is dissolved in the appropriate buffer (described later) while a protein already in solution is dialyzed against the same buffer. As it takes some time for the SDS to reach equilibrium across the dialysis membrane, solid SDS (to make 1% w/v) may be added to the protein solution before dialysis (at least 3 h or overnight). The samples are incubated for 2 min at 95–100°C (to denature the protein and enhance SDS binding) before loading on to the gel. Protein concentration should be about 0.5–1.0 μg/μl and 2–10 μl may be loaded depending on the gel thickness and the width of the sample well. (Much less sample should be used if gels are developed by silver staining.) If the protein concentration is lower than this, a larger volume may be loaded on a gel with a discontinuous buffer system, but for a gel with a continuous buffer system it is better to concentrate the sample solution. This may be possible by ammonium sulfate or trichloroacetic acid precipitation, the precipitated protein being dissolved in the buffer solution to a concentration of 0.5–1.0 $\mu g/\mu l$.

If the protein is difficult to solubilize, e.g. lipoprotein or membrane protein, either 8 M-urea or a non-ionic detergent, e.g. 1 % w/v Triton-X 100, is added to the sample solution or the SDS concentration is increased to 10 % w/v in the sample solution. Potassium or guanidine salts should not be present in buffers or sample solution as the SDS salts of these cations are not very soluble.

Gels are calibrated with proteins of known mol. wt. SDS-mol. wt. calibration kits are available (e.g. Pharmacia, Bio – Rad, BDH, Sigma, Calbiochem).

Continuous buffer system with single polyacrylamide concentration

The electrophoretic method normally used, and described here, contains 0.05 M-phosphate buffer and 0.1 % SDS, and is based on that of Weber and Osborn (1969).

Stock phosphate–SDS buffer contains 0.25 M-phosphate (pH 7), 0.5 % SDS and 0.02 % sodium azide (13 g of $NaH_2PO_4.2H_2O$, 26 g of Na_2HPO_4, 0.2 g of NaN_3, 5.0 g of SDS/l). Dissolve the phosphate by slightly warming the solution and store in a warm place.

Electrode buffer is made by dilution of stock phosphate – SDS buffer with water (1:4, v/v).

The sample is dissolved in a solution consisting of 20 ml of stock phosphate–SDS buffer + 1 g of SDS + 10 g of glycerol (to make the solution denser than electrode buffer) + 0.05 g bromophenol blue (to act as a marker front during electrophoresis) diluted to a final volume of 100 ml. For determination of the mol. wts. of the polypeptide chains after disulfide reduction, the sample solution also contains 1 % (v/v) 2-mercaptoethanol.

The samples after incubation (see earlier) are loaded in the sample wells with a microsyringe or disposable micropipette. Electrophoresis is performed at about 200 V until the marker dye is near the bottom of the gel. The time taken depends on the length of the gel and the geometry of the apparatus. After electrophoresis, the gels are stained, destained and photographed (section 1.2.1.1.).

Discontinuous buffer system with single polyacrylamide concentration

The electrophoretic method normally used and described here contains 0.375 M-Tris and 0.1 % SDS and is based on that of Laemmli (1970). The separating gel contains a sieving acrylamide concentration between 5 % and 25 % while the stacking gel (normally about 20 mm in length) is 4.5 % acrylamide.

Separation gel buffer: 1.5 M-Tris and 0.4 % w/v SDS is titrated to pH 8.8 with HCl.
Stacking gel buffer: 0.5 M-Tris and 0.4 % w/v SDS is titrated to pH 6.8 with HCl.
The sample is dissolved in a solution containing 0.0625 M-Tris, 2.3 % w/v SDS,

10 % w/v glycerol (to make the solution denser) and 0.05 % w/v bromophenol blue (to act as a marker). To examine the mol. wt. of polypeptide chains after disulfide reduction use a sample solution similar to the one above but containing 1 % v/v 2-mercaptoethanol or 100 mM dithiothreitol.

Electrode buffer: 0.025 M-Tris, 0.192 M-glycine and 0.1 % w/v SDS.

Gel ingredients to make 10 ml of separating acrylamide (5 %, 7.5 %, 10 % and 12.5 %) are given below. Degas the solution before adding SDS and again before adding TEMED and persulfate.

	5%	7.5%	10%	12.5%
Stock 30 % acrylamide solution	1.67 ml	2.50 ml	3.33 ml	4.17 ml
Separating gel buffer	2.5 ml	2.5 ml	2.5 ml	2.5 ml
Water	5.83 ml	5.00 ml	4.17 ml	3.33 ml
TEMED	5 μl	5 μl	5 μl	5 μl
Ammonium persulfate	10 mg	10 mg	10 mg	10 mg

After the separating gel has been added to the slab apparatus the gel solution is carefully overlayed with water or isobutanol saturated with water to form a flat surface. When the gel has polymerized the overlay solution is poured off, the separating gel surface first washed with a small amount of stacking gel buffer and the stacking gel buffer added. Ten milliliters of the stacking gel (4.5 %) is prepared by adding 10 mg of ammonium persulfate to a solution containing 1.5 ml of stock 30 % acrylamide solution, 2.5 ml of stacking gel buffer, 6.0 ml of water and 10 μl of TEMED.

After incubating (see earlier) the samples are loaded with a microsyringe or disposable micropipette in the sample wells formed in the stacking gel. Electrophoresis is performed at 100 V until the marker dye has traversed the stacking gel and then at 200 V until the dye is near the bottom of the gel. The time taken depends on the length of the gel and the geometry of the apparatus (often about 3 h).

Continuous and discontinuous buffer systems with gradient polyacrylamide

The procedure uses the same solutions as for the continuous and discontinuous buffer systems with single polyacrylamide concentration described above except that a concentration gradient in acrylamide is formed as the solution is introduced into the slab apparatus. Concentration gradients can be prepared by using two connected vessels (Margolis & Kenrick, 1968) and peristaltic pump or commercially available equipment.

Preformed gradient gels of high quality are currently available from Gradipore and Pharmacia (section 1.2.1.1). These gels do not contain SDS but SDS can be introduced by an initial electrophoresis at 50 mA for 3 h. Electrode buffer is prepared by diluting stock phosphate–SDS buffer with water (1:9, v/v). The proteins are then loaded and electrophoresis is carried out for a further 3.5 h at

50 mA. This is normally sufficient for the proteins to reach their limiting pore position (section 1.2.1).

1.3.1.2. Analysis of results

As described in section 1.2.1, the mol. wt. of a polypeptide chain can be estimated from a calibration curve. However, electrophoresis at a single gel concentration does not eliminate systematic errors introduced by differences in free mobility (Y_0) for standard and unknown proteins (Rodbard, 1976a, b). To avoid this potential source of error, several gel concentrations should be examined and the measured mol. wt. should show no systematic trend with acrylamide concentration (T). The data can be further analyzed according to the empirical equation described by Ferguson (1964):

$$\log(R_f) = -K_R T + \log(Y_0)$$

K_R, the retardation coefficient, for each standard and unknown protein is determined from the slope of the plot $\log(R_f)$ against T, and the mol. wt. of the unknown protein is estimated from the plot of K_R against mol. wt. of the standard proteins. In order to have a reliable measure of K_R and to employ statistical analysis, at least four (e.g. 7%, 8%, 9% and 10%) and preferably seven gel concentrations should be used (Rodbard, 1976a).

1.3.1.3. Gel filtration

The general principles and experimental techniques discussed in section 1.2.2 for gel filtration under non-denaturing conditions apply when denaturing agents are used. Denaturing agents like guanidine HCl, urea and SDS alter the molecular conformation of proteins, often greatly increasing their hydrodynamic volumes. Since separations by gel filtration are based on molecular size, the mol. wt. fractionation range and hence the calibration curve for a particular gel medium changes when the proteins assume extended conformations in denaturing agents. As an example of the change in the fractionation range, Sepharose CL-6B under non-denaturing conditions separates proteins with mol. wt. from 10000 to 4×10^6 while in 6 M-guanidine HCl the range is 3000 to 80000 (Ansari & Mage, 1977).

It has been shown that high concentrations of guanidine HCl denature reduced proteins to linear random coil structures, all of which possess the same gross conformation. Chromatography in 6 M-guanidine HCl on non-cross-linked agarose gel media was recommended as an empirical method for determining mol. wt. (Fish *et al.*, 1969; Mann & Fish, 1972). The major drawbacks are the time required (3–5 days) and the chemical instability of non-cross-linked agarose (Sepharose, Ultrogel and Bio-Gel A) gels in guanidine HCl (Mahuron, 1979). The recent introduction of more stable cross-linked agarose gels (e.g. Sepharose CL-

6B) allows the use of guanidine HCl for at least one year without deterioration of the gel medium (Ansari & Mage, 1977; Mahuron, 1979; Brouwer, 1979) and higher flow rates (approximately 3–4 times) are possible. Guanidine HCl is particularly useful when the protein in question contains carbohydrates since SDS polyacrylamide gel electrophoresis (section 1.3.1.1) gives inaccurate results when there is a significant amount of carbohydrate. It can also be satisfactorily used with Sephacryl and S-200 has been shown to be suitable for estimating mol. wts. of less than 30 000 (Belew *et al.*, 1978a). For good spectral properties high purity guanidine HCl is essential and this may be purchased (e.g. ultra pure grade from Schwarz-Mann), recrystallized from impure material or prepared from guanidine carbonate (Nozaki, 1972) as described below.

Also 8 M-urea or 0.1 % SDS may be used in the elution buffer with many gel media, although the non-cross-linked agarose gels slowly degrade in the presence of high concentrations of urea. The elution buffer is normally about pH 7–8 (0.05 M-phosphate or 0.1 M-ammonium bicarbonate) and when using urea, 0.1 M-sodium chloride is often added to reduce the possibility of ionic interactions between the gel matrix and polypeptide chain. Care should be taken in the use and preparation of urea solutions. The spontaneous formation of cyanate occurs in aqueous solutions of urea (Dirnhuber & Schutz, 1948) as shown.

$$(NH_2)_2CO \rightleftharpoons NH_4NCO$$

The accumulation of cyanate is greatest at pH 6 and above but significant amounts are present in acidic solutions, even at 0°C (Hagel *et al.*, 1971). The rate of reaction of cyanate with amino groups of proteins appears to involve the unprotonated amine and the unionized cyanic acid.

$$\vdash N(H)_2 + HN{=}C{=}O \xrightarrow{H_2O} \left[\vdash N^{+}H_2{-}C(NH_2){=}O + OH^{-} \right] \longrightarrow \vdash NH{-}C(NH_2){=}O + H_2O$$

This is consistent with the difference in reaction rates shown by α-amino ($pK_A \triangleq 8$) and ε-amino groups ($pK_A \triangleq 10.7$) of proteins. At pH 7 α-amino groups react 100 times more rapidly (Stark, 1965). Carbamylation is not a problem below pH 4 and above pH 10 (Thompson & O'Donnell, 1966). Cyanate also reacts with thiols, carbonyl, imidazole, phosphate and reactive hydroxyl groups in proteins (Stark, 1965).

It is important to remember that intact disulfide bonds may exert a certain amount of physical constraint on a polypeptide chain even though the protein is fully denatured in 6 M-guanidine HCl. The effect of this physical constraint is to reduce the hydrodynamic volume of the polypeptide chain in comparison to the same polypeptide chain with its disulfide bonds broken. The effect should become more noticeable with increasing size of the polypeptide chain, increasing number of disulfide bonds and the size of the loops formed by the disulfide bonds. Caution should be exercised in the reliance placed on the mol. wt. A check may be made by chromatographing the constituent chains after reducing and preferably alkylating (section 1.5.1). The sum of the mol. wts. of the constituent chains should agree with the total mol. wt. of the protein. If the proteins are examined in the reduced and non-alkylated form, the elution buffer should be at pH 3 (where the constituent chains are moderately stable at room temperature) (Anfinsen & Haber, 1961) or contain a reducing agent like 2-mercaptoethanol (0.1 M). The addition of a reducing agent often precludes the monitoring of the column effluent at 280 nm due to the absorption at this wavelength of the oxidized form of the reducing agent. It is then necessary to detect the protein by some other method, e.g. radiolabelling the protein before chromatography (see section 1.4.5.1).

Preparation of guanidine HCl (Nozaki, 1972)

Guanidine carbonate (1 kg) is dissolved in 2 l of water at 40°C, followed by the addition of 3 l ethanol. Leave in the cold for several hours. The crystals are filtered and washed first with ethanol–water (2:1, v/v) then ethanol. The recrystallized guanidine carbonate (yield 880 g) and water (500 ml) are cooled in an ice bath. Stir well and slowly add to the slurry 6 M-HCl until the pH is stable at pH 4 after standing overnight. The product is then crystallized in stages. A portion of the reaction mixture is flash evaporated at 40°C until crystallization begins, then stored in the cold to complete the process. After filtering, the mother liquor is added to another portion of the reaction mixture and the procedure repeated, and so on until all of the crystals have been obtained from the reaction mixture. The total product (approximately 780 g) may be recrystallized from water (as above) or methanol, using activated charcoal to decolorize if necessary. A 6 M solution at 225 nm gives $\varepsilon = 0.1$.

Method: preparation of urea solutions

Cyanate ions are produced in urea solutions on standing, the extent of formation varying with the pH of the solution (Hagel *et al.*, 1971). Just prior to use, the solution is passed through a mixed-bed deionizing resin (e.g. Elgalite). The solution should have a conductivity of less than 2 μS.

Amines such as Tris (Henschen-Edman, 1977) and ethanolamine (Crewther *et al.*, 1978) have been added to the solution to compete with the protein for the cyanate.

1.3.1.4. Quantitative N-terminal analysis

N-terminal analysis of either the purified oligomer, monomer or constituent chain is usually made to confirm its purity. When carried out quantitatively on the monomer it can also provide an indication of the mol. wt. of the sample and whether it has cross-linked polypeptide chains (which could yield either twice the expected amount of a derivative or equivalent amounts of two different derivatives). It rests on the assumption that the *N*-terminal amino acid is not blocked, e.g. by carbamylation in urea solution. Some *N*-terminal residues are unstable under the conditions required for cleavage and conversion, e.g. asparagine, glutamine and serine, when correction factors are necessary.

The Edman degradation can be used directly and the quantitation of the derivative made by measuring the absorbance in alcohol at 269 nm (Edman & Henschen, 1975). Preferably, the determination of the amount of derivative formed should be made after further purification of the protein and subsequent hydrolysis.

The use of an isotope dilution procedure with the Edman degradation procedure was well documented by Banks *et al.* (1976). In a variation of this approach Kubota and Tsugita (1980) first made an Edman degradation step, which blocked all lysine side chains, then coupled the new *N*-terminal amino acid with radiolabelled reagent. No separation of the *N*-terminal residue was required with this procedure. The authors used amino acid analysis to determine the amount of protein in the sample and took aliquots of the hydrolysate for determination of radioactivity. A pure sample of a peptide is radiolabelled in an identical manner to the protein to obtain a value for the specific radioactivity of the ^{14}C-PITC.

Sulfophenylisothiocyanate (Braunitzer *et al.*, 1971) can be used to produce a more soluble protein for the second step than that produced by PITC, and addition of SDS (1%) is advantageous. Also, in removing excess radioactive reagent after the reagent, adding n-octanol eliminates any problems caused by SDS with the benzene extractions.

Quantitative *N*-terminal determination (Banks *et al.*, 1976)

Amino acid analysis should be used to quantify the amount of protein if possible. Accurately known weights (W mg) of dry, salt-free protein are coupled with radioactive PITC of known specific activity and after removal of excess reagent, cleavage and conversion are carried out to yield the labelled PTH *N*-terminal amino acid. A known quantity of unlabelled derivative of the *N*-terminal amino acid is added after which a sample of the derivative is isolated and its specific radioactivity determined. If the specific radioactivities of the PITC and of the PTH derivative are S_1 $Ci\,\mathrm{mol}^{-1}$ and S_2 $Ci\,\mathrm{mol}^{-1}$, respectively, and N moles of unlabelled derivative are used for dilution then the subunit molecular weight (M_r) is given by:

$$M_r = W\frac{S_1 - S_2}{NS_2} \times 10^3$$

Edman degradation. To an aliquot (400 μl) of protein solution (approximately 10 mg/ml) of accurately known concentration add pyridine (600 μl), *N*-dimethylallylamine (47 μl) and enough anhydrous trifluoracetic acid (approximately 3 μl) to give an apparent pH of 9.5. Phenyl-^{14}C-isothiocyanate (10 μl) of accurately known specific radioactivity is added and the mixture incubated under nitrogen at 40°C for 1.5 h. Excess reagents are then removed with benzene (3 × 2 ml) and the sample lyophilized. The PTH derivative is formed by addition of water (1 ml) and glacial acetic acid saturated with dry HCl gas followed by incubation at 40°C for 2 h. See also Chapter 12.

Isolation and determination of the N-terminal PTH derivative. After formation of the PTH derivative as described, a known amount of the unlabelled PTH derivative of the *N*-terminal amino acid is added to the reaction mixture. The mixture is then extracted with ethyl acetate (2 × 2 ml) and the combined extracts taken to dryness. The PTH derivative is purified by chromatography using Whatman SG81 paper in two different solvent systems. The first dimension is developed with chloroform–methanol (9 : 1, v/v) after which the PTH derivative is eluted with ethyl acetate. Re-chromatograph using chloroform as solvent and elute with ethanol. HPLC can also be used for PTH purification (see Chapter 13).

The concentration of the ethanolic solution is determined by measurement of absorption at 269 nm against ethanol as a blank (ε 16000 M^{-1} cm^{-1}) and the purity of the sample confirmed from the ratio of absorbance 245 nm/269 nm (0.39). An aliquot (3 ml) of the solution is then counted for radioactivity.

Measurement of the specific radioactivity of the phenylisothiocyanate. The radiolabelled PITC (0.5 mCi) is diluted with unlabelled compound to an approximate specific radioactivity of 0.4 Ci mol^{-1} Aliquots (2 μl) are added to samples of *N*-terminal amino acid (~ 2 mg) dissolved in aqueous pyridine (50 % v/v, 1 ml) and the solutions incubated at 50°C for 35 min under nitrogen. Excess PITC is removed by extraction with benzene (4 × 2 ml) and the samples lyophilized. Cyclization and conversion to the PTH derivative is carried out by incubation in 1 M-HCl (500 μl) at 80°C for 10 min. The PTH derivative is extracted with ethyl acetate (2 × 2 ml) and purified by chromatography on Whatman SG81 paper as described above or by HPLC. The specific radioactivity of the derivative is determined by counting an aliquot of a solution of known concentration (from the absorbance at 269 nm). Four samples of the PTH derivative are prepared and the mean specific radioactivity is taken as the value for the sample of phenylisothiocyanate.

Liquid scintillation counting. The scintillator solution used consists of 5 g of 2-(4-*t*-butyl-phenyl)-5-(4-biphenyl)-oxa-3,4-diazole (butyl-PBD) dissolved in 1 l of toluene (dried over sodium wire). Samples are added in ethanol (3 ml) to 15 ml of scintillator solution and counted in a liquid scintillation system until 40000 counts have been accumulated. For each sample, the true radioactivity is calculated from the observed count and the counting efficiency. The latter is read from a standard graph of counting efficiency against the quenching factor. Efficiencies are generally in the range 92–95 %.

1.3.2. Resolution methods

There may be occasions when it is necessary to isolate and purify monomers for further characterization, e.g. amino acid analysis, peptide mapping or amino acid sequence determination. This section describes methods available for larger scale separation of proteins.

If the available quantity of protein precludes the scaling up, recourse must usually be made to radiolabelling of the protein and analytical scale methods would then be used. Radiolabelling may be intrinsic (Ballou *et al.*, 1976; Schreiber *et al.*, 1979) or effected by reaction with amino acid side chains as discussed in section 1.4.5.1.

1.3.2.1. Size separation

The separation of subunits of sufficiently different size can often be simply accomplished, if the mol. wts. are sufficiently different, using gel filtration in the presence of denaturants. Analytical scale methods were described in section 1.3.1.3, and the principles and methods in that section apply equally to preparative scale gel filtration. Obviously larger size columns are used, the size depending on the amount of protein to be loaded. Columns with 2.5–5.0 cm diameter and 100 cm length are sufficient for most experiments (500 mg to several g).

Volatile buffers, dilute ammonia, acetic acid or formic acid solutions are usually used so that the proteins may be recovered by lyophilization. Gel filtration in 50–75 % formic acid is particularly recommended for the preparation of samples for amino acid sequence determination. Formic acid, a strong denaturing and solubilizing agent, degrades many of the gel filtration media, e.g. Sephadex should be used for no longer than three weeks in 75 % formic acid.

Use an all-glass system and teflon tubing (Fig. 1.3). Proteins are recovered by lyophilization using an all-glass vacuum system after the formic acid concentration has been reduced to about 10 % by dilution with water. Formic acid is generally believed not to degrade or split chemical bonds, but labile bonds, e.g. Asp-Pro may be cleaved.

HPLC (usually 2–3 mg and possibly up to 100 mg of protein) has a distinct advantage over conventional gel filtration chromatography because the time required is much shorter and this more than compensates for the normally lower capacity of the system. Larger capacity columns will certainly become available as the technology develops, but undoubtedly these will be very expensive.

Monomers of different size may also be separated by preparative

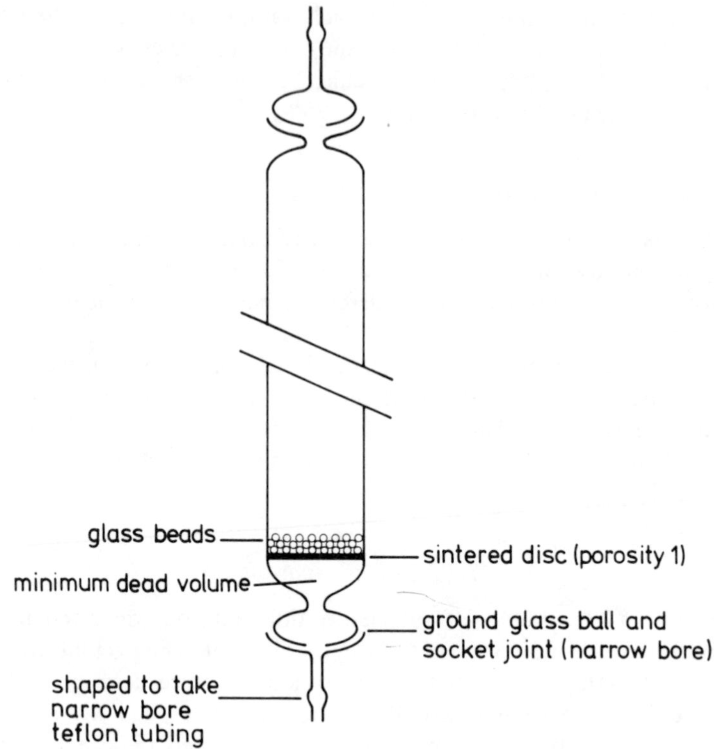

Fig. 1.3. Diagram of all glass column and fittings for chromatography with formic acid.

SDS–polyacrylamide gel electrophoresis. The capacity of this method (e.g. 50 mg of protein) is much lower than that used for filtration but has the advantage that it can separate proteins with similar mol. wts. (Steinert & Idler, 1975). There are basically two different methods. Firstly there is the scaling up of the analytical system using thicker gels or loading the sample across a whole slab gel formed without slots. The proteins are located in the following ways:

1. Staining the whole gel with a weaker stain (0.05% Coomassie Blue) or for a shorter time than normal (section 1.2.1.1).
2. Staining guide strips cut from the gel (remember to allow for the change in gel size after methanol–acetic acid destaining).
3. Immersion of gel in 1 M-KCl when white bands of potassium dodecyl sulfate protein precipitate out (Walker *et al.*, 1980).
4. Using 4 M-sodium acetate to reveal clear bands against a light-scattering background (detection limit 0.1 ng/mm^3, Higgins & Dahmus, 1979).

Bands of interest are then cut out and the protein recovered from the polyacrylamide gel by extraction (Gibson & Gracy, 1979; Djondjurov & Holtzer, 1979) or electrophoretic elution (Tuszynski *et al.*, 1977; Ahmadi, 1979). For accurate determination of amino acid composition of proteins eluted from polyacrylamide gel, the contribution of the gel itself to the composition needs to be taken into account (Brown & Howard, 1980). Another preparative polyacrylamide gel electrophoretic method involves running the proteins off the gel into an elution chamber from where they are flushed out by a constant stream of elution buffer. Elution of protein is followed with a u.v. monitor and protein collected using a fraction collector. Recently the recovery of polypeptide from SDS gels using concentration into a stacking gel by upward electrophoresis was described (Mendel-Hartvig, 1982). Recoveries are high but contamination from buffers and gels must be carefully monitored. Gels may be either rod-shaped (LKB) or slab-shaped (Bio-Rad). Commercial apparatus is available for preparing gels.

1.3.2.2. Ion-exchange chromatography

Subunits may also be separated according to their net charge by ion-exchange chromatography, electrophoresis (section 1.3.2.3) or isoelectric focusing and chromatofocusing (section 1.3.2.4).

In ion-exchange chromatography, the sample is applied to a column of an inert support carrying either positive (anion exchanger) or negative fixed charges (cation exchanger). If the support carries positive charges, those components of the sample which are positively charged (pH of the solution is below the isoelectric point of the protein component) will pass through the column, whereas negatively charged components of the sample will be attracted to the fixed positive charges. Weakly bound components may be eluted with continued flow of starting buffer but for strongly bound components it will be necessary to displace them by applying a salt gradient of increasing ionic strength or by changing the pH. Components with lowest net charge are released first followed by those with higher net charge.

Ionizable groups can be attached to different inert support materials, the most common being cellulose in granular form (e.g. Whatman, Serva, Bio-Rad, Eastman) or more recently in beaded form (Pharmacia). Other support media are cross-linked agarose and cross-linked dextran. The latter has a high capacity and is particularly suited for smaller biological molecules but suffers from shrinkage or swelling with ionic strength or pH gradients. A cation-exchange column (TSK-Gel LS-212) suitable for the use with HPLC was found to be satisfactory in separating multiple forms of soybean lipoxygenase-1 (Aoshima, 1979). Pharmacia have recently introduced large pore HPLC ion-exchange packings which give extremely rapid high resolution separations. Cost may be a deterrent for their widespread use (see Chapter 6).

The most commonly used anion exchangers contain diethylaminoethyl (DEAE) or diethyl-(2-hydroxypropyl)aminoethyl (QAE) functional groups, while for cation exchangers there are carboxymethyl (CM), phospho-(P) or sulfopropyl-(SP) groups. Strong ion exchangers, e.g. QAE, P and SP, are preferred when separating weakly acidic or basic substances or using extremes of pH. Remember that separation of protein components on ion-exchange resins may depend on characteristics other than charge, e.g. proteins containing large numbers of aromatic residues were separated on the basis of their content of these residues at pH 10.5 on QAE-cellulose (Gillespie & Frenkel, 1974).

Choice of ion exchanger depends on the protein components to be separated. Normally chromatography is carried out at a pH where the components of interest are likely to carry the largest charge difference, and the choice of exchanger is then made on the basis of the net charge carried at this pH. To assist in the choice of ion exchanger, monomers can be examined by electrophoresis at different pH values or by isoelectric focusing to determine isoelectric points of the components (section 1.3.2.4). An anion exchanger is chosen if the protein components carry a net negative charge or a cation exchanger if the components carry a net positive charge. Very often the isoelectric points are not known and it is necessary to perform a simple test to determine which type of ion exchanger to use.

Selection of ion-exchange type

The protein is mixed with both a cation and an anion exchanger over a range of pH values, and the amount of protein bound is determined by measuring the absorbance in the supernatant. From the absorbance measurements, it will be clear which ion exchanger with corresponding pH has taken up most protein. The method involves weighing out 0.05 g of each ion exchanger, e.g. DEAE-cellulose and CM-cellulose, into separate test tubes. To each is added about 30 ml of the appropriate buffer, e.g. 0.05 M-Tris-HCl (pH 8.5) for anion exchanger and 0.05 M-sodium acetate (pH 5.0) for cation exchanger, each containing 8 M-urea (to dissociate oligomer into monomers). After equilibration with two more aliquots of buffer, the ion-exchange resin is allowed to sediment and the supernatant removed until the total volume (resin + supernatant) is 10 ml. After adding equal amounts of protein to each test tube, the suspension is shaken and then allowed to sediment. Absorbance due to protein in each supernatant is measured with a spectrophotometer. Some proteins can be absorbed to both anion and cation exchangers, and in such cases separation should be explored on both types of exchangers. On the other hand, if the protein is not absorbed to either DEAE-cellulose or CM-cellulose the ionic strength of the buffer should be reduced or strong exchangers tried.

Ion-exchange procedure

Full details of the properties, physical characteristics and methods of handling for different ion exchangers are available from the manufacturers. For an excellent review of cellulosic ion exchangers refer to Peterson (1970).

Dry ion-exchange resins are normally precycled through acid and alkaline treatments before equilibration with the column buffer. Anion exchangers, e.g. DEAE-cellulose are stirred into 0.5 M-HCl (15 ml/g) and left for 30 min. Supernatant liquor is decanted or filtered off. Wash with water. Add an equal volume of 0.5 M-NaOH with stirring for 30 min. Remove supernatant and wash with water. Equilibrate the resin by stirring for 10 min with buffer, e.g. 20 ml/g. Decant or filter. This treatment is repeated at least three times until the pH and conductivity of the supernatant are the same as the buffer. Cation exchangers, e.g. CM-cellulose are treated similarly except that the first treatment is with 0.5 M-NaOH and the second treatment with 0.5 M-HCl. Pre-swollen ion exchangers can be used without an initial precycling but must be equilibrated completely with the column buffer as described above.

Next the fines are removed. The slurry is allowed to settle in a suitable measuring cylinder for approximately 1 h and the supernatant buffer solution containing fines is removed.

The resin is mixed with more buffer to give a slurry just thin enough to allow air bubbles to rise through it easily. After degassing the slurry using a water aspirator, the column is packed in the usual way with an extension tube. A column 15–25 mm × 500–1000 mm is normally sufficient. After packing, one or two bed volumes of starting buffer are run through at the flow rate to be used in the subsequent chromatographic step. The column bed height should then be constant.

The sample is dissolved in the starting buffer and, if necessary, dialyzed against this buffer overnight. If all components of interest are bound to the resin the sample volume is of little consequence, but if a component is only retarded the sample volume should be kept to 1–2 % of the column volume. The amount of sample applied should not normally exceed 10 % of the available bed capacity.

The eluent is monitored continuously at 280 nm (or as described in section 1.2.2.1) and fractions collected. Bound substances are eluted by changing the ionic strength or the pH of the eluent in a continuous or stepwise fashion. Stepwise elution may give misleading results and should be avoided.

Formation of gradients

Gradient elution can be accomplished using two connected vessels, commercially available gradient formers or a peristaltic pump.

A system for the formation of gradients with two connected vertical-walled

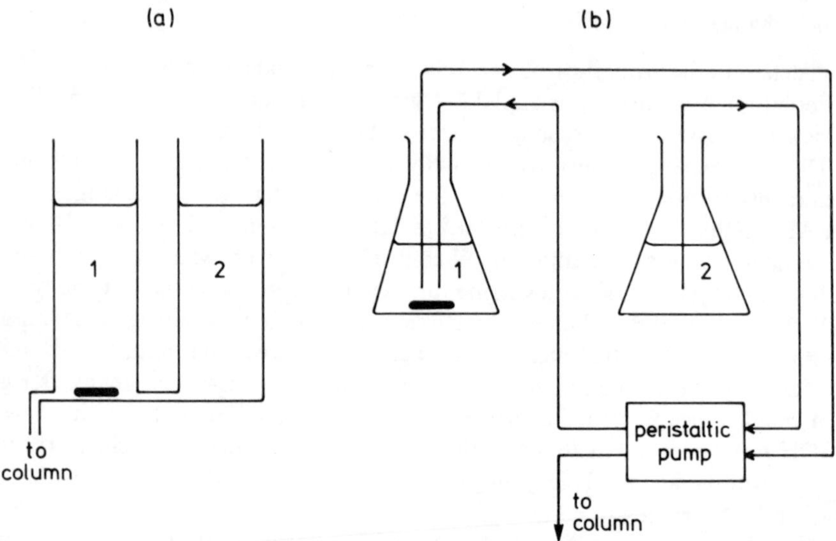

Fig. 1.4. Diagrammatic representation of methods to produce gradients using (a) two connected containers, (b) a peristaltic pump. Starting buffer is in container 1 and limiting buffer in container 2. Shape of gradient depends on cross-sectional areas of containers (a) or diameters of pumping tubes (b) See text for details.

containers is illustrated in Fig. 1.4(a). Container 1 must be stirred continuously. Cross-sectional areas of the containers are A_1 and A_2 while starting and limiting buffer concentrations are C_1 and C_2. When the densities of the two solutions are similar, this system delivers a solution of concentration C, where

$$C = C_2 - (C_2 - C_1)(1 - v/V)^{A_2/A_1}$$

where v is the volume of eluent passed and V is the total volume in containers 1 and 2. A linear gradient is formed when $A_1 = A_2$ while a convex gradient occurs if $A_2 > A_1$ and a concave gradient if $A_2 < A_1$ (Bock & Ling, 1954).

A multichannel peristaltic pump can be used to produce gradients as shown in Fig. 1.4(b). The form of the gradients is specified by the flow rates R_1 (from container 2 to container 1) and R_2 (from container 1 to column), concentrations C_1 and C_2 of starting and limiting buffers, and volume V_1 in container 1. After time t, the concentration in the gradient is given by (Lakshmanan & Lieberman, 1954):

$$C = (C_2 - C_1)\left[1 - \left[\frac{V_1 - t(R_2 - R_1)}{V_1}\right]^{R_1/(R_2 - R_1)}\right]$$

If $R_2 = 2R_1$, this is the condition for the formation of a linear gradient whose slope C/t is

$$\frac{(C_2 - C_1)R_1}{V_1}$$

while a concave gradient is obtained if $R_2 > 2R_1$ and a convex gradient if $R_2 < 2R_1$. If $R_1 < R_2$, then to have sufficient liquid in container 1 to form a gradient of total volume V

$$V_1 \geqslant \frac{R_2 - R_1}{R_2} V$$

R_1 and R_2 can be simply varied with most peristaltic pumps by changing diameters of the pumping tubes. Some peristaltic pumps give a large range of relative flow rates. If the peristaltic pump only accepts a limited range of tubing sizes, linear gradients can be easily generated by fitting three channels of the pump with the same size pump tubing, one channel pumping liquid from container 2 to container 1 and the other two channels pumping liquid from container 1 to the column.

1.3.2.3. Polyacrylamide gel electrophoresis

Analytical scale electrophoresis in 8 M-urea detects monomers with different charge/mass ratios at a particular pH, and is also of use in choosing the pH and type of ion-exchange resin as described in the previous section. Preparative electrophoresis may also be carried out by scaling up an analytical system or electrophoresing the protein off the gel into an elution chamber (section 1.3.2.1).

Optimization of experimental conditions for the best resolution of proteins has recently been discussed by Chrambach (1980). A computer output describing nearly 5000 discontinuous buffer systems covering the pH range 2.5–11 for positively and negatively charged species is available (Chrambach *et al.*, 1976) but normally systems at only a few representative pH values are necessary (e.g. see Maurer, 1971). A frequently used system but incorporating 8 M-urea is one first described by Davis (1964) and is given below.

It is often of interest to know the mol. wt. of components separated by charge at a particular pH. This can be simply accomplished using two-dimensional electrophoresis in which the proteins are separated in the first dimension in a gel rod by electrophoresis at a particular pH in 8 M-urea, and in the second dimension by electrophoresis in the presence of SDS. The use of two-dimensional gels is described in chapter 7, section 7.3.2.

Electrophoresis at pH 8.9 in 8 M-urea

General procedures, equipment, chemicals, staining and photography of gels have been considered in sections 1.2.1.1 and 1.3.1.1. The following method is based on that of Davis (1964) except that 8 M-urea is incorporated into the gel.

Separating gel buffer: 3 M-Tris is titrated to pH 8.9 with HCl.
Stacking gel buffer: 0.5 M-Tris is titrated to pH 6.8 with phosphoric acid.
Sample solution: 2 ml of stacking gel buffer + 6 g of urea + 5 ml of water + 5 mg of bromophenol blue.
Electrode buffer: 0.005 M-Tris/0.04 M-glycine.
Stock 30% acrylamide solution: 29.2% (w/v) acrylamide, 0.8% (w/v) *N,N'*-methylenebisacrylamide (as in section 1.3.1.1 p. 14).

Gel ingredients to make 10 ml of separating acrylamide (7.5%, 10%, 12.5%) are given below.

	7.5%	10%	12.5%
Stock 30% acrylamide solution	2.50 ml	3.33 ml	4.17 ml
Separating gel buffer	1.25 ml	1.25 ml	1.25 ml
Urea	4.8 g	4.8 g	4.8 g
Water	2.65 ml	1.82 ml	0.98 ml
TEMED	5 μl	5 μl	5 μl
Ammonium persulfate	5 mg	5 mg	5 mg

After adding the ammonium persulfate, the solution is quickly degassed using a water aspirator, and transferred to the electrophoresis apparatus. Overlay with water or isobutanol saturated with water. After polymerization pour off the overlaying solution, wash the gel surface with a small amount of stacking gel solution and then add degassed stacking gel solution. Ten milliliters of stacking gel are prepared by adding 5 mg of ammonium persulfate to a solution containing 1.3 ml of stock 30% acrylamide solution, 1.25 ml of stacking gel buffer, 4.8 g of urea, 3.8 ml of water and 5 μl of TEMED.

Samples are loaded with a microsyringe or disposable micropipette, and electrophoresis is performed at 100 V until the marker dye has traversed the stacking gel and then at 200 V until the dye is near the bottom of the gel.

Two-dimensional electrophoresis

Carry out the separation in the first dimension at pH 8.9 in 8 M-urea but other pH values may be used. In the second dimension use electrophoresis in the presence of SDS. In order to sharpen the protein zones before entering the second dimension separating gel, a discontinuous buffer system is used. The gel rod in which the first dimension has been carried out is set in place above the SDS stacking gel using agarose (1% w/v in SDS sample buffer: section 1.3.1.1) (Marshall, 1981).

Instead of using a discontinuous buffer system, a continuous buffer system can be used when the second dimension contains a gradient in polyacrylamide since sharpening of protein bands occurs as the protein nears its equilibrium position. The first dimension gel rod (3.5 mm diameter), after rocking for 30 min in 100 ml of the electrode buffer solution (0.15% SDS and 0.025 M-phosphate, pH 7), is pushed firmly onto the top of a 4–27% commercially prepared continuous gradient polyacrylamide gel (section 1.3.1.1) which has been pre-run at 50 mA for 3 h. Electrophoresis of the proteins is carried out at an initial current of 30 mA for 15 min and then at 50 mA for a further 3.5 h (Marshall, 1980).

1.3.2.4. Isoelectric focusing

In isoelectric focusing, a solution containing a mixture of carrier ampholytes (aliphatic polyamino-polycarboxylic acids) is subjected to an electric field. The different ampholytes move until their net charge is zero, the more acidic ones (lower isoelectric points) migrating towards the anode. As a consequence a pH gradient is formed between the electrodes. Ampholytes are commercially available (LKB, Serva, Bio-Rad, Pharmacia, Pierce).

When a protein is applied to a pH gradient formed by ampholytes, the protein migrates by electrophoresis to the position in the pH gradient corresponding to its isoelectric point. Isoelectric focusing is the most sensitive way of detecting charge heterogeneity within a protein, and if multiple bands are observed it is sometimes difficult to know whether these bands are due to different types of protein components or result from chemical modification of one protein. Isoelectric focusing is sensitive to modifications such as deamidation, different degrees of phosphorylation and cyanate reaction at amino terminal and lysine residues. Hence a protein should be initially characterized by electrophoresis before embarking on an isoelectric focusing study. It is noteworthy that, of 500 proteins studied (Gianazza & Righetti, 1980), 70% have acidic pI values, while 38% have pI values in the range pH 4.5–6.0.

Analytical isoelectric focusing in the presence of 8 M-urea is normally carried out in polyacrylamide gel although Sephadex may also be used as the support medium. Preparative isoelectric focusing may be carried out in a density gradient but in recent years this has been largely superseded by focusing in thick layers of Sephadex or similar solid support (see below and in the commercial literature of LKB and Pharmacia).

Initial experiments usually use wide pH range ampholytes, e.g. pH 3–10, while later experiments may use a narrow range, e.g. pH 5–7. Ampholytes are expensive and in order to minimize cost analytical experiments are conducted in gel rods. Preparative isoelectric focusing is conducted only when conditions for analytical isoelectric focusing have been thoroughly worked out.

Analytical method

Ten milliliters of solution are prepared by adding 5 mg of ammonium persulfate to a solution containing 1.65 ml of stock 30% acrylamide solution (p. 14), 4.8 g of urea, 3.75 ml of water and 1.0 ml of ampholyte. The anode buffer consists of 8 M-urea–0.03 M-phosphoric acid and the cathode solution 8 M-urea–0.05 M-NaOH. Samples are dissolved in 8 M-urea–0.03 M-phosphoric acid–20% glycerol (pH 4.0) or 8 M-urea–0.05 M-Tris–20% glycerol (pH 8.0) for loading at the acidic or basic end of the gel respectively. Gels are pre-focused for 2 h at 400 V, samples applied, and the focusing continued for 17 h at 400 V. Before using the staining procedure described on p. 6, it is necessary to remove the ampholytes by thorough washing with 10% (w/v) trichloroacetic acid. Alternatively, the gels can be stained directly (Righetti & Drysdale, 1974) for 3 h with 0.01% Coomassie Blue R-250 in 25% ethanol–10% acetic acid–0.1% copper sulphate and destained in 10% ethanol – 10% acetic acid.

The pH gradient can be determined on an unstained gel using a surface electrode (e.g. type LoT 403–30 from Ingold). Pieces of wire are inserted into several gels at about 1 cm intervals, the pH measured at these positions and the gels subsequently stained and photographed to locate the protein bands accurately (Marshall & Blagrove, 1979). Results from at least three gels should be averaged to obtain final apparent isoelectric points. These isoelectric points refer to particular protein components in 8 M-urea because urea affects the pK_a of ionizable groups (Bull *et al.*, 1964; Ui, 1971).

The mol. wts. of the separated components can be determined by two-dimensional electrophoresis where the second dimension electrophoresis is in the presence of SDS (O'Farrell, 1975). The first-dimension gel rod is set in place above the SDS stacking slab gel using agarose (as in section 1.3.2.3).

Preparative method

Commercial apparatus is available although it is only necessary to have an efficiently cooled horizontal plate, electrode reservoirs and a power supply.

Fifteen grams of Ultrodex or Sephadex IEF (specially prepared grade of Sephadex G–75) is swollen in 225 ml of 8 M-urea. Add 12 ml of the appropriate ampholyte pH range and mix thoroughly. Deaerate and pour the gel slurry into a gel trough (e.g. 200 × 200 × 5.0 mm) and spread evenly by means of a glass rod. Excess moisture is allowed to evaporate or dry Sephadex is sprinkled onto the surface (cover the mouth of the Sephadex bottle with a fine gauze and hold in place with an elastic band) until the correct consistency is reached, i.e. the gel bed does not move when the trough is tilted to an angle of 45°.

The sample may be applied as a streak directly on the surface of the gel layer or it may be mixed with dry Sephadex (1 ml of sample in 8 M-urea per 60 mg of Sephadex) and poured into a slot formed by cutting out some of the gel. Application of the sample may be at any position in the pH gradient. Load capacity is about 1 mg of protein per ml gel suspension.

Contact between the electrode buffers and the gel layer is by means of Whatman 3 MM paper protected from direct contact by means of dialysis tubing or cellophane sheet. Depending on the distance between electrodes and efficiency of cooling, focusing may be carried out at 300 V for 16 h followed by 600 V for 4 h.

After focusing, proteins are detected by the print technique. Whatman 3 MM is cut to size and rolled onto the gel bed (no air should be trapped between gel and paper). After 2 min the paper is removed, dried in an oven, washed (10 min) twice with 10% TCA to remove ampholytes, washed (5 min) with destain solution to remove TCA, stained with Coomassie Blue G 250 (section 1.2.1.1) for 10 min and then destained.

Gel zones containing substances of interest are cut out and transferred to a small column (e.g. disposable syringe fitted with glass wool). Buffer is added to resuspend the gel and the protein is then eluted. Ampholytes are removed from the protein by dialysis or gel chromatography on Sephadex G–10.

It is sometimes difficult to remove traces of ampholytes and the substitute system marketed by Pierce (Buffalyte) as described by Prestidge & Hearn (1979) should be evaluated where problems are encountered, e.g. in biological assays.

Further experimental details were given by Radola (1973, 1974) and are available from the manufacturers of commercial apparatus.

1.3.2.5. Fractional precipitation

Neutral salts exert striking and specific effects on the conformational stability of proteins (Von Hippel & Schleich, 1969), with anions and cations essentially being independently effective in their abilities to promote unfolding of native structures. Thus, while small amounts of salt can increase the solubility of a protein ('salting-in' effect), at higher concentrations, usually several molar, a 'salting-out' effect occurs which causes precipitation of the protein. The relative effectiveness of various neutral salts in salting-out was first studied by Hofmeister (1888). The lyotropic series for some cations is $Li^+ < Na^+ < K^+ < NH_4^+ < Mg^{2+}$ and for anions is $SO_4^{2-} < CH_3COO^- < Cl^- < NO_3^- < CNS^-$. The physicochemical basis of the precipitation in concentrated salt solutions is complex (Scopes, 1978), one factor being the decrease in activity of protein-associated water molecules which diminishes their solubilizing interactions with polar protein groups. Bull and Breese (1980) suggested that the interfacial tensions between protein molecules and protein–water are the crucial factors in the solubility of proteins and the lyotropic series arises from the effect of the ions on the interfacial tensions. The solubility (S) for many proteins (at high salt concentration) decreases logarithmically as the salt concentration increases:

$$\text{Log } S = \beta - K_s\,\omega$$

where β = solubility of protein in pure water (generally obtained by extrapolation to $\omega = 0$)

K_s = the salting out constant

and ω = ionic strength of the salt solution.

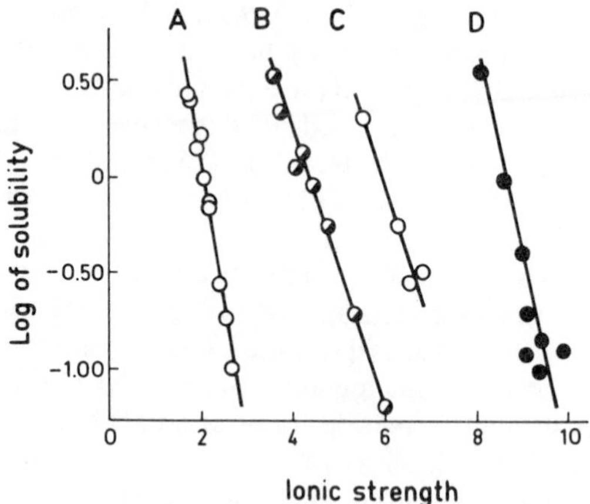

Fig. 1.5. The solubility of proteins (g/l) in ammonium sulfate solutions for fibrinogen (A), hemoglobin (B), serum albumin (C) and myoglobin (D) (Cohn & Edsall, 1943).

Fig. 1.5 (Cohn & Edsall, 1943) shows decreasing protein solubilities with increasing ionic strength of ammonium sulfate solutions. In general, precipitation of protein is most complete near its isoelectric point but this is not universally so.

Precipitation with ammonium sulfate has been used most often as a preliminary purification step, serving to concentrate the protein and remove it from the non-protein material. It may offer a suitable technique for separation of subunits if they have sufficiently different solubilities in ammonium sulfate (or other salt) solutions. Even in favorable situations, however, it is usually necessary to repeat the precipitation step because co-precipitation of other subunits can occur. Ammonium sulfate is usually chosen because of a number of favorable properties. Thus, it has a high salting-out constant, does not denature the protein, is cheap and available in pure form and has a high solubility and a small heat of solution. Also, a saturated solution has a low viscosity and density which is important if the subsequent separation of the protein is by centrifugation.

In practice there is a certain amount of trial and error in obtaining optimum separation conditions. Initially a solubility curve should be obtained for a range of concentrations of ammonium sulfate, bearing in mind that ammonium sulfate lowers the pH of an unbuffered solution. The 'per cent saturation' value for the ammonium sulfate (100% saturation corresponds to a 4.05 M solution at 20°C) required varies from protein to protein and with the pH (usually 5.5–7.5), temperature (usually room temperature) and protein concentration. Dixon (1953) prepared a nomogram for ammonium sulfate solutions. This makes it easy to relate the amount of solid ammonium sulfate to be added to a given volume to obtain the desired degree of saturation. Lowering the pH below 5 can be useful in precipitating the protein at lower salt concentrations by maintaining solvent densities that are lower and more suitable for separation of the protein by centrifugation.

Organic solvent fractionation

Proteins can also be fractionally precipitated with organic solvents such as ethanol and acetone. Propanol was used in conjunction with ammonium sulfate for proteins that tended to aggregate in solution (Dowling & Crewther, 1974). The organic solvents decrease the solubility by lowering the dielectric constant of the medium. Since precipitation occurs by aggregation originating from electrostatic interactions, similar to isoelectric precipitation, the pH for organic solvent fractionation is more important than for aqueous salt precipitations. Hence structurally very similar proteins with different isoelectric points may be separated using organic solvent precipitation. For example, parvalbumins II, III and IV were separated using increasing concentrations of acetone at pH 6.5 (Scopes, 1978). With 55–60% (v/v) acetone, protein IV was precipitated, 60–65% acetone precipitated III and IV, 65–70% acetone precipitated III and parvalbumin II, the most acidic with an isoelectric point of about 4, was precipitated with 70–80% acetone.

As a general rule, the lower the mol. wt. of the protein, the higher the solvent concentration required for precipitation. If too much solvent is added, protein–protein interactions may occur and prevent fractionation. Such problems may be overcome by using a dipolar ion such as glycine to act as a buffer and control the dielectric constant. If denaturation of the protein occurs at room temperature, the procedure can be carried out at 0°C or less. Addition of solvent causes a temperature rise initially (until a concentration of about 20% v/v is reached) and some contraction in volume also occurs. Temperature control of the reaction is important because the solubility of proteins in organic solvents decreases markedly with decrease in temperature, in contrast to salt fractionations where the effect of temperature upon solubility is highly variable.

1.3.2.6. Other methods

Affinity chromatography is considered in Chapter 5.

1.4. MONOMER STOICHIOMETRY WITHIN OLIGOMER

There are several ways of determining the monomer stoichiometry within an oligomeric protein. The method used will depend partly on the techniques available and partly on the complexity and size of the oligomer. Because of difficulties and uncertainties in different methods, stoichiometry should be determined by more than one method. Monomer stoichiometry within an oligomeric protein was often deduced from the mol. wts. of the intact oligomer and component monomers, but a procedure involving the determination of cross-linked products is being increasingly used. Reassociation of monomers, either chemically modified or from different animal species, to form hybrid oligomers followed by polyacrylamide gel electrophoresis may also be used under favorable circumstances. Hybridization and cross-linking methods are usually limited to smaller oligomers containing relatively few monomers. For oligomers containing more than one type of subunit, the ratio of monomer components can be determined after dissociation and polyacrylamide gel electrophoresis, either by densitometry of the Coomassie Blue-stained proteins, or by radioactivity measurements. The absolute number of subunits can then be calculated if estimates of the mol. wts. of oligomer and monomers are available. Amino acid analysis may also be used to estimate the ratio of different types of monomers within an oligomer.

In view of the large errors possible in mol. wt. determinations, the data should be assessed very carefully. Symmetry is important. Dimeric and tetrameric structures are the most common but there are a significant number of examples of hexamers, octamers and dodecamers. There have only been a few reports of trimeric, pentameric or heptameric oligomers (Gianazza & Righetti, 1980). Thus, if calculations suggest that there are eleven monomers this is most likely to be wrong. For larger structures, evidence may often have to be drawn from other sources such as X-ray diffraction analysis and electron microscopy (Baker *et al.*, 1975; Perham, 1975).

1.4.1. Molecular weight analysis

If the mol. wts. of the oligomer and monomer(s) are known, probably by methods outlined in sections 1.2 and 1.3.1, it is then a simple mathematical problem to determine the monomeric composition of the oligomer. For example, if an oligomer with mol. wt. M_r contains two types of monomers α and β (with mol. wt. M_α and M_β respectively) then

$$n_\alpha M_\alpha + n_\beta M_\beta = M_r$$

where n_α and n_β are the number of α and β type monomers, respectively.

It must be remembered that even with the most careful measurements, the mol. wts. of the oligomer and monomer may be in error by 5–10%. Apart from the error of the measurement there is uncertainty in the absolute accuracy especially when empirical methods (e.g. SDS gel electrophoresis, gel filtration) are used. The protein under investigation may not have the same hydrodynamic shape as standard proteins or an unusual composition may affect its behavior.

Consider how 5% or 10% errors affect the estimation of the number of subunits. For an oligomer with a mol. wt. of 300 000 containing 6 subunits of mol. wt. 50 000, a 5% error in the molecular weights gives a range of 5.4–6.6 in the number of monomers and a hexamer could be anticipated with reasonable certainty. However, a 10% error in the mol. wt. leads to a range of 4.9–7.3 and on symmetry requirements the number of monomers could be 4, 6 or 8. The uncertainty in the stoichiometry increases if there are two or more different types of monomers.

1.4.2. Cross-linking with bifunctional reagents

A method that is becoming increasingly popular for determining the number of monomers in an oligomer involves cross-linking with bifunctional reagents, SDS polyacrylamide gel electrophoresis and staining (Davies & Stark, 1970). A set of protein bands is obtained with mol. wts. representing integral multiples of the monomer mol. wt. up to the mol. wt. of the oligomer. With oligomers composed of identical, or near-identical monomers, the number of principal bands observed should be identical to the number of monomers in the oligomer.

Apart from determining the number of subunits, the quaternary structure of the oligomeric protein can often be established from cross-linking patterns (Hucho *et al.*, 1975; Hadju *et al.*, 1976), but this will not be discussed here. Nevertheless, when dealing with a simple system like a tetramer, or maybe a hexamer or an octamer, it would be worthwhile to attempt to establish the arrangement of the monomers in the oligomer.

Most chemical cross-linking has been with bisimidoesters of various chain lengths and in particular dimethyl suberimidate ($n = 6$ in the equation below). These are highly specific modifying agents of proteins which react with α-NH_2 groups and the ε-amino groups of lysine residues to form amidines ($pK_a > 11$).

$$2\,RNH_2 + CH_3\!-\!O\!-\!\overset{\overset{+}{N}H_2}{\overset{\|}{C}}\!-\!(CH_2)_n\!-\!\overset{\overset{+}{N}H_2}{\overset{\|}{C}}\!-\!O\!-\!CH_3 \longrightarrow RNH\!-\!\overset{\overset{+}{N}H_2}{\overset{\|}{C}}\!-\!(CH_2)_n\!-\!\overset{\overset{+}{N}H_2}{\overset{\|}{C}}\!-\!NHR + 2\,CH_3OH$$

Other compounds include acyl azides, sulfhydryl reagents and glutaraldehyde (Peters & Richards, 1977). Relatively stable disuccinimidyl esters, forming either cleavable or non-cleavable cross-links were described (Hill *et al.*, 1979).

Some of the bisimidoester molecules are removed from the system by hydrolysis (Peters & Richards, 1977) of both functional groups, or at one only, the other one reacting with an amino group. This results in only an insignificant increase in the mol. wt. of a product. Other reagent molecules react with two protein amino groups forming cross-links which may be either intra- or intermolecular.

Only the intermonomer cross-links have a marked effect on the mol. wt. of the final product when examined under denaturing conditions. Intramonomer cross-links may exert a physical constraint on a denatured SDS-monomer complex thereby reducing its hydrodynamic volume and giving a smaller apparent mol. wt. (cf. effect of disulfide bonds).

Some of the products from a limited cross-linking reaction of tetramers of the type α_4 and $\alpha_2\beta_2$ are shown in Fig. 1.6. The various products are separated by increasing apparent mol. wt. distribution, which is dependent on the number and position of the cross-links.

However, this range of apparent mol. wts. and hence overlap of the cross-linked product will still permit the determination of the number of monomers in a simple

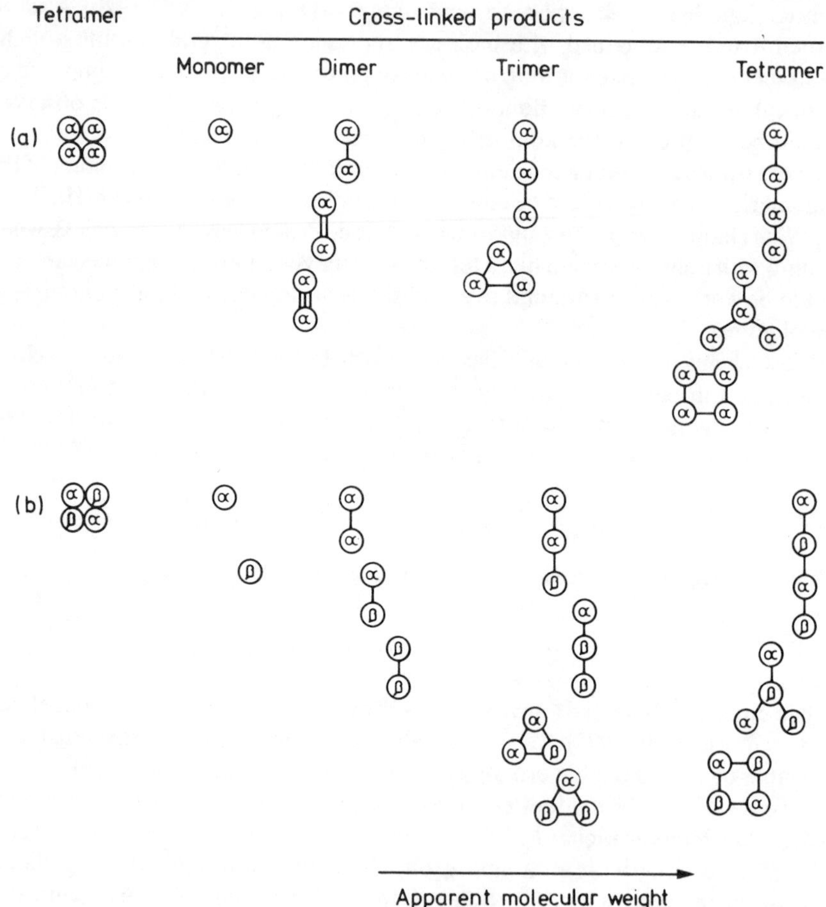

Fig. 1.6. Schematic diagram of some of the products from a limited cross-linking reaction with a bifunctional reagent for two types of tetramers (α_4 and $\alpha_2\beta_2$). Different arrangements of intermonomer cross-links lead to different products with apparent different mol. wts.

oligomer. Oligomers containing monomers of different mol. wts. (i.e. $\alpha_2\beta_2$) will give a more complex range of products, and the interpretation of results will be difficult.

1.4.2.1. Variable factors involved in cross-linking

Cross-linking involves consideration of pH, length and concentration of bifunctional reagent and protein concentration.

pH of reaction

Bisimidoesters react with ε-amino groups of lysine residues (pK_a 9.5–10) and the α-amino group in proteins above pH 8 when some of the groups are unprotonated. The reaction product is an amidine with pK_a > 11, and thus below pH 8 the positive charge of the original protein amino group is retained. At pH 8 or lower, side reactions can occur (Peters & Richards, 1977; Siezen, 1979), while for maximum formation of amidine the reaction should be carried out at or above pH 10. However, many oligomers dissociate at this pH, and it is necessary to compromise with a reaction pH 8–10. If the oligomer can tolerate high pH for only 1–2 min, then as the formation of the amidine is fast, it may be possible to carry out the cross-linking at high pH. It is generally worthwhile to study the effect of pH on the products from cross-linking (Sutoh *et al.*, 1978; Schwam *et al.*, 1979).

Length of bifunctional reagent

The proximity of amino groups in the native oligomer determines the length of bifunctional reagent required to form a cross-link. Maximum lengths of some reagents are: dimethyl malonimidate 0.5 nm; dimethyl succinimidate 0.6 nm; dimethyl adipimidate 0.9 nm; dimethyl suberimidate 1.1 nm; dimethyl dodecylimidate 1.5 nm (Peters & Richards, 1977; De Abreu *et al.*, 1979). The use of a single bifunctional reagent might lead to ambiguous or even erroneous conclusions and ideally a homologous series of reagents covering a reasonable range in length should be employed.

Concentration of bifunctional reagent

The concentration of the bifunctional reagent affects the extent of cross-linking, and in general the higher the initial concentration the greater the extent of cross-linking. However, Peters and Richards (1977) have pointed out that increasing the initial excess of reagent to oligomer will not speed up the cross-linking and large reagent excesses are only likely to provide unwanted side reactions. It is better to sequentially add small increments of reagent. If the extent of cross-linking cannot be increased beyond an intermediate level by manipulating the reagent concentration, then there may be a preference for intramonomer rather than intermonomer cross-linking.

Protein concentration

Protein concentration should be kept low to avoid interoligomer cross-linking. If the mol. wt. of the oligomer is known, an indicator of the formation of interoligomer cross-links is the appearance of species with mol. wts. higher than the oligomer. Cross-linking over a protein concentration range is a test by which cross-linking within and between oligomers can be distinguished. To reduce interoligomer cross-linking, the oligomer may be initially coupled to an insoluble matrix (Pillai & Bachhawat, 1979), e.g. by disulfide groups to Sepharose 4B (see section 3.2.5) before cross-linking with bisimidoesters (Chantler & Gratzer, 1978).

1.4.2.2. Methods

Preparation of bisimidoesters

Bisimidoester hydrochlorides can be readily prepared from corresponding commercially available dinitriles according to McElvain and Schroeder (1949). Dimethyl adipimidate and suberimidate may be purchased.

Dissolve suberonitrile (0.5 g) in an ice-cold mixture of 2 ml of methanol (dried over Linde 3A molecular sieve) and 15 ml of anhydrous ether. Pass dry HCl gas through the solution for 30 min. Allow the reaction mixture to stand at 4°C for 24 h. Add 10 ml of dry ether to precipitate the product, which is washed extensively with dry methanol/ether (1:3, v/v) under anhydrous conditions. The bisimidoester hydrochloride (0.81 g, 81 %) is used without recrystallization.

Amidination of proteins (Davies & Stark, 1970)

The protein (final concentration about 0.5 mg/ml) is dissolved in, or thoroughly dialyzed against, 0.2 M-triethanolamine HCl at pH 8.5. Immediately before use, the bisimidoester HCl is dissolved in the same buffer to a concentration of about 10 mg/ml, and the pH readjusted to about 8.5 with NaOH. The required volume of bisimidoester solution, to give a final concentration of 1 mg/ml, is added to the protein solution, and the reaction is allowed to proceed at room temperature for 3 h. If the bisimidoester is to be sequentially added, allow the reaction to proceed for 30 min before making the next addition.

When studying concentration effects, the protein concentration can be varied between 0.1 and 5.0 mg/ml and the bisimidoester final concentration between 0.2 and 5.0 mg/ml. The pH of the cross-linking reaction can be varied between 8 and 10.5, using 0.2 M-triethanolamine buffer between pH 8 and 8.5 and 0.2 M-borate for pH values above 8.5. Relatively concentrated buffers should be used to maintain constant pH during the course of the reaction and buffers must not contain amino groups.

After a reaction time 3 h, it is not necessary to terminate the reaction since any bisimidoester which has not reacted with the protein is hydrolysed. However, if the time course for cross-linking is being followed, take aliquots and stop the reaction by the addition of a two-fold excess of hydroxylamine relative to initial bisimidoester concentration.

SDS gel electrophoresis

The protein products from the cross-linking experiment are examined most easily by SDS gel electrophoresis, although gel filtration or ultracentrifugation in the presence of denaturants may be used. The methods for SDS gel electrophoresis are described in section 1.3.1.1. Since there are likely to be cross-linked products with mol. wts. above 100 000, 5 % polyacrylamide gels are normally used and about 50 μg of protein is loaded. For a simple oligomer containing identical or near identical monomers, the number of monomers forming the oligomer is indicated by the number of principal species detected. If only a small number of cross-links are present, the mol. wts. of the cross-linked products, which should be integral multiples of the monomer mol. wt., can be determined from a calibration curve. A large number of cross-links may lead to anomalous behavior of the SDS-complexes during electrophoresis.

If the protein concentration is sufficiently high, the reaction mixture can be loaded without dialysis on a discontinuous SDS gel. Most likely it will be necessary to concentrate the protein and this can be conveniently done by precipitation with trichloroacetic acid. The precipitate can then be dissolved in the SDS sample buffer before analysis by the continuous or discontinuous SDS gel method (section 1.3.1.1).

1.4.2.3. Limitations

The cross-linking method has been used to investigate a number of oligomeric proteins of known composition. In principle the method can be used to determine the number of subunits in any oligomeric protein but at times it may fall short of providing the required information. Cross-linking may not occur because amino groups of the protein are scarce, not suitably disposed, or blocked or the bifunctional reagent is not long enough to cross-link the amino groups or it is contaminated with monofunctional reagent.

Some oligomeric proteins do not form intermonomer cross-links for unknown reasons (Hucho *et al.*, 1975), although in some cases the absence of a cross-link may imply unsuitable local protein geometry (Pillai & Bachhawat, 1979). Caution must be exercised in the use of such negative cross-linking results (Peters & Richards, 1977).

The pH may affect association or dissociation of the oligomeric protein, and this could produce unexpected results. It was observed (Sutoh *et al.*, 1978) that pH may also influence the conformation of an individual monomer to such an extent that the cross-linking reaction does not occur.

The cross-linking method is limited to oligomers with mol. wts. of less than about 500 000 because above this molecular weight SDS gel electrophoresis is not capable of analyzing protein species.

1.4.3. Hybridization

Knowledge of the subunit composition of oligomers may be obtained from studies of the hybridization *in vitro* of two different purified forms of the single protein. These may be variants, isozymes or oligomers chemically modified so that the properties of the two forms are sufficiently different to allow easy detection of the products (Klotz *et al.*, 1975). The number of hybrids formed can be used to determine the number of monomers in the oligomer. For example, for a tetramer with identical subunits one would expect to obtain five products, $\alpha\alpha\alpha\alpha$, $\alpha\alpha\alpha\beta$, $\alpha\alpha\beta\beta$, $\alpha\beta\beta\beta$ and $\beta\beta\beta\beta$.

The situation becomes more complicated when the monomers are not identical (e.g. aspartate transcarbamylase with 6 catalytic and 6 regulatory subunits). If they are different but structurally and functionally equivalent, the number of hybrids obtained is specified by the equation

$$N = \frac{(m+s-1)!}{s!\,(m-1)!}$$

where s is the number of subunits and m is the total number of different types of subunit.

Often the required variants (or isozymes) are not readily available so the chemical modification procedure can be more universally applicable. Succinic anhydride was used by Meighen and Schachman (1970) in a study of the subunit structure of aldolase.

$$\begin{matrix} CH_2-C(=O) \\ | \quad\quad\quad\; >O \\ CH_2-C(=O) \end{matrix} + NH_2-R \longrightarrow \begin{matrix} CH_2-C(=O)-OH \\ | \\ CH_2-C(=O)-NH-R \end{matrix}$$

In this reaction (irreversible) the positively charged lysine side chains are converted to negatively charged side chains. Other dicarboxylic acid anhydrides may be used which give a reversible reaction. This type of reaction can lead to dissociation of the oligomer (see below) because of the repulsion effect of many of these like-charged groups. The problem of dissociation may be overcome by partial succinylation but this in turn could cause problems of obtaining homogeneous preparations. The method will fail if the quaternary structures of the modified and unmodified proteins are not identical or will not reform after dissociation into the monomers.

1.4.3.1. Methods

The following methods are examples of two different procedures described for forming hybrid proteins, the first using a variant, the second a chemically modified protein.

Hybridization of hemoglobin (Vinograd *et al.*, 1959)

A 1:1 mixture of hemoglobin S and radioactively labelled hemoglobin A is dialyzed for 48 h at 3°C against 0.1 M-sodium acetate buffer pH 5 to dissociate the α- and β-chains. Dialyze against water for 24 h to reassociate the chains. Chromatographic separation yields radioactive hemoglobin S, the mixed hybrids being formed by exchange of radioactive α-chains of hemoglobin A with α-chains of hemoglobin S.

Hybridization of aldolase (Meighen & Schachman, 1970)

Unmodified aldolase and succinylated aldolase (initial protein concentration 0.4 %) are mixed (in varying molar ratios) in 0.02 M-potassium phosphate buffer (pH 6.5) containing 4.0 M-urea, 0.002 M-EDTA and 0.1 M-dithiothreitol at 4°C. After 30 min the solutions are dialyzed overnight against buffer containing 0.5 M-NaCl and 0.2 mM-dithiothreitol to remove the urea, then two or three changes of the buffer without NaCl.

1.4.4. Dissociation and reassembly

The disruption of tertiary and quaternary structures of oligomers can occur in urea, guanidine HCl and SDS solutions (section 1.3), and by chemical modification as described above (section 1.4.3). Characterization of the products in these systems can indicate the relationship of the monomers to the oligomer. The pH can also be an important parameter in the observation. Both wheat germ agglutinin and the chemically modified succinylated lectin are dimeric at pH values above 5, whereas dissociation occurs when the pH is less than 5 (Monsigny *et al.*, 1979). The difference in stability of proteins at different pH values affects the conditions for refolding and reassociation. Hemocyanin, an assemblage of 8 hexamers, is dissociated into monomers and dimers at pH 8.9 (in the presence of EDTA to complex the essential calcium ions). Alternatively, changes in ionic strength, pH and calcium ion concentration cause hemocyanin molecules to dissociate into a number of distinct states of aggregation of hexamers (Bijlholt *et al.*, 1979).

An extension of the chemical modification approach was demonstrated by Wieland *et al.* (1979) on the multi-enzyme complex of yeast fatty-acid synthetase. Instead of using succinic anhydride, which gives irreversible modification of lysine side chains, they used other anhydrides which gave cleavable groups. Dimethylmaleic anhydride was preferred to citraconic anhydride. Dimethylmaleic acid monoamides can be hydrolyzed under much milder conditions than the corresponding citraconyl amide linkages. After acetylation of the complex with dimethylmaleic anhydride, the modified polypeptide chains were separated into the α- and β-subunits by raising the pH and decreasing the ionic strength. The acetylation reaction is reversed by mild acid treatment (pH 4.6) thus raising the possibility of reassembling the oligomer from the subunits.

$$\vdash NH_2 + \text{dimethylmaleic anhydride } (CH_3\text{–}C\text{–}C(=O)\text{–}O\text{–}C(=O)\text{–}C\text{–}CH_3,\ C{=}C) \underset{\text{pH 4.6}}{\overset{\text{pH 7.5}}{\rightleftharpoons}} \vdash NH\text{–}C(=O)\text{–}C(CH_3){=}C(CH_3)\text{–}COOH$$

Just as the native conformation of monomeric enzymes can be relatively easily regenerated from the constituent polypeptide chains (section 1.5.1.3), apparently because of a favorable thermodynamic configuration, oligomeric enzymes *in vitro* may be shown to have the same specificity after they have been subjected to reversible denaturation and refolding procedures (Cook & Koshland, 1969).

Some parameters such as pH may be more important than others for individual oligomers. After dissociation of fumarase in urea solution (Yamato & Murachi, 1979) dialysis against water did not regenerate the original active tetramer but it did against buffer without urea (see Section 1.4.4.1). The initial product formed is only partially active and is largely a dimer although fluorescence and circular dichroic measurements are virtually identical with the native tetrameric enzyme.

1.4.4.1. Methods

Conditions for dissociation and reassembly of enzyme structures are given. The first example effects dissociation by using a denaturant and the second by chemical modification.

Method: Yamato & Murachi, 1979

Swine heart fumarase (0.5–5 μg/ml) is inactivated after 30 min at 25°C in the presence of 6 M-urea (or guanidine HCl), 50 mM-sodium phosphate buffer (pH 7.3), 10 mM-dithiothreitol, 0.1 mM-EDTA, 10 % glycerol and 200 mM-KCl.

To restore activity the protein subunits are first dialyzed against reactivation buffer (reaction medium above minus the denaturant) for 24 h at 4°C, then incubated at 25°C for a further 3 h.

Method: Wieland *et al.*, 1979

To a solution of fatty acid synthetase (5 to 6 mg/ml in 0.3 M-potassium phosphate buffer (pH 7.5) containing 10 mM-2-mercaptoethanol) is added 13 μl of a 0.8 M-solution of dimethyl maleic anhydride in dry tetrahydrofuran (freshly prepared). After 30 min at 0°C the mixture is equilibrated with 100 mM-Tris-HCl buffer (pH 8.0) containing 1 mM-dithiothreitol. Remove salts by gel filtration on Sephadex G-50.

Reactivation of the enzyme

The modified side chains must first be hydrolyzed. Subunits are diluted with one vol. of 2.0 M-Tris-acetate buffer (pH 8.1). If the resulting protein concentration is lower than 1 mg/ml, bovine serum albumin is added to make an end concentration of 1 mg total protein/ml. This solution is carefully titrated at room temperature with a 100% saturated (23°C) ammonium sulfate solution adjusted to pH 2.0 with conc. sulfuric acid. Protein begins to precipitate at about pH 6.5. Allow the suspension to stand at pH 4.6 for 30 to 60 min at room temperature and then centrifuge for 30 min at 28 000 × g at 4°C. After centrifugation resuspend the pellet in ice-cold reactivation buffer containing 100 mM-potassium phosphate (pH 7.5), 10 mM-dithiothreitol, 0.5 mM-potassium EDTA and 10 μM-flavin mononucleotide. Occasional addition of 0.1 mg/ml bovine serum albumin facilitates the solubilization of the pellet. Residual ammonium sulfate is removed by gel filtration on Sephadex G-50 in reactivation buffer. The desalted protein solutions (1–2 mg/ml) are incubated in reactivation buffer at 25°C.

1.4.5. Polyacrylamide gel electrophoresis

If there are two or more different types of monomers (or polypeptide chains) and the species can be separated by polyacrylamide gel electrophoresis, the ratio of the separated species may be determined by radiolabelling or densitometry. Both methods depend on the isolation of purified species at some stage because for the radiolabelling method a knowledge of the number of radiolabelled residues per molecule is required, while for densitometry the determination of dye binding curves for each species is necessary.

1.4.5.1. Radiolabelling

Each species under investigation must contain at least one amino acid residue capable of being radiolabelled although it is better to have many potential modified residues so that the sensitivity is enhanced. Radiolabelling must be under strongly denaturing conditions so that all components are modified to the same extent. Furthermore, the modification reaction must be specific for a particular amino acid residue and the reaction must go to completion. It is best that no change of charge at the modified residue occurs or if a charge change occurs the electrophoretic separation of the modified non-radioactive species has been investigated.

Most proteins contain a large number of lysine residues which can be specifically labelled by amidination without a change of charge. Reaction conditions can be chosen so that the *N*-terminus is not modified. Another possible site of specific modification is a cysteinyl residue. Cysteine modification reactions are discussed in section 1.5.1 but those of interest here are alkylation with iodoacetate or iodoacetamide; the former increases the negative charge of the protein while the latter does not alter the charge. Labelling may also be effected by reduction of disulfide bonds followed by alkylation but clearly this method will give complications if there are disulfide-bonded polypeptide chains within a monomer. It may be possible to investigate the number of inter- and intrachain disulfide bonds by radiolabelling with and without the presence of a denaturant. The inter- and not the intrachain disulfide bonds of human serum cholinesterase are reduced and alkylated in the absence of denaturant (Lockridge *et al.*, 1979).

If the extent of the modification reaction is assumed to be the same for each species, the radioactivity (D_i in dpm) in species i is given by

$$D_i \propto L_i \cdot X_i$$

where X_i is the number of moles of species i and L_i is the number of potentially modified groups per mole of species i. To determine L_i, amino acid analysis of the purified species i needs to be carried out as well as estimation of the molecular weight. For any two species i and j

$$\frac{X_i}{X_j} = \frac{D_i \cdot L_j}{L_i \cdot D_j}$$

Radiolabelling of proteins

Alkylation of cyteinyl residues with iodoacetate is considered in Chapter 2, section 2.2.1.1. Radioactive ^{14}C-iodoacetate may be incorporated with cold carrier.

A sensitive method of peptide mapping was described which involved amidination with methyl-l-^{14}C-acetimidate, thin layer electrophoresis, chromatography and autoradiography (Bates *et al.*, 1975b).

Amidination of proteins (Bates *et al.*, 1975a)

Protein samples of up to 1 mg are dialysed against 1 mM-EDTA, pH 7.0 and lyophilized. The samples are dissolved in 0.1 ml of 0.2 M-sodium borate buffer (pH 8.5) containing 5 M-guanidine HCl, 2 mM-EDTA and 0.01 % (w/v) sodium azide. A solution of 1 M-methyl-1-^{14}C-acetimidate is prepared and 0.01 ml of this solution is added to the protein samples (final reagent concentration 0.1 M). Reaction is allowed to continue for at least 5 h at room temperature (20°C).

$$CH_3\text{-}\overset{\overset{+NH_2}{\|}}{\underset{*}{C}}\text{-}OCH_3 + NH_2-R \rightarrow CH_3\text{-}\overset{\overset{+NH_2}{\|}}{\underset{*}{C}}\text{-}NH-R + CH_3OH$$

After radiolabelling, the sample is diluted to an appropriate concentration for polyacrylamide gel electrophoresis. The sample may be dialyzed against the electrophoresis sample solution to remove excess radiolabel.

Determination of radioactivity D_i (Bates *et al.*, 1975a)

After polyacrylamide gel electrophoresis, the proteins are visualized by staining, excised and the pieces of gel dried under vacuum in plastic scintillation vials. The dried gels are dissolved by incubation for about 6 h at 60°C with 0.3 ml of 100 vol hydrogen peroxide. Control experiments showed that there was no loss of radiolabel caused by this treatment. Scintillant (3.0 ml of toluene-triton (2:1, v/v) containing 2,5-diphenyloxazole (5 g/l) is added and the clear samples are counted in a scintillation counter. The ^{133}Ba external standard calibrated against a series of sealed quenched samples (Amersham-Searle) is used to correct for quenching. The efficiency is generally 70–75 % and blank gel slices put through the entire procedure are counted for background radioactivity.

1.4.5.2. Densitometry

Coomassie Blue is often used to detect proteins in polyacrylamide gel but as with all protein stains marked deviations from Beer's law at higher protein concentrations are observed. Linearity is in the range of about 1–50 μg of protein, depending on the type of protein, gel dimensions and migration distance of the protein. Densitometry of stained bands cannot be used for quantitation in which bands have widely different amounts of protein. Dye binding characteristics of proteins vary markedly. It is often assumed that dye-protein binding involves an electrostatic association between the dye and basic amino acid residues of the protein (i.e. basic proteins stain more intensely than acidic proteins) but there may be hydrophobic association of dye with protein. It is thus usually necessary to determine dye binding characteristics of each separated protein species.

Method

Using a slab gel, a range of weights of purified protein (e.g. 1, 2, 5, 10, 20, 30, 40 μg; the initial protein concentration having been determined by refractometer measurement, dry wt. analysis, spectrophotometer determination or amino acid analysis) is loaded in different tracks, and the gel stained by the procedure described in section 1.2.1.1. Each track is scanned by a densitometer, the maximum absorption for the dye-protein complex being 590–595 nm with minimum absorption around 400 nm. The area under a peak is obtained using an integrator or by cutting out the peak from the paper and weighing. It was reported (Fishbein, 1972) that the range over which linearity in Beer's law plot is followed is dependent on the migration distance of the protein. Staining and destaining of gels (section 1.2.1.1) must be carried out reproducibly otherwise the peak area per unit weight of protein may vary from gel to gel. If possible the dissociated oligomer should be on the same gel. From an examination of the densitometric scan of the dissociated oligomer the ratio (by weight) of the protein species can be determined. Once this weight ratio of the monomers is obtained the molar ratio may be calculated using the monomer mol. wt.

1.4.6. Amino acid analysis

An estimate for the ratio of different monomers (or polypeptide chains) in an oligomer can be obtained from amino acid composition data if the compositions of the oligomer and separated monomers are known. This method is particularly suited when the amino acid compositions of the monomers are quite different in some amino acids (Burgess, 1969; Wright & Boulter, 1974).

For an oligomer composed of two subunits α and β, then for any particular amino acid

$$X = aX_\alpha + bX_\beta$$

where X, X_α and X_β represent the mole percentage of that amino acid in the oligomer, subunit α and subunit β, and a and b are coefficients representing the proportions (by weight) of α and β. Best fits for a and b can be obtained by regression analysis, and these values can then be expressed on a molar basis.

1.5. OLIGOMER AND MONOMER CROSS-LINKS

An end-group determination on the purified subunit establishes the presence of cross-links between protein subunits when it yields more than one *N*-terminal amino acid residue. A quantitative yield of more than one mole of an *N*-terminal amino acid per mole of protein would suggest that there are two chains both beginning with the same amino acid; likewise a yield of only 50% of a stable amino acid (not serine, for example) could indicate that there are cross-links but one of the chains has a blocked *N*-terminal amino acid. Because the yield of amino terminal residue can be affected by many different conditions, great care should be exercised in the interpretation of the results.

A variety of cross-links has been found in proteins, including disulfides, aldols and aldimines. The disulfide cross-link is the only one that may be cleaved easily and for that reason is the one commonly selected. Methods of separation for the monomers (section 1.3.2) may be applied in many instances to the preparations of the constituent chains.

1.5.1. Disulfide bonds

Cleavage of the disulfide bonds of cystine is usually achieved by oxidation, reduction or other nucleophilic displacement reactions with reagents such as HSO_3^-, CN^-, BH_4^-, H^- (Crewther, 1976; Liu, 1977; see also Chapter 2). Oxidation of a cystinyl residue yields two stable cysteic acid residues (Schram *et al.*, 1954). The sulfonic acid groups confer more hydrophilicity on the molecule and increase the water solubility, particularly at low pH. Disadvantages of oxidation are that it can lead to modifications of amino acids besides cystine and the acidic conditions used may cause partial hydrolysis of labile peptide bonds. If the introduction of sulfonic acid groups is mandatory, reduction of the disulfide bond followed by oxidation with a sultone reagent is preferable (see p. 51 & 75).

Usually disulfide bonds are cleaved by reduction and the extent of reduction may be estimated with Ellman's reagent (Habeeb, 1972; Leighton *et al.*, 1980; see also p. 76). Reduction is usually followed by alkylation of the thiol groups formed to prevent reoxidation to cystine. Various alkylating reagents have been employed, the particular one selected depending on some secondary purpose, for example, enhancement of the solubility properties in aqueous media, production of a site for trypsin attack or the provision of a reversible protecting group for the thiol. However, the alkylation reaction may not always be essential because the reduced protein is stable in cold acetic acid at pH 3 (Anfinsen & Haber, 1961).

1.5.1.1. Cleavage of disulfides

Oxidation

The method of Hirs (1967) is recommended for the oxidation of disulfide bonds or of cysteine. The method is described on p. 78.

Amino acid analysis of a portion of the material is performed after oxidation to ensure that the reaction has proceeded satisfactorily. The yield of cysteic acid should be greater than 85% of theoretical and that of methionine *S,S*-dioxide (from methionine) greater than 95%. Tryptophan is destroyed. Halides must be absent or modification of tyrosine may occur. Prior removal of halide ions is therefore recommended. A column of Dowex 2 (100 mesh) is prepared in the acetate form by washing with 3 M-sodium acetate until the eluate is devoid of halide ions (monitored using acidified silver nitrate). The protein solution is passed slowly through the resin bed (1 ml to 0.1 g protein) which is flushed with solvent to complete the recovery of proteins. Take care not to lose any of the protein on the column.

Reduction

2-Mercaptoethanol has been used extensively. It can form mixed disulfides if not used in sufficient excess and has an unpleasant odor. Dithioerythritol and dithiothreitol are suitable alternatives requiring only a small excess for quantitative reduction. More recently tributylphosphine, has been used extensively by Rüegg & Rudinger (1974, 1977a,b) and Rüegg *et al.* (1979). See Chapter 2, section 2.2.1 for discussion and methods.

Disulfide bridges linking two or more polypeptide chains are often accessible to solvents and reagents without unfolding of the polypeptide chain whereas intrachain disulfide bridges often are not. This difference makes it possible to reduce the exposed disulfide bonds selectively without affecting those in the interior. Complete reduction of disulfides is usually best performed under slightly alkaline conditions in the presence of urea or guanidine HCl. Neither denaturant has been free from criticism. Care must also be taken to remove urea entirely because any urea adsorbed to the proteins could adversely affect sequencing by decomposing to cyanate during the coupling step of the Edman degradation reaction. Our experience with guanidine HCl is less extensive but there have been no sequencing problems encountered after using pure reagent.

Edelman *et al.* (1968) described a method for reducing interchain and not intrachain disulfide bridges in γ_G-immunoglobulin.

The fragments obtained by reductive fission can be analyzed for sulfhydryl content or separated as such by maintaining a reducing medium throughout or by working at low pH. It is often more convenient to convert to more stable and more soluble derivatives. S-carboxymethylation (Gurd, 1972) is the most widely studied alkylation reaction but it is not without pitfalls. It should also be noted that *S*-carboxymethylcysteine decomposes considerably in the Edman degradation procedure unless converted to the phenylthiohydantoin under mild conditions (Inglis, 1976). Radiolabelling with ^{14}C-iodoacetate is helpful in this context but the use of alternative reagents might have advantages. If renaturation studies are to be made subsequently it is possible to prepare derivatives that allow a reversible modification of the thiol. Several of the reagents recommended should be handled with care, e.g. iodoacetate is an allergen and precautionary measures such as working with gloves in a fume hood should be taken as a general rule. See Chapter 2 for methods.

1.5.1.2. Reversible modification

S-sulfoproteins can be prepared by the action of sulfite on cystinyl residues. See also discussion in Chapter 2, section 2.2.4.

$$|\text{-S-S-}| + SO_3^{2-} \rightarrow |\text{-S}^- + {}^-O_3\text{S-S-}|$$

The disulfide bonds may be reformed by reduction and then renaturation of the protein usually by atmospheric oxidation. Oxidative sulfitolysis converts a disulfide into *S*-sulfonated derivatives (Bunte salts).

$$\vdash\!S\!-\!S\!\dashv + 2SO_3^{2-} + H_2O \xrightarrow{[O]} 2\vdash\!SSO_3^- + 2OH^-$$

This reaction proceeds to completion in the presence of cupric ions (Cole, 1967). Chan (1968) treated the protein with sodium sulfite and catalytic amounts of cysteine in the presence of oxygen and 8 M-urea. Treatment of cystine with DTT followed by sodium tetrathionate also gave quantitative conversion to *S*-sulfocysteine (Inglis & Liu, 1970). The *S*-sulfocysteinyl residues are stable in neutral solutions and survive the conditions of the Edman degradation but under reducing conditions they are converted to cysteinyl residues. Under normal hydrolysis conditions with 6 M-HCl they are converted to cystine (Cole, 1967). Measurement of the extent of the reaction is a problem with this procedure but incorporation of radioactivity may be studied if ^{35}S-labelled sulfite is used (Chan, 1968).

A stable derivative, S-2-sulfobenzylcysteine, may be obtained from the reaction of ω-toluene-sultone with the reduced protein at pH 8.3 (Rüegg *et al.*, 1979), as shown.

$$C_6H_4(CH_2OSO_2) + {}^-S\!\dashv \longrightarrow C_6H_4(CH_2\!-\!S\!\dashv)(SO_3^-)$$

The reversal of this reaction is not easy and requires sodium in liquid ammonia. Sodamide should be added to prevent cleavage of threonyl-proline bonds. Both the sulfitolysis and sulfobenzylation reactions confer good solubility on the products.

Oxidative sulfitolysis (Chan, 1968)

The method was applied successfully to aldolase, lactate dehydrogenase and pepsinogen. Complete sulfonation only occurs in the range pH 7.0–8.5, suggesting that the protonated amino group of cysteine plays a role in the reaction mechanism. The protein (2 mg/ml) is incubated with 0.05 M-sodium sulfite and 0.2 mM-cysteine in 8 M-urea or 6 M-guanidine HCl, 0.1 M-Tris-HCl (pH 8.4) for 1 h in an open vessel at 25°C. Dialyze and recover by lyophilization.

Fully active enzyme is obtained by the following treatment of aldolase, *S*-sulfonated as above. Dissolve the protein (0.5–1.0 mg/ml) in 1 M-Tris-HCl (pH 7.5) containing 4 M-urea. After addition of 2-mercaptoethanol (final concentration 0.7 M) the mixture is incubated for 2 h at 25°C. The protein subunits are allowed to reassociate by diluting the mixture ten-fold with 1 M-Tris-HCl (pH 7.5) containing BSA (2.5 mg/ml), EDTA (25 mM) and 2-mercaptoethanol (10 mM). The activity is determined after 15 min.

2-Sulfobenzylation (Rüegg *et al.*, 1979)

The protein (100 μmol) in 100 ml of 0.5 M-sodium bicarbonate/propan-1-ol (1:1, v/v) is treated under nitrogen with tributylphosphine (330 μmol) and ω-toluenesultone (204 mg; 1200 μmol). After 14 h, undissolved ω-toluenesultone is filtered off and the solution volume decreased to about 40 ml. It is then dialyzed against water. 2-Sulfobenzylcysteine is eluted just after cysteic acid on the amino acid analyzer. The yield was quantitative for insulin.

Removal of sulfobenzyl groups (Katsoyannis & Schwartz, 1977)

The protein (300 mg) is thoroughly dried and dissolved in anhydrous liquid ammonia (300 ml) in a round-bottomed flask immersed in a cooling bath of dry ice–acetone. Approximately 120 mg of solid sodamide is also added if labile proline bonds are likely to be present. The subsequent reaction is carried out at the boiling point of the solution. A solution of sodium in liquid ammonia (preferably prepared in the special apparatus described) is added dropwise until a faint blue color is imparted to the reaction mixture and persists for 30 s. Drops of glacial acetic acid (2–3) are added to discharge the color, and the solution is allowed to evaporate until only about 10 ml remains. The flask is evacuated with a pump while swirling until the solution is frozen. It is then lyophilized using the water pump.

1.5.1.3. Re-oxidation

Re-formation of a large proportion of the original disulfide links of some proteins, ribonuclease for example, can be effected by simple aerial oxidation of the reduced protein (Haber & Anfinsen, 1962). Apparently the native molecule possesses the most stable thermodynamic configuration and so the major force in the 'pairing' of the thiol groups to form a disulfide link is the concerted interaction of side chain functional groups distributed along the chain. Proteins in which the arrangement of the disulfide bonds is presumably influenced by the conformation of the precursor form (proinsulin → insulin, for example) may therefore provide more difficulties in renaturation. The conditions described below for the re-oxidation should have some general applicability, although for optimum results, variations of those conditions may be necessary for individual proteins with their unique primary sequences and three-dimensional configurations. For example, addition of a reductant such as mercaptoethanol and an increase in the temperature to 38°C are beneficial for re-oxidation of lysozyme but not for ribonuclease (Epstein & Goldberger, 1963). The appropriate amount of mercaptoethanol required for other proteins should be determined by experiment. Both the low concentration of protein and the presence of mercaptoethanol can diminish aggregation of the units and the formation of intermolecular disulfide bridges. Mercaptoethanol presumably catalyzes the disulfide interchange necessary for cysteinyl residues initially linked incorrectly to revert to the original conformation.

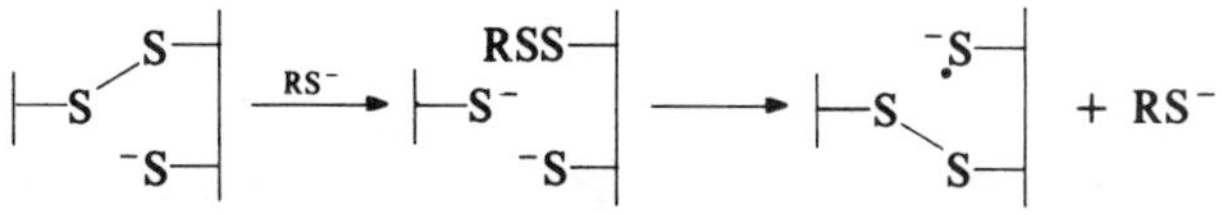

Metal ions which can react with thiols inhibit the reactivation. Slow removal of a denaturant such as urea could also be of use. The solutions are only exposed to air in open vessels during oxidation because bubbling of air through the solution can cause denaturation.

Method: Epstein & Goldberger, 1963

Re-oxidation of reduced lysozyme is carried out in 0.1 M-Tris buffer pH 8.5 for 2 h at 38°C at low protein concentration (0.025 mg/ml). The solution contains 2-mercaptoethanol (180:1 on a molar basis to protein thiol groups). Over 70% reactivation occurs.

Method: Du *et al.*, 1965

The reduced chains of insulin (5 mg protein/ml) are precipitated at pH 3.8 and well washed with de-aerated buffer in the cold. Atmospheric oxidation is carried out in glycine buffer, pH 10.6 at 35°C.

The final re-oxidation product has an activity of about 50% of that of crystalline insulin. The activity varies with the molar ratio of the chains, being highest for a ratio of A to B chains of 1.5 : 1.

1.5.2. Cross-links other than disulfide bonds

Proteins may possess intermolecular bonds that are more intractable to analysis and specific cleavage than disulfide bonds. For example, there can be aldimines (I) in collagen (Tanzer, 1976), desmosine (II) and isodemosine in elastin (which can link 4-polypeptide chains) (Guay & Lamy, 1979), isopeptide bonds formed from enzyme linking of the ε-amino group of lysine to the amide side chain of glutamine (III) in extracellular structural proteins (Asquith *et al.*, 1974; Matacic & Loewy, 1979) and possibly di-(IV) and tri-tyrosine links (Maclaren & Milligan, 1981).

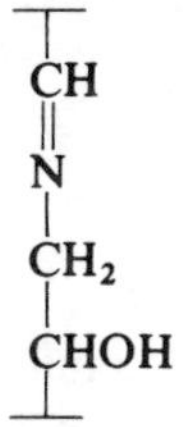

I dehydrohydroxy-lysinonorleucine

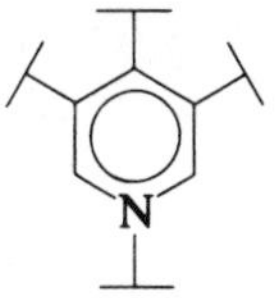

II desmosine

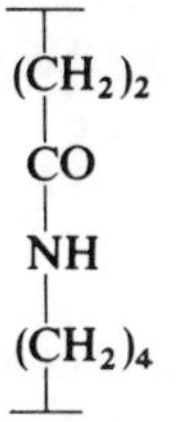

III ε-(γ-glutamyl) lysine

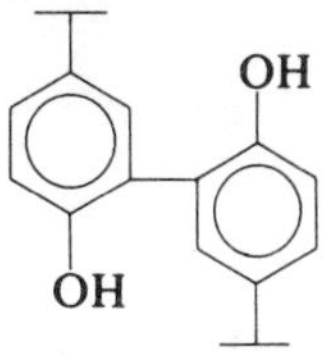

IV dityrosine

1.5.2.1. Aldimines

The initial stage in the formation of the intermolecular bonds in collagen is the oxidative deamination by a copper-dependent enzyme lysyloxidase of specific lysine and hydroxylysine residues at the amino- and carboxy-terminal ends of the molecules. The aldehydes formed, allysine and hydroxyallysine (|–$(CH_2)_3CHO$ and |–$CH_2CHOHCH_2CHO$), condense predominantly with ε-amino groups of hydroxylysine to form either the dehydrohydroxylysinonorleucine (I) or the dehydrodihydroxylysinonorleucine, respectively. The double bonds of both aldimines are reducible with potassium borohydride which renders cross-links stable to acid hydrolysis. However, the dihydroxy compound is labile *in vivo* and rearranges to the ketoimine, hydroxylysino-5-ketonorleucine (V) (Light & Bailey, 1979).

$$\vdash (CH_2)_2CHOH\,CH_2N{=}CHCHOH(CH_2)_2 \dashv$$

$$\longrightarrow \vdash (CH_2)_2\,CHOHCH_2NHCH_2CO(CH_2)_2 \dashv$$

V

Galactosyl and glucosylgalactosyl derivatives of the cross-links have also been characterized after their isolation from alkaline hydrolyzates of the borohydride-reduced collagen (Robins & Bailey, 1974).

Usually the cross-links of collagen are stabilized by reduction (preferably with sodium cyanoborohydride, Robins & Bailey, 1977) and then identified after enzymic digestion and isolation of the cross-linked peptides. These are easily detected if the reduction is made with tritiated reagent. After the reduction step and acid hydrolysis, hydroxylysinonorleucine, dihydroxylysinonorleucine, hydroxymerodesmosine, hydroxylaldolhistidine and histidinohydroxymerodesmosine have been isolated from various tissues (Tanzer, 1976). Housley *et al.* (1975) isolated hydroxyaldolhistidine (VI) (formed from allysine, hydroxyallysine and histidine) which is not reducible by borohydride but stable to acid hydrolysis.

$$\vdash (CH_2)_2\,CHOHCH_2{-}\underset{\underset{\perp}{|}}{\underset{(CH_2)_2}{\underset{|}{C}}}{=}CH{-}N\langle\text{imidazole ring: }{-}CH_2{\dashv},\ {=}N\rangle$$

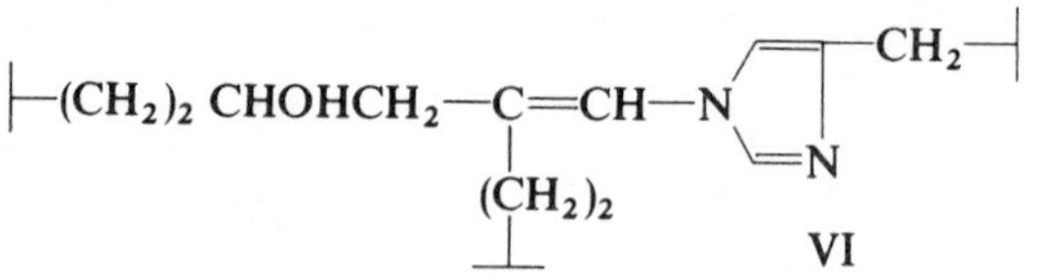

VI

Whether this compound (and others reported) are indeed genuine stable cross-links and not artifacts of the isolation procedures, and how the aldimines are converted to non-reducible stable cross-links during ageing are unsolved problems (Light & Bailey, 1979).

Cleavage of the aldimine (Fessler & Bailey, 1966)

In gelatin the cross-links are severed by the lathyrogen, β-aminopropionitrile.

$$\vdash CH=N\dashv \xrightarrow[R-NH_2]{} \vdash CH=N-R+NH^2\dashv$$

β-Aminopropionitrile is approximately neutralized with glacial acetic acid, then buffer and gelatin are added and the solution is adjusted to pH 7.6 with acetic acid (final concentrations 1–2 mg/ml protein, 1.9 M-β-aminopropionitrile, 1.3 M-acetic acid, 1.6 mM-Tris, 0.8 mM-$MgCl_2$). A few drops of toluene are added and the stoppered tubes are placed in an incubator at 38°C for 2 days.

Reduction with sodium cyanoborohydride (Henkel *et al.*, 1979)

The protein is equilibrated for at least 4 h in 0.1 M-sodium phosphate buffer at pH 4.4. Sodium cyanoborohydride (30:1, protein:$NaBH_3CN$ dry wt ratio) is added and the mixture is left at room temperature for 24 h. Dialyze and lyophilize.

Reaction with ^{14}C-sodium cyanide and ammonia (Pereyra *et al.*, 1974)

Reduction with sodium borohydride, usually at pH 8, is neither quantitative nor reproducible when applied to fibrin. The alternative method proposed for determination of the cross-linkages in this fibrous protein uses ^{14}C-sodium cyanide and ammonia (see equation). The reaction yields either an aminonitrile from the aldehyde or a substituted amino acid characteristic of the aldimine. The products are stable to hydrolysis and can be separated and quantified on the amino acid analyzer.

$$\vdash C(=O)H + Na\overset{*}{C}N + NH_3 \longrightarrow \vdash CH(\overset{*}{C}N)(NH_2) \xrightarrow{H^+} \vdash CH(\overset{*}{C}OOH)(NH_2)$$

$$\vdash CH=N\dashv + Na\overset{*}{C}N \longrightarrow \vdash CH(\overset{*}{C}N)-NH\dashv \xrightarrow{H^+} \vdash CH(\overset{*}{C}OOH)(NH\dashv)$$

Finely divided elastin (10 mg) is suspended in water (2 ml) and 2 mg of ^{14}C-sodium cyanide (specific activity 1.1 mCi/μmol) and 30% ammonia (2 ml) are added. The solution should be at pH 11.5. After 1 h at room temperature the solution is acidified to pH 1 and dialyzed against 0.1 M-HCl at 4°C. The ammonia is not necessary for modification of aldimines. The aldehydes become doubly labelled if tritiated methylamine is used instead of ammonia. Since the specific activity of the $Na^{14}CN$ is known quantitative results are possible. The pyridinium forms of the desmosines do not incorporate the label.

Periodate degradation of cross-links (Robins & Bailey, 1975)

The reduced cross-links of collagen may be cleaved into two products by oxidation with periodate. Reduction with potassium borohydride destroys excess periodate and yields proline and lysine which may be quantified by amino acid analysis. Only 5 min at room temperature is sufficient for the oxidation reaction. If tritiated reagent, KB^3H_4, is used for reduction of the cross-link the labels will appear as shown in the equation by * and **.

$$\rangle CHCH_2CH_2\overset{*}{C}HOHCH_2NH\overset{**}{C}H_2(CH_2)_3CH\langle$$

$$\downarrow \text{1. } NaIO_4 \quad \text{2. } KBH_4$$

$$\begin{array}{l} COOH \\ \;\;\backslash \\ \;\;CH{-}CH_2 \\ \;\;|\qquad\;\;\backslash \\ \;\;|\qquad\quad CH_2 \\ \;\;|\qquad\;\;/ \\ \;\;NH{-}\underset{*}{C}H_2 \end{array} \quad + \quad NH_2\overset{**}{C}H_2(CH_2)_3CH\begin{array}{l} \diagup COOH \\ \diagdown NH_2 \end{array}$$

The protein with reduced cross-link in citrate buffer, pH 5.3 (1 ml) is treated with 0.01 M-$NaIO_4$ in the dark at room temperature. After 5 min the solution is adjusted to pH 7.5 with 2 M-NaOH and KBH_4 (2.5 mg) is added. The reaction is continued for 30 min, at which time the solution is adjusted to pH 2 by the addition of 6 M-HCl.

Separation of cross-linked from non-cross-linked peptides (Stimler & Tanzer, 1974)

Chromatography on hydroxyapatite has been useful in separating cross-linked peptides from non-cross-linked peptides in collagen. The peptides containing the cross-links are more likely to assume the native conformation and preferentially adhere to the hydroxyapatite. Cyanogen bromide peptides from reduced or unreduced insoluble calf bone or skin collagen are dissolved in starting buffer (0.001 M-Na_2HPO_4, pH 6.8, containing 1 M-urea and 0.15 M-NaCl) and dialyzed against starting buffer. They are then adsorbed onto a column (15 × 230 mm) of hydroxyapatite (Hypatite C) equilibrated in starting buffer. Elution is carried out with a linear gradient of 450 ml each of starting buffer and gradient buffer (0.8 M-Na_2HPO_4, pH 6.8, containing 1 M-urea and 0.15 M-NaCl). After a total of 800 ml of buffer has passed through the column the gradient is converted to exponential form by the addition of 450 ml of buffer (0.2 M-Na_2HPO_4, pH 6.8, containing 1 M-urea and 0.15 M-NaCl) to the gradient vessel from a separatory funnel in a system which is closed to atmospheric pressure. Fractions of 5.5 ml are collected at a flow rate of 60 ml/h. Appropriate fractions are pooled, desalted by dialysis against deionized water and then lyophylized.

Two major peaks and one minor peak are obtained from calf bone collagen reduced with NaB^3H_4. The material in the second major peak contains the highest mol. wt. material, is the most helical and contains the cross-linked peptides.

1.5.2.2. Desmosines

The characterization of desmosine is best effected after reduction of elastin with sodium borodeuteride in water or D_2O. The reduced material is subjected to acid hydrolysis. The amino acids are separated chromatographically, derivatized as their *N*-trifluoroacetyl methyl esters and analyzed by mass spectrometry (Paz *et al.*, 1974).

1.5.2.3. Isopeptides

The most common of the isopeptide cross-links in proteins is the N^{ε}-(γ-glutamyl)lysine cross-link found in structural proteins (Matacic & Loewy, 1979; Loewy, 1984). It can be detected by elution in the isoleucine region during amino acid analysis after complete enzymic hydrolysis. However, it may be resolved with special conditions, for example, using an elution buffer at pH 4.8. See also Klostermeyer (1984) for a discussion of the N^{ε}-(β-aspartyl) lysine cross-link.

1.6. MONOMER PREPARATIONS FOR AMINO ACID SEQUENCING

Ideally, preparations for primary structure analysis should give a molar yield of the *N*-terminal amino acid on Edman degradation, i.e. they should be free from salts and other contaminating material and be completely unblocked. Side chain modification should not have occurred during preparation because high yields after each degradation cycle are essential for an extensive automatic sequence determination. Warnings against blocking the *N*-terminal amino acid and producing heterogeneous mixtures have already been given earlier. We include here additional methods that have been found useful for preparation of the protein including methods for removal of lipid and carbohydrate. Removal of carbohydrate from highly glycosylated glycoproteins is helpful before sequencing.

1.6.1. Removal of pyroglutamic acid from amino terminus

The availability of a reliable procedure for removing cyclized glutamyl or glutaminyl residues from the amino terminus of proteins (Podell & Abraham, 1978) has decreased substantially the problems of primary structure determination of proteins blocked in this way. Since the 'deblocking reaction' can generate a free α-amino group of the adjacent amino acid in high yield, the cyclization reaction might now be regarded as an advantage by acting as a protecting group during the protein purification steps. The double incubation method with calf liver pyroglutamate amino peptidase was originally applied to light and heavy chains of immunoglobulins and has recently been successful with the heavy chain of the hemagglutinin from influenza virus (Ward & Dopheide, 1980).

1.6.1.1. Method for removal of pyroglutamic acid from N-terminus

The method of Podell and Abraham (1978) is described. Pyroglutamate amino peptidase from calf liver (*L*-pyroglutamyl peptide hydrolase, EC 3.4.11.8) is stored after lyophilization at −20°C. Ten milliliters of totally reduced and alkylated protein (1 mg/ml) is first dialyzed at 4°C against 0.1 M-phosphate buffer (pH 8.0) containing 5 mM-DTT, 10 mM-EDTA and 5% (v/v) glycerol. The protein preparation is transferred to a screw-top vial and approximately 0.025 mg of active enzyme is added, the vial flushed with nitrogen, capped and mixed. Protein and enzyme are allowed to incubate at 4°C for 9 h with occasional mixing. Another 0.025 mg of active enzyme is added and incubation is continued under nitrogen at room temperature for 14 h with occasional mixing. The treated protein is then dialyzed against 0.05 M-acetic acid, lyophilized and used for automated amino acid sequence analysis.

A chemical procedure for opening the ring has been applied successfully (Muranova & Muranov, 1979). The peptide is incubated at 37°C with a 17% aqueous solution of methylamine for 14 h or longer. N^{γ}-methylglutamine is formed: after dansylation and acid hydrolysis of the peptide, this decomposes to yield glutamic acid.

1.6.2. Reduction of sulfoxides with *N*-methylmercaptoacetamide

The extent of oxidation of methionine during protein preparation is not usually apparent on amino acid analysis because reversal can occur on acid hydrolysis (Savige & Fontana, 1977b) but it can be indicated by low cleavage yields from the CNBr reaction (Gross, 1967). Low values for the cystine analysis may be caused by sulfoxide formation with either cysteinyl (Inglis & Liu, 1970) or alkylated cysteinyl derivatives (Inglis *et al.*, 1976). The disadvantage of partial oxidation is that it leads to difficulties of identification after many degradation cycles. A study of various reagents for the reduction of the methionine *S*-oxide showed that *N*-methylmercaptoacetamide is the most suitable (Houghten & Li, 1979). React methionine *S*-oxide (1 mg/ml) with 0.7–2.8 M-*N*-methylmercaptoacetamide at pH 2.6–8.5 at 37°C for 12–14 h under nitrogen.

The sulfoxide form of *S*-carbamoylmethylcysteine in peptides or proteins takes approximately five times longer to reduce than methionine *S*-oxide.

In some cases, such as after the oxidation of the protein with performic acid to form cysteic acid residues, the methionine will be in the stable *S,S*-dioxide form. Prior oxidation treatments (Inglis *et al.*, 1979) to yield a more stable tryptophyl residue have been used for Edman degradations, one yielding oxindolyl-3-alanine (Savige & Fontana, 1977a) and the other kynurenine (Atassi, 1967). The method of Savige & Fontana (1977a) is more specific. (See Chapter 2, section 2.5.3). Both methods produce methionine *S*-oxide which can be reduced specifically to methionine (Savige & Fontana, 1977b).

1.6.3. Removal of lipid

Apoproteins are commonly prepared from lipoproteins by simple extraction of the lipid with organic solvents such as ethanol, ethanol–ether or methanol–chloroform. However, some apoproteins are at least partially soluble in organic solvents. Serum lipoproteins contain low mol. wt. peptides that are soluble in ethanol and to a lesser extent in ethanol–ether mixtures whereas the higher mol. wt. apoproteins are insoluble (Scanu & Edelstein, 1971). The solubility of the apoproteins decreases with decreasing temperature of extraction. Covalently attached lipids were reported for certain membrane proteins (Schlesinger, 1979).

1.6.3.1. Methods for lipoproteins

Method for very low density lipoproteins

The method of Scanu and Edelstein (1971) is described. Extractions are carried out at -10°C. Protein (2–5 mg/ml) after dialysis against 0.15 M-NaCl containing 1 mM-EDTA, is added dropwise to 50 ml of ethanol (95%)/ether mixture (3:1, v/v) in a 50 ml round-bottom centrifuge tube and extracted for 2 h using equipment providing a slow rotation (15 rev/min). The supernatant is separated (and retained) from the precipitate after centrifuging for 10 min at 2000 rev/min. The precipitate is extracted with a further volume of ethanol–ether for 2 h and separated as above. Finally it is extracted with ether overnight. The precipitate is washed three times with ether and dried under nitrogen.

An additional amount of protein is obtained from the solvent used in the first extraction by adjusting the ethanol–ether ratio to 3:5 (v/v) and extracting overnight at -10°C.

Method for egg yolk lipoprotein

The method of Burley (1973) is described. It does not involve the precipitation of protein. A solution of the lipoprotein in 1.0 M-NaCl (about 6% w/v) is dialyzed against several changes of 6 M-urea – 0.05 M-HCl (pH 3.3) at 20°C to remove the

NaCl almost completely. Alternatively, salt is first removed by dialysis into water at 1°C. To the urea solution (17 ml) a mixture of chloroform (20 ml) and methanol (40 ml) are added and the one-phase solution is allowed to stand for 30 min at 20–25°C. More chloroform (20 ml) is then added. After settling, the lower chloroform layer is discarded and the upper solution treated with another 10 ml of chloroform. If necessary the mixture is centrifuged. Chloroform is then removed and the solution left for 24 h at 20°C. The solution is then dialyzed into 6 M-urea.

1.6.4. Removal of carbohydrate

As a first step it is convenient to use the commercially available HF-pyridine reagent (90 min, room temperature) to remove carbohydrates from glycoproteins. However, it is not always as effective as anhydrous HF which cleaves all the linkages of neutral and acidic sugars of glycoproteins in 1 h at 0°C (Mort & Lamport, 1977). The treatment can adversely affect the physical properties of the protein although it is suitable for preparing material for amino acid sequence determinations.

1.6.4.1. Method for removal of carbohydrate

The method of Mort and Lamport (1977) is described. Commercial apparatus was used (Peninsula Laboratories Inc.). A thoroughly dried sample (up to 100 mg) and anhydrous methanol (10% by volume) are added to the Kel-F reaction vessel. The entire HF line is evacuated except for the HF reservoir. The reaction vessel is cooled in a dry ice–acetone bath and then about 10 ml of HF is distilled over from the reservoir (which also contains cobalt trifluoride as a drying agent), stirring both vessels continuously. The reaction vessel is allowed to warm up to an appropriate temperature (vapor pressure of HF is 364 mmHg at 0°C and 850 mmHg at 23°C). Time the reaction from the appropriate point. If performing the reaction at 0°C an ice bath may be used. At the end of the reaction the vessel is evacuated via a calcium oxide trap. (To prevent excessive frothing of the reaction mixture vacuum must be applied slowly but should be sufficiently fast for rapid HF removal and prevention of oligomerization.) To ensure complete removal of HF the line is kept evacuated for an additional hour after removal of visible HF. After complete HF removal, generally shown by a change from light brown through a transient reddish color to colorless, the sample is taken up in 0.1 M-acetic acid or 50% acetic acid in water if the residue is difficult to solubilize. The product is immediately chromatographed on an appropriate Sephadex column or dialyzed.

More severe treatment with anhydrous HF (3 h at 23°C) cleaves the *O*-glycosidic linkages of amino sugars, but peptide bonds and the *N*-glycosidic linkage between asparagine and *N*-acetylglucosamine still remain intact.

1.6.5. Desalting

Usually the monomer or constituent chain can be freed from salt either by dialysis or gel filtration. The importance of salt removal cannot be over-emphasized. Sometimes dialysis against solutions containing small ions and possibly buffer to keep the proteins in solution is desirable initially, before dialysing against water. Small volume dialysis equipment is commercially available.

Gel filtration (Section 1.2.2.1) is quicker, particularly if a column is already packed. The eluent and gel size used will depend on the solubility and size of the proteins. Poorly soluble hydrophobic proteins can be dissolved in 98 % formic acid which is diluted to 75 % for chromatography (section 1.3.2.1). Sephadex G-25 should be used for proteins with mol. wts. in the range 5 000 to 30 000. Use G-50 for larger molecules. The bed volume should be 3–4 times the sample volume. Prepacked disposable columns containing Sephadex G-25 are available from Pharmacia. Samples of up to 2.5 ml can be desalted in less than 5 min.

1.6.5.1. Method of desalting

The method of Hunkapiller and Hood (1980) is described. Small peptides are desalted on a small column of Bio-Gel P-2 with 1 M-acetic acid or Sephadex G10 with 1 M-ammonium hydroxide followed by lyophilization of the eluate.

Larger peptides and proteins are dialyzed for 12 h against 0.15 M-NaCl containing 0.1 % SDS. This is repeated two more times, followed by three dialysis periods of 12 h against 0.02 % SDS. If the proteins were eluted from SDS-gels the elute is centrifuged at 20 000 rev/min for 20 min to remove particulate matter before dialysis. Coomassie Blue does not interfere with subsequent sequenator analysis. However, it must be remembered that the final concentration of SDS in the sample will depend on the time of dialysis since equilibration of SDS across dialysis membrane may take many days.

Electrodialysis may also be used. See Chapter 17, section 17.3.1.

1.7. REFERENCES

Ackers, G. K. (1975). In *The Proteins*, I, 3rd edn, Neurath, H. & Hill, R. L. (Eds.), Academic Press, New York, pp. 2–94.

Ahmadi, B. (1979). *Anal. Biochem.*, **97,** 229–231.

Allen, R. C. & Maurer, H. R. (Eds.) (1974). *Electrophoresis and Isoelectric Focusing in Polyacrylamide Gels*, Walter de Gruyter, Berlin.

Amos, L. A. (1978). In *Techniques in Protein and Enzyme Biochemistry*, B 111, Elsevier/North-Holland, Shannon, pp. 1–26.

Andersson, L. O., Borg, H. & Mikaelsson, M. (1972). *FEBS Lett.*, **20,** 199–202.

Andrews, P. (1964). *Biochem. J.*, **91,** 222–233.

Anfinsen, C. B. & Haber, E. (1961). *J. Biol. Chem.*, **236,** 1361–1363.
Ansari, A. A. & Mage, R. G. (1977). *J. Chromatogr.*, **140,** 98–102.
Aoshima, H. (1979). *Anal. Biochem.*, **95,** 371–376.
Asquith, R. S., Otterburn, M. S. & Sinclair, W. J. (1974). *Angew. Chem. Int. Edn.*, **13,** 514–520.
Atassi, M. Z. (1967). *Arch. Biochem. Biophys.*, **120,** 56–59.
Axelsson, I. (1978). *J. Chromatogr.*, **152,** 21–32.
Baker, T. S., Eisenberg, D., Eiserling, F. A. & Weissman, L. (1975). *J. Mol. Biol.*, **91,** 391–399.
Ballou, B., McKean, D. J., Freedlender, E. F. & Smithies, O. (1976). *Proc. Natl. Acad. Sci. U.S.A.*, **73,** 4487–4491.
Banks, B. E. C., Doonan, S., Flogel, M., Porter, P. B., Vernon, C. A., Walker, J. M., Crouch, T. H., Halsey, J. F., Chiancone, E. & Fasella, P. (1976). *Eur. J. Biochem.*, **71,** 469–473.
Bates, D. L., Harrison, R. A. & Perham, R. N. (1975a). *FEBS Letts.*, **60,** 427–430.
Bates, D. L., Perham, R. N. & Coggins, J. R. (1975b). *Anal. Biochem.*, **68,** 175–184.
Belew, M., Fohlman, J. & Janson, J.-C. (1978a). *FEBS Letts.*, **91,** 302–304.
Belew, M., Porath, J., Fohlman, J. & Janson, J.-C. (1978b). *J. Chromatogr.*, **147,** 205–212.
Bijlholt, M. M. C., Van Bruggen, E. F. J. & Bonaventura, J. (1979). *Eur. J. Biochem.*, **95,** 399–405.
Birdi, K. S. (1976). *Anal. Biochem.*, **74,** 620–622.
Bock, R. M. & Ling, N.-S. (1954). *Anal. Chem.*, **26,** 1543–1546.
Braunitzer, G., Schrank, B., Ruhfus, A., Peterson, S. & Peterson, U. (1971). *Hoppe-Seyler's Z. Physiol. Chem.*, **352,** 1730–1732.
Broadmeadow, A. C. C. & Wilce, P. A. (1979). *Anal. biochem.*, **100,** 87–91.
Brouwer, J. (1979). *J. Chromatogr.*, **179,** 342–345.
Brown, W. E. & Howard, G. C. (1980). *Anal. Biochem.*, **101,** 294–298.
Bull, H. B. & Breese, K. (1980). *Arch. Biochem. Biophys.*, **202,** 116–120.
Bull, H. B., Breese, K., Ferguson, G. L. & Swenson, C. A. (1964). *Arch. Biochem. Biophys.*, **104,** 297–304.
Burgess, R. R. (1969). *J. Biol. Chem.*, **244,** 6168–6176.
Burley, R. W. (1973). *Biochemistry*, **12,** 1464–1470.
Chan, W. W. -C. (1968). *Biochemistry*, **7,** 4247–4253.
Chantler, P. D. & Gratzer, W. B. (1978). *Anal. Biochem.*, **91,** 626–635.
Chrambach, A. (1980). *Molec. Cellular Biochem.*, **29,** 23–46.
Chrambach, A. & Rodbard, D. (1971). *Science*, **172,** 440–451.
Chrambach, A., Jovin, T. M., Svendsen, P. J. & Rodbard, D. (1976). In *Methods of Protein Separation*, vol. 2, Catsimpoolas, N. (Ed.), Plenum Press, New York, London, pp. 27–144.
Cohn, E. J. & Edsall, J. T. (1943). In *Proteins, Amino Acids and Peptides*, Reinhold, New York, p. 602.
Cole, R. D. (1967). *Methods Enzymol.*, **11,** 206–208.
Cole, E. G. & Mecham, D. K. (1966). *Anal. Biochem.*, **14,** 215–222.
Cook, R. A. & Koshland, D. E. (1969). *Proc. Natl. Acad. Sci. U.S.A.*, **68,** 247–254.
Crewther, W. G. (1976). *Proc. 5th Wool Text. Res. Conf. Aachen, 1975*, **1,** 1–101.
Crewther, W. G., Inglis, A. S. & McKern, N. M. (1978). *Biochem. J.*, **173,** 365–371.
Crewther, W. G., Dowling, L.M., Gough, K. H. Marshall, R. C. & Sparrow, L. G. (1980). In *Fibrous Proteins: Scientific, Industrial and Medical Aspects*, vol. 2, Parry, D. A. D. & Creamer, L. K. (Eds.), Academic Press, London, pp. 151–159.
Davis, B. J. (1964). *Ann. N. Y. Acad. Sci.*, **121,** 404–427.
Davies, G. E. & Stark, G. R. (1970). *Proc. Natl. Acad. Sci. U.S.A.*, **66,** 651–656.
De Abreu, R. A., De Vries, J., De Kok, A. & Veeger, C. (1979). *Eur. J. Biochem.*, **97,** 379–387.

De Jong, W. W., Zweers, A. & Cohen, L. H. (1978). *Biochem. Biophys. Res. Commun.*, **82,** 532–539.
Dirnhuber, P. & Schutz, F. (1948). *Biochem. J.*, **42,** 628–632.
Dixon, M. (1953). *Biochem. J.*, **54,** 457–458.
Djondjurov, L. & Holtzer, H. (1979). *Anal. Biochem.*, **94,** 274–277.
Dohnal, J. C. & Garvin, J. E. (1979). *Biochim. Biophys. Acta.*, **576,** 393–403.
Dowling, L. M. & Crewther, W. G. (1974). *Preparative Biochemistry*, **4,** 203–226.
Du, Y. C., Jiang, R. Q. & Tsou, C. L. (1965). *Sci. Sin.*, **14,** 229–236.
Edelman, G. M., Gall, W. E., Waxdal, M. J. & Konigsberg, W. H. (1968). *Biochemistry*, **7,** 1950–1958.
Edman, P. & Begg, G. (1967). *Eur. J. Biochem.*, **1,** 80–91.
Edman, P. & Henschen, A. (1975). In *Protein Sequence Determination*, Needleman, S. B. (Ed.), 2nd edn, Springer, Berlin, Heidelberg, New York, pp. 232–279.
Epstein, C. J. & Goldberger, R. F. (1963). *J. Biol. Chem.*, **238,** 1380–1383.
Ferguson, K. A. (1964). *Metabolism*, **13,** 985–1002.
Fessler, J. H. & Bailey, A. J. (1966). *Biochim. Biophys. Acta.* **117,** 368–378.
Fischer, L. (1980). In *Laboratory Techniques in Biochemistry and Molecular Biology*, vol. 1, part II, Work T. S. & Burdon, R. H., (Eds.), 2nd edn., North-Holland, Amsterdam, pp. 157–396.
Fish, W. W., Mann, K. G. & Tanford, C. (1969). *J. Biol. Chem.*, **244,** 4989–4994.
Fishbein, W. N. (1972). *Anal. Biochem.*, **46,** 388–401.
Frieden, C. & Nicol, L. W. (Eds.) (1981). *Protein-Protein Interactions*, Wiley, New York.
Gianazza, E. & Righetti, P. G. (1980). *J. Chromatogr.*, **193,** 1–8.
Gibson, D. R. & Gracy, R. W. (1979). *Anal. Biochem.*, **96,** 352–354.
Gillespie, J. M. & Frenkel, M. J. (1974). *Aust. J. Biol. Sci.*, **27,** 617–627.
Gross, E. (1967). *Methods Enzymol.*, **11,** 238–255.
Guay, M. & Lamy, F. (1979). *Trends Biochem. Sci.*, 160–164.
Gurd, F. R. N. (1972). *Methods Enzymol.*, **25,** 424–438.
Habeeb, A. F. S. A. (1972). *Methods Enzymol.*, **25,** 457–464.
Haber, E. & Anfinsen, C. B. (1962). *J. Biol. Chem.*, **237,** 1839–1844.
Hagel, P., Gerding, J. J. T., Fieggen, W. & Bloemendal, H. (1971). *Biochim. Biophys. Acta.*, **243,** 366–373.
Hajdu, J., Bartha, F. & Friedrich, P. (1976). *Eur. J. Biochem.*, **68,** 373–383.
Hames, B. D. & Rickwood, D. (Eds.) (1981). *Gel Electrophoresis of Proteins*, IRL Press, London.
Henkel, W., Rauterberg, J. & Glanville, R. W. (1979). *Eur. J. Biochem.*, **96,** 249–256.
Henschen-Edman, A. (1977). In *Solid Phase Methods in Protein Sequence Analysis*, Proc. Second Internat. Conf., Previero, A. & Coletti-Previero, M.-A. (Eds.), North-Holland, Amsterdam, New York, Oxford, pp. 7–18.
Higgins, R. C. & Dahmus, M. E. (1979). *Anal. Biochem.*, **93,** 257–260.
Hill, M., Bechet, J. & d'Albis, A. (1979). *FEBS Letts.*, **102,** 282–286.
Hirs, C. H. W. (1967). *Methods Enzymol.*, **11,** 197–199.
Hofmeister, F. (1888). *Arch. Exptl. Pathol. Pharmakol.*, **24,** 247–260.
Houghten, R. A. & Li, C. H. (1979). *Anal. Biochem.*, **98,** 36–46.
Housley, T. J., Tanzer, M. L. Henson, E. & Gallop, P. M. (1975). *Biochem. Biophys. Res. Commun.*, **67,** 824–830.
Hucho, F., Mullner, H. & Sund, H. (1975). *Eur. J. Biochem.*, **59,** 79–87.
Hunkapiller, M. W. & Hood, L. E. (1980). Science, **207,** 523–525.
Inglis, A. S. (1976). *J. Chromatogr.*, **123,** 482–485.
Inglis, A. S. & Liu, T. Y. (1970). *J. Biol. Chem.*, **245,** 112–116.
Inglis, A. S., McMahon, D. T. W., Roxburgh, C. M. & Takayanagi, H. (1976). *Anal. Biochem.*, **72,** 86–94.

Inglis, A. S., Rivett, D. E. & McMahon, D. T. W. (1979). *FEBS Lett.*, **104,** 115–118.
Jeffrey, P. D. (1981). In *Protein-Protein Interactions*, Frieden, C. & Nicol, L. W. (Eds.), Wiley, New York, pp. 213–257.
Johansson, I. & Lundgren, H. (1979). *J. Biomed. Biophys. Methods*, **1,** 37–44.
Katsoyannis, P. G. & Schwartz, G. P. (1977). *Methods Enzymol.*, **47,** 501–578.
Kirschner, K. & Bisswanger, H. (1976). *Ann. Rev. Biochem.*, **45,** 143–166.
Klostermeyer, H. (1984). *Methods Enzymol.*, **107,** 258–261.
Klotz, I. M., Darnall, D. W. & Langerman, N. R. (1975). In *The Proteins*, I, 3rd edn, Neurath, H. & Hill, R. L. (Eds.), Academic Press, New York, pp. 293–411.
Kubota, I. & Tsugita, A. (1980). *Eur. J. Biochem.*, **106,** 263–268.
Laemmli, U. K. (1970). *Nature (Lond.)*, **227,** 680–685.
Lakshmanan, T. K. & Lieberman, S. (1954). *Arch. Biochem. Biophys.*, **53,** 258–281.
Lambin, P. (1978). *Anal. Biochem.*, **85,** 114–125.
Lambin, P. & Fine, J. M. (1979). *Anal. Biochem.*, **98,** 160–168.
Lambin, P., Rochu, D. & Fine, J. M. (1976). *Anal. Biochem.*, **74,** 567–575.
Landon, M. (1977). *Methods Enzymol.*, **47,** 145–149.
Lasky, M. (1978). In *Electrophoresis '78*, Catsimpoolas, N. (Ed.) Elsevier, Amsterdam, pp. 195–210.
Leighton, P. H., Fisher, W. K., Moon, K. E. & Thompson, E. O. P. (1980). *Aust. J. Biol. Sci.*, **33,** 513–520.
Light, N. D. & Bailey, A. J. (1979). In *Fibrous Proteins: Scientific, Industrial and Medical Aspects*, Parry, D. A. D. & Creamer, L. K. (Eds.), vol. 1, Academic Press, New York, pp. 151–177.
Lin, A. W. M. & Castell, D. O. (1978). *Anal. Biochem.*, **86,** 345–356.
Liu, T. Y. (1977). In *The Proteins, III*, 3rd edn, Neurath, H. & Hill, R. L. (Eds.), Academic Press, New York, pp. 239–402.
Lockridge, O., Eckerson, H. W. & La Du, B. N. (1979). *J. Biol. Chem.*, **254,** 8324–8330.
Loewy, A. G. (1984). *Methods Enzymol.*, **107,** 241–257.
McElvain, M. & Schroeder, J. P. (1949). *J. Am. Chem. Soc.*, **71,** 40–46.
Maclaren, J. A. & Milligan, B. (1981). *Wool Science: The Chemical Reactivity of the Wool Fibre*, Science Press, Sydney.
Mahuron, D. (1979). *J. Chromatogr.*, **172,** 394–396.
Mann, K. G. & Fish, W. W. (1972). *Methods Enzymol.* **26,** 28–42.
Manwell, C. (1977). *Biochem. J.*, **165,** 487–495.
Margolis, J. & Kenrick, K. G. (1968). *Anal. Biochem.*, **25,** 347–362.
Marshall, R. C. (1980) *J. Invest. Derm.*, **75,** 264–269.
Marshall, R. C. (1981). *Text. Res. J.*, **51,** 106–108.
Marshall, R. C. & Blagrove, R. J. (1979). *J. Chromatogr.*, **172,** 351–356.
Matacic, S. S. & Loewy, A. G. (1979) *Biochim. Biophys. Acta*, **576,** 263–268.
Matheka, H. D., Enzmann, P. J., Bachrach, H. L. & Migl, B. (1977). *Anal. Biochem.*, **81,** 9–17.
Maurer, H. R., (Ed.) (1971). In *Disc Electrophoresis and Related Techniques of Polyacrylamide Gel Electrophoresis*, 2nd edn, Walter de Gruyter, Berlin.
Meighen, E. A. & Schachman, H. K. (1970). *Biochemistry*, **9,** 1163–1176.
Mendel-Hartvig, I. (1982). *Anal. Biochem.*, **121,** 215–217.
Monsigny, M., Seine, C., Obrenovitch, A., Roche, A., Delmotte, F. & Boschetti, E. (1979). *Eur. J. Biochem.*, **98,** 39–45.
Morgan, G. & Ramsden, D. B. (1978). *J. Chromatogr.*, **161,** 319–323.
Mort, A. J. & Lamport, D. T. A. (1977). *Anal. Biochem.*, **82,** 289–309.
Muranova, T. A. & Muranov, A. V. (1979). *Sov. J. Bioorg. Chem.*, **5,** 747–750.
Noel, D., Nikaido, K. & Ames, G. F. (1979). *Biochemistry*, **18,** 4159–4165.

Nozaki, Y. (1972). *Methods Enzymol.*, **26,** 43–50.
O'Farrell, P. H. (1975). *J. Biol. Chem.*, **250,** 4007–4021.
Paz, M. A., Pereyra, B., Gallop, P. M. & Seifter, S. (1974). *J. Mechanochem. Cell Motility*, **2,** 231–239.
Pereyra, B., Blumenfeld, O. O., Paz, M. A., Henson, E. & Gallop, P. M. (1974). *J. Biol. Chem.*, **249,** 2212–2219.
Perham, R. N. (1975). *Phil. Trans, R. Soc. Lond. B,* **272,** 123–136.
Peters, K. & Richards, F. M. (1977). *Ann. Rev. Biochem.*, **46,** 523–551.
Peterson, E. A. (1970). In *Laboratory Techniques in Biochemistry and Molecular Biology*, vol. 2, Work, T. S. & Work, E. (Eds.), North-Holland, Amsterdam, Oxford, pp. 223–396.
Pharmacia (1979). *Gel Filtration Theory and Practice*, Uppsala, Sweden.
Pillai, S. & Bachhawat, B. K. (1979). *J. Mol. Biol.*, **131,** 877–881.
Podell, D. N. & Abraham, G. N. (1978). *Biochem. Biophys. Res. Commun.*, **81,** 176–185.
Poduslo, J. F. & Rodbard, D. (1980). *Anal. Biochem.*, **101,** 394–406.
Poehling, H.-M. & Neuhoff, V. (1981). *Electrophoresis*, **2,** 141–147.
Prestidge, R. L. & Hearn, M. T. W. (1979). *Anal. Biochem.*, **97,** 95–102.
Radola, B. J. (1973). *Biochim. Biophys. Acta*, **295,** 412–428.
Radola, B. J. (1974). *Biochim. Biophys. Acta*, **386,** 181–195.
Reid, M. S. & Bieleski, B. L. (1968). *Anal. Biochem.*, **22,** 374–381.
Reynolds, J. A. & Tanford, C. (1970). *J. Biol. Chem.*, **245,** 5161–5165.
Righetti, P. G. & Drysdale, J. W. (1974). J. Chromatogr. **98,** 271–321.
Robins, S. P. & Bailey, A. J. (1974). *FEBS Lett.*, **38,** 334–336.
Robins, S. P. & Bailey, A. J. (1975). *Biochem. J.*, **149,** 381–385.
Robins, S. P. & Bailey, A. J. (1977). *Biochem. J.*, **163,** 339–346.
Rodbard, D. (1976a). In *Methods of Protein Separation*, vol. 2, Catsimpoolas, N., (Ed.), Plenum Press, New York, pp. 145–179.
Rodbard, D. (1976b). In *Methods of Protein Separation*, vol. 2, Catsimpoolas, N. (Ed.), Plenum Press, New York, pp. 181–218.
Rüegg, U. T. & Rudinger, J. (1974). *Int. J. Pept. Prot. Res.* **6,** 447–456.
Rüegg, U. T. & Rudinger, J. (1977a). *Methods Enzymol.*, **47,** 111–116.
Rüegg, U. T. & Rudinger, J. (1977b). *Methods Enzymol.* **47,** 116–122.
Rüegg, U. T., Jarvis, D. & Rudinger, J. (1979). *Biochem. J.*, **179**, 127–134.
Sammons, D. W., Adams, L. D. & Nishizawa, E. E. (1981). *Electrophoresis*, **2,** 135–141.
Savige, W. E. & Fontana, A. (1977a). *Methods Enzymol.*, **47,** 442–453.
Savige, W. E. & Fontana, A. (1977b). *Methods Enzymol.*, **47,** 453–459.
Scanu, A. M. & Edelstein, C. (1971). *Anal. Biochem.*, **44,** 576–588.
Schirmer, R. H., Untucht-Grau, R., Pai, E. F. & Schulz, G. E. (1977). In *Solid Phase Methods in Protein Sequence Analysis*, Proc. Second Internat. Conf., Previero, A. & Coletti-Previero, M.-A. (Eds.), North-Holland, Amsterdam, New York, Oxford, pp. 265–276.
Schlesinger, M. (1979). *Cell*, **17,** 813–819.
Schram, E., Moore, S. & Bigwood, E. J. (1954), *Biochem. J.*, **57,** 33–37.
Schreiber, G., Dryburgh, H., Millership, A., Matsuda, Y., Inglis, A., Phillips, J., Edwards, K. & Maggs, J. (1979). *J. Biol. Chem.*, **254,** 12013–12019.
Schwam, H., Michelson, S., Randall, W. C., Sondey, J. M. & Hirschmann, R. (1979). *Biochemistry*, **18,** 2828–2833.
Scopes, R. K. (1978). In *Techniques in Protein and Enzyme Biochemistry* B101, Elsevier/North-Holland, Shannon, pp. 1–42.
Segrest, J. P. & Jackson, R. L. (1972). *Methods Enzymol.*, **28,** 54–63.
Siezen, R. J. (1979). *FEBS. Lett.*, **100**, 75–80.

Stark, G. R. (1965). *Biochemistry*, **4,** 1030–1036.
Stegemann, H. (1979). In *Electrokinetic Separation Methods*, Righetti, P. G., Van Oss, C. J. & Vanderhoff, J. W. (Eds.), Elsevier/North-Holland, Amsterdam, pp. 313–336.
Steinert, P. M. & Idler, W. W. (1975). *Biochem. J.*, **151,** 603–614.
Stimler, N. P. & Tanzer, M. L. (1974). *Biochim. Biophys. Acta,* **365,** 425–433.
Studier, F. W. (1973). *J. Mol. Biol.*, **79,** 237–248.
Sutoh, K., Chiao, Y.-C. C. & Harrington, W. F. (1978). *Biochemistry*, **17,** 1234–1239.
Tanzer, M. L. (1976). In *Biochemistry of Collagen*, Ramachandran, G. N. & Reddi, A. H., (Eds.), Plenum Press, New York, pp. 137–162.
Thomas, J. O. (1978). In *Techniques in Protein and Enzyme Biochemistry* B106, Elsevier/North-Holland, Shannon, pp. 1–22.
Thompson, E. O. P. & O'Donnell, I. J. (1966). *Aust. J. Biol. Sci.*, **19,** 1139–1151.
Tuszynski, G. P., Damsky, C. H., Fuhrer, J. P. & Warren, L. (1977). *Anal. Biochem.*, **83,** 119–129.
Ui, N. (1971). *Biochim. Biophys. Acta*, **229,** 567–581.
Van Holde, K. E. (1975). In *The Proteins*, *I*, 3rd edn, Neurath, H. & Hill, R. L. (Eds.), Academic Press, New York, pp. 225–291.
Vinograd, J. R., Hutchinson, W. D. & Schroeder, W. A. (1959). *J. Am. Chem. Soc.*, **81,** 3168–3169.
Von Hippel, P. H. & Schleich, T. (1969). In *Structure and Stability of Biological Macromolecules*, Timasheff, S. N. & Fasman, G. D. (Eds.), Marcel Dekker, New York, pp. 417–454.
Walker, J. E., Auffret, A. D., Carne, A. F., Naughton, M. A. & Runswick, M. J. (1980). In *Methods in Peptide and Protein Sequence Analysis*, Proc. Third Internat. Conf., Birr, C. (Ed.), Elsevier/North-Holland, Amsterdan, New York, Oxford, pp. 257–265.
Ward, C. W. & Dopheide, T. A. A. (1980). *Virology*, **103,** 37–53.
Weber, K. & Osborn, M. (1969). *J. Biol. Chem.*, **244,** 4406–4412.
Weber, K. & Osborn, M. (1975). In *The Proteins*, *I*, 3rd edn, Neurath, H. & Hill, R. L. (Eds.), Academic Press, New York, pp. 179–223.
Wieland, F., Renner, L., Verfurth, C. & Lynen, F. (1979). *Eur. J. Biochem.*, **94,** 189–197.
Wright, D. J. & Boulter, D. (1974). *Biochem. J.*, **141,** 413–418.
Yamato, S. & Murachi, T. (1979). *Eur. J. Biochem.*, **93,** 189–195.

Practical Protein Chemistry—A Handbook
Edited by A. Darbre
© 1986, John Wiley & Sons Ltd.

ANGELO FONTANA and the late ERHARD GROSS
Institute of Organic Chemistry (Biopolymer Research Centre, C. N. R.), University of Padova, Padova, Italy (A. F.), and National Institute of Child Health and Human Development, Endocrinology and Reproduction Research Branch, Section on Molecular Structure, Bethesda, Maryland, U. S. A. (E. G.).

2

Fragmentation of Polypeptides by Chemical Methods

CONTENTS

2.1. INTRODUCTION

A fundamental step in the elucidation of the covalent structure of a protein is the degradation of the macromolecule to low molecular weight peptides which can be completely characterized with respect to composition and sequence. In general, the primary structure of a protein cannot be determined simply by sequential degradation from either terminus of the entire molecule. Instead one must resort to the examination of fragments obtained from selective enzymatic or chemical cleavage of the polypeptide chain. This chapter deals with the fragmentation of polypeptide chains by chemical methods, whereas enzymic methods are discussed by Wilkinson (Chapter 3) and Winter (Chapter 10).

Chemical cleavage of a polypeptide chain exploits the unique reactivity of chemically modified side chains of particular amino acids in the labilization of adjacent peptide bonds by neighboring group participation (Capon, 1964; Capon

& McManus, 1976). Over the years several methods have been explored to effect protein cleavage at specific amino acid residues. A good procedure for the selective chemical cleavage of polypeptide chains must meet the criteria of high specificity and high cleavage yield. Ideally it ought to be free of side reactions. However, many of the procedures devised suffer from the shortcomings of poor yield of peptide bond fission and the occurrence of undesired reactions in side-chain groups.

High specificity is an important requirement for a fragmentation technique to be useful because with few peptide bonds cleaved the resulting mixture of peptide fragments is easily fractionated. Given the potential to determine relatively long amino acid sequences in the sequencer the protein is best cleaved into a small number of large fragments which, after separation, are directly subjected to automated sequence analysis. Side-chain functions of relatively rare amino acids such as methionine, tryptophan and cysteine are potential sites of attack to achieve limited fragmentation.

The demand for cleavage procedures that yield large fragments has stimulated more and more interest in the development of specific chemical methods for polypeptide cleavage. The methods to be presented here are restricted to those which have been proved to be of practical utility and are most often used at present. Besides the methods available for cleavage at methionine, tryptophan, tyrosine and cysteine residues, partial acid hydrolysis and reductive and oxidative cleavage of disulfide bonds of cystine residues are also discussed. There is easy access to the literature on the chemical fragmentation of peptides and proteins and no attempt is made in this article to cover all aspects of the chemical degradation of proteins. Detailed reviews (Witkop, 1961; Spande *et al.*, 1970; Fontana & Scoffone, 1972; Heinrikson & Kramer, 1974; Atassi, 1977; Konigsberg & Steinman, 1977) provide ample documentation for the reader who is interested in greater insights into the procedures here described. Specific experimental procedures have already been reported (Hirs 1967a; Hirs & Timasheff, 1972, 1977, 1983).

2.2. CLEAVAGE OF DISULFIDE BONDS

Cystine residues may form interchain links and/or intrachain links. Disulfide bonds of a protein may need to be cleaved before any sequence work in order to separate the peptide chains linked through the S-S-bridges. In addition, disulfide bonds are reactive groups which can cause complications during structural studies, so that it is necessary to stabilize them as suitable derivatives. Finally, proteins in which the cystine residues are split are more susceptible to attack by proteolytic enzymes. Methods available for cleaving disulfide bonds are herein discussed. There are several reviews on this subject (Cecil, 1963; Jocelyn, 1972;

Friedman, 1973; Liu, 1977). Only selected methods more frequently employed are given here.

2.2.1. Reduction and *S*-carboxymethylation

Reductive methods are by far the most selective and useful procedures for splitting the disulfide bonds of proteins. The traditional method of reduction involves treatment of a protein with high concentrations of a low molecular weight thiol. The reaction may be represented by equation 2.1, from which it is obvious that in order to drive the reaction in the desired direction a considerable excess of thiol has to be added. Thiol reagents that have been used are cysteine, reduced glutathione, 2-mercaptoethylamine, thioglycollic acid, 2-mercapto-ethanol (Hirs, 1967; White, 1972) and dithiothreitol (DTT) (Cleland, 1964).

$$\text{protein}(\text{S—S}) + 2\,\text{RSH} \rightleftharpoons \text{protein}(\text{SH}\ \ \text{SH}) + \text{RSSR} \qquad (2.1)$$

Reduction is usually performed in the absence of oxygen, at a slightly alkaline pH and under conditions in which the protein is unfolded, i.e. in the presence of 8 M-urea or 6 M-guanidine HCl. When urea is used as an unfolding agent, care must be taken to avoid the presence of cyanate ion, which is formed on standing in urea solutions and which causes protein modification (Stark, 1967). It is advantageous to use recrystallized urea and preferably freshly prepared solutions which have been deionized using a mixed bed resin.

Upon removal of the reducing agent, the protein thiol groups readily oxidize in air to reform disulfide bonds. This can be prevented by the modification of the thiol groups to stable derivatives and is generally accomplished in the same reaction mixture that is used for the reduction. The most frequently used -SH blocking reagent is iodoacetic acid. Cysteine residues are thus converted to *S*-carboxymethylcysteine that will remain stable during further study of the protein and in particular during protein sequence determination. The extent to which complete reduction and *S*-carboxymethylation have been attained is determined by amino acid analysis. S-carboxymethylcysteine is stable to acid hydrolysis and is determined with an amino acid analyzer. This also serves as a check on the extent of side reactions involving other amino acids (methionine, tyrosine, histidine) (Gurd, 1972).

It has been shown that *S*-carboxymethylcysteine (**1**) can undergo intramolecular cyclization under acid conditions to yield a thiazine derivative (**2**). This derivative escapes detection by amino acid analysis because it lacks a free amino group (Bradbury & Smyth, 1973). The rate and extent of this reaction depend upon the temperature and pH. Cyclization to the thiazine derivative occurs most rapidly in citrate buffer, pH 3.0, at 110°C, whereas under the usual conditions of acid

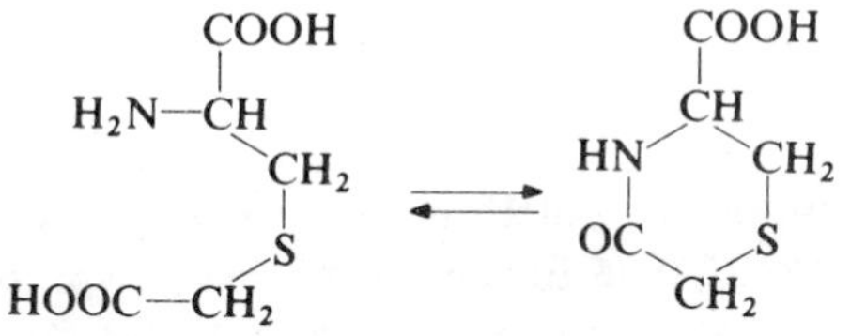

$$\underset{\mathbf{1}}{HOOC{-}CH_2{-}S{-}CH_2{-}CH(NH_2){-}COOH} \rightleftharpoons \underset{\mathbf{2}}{\text{cyclo}[HN{-}CH(COOH){-}CH_2{-}S{-}CH_2{-}CO]}$$

hydrolysis for amino acid analysis (6 M-HCl, 110°C, 22 h) *S*-carboxymethylcysteine is stable. In order to avoid this complication, Bradbury & Smyth (1973) reacted cysteine SH-groups with 3-bromopropionic acid, which yields a stable *S*-carboxyethylcysteine derivative.

A general procedure for reduction of disulfide bonds in ribonuclease using 2-mercaptoethanol and their stabilization by carboxymethylation was reported (White, 1960; Anfinsen & Haber, 1961; Crestfield *et al.*, 1963). At present, the reagent of choice for the reduction of disulfide bonds of proteins is dithiothreitol (DTT) (3) (see review Konigsberg, 1972). The use of DTT as a reducing agent for disulfides was first described by Cleland (1964), and the reagent bears his name. It is commercially available.

$$RSSR + \underset{\mathbf{3}}{HSCH_2CH(OH)CH(OH)CH_2SH} \longrightarrow \underset{\mathbf{4}}{RSSCH_2CH(OH)CH(OH)CH_2SH} + RSH \quad (2.2)$$

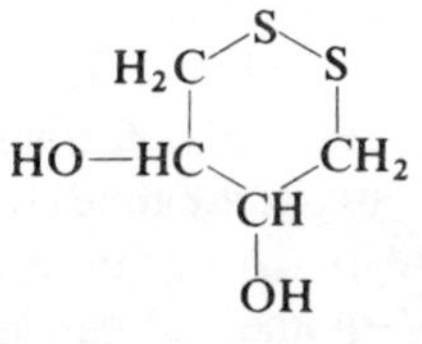

$$\mathbf{4} \longrightarrow \underset{\mathbf{5}}{\text{cyclo}[S{-}S{-}CH_2{-}CH(OH){-}CH(OH){-}CH_2]} + RSH \quad (2.3)$$

The reaction (equations 2.2 and 2.3) proceeds nearly to completion because the formation of a 6-membered ring (5) containing a disulfide bridge is energetically favored over the mixed disulfide. The redox potential of DTT at pH 7 and 25°C is -0.33 V compared to -0.22 V for cysteine. These two values allow one to calculate an overall equilibrium constant of 10^4 for the reduction of cysteine by DTT. This reagent can be used at a much lower concentration than mercaptoethanol by virtue of its lower redox potential and its resistance to air oxidation. An additional advantage of DTT is its relatively low level of unpleasant odor.

2.2.1.1. *Method of reduction and S-carboxymethylation*

Method: after Waxdal *et al.* (1968)

A 10–20 mg/ml solution of protein in buffer containing 6 M-guanidine HCl, 0.2–0.5 M-Tris-HCl and 2 mM-EDTA, pH 8.0, in a polypropylene tube is flushed with nitrogen, capped and incubated at 37–50°C for 30 min to aid denaturation. DTT is added to give a 50-fold molar excess (or add 300 mol/mol of protein if the disulfide content is unknown), the tube flushed with nitrogen and incubated at 37°C for 4 h.

Cool the solution in ice and add recrystallized iodoacetic acid or iodoacetamide to give a slight excess over the amount of SH groups in the mixture. Keep the tube in the dark. Check the pH and adjust to pH 8 if necessary. After 1 h remove excess reagents by adding mercaptoethanol to 'mop-up' excess alkylating reagent and dialyze at alkaline pH in the cold, to prevent modification of methionine residues, which can take place during dialysis against an acidic buffer. Rapid desalting with a buffer or solution in which the protein is soluble can also be used (e.g. Sephadex G-10 or Biogel P-10 in 0.01 M-NH_4HCO_3).

2.2.1.2. *Method using tributylphosphine*

Trialkyl phosphines have been shown to be potent and specific agents for the cleavage of disulfide bonds in wool (MacLaren & Sweetman, 1966). Rüegg and Rudinger (1977a) studied tributylphosphine reduction with proteins of known structure.

$$R_1SSR_2 + Bu_3P + H_2O \rightarrow R_1SH + R_2SH + Bu_3PO \qquad (2.4)$$

Equation 2.4 shows the particular advantage of the stoichiometry of the reaction. Only a 5–20% molar excess of tributylphosphine over protein disulfide groups is used. In addition, the reaction is rapid, specific for disulfides, and alkylation of the resultant thiol groups can be performed in the presence of the reducing agent. The reaction is usually carried out under alkaline conditions. Under acid conditions the reaction is slower (24–48 h). To ensure a one-phase system the solution contains up to 50% propan-1-ol and this can assist in solubilizing the protein. Tributylphosphine has an unpleasant smell and is *toxic*. Work in the hood. Store the reagent and all solutions under nitrogen. A 2% (v/v) solution in propan-1-ol has a reducing capacity of approximately 55–70 μmol/ml.

Method: Rüegg and Rudinger (1977a)

Bovine pancreatic ribonuclease (64 mg, 5 μmol) is dissolved in 7.5 ml of 7 M-urea in 0.1 M-Tris-HCl (pH 8.2)-propan-1-ol (2:1, v/v). Under nitrogen and with stirring add 50 μl of tributylphosphine (130–180 μmol, 6- to 9-fold excess). After 2 h add 140 μl of 1 M-iodoacetate (1 M-iodoacetic acid in 1 M-NaOH, 3.5-fold excess) in the dark. The solution is adjusted after 5 min to pH 8.2 with 1 M-Na_2CO_3 and after another 30 min 0.5 ml of glacial acetic acid is added. The solution is then dialyzed in the cold against three changes of 2 l of 0.1 % acetic acid and the material is recovered by lyophilization.

Several other methods for the reductive cleavage of disulfide bonds have been employed in protein chemistry. For sequence studies reduction with sodium borohydride at pH 7 to 10 is no longer used because the reagent was found to produce cleavage of the peptide bond (Crestfield *et al.*, 1963). Diborane, which reduces carboxyl (but not carboxylate) groups to their corresponding alcohols, also affects the reduction of disulfide bonds (Atassi & Rosenthal, 1969; Atassi *et al.*, 1975). It has been shown that diborane may be employed to achieve disulfide bond reduction in non-protonated proteins without effect on the carboxyl groups. Electroreduction of disulfide bonds by use of a dropping mercury electrode has also been employed successfully, although rarely with proteins (Cecil & Weitzman, 1964; Leach *et al.*, 1965).

2.2.2 Other *S*-protecting groups

Alternatives to the *S*-carboxymethylation reaction are available. The method employed should be chosen to suit the purpose of the investigation. It may be desirable to have a reversible SH-protecting group, or a highly soluble derivative, or a chromophoric or fluorescent label attached to the protein, or the ability to cleave subsequently by trypsin at the cysteine residues. The sulfydryl group is the most reactive function in amino acid side chains of proteins. It may be readily alkylated, acylated, arylated, oxidized and form complexes with heavy metals (Jocelyn, 1972; Freedman, 1973; Torchinskii, 1974; Kenyon & Bruice, 1977).

2.2.2.1. Ethyleneimine

The positively-charged derivative *S*-aminoethylcysteine is prepared by reacting an SH-protein with ethyleneimine and this is often used to introduce additional sites for the tryptic cleavage of proteins (equation 2.5) (Lindley, 1956; Raftery & Cole, 1963, 1966). Trypsin cleaves peptide bonds involving residues of *S*-(*β*-aminoethyl) cysteine, an analogue of lysine. Proteolysis, however, is slower than for bonds involving lysyl (Cole, 1967b) or arginyl residues. The *S*-alkylation reaction with ethyleneimine can give rise to side reactions, such as alkylation of methionine (Schroeder *et al.*, 1967) or of the α-amino group of the peptide chain (Shotton & Hartley, 1973), particularly if the alkylation reaction is allowed to

$$\text{protein—SH} + \underset{\underset{\displaystyle\text{H}}{\displaystyle\text{N}}}{\text{CH}_2\text{—CH}_2} \rightarrow \text{protein—S—CH}_2\text{—CH}_2\text{—NH}_2 \qquad (2.5)$$

proceed for a prolonged time in the presence of a large excess of ethyleneimine. *S*-aminoethylcysteine is eluted with the basic amino acids on the amino acid analyzer (Schroeder *et al.*, 1967).

Reduction with tributylphosphine and alkylation with ethyleneimine

Contribution by Marshall and Inglis. Dissolve the protein (2–10 mg/ml) in 0.5 M-$NaHCO_3$-propan-1-ol (1:1,v/v). Adjust to pH 8.0–8.3 and add with stirring under nitrogen tributylphosphine (1–5 μmol per mg of protein) and ethyleneimine (4–40 μmol per mg of protein). React for 2-4 h.

Reduction with mercaptoethanol and alkylation with ethyleneimine

This is the method of Slobin and Singer (1968). The protein (25 mg of γ-globulin) is dissolved in 5 ml of fresh 8 M-urea or 6 M-guanidine HCl in 1 M-Tris-HCl buffer, pH 8.9, containing 75 μl of 2-mercaptoethanol. After 2 h at room temperature under nitrogen 150 μl of ethyleneimine is added, and the alkylation allowed to proceed for 2 h. Th reaction mixture is then dialyzed, first against 100 vol. of 1% NH_4HCO_3 and then against distilled water. The protein derivative is recovered by lyophilization.

2.2.2.2. 4-Vinylpyridine

4-Vinylpyridine was used to alkylate the –SH group in proteins (equation 2.6) (Cavins & Friedman, 1970; Friedman *et al.*, 1970).

$$\text{protein—SH} + CH_2{=}CH\text{—}C_5H_4N \longrightarrow \text{protein—S—}CH_2\text{—}CH_2\text{—}C_5H_4N \quad (2.6)$$

This group introduces a polar nature to the side chain, which increases solubility and facilitates purification of the protein. The derivative is stable to acid hydrolysis and has a characteristic absorption at 260 nm. The PTH derivative is easily identified during protein sequencing (Hermodson *et al.*, 1973).

Alkylation with 4-vinylpyridine

This is the method of Inglis *et al.* 1976.
Dissolve the protein (300–500 mg) in 20–30 ml of Tris-HCl buffer, pH 7.5 containing 8 M-urea, under nitrogen. Add 4-vinylpyridine (20 M excess over –SH groups) and stir for 2–3 h. Acidify to pH 3 with glacial acetic acid. Dialyze against 0.01 M-acetic acid and lyophilize.

Reduction with DTT and alkylation with 4-vinylpyridine

This is the method of Hermodson *et al.* (1973).
Porcine trypsin (20 mg) was dissolved in a solution of 1 ml of 6 M-guanidine HCl, 0.13 M-Tris and 0.1 mg of EDTA, pH 7.6. DTT was added to yield a 20-fold molar excess over the concentration of protein disulfide groups. After 3 h at room temperature, the mixture was treated for 90 min with 3.0 mol 4-vinylpyridine/mol DTT. The solution was then acidified to pH 2.0 with 88 % formic acid and the alkylated protein separated from reagents and salts on a column of Sephadex G-75 equilibrated and eluted with 9 % aqueous formic acid. Three fractions were obtained corresponding to the single chain of β-trypsin and to the two fragments of α-trypsin.

2.2.2.3. Other S-alkylating groups

Three new SH-alkylating agents have been developed to introduce charged groups at cysteinyl residues in proteins, sodium 2-bromoethane-sulfonate (**6**) (Niketic *et al.*, 1974), 1,3-propane sultone (**7**) (Rüegg & Rudinger, 1974, 1977b) and 2-bromoethyl-trimethylammonium bromide (**8**) (Itano & Robinson, 1972). Reagents **6** and **7** generate protein-$S(CH_2)_nSO_3^-$ and reagent **8** protein-$SCH_2CH_2N^+(CH_3)_3$. These charged derivatives are potentially useful in electrophoretic studies, maintaining their respective charges over a broad range of pH.

$$Br{-}CH_2CH_2{-}SO_3^- Na^+ \quad (\mathbf{6})$$

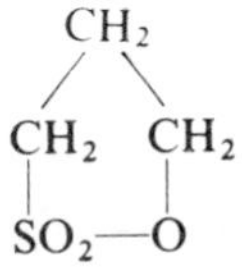

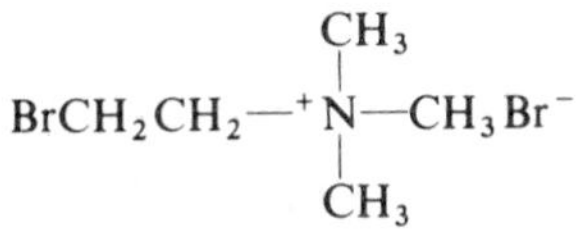

$$\begin{array}{ccc} & CH_2 & \\ / & & \backslash \\ CH_2 & & CH_2 \\ | & & | \\ SO_2 & - & O \end{array} \quad (\mathbf{7})$$

$$BrCH_2CH_2{-}{}^+N(CH_3)_3\,Br^- \quad (\mathbf{8})$$

Reduction and sulfopropylation with 1,3-propane sultone

This is the method of Rüegg and Rudinger (1977). Dissolve ribonuclease (64 mg, 5μ mol) in 7.5 ml of 7 M-urea in 0.1 M-Tris-HCl, pH 8.2/propan-1-ol (2 : 1, v/v). De-aerate and saturate with nitrogen. Add tributyl phosphine (22–40 μmol, 1.1- to 2-fold excess) as a 5 % titrated solution in propan-1-ol and 320 μl (80 μmol) of a 250 mM solution of 1,3-propane sultone in propan-1-ol/water (1:1, v/v). Allow to stand at room temperature for 16 h. Acidify with acetic acid and dialyze in the cold against three changes of 0.1 % acetic acid. Recover the *S*-sulfopropylated protein by lyophilization.

2.2.2.4. Reversible protection of thiol group

In some cases it could be advantageous to reversibly protect the thiol group of cysteine, i.e. with a group that can be eventually removed. Reversible protection of thiol groups of cysteine residues in proteins can be achieved via mixed disulfides, which can be easily cleaved by reduction. A class of mixed disulfide forming -SH reagents are the arylsulfenyl halides (Fontana *et al.*, 1968; Fontana & Scoffone, 1972). 2-Nitro-phenylsulfenyl chloride reacts in aqueous acetic acid with a cysteine residue to form the mixed disulfide (**9**). This reagent reacts also with tryptophan residues which are converted to 2-thioether-indole derivatives (**10**).

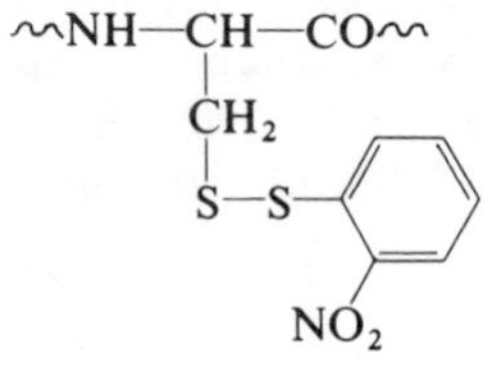

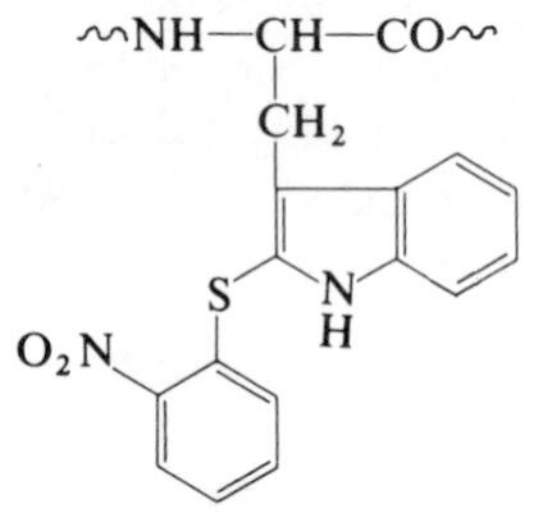

9

10

Azobenzene-2-sulfenyl bromide, on the other hand, was found to react specifically with -SH groups (equation 2.7). The lack of reactivity towards the indole nucleus was related to the special structure of the sulfenyl halide, which is a 2-phenylbenzo-1-thio-2,3-diazonium ion (**11**).

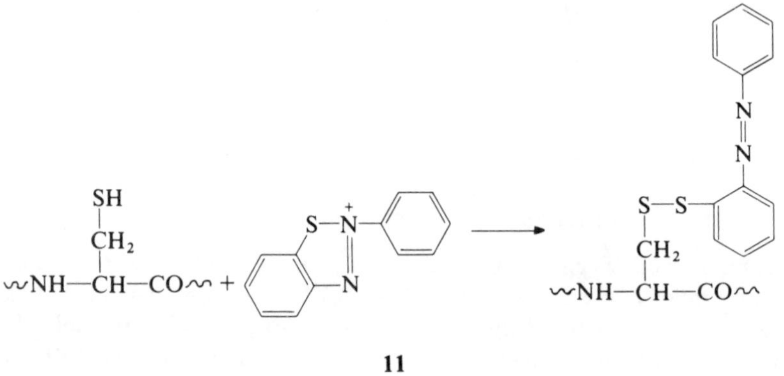

(2.7)

11

A similar alkyl-aryl mixed disulfide can be obtained also by reacting the cysteine-containing protein with 5,5′-dithiobis-(2-nitrobenzoic acid) (**12**) (DTNB; Ellman's reagent; Ellman, 1959) (equation 2.8). DTNB has a higher redox potential than aliphatic disulfides and will react with aliphatic thiols by an exchange reaction to form a mixed disulfide (**13**) and the colored 2-nitro-5-thiobenzoate (**14**). Ellman's reagent is widely used for the quantitative determination of cysteine content in proteins (Habeeb, 1972).

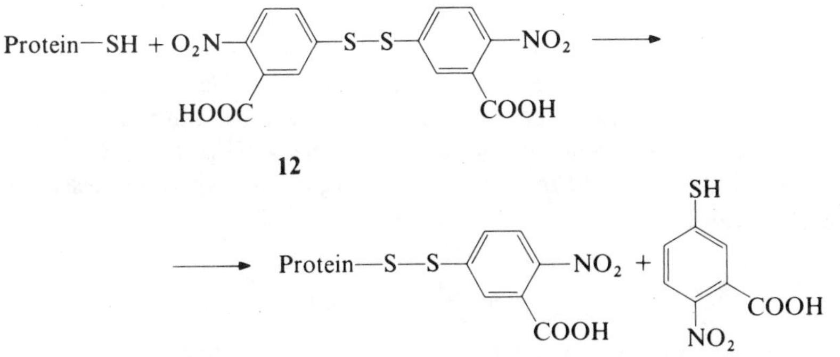

Protein—SH + O_2N–C_6H_3(COOH)–S–S–C_6H_3(COOH)–NO_2 ⟶

HOOC COOH

12

(2.8)

⟶ Protein—S—S–C_6H_3(COOH)–NO_2 + HS–C_6H_3(COOH)–NO_2

SH

COOH COOH

NO_2

13 **14**

Alkyl alkanethiolsulfonates were found to be specific reagents for -SH-groups to form mixed disulfides (equation 2.9). The reaction occurs

$$\text{Protein-SH} + \text{R-S-SO}_2\text{-R} \rightarrow \text{Protein-S-S-R} + \text{R-SO}_2\text{H} \qquad (2.9)$$

rapidly under mild conditions without a large excess of reagent (Smith *et al.*, 1975; Kenyon & Bruice, 1977). The most widely used reagent is methyl methanethiolsulfonate, which introduces the small uncharged CH_3S-group into a thiol-containing protein.

2.2.3. Oxidation

Performic acid oxidation is the most widely used procedure for oxidative cleavage of disulfide bonds in proteins. Treating a protein with performic acid leads to the oxidation of cysteine and cystine residues to cysteic acid residues (**15**) in yields of more than 90%. Cysteic acid is stable to acid hydrolysis and elutes very rapidly before aspartic acid on the amino acid analyser (Spackman *et al.*, 1958). The reaction is useful for the determination of the combined content of cystine and cysteine residues in proteins. Simultaneously, methionine is converted to methionine *S*, *S*-dioxide (**16**), tryptophan to N-formylkynurenine (**17**) and to other unknown substances (Toennies & Homiller, 1942; Benassi *et al.*, 1965) and

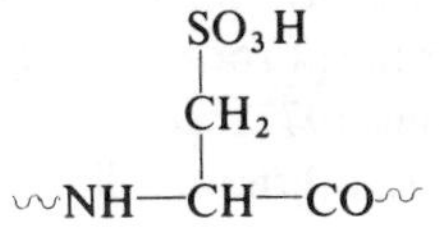

SO_3H–CH_2–(～NH—CH—CO～)

15

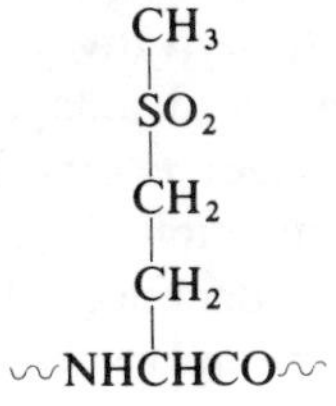

CH_3–SO_2–CH_2–CH_2–(～NHCHCO～)

16

HCO—NH–C_6H_4–CO–CH_2–(～NH—CH—CO～)

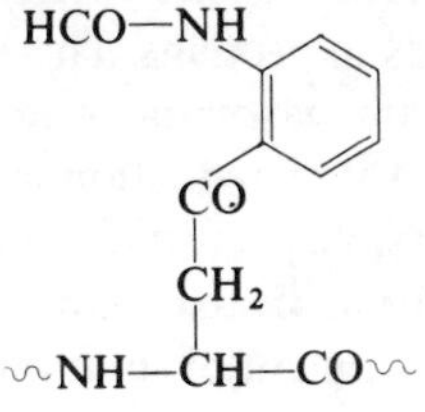

17

tyrosine, in part, to halogenated derivatives (Hirs, 1956; Sanger & Thompson, 1963). Attempts to find conditions which would either minimize side reactions with amino acids other than cysteine and cystine residues, or lead to single well-defined products have been unsuccessful. Performic acid oxidation of proteins was first applied to insulin (Sanger, 1949) and ribonuclease (Hirs, 1956; Hirs *et al.*, 1956). The method of Hirs (1967b) is described below (see also this reference for a detailed discussion).

2.2.3.1. Performic acid oxidation

The performic acid reagent is made by mixing 5 vol. of 30% hydrogen peroxide and 95 vol. of 99% formic acid and the closed container is allowed to stand at room temperature for 2 h. The titer of the peracid reaches a maximum about this time and decreases slowly thereafter. The reagent (1 ml for 5 mg of protein) is then added to the protein solution at −5°C in the ratio of 2:1 (v/v). Dissolve the protein in 99% formic acid and add methanol to 20% concentration to prevent freezing. The final protein concentration is about 0.5%. After 2–3 h at −5 to −10°C in a salt–ice bath, the sample is diluted with water and lyophilized. The oxidized protein may also be recovered by precipitation with trichloroacetic acid or an organic solvent. Alternatively, the reaction mixture is diluted with an equal vol. of ice-cold water and dialyzed against two changes of water in the cold and finally against 1 mM-2-mercaptoethanol. The protein is then recovered by lyophilization. See also Chapter 1, section 1.5.1.1.

2.2.4 Sulfitolysis

When disulfides are treated with alkali sulfite, S-sulfonated derivatives and thiols are formed (equation 2.10). The thiols must be re-oxidized to disulfides which are reacted with more sulfite and the reaction continues until conversion of disulfides into *S*-sulfonate is complete (Cecil, 1963; Cole, 1967a; Liu, 1977).

$$RSSR + SO_3^{2-} \rightleftharpoons RS^- + RSSO_3^- \quad \xrightarrow{[O]} \text{(RS}^- \text{ back to RSSR)} \qquad (2.10)$$

Sulfitolysis has been employed as a mild procedure for the cleavage of disulfide bonds in proteins. The *S*-sulfonate derivatives are stable at neutral and acid pH and are reasonably water soluble. An advantage of the *S*-sulfo group is its ready removal by reduction with excess thiol reagent. This group has been used for the reversible protection of the thiol function of cysteine for the purposes of peptide synthesis (Katsoyannis, 1969; Brandenburg, 1972; Rüegg, 1977).

Treatment of solutions of cystine and cysteine with dithiothreitol followed by treatment with an excess of sodium tetrathionate ($Na_2S_4O_6$) is the basis of the method for the determination of the half-cystine residues in protein hydrolysates as *S*-sulfocysteine (Inglis & Liu, 1970; Liu & Inglis, 1972). It was suggested (Pihl & Lange, 1962) that the reaction of protein -SH groups with tetrathionate yields a *S*-sulfosulfenyl derivative rather than the *S*-sulfocysteine derivative, but under the experimental conditions described by Inglis and Liu (1970) it appears that the reaction of tetrathionate with free cysteine produces *S*-sulfocysteine only. This elutes close to cysteic acid on the amino acid analyzer (Liu & Inglis, 1972).

See Chapter 1, Section 1.5.1.2 for oxidative sulfitolysis, discussion and method.

2.3. PARTIAL ACID HYDROLYSIS

Preferential peptide bond fission of proteins by acids under milder conditions than those required for the complete hydrolysis to their constituent amino acids is useful for fragmenting peptide chains (Kasper, 1975). Basically, two methods are employed, partial acid hydrolysis with conc. HCl and hydrolysis in dilute acid solution.

The relative rate of hydrolysis of a given peptide bond depends on various steric and electrostatic factors. Valyl and leucyl peptides were found to be the most stable (Synge, 1945). Peptide bonds formed between two β-branched amino acids (e.g. Val-Ile, Ile-Ile) are extremely difficult to hydrolyze and probably form the most acid stable type of peptide bond. Positively charged basic groups might be expected to oppose the approach of hydrogen ions and this would explain the stability of dipeptides which accumulate towards the end of a hydrolysis.

The most frequently used acid is HCl although for particular problems acetic acid in combination with other acids has been used. Some experimental conditions that have been employed for partial acid hydrolysis of proteins are presented in Table 2.1

Table 2.1. Examples of partial acid hydrolysis of proteins

Protein	Conditons	Reference
Insulin	12 N HCl, 37°C, 3 days	Sanger & Thompson, 1953
Lysozyme	12 N HCl, 37°C, 7 days	Thompson, 1955
Trypsin	5.7 N HCl, 100°C, 20–30 min	Naughton *et al.*, 1960
Phosphoglucomutase	2 N HCl, 100°C, 30 min	Milstein & Sanger, 1961
Ribonuclease, insulin, glucagon	0.03 N HCl, 105°C, 4–48 h	Grannis, 1960 Schultz *et al.*, 1962

Since the specificity of partial acid hydrolysis is not high, extensive peptide bond cleavage occurs at numerous sites along the polypeptide chain. For example, the separated A and B chains of insulin (Sanger & Thompson, 1953) and lysozyme (Thompson, 1955) yielded numerous di- and tripeptides which were purified and characterized. Although determining the sequence of small peptides is easy (when they have been isolated), partial acid hydrolysis has rarely been used since fast and accurate methods for determining the sequences of relatively large peptide fragments became available.

2.3.1. Cleavage at aspartyl-prolyl peptide bonds

The Asp-Pro peptide bond appears to be particularly sensitive to hydrolysis under mild acid conditions (Fraser *et al.*, 1972; Hass *et al.*, 1975; Jauregui-Adell & Marti, 1975; see review by Landon, 1977). Selective cleavage of such bonds has been observed even during the acidic conditions employed in the proteolytic digestion of a protein with pepsin, in the chemical cleavage using CNBr in the presence of 70% formic acid, or during the purification of peptides by gel filtration using columns equilibrated with aqueous acetic or formic acid. Piszkiewicz *et al.* (1970) observed partial hydrolysis of an Asp-Pro bond in glutamic dehydrogenase treated with CNBr in 70% formic acid. These authors discussed the likely mechanism of the preferential splitting of the Asp-Pro bond.

This selective acid-catalyzed cleavage of the Asp-Pro bond is limited by the relative paucity of Asp-Pro bonds in proteins, but the procedure can provide relatively large fragments amenable to automated Edman procedures.

2.3.1.1. Selective cléavage

This is the method of Nute and Mahoney (1979).

S-pyridylethylated γ-chain of fetal hemoglobin (51 mg) was dissolved in 2 ml of 70% formic acid-7 M-guanidine HCl solution which was incubated at 40°C for 24 h. The reaction was stopped by the addition of 2 ml of cold, deionized water and the solution was immediately applied to a 20 × 950 mm column of Sephadex G-75 superfine, equilibrated and subsequently eluted with 9% formic acid. The carboxyl terminal fragment (residues 100 to 146) was recovered by lyophilization and further purified by passage through a 20 × 950 mm column of Sephadex G-50 superfine, using 9% formic acid as the eluent.

2.3.2. $N \rightarrow O$ acyl shift

Peptide bonds involving the amino groups of serine and threonine are quite susceptible to hydrolysis by conc. acid at room temperature. Cleavage is especially favorable for serine peptides, whereas threonine peptides cleave to a lesser extent (see review Iwai & Ando, 1967). The mechanism of this preferential cleavage seems to involve an $N \rightarrow O$ peptidyl shift, followed by the specific hydrolysis of the ester bond, as shown for a serine containing peptide (equation 2.11).

The $N \rightarrow O$ acyl shift takes place in conc. anhydrous acids such as sulfuric, phosphoric, hydrofluoric or formic. Conc. sulfuric acid is a classical reagent used for the reaction (Desnuelle & Bonjour, 1951; Elliott, 1952). Little non-specific splitting of peptide bonds appears to take place in sulfuric acid and indeed it has

$$\sim\!\mathrm{NHCH(R)\!-\!C(\!=\!O)\!-\!NHCH(CH_2OH)CO}\!\sim\; \rightleftharpoons\; \sim\!\mathrm{NHCH(R)\!-\!C(\leftarrow OH_2)(O\!-\!)\!=\!\overset{+}{N}HCH(\!-\!CH_2\!-\!)CO}\!\sim\; \longrightarrow \qquad (2.11)$$

$$\sim\!\mathrm{NHCH(R)\!-\!CO\!-\!O\!-\!CH_2\!-\!CH(NH_2)CO}\!\sim\; \xrightarrow{H_2O}\; \sim\!\mathrm{NHCH(R)COOH} + \mathrm{H_2NCH(CH_2OH)CO}\!\sim$$

been found that insulin retains its biological activity after sulfuric acid treatment (Glendenning *et al.*, 1947). However, some sulfonation of tyrosine may occur and tryptophan may be partly destroyed.

Chemical fragmentation at serine and threonine residues by the $N \rightarrow O$ acyl shift reaction requires particular conditions for the selective hydrolysis of the ester bonds (*O*-peptide bonds) formed at serine and threonine residues. These ester bonds are hydrolyzable by acid or alkali. However, since the ester, in the presence of alkali, reverts to the original amide, protection of the amino groups by acylation is necessary to prevent the reverse shift and thus improve yields of cleavage. The reversible nature of the reaction is explained by the formation of a cyclic hydroxyoxazolidine from the ester derivative in alkaline media (see Iwai & Ando, 1967).

N-acylation of the *O*-peptide derivative can be achieved with acetic or formic anhydride in great excess. Removal of the acyl groups is subsequently needed for sequence studies of the peptide fragments. Formyl groups may be removed by treatment with methanolic HCl at room temperature for a short time, whereas the more resistant acetyl groups are cleaved by 0.01 M-potassium hydroxide. Among the other methods employed to cleave the *O*-peptide bonds, mention should be made of the use of hydroxylamine or hydrazine to form hydroxamic acids or acid hydrazides, respectively, with the residue preceding the hydroxyamino acid. Piperidine (1 M) was also used (Goldberger & Anfinsen, 1962). See Chapter 8, Section 8.13 for determination of acetyl and formyl groups.

Acid hydrolysis in the presence of 6 M-HCl cleaves the *O*-peptide bonds; however, since non-specific cleavage of peptide bonds generally tends to increase with prolongation of the hydrolysis time, it is important to limit the reaction to the shortest time possible.

2.3.3. Side reactions

Partial acid hydrolysis of a protein may induce several acid-catalyzed reactions, such as disulfide interchange, diketopiperazine formation, cyclization of amino terminal glutamine to pyrrolidone carboxylic acid, α-β-shift at the level of aspartyl residues, hydrolysis of asparagine and glutamine to the corresponding carboxylic acids and partial destruction of tryptophan residues.

Disulfide interchange reactions (equation 2.12) occur in strongly acidic solutions, presumably by a mechanism involving sulfenium

$$R_1{-}SS{-}R_1 + R_2{-}SS{-}R_2 \rightleftharpoons 2\,R_1{-}SS{-}R_2 \qquad (2.12)$$

ions (RS^+) as reactive species (Ryle & Sanger, 1955; Benesch & Benesch, 1958). In fact, after hydrolysis of insulin with conc. HCl more cystine peptides were obtained than could be accounted for by a unique structure of the protein (Ryle & Sanger, 1955). Such disulfide interchange during partial acid hydrolysis would complicate the task of determining which half-cystine residues are joined together in the original protein.

If a dipeptide is isolated from a partial acid hydrolysate of a protein, the possibility exists that its sequence is not correct as the result of an inversion of the order of the amino acid residues. The cyclization of a dipeptide to a diketopiperazine is favored in dilute acid solution at high temperature (equation 2.13). Hydrolysis of this cyclic derivative may yield either the parent peptide or a new, isomeric dipeptide with an inverted sequence. Diketopiperazines are less stable to acid hydrolysis than the corresponding dipeptides. Inversion of sequence was clearly demonstrated for glycylvaline in 0.1 M-HCl (Sanger & Thompson, 1952).

$$H_2N{-}CH(R_1){-}CO{-}NH{-}CH(R_2){-}COOH \underset{+H_2O}{\overset{-H_2O}{\rightleftharpoons}} \text{cyclo}[NH{-}CH(R_1){-}CO{-}NH{-}CH(R_2){-}CO] \xrightarrow{+H_2O} H_2NCH(R_2)CO{-}NHCH(R_1)COOH \qquad (2.13)$$

dipeptide — diketopiperazine

An α,β-transpeptidation reaction can occur with an aspartyl residue in acid solution through the intermediate formation of a succinimido derivative (Swallow & Abraham, 1958; Naughton *et al.*, 1960), as shown in equation 2.14.

$$\sim NH{-}CH(CH_2COOH){-}CO{-}NH\sim \;\longrightarrow\; \sim NH{-}CH{-}CO{-}N(\sim){-}CO{-}CH_2 \text{ (ring)} \;\longrightarrow\; \sim NH{-}CH(COOH){-}CH_2{-}CO{-}NH\sim \qquad (2.14)$$

α-aspartyl-peptide — α,β-aspartyl-peptide — β-aspartyl-peptide

The newly formed β-aspartyl structure in a peptide chain renders the peptide bond resistant to hydrolysis by exopeptidases such as leucine aminopeptidase and carboxypeptidases. Furthermore, the Edman degradation is blocked at the cyclization stage by the imide of the β-aspartyl structure.

Cyclization of amino terminal glutamine (**18**) to the pyrrolidone carboxylyl residue (**19**) occurs easily (Hirs *et al.*, 1956; Smyth *et al.*, 1962). Therefore, *N*-terminal glutaminyl-peptides yield ninhydrin negative peptides and the Edman degradation cannot be employed. Glutamic acid may also form the pyrrolidone ring, but this conversion requires more vigorous conditions. See Chapter 1 section 1.6.1, for removal of amino terminal pyrrolidone carboxylic acid.

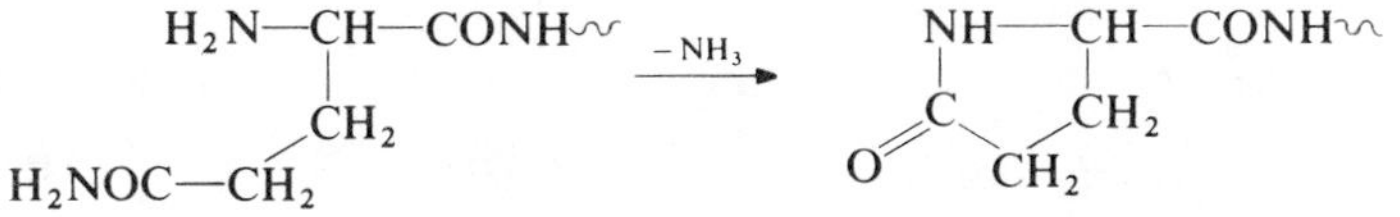

H_2N—CH—CONH~ (CH: —CH_2—CH_2—CONH$_2$, i.e. H_2NOC—CH_2—CH_2) $\xrightarrow{-NH_3}$ NH—CH—CONH~ (ring: NH—C(=O)—CH_2—CH_2—CH)

18 **19**

2.4. CLEAVAGE AT METHIONINE

2.4.1. Cyanogen bromide

Gross and Witkop (1961) have shown that cyanogen bromide reacts with the sulfur of the thioether side chain of methionine to yield a mixture of homoserine and homoserine lactone plus methylthiocyanate. When the reaction is applied to methionine-containing peptides, the bond involving the carboxyl group of methionine is cleaved. The mechanism of the cleavage reaction is shown below.

The cyanogen group of CNBr electrophilically attacks the sulfur of methionine, forming an intermediary cyano sulfonium salt (**20**). Nucleophilic attack by the carbonyl oxygen of the methionyl moiety on the γ-carbon of the cyanosulfonium brings about the release of methyl thiocyanate and the formation of an iminolactone (**21**). The iminolactone hydrolyses to give peptidyl homoserine lactone (**22**) and the corresponding peptidyl homoserine and the amino peptide (**23**) (Gross & Witkop, 1961, 1962; Gross, 1967).

Following its first application to the cleavage of methionyl peptide bonds in ribonuclease (Gross & Witkop, 1962), the reaction has become the most widely used chemical reaction for protein fragmentation (see Gross, 1967; Spande *et al.*, 1970, for numerous applications). Under appropriate conditions many proteins have been cleaved selectively at methionine peptide bonds, often in quantitative yields. The longest protein sequence established so far by CNBr cleavage is the single chain of 1021 amino acids of β-galactosidase of *Escherichia coli* (Fowler & Zabin, 1977). The convenience of this cleavage reaction is also related to the fact that the reagent and its by-products are volatile and easily removed by

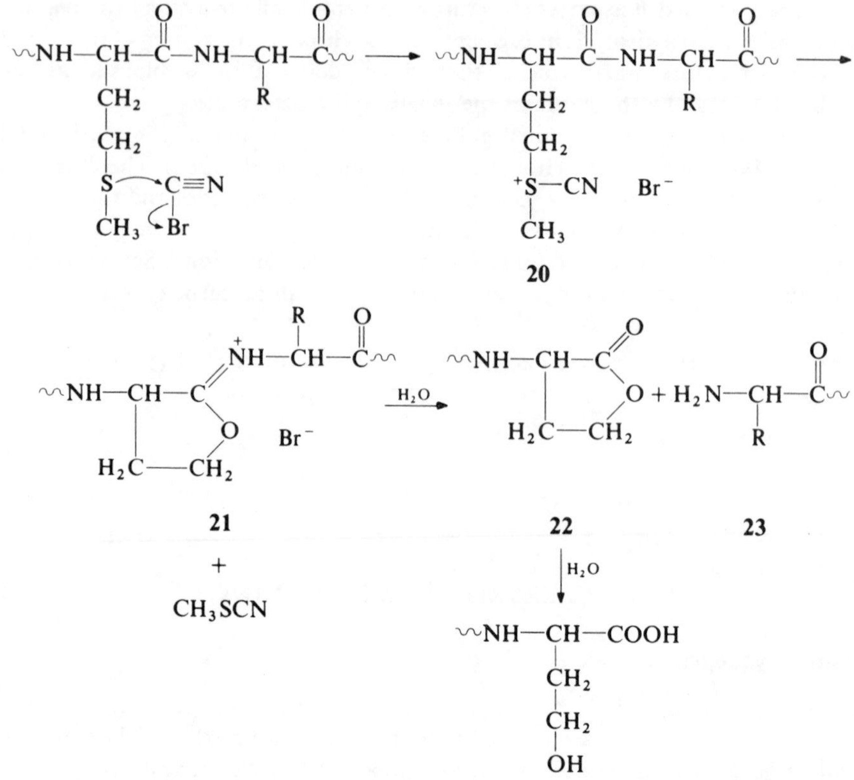

$$\sim NH-CH(CH_2CH_2S(CH_3))-C(=O)-NH-CH(R)-C(=O)\sim \; + \; Br-C\equiv N \longrightarrow \sim NH-CH(CH_2CH_2\overset{+}{S}(CH_3)-CN)-C(=O)-NH-CH(R)-C(=O)\sim \;\; Br^-$$

20

$$\longrightarrow$$

$$\sim NH-CH-C(=\overset{+}{N}H-CH(R)-C(=O)\sim)-O \;\; (\text{ring: } H_2C-CH_2) \;\; Br^-$$

21 + CH_3SCN

$$\xrightarrow{H_2O} \sim NH-CH-C(=O)-O \;\; (\text{ring: } H_2C-CH_2) \;+\; H_2N-CH(R)-C(=O)\sim$$

22 **23**

$$\mathbf{22} \xrightarrow{H_2O} \sim NH-CH(CH_2CH_2OH)-COOH$$

lyophilization. In addition, since methionine is a relatively rare amino acid in proteins, the CNBr reaction can be usefully employed to generate large fragments to be sequenced with the aid of a sequenator.

All fragments produced by the action of CNBr will contain *C*-terminal homoserine or its lactone except for the *C*-terminal peptide of the protein. Homoserine lactone elutes from the columns of the amino acid analyzer with the basic amino acids, usually between the ammonia and arginine peaks, while homoserine coelutes with glutamic acid. However, conditions may be found to separate these two amino acids (Gross, 1967). Homoserine lactone may be hydrolyzed to the free acid by dissolving it in 0.1 M-NaOH for 1 h (Shechter *et al.*, 1976) or in pyridine–acetic acid, pH 6.5 and heating at 105°C for 1 h (Ambler, 1965).

Quantitation of homoserine (Gross, 1967) or of methyl thiocyanate (Inglis & Edman, 1970) during CNBr cleavage may give misleading information about the degree of cleavage of peptide bonds. In fact, it has been demonstrated that homoserine can be formed by a mechanism which does not cleave the methionyl peptide bond. A side reaction has been claimed to take place through a six-

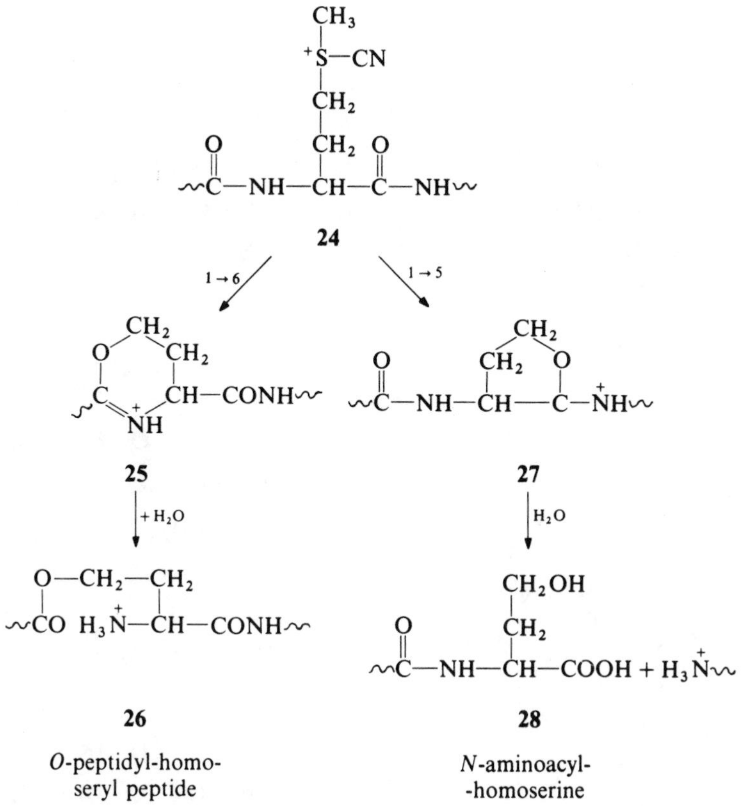

$$\sim C(=O)-NH-CH(CH_2CH_2S^{+}(CH_3)-CN)-C(=O)-NH\sim$$

24

$1 \rightarrow 6$ $\quad$ $1 \rightarrow 5$

25 $\qquad$ **27**

$+H_2O$ $\qquad$ H_2O

$$\sim CO\!-\!O-CH_2-CH_2-CH(H_3\overset{+}{N})-CONH\sim$$

$$\sim C(=O)-NH-CH(CH_2CH_2OH)-COOH + H_3\overset{+}{N}\sim$$

26

O-peptidyl-homo-seryl peptide

28

N-aminoacyl--homoserine

membered ring (**25** → **26**) which competes with the forward reaction through a five-membered iminolactone ring (**27** → **28**) (Carpenter & Shiigi, 1974).

Upon neutralization of the acidic reaction mixture the *O*-peptidyl group migrates from oxygen to nitrogen of the homoserine residue (see section 2.3.2). The degree of this side reaction seems to vary with the nature of the two amino acids located on both sides of the methionine residue.

In some instances, low yields of cleavage have been attributed to the presence of particularly resistant sequences such as Met-Thr or Met-Ser (Schroeder *et al.*, 1969; Titani *et al.*, 1972). In these cases the intermediate iminolactone is attacked by the neighboring hydroxyl group of the threonine or serine to yield homoseryl-*O*-threonyl or homoseryl-*O*-seryl peptides (**29** → **33**). The bicyclic structure (**30**) releases a homoserine side chain in (**31**) and by additional ring opening gives either an ester linkage (**32**) or a peptide linkage (**33**).

It has been found that the use of aqueous trifluoroacetic acid instead of formic acid for the cleavage reaction is advantageous when Met-Thr or Met-Ser sequences are involved. Perhaps the increased acidity of trifluoroacetic acid may

Met-Thr-Peptide

29 → 30 → 31 → 32 / 33

$\sim NH-CH(CH_2CH_2O)-C(=NH^+)-CH(CH(OH)CH_3)-CO\sim$ (29) → $\sim NH-CH-C(O)(NH-CH-CO\sim)$ (30) → $\sim NH-CH(CH_2CH_2OH)-C$ iminolactone NH^+ $CH-CO\sim$ (31) → $\sim NH-CH(CH_2CH_2OH)-C(=O)-O-CH(CH_3)-CH(NH_3^+)-CO\sim$ (32) and $\sim NH-CH(CH_2CH_2OH)-C(=O)-NH-CH(CH(OH)CH_3)-CO\sim$ (33)

have the effect of preventing the side reaction described above by protonation of the hydroxyl groups as well as the iminolactone (Schroeder *et al.*, 1969; Titani *et al.*, 1972).

Peptide bonds involving methionine *S*-oxide or *S,S*-dioxide are not cleaved by CNBr (Gross, 1967). An interesting observation was recently reported by Fontana *et al.* (1980) that methionine by reaction with a great excess (1000 equiv.) of CNBr in 70% formic acid was converted to methionine *S*-oxide. This indicates that excess of reagent in the cleavage reaction should be avoided and may explain the particular resistance of methionyl peptide bonds when proteins are treated with a great excess of reagent (Ozols & Gerard, 1977a, b). Cyanogen bromide oxidizes cysteine to cystine (Foye *et al.*, 1967) and more slowly to cysteic acid

(Gross, 1967). Excess of CNBr in acidic solution causes bromination of tyrosine residues (Fontana *et al.*, 1980) and tryptophyl peptide bond cleavage (see section 2.4.2).

The cleavage reaction has been performed in 0.1 M-HCl, aqueous formic or trifluoroacetic acid. At present, the most generally used solvent is 70 % formic acid. The acidic solvent protonates ε-amino groups and other basic groups and thereby prevents reactions with CNBr. In addition, aqueous formic acid is a strong protein unfolding agent, causing exposure of methionine residues.

The CNBr reaction, besides its extreme value for sequence studies, has found additional important applications. Peptide fragments, obtained from proteins via CNBr cleavage and bearing a *C*-terminal homoserine lactone residue, have been successfully coupled to amino-polymers in high yields, as shown below (**34** → **36**). The procedure offers an easy way to attach peptides selectively *via* their *C*-terminus to a solid support for subsequent solid phase sequential degradation (Horn & Laursen, 1973).

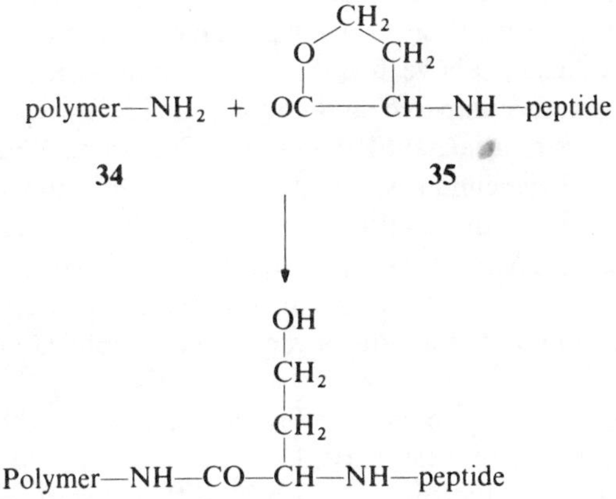

$$\text{polymer—NH}_2 \;+\; \text{OC(—O—CH}_2\text{—CH}_2\text{—)CH—NH—peptide}$$

34 **35**

↓

$$\text{Polymer—NH—CO—CH(CH}_2\text{CH}_2\text{OH)—NH—peptide}$$

36

It was shown that mixing of fragments 1–65 and 66–104 obtained by cleavage of horse heart cytochrome *C* with CNBr, in aqueous solution, resulted in their rejoining through restoration of the peptide bond originally cleaved (Corradin & Harbury, 1974). Analogously, fragment 1–52 of basic pancreatic trypsin inhibitor was recoupled to the peptide fragment 53–58 (Dyckes *et al.*, 1974). Thus, the *C*-terminal homoserine lactone ring is sufficiently activated to undergo smooth and selective aminolysis by the single α-amino group of the terminal amino acid residue. These recouplings allowed for the possibility of achieving cytochrome C and trypsin inhibitor semi-synthesis (Sheppard, 1979). The usefulness of the CNBr reaction has been most dramatically demonstrated in the genetic syntheses of important hormones. The reaction was used to cleave the tetradecapeptide somatostatin from methionyl-somatostatin linked to β-galactosidase produced in *E. coli* (Itakura *et al.*, 1977).

2.4.1.1. Method for cyanogen bromide cleavage

Dissolve the protein (10–20 mg/ml) in 70% formic acid and add solid CNBr (2 mg reagent/mg protein) directly to the protein solution. CNBr is usually present at 20 to 100-fold molar excess with respect to methionine. After 18–20 h at room temperature in the dark the reaction mixture is diluted with 5–10 vol. of water and excess reagent and by-products are removed by lyophilization. The products are then separated by gel filtration.

Waxdal *et al.* (1968) treated 0.5 μmol of peptide/ml with CNBr (5mg/ml) for 4 h at room temperature. See also Kasper (1975).

2.4.2 Other reagents

Cleavage of methionine peptides can also be achieved by reaction of the thioether group of methionine with alkylating agents, the mechanism of the reaction being similar to that discussed above for the cyanogen bromide reaction. Several alkylating agents were compared and iodoacetamide was the most effective cleaving agent (Lawson *et al.*, 1961; Lawson & Schramm, 1962). Treatment of apocatalase with ethyleneimine resulted in aminoethylation of the thioether group of methionine and the resulting aminoethylsulfonium derivative allowed peptide bond cleavage (Schroeder *et al.*, 1967). The reaction of methionyl peptides with iodoacetamide has been used in the diagonal electrophoretic method for the selective identification and isolation of methionine peptides (Tang & Hartley, 1967, 1970).

Hydrogen fluoride was shown to effect cleavage at methionine in model peptides in high yields (Lenard & Hess, 1964; Lenard *et al.*, 1964). The acid also causes cleavage at serine and threonine via $N \rightarrow O$ acyl shift (see section 2.3.2). This complication precluded a useful application of the HF-cleavage at methionine for protein fragmentation.

2.5. CLEAVAGE AT TRYPTOPHAN

The great reactivity of the indole nucleus of tryptophan has evoked much interest in finding techniques for the selective cleavage of tryptophyl bonds. The usually low content of tryptophan in a protein should result in some large fragments suitable for sequencing. However, in spite of the variety of reagents and procedures introduced over the years for this purpose, only a few have been proved to be of practical utility (see reviews Spande *et al.*, 1970; Fontana & Toniolo, 1976). In the following, mention is made of the various procedures explored to cleave tryptophyl peptide bonds, while the most useful and frequently used are discussed in greater detail.

2.5.1. Oxidative halogenation

Oxidative halogenation of the indole nucleus of tryptophan residues with *N*-bromosuccinimide (NBS) was the first procedure proposed for the modification and cleavage of the tryptophyl peptide bond. Other brominating agents employed included *N*-bromoacetamide, *N*-bromophthalimide and bromine (Patchornik *et al.*, 1960; see reviews Witkop, 1961; Ramachandran & Witkop, 1967; Spande *et al.*, 1970, Fontana & Toniolo, 1976).

The major shortcoming of the NBS-cleavage of the tryptophyl peptide bond is that the yields obtained with proteins may be as low as 5 %, seldom exceeding 50 %. Moreover, undesirable side reactions may be extensive. In fact, NBS is an extremely reactive agent which can cause modification, as well as peptide bond fission, not only of tryptophan, but also tyrosine (see section 2.6.1) and histidine (see section 2.8.3) residues. However, tyrosine cleavage can be prevented by *O*-acetylation of the phenolic hydroxyl group (Shaltiel & Patchornik, 1961, 1963a, b) and histidine cleavage can be minimal since it occurs only after heating. Methionine is extensively oxidized to methionine *S*,*S*-dioxide and cysteine and cystine to cysteic acid. Funatsu *et al.* (1964) showed that in 8 M-urea solution NBS cleaved tryptophyl bonds without affecting tyrosyl bonds. In view of these disadvantages of NBS, several other reagents of the positive halogen type have been subsequently explored to obtain higher yields and more selective peptide bond cleavage at tryptophan (see below).

The cleavage with NBS is conducted in an acidic medium with enough NBS to allow for consumption by sulfur-containing amino acids, tyrosine and histidine. Two to three moles of NBS are required to cleave simple tryptophyl peptides. Considering the results obtained by reacting with NBS 3-substituted indoles and simple tryptophyl derivatives (Witkop, 1961) and the isolation of some intermediates in the cleavage reaction (Savige & Fontana, 1977a, b; Fontana *et al.*, 1980), the most plausible mechanism of cleavage of tryptophyl peptide bonds via oxidative halogenation is shown (**37** → **43**).

According to this mechanism, two equivalents of positive halogen participate in the modification–cleavage reaction. With the very reactive NBS, halogenation in the 5-position of the benzene ring is also observed (Patchornik *et al.*, 1960). 3-Haloindolenine (**38**) appears to be the first initial intermediate, which undergoes conversion to oxindole (**39**) by hydrolysis and elimination reactions. The oxindole is then converted by a second equivalent of positive halogen to a 3-halo-oxindole moiety (**40**). An iminolactone (**42**) is subsequently formed by nucleophilic attack of the carbonyl oxygen of the former tryptophyl (**41**) moiety on the carbon bearing the halogen. The iminolactone should undergo ready hydrolysis to the N-acyl-dioxindolylalanine lactone (**43**) (in equilibrium with the open form of the amino acid), which becomes the *C*-terminal amino acid of a peptide fragment.

That the oxindole function was an intermediate in the cleavage reaction was clearly shown with horse heart cytochrome C (Savige & Fontana, 1977b; Fontana *et al.*, 1980). The single tryptophan residue of the protein was first selectively oxidized to the oxindole derivative and then cleavage effected by a subsequent bromination with a mixture of dimethyl sulfoxide (DMSO)/HBr (see p. 9).

The reaction scheme (**37** → **43**) involves also a pathway in which a 3-halo-oxindolylalanine derivative (**40**), besides cyclization to (**42**), is hydrolyzed to a dioxindolylalanine moiety (**41**). Selective (or preferential) splitting of the peptide bond involving the dioxindolylalanine residue can be obtained by mild acid treatment (Savige & Fontana, 1977a). This last reaction could presumably occur via a carbonium ion intermediate leading to the iminolactone (**42**). The anchimeric assistance of a γ-hydroxy group in the acid catalyzed hydrolysis of aliphatic amides is well documented (Capon, 1964), as exemplified by the easy and selective peptide bond fission in the toxic cyclopeptide phalloidin, in which a γ-hydroxy group similarly facilitates peptide bond splitting (Wieland & Schön, 1955; Wieland *et al.*, 1966).

It has been found that cleavage at tryptophan mediated by oxidative

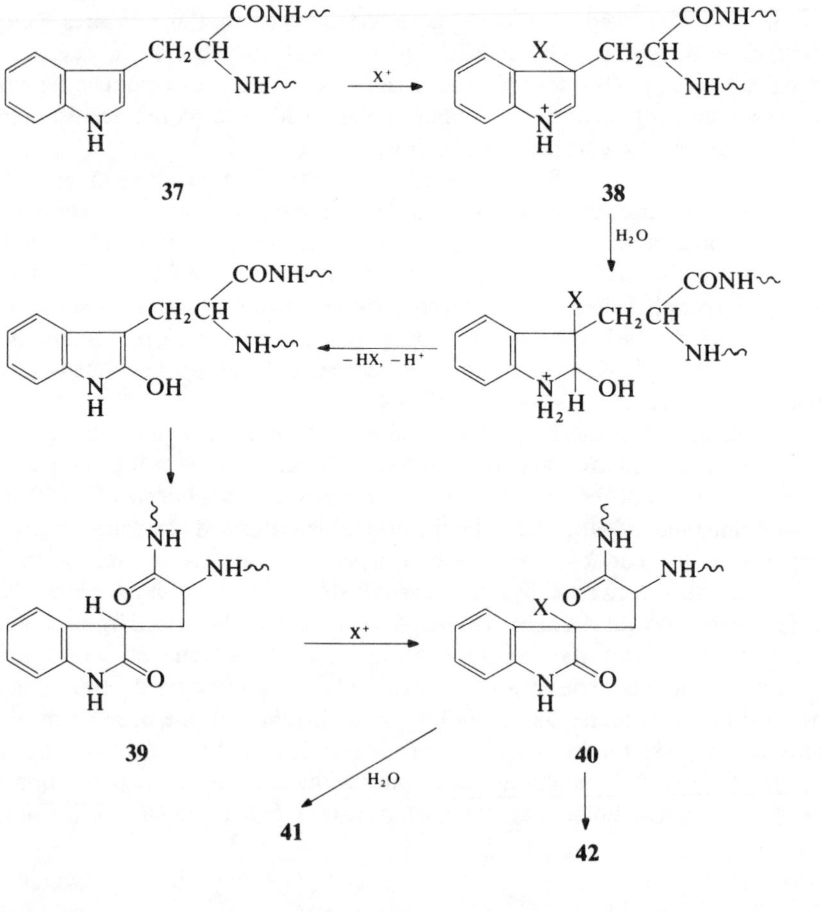

CONH~ CH$_2$CH NH~ (indole, N H) **37** —X^+→ X CH$_2$CH CONH~ NH~ ($\overset{+}{N}$ H) **38**

38 —H_2O→ X CH$_2$CH CONH~ NH~ $\overset{+}{N}H_2$ H OH —(−HX, −H$^+$)→ CONH~ CH$_2$CH NH~ N H OH

→ **39** (NH~, NH~, H, O, N H, O) —X^+→ **40** (NH~, NH~, X, O, N H, O)

40 —H_2O→ **41**

40 → **42**

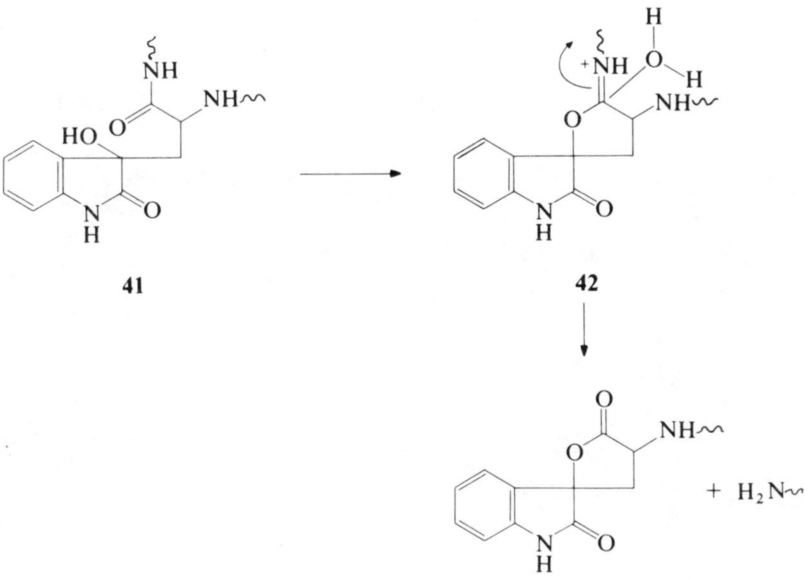

41

42

43

$+ H_2N\sim$

halogenation can occur also at the amino group of this residue, although in low yields (Martenson *et al.*, 1977; Savige & Fontana, 1977a, b). When glycyl-tryptophan was incubated with BNPS-skatole (see 2.5.2) in 75% acetic acid at 37°C for 24h free glycine was liberated in 20% yields (Martenson *et al.*, 1977). Cleavage of the pentapeptide Phe-Val-Gln-Trp-Leu with DMSO/haloacid (see 2.5.3) in acetic acid and subsequent acid treatment at pH 2 (60°C, several h), in addition to the expected leucine, free dioxindolylalanine (and/or its lactone) was identified in the reaction mixture. No other amino acid was detected on the amino acid chromatogram (Savige & Fontana, 1977b).

The mechanism of cleavage at the amino group of tryptophan can be explained by a reaction pathway involving a 6-membered iminolactone intermediate formed at the level of the *C*-terminal dioxindolylalanine moiety (**44**) originated by oxidative cleavage of the former tryptophyl peptide bond (**44** → **48**).

The peptide bond fission at the amino group of tryptophan can also be explained by a 1–6 interaction, the mechanism being similar to that proposed for the CNBr reaction of methionine peptides without cleavage (Carpenter & Shiigi, 1974) (see 2.4.1). With these peptides an *O*-aminoacyl-homoserine derivative is formed, which should be relatively stable to the aqueous acidic solvent used for the reaction (70% formic acid). With tryptophyl peptides the corresponding *O*-aminoacyl derivative involves a tertiary alkyl ester (**51**) which is expected to be much more labile to acidic solvents. The possible mechanism of the 1–6 interaction is shown below (**49** → **52**).

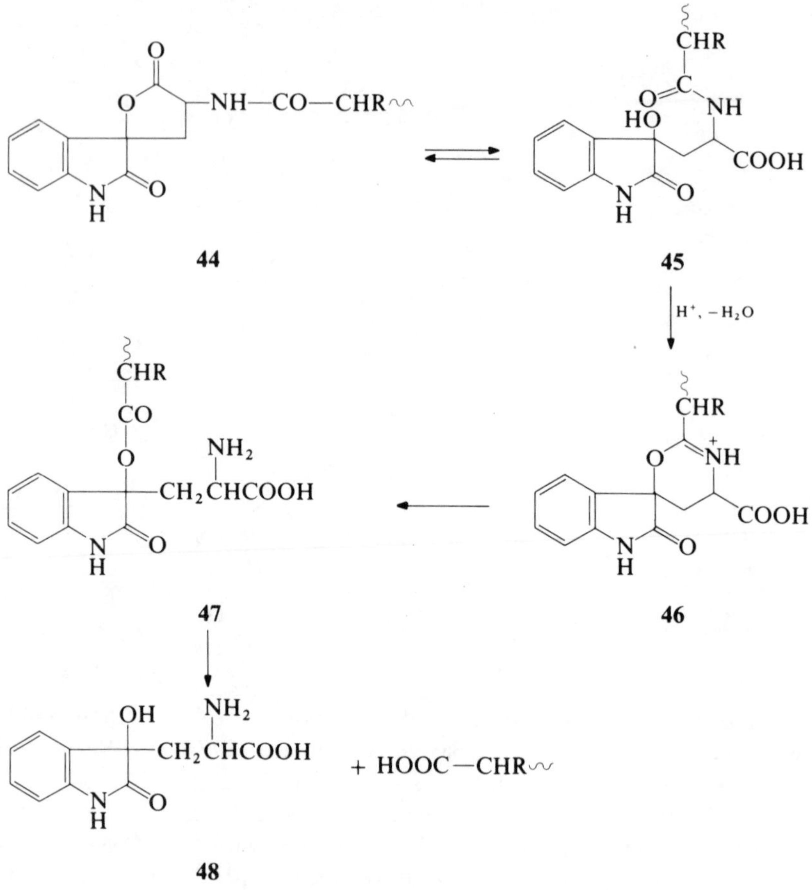

44 ⇌ 45 —(H^+, $-H_2O$)→ 46 → 47 → 48 + HOOC—CHR~

In this context it is of interest to mention that Veronese *et al.* (1969) reported cleavage both at the amino and carboxyl groups by mild acid treatment of γ-phenyl-homoserine peptides (**53**), which are structurally related both to homoserine (**54**) as well as to dioxindolylalanine peptides (**55**).

N-terminal tryptophyl peptides are not usually cleaved by halogenating agents (see Spande *et al.*, 1970). The reason for the failure of the cleavage reaction may be related to the possibility that the free amino group of tryptophan derivatives and not the carboxamide function interact preferentially with the bromoindolenine (see formula **37**), leading to the formation of tricyclic pyrrolo-indole derivatives without cleavage (Ohno *et al.*, 1970). Similar amino group participation has been found with *N*-terminal tyrosyl peptides, which are not cleaved by NBS (Wilchek *et al.*, 1968) see 2.6).

Ultraviolet absorption spectra of the peptide fragments obtained by oxidative

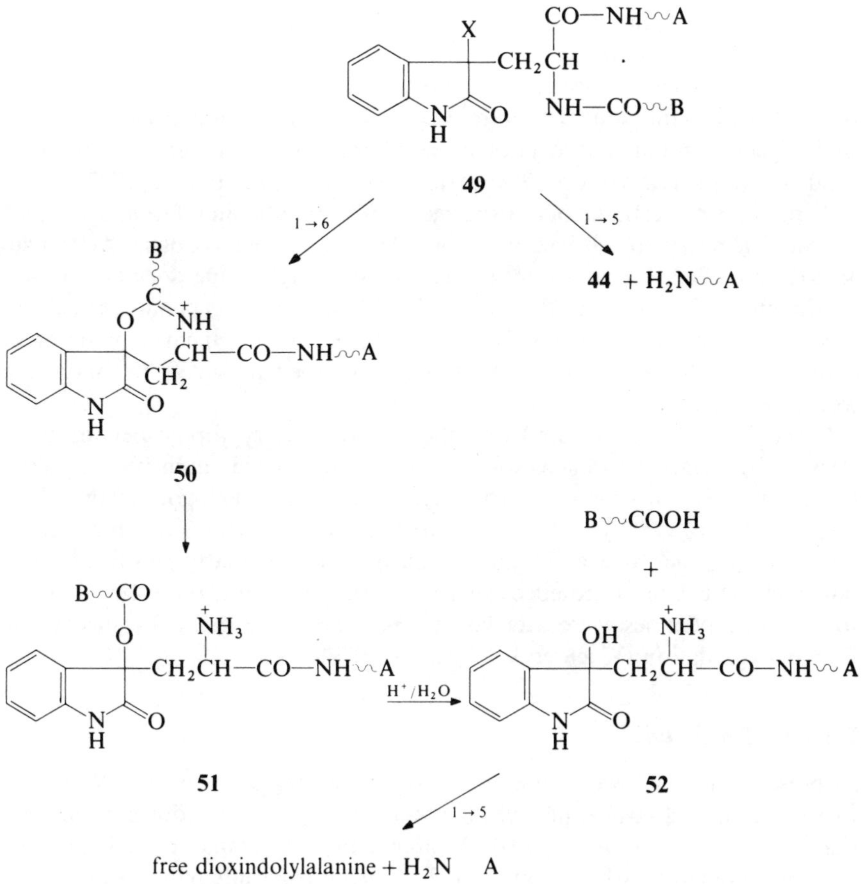

X
CO—NH∿A
—CH_2CH
N O
H
NH—CO∿B

49

1 → 6

1 → 5

44 + H_2N∿A

B
C
O $\overset{+}{N}$H
CH—CO—NH∿A
CH_2
N O
H

50

B∿CO
O
$\overset{+}{N}H_3$
—CH_2CH—CO—NH∿A
N O
H

51

H^+/H_2O

B∿COOH
+
OH $\overset{+}{N}H_3$
—CH_2CH—CO—NH∿A
N O
H

52

1 → 5

free dioxindolylalanine + H_2N A

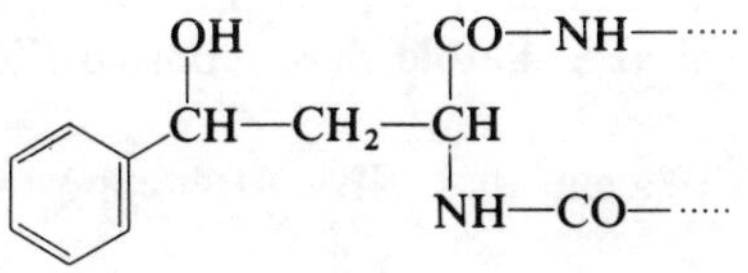

OH CO—NH— .....
CH—CH_2—CH
NH—CO— .....

53

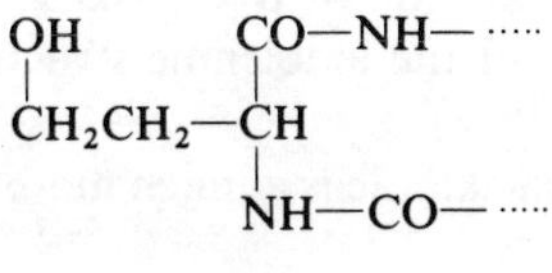

OH CO—NH— .....
CH_2CH_2—CH
NH—CO— .....

54

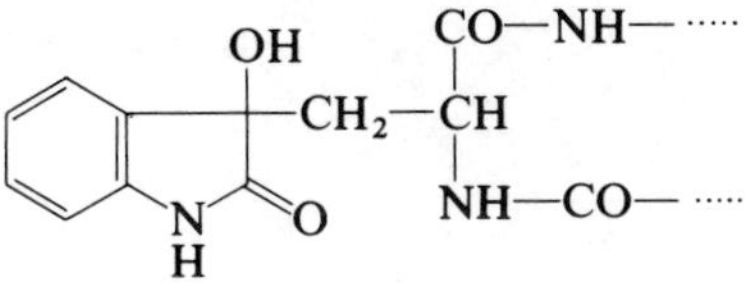

OH CO—NH— .....
—CH_2—CH
N O
H
NH—CO— .....

55

halogenation are indicative of the presence of *C*-terminal dioxindolylalanine or its lactone (see reaction pathway of cleavage, **37** → **43**). The dioxindolylalanine moiety in water has absorption maxima near 253 nm (ε 4 070) and near 287 nm (ε 1 220), with minima at 230 and 277 nm. The corresponding lactone shows similarly a maximum of absorption near 253 nm, but the other maximum occurs at 300 nm (Green & Witkop, 1964; Savige, 1975; Savige & Fontana, 1977b). With NBS or other powerful halogenating agents 5-bromo-dioxindolylalanine (λ max 261 and 302 nm) and/or its lactone (λ max 261 and 310 nm) is obtained (Hinman & Bauman, 1964; Martenson *et al.*, 1977). Oxindolylalanine does not show a maximum of absorption in the region 270–300 nm, but a maximum at 250 nm and a shoulder near 280 nm. Of course, if the peptide fragments contain other chromophores (tyrosine) the actual spectrum observed in the 250–300 nm region would be altered.

It should be borne in mind that the methods of tryptophyl peptide bond cleavage by oxidative halogenation are customarily carried out in 50–80 % acetic (or formic) acid, for several h at room temperature and analogous to the CNBr reaction (see section 2.4.1). These are rather drastic conditions which can cause side reactions, such as deamidation or splitting of particularly sensitive peptide bonds (Asp-Pro). After protein cleavage of tryptophyl peptides the dioxyindolyl-spirolactone peptides were attached to 3-aminopropyl glass for solid-phase Edman degradation (Wachter & Werhahn, 1979).

2.5.2 BNPS-skatole

At present, the most widely used method for cleaving tryptophyl peptides and proteins utilizes 2-(2-nitrophenylsulfenyl-3-bromo-3-methylindolenine (BNPS-skatole) (**57**) (Omenn *et al.*, 1970; Fontana, 1972; Fontana *et al.*, 1973). The reagent is prepared by bromination of 2-(2-nitrophenylsulfenyl)-3-methylindole (**56**) with NBS in glacial acetic acid (Omenn *et al.*, 1970). A modified procedure for the preparation of BNPS-skatole and a detailed spectroscopic investigation in support of the indolenine structure of BNPS-skatole were reported (Zeitler & Eulitz, 1978).

BNPS-skatole is a much more selective agent than NBS. After exposure of an

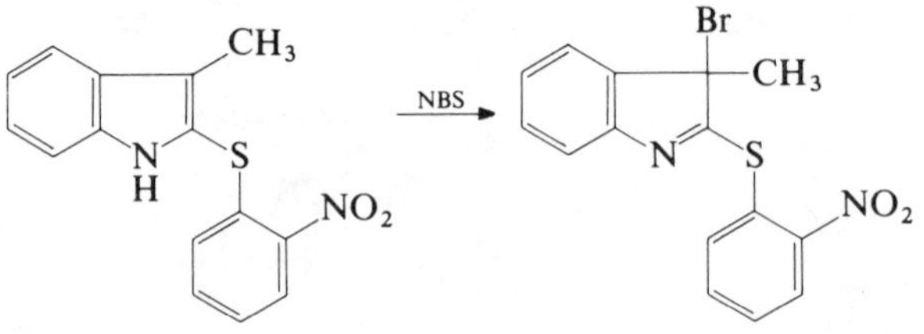

56 **57**

amino acid mixture to 10 equivalents of BNPS-skatole in 50 % acetic acid for 30 min at room temperature, tryptophan was completely absent, methionine was converted to methionine *S*-oxide, and the other amino acids were recovered quantitatively. By contrast, in an amino acid mixture oxidized with NBS, not only tryptophan, but also methionine, tyrosine, histidine and cystine were completely absent. Cysteine, as expected, is easily oxidized by BNPS-skatole to cystine and to some extent, by excess reagent, to cysteic acid. If present, sulfhydryl groups must be protected, using suitable reagents (Omenn *et al.*, 1970).

Since cleavage is time-dependent and tryptophan reacts rapidly with the reagent, tryptophan modification without any significant cleavage was observed when the reaction was carried out with a low excess of reagent and for shorter times. Much longer reaction times and excess reagent (20–100 fold) are needed for cleavage, as is clearly indicated by the studies with nuclease (Omenn *et al.*, 1970).

Numerous applications of the reagent were published (see Fontana *et al.*, 1980, for references) and the yields of specific cleavage were 40–80 %. It seems that such variability in yields of cleavage may depend on the quality of the reagent employed, which is commercially available. BNPS-skatole is light sensitive, decomposes at room temperature, becoming dark and releasing bromine. On the other hand it has been found that a highly pure sample of the crystalline reagent (pale yellow stars from ligroin; Omenn *et al.*, 1970) can be stored in closed vials at −20°C for prolonged periods of time.

2.5.2.1. Cleavage with BNPS-skatole

This is the method of Martenson *et al.* (1977).

Peptide 89-169 of bovine myelin basic protein (1.7 μmol/ml) and BNPS-skatole (17 μmol/ml) in 75 % acetic acid were incubated at 37°C for 24h. Subsequently excess reagent and its by-products were removed from the reaction mixture by repeated extraction with ethyl acetate and the aqueous solution was lyophilized. Peptide fragments were isolated after gel filtration.

2.5.3. Dimethyl sulfoxide/halogen acids

Tryptophyl peptides and proteins can be selectively cleaved by treatment with a mixture of DMSO and hydrobromic acid in acetic acid solution. The yield of specific cleavage is comparable to that obtained with BNPS-skatole; however, the particular advantages of the procedure include availability of reagents and experimental convenience (Savige & Fontana, 1977b; Fontana *et al.*, 1980). The mechanism of the cleavage reaction involves oxidative halogenation of the indole nucleus of tryptophan. In fact, the interaction between a sulfoxide and a halogen acid leads to several equilibrium reactions, involving the halosulfonium ion $R_1R_2SX^+$ as the principal reactive species (Bovio & Miotti, 1978, and references

$$
\begin{array}{ccc}
 & & R_1R_2SX^+ + X^- \\
 & & \updownarrow \\
R_1R_2SO + 2HX & \rightleftharpoons & R_1R_2SX_2 + H_2O \\
 & & \updownarrow \\
 & & R_1R_2S + X_2
\end{array}
\tag{2.15}
$$

cited therein). Any of the three species, indicated in equation 2.15, sulfide dihalide, halosulfonium halide or free halogen can act as halogenating agent of the indole nucleus of tryptophan.

Whereas a mixture of DMSO/HCl causes a clean conversion of tryptophan to oxindolylalanine (Savige & Fontana, 1977a, 1980) the use of DMSO/HBr causes oxidation of tryptophan to dioxindolylalanine and the formation of the spirolactone (Fontana *et al.*, 1980). Tryptophyl peptides are also oxidized by DMSO/HBr to the dioxindolylalanine level, but in this case oxidation is accompanied by peptide bond cleavage. It was also found that oxindolylalanine peptides, obtained by reaction of the corresponding tryptophan peptides with DMSO/HCl could be cleaved by subsequent treatment with DMSO/HBr, indicating that the oxindole function is an intermediate in the cleavage reaction (Savige & Fontana, 1977a, b; 1980). Optimum cleavage yields are best obtained by initial treatment of the protein with DMSO/HCl, followed by the subsequent addition of DMSO/HBr *in situ*. This double treatment is recommended since it has been found that direct treatment of a tryptophan peptide with DMSO/HBr leads to the formation of a by-product of tryptophan, identified as 2-dimethylsulfonium-tryptophan (**58**) (Savige & Fontana, unpublished results). The reaction appears to involve electrophilic substitution at the 2-position of the indole nucleus of tryptophan by protonated DMSO, analogous to the reaction observed with phenols and phenol ethers (Goethals & Radzitski, 1964).

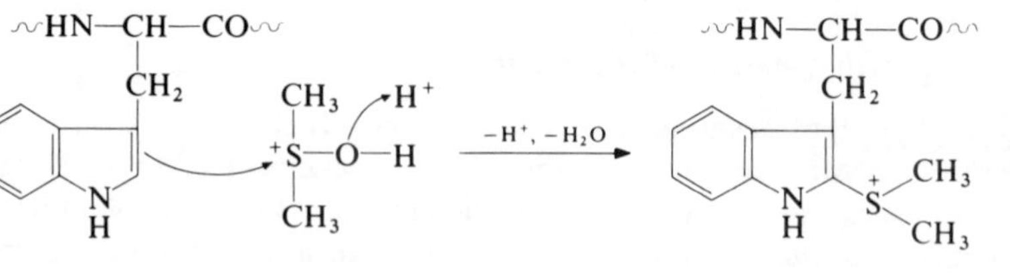

~HN—CH—CO~ | CH_2 (indole, N–H) + CH_3–$^+$S(–O–H, H^+)–CH_3 $\xrightarrow{-H^+,\ -H_2O}$ ~HN—CH—CO~ | CH_2 (indole, N–H, 2-position $\overset{+}{S}(CH_3)_2$)

58

Cleavage caused by the DMSO/haloacid procedure is selective for the tryptophyl peptide bond. Methionine is also oxidized to methionine *S*-oxide and cysteine to cystine. In practice, it has been found convenient to add to the reaction mixture a small amount of phenol to prevent tyrosine modification. It is also advisable to add some phenol to the 48 % HBr solution to destroy traces of free bromine. With these precautions, it appears that tyrosine does not react to any significant extent (Savige & Fontana, 1977b) but in one case tyrosine destruction was reported (Frank *et al.*, 1978).

2.5.3.1. Cleavage with DMSO/HCl and DMSO/HBr

The method of Savige and Fontana (1977b) is described.
Cytochrome C (17 mg) was dissolved at room temperature, together with 20 mg of phenol, in a preformed mixture of glacial acetic acid (600 μl), 12 M-HCl (300 μl) and DMSO (25 μl). After 30 min, 48 % HBr (100 μl) and DMSO (25 μl) were added to the reaction mixture. After a further 30 min at room temperature, water (2 ml) was added and the solution was extracted several times with ethyl acetate to remove the heme cleaved from the protein and its decomposition products. The aqueous layer was concentrated *in vacuo* and applied to a column (20 × 1440 mm) of Sephadex G-50 SF equilibrated and eluted with 10 % formic acid. Three peaks of peptide material were eluted almost completely separated. The first one, preceded by traces of presumably aggregated material, consisted of oxidized but uncleaved apocytochrome C, with the tryptophan-59 converted to the dioxindolylalanine residue. The second and third peak were fragments 1–59 and 60–104, respectively. The yields of cleavage were estimated on the basis of the weighed amount of fragments obtained after lyophilization and found to be approximately 60 %.

The identity of the material eluted from the column was established by amino acid analysis of the hydrolyzate with 3N-*p*-toluene-sulfonic acid (Liu & Chang, 1971). The amino acid recovery on the analyzer was in agreement with the theory for apocytochrome C and the fragments. The methionine residues in positions 65 and 80 of the sequence were oxidized to methionine *S*-oxide. The heme was completely cleaved from the protein by DMSO/haloacid, similar to the action of BNPS-skatole on cytochrome C (Fontana *et al.*, 1973).

2.5.3.2. Cleavage with DMSO/HCl and CNBr

This is the method of Huang *et al.* (1983).
If cysteine is present in the protein, it should first be reduced and alkylated. Reagent grade chemicals may be used without further purification.

Place a small vol. of the sample containing 2–3 nmol of protein in a 1.5 ml Eppendorf microtest tube and lyophilize. Add 4.9 μl of oxidizing solution. This is freshly prepared by mixing 300 μl of glacial acetic acid + 150 μl of 9 M-HCl + 40 μl of DMSO. Allow the reaction to proceed for 30 min at room temperature or 2 h at 4°C. Partially neutralize the sample by cooling in an ice bath and adding 4.4 μl of ice-cold NH_4OH (15 M as supplied). The protein may precipitate at this stage, but should redissolve in the next step. Add 40 μl of a solution of CNBr (0.3g/ml) in either 5 M-acetic acid or 60 % formic acid. Cap the tube tightly and seal with Parafilm. Allow the reaction to proceed for 12–15 h at room temperature or 30 h at 4°C in the dark.

Cleavage occurs nearly quantitatively, irrespective of the amino acids attached to tryptophan. The method neither derivatizes nor cleaves at any other amino acid. Small amounts of salts, ionic detergent or Coomassie Blue do not interfere.

2.5.4 Cyanogen bromide/heptafluorobutyric acid

The interaction of CNBr with tryptophan residues in proteins can lead to the cleavage of the tryptophyl peptide bonds. Moderate yields of cleavage (~20 %) have been observed with β-lactoglobulin (Braunitzer & Aschauer, 1975) and *Neurospora crassa* glutamate dehydrogenase (Wootton *et al.*, 1974) Subsequently, Ozols and Gerard (1977a,b,c) were able to show that the yield of CNBr cleavage at tryptophan can be greatly improved using a large excess (up to 10 000 molar excess over protein to be cleaved) and working in the more acidic medium provided by heptafluorobutyric acid instead of the commonly used 70 % formic acid.

In a recent study (Fontana *et al.*, 1980), it was shown that the mechanism of cleavage of tryptophyl peptide bonds by a large excess of CNBr in the presence of heptafluorobutyric acid involves oxidative halogenation of the indole nucleus of tryptophan. The model compound, *N*-acetyltryptophan amide, was converted by excess CNBr to *N*-acetyl-5-bromo-dioxindolylalanine lactone (among other products), in equilibrium with its open form, much the same as treatment of this compound with NBS (Patchornik *et al.*, 1960). The CNBr/heptafluorobutyric acid mixture partly converted free tyrosine to bromotyrosine and free methionine to methionine *S*-oxide. It was proposed (Fontana *et al.*, 1980) that in the presence of acid, a nucleophile (indolyl, phenolic or thioether) approaches the bromine end of the dipole of the pseudohalogen CNBr to form HCN and a bromo-derivative, which eventually rearranges to other derivatives (oxi- and dioxindole, sulfoxide), as shown.

$$\mathrm{Nu}\!:\curvearrowright \mathrm{Br{-}C{\equiv}N}\!:\curvearrowright \mathrm{H}^+$$

However, bromination may be also due to some contaminant bromine pre-existent in the CNBr sample.

2.5.4.1. Cleavage with CNBr/heptafluorobutyric acid

This is the method of Ozols and Gerard (1977b).
Apocytochrome (0.1–0.5 μmol) is dissolved in 1 ml each of 88 % formic acid and heptafluorobutyric acid. After addition of 700 mg of solid CNBr, the reaction mixture is held for 24 h in the dark. The reagent and solvent are removed with a stream of nitrogen, and the remaining material dissolved in 10 ml of water and

lyophilized. The dried residue is then dissolved in 9 M-acetic acid and applied to a column of Sephadex G-50 equilibrated and eluted with 9 M-acetic acid.

2.5.5. *o*-Iodosobenzoic acid.

A new procedure to cleave tryptophyl peptide bonds was described by Mahoney and Hermodson (1979). This involves treatment of a protein in 80% aqueous acetic acid with *o*-iodosobenzoic acid in the presence of 4 M-guanidine HCl. The procedure is selective for tryptophan with 70–100% yields of cleavage. However, subsequent reports (Wachter & Werhahn, 1980; Fontana *et al.*, 1980, 1981, 1982, 1983; Johnson & Stockmal, 1980) have indicated that the procedure is not as specific as originally reported, since cleavage occurs also at tyrosine and the yields of cleavage at tryptophan compare with those obtained with other reagents of the positive halogen type. Fontana *et al.* (1981, 1982) have reinvestigated in detail the procedure of protein fragmentation using *o*-iodosobenzoic acid and have shown that the reagent, under the experimental conditions of 80% aqueous acetic acid containing 4 M-guanidine HCl mediates the oxidative halogenation of the indole nucleus of tryptophan and of the phenol ring of tyrosine with concomitant peptide bond fission. Cleavage does not occur when halides are carefully excluded from the reaction mixture. The mechanisms of the peptide bond fissions are probably similar to those already proposed for other halogenating agents (e.g. NBS). It was proposed that incubation of *o*-iodosobenzoic acid under acid conditions in the presence of halides involves a series of equilibrium reactions leading to the formation of halogenating species, including free halogen, as shown in equation 2.16. Therefore, oxidative halogenation, causing cleavage at both tryptophan and tyrosine, is an intrinsic property of the *o*-iodosobenzoic acid/4 M-guanidine HCl mixture and not due to impurities of the reagent employed, such as *o*-iodoxybenzoic acid (Mahoney & Hermodson, 1980; Mahoney *et al.*, 1981). In fact, the purity of the reagent can be established by a number of analytical criteria (Fontana *et al.*, 1981, 1982). In order to minimize cleavage at tyrosine, it has been found that it is convenient to add *p*-cresol to the reaction mixture as a scavenger for tyrosine modification and cleavage (Fontana *et al.*, 1981; Mahoney *et al.*, 1981). With this modification, the procedure using *o*-iodosobenzoic acid can be successfully employed to cleave proteins at tryptophan only.

$$\begin{array}{ccc} & & HOOC{-}C_6H_4{-}ICl^+ \;+\; Cl^- \\ & & \updownarrow \\ HOOC{-}C_6H_4{-}IO + 2H^+ + 2Cl^- & \rightleftharpoons & HOOC{-}C_6H_4{-}ICl_2 \;+\; H_2O \\ & & \updownarrow \\ & & HOOC{-}C_6H_4{-}I \;+\; Cl_2 \end{array} \qquad (2.16)$$

2.5.5.1. *Cleavage with o-iodosobenzoic acid*

This is the method of Mahoney *et al.* (1981).
o-Iodosobenzoic acid (10 mg) is dissolved in 80% acetic acid containing 4 M-guanidine HCl and 20 μl of *p*-cresol. After 2 h incubation of this mixture at room temperature, the protein (5 mg) is added and the reaction allowed to proceed for an additional 24 h at room temperature in the dark. The peptide fragments are obtained after gel filtration on a Sephadex column.

2.5.6. Other reagents

Cleavage at tryptophan can be achieved via oxidative halogenation by 'active iodine' that is generated with H_2O_2, iodide and lactoperoxidase or horseradish peroxidase. Simple tryptophan peptides are cleaved in yields of 30% to 40% in 10 min at pH 5.0. All three constituents (iodide, H_2O_2 and peroxidase) are required for peptide bond fission to proceed. Other iodinating agents, such as I_2, I_3^-, ICl, and chloramine-T/KI also cause cleavage at tryptophan in similar yield (Alexander, 1974). It remains to be established whether the peroxidase-mediated halogenation is applicable to the cleavage at tryptophan in proteins. The procedure seems promising.

2.5.6.1. *Cleavage with N-chlorosuccinimide*

Tryptophyl peptide bonds are selectively cleaved by *N*-chlorosuccinimide (NCS) under acidic conditions. All other peptide bonds are resistant to cleavage by this reagent. The results obtained with model compounds indicate that the yields of cleavage with NCS are somewhat lower than those with NBS and that the reaction is slower. However, the cleavage with NCS is very selective and not a trace of cleavage of tyrosyl and histidyl peptide bonds is observed. In α-lactalbumin, Kunitz trypsin inhibitor, myoglobin and glucagon, all expected tryptophyl peptide bonds were cleaved with NCS in aqueous acetic acid in 19–58% yield (Shechter *et al.*, 1976). Lischwe and Sung (1977) incorporated urea in the reaction mixture with NCS.
This is the method of Lischwe and Sung (1977).
The NCS/urea cleavage is performed at a protein concentration of 0.286 mM with varying molar ratios of NCS. The cleavage buffer consists of 27.5% acetic acid and 4.68 M-urea. The reaction is carried out at room temperature for 30 min and stopped by the addition of *N*-acetyl-L-methionine (Shechter *et al.*, 1976). For the routine procedure the protein is dissolved at a concentration of 1 nmol/ml H_2O and NCS added in buffer (1 ml of glacial acetic acid, 1 g of urea, 1 ml of H_2O). The protein solution (4 vol.) is mixed with NCS buffer solution (10 vol.) to give the final ratios of reactants and buffer components described above.

Cleavage of the tryptophyl peptide bonds was optimal with about 50% yield in several species of cytochrome C.

2.5.6.2. Cleavage with tribromocresol

Burstein and Patchornik (1972) reported the use of 2,4,6-tribromo-4-methylcyclohexadienone (tribromocresol) (**59**) for the selective cleavage of

Br O Br CH$_3$ Br

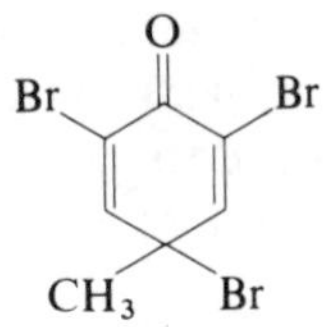

59

tryptophyl peptide bonds in peptides and proteins. Tyrosyl and histidyl peptide bonds, which usually are cleaved by other brominating agents such as NBS, are sensitive to modification but are not cleaved by tribromocresol. Tyrosine was converted by tribromocresol to 3, 5-dibromotyrosine, cysteine to cysteic acid, and methionine to methionine *S*-oxide. Optimal conditions for the cleavage of the tryptophyl peptide bond were found to be 3 equivalents of tribromocresol at pH 3 for 5–15 min at room temperature. In lysozyme, selective cleavage of the expected tryptophyl peptide bonds (yields 5–60%) was obtained. The mechanism of the cleavage reaction of tryptophyl peptide bonds was thought to be similar to the one proposed with NBS.

The method of Burstein and Patchornik (1972) is described.

To a solution of 0.5 μmol of glucagon in 2 ml of 70% acetic acid add 25 μmol of tribromocresol in 50 μl of dioxane. The reaction is allowed to proceed at room temperature with constant stirring for 15 min. The mixture is applied to a Sephadex G-25 column and eluted with 0.1M-acetic acid to separate the peptide fragments.

2.5.7. Ozonization

Tryptophan peptides are converted to *N*-formyl-kynurenine peptides (**60**) by ozonization in anhydrous formic acid (Previero *et al.*, 1963, 1964; Veronese *et al.*, 1967, 1969; Morishita & Sakiyama, 1970). Mild acid treatment can subsequently be employed to cleave the *N*-formyl group. When kynurenine or formyl-kynurenine peptides are heated in bicarbonate buffer for periods of up to 4 h the

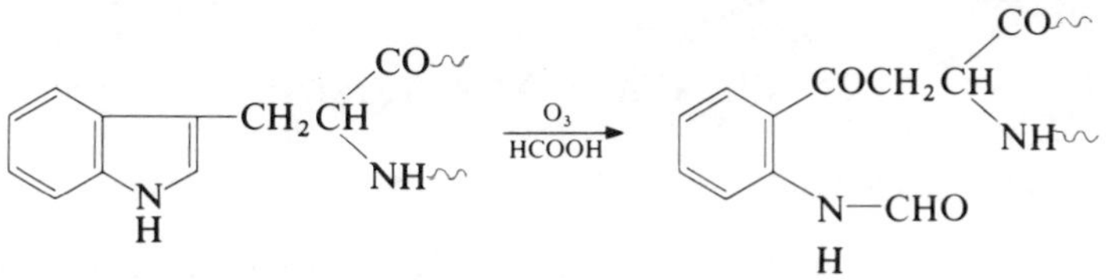

CO~ CH$_2$CH NH~ N H $\xrightarrow[\text{HCOOH}]{O_3}$ CO~ COCH$_2$CH NH~ N—CHO H

60

peptide bond is selectively hydrolyzed in yields of 49–59 %. Intramolecular catalysis involving the neighboring γ-keto group of the formyl-kynurenine residue was thought to be responsible for the labilization of the peptide bond (Previero *et al.*, 1966a).

N-formyl-kynurenine peptides in 0.2 M-hydrazine-acetate buffer, pH 3.6 at 100°C for 2 h form hydrazones (**61**) which undergo ring closure to tetrahydropyridazones (**62**) with concomitant cleavage. Yields of cleavage were 35–68 %, but some non-specific peptide bond cleavage by the action of hydrazine was observed (Morishita *et al.*, 1967a, b; Morishita & Sakiyama, 1970).

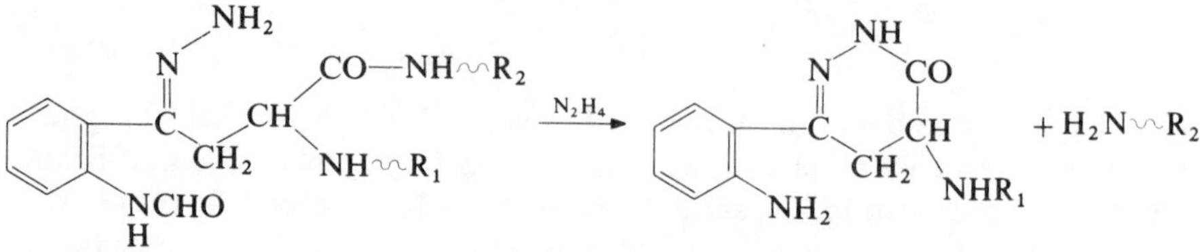

61 $\xrightarrow{N_2H_4}$ **62** $+ H_2N\sim R_2$

Reduction of kynurenine peptides with $NaBH_4$ (Previero *et al.*, 1966b) or electrolytic reduction at controlled potential (Veronese *et al.*, 1969) to γ-(*o*-aminophenyl) homoserine derivatives (**63**) and subsequent hydrolysis allows cleavage in 20–40 % yields of both the *N*- and *C*-peptide bonds of the former tryptophyl residue.

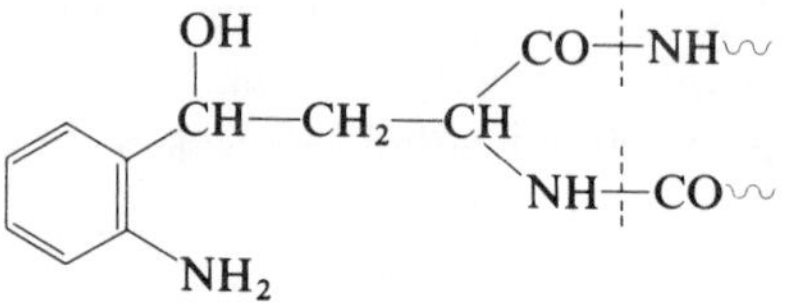

63

Tryptophyl peptide bond cleavage via ozonolysis has not yet been employed to cleave proteins. Yields of cleavage are low and the procedure converts cystine quantitatively to cysteic acid, methionine to its *S*, *S*-dioxide and tyrosine is slowly oxidized.

2.6. CLEAVAGE AT TYROSINE

2.6.1. *N*-bromosuccinimide

The tyrosyl peptide bond can be selectively cleaved by NBS, if tryptophan is absent (Corey & Haefele, 1959; Schmir *et al.*, 1959; Ramachandran & Witkop,

1967). Since tryptophyl peptide bonds when present cleave most rapidly, tyrosyl bonds will be split only with a larger excess of reagent. The cleavage of tyrosyl peptide bonds with NBS in mildly acidic medium probably proceeds by the mechanism shown below. Tyrosyl peptides are converted by two equivalents of NBS to *o*, *o*'-dibromotyrosyl derivatives and then by a third equivalent of NBS to a tribromodienone intermediate (**65**), which invites 1,5-interaction of the carboxyl to form a spiro-γ-iminolactone. This reactive lactone (**66**) hydrolyzes easily to a dibromodienone spirolactone (**67**) and an amine component.

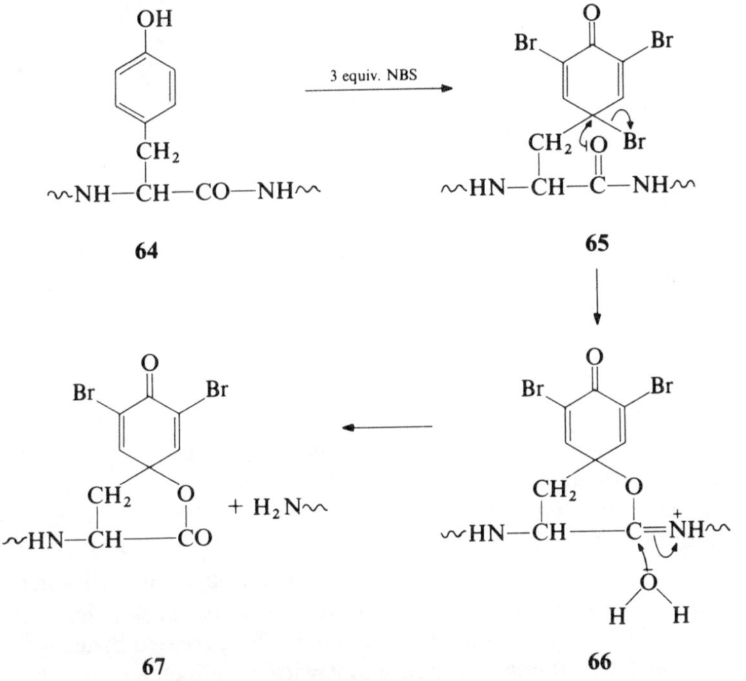

OH
CH₂
~NH—CH—CO—NH~
64

3 equiv. NBS →

O
Br Br
CH₂ O Br
~HN—CH—C—NH~
65

O
Br Br
CH₂ O
~HN—CH—C=NH⁺~
O
H H
66

←

O
Br Br
CH₂ O
~HN—CH—CO + H₂N~
67

The dienone spirolactone, which has been isolated and characterized (Corey & Haefele, 1959; Schmir *et al.*, 1959), is a strong chromophore at 260 nm (ε 10 000–11 000). With simple peptides, the changes in absorption at 260 nm may be used to follow the course of the cleavage reaction.

N-terminal tyrosyl peptides fail to cleave with excess NBS under the acidic conditions usually employed. Wilchek *et al.* (1968) studied the reaction of NBS with model tyrosyl compounds bearing a free α-amino group. It was shown that the free amino group of tyrosine participates in a reaction alternative to the cleavage mechanism. This alternative reaction involves an internal Michael-type addition to a tribromodienone intermediate, leading ultimately to a dibromo-6-hydroxyindole-derivative (**73**), as shown below (**68** → **73**).

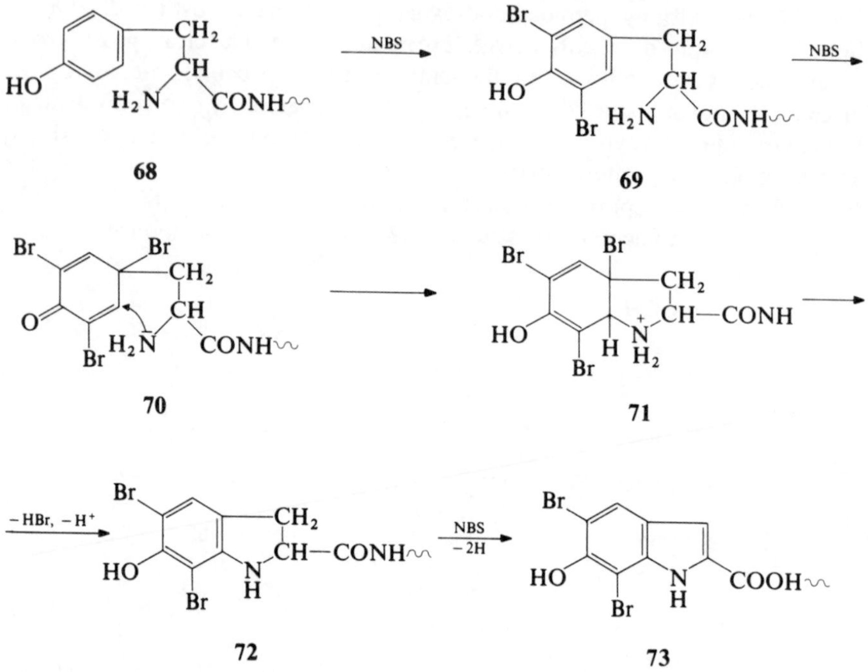

The NBS-cleavage of tyrosyl peptide bonds was successfully employed with peptides and proteins. All six tyrosyl bonds in *S*-carboxymethyl-ribonuclease, which contains no tryptophan, were cleaved in yields of 30–65 % (Ramachandran & Witkop, 1967; Spande *et al.*, 1970).

Since all histones contain tyrosine but no tryptophan, the NBS-method has been used widely in histone sequence determination (Bustin & Cole, 1969; Rall & Cole, 1970). The yield is not quantitative, but can be increased by using an excess of reagent, with concomitant side reactions with histidine and the sulfur amino acids, if present.

2.6.1.1. Cleavage with N-bromosuccinimide

This is the method of Hurley & Stout (1980).

The cleavage is performed by adding NBS to a 1 mg/ml solution of histone H 1 in 50 % acetic acid to give a final concentration of 0.09 mg of NBS per mg of protein. After incubation at room temperature for 2 h an equal amount of fresh NBS is added to the reaction mixture and the incubation continued for two additional 4-h periods. The mixture is then diluted 10-fold with water and lyophilized. The peptide fragments are separated by gel filtration on a Sephadex G-100 column equilibrated with 0.01 M-HCl.

2.6.2 Other reagents

N-iodosuccinimide is as effective as NBS in the oxidative cleavage of the tyrosyl peptide bond. The reaction is conducted at pH 4.5 and yields of cleavage with model compounds and simple tyrosyl peptides are 25–95%. The mechanism of cleavage is the same as with NBS (Junek *et al.*, 1969). Electrolytic oxidation at a platinum anode was shown to cleave tyrosyl peptide bonds in moderate yields. This procedure does not cause cleavage at tryptophan and histidine, as does NBS. However, other functional groups in peptides and proteins (imidazole, thioether, disulfide, amine) are subject to electrolytic oxidation, but at a lower rate than the phenolic group of tyrosine. Angiotensin, insulin and ribonuclease were specifically cleaved to varying degrees (see review Cohen & Farber, 1967).

2.7. CLEAVAGE AT CYSTEINE

Conversion of the -SH group of cysteine residues in proteins to the thiocyano function is a method for selective cleavage at the amino group of the modified cysteine residue (Stark, 1977). Catsimpoolas & Wood (1964, 1966) first reported that peptide bond fission occurred when cystine-containing proteins were incubated in the presence of cyanide. The formation of thiocyanoalanine residues (**74**) which cyclized to an acyliminothiazolidine (**75**) was followed by rapid hydrolysis to an amino-terminal peptide and a COOH-terminal 2-iminothiazolidine-4-carboxyl fragment (**76**) in good yield. Since the scission of the disulfide bonds of cystine peptides by cyanide produces partial, random formation of thiocyano groups from both members of the disulfide bond, a complex mixture of peptides is expected. In fact, bovine ribonuclease released nine peptides when treated with a 1000-fold excess of cyanide at pH 8, 37°C for 48 h. The peptides showed a composition consistent with the known sequence of

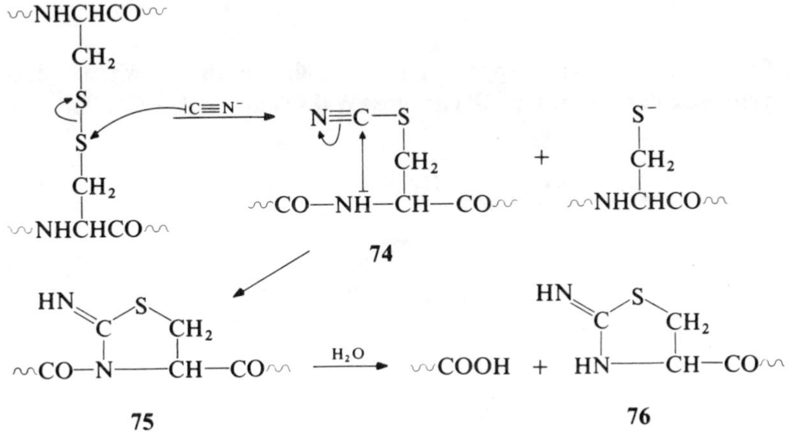

~NHCHCO~ / CH_2 / S—S / CH_2 / ~NHCHCO~ $\xrightarrow{C\equiv N^-}$ N≡C—S—CH_2 (~CO—NH—CH—CO~) + S^-—CH_2 (~NHCHCO~)

74

HN=C(S—CH_2) ~CO—N———CH—CO~ $\xrightarrow{H_2O}$ ~COOH + HN=C(S—CH_2) HN———CH—CO~

75 **76**

ribonuclease and corresponded to peptides resulting from cleavage at the *N*-acyl peptide bond at each of the eight half-cystine residues of the protein (Catsimpoolas & Wood, 1966).

A method was reported for the direct and quantitative conversion of -SH groups in proteins to the corresponding thiocyanate derivatives using 2-nitro-5-thiocyanobenzoic acid (NTCB) (**78**) (Degani & Patchornik, 1971, 1974) (equations 2.17 and 2.18). The reagent was synthesized by treatment of 5,5′-dithiobis-(2-nitrobenzoic acid) (see p. 76) with NaCN (equation 2.17). NCTB is a remarkably mild and selective cyanylating agent for -SH groups. Its reactivity is related to the easy displacement by the sulfur nucleophile of the good leaving group *p*-nitrothiophenolate (equation 2.18). Radioactive NCTB, prepared by cleaving Ellman's reagent with $Na^{14}CN$, allows introduction of a radioactive label into the thiocyanoalanine side chains of modified proteins in order to follow the appearance of radioactive 2-iminothiazolidine peptide derivatives.

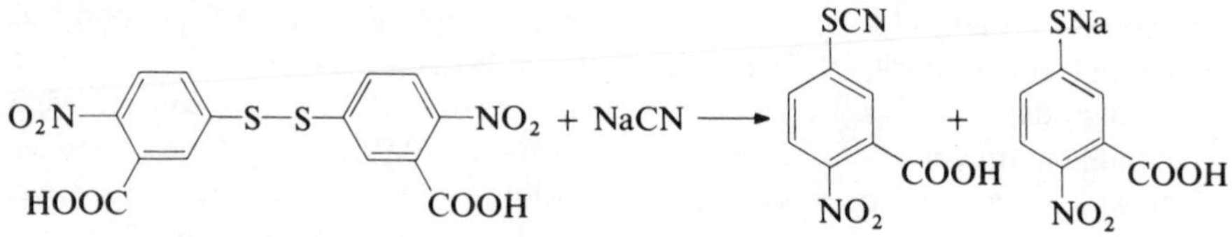

O_2N–C₆H₃(COOH)–S–S–C₆H₃(COOH)–NO_2 + NaCN ⟶ SCN–C₆H₃(COOH)–NO_2 + SNa–C₆H₃(COOH)–NO_2

77 **78** **(2.17)**

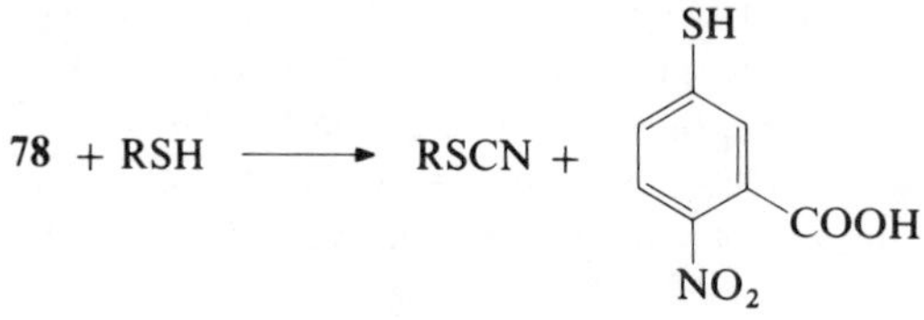

78 + RSH ⟶ RSCN + SH–C₆H₃(COOH)–NO_2 **(2.18)**

79

1-Cyano-4-dimethylamino pyridinium salts (**80**) were also shown to be effective cyanylating agents for protein -SH groups (Wakselman *et al.*, 1976). The reaction

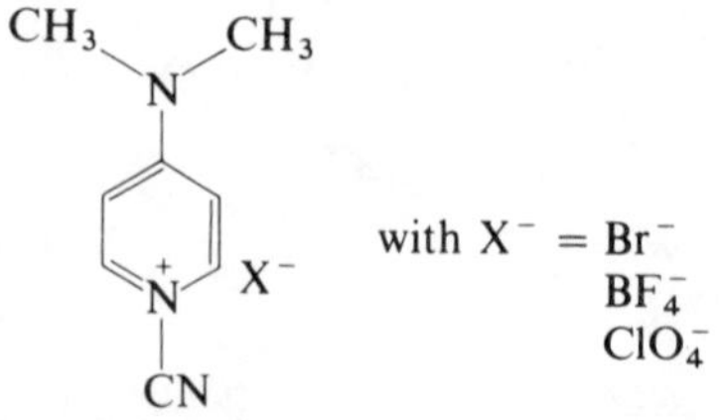

$(CH_3)_2N$–C₅H₄$\overset{+}{N}$–CN X^- with $X^- = Br^-$, BF_4^-, ClO_4^-

80

occurs in neutral or acidic media (7 M-urea, 0.1 M acetate, pH 2–7, 11 min, 25C°, 3-fold molar excess reagent over -SH groups) to generate protein-SCN.

A two-step procedure of *S*-cyanylation involves treatment of a cysteine-containing protein with Ellman's reagent under mildly alkaline conditions (see Section 2.2.2.4). The protein thionitrobenzoate mixed disulfide is then cleaved by cyanide to the thiocyanoalanine derivative of the protein and 2-nitro-5-thiobenzoate anion (Vanaman & Stark, 1970). Also, it was found (A. Fontana, unpublished) that a similar two-step cyanylation reaction occurred by clearing the mixed disulfide obtained by reaction of a cysteine-containing protein with 2-nitro-phenylsulfenyl chloride in aqueous acetic acid (Fontana *et al.*, 1968; Fontana & Scoffone, 1972).

Selective cleavage of the polypeptide chain at the thiocyanoalanine residues may be achieved by incubation at alkaline pH, e.g. pH 9.0 for 24 h at 37°C. Thiocyanoalanine residues can undergo a β-elimination reaction to form thiocyanate and dehydroalanine (see section 2.5.2). Since both cleavage and β-elimination reactions occur under mildly alkaline conditions, the two reactions might be competitive resulting only to a moderate extent in cleavage of the peptide bond. However, the results so far obtained with several proteins (Stark, 1977) indicate that cleavage goes nearly to completion. Low cleavage yields may result from incomplete *S*-cyanylation.

The observation was made that the mono-thiocyano derivatives of papain (Klein & Kirsch, 1969), of the catalytic subunit of *E. coli* aspartate trans-carbamylase (Vanaman & Stark, 1970) and of *Azotobacter vinelandii* isocitrate dehydrogenase (Chung *et al.*, 1971) do not cyclize with the protein in its native state. Peptide bond fission is achieved only after denaturation of the protein derivative (Vanaman & Stark, 1970). Therefore, the cleavage step of the *S*-cyanylated protein is carried out in the presence of an unfolding agent such as guanidine HCl.

2.7.1 Cyanylation with 2-nitro-5-thiocyanobenzoic acid

The method of Jacobson *et al.* (1973) is described.

The protein is dissolved in 6 M guanidine HCl-0.2 M-Tris-acetate buffer, pH 8. If there are no disulfides, add a slight excess of DTT and allow the solution to stand at room temperature for 30 min. To reduce disulfides, add DTT to 10 mM and heat the solutions to 37°C for 1-2 h. Add NCTB in five-fold molar excess over total thiol, readjust to pH 8 with NaOH if necessary and react for 15 min at 37°C. Acidify the reaction mixture of pH 4 or below, cool to 4°C and separate the modified protein from small molecules by dialysis or by gel filtration with 50% acetic acid or another volatile solvent of low pH. The sample may be stored at −20°C in 50% acetic acid without cleavage. For cleavage, take the protein to dryness to remove the acidic solvent thoroughly and redissolve it in 6 M-guanidine HCl-0.1 M-sodium borate buffer, pH 9, for 12–16 h at 37°C.

A drawback to the more extensive use of the cysteine cleavage method to produce large fragments for sequence analysis is the fact that the *C*-terminal peptide is blocked at its amino-terminus by an iminothiazolidinyl residue. Because attempts to specifically remove the iminothiazolidinyl residue by application of the classical Edman degration technique failed (Jacobson *et al.*, 1973), the cleaved peptides, with the exception of the peptide fragment which contained the original amino-terminus, could not be used for sequential degradation analysis. Recently a method was reported to unmask the blocked *N*-terminal group with Raney nickel (Otieno, 1978). Upon treatment of acyliminothiazolidine-peptides with the very active (neutral) W-6 catalyst (Billica & Adkins, 1967) in neutral solution at 50°C for 6 to 10 h smooth desulfurization to the expected *N*-terminal alanine peptides occurred in 90% yield. Cysteine and methionine were rapidly converted to alanine and α-aminobutyric acid, respectively.

2.7.1.1. Desulfurization with Raney nickel

The method of Otieno (1978) is described.
Desulfurization of the acyliminothiazolidine peptide was carried out 50°C in 0.05 M Tris-HCl buffer, pH 7. The W-6 Raney nickel catalyst was prepared as described by Billica and Adkins (1967). A mixture of 20 mg of the peptide and 200 mg of the catalyst in 50 ml of Tris buffer was refluxed for 7 h under nitrogen. The solution was cooled and then centrifuged at 4°C. The supernatant was passed through a column (6.0 × 50 mm) of Chelex-100 chelating resin. The peptide was eluted with 1 M-ammonium hydroxide and recovered by lyophilization.

In view of the high yields of specific cleavage of cysteine peptide bonds after *S*-cyanylation and the successful unblocking by Raney nickel desulfurization of the resulting iminothiazolidine peptides, cleavage at cysteine residues should be useful for the generation of large peptides.

2.8. MISCELLANEOUS CLEAVAGES

2.8.1. Asparaginyl-glycyl peptide bond

The ability of hydroxylamine to cleave specifically Asn-Gly peptide bonds (**81**) has been performed successfully in a substantial number of cases in both proteins and large peptide fragments. Cleavage results from the tendency of the asparaginyl side chain to cyclize forming a substituted succinimide (**82**) that is susceptible to nucleophilic attack by hydroxylamine (see review Bornstein & Balian, 1977). The increased susceptibility of Asn-Gly bonds in comparison with other asparaginyl bonds appears to be a function of the relative ease with which the asparaginyl side chain can cyclize in the absence of steric hindrance imposed by a side chain on the next amino acid. The mechanism of cleavage of the Asn-Gly peptide bond is shown

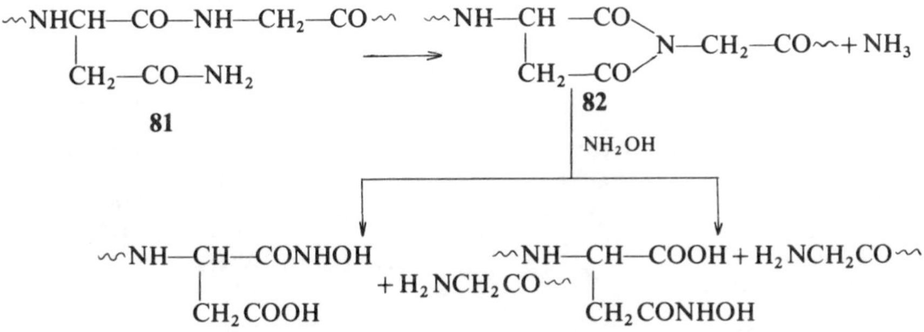

$$\sim NHCH(CH_2{-}CO{-}NH_2){-}CO{-}NH{-}CH_2{-}CO\sim \longrightarrow \sim NH{-}CH{-}CO(CH_2{-}CO){>}N{-}CH_2{-}CO\sim + NH_3$$

81 → **82**

NH_2OH

$$\sim NH{-}CH(CH_2COOH){-}CONHOH + H_2NCH_2CO\sim \qquad \sim NH{-}CH(CH_2CONHOH){-}COOH + H_2NCH_2CO\sim$$

above. Although some γ-glutaminyl esters in peptides are also known to form cyclic compounds, cleavage of Gln-Gly bonds by hydroxylamine has not been observed.

The original method for hydroxylamine cleavage was described by Blumenfeld *et al.* (1965). Modifications that have been subsequently introduced include a higher hydroxylamine concentration, slightly higher temperature, a lower pH and the use of 6 M-guanidine HCl as a solvent and lithium hydroxide for titration.

2.8.1.1. Cleavage at Asn-Gly

The method of Enfield *et al.* (1980) is described.
A solution of 102 mg of *S*-pyridylethylated light chain of factor X_1 in 25 ml of 6 M-guanidine HCl and 2 % hydroxylamine HCl was prepared at ambient temperature with sufficient 4.5 M-lithium hydroxide added to give pH 9.0. During the reaction, the pH was maintained by the addition of 4.5 M-LiOH. A reaction time of 4 h was found to be optimal to obtain cleavage of the Asn-Gly peptide bond.

2.8.2. Cleavage at proline

Birch *et al.* (1955) reported that sodium in liquid ammonia easily cleaved tertiary amides to aldehydes and secondary amines. This reagent was first observed to cleave the *N*-peptide bond of proline (Hofmann & Yajima, 1961) and prompted several investigators (Wilchek *et al.*, 1965; Benisek *et al.*, 1967) to exploit the use of sodium in liquid ammonia for cleaving prolyl bonds. Reductive cleavage of the *N*-prolyl peptide bond (**83**) converts the amino acid preceding proline to its aldehyde (**84**) or alcohol (**85**) and liberates a peptide with *N*-terminal proline (**86**) (Wilchek *et al.*, 1965, Spande *et al.*, 1970).

The method, however, suffers from several shortcomings, including non-specific cleavage and extensive degradation of alkali-sensitive amino acids. To overcome these difficulties, modifications of the original procedure have been

$$\sim NH\overset{R}{\overset{|}{C}}HCO-N\begin{matrix}CH_2-CH_2\\ | \quad\quad | \\ CH-CH_2\\ | \\ CO-NH\sim\end{matrix} \xrightarrow{Na/NH_3} \sim NH\overset{R}{\overset{|}{C}}H-\overset{O}{\overset{\|}{C}}-H + HN\begin{matrix}CH_2-CH_2\\ | \quad\quad | \\ CH-CH_2\\ | \\ CO-NH\sim\end{matrix}$$

83 **84** **86**

$$+ \quad \sim NH-\overset{R}{\overset{|}{C}}H-CH_2OH$$

85

described. Sodium hydrazide in hydrazine-ether at 0°C; 45–60 min, cleaved model peptides and the insulin B-chain in 70%–100% yields (Kauffmann & Sobel, 1966). Lithium aluminium hydride in anhydrous tetrahydrofuran has been used to cleave model peptides with proline, gramicidin S and tyrocidine C (Ruttenberg *et al.*, 1964, 1965).

2.8.3 Cleavage at histidine

Histidine carries a double bond in the imidazole ring situated in the γ-δ position, which is the position analogous to the double bond present in tryptophyl and tyrosyl residues. *N*-bromosuccinimide, in fact, cleaves also histidyl peptides although in much lower yields than tryptophyl or tyrosyl peptides (Shaltiel & Patchornik, 1963 a,b). The mechanism presumably operative in the NBS-cleavage of histidyl peptide bonds seems to involve a five-membered iminolactone ring which hydrolyzes readily under acidic conditions analogous to the mechanism of cleavage with tryptophan and tyrosine peptides (Spande *et al.*, 1970). Cleavage can be carried out by allowing histidyl peptides to react with NBS in pyridine-aqueous acetic acid (pH 3.3) for 1 h at 100°C. Concomitant cleavage of tryptophyl and tyrosyl peptide bonds, if present, under these conditions can be prevented by *O*-acylation of tyrosine (Shaltiel & Patchornik, 1963a) and by selective alkylation of tryptophan with 2-hydroxy-5-nitrobenzyl bromide (Wilchek & Witkop, 1967). Faced with the low yield of cleavage and the strong oxidizing properties of NBS for several amino acid side-chain functions, in particular all sulfur amino acids, cleavage of the histidine peptide bond by NBS has found little use in peptide and protein fragmentation.

2.8.4. Cleavage via dehydroalanine

When serine and cysteine peptides (**87**) are converted to derivatives in which β-elimination is facilitated, the resulting dehydroalanyl peptide (**88**) can be induced to cleave at the α-carbon atom, as shown.

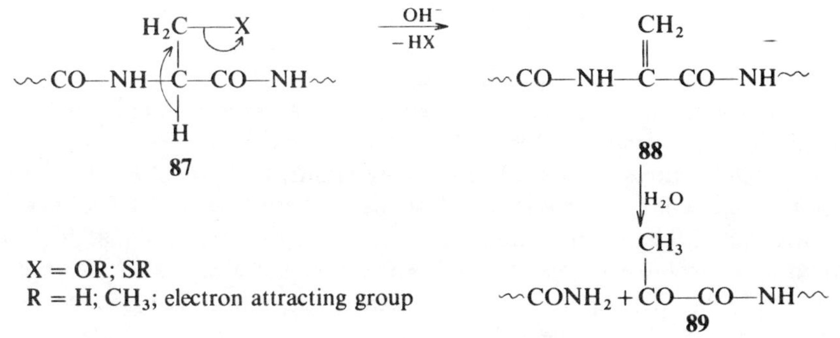

$$\sim CO{-}NH{-}\underset{\underset{H}{|}}{\overset{\overset{H_2C-X}{|}}{C}}{-}CO{-}NH\sim \xrightarrow[-HX]{OH^-} \sim CO{-}NH{-}\overset{\overset{CH_2}{\|}}{C}{-}CO{-}NH\sim$$

87 **88**

$$\mathbf{88} \xrightarrow{H_2O} \sim CONH_2 + \overset{\overset{CH_3}{|}}{CO}{-}CO{-}NH\sim$$

89

X = OR; SR
R = H; CH_3; electron attracting group

Substituting the serine hydroxyl group of a peptide by strong electron-attracting groups, such as the *O*-tosyl (Photaki, 1963) or *O*-diphenyl-phosphoryl-group (Riley *et al.*, 1957) a β-elimination process can easily be induced by base treatment. For example, chymotrypsin was converted to anhydro-chymotrypsin by tosylation at the serine residue of the active site with tosyl fluoride and subsequent base treatment (Strumeyer *et al.*, 1963; Weiner *et al.*, 1966).

Conversion of cysteine to dehydroalanine in proteins may be achieved by *S*-nitroarylation with fluorodinitrobenzene followed by base treatment in 0.1 M-alkali (Sokolovsky & Patchornik, 1964). *S*-methylsulfonium derivatives form dehydrolanine at a lower pH (8.5–9) than *S*-dinitrophenyl derivatives (Sokolovsky *et al.*, 1964).

Dehydroalanyl-peptides (**91**) are also formed directly from cystine peptides (**90**) by the action of alkali. The breakdown of the disulfide group is initiated by hydrogen abstraction via the attack of hydroxyl ion followed by the release of a *S*-thiocysteine residue (**92**). The latter decomposes further into cysteine and free sulfur (Tarbell & Harnisch, 1951).

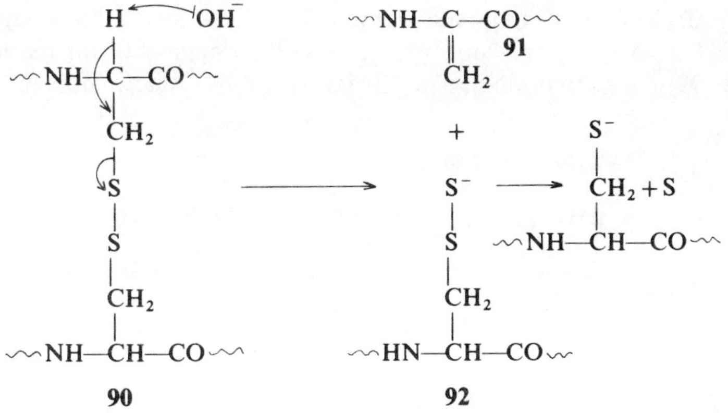

$$\underset{\mathbf{90}}{\sim NH{-}\overset{\overset{H \leftarrow OH^-}{|}}{\underset{\underset{\underset{\underset{\underset{\sim NH-CH-CO\sim}{|}}{CH_2}}{\underset{|}{S}}}{\underset{|}{S}}}{\underset{|}{CH_2}}}{C}}{-}CO\sim} \longrightarrow \underset{\mathbf{91}}{\sim NH{-}\underset{\underset{CH_2}{\|}}{C}{-}CO\sim} + \underset{\mathbf{92}}{\underset{\underset{\underset{\sim HN-CH-CO\sim}{|}}{CH_2}}{\underset{|}{\underset{S}{\overset{S^-}{|}}}}} \longrightarrow \sim NH{-}\overset{\overset{S^-}{|}}{\overset{CH_2}{|}}{CH}{-}CO\sim + S$$

The initial *C–S* bond breakage can occur on both sides of the disulfide bridge, so that a complex mixture of dehydroalanine peptides is expected when a cystine-containing protein is exposed to alkaline treatment.

The dehydroalanine peptide can be cleaved either by hydrolysis or by oxidation (Sokolovsky & Patchornik, 1964). Performic acid oxidizes dehydroalanyl peptides to α,β-dihydroxyalanine which is subsequently hydrolyzed at pH 10 to a peptide amide and a hydroxypyruvyl peptide. Alternatively, cleavage may be performed by bromination in aqueous acetic acid, presumably via a labile α,β-dibromoalanine intermediate. Use of bromide would cleave also tryptophyl, tyrosyl, and histidyl-peptide bonds (Spande *et al.*, 1970). The peptide bond cleavage via dehydroalanine suffers from several shortcomings. The dehydroalanyl residue (**93**) can react with nucleophiles, among them, the ε-amino groups of lysine residues (**94**) of the protein to give DL-α-amino-β-(ε-N-L-lysine) propionic acid residues (lysinoalanine) (**95**) (Bohak, 1964). Hydrogen peroxide under the alkaline conditions of cleavage will oxidize several amino acid side chain groups, resulting in peptides with extensive modifications. Treatment with alkali may also give rise to several undesired reactions, e.g. transpeptidation, hydrolysis of amide bonds and non-specific peptide bond cleavage.

$$\underset{\mathbf{93}}{\sim\text{NH—}\overset{\overset{\text{CH}_2}{\|}}{\text{C}}\text{—CO}\sim} \quad \underset{\mathbf{94}}{\sim\text{NH}\overset{\overset{\text{H}_2\text{N—(CH}_2)_4}{|}}{\text{C}}\text{HCO}\sim} \longrightarrow \underset{\mathbf{95}}{\sim\text{NH—}\overset{\overset{\text{CH}_2\text{—NH——(CH}_2)_4}{|}}{\text{C}}\text{H—CH}\sim\sim\text{NHCHCO}\sim}$$

2.8.5. Other cleavages

S-methyl-cysteine (**96**), the lower homolog of methionine, does not react with CNBr with cleavage of the peptide bond at its carboxyl group. The β-lactone (**97**) that would result from a reaction analogous to that of methionine with CNBr, is not formed. However, if the amino group of *S*-methyl-cysteine (**96**) is acylated or aminoacylated (**98**) the reaction proceeds with formation of an oxazolinium bromide (**99**) and methylthiocyanate (Gross *et al.*, 1967). At low temperature **99** is

$$\underset{\mathbf{96}}{\begin{array}{l}\sim\text{NH—CH—CO—NH}\sim\\ \qquad\;\;|\\ \qquad\text{CH}_2\\ \qquad\;\;|\\ \qquad\text{S}\rightarrow\text{CN}\\ \qquad\;\;|\\ \qquad\text{CH}_3\;\;\text{Br}^-\end{array}} \quad \xrightarrow[\times]{-\text{CH}_3\text{SCN}} \quad \underset{\mathbf{97}}{\begin{array}{l}\sim\text{NH—CH——CO}\\ \qquad\;\;|\qquad\;\;|\\ \qquad\text{CH}_2\text{—O}\end{array}}$$

$CH_3CO{-}NH{-}CH(R){-}CO{-}NH{-}CH(CH_2{-}S(CH_3){-}C{\equiv}N \cdot Br){-}CONH\sim$ **96** $\longrightarrow$ $CH_3CO{-}NH{-}CH(R){-}CO{-}NH{-}CH(CH_2{-}{}^{+}S(CH_3){-}CN){-}CONH\sim + Br^-$ **98**

98 $\xrightarrow{\text{A, } 0°\text{, 6 days}}$ $CH_3CO{-}NH{-}CH(R){-}C(=\overset{+}{N}H){-}O{-}CH_2{-}CH{-}CONH\sim$ (ring) **99** $\xrightarrow{H_2O}$ $CH_3CO{-}NH{-}CH(R){-}CO{-}O{-}CH_2{-}CH(\overset{+}{N}H_3){-}CONH$ **100**

98 $\xrightarrow{\text{B}}$ $CH_3CO{-}NH{-}CH(R){-}CO{-}NH{-}C(=CH_2){-}CONH\sim$ $\xrightarrow{H_2O}$ $CH_3CO{-}NH{-}CH(R){-}CONH_2$ + $CH_3{-}CO{-}CONH\sim$

readily hydrolyzed to the *O*-acyl-derivative (*O*-seryl-peptide) (**100**) (mechanism A). To complete the cleavage the ester has to be hydrolyzed to produce the two expected fragments. At high temperature the dehydroalanine derivative is formed via β-elimination (mechanism B). The reaction was successfully applied with several *S*-methyl-cysteine peptides.

Cleavage at serine and threonine peptide bonds (**101**) preceded by an oxidation of the hydroxyl group using dicyclohexylcarbodiimide and phosphoric acid in dimethylsulfoxide (**102**) can be obtained by reaction with phenylhydrazine for a few hours at room temperature or for 20–40 min at 40–70°C. The intermediate phenylhydrazone (**103**) cyclizes to a pyrazolone (**104**) with release of the adjacent amino terminus. Alternatively, hydroxylamine yields an intermediate hydrazone (**105**) which cyclizes to an iso-oxazolinone (**106**) and the amine moiety (D'Angeli *et al.*, 1966, 1967). Experiments carried out with model peptides gave satisfactory yields of cleavage, but the method does not seem suitable for application to proteins.

Editor's Note Inglis (1983) reported on the cleavage of proteins at aspartic acid residues using dilute HCl (pH 2) or dilute formic acid under vacuum for 2 h at 108°C. Most of the aspartyl bonds in a number of proteins were cleaved under these conditions. Some deamidation of asparagine in insulin and deacetylation of *N*-terminal glycine in cytochrome C occurred.

~CONH—CH—CONH~ (CH—OH, R) **101** $\xrightarrow[H_3PO_4]{DMSO/DCC}$ ~CONH—CH—CONH~ (CO, R) **102**

102 $\xrightarrow{C_6H_5NHNH_2}$ **103**; **102** $\xrightarrow{NH_2OH}$ **105**

~CONH—CH—CO—NH~ (C, R, N, NH, C_6H_5) **103** → ~CO—NH—CH—CO (C, R, N, N—C_6H_5) + H_2N~ **104**

~CONH—CH—CO—NH~ (C, R, N, O, H) **105** → ~CONH—CH—CO (C, R, N, O) + H_2N~ **106**

R = H, CH_3

2.9. CONCLUSIONS

There are available today a number of chemical procedures to cleave polypeptide chains at specific residues. Principles of selectivity and high yield of cleavage determine the utility of a particular reagent or procedure. In this article the authors have discussed those non-enzymic methods of protein cleavage that have been, and promise to be, most useful in terms of specificity and extent of peptide bond rupture.

The cyanogen bromide reaction at methionine residues, because of its high selectivity and nearly quantitative yield of cleavage, is widely applied. Suitable methods to cleave at tryptophyl and cysteinyl peptide bonds have been developed and their utility demonstrated.

No single technique is sufficient to fragment a protein in such a fashion that the study of the isolated fragments will allow the deduction of the full primary structure. Frequently the proper combination of supplementary chemical and enzymic techniques is required to attain this objective.

Chemical methods of peptide bond cleavage are supplementary to other domains of peptide chemistry, such as forming peptide bonds in novel and unique ways. It has been demonstrated that non-enzymic fragmentation techniques are applicable to the synthesis of biologically active peptides via recombinant DNA techniques.

The recognition of the neighboring group effect in a 1,5-type of interaction as the underlying cause of facile cleavage of the peptide bond of γ-substituted amino acids, e.g. γ,δ-dihydroxyleucine in phalloidin (Wieland & Schön, 1955), was followed by the systematic exploration of this principle in other amino acids. While numerous agents and reagents have been investigated for the purpose, many more await testing to enlarge the arsenal of useful compounds.

2.10. REFERENCES

Alexander, N. M. (1974). *J. Biol. Chem.*, **249,** 1946–1952.

Ambler, R. P. (1965). *Biochem. J.*, **96,** 32P.

Anfinsen, C. B. & Haber, E. (1961). *J. Biol. Chem.*, **236,** 1361–1363.

Atassi, M. Z. (1977). In *Immunochemistry of Proteins*, Atassi, M. Z. (Ed.), vol. 1, Plenum Press, New York and London, pp. 114–161.

Atassi, M. Z. & Rosenthal, A. F. (1969). *Biochem. J.*, **111,** 593–601.

Atassi, M. Z., Suliman, A. M. & Habeeb, A. F. S. A. (1975). *Biochem. Biophys. Acta*, **405,** 452–463.

Benassi, C. A., Scoffone, E. & Veronese, F. M. (1965). *Tetrahedron Lett.*, **49,** 4389–4393.

Benisek, W. F., Raftery, M. A. & Cole, R. D. (1967). *Biochemistry*, **6,** 3780–3790.

Billica, R. M. & Adkins, H. (1967). *Org. Syn. Coll.*, **3,** 176.

Birch, A. J., Cymerman-Craig, J. & Slaytor, M. (1955). *Austr. J Chem.*, **8,** 512–518.

Blumenfeld, O. O., Rojkind, M & Gallop, P. M. (1965). *Biochemistry*, **4,** 1780–1788.

Bohak, Z. (1964). *J. Biol. Chem.*, **239,** 2878–887.

Bornstein, P. & Balian, G. (1977). *Methods Enzymol.*, **47,** 132–145.

Bovio, A. & Miotti, U. (1978). *J. Chem. Soc. Perkin*, **II,** 172–177.

Bradbury, A. F. & Smyth, D. G. (1973). *Biochem. J.*, **131,** 637–642.

Brandenburg, D. (1972). *Hoppe Seyler's Z. Physiol. Chem.*, **353,** 263–267.

Braunitzer, G. & Aschauer, H. J. (1975). *Hoppe Seyler's Z. Physiol. Chem.*, **356,** 473–474.

Burstein, Y. & Patchornik, A. (1972). *Biochemistry*, **11,** 4641–4650.

Bustin, M. & Cole, R. D. (1969). *J. Biol. Chem.*, **244,** 5291–5294.

Capon, B. (1964). *Quarterly Reviews*, **18,** 45–111.

Capon, B. & McManus, S. P. (1976). *Neighboring Group Participation*, Plenum Press, New York, London.

Carpenter, F. H. & Shiigi, S. M. (1974). *Biochemistry*, **13,** 5159–5164.

Catsimpoolas, N. & Wood, J. L. (1964). *J. Biol. Chem.*, **239,** 4132–4137.

Catsimpoolas, N. & Wood, J. L. (1966). *J. Biol. Chem.*, **241,** 1790–1796.

Cavins, J. F. & Friedman, M. (1970). *Anal. Biochem.*, **35,** 489–493.

Cecil, R. (1963). In *The Proteins*, Neurath, H. (Ed.), 2nd edn, vol. 1, Academic Press, New York, London, pp. 379–476.

Cecil, R. & Weitzman, P. D. J. (1964). *Biochem. J.*, **93,** 1–11.

Chung, A. E., Franzen, J. S. & Braginski, J. E. (1971). *Biochemistry*, **10,** 2872–2876.

Cleland, W. W. (1964). *Biochemistry*, **3,** 480–482.

Cohen, L. A. & Farber, L. (1967). *Methods Enzymol.*, **11,** 299–308.

Cole, R. D. (1967a). *Methods Enzymol.*, **11,** 206–208.

Cole, R. D. (1967b). *Methods Enzymol.*, **11,** 315–317.

Corey, E. J. & Haefele, L. F. (1959). *J. Am. Chem. Soc.*, **81,** 2225–2228.

Corradin, G. & Harbury, H. A. (1974). *Biochem. Biophys. Res. Commun.*, **61,** 1400–1406.
Crestfield, A. M., Stein, W. H. & Moore, S. (1963). *J. Biol. Chem.*, **238,** 2413–2420.
D'Angeli, F., Scoffone, E., Filira, F. & Giormani, V. (1966). *Tetrahedron Lett.*, **24,** 2745–2748.
D'Angeli, F., Giormani, V., Filira, F. & Di Bello, C. (1967). *Biochem. Biophys. Res. Commun.*, **28,** 809–814.
Degani, Y. & Patchornik, A. (1971). *J. Org. Chem.*, **36,** 2727–2728.
Degani, Y. & Patchornik, A. (1974). *Biochemistry,* **13,** 1–11.
Desnuelle, P. & Bonjour, G. (1951). *Biochim. Biophys. Acta*, **7,** 451–459.
Dyckes, D. F., Creighton, T. E. & Sheppard, R. C. (1974).*Nature* (*Lond.*), **247,** 202–204.
Elliott, D. F. (1952). *Biochem. J.*, **50,** 542–550.
Ellman, G. L. (1959). *Arch. Biochem. Biophys.*, **82,** 70–77.
Enfield, D. L., Ericsson, L. H., Fujikawa, K., Walsh, K. A., Neurath, H. & Titani, R. (1980). *Biochemistry*, **19,** 659–667.
Fontana, A. (1972). *Methods Enzymol.*, **25,** 419–423.
Fontana, A. & Scoffone, E. (1972). *Methods Enzymol.*, **25,** 482–494.
Fontana, A. & Toniolo, C. (1976). *Progr. Chem. Organic Natl. Compounds*, **33,** 309–409.
Fontana, A., Scoffone, E. & Benassi, C. A. (1968a). *Biochemistry*, **7,** 980–986.
Fontana, A., Veronese, F. M. & Scoffone, E. (1968b). *Biochemistry*, **7,** 3901–3905.
Fontana, A., Vita, C. & Toniolo, C. (1973). *FEBS Lett.*, **32,** 139–142.
Fontana, A., Savige, W. E. & Zambonin, M. (1980). In *Methods in Peptide and Protein Sequence Analysis*, Third Internat. Conf., Birr, C. (Ed.), Elsevier/North Holland, Amsterdam, New York, Oxford, pp. 309–322.
Fontana, A., Dalzoppo, D., Grandi, C. & Zambonin, M. (1981). *Biochemistry*, **20,** 6997–7004.
Fontana, A., Dalzoppo, D., Grandi, C. & Zambonin, M. (1982). In *Methods in Protein Sequence Analysis, Proc.* Fourth Internat. Conf., Elzinga, M. (Ed.), Humana Press, New Jersey, pp. 325–334.
Fontana, A., Dalzoppo, D., Grandi, C. & Zambonin, M. (1983). *Methods Enzymol.*, **91,** 311–318.
Fowler, A. V. & Zabin, I. (1977). *Proc. Natl. Acad. Sci. U.S.A.*, **74,** 1507–1510.
Foye, W. O., Helb, A. M. & Mickles, J. (1967). *J. Pharm. Sci.*, **56,** 292–293.
Frank, G., Sidler, W., Widmer, H. & Zuber, H. (1978). *Hoppe-Seyler's Z. Physiol. Chem.*, **359,** 1491–1507.
Fraser, J. F., Poulsen, K. & Haber, E. (1972). *Biochemistry*, **11,** 4974–4977.
Friedman, M. (1973). *The Chemistry and Biochemistry of the Sulfhydryl Group in Amino Acids, Peptides and Proteins*, Pergamon Press, New York.
Friedman, M., Krull, L. H. & Cavins, J. F. (1970). *J. Biol. Chem.*, **245,** 3868–3871.
Funatsu, M., Green, N. M. & Witkop, B. (1964). *J. Am. Chem. Soc.*, **86,** 1846–1848.
Glendenning, M. B., Greenberg, D. M. & Fraenkel-Conrat, H. (1947). *J. Biol. Chem.*, **167,** 125–128.
Goethals, E. & Radzitski, P. De (1964). *Bull. Soc. Chim. Belges*, **73,** 546–559.
Goldberger, R. F. & Anfinsen, C. B. (1962). *Biochemistry*, **1,** 401–405.
Grannis, G. F. (1960). *Arch. Biochem. Biophys.*, **91,** 255–265.
Green, N. M. & Witkop, B. (1964). *Trans. N. Y. Acad. Sci.*, Ser. II **26,** 659–669.
Gross, E. (1967). *Methods Enzymol.*, **11,** 238–255.
Gross, E. & Witkop, B. (1961). *J. Am. Chem. Soc.*, **83,** 1510–1511.
Gross, E. & Witkop, B. (1962). *J. Biol. Chem.*, **237,** 1856–1860.
Gross, E., Morell, J. L. & Lee, P. Q. (1967). *Int. Congress of Biochemistry (Tokyo)*, Abstr. p. IX–535.
Gurd, F. R. N. (1972). *Methods Enzymol.*, **25,** 424–438.

Habeeb, A. F. S. A. (1972). *Methods Enzymol.*, **25,** 457–464.
Hass, G. M., Nau, H., Biemann, K., Grahn, D. T., Ericsson, L. H. & Neurath, H. (1975). *Biochemistry*, **14,** 1334–1342.
Heinrikson, R. L. & Kramer, K. J. (1974). *Progress in Bioorganic Chemistry*, Kaiser, E. T. & Kézdy, F. J. (Eds.), vol. 3, Wiley, New York, pp. 141–250.
Hermodson, M. A., Ericsson, L. H., Neurath, H. & Walsh, K. A. (1973). *Biochemistry*, **12,** 3146–3153.
Hinman, R. L. & Bauman, C. P. (1964). *J. Org. Chem.*, **29,** 2431–2437.
Hirs, C. H. W. (1956). *J. Biol. Chem.*, **219,** 611–621.
Hirs, C. H. W. (Ed.) (1967a). *Methods Enzymol.*, **11,** 1–988.
Hirs, C. H. W. (1967b). *Methods Enzymol.*, **11,** 197–199.
Hirs, C. H. W. (1967c). *Methods Enzymol.*, **11,** 199–203.
Hirs, C. H. W. & Timasheff, S. N. (Eds.) (1972). *Methods Enzymol.*, **25,** 1–724.
Hirs, C. H. W. & Timasheff, S. N. (Eds.) (1977). *Methods Enzymol.*, **47,** 1–668.
Hirs, C. H. W. & Timasheff, S. N. (Eds.) (1983). *Methods Enzymol.*, **91,** 1–693.
Hirs, C. H. W., Stein, W. H. & Moore, S. (1956). *J. Biol. Chem.*, **221,** 151–169.
Hofmann, K. & Yajima, H. (1961). *J. Am. Chem. Soc.*, **83,** 2289–2293.
Horn, M. J. & Laursen, R. A. (1973). *FEBS Lett.*, **36,** 285–288.
Huang, H. V., Bond, M. W., Hunkapiller, M. W. & Hood, L. E. (1983). *Methods Enzymol.*, **91,** 318–324.
Hurley, C. K. & Stout, J. T. (1980). *Biochemistry*, **19,** 410–416.
Inglis, A. S. (1983). *Methods Enzymol.*, **91,** 324–332.
Inglis, A. S. & Edman, P. (1970). *Anal. Biochem.*, **37,** 73–80.
Inglis, A. S. & Liu, T.-Y. (1970). *J. Biol. Chem.*, **245,** 112–116.
Inglis, A. S., McMahon, D. T. W., Roxburgh, C. M. & Takayanagi, H. (1976). *Anal. Biochem.*, **72,** 86–94.
Itakura, K., Hirose, T., Crea, R., Riggs, A. D., Heyneker, H. L., Bolivar, F. & Boyer, H. W. (1977). *Science*, **198,** 1056–1063.
Itano, H. A. & Robinson, E. A. (1972). *J. Biol. Chem.*, **247,** 4819–4824.
Iwai, K. & Ando, T. (1967). *Methods Enzymol.*, **11,** 263–282.
Jacobson, G. R., Schaffer, M. H., Stark, G. R. & Vanaman, T. C. (1973). *J. Biol. Chem.*, **248,** 6583–6591.
Jauregui-Adell, J. & Marti, J. (1975). *Anal. Biochem.*, **69,** 468–473.
Jocelyn, P. C. (1972). *Biochemistry of the -SH Group*, Academic Press, New York.
Johnson, P. & Stockmal, V. B. (1980). *Biochem. Biophys. Res. Commun.*, **94,** 697–703.
Junek, H., Kirk, K. L. & Cohen, L. A. (1969). *Biochemistry*, **8,** 1844–1848.
Kasper, C. B. (1975). In *Protein Sequence Determination*, Needleman, S. B. (Ed.), 2nd edn, Springer, Berlin, Heidelberg, New York, pp. 114–161.
Katsoyannis, P. G. (1969). In *Diabetes*, Proc. Congr. Internat. Diabetes Fed., 6th 1967, Ostman, J. (Ed.), Exerpta Medica, Amsterdam, pp. 379–394.
Kauffmann, T. & Sobel, J. (1966). *Liebig's Ann. Chem.*, **698,** 235–241.
Kenyon, G. L. & Bruice, T. W. (1977). *Methods Enzymol.*, **47,** 407–430.
Klein, I. B. & Kirsch, J. F. (1969). *J. Biol Chem.*, **244,** 5928–5935.
Konigsberg, W. (1972). *Methods Enzymol.*, **25,** 185–188.
Konigsberg, W. & Steinman, H. M. (1977). In *The Proteins*, Neurath, H. & Hill, R. L. (Eds.), 3rd edn, vol. 3, Academic Press, New York, pp. 1–178.
Landon, M. (1977). *Methods Enzymol.*, **47,** 145–149.
Lawson, W. B. & Schramm, H. J. (1962). *J. Am. Chem. Soc.*, **84,** 2017–2018.
Lawson, W. B., Gross, E., Foltz, C. M. & Witkop, B. (1961). *J. Am. Chem. Soc.*, **83,** 1509–1510.
Leach, S. I., Meschers, A. & Swanepoel, O. A. (1965). *Biochemistry*, **4,** 23–27.

Lenard, J. & Hess, G. P. (1964). *J. Biol. Chem.*, **239,** 3275–3281.
Lenard, J., Schally, A. V. & Hess, G. P. (1964). *Biochem. Biophys. Res. Commun.*, **14,** 498–502.
Lindley, H. (1956). *Nature (Lond.)*, **178,** 647–648.
Lischwe, M. A. & Sung, M. T. (1977). *J. Biol. Chem.*, **252,** 4976–4980.
Liu, T.-L. (1977). In *The Proteins*, Neurath, H. & Hill, R. L. (Eds.), 3rd edn, vol. 3, Academic Press, New York, pp. 239–402.
Liu, T.-Y. & Chang, Y. H. (1971). *J. Biol. Chem.*, **246,** 2842–2848.
Liu, T.-Y. & Inglis, A. S. (1972). *Methods Enzymol.*, **25,** 55–60.
MacLaren, J. A. & Sweetman, B. J. (1966). *Austr. J. Chem.*, **19,** 2355–2360.
Mahoney, W. C. & Hermodson, M. A. (1979). *Biochemistry*, **18,** 3810–3814.
Mahoney, W. C. & Hermodson, M. A. (1980). In *Methods in Peptide and Protein Sequence Analysis, Proc. Third Internat. Conf.*, Birr, C. (Ed.), Elsevier/North Holland, Heidelberg, Amsterdam, New York, Oxford, pp. 323–328.
Mahoney, W. C., Smith, P. K. & Hermodson, M. A. (1981). *Biochemistry*, **20,** 443–448.
Martenson, R. E., Deibler, G. E. & Kramer, A. J. (1977). *Biochemistry*, **16,** 216–221.
Milstein, C. & Sanger, F. (1961). *Biochem. J.*, **79,** 456–469.
Morishita, M. & Sakiyama, F. (1970). *Bull. Chem. Soc. Japan*, **43,** 524–530.
Morishita, M., Sakiyama, F. & Narita, K. (1967a). *Bull. Chem. Soc. Japan*, **40,** 433.
Morishita, M., Sowa, T., Sakiyama, F. & Narita, K. (1967b). *Bull. Chem. Soc. Japan*, **40,** 632–639.
Naughton, M. A., Sanger, F., Hartley, B. S. & Shaw, D. C. (1960). *Biochem. J.*, **77,** 149–163.
Niketic, V., Thomsen, J. & Kristiansen, K. (1974). *Eur. J. Biochem.*, **46,** 547–551.
Nute, P. E. & Mahoney, W. C. (1979). *Biochemistry*, **18,** 467–472.
Ohno, M., Spande, T. F. & Witkop, B. (1970). *J. Am. Chem. Soc.*, **92,** 343–348.
Omenn, G. S., Fontana, A. & Anfinsen, C. B. (1970). *J. Biol. Chem.*, **245,** 1895–1902.
Otieno, S. (1978). *Biochemistry*, **17,** 5468–5474.
Ozols, J. & Gerard, C. (1977a). *J. Biol. Chem.*, **252,** 5986–5989.
Ozols, J. & Gerard, G. (1977b). *J. Biol. Chem.*, **252,** 8549–8553.
Ozols, J. & Gerard, C. (1977c). *Proc. Natl. Acad. Sci. U.S.A.*, **74,** 3725–3729.
Patchornik, A., Lawson, W. B., Gross, E. & Witkop, B. (1960). *J. Am. Chem. Soc.*, **82,** 5923–5927.
Photaki, I. (1963). *J. Am. Chem. Soc.*, **85,** 1123–1126.
Pihl, A. & Lange, R. (1962). *J. Biol. Chem.*, **237,** 1356–1362.
Piszkiewicz, D., Landon, M. & Smith, E. L. (1970). *Biochem. Biophys. Res. Commun.*, **40,** 1173–1178.
Previero, A., Scoffone, E., Benassi, C. A. & Pajetta, P. (1963). *Gazz. Chim. Ital.*, **93,** 849–858.
Previero, A., Coletti, M.-A. & Galzigna, L. (1964). *Biochem. Biophys. Res. Commun.*, **16,** 195–198.
Previero, A., Coletti-Previero, M.-A. & Jollès, P. (1966a). *Biochem. Biophys. Res. Commun.*, **22,** 17–21.
Previero, A., Coletti-Previero, M.-A. & Jollès, P. (1966b). *Biochim. Biophys. Acta*, **124,** 400–402.
Raftery, M. A. & Cole, R. D. (1963). *Biochem. Biophys. Res. Commun.*, **10,** 467–472.
Raftery, M. A. & Cole, R. D. (1966). *J. Biol. Chem.*, **241,** 3457–3461.
Rall, S. C. & Cole, R. D. (1970). *J. Am. Chem. Soc.*, **92,** 1800–1801.
Ramachandran, L. K. & Witkop, B. (1967). *Methods Enzymol.*, **11,** 283–299.
Riley, G., Turnbull, J. & Wilson, W. (1957). *J. Chem. Soc.*, 1373–1379.
Rüegg, U. T. (1977). *Methods Enzymol.*, **47,** 123–126.

Rüegg, U. T. & Rudinger, J. (1974). *Int. J. Peptide Protein Res.*, **6,** 447–456.
Rüegg, U. T. & Rudinger, J. (1977a). *Methods Enzymol.*, **47,** 111–116.
Rüegg, U. T. & Rudinger, J. (1977b). *Methods Enzymol.*, **47,** 116–122.
Ruttenberg, M. A., King, T. P. & Craig, L. C. (1964). *Biochemistry*, **3,** 758–764.
Ruttenberg, M. A., King, T. P. & Craig, L. C. (1965). *Biochemistry*, **4,** 11–18.
Ryle, A. P. & Sanger, F. (1955). *Biochem. J.*, **60,** 535–540.
Sanger, F. (1949). *Biochem. J.*, **44,** 126–128.
Sanger, F. & Thompson, E. O. P. (1952). *Biochim. Biophys. Acta*, **9,** 225–226.
Sanger, F. & Thompson, E. O. P. (1953). *Biochem. J.*, **53,** 353–366.
Sanger, F. & Thompson, E. O. P. (1963). *Biochim. Biophys. Acta*, **71,** 468–471.
Savige, W. E. (1975). *Austr. J. Chem.*, **28,** 2275–2287.
Savige, W. E. & Fontana, A. (1977a). *Methods Enzymol.*, **47,** 442–453.
Savige, W. E. & Fontana, A. (1977b). *Methods Enzymol.*, **47,** 459–469.
Savige, W. E. & Fontana, A. (1980). *Int. J. Peptide Protein Res.*, **15,** 285–297.
Schmir, G. L., Cohen, L. A. & Witkop, B. (1959). *J. Am. Chem. Soc.*, **81,** 2228–2233.
Schroeder, W. A., Shelton, J. R. & Robberson, B. (1967). *Biochim. Biophys. Acta*, **147,** 590–592.
Schroeder, W. A., Shelton, J. B. & Shelton, J. R. (1969). *Arch. Biochem. Biophys.*, **130,** 551–556.
Schultz, J., Allison, H. & Grice, M. (1962). *Biochemistry*, **1,** 694–698.
Shaltiel, S. & Patchornik, A. (1961). *Bull. Res. Council Israel*, **10A,** 48.
Shaltiel, S. & Patchornik, A. (1963a). *J. Am. Chem. Soc.*, **85,** 2799–2806.
Shaltiel, S. & Patchornik, A. (1963b). *Israel J. Chem.*, **1,** 187–190.
Shechter, Y., Patchornik, A. & Burstein, Y. (1976). *Biochemistry*, **15,** 5071–5075.
Sheppard, R. C. (1979). In *Peptides: Structure and Biological Function*, Gross, E. & Meienhofer, J. (Eds.), Pierce, Rockford, Illinois, pp. 577–585.
Shotton, D. & Hartley, B. S. (1973). *Biochem. J.*, **131,** 643–675.
Slobin, L. I. & Singer, S. J. (1968). *J. Biol. Chem.*, **243,** 1777–1786.
Smith, D. J., Maggio, E. T. & Kenyon, G. L. (1975). *Biochemistry*, **4,** 766–771.
Smyth, D. G., Stein, W. H. & Moore, S. (1962). *J. Biol. Chem.*, **237,** 1845–1850.
Sokolovsky, M. & Patchornik, A. (1964). *J. Am. Chem. Soc.*, **86,** 1859–1860.
Sokolovsky, M., Sadeh, T. & Patchornik, A. (1964). *J. Am. Chem. Soc.*, **86,** 1212–1217.
Spackman, D. H., Stein, W. H. & Moore, S. (1958). *Anal. Chem.*, **30,** 1190–1206.
Spande, T. F., Witkop, B., Degani, Y. & Patchornik, A. (1970). *Adv. Protein Chem.*, **24,** 97–260.
Stark, G. R. (1967). *Methods Enzymol.*, **11,** 125–138.
Stark, G. R. (1977). *Methods Enzymol.*, **47,** 129–132.
Strumeyer, D. H., White, W. N. & Koshland, D. E., Jr. (1963). *Proc. Natl. Acad. Sci. U.S.A.*, **50,** 931–935.
Swallow, D. L. & Abraham, E. P. (1958). *Biochem. J.*, **70,** 364–373.
Synge, R. L. M. (1945). *Biochem. J.*, **39,** 351–355.
Tang, J. R. & Hartley, B. S. (1967). *Biochem. J.*, **102,** 593–599.
Tang, J. R. & Hartley, B. S. (1970). *Biochem. J.*, **118,** 611–623.
Tarbell, D. S. & Harnisch, D. P. (1951). *Chem. Rev.*, **49,** 1–90.
Thompson, A. R. (1955). *Biochem. J.*, **60,** 507–515.
Titani, K., Hermodson, M. A., Ericsson, L. H., Walsh, K. A. & Neurath, H. (1972). *Biochemistry*, **11,** 2427–2435.
Toennies, G. & Homiller, R. P. (1942). *J. Am. Chem. Soc.*, **64,** 3054–3056.
Torchinskii, Y. M. (1974). *Sulfhydryl and Disulfide Groups of Proteins*, Plenum Press, New York.
Vanaman, T. C. & Stark, G. R. (1970). *J. Biol. Chem.*, **245,** 3565–3573.

Veronese, F., Fontana, A., Boccù, E. & Benassi, C. A. (1967). *Gazz. Chim. Ital.*, **97,** 321–331.
Veronese, F. M., Boccù, E., Benassi, C. A. & Scoffone, E. (1969). *Z. Naturforsch.*, **24,** 294–300.
Wachter, E. & Werhahn, R. (1979). *Anal. Biochem.*, **97,** 56–64.
Wachter, E. & Werhahn, R. (1980). In *Methods in Peptide and Protein Sequence Analysis, Proc. Third Internat. Conf.*, Birr, C. (Ed.), Elsevier/North Holland, Heidelberg, Amsterdam, New York, Oxford, pp. 21–33.
Wakselman, M., Guibejampel, E., Raoult, A. & Busse, W. D. (1976). *J. Chem. Soc. Chem. Commun.*, 21–22.
Waxdal, M. J., Konigsberg, W. H., Henley, W. L. & Edelman, G. M. (1968). *Biochemistry*, **7,** 1959–1966.
Weiner, H., White, W. N., Hoare, D. G. & Koshland, D. E. (1966). *J. Am. Chem. Soc.*, **8,** 3851–3859.
White, F. H., Jr. (1960). *J. Biol. Chem.*, **235,** 383–389.
White, F. H., Jr. (1972). *Methods Enzymol.*, **25,** 387–392.
Wieland, T. & Schön, W. (1955). *Liebig's Ann. Chem.*, **593,** 157–178.
Wieland, T., Lamperstorfer, C. & Birr, C. (1966). *Makromol. Chem.*, **92,** 277–286.
Wilchek, M. & Witkop, B. (1967). *Biochem. Biophys, Res. Commun.*, **26,** 296–300.
Wilchek, M., Sarid, S. & Patchornik, A. (1965). *Biochim. Biophys Acta*, **104,** 616–618.
Wilchek, M., Spande, T., Milne, G. & Witkop, B. (1968). *Biochemistry*, **7,** 1777–1786.
Witkop, B. (1961). *Adv. Protein Chem.*, **16,** 221–321.
Wootton, J. C., Chambers, G. K., Holder, A. A., Baron, A. J., Taylor, J. G., Fincham, J. R. S., Blumenthal, K. M., Moon, K. & Smith, E. L. (1974). *Proc. Natl. Acad. Sci. U.S.A.*, **71,** 4361–4365.
Zeitler, H.-J. & Eulitz, M. (1978). *J. Clin. Chem. Clin. Biochem.*, **16,** 669–674.

Practical Protein Chemistry—A Handbook
Edited by A. Darbre
© 1986, John Wiley & Sons Ltd

J. M. WILKINSON
Department of Biochemistry, Royal College of Surgeons of England, 35/43, Lincoln's Inn Fields, London WC2A 3PN, U.K.

3

Fragmentation of Polypeptides by Enzymic Methods

CONTENTS

3.1. INTRODUCTION

By virtue of the broad range of specificities available, the use of proteolytic enzymes offers a very wide choice of methods for polypeptide chain cleavage and the preparation of defined peptides. The subject has been extensively reviewed in the past, some reviews dealing particularly with polypeptide chain cleavage (Hill, 1965; Smyth, 1967; Kasper, 1975; Konigsberg & Steinman, 1977) while others have been more concerned with the properties and activities of the enzymes (Perlman & Lorand, 1970; Boyer, 1971; Lorand, 1976). Some information on more recently available enzymes was reviewed by Keil (1977).

The development of automated methods of sequencing has led to an emphasis on the production of a few large and easily fractionated fragments from a given protein. As a result there is an increased interest in the use of highly specific enzymes. A number of these have recently come into wide use and it is to be hoped that there will shortly be others. There is still, however, a requirement for a variety of enzymes of different specificity to complete the armoury of the protein chemist so that overlapping peptides of all kinds may be prepared.

The aim of this chapter is to bring together information on the most useful enzymes which are now available. A summary of the specificities of enzymes is given in Tables 3.1 and 3.2. Some practical details of their use are given together with recent references. The preliminary steps to be taken in preparing a protein for digestion are reviewed, together with methods of chemical modification which may restrict or enhance enzyme specificity.

A proteolytic enzyme is only of real use if it is readily available in a pure form, as it is seldom worth the time and effort of preparing one's own. Sources of enzymes commercially available are therefore given where possible. Unfortunately several potentially useful enzymes are not yet readily available. Details of the use of some of these are given here in the hope that they will become available in the near future.

3.2. PREPARATION OF SUBSTRATE

The physical condition of the substrate is of considerable importance in determining the extent of digestion. If a native protein structure is digested it is probable that cleavage will be restricted. This may be an advantage in isolating fragments with a particular biological activity and will be discussed below (see under trypsin, pepsin and papain). In general, however, proteins require both to be denatured and to have their disulphide bonds cleaved before enzymic digestion can go to completion.

Denaturation may be accomplished simply by heating or by acid precipitation. Boiling for 5 to 10 min at neutral pH or precipitation with 5 % trichloroacetic acid should be adequate treatment in many cases. Disulphide bond cleavage in a denaturing solvent, such as 8 M-urea or 6 M-guanidine HCl followed by alkylation of the resulting cysteine residues is probably the method of choice. Methods of reduction and alkylation are outlined in the next section and given in detail in Chapter 2.

Denatured proteins are frequently insoluble and although an insoluble substrate does not necessarily present a problem it may sometimes do so. In attempting to digest an insoluble protein it is necessary that the precipitate should be as finely divided as possible. This may be achieved by dissolving the protein at an extreme pH, if this is possible, and then titrating back to the desired pH, or by dissolving in some solvent with 6 M-guanidine HCl which may then be removed by dialysis. If this is not successful the protein may in many cases be solubilized by the reversible modification of lysine residues, citraconylation being the method of choice (see below). This gives the protein an overall net negative change and leads to disaggregation.

Some proteolytic enzymes retain some or all of their activity under denaturing conditions and advantage may be taken of this in particularly difficult circumstances. Use has been made of low concentrations of urea or guanidine HCl or even of sodium dodecyl sulphate (SDS). Alternatively it may be possible to solubilize a protein by increasing the salt concentration of the buffer or by

Table 3.1. Highly specific proteases.

Enzyme	Optimum pH	Major cleavages	Minor cleavages	Exceptions
Trypsin	7–9	–Lys↓X–; –Arg↓X; –Aec↓X–		–Lys↓̸Pro–
Thrombin	8	–Arg↓X–		
		(X = Gly, Ala, Val, Asp, Cys, Arg)		
S. aureus V8 protease	4 & 7.8	–Glu↓X–	–Asp↓X–	–Glu↓̸Pro–; –Glu↓̸Glu–
Clostripain	7.7	–Arg↓X–	–Lys↓X–	
Mouse submaxillary protease	7.5–8	–Arg↓X–		
A. mellea protease	8.0	–X↓Lys–; –X↓Aec–	–Arg↓X–	–X↓̸Lys–Pro–
Myxobacter AL1 protease	9.0	–X↓Lys–		
Post-proline cleaving enzyme	7.5–8	–Pro↓X–		–Pro↓̸Pro–

Aec: aminoethylcysteine

Table 3.2. Less specific proteases

Enzyme	Optimum pH	Cleavage sites	Exceptions
Chymotrypsin	7–9	–H↓X– H = Tyr, Phe, Trp, Leu	–H↓Pro–
Thermolysin	7–8	–X↓H– H = Val, Leu, Ile, Phe, Tyr, Trp	–X↓H–Pro–
Crotalus atrox α–protease	7.5–8	–X↓H– (similar to thermolysin)	
Pepsin	2	–X↓H↓Y–, –X↓Glu↓Y– H is aromatic or large aliphatic	
Papain	5–7.5	–Phe–X↓Y– Many other bonds cleaved	
Elastase	7–9	–N↓Y– N = small neutral or hydrophobic residues (Ser, Ala, Gly, Val, Leu)	
α–Lytic protease	7–8	–N↓Y– (Similar to elastase)	

increasing the temperature of digestion. Details of those enzymes which retain their activity under denaturing conditions or at an elevated temperature are given below and reference should be made to trypsin, clostripain, *S. aureus* V8 protease, chymotrypsin, thermolysin and papain.

3.3. GENERAL CONDITIONS OF DIGESTION

For any particular enzymic digest attention must be paid to four variables: buffer, enzyme-substrate ratio, temperature and time. Suggested conditions are given in the section on each enzyme, but these must be regarded as a guide only and some general comments may be made here.

3.3.1. Buffer

It is generally preferable to use a volatile buffer, especially if the peptides are to be separated by ion-exchange chromatography or paper electrophoresis. The salt concentration may not be a problem if HPLC is used (Chapter 6). Most of the enzymes described here have pH optima in the range of 7.5 to 8.5. The most generally useful buffers in this range are ammonium bicarbonate, at concentrations of 1% (w/v) or 100 mM, giving a pH of 7.9, and 100 mM-*N*-ethylmorpholine acetate. For peptic digestions at pH 2.0, 10 mM-HCl or 5% acetic acid may be used. Some of the enzymes described require specific additions to the buffer for enzymic activity, either a sulphydryl reagent (see clostripain and papain) or Ca^{2+} ion (see thermolysin).

3.3.2. Enzyme–substrate ratio

The usual ratio used to obtain complete digestion is of the order of 1 or 2% (w/w). If only a limited degree of digestion is required this may be reduced considerably. On the other hand it may sometimes be necessary to increase the amount of enzyme to obtain complete digestion.

3.3.3. Temperature

The temperature at which most digests are performed is 37°C. Reducing the temperature to 20°C will, of course, reduce the rate of reaction and this is useful if a limited digest is required. Thermolysin and myxobacter protease are the only enzymes described here which retain a significant degree of activity at temperatures above 40°C.

3.3.4. Time

Quoted digestion times are generally of the order of 2 to 4 h to obtain complete digestion. A shorter time, of the order of 10 min, may result in the cleavage of only a few specific bonds. Longer times of 16 or 24 h may be necessary with some enzymes or with particularly resistant substrates. Care must be exercised in this latter case to eliminate bacterial contamination and it may be helpful to add a drop of toluene.

3.3.5. Monitoring of digestion

The majority of enzyme digests are required to go to completion and it is not therefore necessary to monitor their progress. It may sometimes be useful if sufficient material is available to follow the rate of reaction by the use of an automatic titrator which can follow base uptake in an unbuffered solution.

If, however, it is desired to limit digestion to a few specific cleavage points it is necessary to select appropriate conditions of time and temperature. The progress of the digest may be monitored by gel electrophoresis as the digestion products will generally be large peptides. An example of this type of monitoring is given by Hale & Perham (1979).

3.3.6. Termination of digestion

Digests may be terminated in a number of ways, the simplest being by freezing followed by freeze-drying. Alternatively the pH of the digest may be altered to one at which the enzyme is no longer active, generally by acidification.

It may be desirable, however, to stop the reaction by specific inhibition and this is particularly the case for a limited digest. Details of specific inhibitors are given under each enzyme. In general the proteases fall into three classes, serine, sulphydryl and metalloenzymes. The serine enzymes may be inactivated specifically by the inhibitors diisopropylphosphorofluoridate (DFP) and phenylmethylsulphonyl-fluoride (PMSF). The latter is preferred as DFP is highly toxic. DFP may be dissolved in anhydrous propanediol and PMSF in propan-2-ol, either being made up as a 1 M solution. The digest should be made 1 mM in inhibitor to prevent further digestion. Sulphydryl proteases may be inhibited by alkylating agents such as iodoacetic acid which should be added in a concentration sufficient to be in a slight excess over the sulphydryl reagent added to activate the enzyme. Metalloenzymes are inhibited by chelating agents such as EDTA (10 mM).

3.4. PROTEIN MODIFICATION REACTIONS

Most proteins will require some sort of chemical modification before proteolytic digestion. In many cases this will be restricted to cleavage of disulphide bonds and, in the case of reduction, alkylation of the resulting cysteine residues to give stable derivatives. In the case of aminoethylation the derivative also provides an extra cleavage point for trypsin and the *A. mellea* protease. A modification reaction for aspartic acid residues that gives rise to an extra point of tryptic cleavage is also included.

Other modification reactions described are designed to modify lysine and arginine residues in order to render adjacent peptide bonds resistant to attack, particularly by trypsin. The emphasis here is on reversible reactions so that the original residue may be regenerated after digestion.

3.4.1. Disulphide bond cleavage

3.4.1.1 Oxidation

Cleavage of disulphide bonds by oxidation with performic acid gives rise to the stable derivative cysteic acid and introduces extra negative charges into the polypeptide chain. See Chapter 2 (section 2.2.3) for further discussion and a practical method.

3.4.1.2. Reduction

Cleavage of disulphide bonds by reduction is fully discussed in Chapter 2. See also Konigsberg (1972).

3.4.2. Alkylation of sulphydryl groups

A number of alkylating reagents are available and the choice will depend on the type of group it is desired to introduce into the protein. Three main types of reagent are given here.

3.4.2.1. Carboxymethylation

This is the most commonly used modification. The reagents iodoacetic acid and iodoacetamide may be used to introduce an acidic or neutral substituent, respectively. Both reagents should be recrystallized before use to remove free iodine, the acid from hexane and the amide from petroleum ether. A solution of either reagent should be added to the reduced protein to give a slight excess over all sulphydryl groups. The iodoacetic acid solution should be neutralized before use with Tris or NaOH. If the reaction is carried out at 0°C for 1 h there should be no danger of alkylating other functional groups in the protein. A discussion of this may be found in Gurd (1972) and in Chapter 2 (section 2.2.1).

A major advantage of carboxymethylation is that both reagents may be obtained radioactively labelled. In order to obtain maximum labelling of sulphydryl groups the radioactive reagent should be added first, followed by a slight excess of cold reagent. Protein sulphydryl groups are found to react more rapidly than those of the reducing agent and hence a high proportion of the label is incorporated into protein.

3.4.2.2. Aminoethylation

Ethyleneimine may be used to introduce basic aminoethyl-cysteine residues into proteins (Raftery & Cole, 1963, 1966). The importance of this reaction is due to the fact that aminoethyl-cysteine is an analogue of lysine and creates extra points of tryptic cleavage (see section 3.5.1). Cole (1967) suggested the use of an eight-fold excess of ethyleneimine over sulphydryl groups. The only side reaction noted was the modification of methionyl residues at low pH (Schroeder *et al.*, 1967). Thus lowering the pH before all the excess reagent has been removed should be avoided. See Chapter 2 for practical details.

3.4.2.3. Other alkylation reactions

4-Vinylpyridine was used to form the acid-stable basic derivative *S*-pyridylethyl cysteine (Friedman *et al.*, 1970). 2-Bromoethyltrimethylammonium bromide was introduced by Itano & Robinson (1972). This gives the basic cysteine derivative 4-thialaminine, which is also claimed to be stable to acid hydrolysis and to improve the handling characteristics of proteins in the automated amino acid sequencer. See Chapter 2 for further details.

3.4.3. Modification of lysine residues

A number of reactions have been described which will reversibly modify lysine residues in proteins. Three of these, citraconylation, maleylation and trifluoroacetylation are described here. The choice between them depends on the particular stability required. All three modified residues are stable at pH 8 to 9 but while the first two are labile at low pH the trifluoroacetyl groups are labile at high pH. To these three may be added tetrafluorosuccinylation (Braunitzer *et al.*, 1968), which gives an alkali-labile derivative, but it has not been much used. Atassi & Habeeb (1972) investigated the stability of the various blocking groups, the extent of the regeneration of amino groups and the recovery of native structure and biological activity. Citraconylation is the method of choice due to the mild conditions required for unblocking, the completeness of unblocking and the high recovery of activity. Citraconylation suffers from one problem in that the blocking groups are appreciably labile at 40°C even at pH 8. If the protein needs to be handled under these conditions for an extended period, maleylation may be a better choice. A further problem with both citraconylation and maleylation is that free sulphydryl groups may become alkylated during the reaction (Butler & Hartley, 1972). If it is necessary to regenerate the sulphydryl groups after unblocking, they should be protected with sulphite or by forming a mixed disulphide with cysteine.

3.4.3.1. Citraconylation

The use of citraconic anhydride as a reversible blocking reagent was introduced by Dixon & Perham (1968). The reaction is performed at room temperature and is best followed by using a pH electrode in the stirred solution. The protein (15 mg/ml) is dissolved or suspended in water and the pH adjusted to 8.0. Citraconic anhydride (1 μl/mg protein) is added in aliquots to the solution, while maintaining pH 8 by the addition of 5 M-NaOH. The reaction is complete in about 30 min at which point the pH remains stable. The protein solution may be desalted either by dialysis or by gel filtration using 0.1 M-ammonium bicarbonate as solvent. Most proteins will become solubilized during modification, but the reaction may, if necessary, be performed in urea or guanidine HCl solution.

The blocking groups may be removed by incubation at pH 3.5 at 20°C overnight. The unblocking reaction is faster with pH < 3.5. Gibbons and Perham (1970) suggested dialysis against 10 mM-HCl, pH 2.0 for 6 h at 20°C.

3.4.3.2. Maleylation

Butler & Hartley (1972) discussed the reaction in detail and suggested that the reaction may be carried out in carbonate or phosphate buffer and that a temperature of 2°C should be used to avoid pH fluctuations. The unblocking reaction requires rather more stringent conditions than those used for citraconylated proteins. The half-life of the maleyl groups at pH 3.5, 37°C is 11 to 12 h. Freedlender & Haber (1972) used pH 2.5, 40°C for 120 h.

Method

The method of Butler *et al.* (1969) is described.
Maleic anhydride should be sublimed before use.
Dissolve 20 mg of bovine chymotrypsinogen in 2.0 ml of 0.1 M-sodium pyrophosphate buffer at pH 9.0. Treat at 2°C with 300 μl of 1.0 M-maleic anhydride in redistilled dioxane (added in portions 50 μl × 6). Maintain pH 9.0 with 0.1 M-NaOH. Desalt by passing through Sephadex G-25 column (400 × 30 mm) in 0.01 M-ammonia.

To unblock, the protein is taken to pH 3.5 with formic acid and ammonia and incubated for 30 h at 37°C and then for 60 h at 60°C.

See p. 156 for a method of partial maleylation.

3.4.3.3. Trifluoroacetylation

The use of ethylthioltrifluoroacetate as a reversible blocking reagent was introduced by Goldberger & Anfinsen (1962) and its use was reviewed by Goldberger (1967). The protein is dissolved in water, or urea solution and the pH adjusted to 10 with 1 M-KOH. The solution is stirred vigorously and ethylthioltrifluoroacetate (10 μl/mg protein) added and the solution maintained at pH 10 by the addition of 1 M-KOH at room temperature. Excess reagents may be removed by dialysis.

The trifluoroacetyl groups may be removed by treatment with 1 M-piperidine at 0°C for 2 h. Afterwards, adjust to pH 6 by addition of 0.5 M acetic acid and remove excess reagent by freeze drying.

3.4.4. Modification of arginine residues

Arginine residues in proteins have been modified by the use of a number of reagents, among them being malonaldehyde (King, 1966), cyclohexanedione (Toi *et al.*, 1967), 2,3-butanedione (Yankeelov & Acree, 1971) and nitromalondialdehyde (Signor *et al.*, 1971). Some of these methods were reviewed by Yankeelov (1972). Unfortunately most of the reactions require extreme conditions of pH which may result in undesirable side reactions and none of them are reversible. Patthy & Smith (1975) described a modification of the reaction with 1,2-cyclohexanedione using pH 8–9. This is given below. The resulting arginine derivative is stable at acid pH and forms a complex with borate which is stable at pH 8. The method was reviewed by Smith (1977).

Modification of arginine with 1,2-cyclohexanedione

The protein is dissolved in 0.2 M-sodium borate buffer, pH 9.0 (10 mg/ml) and reacted with 0.15 M-1,2-cyclohexanedione at 35°C for 2 h. Remove excess reagent by acidification with an equal volume of 30% acetic acid followed by dialysis against dilute acetic acid and freeze drying. Alternatively the protein may be dialysed directly against 0.1 M-sodium borate, pH 8.0, when it will be in a suitable state for digestion.

The arginine residues may be regenerated from the complex by incubation at pH 7.0 in 0.5 M-hydroxylamine at 37°C for 12 h in an inert atmosphere.

3.4.5. Modification of carboxyl groups

The modification of carboxyl groups in small peptides by use of a water-soluble carbodiimide in order to convert aspartic acid residues to a lysine analogue was suggested by Wang & Young (1978). Either 1,2-diaminoethane or diaminomethane was used for the modification. Peptides were treated with 2 M-diamine at pH 4.75 and 0.4 M-1-ethyl-3-(3-diethyl-aminopropyl) carbodiimide HCl. After 1 h at room temperature a second aliquot of carbodiimide was added and allowed to react for 5 h. The modified peptides may be freed of reaction products by gel filtration. This reaction is not reversible and will result in the modification of both α- and side chain carboxyl groups. It has not been tested with proteins, the largest polypeptide modified being glucagon, which has 29 amino acids.

3.5. PROTEASES OF HIGH SPECIFICITY

The following section describes enzymes whose specificity is restricted to the cleavage of a limited number of peptide bonds. They are listed in Table 3.1. Trypsin is probably the most important and most widely used. Its usefulness is enhanced by the modification reactions described in the previous section which means that it can be used to cleave specifically at either arginine or lysine residues. Greatly increased use is now being made of the V8 protease from *S. aureus*. This enzyme gives reliable and specific cleavage at acidic residues, particularly glutamic acid, and therefore provides an important alternative to trypsin. A recent report (Drapeau, 1980) has described a protease isolated from a mutant of *Pseudomonas fragi*. Preliminary data suggest that it cleaves on the amino terminal side of aspartic or cysteic acid residues. It is too early to assess its usefulness but it may provide a valuable alternative to the V8 protease. The other enzymes described are, in general, specific for either arginine or lysine but have not been as widely used as trypsin. On the whole they attack less than the expected number of bonds in any given protein and this may be a considerable advantage in some investigations.

3.5.1. Trypsin

Trypsin is one of the most specific proteolytic enzymes available and certainly the most widely used. It is available from a wide variety of sources but may be contaminated with small amounts of chymotrypsin. The chymotryptic activity may be inactivated by treatment with L-(1-tosylamido-2-phenyl) ethylchloromethyl ketone (TPCK) (Kostka & Carpenter, 1964). TPCK trypsin is available commercially (Worthington) and enzyme of this grade should always be used.

3.5.1.1. Specificity and conditions

Specificity

Trypsin cleaves specifically peptide bonds on the *C*-terminal side of lysine and arginine residues, bonds of the type -Lys-X- and -Arg-X- being susceptible. There are some restrictions on this specificity. When X is proline the bond is almost completely resistant to tryptic cleavage. The presence of acidic residues on either side of a potentially susceptible bond may lead to considerably reduced rates of cleavage or even total resistance. The presence of positively charged groups may also reduce the rate of cleavage and in particular adjacent lysine or arginine residues, or peptides in which these residues are *N*-terminal, may only give partial cleavage.

Despite its very high specificity there are numerous instances in the literature of non-specific cleavage but in any particular protein non-specific cleavage is likely to be rare. In general susceptible bonds are adjacent to aromatic or hydrophobic residues. Such cleavage is not thought to be the result of chymotryptic activity but rather to be intrinsic to trypsin and to be due to the presence of small amounts of ψ-trypsin (Keil-Dlouhá *et al*., 1971). There are also examples of cleavage of -Arg-Pro- bonds in the sequences -Trp-Arg-Pro-Ala- (Nyman *et al*., 1966) and -Ala-Arg-Pro-Ala- (Grand *et al*., 1976). Caution should therefore be exercised in interpreting the results of a tryptic digest. The number of peptides obtained may not add up to the total of arginine and lysine residues plus one and it is possible to obtain peptides which contain a single lysine or arginine residue which is not, in fact, *C*-terminal, although it should be stressed that this situation is rare.

Conditions

Trypsin is a serine protease which is maximally active in the pH range 7 to 9. It is reversibly inactivated below pH 4. Trypsin solutions should be made up in 10 mM-HCl to prevent autodigestion and can be stored, frozen, for a few weeks. Ammonium bicarbonate (0.1 M) is widely used as a buffer for tryptic digestion, using an enzyme–substrate ratio of 1% or 2% at 37°C for between 1 and 4 h. Tryptic digestion may be limited by using native protein as substrate and reducing both the time and temperature. Thus, Leavis *et al*. (1978a) used digestion at 25°C for 30 min to limit cleavage to six out of a possible 15 sites in troponin C.

Tryptic activity is resistant to mild denaturing conditions. Harris (1956) has shown that 48 % of the activity is retained in 4 M-urea and Spackman *et al.* (1960) used buffers containing 2 M-guanidine HCl. Trypsin is rapidly and irreversibly inactivated by both DFP and PMSF as described above. There are also a number of specific protein inhibitors available such as soybean trypsin inhibitor (Birk, 1976), which form a stoichiometric complex with trypsin and may be used to stop the reaction (Leavis *et al.*, 1978a). These authors add 4 mg of soybean inhibitor for 1 mg of trypsin. If the reaction is terminated by acidification it must be remembered that the trypsin will once again become active if the pH is raised.

Specific cleavage at arginine residues

The specificity of trypsin may be restricted to cleavage at arginine residues by reversible blocking of the lysine residues (see section 3.4.3). Citraconylation is the method of choice unless the digest is to be a very long one in which case maleylation should be used. Peptides are generally unblocked by reducing the pH to below 3.5 before attempting their separation. If this is done the trypsin will be inactivated but it must be remembered that reactivation will occur at high pH unless it is removed or irreversibly inactivated. If for any reason it is necessary to handle the peptides at a low pH before unblocking, trifluoroacetylation may be chosen.

Specific cleavage at lysine residues

The reversible arginine blocking method of Patthy & Smith (1975) may be used to restrict tryptic cleavage to lysine residues (see 3.4.4). These authors separated the peptides in the digest at low pH before removing the blocking group. The method was used by Blumenthal *et al.* (1975) and Yoshida *et al.* (1976), and seems to work well although there may be some problems with the solubility of the protein after blocking, especially if it is freeze dried (R.J.A. Grand, personal communication).

Specific cleavage at cysteine residues

Raftery & Cole (1963) showed that alkylation of cysteine residues to give the lysine analogue, aminoethylcysteine, gives rise to an extra point of tryptic cleavage in a polypeptide chain. The rate of cleavage at aminoethylcysteine is not as great as at arginine or lysine (Cole, 1967; Plapp *et al.*, 1967) and a longer digestion time or higher enzyme–substrate ratio may be required for complete cleavage. The possibility of reversibly blocking both lysine and arginine residues means that it should be possible to restrict cleavage with trypsin to cysteine residues. The lysine and arginine residues would have to be blocked before aminoethylation and it would also be necessary to protect the sulphydryl groups if citraconylation was used. The acidic nature of the lysine blocking groups and the bulkiness of the arginine blocking groups make it unlikely for tryptic cleavage to occur if either of these residues are adjacent to an aminoethylcysteine residue.

Specific cleavage at aspartic acid residues

The recent method of Wang & Young (1978) for the modification of aspartyl carboxyl groups with diamines brings the possibility of specific cleavage at aspartic acid residues. The only other method for such cleavage is partial acid hydrolysis (Schultz, 1967). As the carboxyl groups of glutamic acid and the α-carboxyl group are also modified in the reaction, it is unlikely that this method will be generally applicable to proteins. It may, however, be useful for smaller peptides, particularly those containing little or no glutamic acid.

3.5.2. Thrombin

Thrombin is one of the most specific proteolytic enzymes available and it is somewhat surprising that it has not found greater use. This may be due to the fact that different batches of the enzyme differ in their purity and activity. The enzyme is available commercially from a number of suppliers (e.g. Calbiochem., Miles Laboratories, Sigma). Uehara *et al.* (1979) purified commercial thrombin by chromatography on SP Sephadex G50 as described by Lundblad *et al.* (1976). For a discussion of the methods of assay of thrombin activity and the different molecular forms in which it may be found, see Lundblad *et al.* (1976).

3.5.2.1. Specificity and conditions

Specificity

Cleavage is restricted to bonds on the *C*-terminal side of arginine residues of the type -Arg-X-. In fibrinogen, the natural substrate, X, is glycine but other bonds have been shown to be cleaved, where X is alanine, arginine, aspartic acid, cysteine and valine (Lundblad *et al.*, 1976). Wall *et al.* (1981) have reported the cleavage of an -Arg-His- bond in calmodulin. In any given protein it seems likely that only a very limited number of arginyl bonds will be cleaved and the method should be ideal for preparing large fragments suitable for automatic sequence determination. In some cases only a single bond will be cleaved, even under rigorous conditions, while in others the extent of cleavage will depend on the time and temperature used. An example of this is given by Uehara *et al.* (1979)

Conditions

Thrombin is a serine protease active near pH 8.0. Digestion conditions are similar to those used for trypsin, i.e. a buffer of 100 mM-NH_4HCO_3 at 37°C for 4 to 8 h, but the temperature and time used by various workers differ considerably. The activity of thrombin preparations is generally quoted in NIH units based on fibrinogen clotting activity (Lundblad *et al.*, 1976). The enzyme–substrate ratio used varies widely. On a weight basis variations are found from 1 % (Morgan *et al.*, 1975) to 20 % (Iwanaga *et al.*, 1969) and on an activity basis from 1 unit per mg substrate (Leavis *et al.*, 1978b) to 60 units per mg substrate (Winstanley & Trayer, 1979). In view of this variation it is particularly desirable to monitor thrombin digests by gel electrophoresis in order to find the most suitable reaction conditions. An example of this method is given by Muszbek & Laki (1974). Thrombin may be inhibited by both DFP and PMSF.

3.5.3. *Staphylococcus aureus* V8 protease

The protease isolated by Houmard & Drapeau (1972) from the V8 strain of *Staphylococcus aureus* has been used for the preparation of specific proteolytic fragments. This enzyme is commercially available (Miles). Proteases of similar specificity have been isolated from the 8325N strain of *S. aureus* by Rydén *et al.* (1974) and from sorghum by Garg & Virupaksha (1970a, b), but neither of these is commercially obtainable.

3.5.3.1. Specificity and conditions

Specificity

The major specificity is for cleavage of peptide bonds *C*-terminal to glutamic acid residues, -Glu-X-, unless X is proline or glutamic acid. Drapeau (1977) showed that the substrate specificity depends on the buffer used. In 50 mM-ammonium acetate, pH 4.0, or 50 mM NH_4 HCO_3, pH 7.8, the enzyme is quite specific for glutamic acid residues, whereas in 50 mM-sodium phosphate, pH 7.8, cleavage occurs after both glutamyl and aspartyl residues. Austen & Smith (1976) also investigated the specificity under these conditions and found in some cases an increased rate of hydrolysis in phosphate buffer. They suggested that the observed difference in specificity may be due to the difference in rate of cleavage and that the rate of cleavage of aspartyl bonds in bicarbonate or acetate buffer is generally too small to be observed. Wootton *et al.* (1975) showed that bonds adjacent to *S*-carboxymethylcysteine residues are not cleaved.

Investigations with the V8 protease include those of Emmens *et al.* (1976), Evanberg *et al*, (1977), Walker *et al.* (1977) and Fleer *et al.* (1978). In addition to the expected cleavages these workers reported a number of non-specific cleavages, notably of -Ser-X-bonds. In our hands the only cleavages observed were those expected, involving glutamyl and aspartyl residues (Wilkinson & Grand, 1978). There may be some partial cleavage of bonds especially in a sequence which contains several glutamyl residues.

Conditions

The *S. aureus* V8 protease is active over the pH range 3.5 to 9.5 (Drapeau, 1977) and has pH optima at pH 4.0 and pH 7.8 (Houmard & Drapeau, 1972). It is a serine enzyme and is inhibited by DFP. The enzyme is active in the presence of 0.2 % SDS and retains 50 % activity in 4 M-urea (Drapeau, 1977). It may thus be useful if dissociating conditions are needed to solubilize the substrate, although no information is available as to whether SDS binding to the substrate may limit proteolytic cleavage. At low pH a buffer of 50 mM-ammonium acetate, pH 4.0, is recommended, while at pH 7.8 either 50 mM-NH_4HCO_3 or 50 mM-sodium phosphate may be used, depending on the specificity required. A temperature of 37°C is generally suitable and digestion may be continued for up to 18 h (Houmard & Drapeau, 1972).

3.5.4. Clostripain

Clostripain is a sulphydryl protease isolated from *Clostridium histolyticum* (Mitchell & Harrington, 1968) and its use was reviewed by Mitchell (1977). It is available commercially (Worthington, Boehringer). Gilles *et al.* (1979) published a method of purification by affinity chromatography on ω-amino alkylagarose columns (Sepharose-C_7NH_2). This purification gives a highly active preparation which differs slightly from those described previously. It would seem advantageous to submit commercial preparations to this further purification procedure before use.

3.5.4.1. Specificity and conditions

Specificity

The major specificity is for cleavage at the *C*-terminal side of arginine residues at bonds of the type -Arg-X-, although some -Lys-X-bonds are also cleaved. The extent of cleavage at lysine residues may be limited by varying the digestion conditions and these need to be investigated for each individual protein. Shih and Hash (1971) demonstrated cleavage at two arginine residues, but not at the lysine residue of a 20-residue CNBr fragment from lysozyme. Liu & Putnam (1977) reported cleavage at arginine but also at some lysine residues in human β_2-glycoprotein I. D. Stone (personal communication) observed cleavage at arginine but not lysine residues in a CNBr fragment from dihydrofolate reductase (Stone *et al.* 1977). The extent of cleavage at the arginine residues, however, was not uniform and varied from 40 to 100 %. A similar situation was observed in horse phosphoglycerate kinase (G. Hardy, personal communication). Complete cleavage was found at the three arginine residues in a CNBr fragment with no cleavage at the five lysine residues. Gilles *et al.* (1979) investigated the specificity of their purified enzyme acting on hake parvalbumin. This protein contains one arginine and twelve lysine residues. They only observed cleavage at the single arginine residue after overnight digestion at 37°C. Similar specificity was demonstrated by Moonen *et al.* (1980) and by Drickamer (1981). The specificity is thus similar to that provided by tryptic digestion of proteins whose lysine residues have been blocked. As no blocking or unblocking reactions are required it may be found to be more convenient in some cases.

Conditions

Mitchell & Harrington (1968) reported an optimum pH 7.7 for clostripain. The presence of a sulphydryl reagent is essential. Mitchell & Harrington (1968) suggested that there may be a requirement for Ca^{2+}, although this does not seem essential. Conditions used by D. Stone (personal communication) are 2% enzyme–substrate ratio (w/w) in 100 mM-NH_4HCO_3, 1 mM-DTT at 25°C for 4 h. Gilles *et al.* (1979) activated the enzyme for 1 h in 10 mM-DTT, 20 mM-NH_4HCO_3 and then used 0.5% enzyme-substrate (w/w) in 50 mM-NH_4HCO_3 at 37°C overnight. Time and temperature may thus be varied depending on the extent of digestion required. Shih and Hash (1971) incorporated 2 M-urea in their buffers to help solubilize the substrate.

3.5.5. Submaxillary protease

This enzyme was isolated from the submaxillary glands of male mice by Schenkein *et al.* (1969a, b). It has an optimum pH 7.5 to 8.0 and was further characterized by Boesman *et al.* (1976). Schenkein *et al.* (1979, 1981) showed that it is a serine protease inhibited by DFP and that it possesses sequence homology with thrombin. Similar results were obtained by Thomas *et al.* (1979). It is available commercially (Pierce, Boehringer).

3.5.5.1. Specificity and conditions

Specificity

Schenkein *et al.* (1977) used a variety of synthetic and natural sustrates and showed that cleavage is restricted to bonds *C*–terminal to arginine residues. Using hen egg lysozyme and beef insulin they found no cleavage at lysine residues but cleavage at some arginine residues did not go to completion. Wasserman & Capra (1978) obtained similar results with an immunoglobulin kappa chain. They found cleavage at most arginine residues with the exception of an Arg-Val and an Arg-Arg bond which were resistant to digestion. The specificity is thus similar to that of clostripain. Moonen *et al.* (1980), however, obtained cleavage at only one arginine residue out of ten in the phosphatidylcholine exchange protein from bovine liver.

Conditions

Schenkein *et al.* (1977) used 1 % NH_4HCO_3, pH 8.0 at 37°C for up to 24 h, using an enzyme–substrate ratio of 2 % (w/w). Digests were stopped by acidification to pH 1.5 to 2.0 with HCl.

3.5.6. Armillaria mellea protease

This enzyme was isolated from the basidiomycete *Armillaria mellea* by Walton *et al.* (1972). The enzyme is not available commercially but it can be prepared (Walton *et al.*, 1972, 1978; Lewis *et al.*, 1978). It is a Zn^{2+}-containing enzyme consisting of a single polypeptide chain of 14 000 daltons with optimum pH 6.8.

3.5.6.1. Specificity and conditions

Specificity

The specificity was investigated by Shipolini *et al.* (1974). They showed that cleavage occurs at peptide bonds on the *N*-terminal side of lysine and aminoethylcysteine residues, -X-Lys- and -X-Aec-, including those bonds where X is proline. -X-Lys- bonds are resistant to cleavage when the sequence is -X-Lys-Pro-. The cleavage of a -Pro-Lys- bond in insulin was demonstrated by Gregory (1975). This study was extended by Doonan *et al.* (1975) using a different substrate. They suggested that cleavage may be restricted if adjacent residues on either side of a lysine are acidic. Cleavage was also observed on the *C*-terminal side of arginine residues, -Arg-X-. This appeared to be enhanced if X was leucine or isoleucine, although not all bonds of this type were cleaved.

Studies on the use of *A. mellea* protease for the limited cleavage of immunoglobulins G and D were carried out by Hunneyball & Stanworth (1975) and by Jefferis and Matthews (1977), respectively. The latter demonstrated cleavage *N*-terminal to a lysine residue.

Conditions

Doonan & Fahmy (1975) used 0.2 M-*N*-ethylmorpholine-acetic acid buffer, pH 8.0 and 1 % enzyme-substrate ratio (w/w) at 35°C for 22 h. Gregory (1975) used 0.1 M-NH_4HCO_3 buffer with an enzyme–substrate ratio of between 0.1 % and 0.5 % at 37°C. He also varied the pH and temperature showing that cleavage took place satisfactorily from pH 4–9 and that the rate of cleavage increased up to 50°C. Hunneyball & Stanworth (1975) used a buffer containing 10 mM-Tris-HCl – 0.15M-NaCl-2 mM-Ca acetate, pH 7.0, with a 1 % enzyme–substrate ratio (w/w) at 37°C for 3 h for optimal cleavage. Walton *et al.* (1978) showed that the enzyme is inhibited by EDTA and also by lysine and cysteine at concentrations of 1 mM.

Specific cleavage at cysteine residues

The possibility of specific cleavage at aminoethylcysteine residues led Doonan & Fahmy (1975) to suggest the use of *A. mellea* protease. They alkylated the cysteine residues of both human haemoglobin β-chain and a bee venom with ethyleneimine after the lysine residues had been blocked by trifluoroacetylation. Specific cleavage at aminoethylcysteine was obtained except when cysteine and lysine were adjacent.

3.5.7. Myxobacter ALl protease II.

The purification of the protease II from *Myxobacter* ALl was described by Wingard *et al.* (1972) and its properties by Jackson & Matsueda (1970). It is not available commercially.

3.5.7.1. Specificity and conditions

Specificity

Myxobacter ALl protease II was shown to be specific for peptide bonds *N*-terminal to lysine residues in a similar manner to the *A. mellea* protease (Wingard *et al.*, 1972). Cunningham *et al.* (1976) found cleavage only at lysine residues of penicillopepsin IV using a digestion time of 1 h. Jörnvall (1977), with a digestion time of 4 h, observed the same specificity in a digest of yeast alcohol dehydrogenase including cleavage of a -Pro-Lys- bond. No cleavage was found at arginine residues but some non-specific cleavage occurred at other residues, there being no discernable pattern in this. It is possible that this might have been reduced by using a shorter digestion time. Freisheim *et al.* (1978) obtained cleavage at both -Gly-Lys- bonds in a CNBr fragment of a dihydrofolate reductase but not at the single arginine residue in this peptide.

In a comparative study of the cleavage of penicillopepsin by the *A. mellea* and *Myxobacter* enzymes (T. Hofmann, personal communication), it was found that the *Myxobacter* enzyme was more specific than the *A. mellea* enzyme which gave rise to a number of non-specific cleavages.

Conditions

The *Myxobacter* enzyme has an optimum pH of 8.5 to 9.0. It is resistant to autodigestion, heat stable at 60° C and inhibited by 50 mM-EDTA (Wingard *et al.* 1972). The conditions used by Jörnvall (1977) for digestion are 20 mM-Tris-HCl buffer, pH 9.0 at 37°C for 4 h, using an enzyme-substrate ratio of 1 % (w/w).

3.5.8. Post-proline cleaving enzyme

As the side chain of proline differs markedly from all other amino acid side chains it might be expected that an enzyme that cleaves specifically at this residue has evolved in some organism. Such an enzyme would be of great use as the occurrence of proline is relatively rare in polypeptides and also its proximity to many peptide bonds renders them resistant to proteolytic attack by other enzymes. Only one such enzyme has been isolated to date. This is the post-proline cleaving enzyme of Koida & Walter (1976), isolated by affinity chromatography from lamb kidney. This was tentatively characterized as a serine protease (Yoshimoto *et al.* 1977). Further characterization was carried out (Yoshimoto *et al.*, 1978; Walter & Yoshimoto, 1978). It is not commercially available.

3.5.8.1. Specificity and conditions

Specificity

Koida and Walter (1976) demonstrated cleavage of a number of small peptides of up to 13 residues. All types of -Pro-X- bonds were cleaved with the exception of -Pro-Pro-. The rate of cleavage depended on the nature of X, being fastest when X is hydrophobic and decreasing when X is basic or acidic. Yoshimoto *et al.* (1977) showed cleavage at -Ala-X- bonds but at rates of 1/100 to 1/1000 that of -Pro-X- bonds. Eerd *et al.* (1978) obtained limited cleavage of a -Pro-Thr- bond in an octapeptide, while Fleer *et al.* (1978) obtained only 10–20% cleavage of two peptides, of 33 and 22 residues, during a 36-h digestion period. They suggest that the enzyme may only be useful for cleavage of small peptides.

Conditions

Koida and Walter (1976) showed that maximum activity is obtained between pH 7.5 and 8.0. They used 100 mM-NH_4HCO_3, pH 7.8 with 1 mM-DTT and 1 mM-EDTA at 22°C for 24 h.

3.6. PROTEASES OF LOWER SPECIFICITY

Included in this section are a number of important enzymes of more general specificity. They are listed in Table 3.2. This is not an exhaustive compilation, but contains proteases which will cleave a wide variety of peptide bonds. In most cases these cleave bonds adjacent to hydrophobic residues. Enzymes of rather broader specificity, such as pepsin and papain, have also been included together with examples of their use to provide limited proteolysis of native proteins.

3.6.1. Chymotrypsin

Chymotrypsin is the most specific enzyme described in this section and has been widely used in the preparation of primary digests of proteins (Smyth, 1967; Kasper, 1975; Konigsberg & Steinman, 1977).

3.6.1.1. Specificity and conditions

Specificity.

The major cleavage sites are at peptide bonds *C*-terminal to aromatic or large hydrophobic residues, of the type -H-X-, where H is tyrosine, phenylalanine, tryptophan or leucine. Bonds are resistant to cleavage where X is proline. Chymotrypsin has a less rigid specificity than trypsin and cleavage may occur *C*-terminal to a variety of residues. Cleavage has been reported adjacent to glycine, histidine, isoleucine, methionine, serine, threonine and valine.

Conditions

The conditions for chymotrypsin are similar to those used for trypsin, namely 100 mM-NH_4HCO_3, 2% enzyme–substrate ratio (w/w) at 37°C for 4 h. Spackman *et al.* (1960) used a buffer, pH 7.0, containing 2 M-guanidine HCl for digestion. Chymotrypsin is a serine protease and is inhibited by both DFP and PMSF. Trypsin contamination is not usually a problem in chymotryptic digests but if necessary soybean trypsin inhibitor may be added to the digest (see 3.5.1.1).

3.6.2. Thermolysin

Thermolysin is isolated from the *Bacillus thermoproteolyticus* and has found extensive use in amino acid sequence investigations. It is available from Calbiochem and other sources. See reviews by Matsubara (1970) and Heinrikson (1977).

3.6.2.1. Specificity and conditions

Specificity

The major specificity is for cleavage of bonds on the *N*-terminal side of hydrophobic residues of the type -X-H-, where H is generally valine, leucine, isoleucine, phenylalanine, tyrosine or tryptophan. However, the specificity is rather broad and bonds adjacent to alanine are fairly readily cleaved. Cleavage has also been noted adjacent to asparagine, threonine, histidine and glycine, although these are rare. The specificity was investigated by a number of workers, notably by Ambler and Meadway (1968) and reviewed by Heinrikson (1977). Bonds involving proline of the type -Pro-Ile- are cleaved but in sequences of the type -X-Phe-Pro-, the -X-Phe- bond is resistant to cleavage. Cleavage is also inhibited by an adjacent α-amino or carboxyl group so that thermolysin has no exopeptidase activity. Side chain amino and carboxyl groups do not appear to have any marked effect on specificity.

Conditions

Thermolysin is active between pH 7.0 and 8.0 and has an absolute requirement for Ca^{2+}. The buffer used may thus be 100 mM-NH_4HCO_3 but should contain 5 mM-$CaCl_2$. Most preparations contain sufficient Ca^{2+} to be active without added $CaCl_2$. The enzyme may be inhibited with EDTA. Digestion at 37°C for 4 h using 2 % enzyme–substrate (w/w) may generally be found suitable although the conditions may need to be varied in particular cases. As its name implies, thermolysin is heat stable. It shows no loss of activity after 1 h at 60°C and retains significant activity at 80°C (Matsubara, 1970). The enzyme is stable in 8 M-urea and also in 20 % ethanol or methanol but these conditions enhance thermal denaturation and digestion should then take place at 20°C.

3.6.3. Pepsin

The use of pepsin in proteolysis and its catalytic activity have been extensively reviewed (Fruton, 1971; Clement, 1973). With the exception of the *S. aureus* V8 protease, it is the only well-characterized and generally available acid protease.

3.6.3.1. Specificity and conditions

Specificity

The major specificity is for cleavage of bonds on either side of aromatic or large aliphatic residues. Bonds involving glutamic acid are also susceptible. The specificity, however, is wide and a large number of other cleavages have been observed, and it is difficult to predict the outcome of a particular digest. Use has been made of pepsin in limited cleavage of native proteins. This has proved particularly useful in obtaining large fragments of immunoglobulins (Nissonoff *et al.*, 1960). The pH for this digest was between 4.5 and 5.0.

Conditions

Pepsin is active in the range pH 1 to 5, with optimum pH 2.0. Digests may thus be conveniently carried out in 10 mM-HCl or 5 % acetic acid. The enzyme is irreversibly inactivated above pH 6.0. Pepsin is particularly suited to the investigation of the arrangement of disulphide bridges in proteins. The low pH of digestion minimizes the likelihood of disulphide interchange reactions and enables peptides containing the correct pairing of cysteine residues to be isolated (Spackman *et al.*, 1960). See also Chapter 4, 4.4.2.

3.6.4. Papain

Papain is a sulphydryl enzyme isolated from *Papyra* latex. See review by Glazer & Smith (1971). It is readily available from a number of sources.

3.6.4.1. Specificity and conditions

Specificity

The substrate specificity of papain was extensively investigated by Schechter & Berger (1968). They showed that the enzyme contains an extended binding site, cleavage of a specific bond being affected by the binding of residues on either side. In particular, there is a preference for binding to phenylalanine two residues on the *N*-terminal side of the bond cleaved. Despite this, however, papain has been shown to cleave at a very large number of different types of bond (Glazer & Smith, 1971) and it is difficult to predict its precise action.

Papain has found considerable use in the preparation of large fragments from native proteins such as immunoglobulins (Porter, 1959) and myosin (Lowey *et al.*, 1969) and the cleavage of the extracellular portion of human histocompatibility antigens (Prober *et al.*, 1978).

Conditions

Papain is optimally active between pH 5 and 7.5, and requires activation by a sulphydryl reagent. Scawen *et al.* (1974) used 0.2 M-pyridine acetate buffer, pH 6.5, containing 1% (v/v) 2,3-dimercaptopropan-1-ol for 1 h at 37°C with an enzyme–substrate ratio of 2% (w/w). Papain is rapidly inactivated by oxidation in the presence of low concentrations of cysteine, by heavy metal ions and by cyanate (Sluyterman, 1967b). Papain is particularly resistant to denaturation by urea. Activity is retained even in 8 M-urea (Sluyterman, 1967a), but this must be carefully deionized to prevent inactivation by cyanate.

3.6.5. Elastase

Elastase is a serine protease structurally homologous with trypsin and chymotrypsin. See review by Hartley & Shotton (1971). It is widely available commercially.

3.6.5.1. Specificity and conditions

Specificity

Elastase has a rather broad specificity. Its major cleavage points are peptide bonds *C*-terminal to small hydrophobic side chains such as alanine (Hartley & Shotton, 1971). An investigation of the specificity by Narayanan & Anwar (1969) suggested a specificity for bonds adjacent to neutral amino acids. Using oxidized insulin chains as substrate, bonds involving serine, alanine, glycine, valine and leucine were cleaved.

Conditions

Similar digestion conditions to those used for trypsin and chymotrypsin may be used, i.e. 100 mM-NH_4HCO_3, at 37°C for 1 to 4 h with an enzyme–substrate ratio of 2% (w/w).

3.6.6. α-Protease from *Crotalus atrox*

The α-protease from *Crotalus atrox* venom was isolated and characterized (Pfleiderer & Krauss, 1965; Zwilling & Pfleiderer, 1967). It is commercially available (Pierce). The enzyme has an optimum pH 7.5 to 8.0. It is specific for bonds *N*-terminal to hydrophobic residues (Pfleiderer & Krauss, 1965; Mella *et al*, 1967) and was recently used in sequencing studies (Walker *et al*., 1977). The specificity is thus very similar to that of thermolysin and more work requires to be done to determine whether there is any advantage in using it.

3.6.7. α-Lytic protease

The isolation of the α-lytic protease from *Sorangium sp.* was described by Whitaker (1965), but it is not commercially available. The specificity is somewhat similar to elastase. Whitaker (1970) suggested that it may be more specific, but no comparative work is available. Its use was described recently in a number of investigations (Stone *et al*., 1977; Sodek *et al*., 1978; Stone & Smillie, 1978).

3.7. CONCLUSION

The main emphasis of this chapter is on the production of large peptide fragments and thus on enzymes of high specificity and the use of techniques for limited cleavage. This is of importance both for amino acid sequence determination and for the investigation of structure-function relationships. The most specific proteases cleave bonds which are adjacent to either the basic residues, lysine and arginine, or the acidic residue, glutamic acid. The other enzymes described tend to cleave bonds adjacent to hydrophobic residues and thus lack absolute specificity, but they are nevertheless of great importance in structural studies for further cleavage of large fragments. One example has been given of a protease cleaving adjacent to proline, but, although of great potential use, it does not seem to be of general applicability on present evidence. It is to be hoped that further work will lead to the characterization of other enzymes of differing specificity and that a number of the enzymes described here will become available commercially.

3.8. REFERENCES

Ambler, R. P. & Meadway R. J. (1968). *Biochem. J.*, **108,** 893–895.
Atassi, M. Z. & Habeeb, A. F. S. A. (1972). *Methods Enzymol.* **25,** 546–553.

Austen, B. M. & Smith, E. L. (1976). *Biochem. Biophys. Res. Commun.*, **72,** 411–417.
Birk, Y. (1976). *Methods Enzymol.*, **45,** 700–707.
Blumenthal, K. M., Moon, K. & Smith, E. L. (1975). *J. Biol. Chem.*, **250,** 3644–3654.
Boesman, M., Levy, M. & Schenkein, I. (1976). *Arch. Biochem. Biophys.*, **175,** 463–476.
Boyer, P. D. (Ed.) (1971). *Enzymes*, 3rd edn, vol. 3, Academic Press, New York & London.
Braunitzer, G. von, Beyreuther, K., Fujiki, H. & Schrank, B. (1968). *Hoppe-Seyler's Z. Physiol. Chem.*, **349,** 265.
Butler, P. J. G. & Hartley, B. S. (1972). *Methods Enzymol.*, **25,** 191–199.
Butler, P. J. G., Harris, J. I., Hartley, B. S. & Leberman, R. (1969). *Biochem. J.*, **112,** 679–689.
Clement, G. (1973). In *Progress in Bioorganic Chemistry*, Kaiser, E. T. & Kézdy, F. J. (Eds.), vol. 2, Wiley, New York, London, Sydney, Toronto, pp. 177–238.
Cole, R. D. (1967). *Methods Enzymol.*, **11,** 315–317.
Cunningham, A., Wang, H. -M., Jones, S. R., Kurosky, A., Rao, L., Harris, C. I., Rhee, S. H. & Hofmann, T. (1976). *Can. J. Biochem.*, **54,** 902–914.
Dixon, H. B. F. & Perham, R. N. (1968). *Biochem. J.*, **109,** 312–314.
Doonan, S. & Fahmy, H. M. A. (1975). *Eur. J. Biochem.*, **56,** 421–426.
Doonan, S., Doonan, H. J. Hanford, R., Vernon, C. A., Walker, J. M., Airoldi, L. P. da S., Bossa, F., Barra, D., Carloni, M., Fassela, P. & Riva, F. (1975). *Biochem. J.*, **149,** 497–506.
Drapeau, G. R. (1977). *Methods Enzymol.*, **47,** 189–191.
Drapeau, G. R. (1980). *J. Biol. Chem.*, **255,** 839–840.
Drickamer, K. (1981). *J. Biol. Chem.*, **256,** 5827–5839.
Eerd, J. -P. van, Capony, J. -P., Ferraz, C. & Pechère, J. -F. (1978). *Eur. J. Biochem.*, **91,** 231–242.
Emmens, M., Welling, G. W. & Beintema, J. J. (1976). *Biochem. J.*, **157,** 317–323.
Evanberg, A., Meyer, H., Gaastra, W., Verheij, H. M., & Haas G. H. de (1977). *J. Biol. Chem.*, **252,** 1189–1196.
Fleer, E. A. M., Verheij, H. M. & Haas, G. H. de (1978). *Eur. J. Biochem.*, **82,** 261–269.
Freedlender, E. F. & Haber, E. (1972). *Biochemistry*, **11,** 2362–2370.
Freisheim, J. H., Bitar, K. G., Reddy, A. V. and Blankenship, D. T. (1978). *J. Biol. Chem.*, **253,** 6437–6444.
Friedman, M., Krull, L. H. & Cavins, J. F. (1970). *J. Biol. Chem.*, **245,** 3868–3871.
Fruton, J. S. (1971). In *Enzymes*, 3rd edn, vol. 3, Boyer, P. D. (Ed.), Academic Press, New York & London, pp. 119–164.
Garg, G. K. & Virupaksha, T. K. (1970a). *Eur. J. Biochem.*, **17,** 4–12.
Garg, G. K. & Virupaksha, T. K. (1970b). *Eur. J. Biochem.*, **17,** 13–18.
Gibbons, I. & Perham, R. N. (1970). *Biochem. J.*, **116,** 843–849.
Gilles, A. -M., Imhoff, J. -M. & Keil, B. (1979). *J. Biol. Chem.*, **254,** 1462–1468.
Glazer, A. N. & Smith, E. L. (1971). In *Enzymes*, 3rd edn, vol. 3, Boyer, P. D. (Ed.), Academic Press, New York & London, pp. 501–546.
Goldberger, R. F. (1967). *Methods Enzymol.*, **11,** 317–322.
Goldberger, R. F. & Anfinsen, C. B. (1962). *Biochemistry,* **1,** 401–405.
Grand, R. J. A., Wilkinson, J. M. & Mole, L. E. (1976). *Biochem. J.*, **159,** 633–641.
Gregory, H. (1975). *FEBS Lett.*, **51,** 201–205.
Gurd, F. R. N. (1972). *Methods Enzymol.*, **25,** 424–438.
Hale, G. & Perham, R. N. (1979). *Eur. J. Biochem.*, **94,** 119–126.
Harris, J. I. (1956). *Nature (Lond.)*, **177,** 471–473.
Hartley, B. S. & Shotton, D. M. (1971). In *Enzymes*, 3rd edn, vol. 3, Boyer, P. D. (Ed.), Academic Press, New York & London, pp. 323–373.
Heinrikson,R. L. (1977). *Methods Enzymol.*, **47,** 175–189.

Hill, R. L. (1965). *Adv. Protein Chem.*, **20,** 37–107.
Houmard, J. & Drapeau, G. R. (1972). *Proc. Natl. Acad. Sci. U.S.A.*, **69,** 3506–3509.
Hunneyball, I. M. & Stanworth, D. R. (1975). *Immunology,* **29,** 921–931.
Itano, H. A. & Robinson, E. A. (1972). *J. Biol. Chem.*, **247,** 4819–4824.
Iwanaga, S., Wallén, P., Gröndahl, N. J., Henschen, A. & Blombäck, B. (1969). *Eur. J. Biochem.*, **8,** 189–199.
Jackson, R. L. & Matsueda, G. R. (1970). *Methods Enzymol.,* **19,** 591–599.
Jefferis, R. & Matthews, J. B. (1977). *Immunochemistry*, **14,** 171–178.
Jörnvall, H. (1977). *Eur. J. Biochem.*, **72,** 425–442.
Kasper, C. B. (1975). In *Protein Sequence Determination*, Needleman, S. B. (Ed.), 2nd edn, Springer, Heidelberg, Berlin, New York, pp. 114–161.
Keil, B. (1977). In *Solid Phase Methods in Protein Sequence Analysis*, Proc. Second Internat. Conf., Previero, A. & Coletti-Previero, M. -A. (Eds.), North Holland, Amsterdam, New York, Oxford, pp. 287–292.
Keil-Dlouhá, V., Zylba, N., Imhoff, J. -M., Tong, N. -T. and Keil, B. (1971). *FEBS Lett.*, **16,** 291–295.
King, T. P. (1966). *Biochemistry,* **5,** 3454–3459.
Koida, M. & Walter, R. (1976). *J. Biol. Chem.*, **251,** 7593–7599.
Konigsberg, W. (1972). *Methods Enzymol.*, **25,** 185–188.
Konigsberg, W. H. & Steinman, H. M. (1977). In *The Proteins*, Neurath, H. & Hill, R. L. (Eds.), 3rd edn, vol. 3, Academic Press, New York, pp. 1–178.
Kostka, V. & Carpenter, F. H. (1964). *J. Biol. Chem.*, **239,** 1799–1803.
Leavis, P. C., Rosenfeld, S. S., Gergely, J., Grabarek, Z. & Drabikowski, W. (1978a). *J. Biol. Chem.*, **253,** 5452–5459.
Leavis, P. C., Rosenfeld, S. & Lu, R. C. (1978b). *Biochim. Biophys. Acta*, **535,** 281–286.
Lewis, W. G., Basford, J. M. & Walton, P. L. (1978). *Biochim. Biophys. Acta*, **522,** 551–560.
Liu, Y. -S. & Putnam, F. W. (1977). *Fed. Proc. Fed. Amer. Soc. Exp. Biol.*, **36,** 763.
Lorand, L. (Ed.) (1976). *Methods Enzymol.*, **45,** 1–939.
Lowey, S., Slayter, H. S., Weeds, A. G. & Baker, H. (1969). *J. Mol. Biol.*, **42,** 1–29.
Lundblad, R. L., Kingdon, H. S. & Mann, K. G. (1976). *Methods Enzymol.*, **45,** 156–176.
Matsubara, H. (1970). *Methods Enzymol.*, **19,** 642–651.
Mella K., Volz, M. & Pfleiderer, G. (1967). *Anal. Biochem.*, **21,** 219–226.
Mitchell, W. M. (1977). *Methods Enzymol.*, **47,** 165–170.
Mitchell, W. M. & Harrington, W. F. (1968). *J. Biol. Chem.*, **243,** 4683–4692.
Moonen, P., Akeroyd, R., Westerman, J., Puijk, W. C., Smits, P. & Wirtz, K. W. A. (1980). *Eur. J. Biochem.*, **106,** 279–290.
Morgan, F. J., Birken, S. & Canfield, R. E. (1975). *J. Biol. Chem.*, **250,** 5247–5258.
Muszbek, L. & Laki, K. (1974). *Proc. Natl. Acad. Sci. U.S.A.*, **71,** 2208–2211.
Narayanan, A. & Anwar, R. A. (1969). *Biochem. J.*, **114,** 11–17.
Nissonoff, A., Wissler, F. C., Lipman, L. N. & Woernley, D. L. (1960). *Arch. Biochem. Biophys.*, **89,** 230–244.
Nyman, P. O., Strid, L. & Westermark, G. (1966) *Biochim. Biophys. Acta.*, **122,** 554–556.
Patthy, L. & Smith, E. L. (1975). *J. Biol. Chem.*, **250,** 557–564.
Perlman, G. E. & Lorand, L. (Eds.) (1970). *Methods Enzymol.*, **19.**
Pfleiderer, G. & Krauss, A. (1965). *Biochem. Z.*, **342,** 85–94.
Plapp, B. V., Raftery, M. A., & Cole, R. D. (1967). *J. Biol. Chem.*, **242,** 265–270.
Porter, R. R. (1959). *Biochem. J.*, **73,** 119–126.
Prober, J. S., Guild, B. C. & Strominger, J. L. (1978). *Proc. Natl. Acad. Sci. U.S.A.*, **75,** 6002–6006.

Raftery, M. A. & Cole, R. D. (1963). *Biochem. Biophys. Res. Commun.*, **6,** 467–472.
Raftery, M. A. & Cole, R. D. (1966). *J. Biol. Chem.*, **241,** 3457–3461.
Rydén, A. -C., Rydén, L. & Philipson, L. (1974). *Eur. J. Biochem.*, **44,** 105–114.
Scawen, M. D., Ramshaw, J. A. M., Brown, R. H. & Boulter, D. (1974). *Eur. J. Biochem.*, **44,** 299–303.
Schechter, I. & Berger, A. (1968). *Biochem. Biophys. Res. Commun.*, **32,** 898–902.
Schenkein, I., Boesman, M., Tokarsky, E., Fishman, L. & Levy, M. (1969a). *Biochem. Biophys. Res. Commun.*, **36,** 156–165.
Schenkein, I., Levy, M. & Fishman, L. (1969b). *Methods Enzymol.*, **19,** 672–681.
Schenkein, I., Levy, M., Franklin, E. C. & Frangione, B. (1977). *Arch. Biochem. Biophys.*, **182,** 64–70.
Schenkein, I., Gabor, M., Franklin, E. C. & Frangione, B. (1979). *Fed. Proc. Fed. Amer. Soc. Exp. Biol.*, **38,** 326.
Schenkein, I., Franklin, E. C. & Frangione, B. (1981). *Arch. Biochem. Biophys.*, **209,** 57–62.
Schroeder, W. A., Shelton, J. R. & Robberson, B. (1967). *Biochim. Biophys. Acta*, **147,** 590–592.
Schultz, J. (1967). *Methods Enzymol.*, **11,** 255–263.
Shih, J. W. K. & Hash, J. H. (1971). *J. Biol. Chem.*, **246,** 994–1006.
Shipolini, R. A., Callewaert, G. L., Cottrell, R. C. & Vernon, C. A. (1974). *Eur. J. Biochem.*, **48,** 465–476.
Signor, A., Bonora, G. M., Biondi, L., Nisato, D., Marzotto, A. & Scoffone, E. (1971). *Biochemistry*, **10,** 2748–2752.
Sluyterman, L. A. A. (1967a). *Biochim. Biophys. Acta*, **139,** 418–429.
Sluyterman, L. A. A. (1967b). *Biochim. Biophys. Acta*, **139,** 430–438.
Smith, E. L. (1977). *Methods Enzymol.*, **47,** 156–161.
Smyth, D. G. (1967). *Methods Enzymol.*, **11,** 214–231.
Sodek, J., Hodges, R. S. & Smillie, L. B. (1978). *J. Biol. Chem.*, **253,** 1129–1136.
Spackman, D. H., Stein, W. H. & Moore, S. (1960). *J. Biol. Chem.*, **235,** 648–659.
Stone, D. & Smillie, L. B. (1978). *J. Biol. Chem.*, **253,** 1137–1148.
Stone, D., Phillips, A. W. & Burchall, J. J. (1977). *Eur. J. Biochem*, **72,** 613–624.
Thomas, K. A., Silverman, R. E., Jeng, I., Baglan, N. C. & Bradshaw, R. A. (1979). *Fed. Proc. Fed. Amer. Soc. Exp. Biol.*, **38,** 324.
Toi, K., Bynum, E., Norris, E. & Itano, H. A. (1967). *J. Biol. Chem.*, **242,** 1036–1043.
Uehara, H., Ewenstein, B. M., Martinko, J. M., Nathenson, S. G., Kindt, T. J. & Coligan, J. E. (1979). *Biochemistry*, **19,** 6182–6188.
Walker, J. M., Hastings, J. R. B. & Johns, E. W. (1977). *Eur. J. Biochem.*, **76,** 461–468.
Wall, C. M., Grand, R. J. A. & Perry, S. V. (1981). *Biochem. J.*, **195,** 307–316.
Walter, R. & Yoshimoto, T. (1978). *Biochemistry*, **17,** 4139–4144.
Walton, P. L., Turner, R. W. & Broadbent, D. (1972). Br. Patent 1263956.
Walton, P. L., Jones, C. & Jackson, S. J. (1978). In *Progress in Chemical Fibrinolysis and Thrombolysis*, Davidson, J. F., Rowan, R. M., Samama, M. M. & Desnoyers, P. C. (Eds.), vol. 3, Raven Press, New York, pp. 373–378.
Wang, T. -T. & Young, N. M. (1978). *Anal. Biochem.*, **91,** 696–699.
Wasserman, R. L. & Capra, J. D. (1978). *Immunochemistry*, **15,** 303–305.
Whitaker, D. R. (1965). *Can. J. Biochem.*, **43,** 1935–1954.
Whitaker, D. R. (1970). *Methods Enzymol.*, **19,** 599–613.
Wilkinson, J. M. & Grand, R. J. A. (1978). *Eur. J. Biochem.*, **82,** 493–501.
Wingard, M., Matsueda, G. & Wolfe, R. S. (1972). *J. Bacteriol.*, **112,** 940–949.
Winstanley, M. A. & Trayer, I. P. (1979). *Biochem. Soc. Trans.*, **7,** 703–704.
Wootton, J. C., Baron, A. J. & Fincham, J. R. S. (1975). *Biochem. J.*, **149,** 749–755.

Yankeelov, J. A. (1972). *Methods Enzymol.*, **25,** 566–579.
Yankeelov, J. A. & Acree, D. (1971). *Biochem. Biophys. Res. Commun.*, **42,** 886–891.
Yoshida, N., Sasaki, A., Rashid, M. A. & Otsuka, H. (1976). *FEBS Lett.*, **64,** 122–125.
Yoshimoto, T., Orlowski, R. C. & Walter, R. (1977). *Biochemistry*, **16,** 2942–2948.
Yoshimoto, T., Fischl, M., Orlowski, R. C. & Walter, R. (1978). *J. Biol. Chem.*, **253,** 3708–3716.
Zwilling, R. & Pfleiderer, G. (1967). *Hoppe-Seyler's Z. Physiol. Chem.*, **348,** 519–524.

Practical Protein Chemistry—A Handbook
Edited by A. Darbre
© 1986, John Wiley & Sons Ltd.

RALPH SCHROHENLOHER and J. CLAUDE BENNETT
Division of Clinical Immunology & Rheumatology,
Department of Medicine
and
Department of Microbiology,
University of Alabama in Birmingham,
Schools of Medicine & Dentistry,
Birmingham,
Alabama,
U.S.A.

4

Disulfide Bonds

CONTENTS

4.1. INTRODUCTION

Disulfide bonds comprise the major covalent cross-linkage in proteins and may be intra- or inter-chain in nature. Intra-chain disulfide bonds serve to confer conformational stability on the folded polypeptide chain. Additionally, by limiting or directing this folding these bonds may contribute to the correct orientation of the amino acid residues that form the active sites of enzymes, antibodies and other biologically active proteins. Inter-chain disulfide bonds are functional in maintaining the quaternary structure of multi-chain proteins, serving as the only linkage between subunits or providing covalent stability to structures otherwise maintained by non-covalent forces.

Sanger and his coworkers (Sanger, 1953; Ryle & Sanger, 1955; Ryle *et al.*, 1955), in their pioneering work on the structure of insulin, recognized that an interchange reaction took place with disulfide bonds when the protein was subjected to partial hydrolysis in cold conc. HCl. Subsequent studies using a model system consisting of *N*,*N*′-bis-2,4-dinitrophenyl-L-cystine (bis-DNP-cystine) and cystyl-bis-glycine or oxidized glutathione demonstrated that this reaction occurred in the presence of conc. mineral acids or in slightly alkaline buffers, although by different mechanisms. In acid solution the reaction was found to be inhibited by sulfhydryl compounds, whereas in alkaline buffer the reaction was catalyzed by these compounds and inhibited by sulfhydryl binding reagents (Ryle & Sanger, 1955). These investigators established conditions for the partial cleavage of insulin that prevented rearrangement of the disulfide bonds and thus established the locations of these bonds in the molecule. Spackman *et al.* (1960) further characterized the stability of the disulfide bond by examining the interaction of cystine and oxidized glutathione and found disulfide interchange to be minimal over the range of pH 2 to pH 6.5. Further, isolation of cystine peptides from two of the four disulfide bonds of ribonuclease after proteolytic cleavage under conditions that favored disulfide interchange clearly indicated that disulfide bonds differ in susceptibility to this reaction. In fact there seems to be little tendency for the disulfide bonds of intact proteins to rearrange by this mechanism, although interchange reactions with small disulfide compounds have been reported (Smithies, 1965; Schrohenloher, 1971).

Thus, the design of studies to establish the locations of disulfide bonds in proteins is complicated by the structural diversity of the disulfide bond and its chemical reactivity. These studies must be performed without prior cleavage of the bonds in question and differences in the susceptibility of native proteins to fragmentation is an important factor. Thus there are no methods which are universally applicable. Because of space limitations this chapter will be concerned only with the general approach for establishing the location of disulfide bonds in proteins and will present those methods that have proved most useful.

4.2. STRATEGY

The basic approach to identifying which half-cystine residues in a protein are joined together remains essentially unchanged from that first applied to ox insulin by Ryle *et al.* (1955). The five-step procedure can be summarized as follows:

1. partial hydrolysis of the protein under conditions that prevent rearrangement of disulfide bonds;
2. fractionation of the cystine peptides from one another, also under conditions that stabilize the disulfide bonds;
3. cleavage of the cystine peptides to derivatized half-cystine peptides;
4. fractionation of the half-cystine peptides;
5. identification of the half-cystine peptides from amino acid compositions.

This approach assumes a knowledge of the amino acid sequence of the chain or chains that comprise the protein. It is also essential that the cystine peptides should each contain only one disulfide bridge for unequivocal identification of the bonding.

Additional details of the general strategy are outlined in the flowsheet presented in Fig. 4.1. Normally the amino acid sequence(s) of the isolated polypeptide chain(s) is established prior to initiating studies on disulfide bonding. Information regarding the number and distribution of these half-cystine residues can greatly aid the task of sorting out the cross-linkages. As described below, the number of disulfide bonds can readily be determined in a number of ways. Similarly, information about the extent of intra- and inter-chain bonding might be generated by physicochemical studies on the protein which usually precede or parallel characterization of its covalent structure. For example, it was reported that only 8 mol of ^{14}C-iodoacetamide reacted with a human IgG protein following mild reduction with dithiothreitol (Edelman *et al.*, 1968; Gall *et al.*, 1968). Alkylation without prior reduction resulted in incorporation of only 0.5 mol of the ^{14}C-iodoacetamide. Since the conditions for reduction permitted complete dissociation of the component polypeptide chains, the molecule contained no more than four inter-chain bonds. The remaining 26 half-cystine residues shown to be present by amino acid analysis were therefore involved in intra-chain bonds. Furthermore, following separation of the chains, the distribution of the labeled residues demonstrated that each light chain was attached to

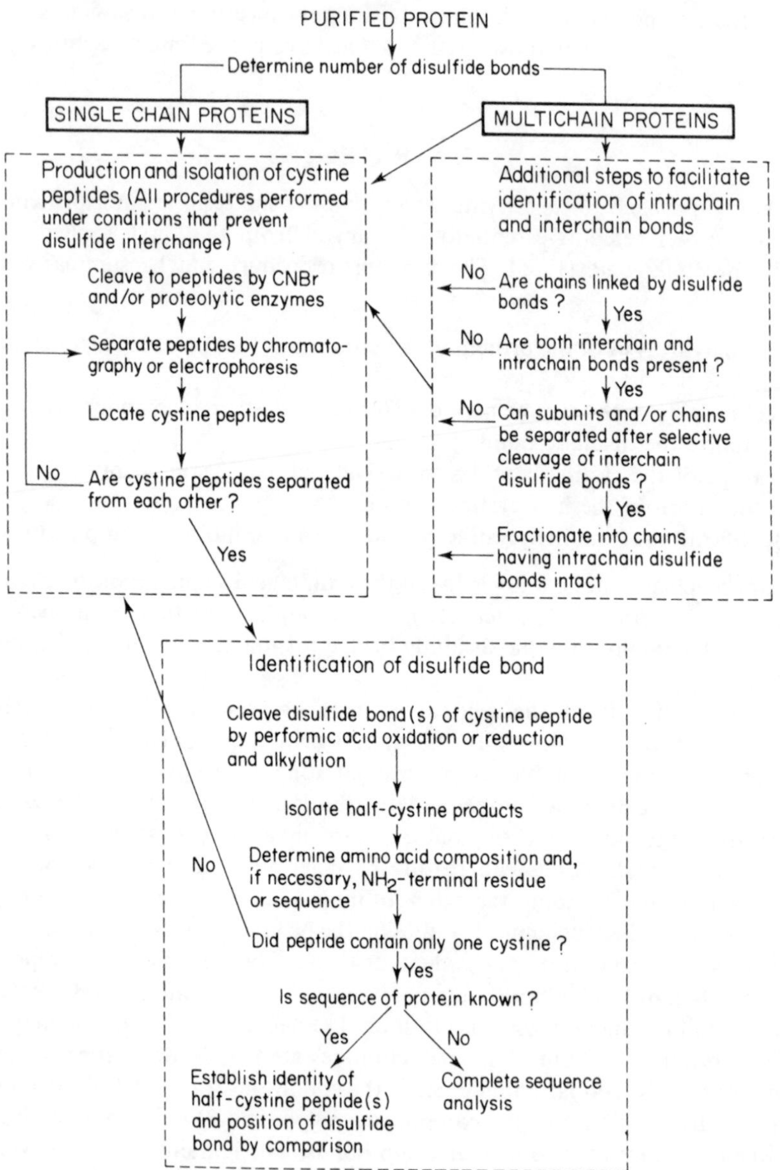

Fig. 4.1. General strategy to determine disulfide bonds in purified proteins and peptides.

a heavy chain by a single bond and that the two heavy chains were joined by two bonds.

Analysis of larger, multi-chain proteins is facilitated by fractionation into smaller units. Dissociation into polypeptide chains by selective cleavage of inter-chain bonds was applied in studies on immunoglobulins (Waxdal *et al.*, 1968a; Strosberg *et al.*, 1975). Cleavage of the native molecule by CNBr or proteolytic enzymes was also used to obtain large, well-defined fragments from identifiable regions of the molecule that were suitable for further cleavage to cystine peptides and subsequent identification of the disulfide bonds (Steiner & Porter, 1967; Gall *et al.*, 1968; Waxdal *et al.*, 1968 a, b; Mestecky *et al.*, 1974).

Quantitative data on the yields of the cystine peptides are essential in evaluating whether they represent the unique arrangement of bonds in the protein. Indeed, recovery of a peptide in high yield provides the strongest evidence that the two half-cystine residues are joined in the manner indicated. While yields in excess of 50% are preferred, mechanical losses and failure to achieve quantitative cleavage may contribute to lower recoveries. Although a low yield may not in itself preclude other bond arrangements, such data might be acceptable as evidence for the indicated structure in the absence of conflicting results, e.g. another cystine peptide containing only one of the half-cystine residues.

4.3. NUMBER OF DISULFIDE BONDS

The number of disulfide bonds can be determined from the carboxymethyl-cysteine content of the protein before and after complete reduction and alkylation. Carboxymethylcysteine is determined by amino acid analysis. The method is given in Chapter 2, p. 72.

Fractionation of multi-chain proteins into component chains prior to analysis may provide additional information on the distribution of the disulfide bonds. Alkylation with a radiolabeled sulfhydryl-blocking reagent permits the number of half-cystine residues released by reduction to be determined from the radioactivity of the product. Satisfactory results have been obtained with iodoacetamide-l-^{14}C (1.2 μCi/μmol) and iodoacetic-l-^{14}C acid (0.5 μCi/μmol) (Gall *et al.*, 1968; Oliveira & Lamm, 1971).

Disulfide bond content can also be estimated by spectrophotometric and fluorometric assays. The most widely applied procedures are based on the spectrophotometric analysis of protein sulfhydryl groups by DTNB (Ellman, 1959) before and after reduction (see Chapter 2, p. 76). Miller and Metzger (1965) applied this procedure to establish the number of inter- and intra-chain bonds in a human IgM protein. Reduced (or untreated) protein was precipitated with cold 5% trichloroacetic acid (TCA), collected by centrifugation and residual reducing agent removed by washing five times with the TCA. The precipitate was redissolved in a small volume of 5 M-guanidine and the protein concentration determined by absorbance at 280 nm after removing trace turbidity by centrifugation. The sulfhydryl content was determined by reacting an aliquot of the redissolved protein with an equal volume of 0.1 M Tris-HCl, pH 9.1, which was 5 M in guanidine and 2×10^{-4} M in DTNB (final pH 8.1). The sulfhydryl content was calculated from the absorbance of the mixture at 412 nm using a molar extinction coefficient of 13 600. The determination of disulfide groups may also be accomplished by fluorescence quenching using fluorescein mercuric acetate (Karush *et al.*, 1964).

4.4. CLEAVAGE TO CYSTINE PEPTIDES

As previously emphasized, care must be taken to avoid disulfide interchange during the fragmentation of proteins. For this reason, it is preferable to use procedures that can be performed at pH 2.0 to 6.5. Nonetheless, it is frequently necessary to use procedures that require somewhat higher pH values to achieve the necessary cleavages. Hydrolyses performed at pH 7.0 to 7.5 with trypsin and thermolysin have been applied successfully (Spackman *et al.*, 1960; Steiner & Porter, 1967; Gall *et al.*, 1968; Narita *et al.*, 1978). However, the wide variation in the stability of protein disulfide bonds requires a certain caution.

Because of its activity at low pH and its ability to hydrolyze native proteins, pepsin has been a popular choice for initial cleavage. Once the structure of an otherwise resistant protein has been disrupted by pepsin, further hydrolysis can be effected by other proteolytic enzymes at a suboptimal pH, as may be required for disulfide stability. Specific chemical cleavage at methionine residues by CNBr in weak acid has also been widely applied for the initial cleavage of proteins in disulfide bond placement studies. It has the advantage of producing well-defined fragments that can be separated prior to further fragmentation, thus simplifying the identification process.

The choice of enzymes for further cleavage is largely dictated by the difficulties encountered in obtaining small cystine peptides that contain only one disulfide. Trypsin, chymotrypsin, staphylococcal protease and thermolysin, used individually or in various combinations, are frequent choices. Occasionally, enzymes of more limited specificity, such as plasmin and elastase, can be used to advantage (Blumenthal & Kem, 1977; Doolittle *et al.*, 1978). Edman degradation has also been applied to effect cleavage of a peptide containing two disulfide bonds separated by only a few residues (Gauldie *et al.*, 1978).

Details of cleavage methods used in studies of disulfide bonds are given here, but see Chapters 3 and 10 for enzymic cleavage and Chapters 1 and 2 for chemical cleavage.

4.4.1. Cleavage by cyanogen bromide

Cyanogen bromide cleavage of proteins is discussed in detail and a method is given by Fontana and Gross in Chapter 2 (section 2.4.1.1).

4.4.2. Hydrolysis by pepsin

As initially applied by Spackman *et al.* (1960) to ribonuclease A, a 1 % solution of the protein was hydrolyzed by pepsin in 0.05 N-sodium citrate buffer at pH 1.0 for 16 h at 25°. The enzyme to substrate ratio was 1:50. Modifications include use of larger quantities of enzyme, incubation at 37°C, shorter reaction times and other solvents, including 5 % formic acid and 0.03 M-HCl (Steiner & Porter, 1967; Strausbauch *et al.*, 1971; Préval & Fougereau, 1972; Narita *et al.*, 1978). Digestion can be stopped by freezing and the products lyophilized if desired. See also Chapter 3, section 3.6.3.

4.4.3. Hydrolysis by trypsin and chymotrypsin

The pepsin-treated protein from 4.4.2 is adjusted to pH 6.5 and incubated with trypsin for 8 h at 25°C using an enzyme to substrate ratio of 1:25 (Spackman *et al.*, 1960). Chymotrypsin is then added (enzyme to substrate ratio of 1:100) and incubation is terminated by bringing the solution to pH 2 by adding a few drops of conc. HCl. The extent of cleavage can be followed by the appearance of free amino groups (ninhydrin or *o*-phthalaldehyde reaction) on aliquots withdrawn periodically from the reaction mixture. Performing the hydrolyses at 40°C may be advantageous (Narita *et al.*, 1978). See also Chapter 3, section 3.6.1.

It is sometimes possible to carry out hydrolysis with trypsin at neutral or slightly alkaline pH, taking care to establish that disulfide interchange does not occur. Gall *et al.* (1968) subjected a CNBr fraction of human IgG to trypsin (2 % by wt.) at pH 7.2 for 4 h. Digestion in the presence of an excess of iodoacetamide-1-^{14}C indicated the absence of unstable disulfide bonds. Similar conditions were used for the tryptic digestion of a peptic fragment from another human IgG by Steiner & Porter (1967) without evidence for disulfide interchange. Narita *et al.* (1978) placed two of the four disulfide bonds of cobra venom cardiotoxin using cystine peptides generated by hydrolysis with trypsin at pH 7.0 for 24 h. See also Chapter 3, section 3.5.1.

4.4.4. Hydrolysis by staphylococcal protease

This protease is isolated from culture filtrates of *Staphylococcus aureus*, strain V8 (Drapeau *et al.*, 1972), and is specific for glutamoyl bonds at pH 4.0 in ammonium acetate buffer (Houmard & Drapeau, 1972).

Sequential degradation of human β-thromboglobulin by CNBr in 70 % formic acid and staphylococcal protease at pH 4.0 gave cystine peptides from which the two disulfide bonds in the molecule were located (Begg *et al.*, 1978). The CNBr-treated protein (20 mg/ml) was digested with the enzyme for 18 h at 37°C in 0.05 M-ammonium acetate buffer, pH 4.0. The enzyme to substrate ratio was 1:40. See also Chapter 3, section 3.5.3.

4.4.5. Hydrolysis by thermolysin

Cystine peptides produced by hydrolysis of peptic peptides by thermolysin were used to establish the disulfide bridges in α-lactalbumin (Vanaman *et al.*, 1970) and mouse 2.5 S nerve growth factor (Angeletti *et al.*, 1973). Hydrolysis of the peptide from α-lactalbumin was performed at 37°C in 0.0025 M-calcium chloride adjusted to pH 7.5 by frequent addition of 1 M-Tris base and digestion terminated by adjusting to pH 3.1 with HCl. The peptide from mouse nerve growth factor dissolved in water was digested at pH 7.0 in a pH-stat for 22 h. High yields of peptides precluded extensive disulfide interchange (Vanaman *et al.*, 1970). Hydrolysis with thermolysin was performed at pH 6.5 and pH 6.9 (Blumenthal & Kem, 1977; Narita *et al.*, 1978). See also Chapter 3, section 3.6.2.

4.4.6. Cleavage after maleylation or succinylation

Occasionally the resistance of native proteins to proteases is not relieved by pretreatment with 8 M-urea, 6 M-guanidine HCl, boiling or acidification. A possible solution was provided by Blumenthal and Kem (1977). Although toxin B-IV (a crustacean-selective axonal neurotoxin) was extremely resistant to proleolysis both in its native form and after denaturation by the methods listed above, maleylation or succinylation rendered the protein susceptible to trypsin.

4.4.6.1. Succinylation

Method: Chu *et al.*, 1969

The protein (1 %) in 1.0 M-sodium bicarbonate buffer, pH 8.0, is reacted at 25°C with approximately 0.8 parts by wt. of succinic anhydride, added gradually over a period of 1 h. The reaction mixture is maintained at pH 8.0 by the addition of 0.1 M -NaOH and stopped after 1 or 2 h by dilution with 2.5 vol. of distilled water, then dialyzed overnight. Milder conditions may be effective for some proteins (Habeeb *et al.*, 1958). The succinylated protein can then be treated simultaneously by trypsin, chymotrypsin and thermolysin (enzyme to substrate ratio of 1:50 each) in 0.1 M-potassium phosphate buffer, pH 6.8 at 40°C. The reaction is terminated after 4 h by adjustment to pH 3.0 and the peptide solution lyophilized.

4.4.6.2. Maleylation

Method: Blumenthal & Kem, 1977

Partial maleylation of a resistant protein may make it sensitive to trypsin. Use a 10-fold excess of maleic anhydride over the amino groups and a reaction time of 5 min at pH 8.0. After adjusting to pH 7.0, the maleylated protein is hydrolyzed by trypsin at 37°C with an enzyme to substrate ratio of 1:33 (w/w). Terminate by adjusting to pH 3.0. The peptides can be demaleylated at pH 3.5, 48 h at 37° (Butler *et al.*, 1969). They can then be subjected to further cleavage by trypsin or other proteases.

Note that the protein is exposed to neutral or slightly alkaline pH conditions during succinylation, maleylation and tryptic hydrolysis and the possible occurrence of disulfide interchange must be considered.

See p. 130 for another method of maleylation.

4.5. FRACTIONATION OF CYSTINE PEPTIDES

The preparative isolation and purification of disulfide-bonded peptides constitutes a special problem for the protein chemist. The difficulties in this area relate to the instability of disulfide bonds and the problem of rapid analysis of these bonds by colorimetric techniques. Under conditions of mild acidity, i.e. pH 2.0–6.5, the disulfide bond is relatively stable. This requirement restricts the choice of chain cleavage and isolation techniques. Because of these limitations the primary structure of the half-cystine derivatized polypeptide is generally determined before taking the steps to determine the nature of the disulfide bond in the native molecule. Once the primary structure is known it is then a much easier problem to design cleavage strategies for the native molecule which allow one to focus only on those peptides containing disulfide linkages.

4.5.1. Methods of fractionation

Keeping in mind the necessity for exposing cystine peptides to only mildly acidic conditions, i.e. pH 2.0 to 6.5, most conventional gel filtration (Canfield & Liu, 1965) and ion-exchange chromatographic technologies (Spackman *et al.*, 1960, Angeletti *et al.*, 1973) are widely applicable for fractionation. Effective purification of cystine-containing polypeptides has been achieved by the use of 25 % acetic acid for gel filtration on Sephadex (Fox *et al.*, 1977) and also by the use of cation exchangers (Vanaman *et al.*, 1970). Although the separation of cystine peptides from one another is a prerequisite for identifying the bonding, their separation from peptides not containing cystine may not be essential (Ryle *et al.*, 1955; Spackman *et al.*, 1960). Non-cystine peptides may be separated with greater ease from the more highly charged cysteic acid peptides generated from cystine peptides. Effective fractionation of cysteic acid peptides has been achieved with a very high degree of efficiency by the utilization of HPLC and ion-pairing (Hearn & Hancock, 1979). The addition of small amounts of phosphoric acid to fully protonate amino groups, while at the same time masking them by pairing with the phosphate ion, provides an ideal separation in organic solvents with appropriately designed HPLC columns (see Chapter 6).

4.5.2. Detection of cystine peptides

In the separation and purification of cystine-containing peptides, it is desirable to select methods which quickly eliminate those peptides that do not contain cystine. One widely used method is to monitor each column fraction by amino acid analysis following performic acid oxidation (Spackman, *et al.*, 1960). This allows the investigator to focus on the components that contain a derivative of cystine. It was shown that certain of the oxidation states of sulfur will bleach an iodoplatinate solution (Fowler & Robins, 1972). This may be used as a very sensitive and rapid analysis for disulfides. However, sulfur present as thiol, disulfide or thioether will react in this assay. A sensitive fluorimetric method involves treatment of a disulfide with an alkaline solution of fluorescein–mercuric acetate (Karush *et al.*, 1964). Quenching of fluorescence is observed in those samples that contain disulfide bonds. Unfortunately, this method may be very difficult to reproduce and may not lend itself to easy analysis.

4.5.2.1. Disulfide analysis

Method: Anderson & Wetlaufer, 1975

The method is reproducible and can be readily automated. A solution of peptide (e.g. column eluate) is mixed with an equal volume of 6 M-NaOH. The alkaline cleavage is allowed to proceed for an appropriate length of time (about 30 min). The reaction is stopped by the addition of 0.5 ml of 2 M-phosphoric acid which is 2×10^{-3} M in EDTA. This must be mixed rapidly to avoid layering of the solutions of different densities. The pH should now be between 6.0 and 7.0 and a color may be developed by the addition of 0.1 ml of DTNB (1.0 mg/ml in 0.02 M-sodium acetate, pH 5.5). The absorbance is read at 412 nm.

The EDTA partially stabilizes the color by inhibiting metal ion-catalyzed oxidation of thiols, but there is a decrease of 1–2 % absorbance over 60 min. Use glutathione for a standard curve with concentrations zero to 2.5×10^{-4} M.

4.6. IDENTIFICATION AND LOCALIZATION OF CYSTINE PEPTIDES

Generally, the precise identification of peptides bonded by disulfide bridges will require cleavage of that bridge and subsequent isolation of the derivatized half-cystine peptides.

4.6.1. Cleavage to half-cystines

The cleavage of disulfide bonds is discussed in Chapters 1 (section 1.5.1) and 2 (section 2.2). The most common methods are those of performic oxidation (Spackman *et al.*, 1960) or reduction and alkylation (Zahler & Cleland, 1968). The resultant half-cystine peptides may then be purified and isolated according to standard peptide techniques involving either gel filtration, chromatography, paper electrophoresis, HPLC, etc.

4.6.2. Identification if polypeptide sequence is known

Inspection of the known amino acid sequence will indicate the proper approach to take for defining the relationships of the disulfide bonds. For unambiguous identification each purified cystine peptide should contain only one disulfide bridge. Therefore, in selection of the appropriate fractionation and cleavage strategy the intact polypeptide chain must be cleaved in at least one position between consecutive half-cystine residues in the linear sequence. CNBr may be very useful in certain instances.

However, if the protein is large and there are a number of such bridges a relatively non-specific proteolytic enzyme, such as thermolysin (Vanaman *et al.*, 1970), may be more effective in producing the appropriate cleavages to deduce the linked peptides.

Although there are many possible strategies, recent work on myotoxin *a* demonstrates an approach with a rather simple protein (Fox *et al.*, 1979). It also illustrates the elegance of a simple scheme that can be used to define disulfide bonds, as follows:

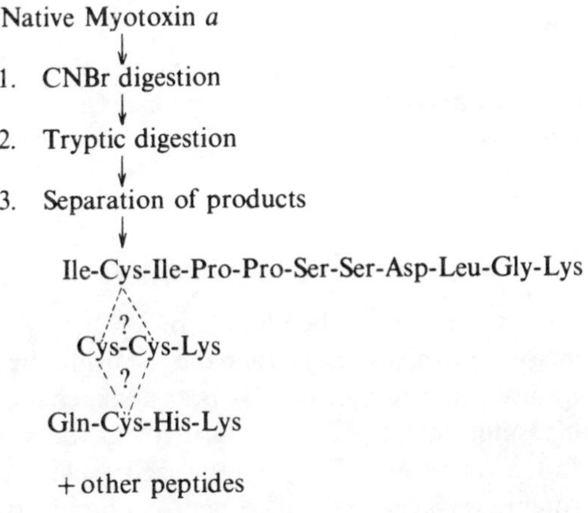

Following CNBr cleavage and tryptic digestion a disulfide-bonded peptide was isolated. This consisted of three different small peptides linked through disulfide bonds. The dashed lines show the possible linkages that could be present. Because of the location of the cleavages relative to the known sequence, they then subjected the peptide to a single-step Edman degradation in the sequencer cup, followed by desalting and separation of the products. This resulted in isolation of the following peptides:

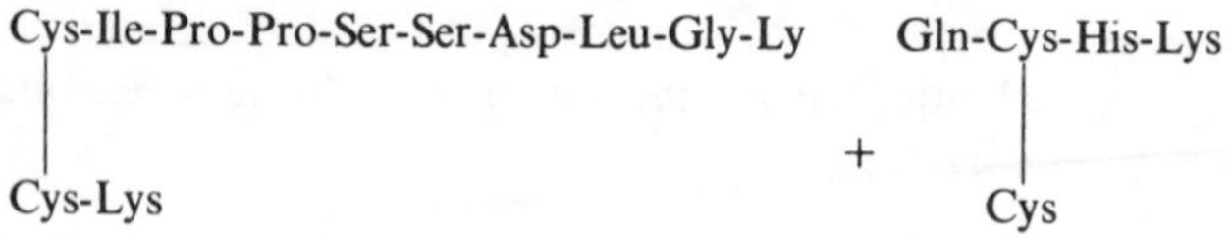

```
Cys-Ile-Pro-Pro-Ser-Ser-Asp-Leu-Gly-Ly      Gln-Cys-His-Lys
 |                                               |
 |                                       +       |
Cys-Lys                                         Cys
```

The Edman degradation removed Ile and one Cys residue but not Gln because it cyclized to the pyrrolidonecarboxylic acid. This unambigously defined the disulfide linkages as follows:

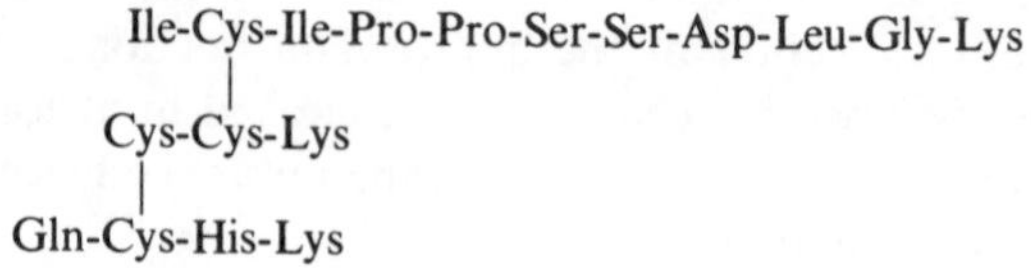

```
        Ile-Cys-Ile-Pro-Pro-Ser-Ser-Asp-Leu-Gly-Lys
             |
        Cys-Cys-Lys
             |
    Gln-Cys-His-Lys
```

The specific strategy selected is clearly unique to each protein.

4.6.3. Identification if polypeptide sequence is unknown

When the sequence of the starting polypeptide is not known the approach must be blind. Generally, thermolysin is a good enzyme to use for producing S-S bonded peptides that are easily discernable, i.e. peptides that consist of only a single disulfide bond (Vanaman *et al.*, 1970). This allows isolation of peptides which then can be sequenced independently before the total sequence or the absolute order of the disulfide bonds within that sequence are known. For certain comparative purposes, such as structural homologies, this approach can be very significant and can generate much information about a molecule before its complete primary structure has been determined.

4.6.4. General comments on locating disulfide bonds

It is obvious that in proteins where the number of S-S bonds are known and the amino acid sequence is known, one may assign the last bridge by simple deduction. Thus, if out of four disulfide bonds three have been assigned unambiguously, then it is legitimate to assume that the final bond involves the remaining two half-cystine residues in the chain. As already emphasized, yields of disulfide-bonded peptides from protein cleavages and subsequent peptide isolations may not be high—in some cases even less than 50%. However, if there is no conflicting evidence then yields even at this level are sufficient to allow unambiguous assignment of the disulfide bridges to particular peptides. This, of course, assumes that there are no instabilities in the disulfide pattern within the protein. Although there is an example of such instability *in vivo* with β-lactoglobulin (McKenzie *et al.*, 1972) this seems to be a rare occurrence and generally should not be encountered.

4.7. SPECIAL PROCEDURES

4.7.1. Stains

Specific stains (Scoffone & Fontana, 1975) can be used to detect cysteine and cystine on chromatograms (see Chapter 8, section 8.14.3 for details.)

4.7.2. Diagonal mapping

The technique of diagonal mapping has been adapted for the location of disulfide bonds (Brown & Hartley, 1966; Bennett, 1967). It can be performed in several ways but it is most often used with paper techniques. For example, an enzymic digest of a protein under study may be subjected to electrophoresis on a sheet of Whatman 3 MM paper (400 × 400 mm). Following electrophoresis the paper is dried and exposed to performic acid vapor for 2 h. This allows oxidation of S-S bonds in those peptides that contain cystine. After drying *in vacuo* over NaOH, electrophoresis is then performed at right angles to the original direction using identical conditions. This would result in all unmodified peptides lying along the 45° diagonal. The cystine peptides, however, which have been modified between the two electrophoretic runs, will be expected to generate a parallel pair of cysteic acid-containing peptides, lying off this line. These are cut from the paper and used in a preparative way. In order to isolate cystine peptides, a test strip may be cut from the original electrophoretogram, exposed to performic acid and then stitched to another sheet of chromatographic paper prior to electrophoresis in the second dimension. Such a technique allows matching of spots on the unoxidized original strip so that cystine peptides may be cut and isolated with their S-S bonds intact for subsequent handling. Modifications of the diagonal technique have also been utilized by producing various derivatives, such as *S*-aminoethylcysteine (Perham, 1967).

4.7.3. Specific cleavage at cysteine residues

The technique of cleaving peptides at cysteine residues using 2-nitro-5-thiocyanobenzoic acid (NTCB) has a high degree of specificity (Jacobson *et al.*, 1973). The significance of this approach is seen in the study of histocompatibility antigens in which disulfide loops that appear to be arranged in tandem can be compared (Ferguson *et al.*, 1979).

This technique gives an indication of the size of that loop by virtue of the size of the fragment released. It may be of considerable importance in the establishment of structural homologies among related molecules. Unfortunately, fragments produced by this method have an *N*-terminal thiazolidine ring and hence cannot be used directly for Edman degradation sequence analysis. It has recently been reported, however, that deblocking of the peptide can be successfully accomplished with Raney nickel (Otieno, 1978) (see Chapter 2, p. 108). This converts the cysteine derivative to alanine which can then be sequenced directly. Methionine in this reaction is also converted to β-aminobutyric acid.

4.7.3.1. Cleavage with 2-nitro-5-thiocyanobenzoic acid

Method: Degani & Patchornik, 1974; modified by Ferguson *et al.*, 1979

Several nmol of lyophilized protein are dissolved in a buffer that contains 6 M-guanidine HCl, 2 mM-dithiothreitol, 1 mM-EDTA and 0.2 M-Tris acetate at pH 8.0. Incubate at 37°C for 2 h. Following reductive cleavage, the sample is diluted with 0.7 ml of the same buffer and 0.5 ml of 30 mM NTCB is added. After incubation at room temperature for 30 min, adjust to pH 9.5 with 1 M-NaOH and incubate at 55° C for 24 h. The reaction may be stopped by adjusting to pH 4 with glacial acetic acid.

4.8. REFERENCES

Anderson, W. L. & Wetlaufer, D. B. (1975). *Anal. Biochem.*, **67,** 493–502.
Angeletti, R. H., Hermodson, M. A. & Bradshaw, R. A. (1973). *Biochemistry*, **12,** 100–115.
Begg, G. S., Pepper, D. S., Chesterman, C. N. & Morgan, F. J. (1978). *Biochemistry*, **17,** 1739–1744.
Bennett, J. C. (1967). *Methods Enzymol.*, **11,** 330–339.
Blumenthal, K. M. & Kem, W. R. (1977). *J. Biol. Chem.*, **252,** 3328–3331.
Brown, J. R. & Hartley. B. S. (1966). *Biochem. J.*, **101,** 214–228.
Butler, P. J. G., Harris, J. I., Hartley, B. S. & Leberman, R. (1969). *Biochem. J.*, **112,** 679–689.
Canfield, R. E. & Liu, A. F. (1965). *J. Biol. Chem.*, **240,** 1997–2002.
Chu, F. S., Crary, E. & Bergdoll, M. S. (1969). *Biochemistry*, **8,** 2890–2896.
Degani, Y. & Patchornik, A. (1974). *Biochemistry*, **13,** 1–11.
Doolittle, R. F., Cottrell, B. A., Strong, D. & Watt, K. W. K. (1978). *Biochem. Biophys. Res. Commun.*, **84,** 495–500.
Drapeau, G. R., Boily, Y. & Houmard, J. (1972). *J. Biol. Chem.*, **247,** 6720–6729.
Edelman, G. M., Gall, W. E., Waxdal, M. J. & Konigsberg, W. A. (1968). *Biochemistry*, **7,** 1950–1958.
Ellman, G. L. (1959). *Arch. Biochem. Biophys.*, **82,** 70–77.
Ferguson, W. S., Terhorst, C. T., Robb, R. J. & Strominger, J. L. (1979). *Mol. Immunol.*, **16,** 23–28.
Fowler, B. & Robins, A. J. (1972). *J. Chromatogr.*, **72,** 105–111.
Fox, J. W., Elzinga, M. & Tu, A. T. (1977). *Fed. Eur. Biol. Scientists Lett.*, **80,** 217–220.
Fox, J. W., Elzinga, M. & Tu, A. T. (1979). *Biochemistry*, **18,** 678–684.
Gall, W. E., Cunningham, B. A., Waxdal, M. J., Konigsberg, W. H. & Edelman, G. M. (1968). *Biochemistry*, **7,** 1973–1982.
Gauldie, J., Hanson, J. M., Shipolini, R. A. & Vernon, C. A. (1978). *Eur. J. Biochem.*, **83,** 405–410.
Gross, E. & Witkop, B. (1961). *J. Am. Chem. Soc.*, **83,** 1510–1511.
Habeeb, A. F. S. A., Cassidy, H. G. & Singer, S. J. (1958). *Biochim. Biophys. Acta*, **29,** 587–593.

Hearn, M. T. W. & Hancock, W. S. (1979). *Trends Biochem. Sci.*, Pers. Edn, N58–N62.
Houmard, J. & Drapeau, G. R. (1972). *Proc. Natl. Acad. Sci. U. S. A.*, **69**, 3506–3509.
Jacobson, G. R., Schaffer, M. H., Stark, G. R. & Vanaman, T. C. (1973). *J. Biol. Chem.*, **248**, 6583–6591.
Karush, J., Klinman, N. R. & Marks, R. (1964). *Anal. Biochem.*, **9**, 100–114.
McKenzie, H. A., Ralston, G. B. & Shaw, D. C. (1972). *Biochemistry*, **11**, 4539–4547.
Mestecky, J., Schrohenloher, R. E., Kulhavy, R., Wright, G. P. & Tomana, M. (1974). *Proc. Natl. Acad. Sci. U.S.A.*, **71**, 544–548.
Miller, F. & Metzger, H. (1965). *J. Biol. Chem.*, **240**, 4740–4745.
Narita, K., Cheng, K.-L., Chang, W.-C. & Lo, T.-B. (1978). *Int. J. Peptide Protein Res.*, **11**, 229–237.
Oliveira, B. & Lamm, M. E. (1971). *Biochemistry*, **10**, 26–31.
Otieno, S. (1978). *Biochemistry*, **17**, 5468–5473.
Perham, R. N. (1967). *Biochem. J.*, **105**, 1203–1207.
Préval, C. de & Fougereau, M. (1972). *Eur. J. Biochem.*, **30**, 452–462.
Ryle, A. P. & Sanger, F. (1955). *Biochem. J.*, **60**, 535–540.
Ryle, A. P., Sanger, F., Smith, L. F. & Kitai, R. (1955). *Biochem. J.*, **60**, 541–556.
Sanger, F. (1953). *Nature* (Lond.), **171**, 1025–1026.
Schrohenloher, R. E. (1971). *Immunochemistry*, **8**, 375–389.
Scoffone, E. & Fontana, A. (1975). In *Protein Sequence Determination*, Needleman, S. B. (Ed.), 2nd edn, Springer, Berlin, Heidelberg, New York, p. 202.
Smithies, O. (1965). *Science*, **150**, 1595–1598.
Spackman, D. H., Stein, W. H. & Moore, S. (1960). *J. Biol. Chem.*, **235**, 648–659.
Steiner, L. A. & Porter, R. R. (1967). *Biochemistry*, **6**, 3957–3970.
Strausbauch, P. H., Hurwitz, E. & Gival, D. (1971). *Biochemistry*, **10**, 2231–2237.
Strosberg, A. D., Margolies, M. N. & Haber, E. (1975). *J. Immunol.*, **115**, 1422–1424.
Vanaman, T. C., Brew, K. & Hill, R. L. (1970). *J. Biol. Chem.*, **245**, 4583–4590.
Waxdal, M. J., Konigsberg, W. A., Henley, W. L. & Edelman, G. M. (1968a). *Biochemistry*, **7**, 1959–1966.
Waxdal, M. J., Konigsberg, W. A. & Edelman, G. M. (1968b). *Biochemistry*, **7**, 1967–1972.
Zahler, W. L. & Cleland, W. W. (1968). *J. Biol. Chem.*, **243**, 716–719.

Practical Protein Chemistry—A Handbook
Edited by A. Darbre
© 1986, John Wiley & Sons Ltd

A. DONNY STROSBERG
Molecular Immunology Laboratory,
IRBM–CNRS and University of Paris VII,
Place Jussieu 2, Paris 75251, France.

5

Affinity Chromatography of Proteins

CONTENTS

5.1. INTRODUCTION

Affinity chromatography, a major method of protein purification, is based on the interaction between biologically active materials, one of which is usually covalently coupled to an inert support. Affinity chromatography is used in the purification of antibodies, enzymes, hormones, receptors and other types of macromolecules such as polysaccharides or nucleic acids, as well as biological particles such as viruses and cells.

In this chapter discussion is restricted to practical details and a number of the difficulties which may arise at different stages of experimentation. A schematic summary of the various procedures to be discussed is presented in Fig. 5.1.

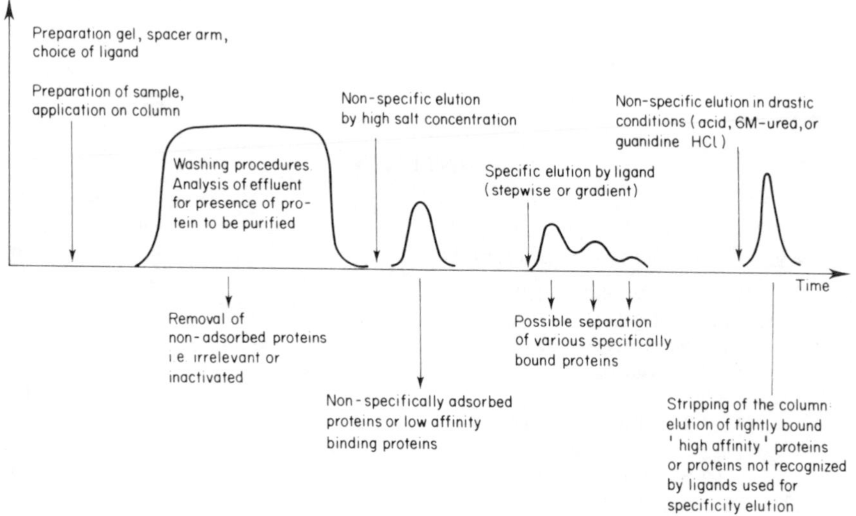

Fig. 5.1. Main steps in affinity chromatography.

For a more complete discussion of the theoretical basis of affinity chromatography, the reader is referred to books and reviews listed in section 5.10.

5.2. WHEN TO USE AFFINITY CHROMATOGRAPHY

At which step in a purification procedure should one make use of affinity chromatography techniques? Various possibilities are discussed below and summarized in Fig. 5.2.

Affinity chromatography may be used as a primary method of purification when the interactions involved are sufficiently strong and specific to allow the

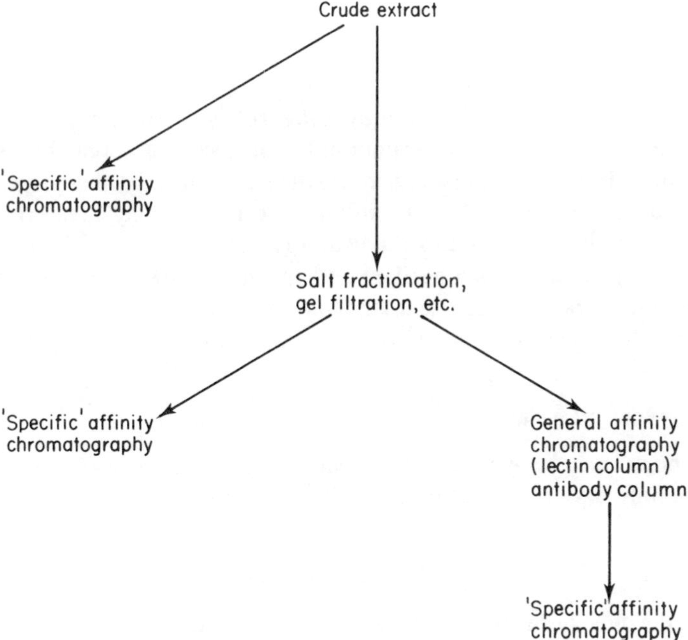

Fig. 5.2. The 'timing' of affinity chromatography.

separation of the desired protein from a large number of contaminating molecules. Examples of this type of use include the extraction of specific antibodies from serum using antigen-containing adsorbants, the extraction of hormones from placental extracts and the separation of hormone receptors from solubilized plasma membranes. Affinity chromatography is especially useful as a primary purification method when the stability of the molecules to be purified is in question, in which case it is important to limit the number of steps involved in the purification. Reasons for instability may include thermal lability of the protein, susceptibility to proteolytic enzymes present in the crude extract or the necessity to conserve the native environment of phospholipids or other substances which may be associated with the protein.

It is often advantageous to use affinity chromatography in the final purification of a molecule, after initial non-specific partial isolation with procedures such as ammonium sulfate precipitation, gel filtration or ion-exchange chromatography. Specific elution from the adsorbent can then also be considered a criterion for purity and test for biological integrity.

Affinity chromatography may also be used in intermediate steps to climinate contaminants, without necessarily reaching complete purification.

5.3. PREPARATION OF THE AFFINITY GEL

5.3.1. Choice of gel

The choice of the affinity chromatography gel is often not given sufficient consideration, although it is of paramount importance. The gel can be considered in three parts: the support, the side arm and the ligand. It is not always necessary to have all three parts, e.g. in lectin affinity chromatography a matrix such as Sephadex G-75 itself constitutes the ligand (glucose, mannose). Similarly, anti-dextran antibodies have been purified on Sephadex G-100. In general, however, the ligand should be attached to a matrix and a side arm is introduced between the ligand and the matrix support to avoid steric hindrance of affinity binding.

5.3.2. Support material

The support must be inert to biological and chemical degradation. Agarose, cellulose, polyacrylamide and glass beads are commonly used.

5.3.2.1. Agarose

Agarose, the most popular support, is a polymer of D-galactose and 3-6, anhydro-L-galactose. This substance has several attractive features. It is widely available in a standardized form and is reasonably stable in normal, neutral conditions.

Ligands and side arms are very easily attached to agarose by cyanogen bromide activation. Several procedures for this reaction are described in the literature. CNBr can be used in a buffer solution or may be dissolved in dimethylformamide. The pH must be maintained between 9 and 10 during activation, either by working in a concentrated buffer, or by constant adjustment of the pH by NaOH. If the laboratory is not equipped to handle the toxic CNBr (a well-ventilated hood and a good vacuum pump are necessary), CNBr-activated agarose is commercially available. Several disadvantages of this support must however be noted:

1. Biodegradation. Agarose is easily degraded by hydrolytic enzymes which may be produced by bacteria present in the biological solutions under study. Sodium azide (0.02%) may be helpful in limiting bacterial proliferation.
2. Chemical degradation. It has recently been observed that the chemical link between the Sepharose and the side arm or ligand may be progressively hydrolyzed by amines, resulting in leakage from the gel. Tris buffers should therefore be avoided. Acidic conditions, sometimes used for the elution of tightly bound proteins, may also contribute to the degradation of the affinity gel.
3. Non-specific adsorption. Another type of problem frequently observed with agarose adsorbents is related to their mode of activation. CNBr treatment results in the activation of many hydroxyl groups and their number generally far exceeds the number of groups available for coupling. Unreacted groups can usually be

saturated by the addition of an excess of a free primary amine, such as lysine or ethylene diamine. Despite this precaution, non-specific coupling may still occur. Far more non-covalent adsorption is observed when affinity chromatography is carried out in the presence of concentrated solutions of proteins, such as serum and cell or seed extracts. To reduce the non-specific adsorption, washing with concentrated salt solutions is often advised, but this may result in the desorption of low affinity specifically bound proteins.

5.3.2.2. Cellulose

This heterogeneous polymer of glucose is frequently used in the preparation of immunoabsorbents since it appears to produce less non-specific interaction with some proteins than agarose and the chemical derivatization for coupling of ligands is quite easy. However, its hydrodynamic properties are considerably less attractive than those of the other supports described here.

5.3.2.3. Polyacrylamide

Polyacrylamide has the advantage of low, non-specific absorption of protein. However, difficulties may be encountered in introducing spacer arms or binding ligands to this matrix.

5.3.2.4. Glass beads

Glass beads are commercially available in various sizes and with varying degrees of porosity. Porous glass has an excellent rigidity and permeability as a matrix, but it lacks chemical stability especially in alkaline solution and may show non-specific adsorption.

5.3.2.5. Protein supports

One of the first types of adsorbents used with much success consisted of the polymer obtained by the addition of ethylchloroformate to the solution of proteins which contained the ligand. The insoluble gel was reduced to fine particles simply by stirring with a magnetic bar (Avrameas & Ternynck, 1967). To improve the hydrodynamic properties of the affinity column, the gel can be mixed with G-25 Sephadex coarse beads.

5.3.3. Spacer arm

Direct binding of the ligand to the matrix may lead to steric hindrance because of the often relative inaccessibility of the active site in the protein under study, in which case the efficiency of the affinity chromatography is strongly diminished. Incorporation of a spacer arm increases the distance between matrix and ligand

and allows a greater accessibility of the ligand. The nature of the ligand and the matrix are both determining factors in deciding on the length and the chemical nature of the spacer arm.

5.3.4. Ligand

Different factors are to be considered when the ligand is chosen. The ligand must be coupled to the matrix directly or via the spacer arm. The bond must be stable. The ligand coupled to the matrix must still bind strongly to the substance to be purified so that washingc is possible, and the binding must be reversible in order to permit the release of the bound substance. No ligand is perfect: the advantages and disadvantages of each must be considered in order to arrive at the best choice.

5.3.5. Coupling reaction

The chemistry involved will depend on the groups involved. It is essential that the reaction should not alter the affinity of the ligand for the substance to be purified.

5.4. ADSORPTION AND ELUTION

Before using the affinity gel, it is best to wash away the 0.02% sodium azide usually added to prevent biodegradation of the agarose support. Any contaminating proteins or released ligand are 'stripped' off by a short wash with 1 M-acetic acid, followed by a thorough wash with the generally neutral application buffer. Before applying the sample to the column, it is best to determine the specific activity of the protein to be purified. This pre-purification concentration may be determined by a quantitative immunochemical determination such as an enzyme-linked or radioimmunoassay.

When starting with solubilized membrane proteins, detergent often remains present in the sample. Used in low concentration, digitonin or SDS does not appear to affect affinity chromatography. However, possible interference should be tested for in each individual case; it may best be evaluated by testing the effect of detergent on the biological test used prior to application of the sample. Use of buffers without detergent may result in non-specific precipitation on the affinity column, in which case little purification would take place.

After sample application, the column is washed, first with neutral buffers, then with 3 M-NaCl. Washings should be carefully recovered and tested for biological activity. It is best to avoid excessive dilution of the washings since they may have to be reapplied to the affinity gel if the column becomes saturated or if different conditions are to be tried.

Elution of the bound substance can now be initiated. One of the essential conditions for success in affinity chromatography is the correct choice of the reagent used to specifically displace the bound proteins from the column. The most obvious compounds for this purpose are the ligands coupled to the inert support. These, however, are not always available for this purpose; poor solubility or excessive cost may prevent their utilization. Also, it may be advantageous to use another ligand with the same specificity but with a different structure. If for one reason or another an unwanted protein interacts with the affinity gel, elution with the same ligand may be misleading. Eluents which are commonly used include enzyme inhibitors or cofactors, haptens corresponding to fragments of the antigen coupled on the gel, antagonists of hormone receptors, etc.

Stepwise or gradient elution with increasing concentration of the eluent may resolve adsorbed protein components of different affinities. A drastic 'stripping' of the affinity column may be effected by the use of conditions which minimize the protein-adsorbent interaction, such as low pH (1 M-acetic acid), or which affect hydrophobic bonds (6 M-urea or 4 M-guanidine HCl), or electrostatic interactions (2 M-ferric thiocyanate). These elution conditions may be necessary to recover tightly bound proteins having strong affinity for the gel. However, they may also remove proteins bound non-specifically. The distinction between these two possibilities is sometimes difficult since the drastic elution conditions may denature the eluted protein and destroy the biological activity. Alternative identification methods, such as immunochemical techniques, may be helpful in this case. Another disadvantage of using strong elution conditions is that they may shorten the useful life of the affinity gel. To limit the effects of low pH, one should apply only a limited amount of acid and recover the eluate in tubes containing concentrated phosphate buffer. Proteins eluted in strong denaturants should be dialyzed immediately.

It is essential to calculate the yield at each step of the affinity purification; therefore, no solution should be discarded before final evaluation. Extensive variations in yield are commonly observed and should not discourage users of this method. Addition of carrier protein may increase the yield of biologically active protein but may also interfere with purification. The release of ligand from the gel may also diminish the amount of adsorbed protein or may lower the apparent yield of purified biologically active material. Leakage of ligand may be evaluated by incorporation of radiolabeled molecules in the affinity gel, or by testing washes of the pure gel (Vauquelin *et al.*, 1977). The addition of protease inhibitors should be considered as a means of preventing degradation during application of the sample on the affinity column. Elution with a specific ligand may interfere in assay procedures for biologically active proteins. It may be necessary to remove the ligand effect by dilution of the sample, or to use exhaustive dialysis, with the consequent risk of protein denaturation.

5.5. GENERAL AFFINITY PURIFICATION METHODS

General affinity chromatography procedures can be applied when the specific ligands to be incorporated in the affinity gel are not available in purified form.

Three main types of procedures have been developed:

1. Lectin affinity chromatography for the purification of glycoproteins in solution or attached to cell walls.
2. Antibody 'immunoadsorption' to obtain pure antigen.
3. Enzyme co-factor columns.

5.5.1. Lectin columns

Lectin affinity chromatography utilizes the property of various proteins to specifically bind particular sugar moieties. By using proteins such as Concanavalin A or lentil lectin, it is possible to retain specifically proteins or cells which possess glucose or mannose moieties. Limulin permits the selection of sialic acid-rich glycoproteins. Peanut agglutinin selects *N*-acetyl-galactosamine-containing proteins and has also been used to retain T lymphocyte populations at specific differentiation stages. Elution from the lectin gels is effected by concentrated carbohydrate solutions corresponding to the specificity of the lectin, e.g. α-glycopyranoside is used for Concanavalin A or lentil lectin and sialic acid for limulin.

5.5.2. Antibody columns

The so-called 'immunoadsorption' technique was an early development of affinity chromatography. Initially the purified antigen used for immunization was linked to an insoluble matrix and the specific antibodies present in the serum were retained. Elution was carried out using concentrated solutions of antigen or fragments thereof ('haptens') and if these were not available or not efficient enough, elution buffers of lower pH or containing guanidine or urea were used. Immunoadsorption used in this way yields heterogeneous populations of antibody molecules, or antibody-bearing lymphocytes. Appreciable amounts of purified antigen are required for preparation of the immunoadsorbent and also for the elution.

More recently, immunoadsorption has been applied to situations in which the antigen is only available in amounts required for the initial immunization. In this procedure, the antibodies are linked to the insoluble matrix and the extract containing the antigen is applied to the affinity column. Elution is generally carried out in the same way as before, but much greater caution must be used in preparing and handling the adsorbent, since the antibody is generally quite susceptible to denaturation.

When immobilized antibodies are used to adsorb a protein, it is essential that the antibodies be mono specific. This condition can only be fulfilled if highly purified antigen is available to raise antibodies in test animals. Recently, a new method has been developed which circumvents this difficulty. In this method, mice are immunized with a mixture of antigens. A number of antibody-producing spleen cells are taken and fused with permanently growing myeloma cells, thereby obtaining antibody-producing hybrids called 'hybridomas'. Each separate clone of cells makes antibodies against only one antigenic determinant of a single antigen and therefore fulfills perfectly the requirement of 'monospecificity'. Monoclonal hybridoma-produced antibodies can then be obtained in large quantities by growing the cells in *vitro*, or *in vivo* in the peritoneal cavities of mice. The easily purified antibodies have been used to purify various lymphocyte cell surface antigens by affinity chromatography.

An example of the use of hybridomas in purification procedures is given in a study carried out by Sunderland *et al.* (1979). In this work a predominant leukocyte-common (L-C) antigen from rat thymocytes was purified by the use of a monoclonal antibody. In the first step, a crude preparation of the 100 000 mol. wt. thymocyte membrane glycoprotein was prepared at 400- to 900-fold purification by lentil lectin affinity chromatography and gel filtration in deoxycholate. Mice were then immunized with this fraction. Spleen cells were removed from antibody-producing animals and hybrid myeloma cell lines secreting antibody to the L-C antibody were verified by indirect radioactivity binding assay and inhibition with the crude antigen preparation. This 'monoclonal' antibody was used for affinity chromatography and gave an electrophoretically pure L-C antigen. The authors explain the good elution yield from the antibody column by the fact that the monoclonal antibody recognizes only one determinant per molecule in contrast to conventional antibodies where several determinants are usually involved. A lower affinity for antigen also distinguishes the monoclonal mouse antibody from the hyperimmune rabbit antibody.

5.5.3. Enzyme substrate inhibitor or cofactor columns

The best ligand for the affinity purification of enzymes is the substrate itself, but its use is generally precluded by the rapid turnover of the enzymic reaction resulting in a very inefficient adsorption and rapid destruction of the adsorbent. However, under conditions where catalytic reactions are much reduced while enzyme–substrate binding remains considerable, substrates may be used. Using a subzero temperature (-2 to $-50°$C), association is still strong enough to allow specific retention of the enzyme, which is easily removed from the column by warming (Douzou, 1976).

More generally used are enzyme inhibitors or other effectors known to bind the enzyme outside the catalytic site. In some cases the coupled substrate itself may be very specific inhibitor. The usefulness and thus the choice of the inhibitor depends entirely on the association–dissociation rates of the ligand-protein complex, i.e. the conditions for adsorption and those subsequently used for recovery of the enzyme must be carefully considered.

To avoid having to synthesize a new adsorbent for the purification of each individual enzyme, a search was made for a 'universal' affinity ligand which would be of more general use. Adenine nucleotide coenzymes such as 5′-AMP or 2′,5′-ADP and NAD^+ or $NADP^+$ analogues constitute good candidates for general applicability of enzyme purification since about one-third of the 2000 listed enzymes require the presence of this type of cofactor. Blue dextran, a structural analogue of a NAD^+-derived gel has been used for this purpose.

Interestingly, the use of 'general' ligands has improved our understanding of the coupling of cofactors, as well as the adsorption and elution conditions and the action of various proteins and salt ions in increasing or decreasing the biospecific retention or release of a particular enzyme. Thus, despite the relatively low specificity of a 'general' ligand, excellent purification of enzymes can be obtained by adequately varying elution conditions.

An elegant example of such a 'cascade' purification was recently provided by Beeckmans and Kanarek (1981) who purified citrate synthase (EC 4.1.3.7.) and fumarase (EC 4.2.1.2). Both enzymes were isolated using the same affinity column containing pyromellitic acid (PMA) coupled to Sepharose-4B through a diaminopropanol spacer arm. When crude tissue extract was applied to the PMA column in 0.01 M-phosphate buffer, pH 7.3, containing 0.01 M-2-mercapto-ethanol, citrate synthase remained bound whereas fumarase binding was completely prevented by the presence of phosphate, which is a competitive inhibitor of this enzyme. The citrate synthase was eluted by the same buffer containing 0.01 M-citric acid. When this eluate from the PMA column was applied to Blue Sepharose-4B gel, this enzyme remained quantitatively bound. It was released from the column by 0.01 M-phosphate buffer, pH 7.3, containing mercaptoethanol but without citric acid. The enzyme fraction not bound to the PMA column when using the phosphate buffer and which contained all the fumarase activity was extensively dialyzed against 0.01 M-Tris-acetate buffer, pH 7.3, containing 0.014 M-2-mercaptoethanol. When the same PMA column was used with this buffer the fumarase remained quantitatively bound. It was eluted with L-malic acid. At this stage, the fumarase was easily purified by using the Blue Sepharose column. In the absence of L-malic acid, fumarase binds to this column. The enzyme is eluted by re-addition of L-malic acid to the Tris-acetate buffer.

5.6. EXAMPLES

5.6.1. Separation and purification of antibodies

Affinity chromatography can be employed with considerable efficiency in the isolation and fractionation of structurally heterogeneous populations of anti-bodies. For example, antibodies specific for antigens of the cell wall of *Pneumococcus* strain VIII were raised in rabbits hyperimmunized with the whole, formalin-killed organisms and various antigen-containing 'immunoadsorbents' were prepared starting from the *Pneumococcus* cell wall polysaccharide. The simplest adsorbent was made by coupling the polysaccharide to Sepharose-4B. This was done by grafting amino groups on either of these two compounds, exclusively built up by sugar moieties. CNBr activation of the non-aminated saccharide and reaction with the poly-amino derivative yielded the final adsorbent. Several antibody populations were separated using various affinity gels derived from this adsorbent either by acid or alkaline treatment.

In our experience, we have observed that if the native antigen is attached directly onto the matrix, it may prove to be too strong an adsorbant and the native antibodies may be extremely difficult to remove (Cheng *et al.*, 1973). In these cases, the use of a strong eluent such as 6 M-guanidine HCl may then result in denaturation or unwanted alteration of the antibodies. To avoid this situation, one can use as starting material cell walls from cross-reacting organisms or one may chemically alter the original saccharides by acid or alkaline treatment, amination, etc.

The adsorbent is contained in an acetone-treated 5 ml syringe barrel. The rubber top of the syringe piston is used as the 'top' of the column and a hole is pierced to introduce a piece of tubing for washing the elution buffers (Fig. 5.3). To avoid disturbing the surface of the gel and to prevent clogging, a small filter disc is laid on top of the agarose. Before introducing 3 ml of the affinity gel, a 2 ml layer of Sephadex G-25 medium size beads is placed on top of the glass wool which closes off the barrel. These beads have several functions: they prevent mixing of the adsorbent and the glass wool, they concentrate the protein effluent, and they serve as a buffering chamber in which the acid-eluted proteins are rapidly renatured.

A new adsorbent should be tested by adsorption of the immunoglobulins prepared by ammonium sulfate precipitation. Then the various solutions

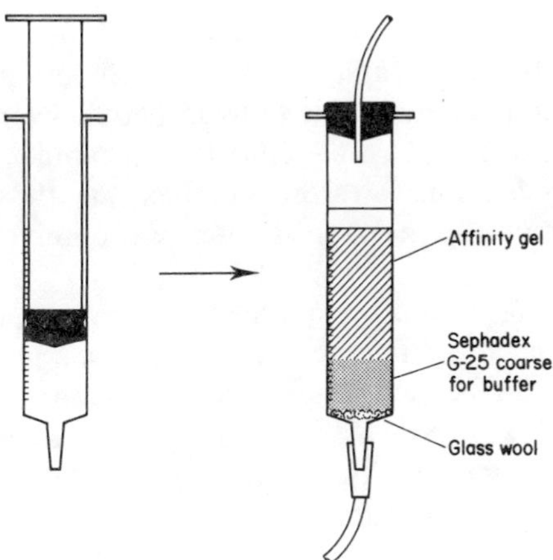

Fig. 5.3. A simple and inexpensive affinity chromatography column.

intended to be used to elute the specifically adsorbed antibodies are added in a stepwise fashion. The yield must be calculated at each step to account for the full amount of applied immunoglobulin and to establish the efficiency of the adsorbent.

To effect the gradual elution of various fractions of antibodies, one resorts to a gradient of increasing concentrations of the salt, the hapten, or both. Final elution usually is carried out using an acid buffer (0.2 M-glycine-HCl, pH 2.2), 6 M-urea-HCl, or 4 M-guanidine-HCl solutions. The utmost care must be taken to neutralize and wash the adsorbent immediately after elution and also to neutralize and dialyze the eluted antibodies.

Although concentrated salt solutions such as 3 M-NaCl do not really constitute 'specific' eluents, they have been found to be of considerable use in affinity chromatography. The reason for this may be the weakening of electrostatic interactions between antibodies and the antigen-containing matrix. More specific eluents include haptens such as single sugars (fucose, mannose), disaccharides (cellobiose, lactose), or even less well-defined oligosaccharides derived by mild acid hydrolysis from whole polysaccharide.

5.6.2. Hormone receptor purification

Hormone receptors are present in the cell plasma membrane in minute quantities and their isolation requires highly specific methodology. Affinity chromatography is the method of choice since it bases the purification on the functional rather than on the overall chemical and physical properties of the receptors. However, a number of important criteria must be fulfilled for a successful affinity purification.

1. Non-specific retention of receptors and unrelated proteins by the spacer arm or the supporting material must be avoided because this will hamper the biospecific elution by free ligand and will reduce the purity of the eluted receptors. Such retention may be due to hydrophobic interactions between the spacer arms and the often detergent-solubilized membrane components. In the example below, introduction of an aromatic moiety into the spacer arm increased non-specific adsorption and the opposite effect was observed by introduction of polar groups.
2. The immobilized ligand must retain the ability to recognize and bind to the specific site on the receptor. This criterion is of particular importance when small molecules (steroids, cholinergic and adrenergic drugs) are used, since they may be coupled to the spacer arm through groups involved in receptor reconition. The coupling may drastically modify the chemical, physical and steric characteristics of the ligand and thus considerably lower the specific binding.
3. An efficient elution procedure should be available to gently remove receptors bound to the affinity adsorbants with association constants which may be as high as 10^9 to 10^{10} M/l. Drastic elution conditions are not acceptable since they would

destroy the biological activity, often the only property by which the eluted receptor can be identified.
4. The determination of the small amounts of affinity-purified receptor must be by a technique easy to repeat in a variety of different conditions.

We will discuss here the methodology developed for the affinity purification of the catecholamine hormone receptor of the turkey erythrocyte plasma membrane. This receptor is linked to the adenylate cyclase enzyme. The agonists are epinephrine or isoprenaline (isoproterenol) and the antagonists are alprenolol or propranolol. Binding in this case is rapid and reversible.

In the first attempts to prepare a convenient affinity column norepinephrine was linked directly or via the amino group through spacer arms to agarose (Lefkowitz *et al.*, 1972). The amino group, however, plays a major role in the pharmacological activity of the agonist, and subsequent purification procedures avoided blocking this group.

Alternatively, catecholamines were linked via the aromatic ring to diazotized spacer arms on glass beads. The bound hormone could still activate the adenylate cyclase systems in tissues, cells, and plasma membranes (Venter *et al.*, 1975). Additional studies suggested, however, that ^{3}H-norepinephrine binding sites were not identical to the functional β-adrenergic receptor (Yong, 1973). Biological activities attributed to the beads could be due to released ligands. The release was shown to be due not only to leakage of unattached, adsorbed free ligand, but also to chemical hydrolysis of the side arm by amino group-containing compounds such as Tris-HCl and serum albumin.

Several compounds were prepared and tested for release of ligand from the beads, efficiency of adsorption of receptors and amount of non-specific absorption (Vauquelin *et al.*, 1978). Three classes of potential affinity adsorbents were synthesized (Table 5.1):

1. The agonist isoproterenol was coupled via the aromatic ring to diazotized spacer arm on agarose (gel I).
2. The agonist norepinephrine was immobilized via the ethanolamine side chain through an amide bond (gel II).
3. The antagonist alprenolol was treated with *N*-bromosuccinimide and the resulting bromohydrin form of alprenolol was linked to a thiolated spacer arm on agarose (gel III).

The adsorption characteristics of the different gels were assayed. The gel containing azo-coupled isoproterenol adsorbed 62 % of the receptor sites (Table 5.1) and 25 % of the solubilized protein. The corresponding spacer arm–agarose conjugate (gel IV) retained 67 % of the receptors and 23 % of the proteins. Thus the observed adsorption process was governed by the structure of the spacer arm rather than by the presence of the immobilized ligand. The same phenomenon was observed when investigating the characteristics of the gel containing amido-linked norepinephrine (gel II). Both this gel and the corresponding spacer arm on agarose (gel V) adsorbed 35 % and 22 % of the receptors, respectively, and 12 % and 9 % of protein. To reduce this considerable amount of non-specific retention of protein by the spacer arm, the adsorption characteristics of various spacer arm–agarose conjugates were investigated. Table 5.1 shows that the spacer arm containing both an aromatic ring and a diaminodipropylamino chain (gel IV) retained the largest amount of (-)-[^{3}H]-dihydroalprenolol binding sites and protein. This retention was markedly reduced by replacing the aromatic moiety by a methyl group (gel V), or by rendering the aliphatic component of the arm more hydrophilic (i.e. by replacing the diaminodipropylamine moiety by a diaminopropan-2-ol group—gel VI). Both the acetamido- and the *N*-acetylcysteamido derivatives of diaminopropan-2-ol-agarose (gels VII and VIII) were nearly inert. These data show that the presence of the aromatic ring and of hydrophobic residues on the spacer arms increased the extent of non-specific retention of β-receptors and protein by the gels. Gel III adsorbed 88 % of the receptor sites. In contrast, the corresponding spacer arm on agarose (gel VIII) retained 4 % of the sites. These gels retained only 9 % and 6 % of the protein, respectively.

From the three classes of potential affinity gel, the alprenolol–agarose conjugate was the only one to show the desired adsorption characteristics. It retained most of the receptor sites, negligible amounts of protein, and the adsorption was biospecific, i.e. due to the presence of immobilized alprenolol rather than to the presence of the spacer arm. By using a radiolabeled spacer arm or radioactive antagonist, it was verified that no measurable free or arm-bound ligand was released from the agarose beads during the incubation step: the supernatants obtained by treatment of the gels with either buffer or digitonin extract did not inhibit the binding of (-)-[^{3}H]-dihydroalprenolol to fresh membranes.

Because the agarose-coupled antagonist alprenolol gave the best available affinity gel, this compound was used to determine optimal conditions for purification, in batch, on a column, or a combination of the two procedures in which adsorption was carried out in a batchwise manner, followed by washing and elution on a column. The mixed assay was found to be the easiest to use since it simplified the washing procedure and also allowed a careful standardization of the elution procedure. This latter procedure is described.

Digitonin-solubilized turkey erythrocyte purified membranes (1.5 mg protein, i.e. 4 pmol receptor in 1 ml) are applied to 0.5 ml of affinity gel packed in a small column which has been extensively washed with 3 ml of 0.01 M-Tris buffer containing 0.09 M-NaCl. After application and adsorption at 30°C in a bench-incubator, the column is washed with a predetermined volume (2 ml) of buffer, supplemented with 1 M-NaCl. Elution is carried out with 5 ml of a relatively concentrated solution of [^{3}H]-dihydroalprenolol containing either 1 M-NaCl or, as used more recently, 0.05 % digitonin. Receptor binding activity is assessed either by a polyethylene glycol precipitation method on glass fiber filters (Vauquelin *et al.*, 1978) or a Sephadex G-50 gel filtration technique by which radiolabel is separated from unbound ligand.

By using the β-adrenergic antagonist–agarose conjugate, a 20 000-fold purification of the turkey erythrocyte plasma membrane receptors with a yield of 25% to 30% was obtained in a single affinity-chromatography step. The procedure was successfully used for purification of frog erythrocyte β-receptor protein (Caron *et al.*, 1979).

GEL N°	GEL STRUCTURE: MATRIX	SPACER ARM	LIGAND	% ADSORPTION: RECEPTOR	% ADSORPTION: PROTEIN
I	agarose	$NH-(CH_2)_3-NH-(CH_2)_3-NH-\overset{O}{\overset{\parallel}{C}}-C_6H_4-N_2-$	$C_6H_2(OH)(OH)-CHOH-CH_2-NH-CH(CH_3)_2$ (IP)	62	25
II	agarose	$NH-(CH_2)_3-NH-(CH_2)_3-NH-\overset{O}{\overset{\parallel}{C}}-CH_2-CH_2-\overset{O}{\overset{\parallel}{C}}-$	$NH-CH_2-CHOH-C_6H_3(OH)(OH)$ (NE)	35	12
III	agarose	$NH-CH_2-CHOH-CH_2-NH-\overset{O}{\overset{\parallel}{C}}-CH(NH-CO-CH_3)-CH_2-CH_2-S-$	$CH_2-CHOH-CH_2-C_6H_4-O-CH_2-CHOH-CH_2-NH-CH(CH_3)_2$ (ALP)	88	9
IV	agarose	$NH-(CH_2)_3-NH-(CH_2)_3-NH-\overset{O}{\overset{\parallel}{C}}-C_6H_4-NO_2$		67	23
V	agarose	$NH-(CH_2)_3-NH-(CH_2)_3-NH-\overset{O}{\overset{\parallel}{C}}-CH_3$		22	9
VI	agarose	$NH-CH_2-CHOH-CH_2-NH-\overset{O}{\overset{\parallel}{C}}-C_6H_4-NO_2$		30	16
VII	agarose	$NH-CH_2-CHOH-CH_2-NH-\overset{O}{\overset{\parallel}{C}}-CH_3$		8	0
VIII	agarose	$NH-CH_2-CHOH-CH_2-NH-\overset{O}{\overset{\parallel}{C}}-CH(NH-CO-CH_3)-CH_2-CH_2-SH$		4	6

Table 5.1. Adsorption of solubilized β-receptors and protein by potential affinity gels and spacer arm agarose derivatives. Digitonin extracts of turkey erythrocyte membranes (0.9 ml) were incubated with 100 mg of each gel for 10 min at 30°C. After centrifugation, the supernatants were collected and assayed for $(-)$-$[^3H]$-dihydroalprenolol binding activity and protein content. The difference between the original digitonin extract and the supernatants was defined as "adsorption". (IP) = isoproterenol, (NE) = norepinephrine and (ALP) = alprenolol. Native agarose adsorbed less than 2% of solubilized receptor and protein (from Vauquelin *et al.*, 1978).

5.7. STRATEGY FOR THE PURIFICATION OF RARE PROTEINS

A survey of the literature and our own experience suggest that the most fruitful strategy for large scale isolation of a rare protein involves the following steps:

1. Preparation of a small amount of the pure protein by specific affinity chromatography using a natural or synthetic ligand, followed by verification of the biological activity.

2. Immunization of a mouse with the affinity-purified material.
3. Preparation of monoclonal antibodies using the splenocytes of the immunized mouse.
4. Large scale immunoaffinity purification of the desired protein on a matrix containing the specific monoclonal antibody.

5.8. ACKNOWLEDGEMENTS

The original research reviewed herein was supported by grants from the CNRS (79.7.054), INSERM (80.1017), DGRST (79.7.0780), University Paris VII ('Credits C, 1981'), Fonds de la Recherche Medicale Française and Association pour le Dévéloppement de la Recherche sur le Cancer.

5.9. REFERENCES

Avrameas, S. & Ternynck, T. (1967). *J. Biol. Chem.*, **242,** 1651–1659.
Beeckmans, S. & Kanarek, L. (1981). *Eur. J. Biochem.*, **117,** 527–535.
Caron, M. G., Srinivasa, Y., Pitha, J., Kocolek, K. & Lefkowitz, R. J. (1979). *J. Biol. Chem.*, **254,** 2923–2927.
Cheng, W. C., Fraser, K. J. & Haber, E. (1973). *J. Immunol.*, **111,** 1677–1689.
Douzou, P. (1976). *Trends Biochem. Sci.*, **1,** 25–27.
Lefkowitz, R. J., Haber, E. & O'Hara, D. (1972). *Proc. Natl. Acad. Sci. U.S.A.*, **69,** 2828–2832.
Sunderland, C. A., McMaster, W. R. & Williams, A. F. (1979). *Eur. J. Immunol.*, **9,** 155–159.
Vauquelin, G., Geynet, P., Hanoune, J. & Strosberg, A. D. (1977). *Proc. Natl. Acad. Sci. U.S.A.*, **74,** 3710–3714.
Vauquelin, G., Geynet, P., Hanoune, J. & Strosberg, A. D. (1978). *Life Sci.*, **23,** 1791–1796.
Venter, J. C., Ross, J. Jr. & Kaplan, N. O. (1975). *Proc. Natl. Acad. Sci. U.S.A.*, **72,** 824–828.
Yong, M. S. (1973). *Science*, **182,** 157–158.

5.10. FURTHER READING

Hoffman-Osterhof, O., Breitenbach, M., Koller, F., Kraft, D. & Scheiner, A. (Eds.) (1978). *Affinity Chromatography*, Pergamon, Oxford.
Jacoby, W. B. & Wilchek, M. (Eds.) (1974). *Methods in Enzymology*, vol. 34, *Affinity Techniques, Enzyme Purification*, part B, Academic Press, New York.
Lowe, C. R. (1979). *An Introduction to Affinity Chromatography*, North-Holland, Amsterdam.
Lowe, C. R. & Dean, P. D. G. (1974). *Affinity Chromatography*, Wiley, London.
Schott, H. (1984). *Affinity Chromatography, Chromatographic Science Series*, vol. 27, Marcel Dekker, New York and Basel.
Scouten, W. H. (1981). *Affinity Chromatography*, Wiley, New York, Chichester, Brisbane, Toronto.
Wilchek, M. & Hexter, C. S. (1976). In *Methods of Biochemical Analysis*, vol. 23, Glick, D. (Ed.), Wiley, New York, pp. 347–385.

Practical Protein Chemistry–A Handbook
Edited by A. Darbre
© 1986, John Wiley & Sons Ltd

M. D. WATERFIELD
Protein Chemistry Laboratory,
Imperial Cancer Research Fund,
Lincoln's Inn Fields,
London WC2A 3PX, U. K.

6

Separation of Mixtures of Proteins and Peptides by High Performance Liquid Chromatography

CONTENTS

6.1. INTRODUCTION

The development of instrumentation for high performance liquid chromatography (HPLC) has resulted in the introduction of many new separation methods which give significant improvements in the capability to purify proteins and peptides for functional and structural analysis. Progress in separation technology is related very closely to the development of suitable columns for fractionation where advantage can be taken of size, charge, hydrophobicity or affinity differences in the solute. The chromatographic techniques employed include:

1. *size exclusion*, where physical interactions alone should govern separation;
2. *adsorption*, where competition between solute and solvent molecules for a place on the surface of the absorbent and the selective disruption of multiple solute-support interactions are exploited;
3. *partition*, where a stationary phase is either non-covalently or covalently attached to the column packing and fractionation can be achieved using normal phase (polar interactions) or reverse phase (non-polar interactions) techniques.

The classification of chromatographic methods within this framework can often be difficult. For example, current gel permeation columns do not have completely inert packings and in addition to their use in size fractionation they can be used for both adsorption and partition based separations. The development of reverse phase techniques has made it possible to perform extremely rapid high resolution peptide separations and reverse phase techniques applicable to larger polypeptides are beginning to appear in the literature. It is likely that the most significant developments in this field will concern column packings and we can anticipate the introduction of methods which will give increased resolution capabilities for larger peptides and proteins. The use of affinity techniques in peptide and protein separations will undoubtedly take advantage of the chromatographic properties of HPLC particles and we can expect the introduction of HPLC-based methods within a short time. Adequate instrumentation for preparative and large scale HPLC of peptides is available commercially and we can expect to see the development of high sensitivity microbore systems to cover the opposite end of the spectrum. Most current instruments can probably not cope with the low flow rates, minimum dead volume and micro flow cells needed to handle microbore analysis. However, suitable instruments are being introduced and no doubt the separation and analysis of pmol amounts of peptides and proteins will be practical very soon. The other aspect of instrumentation development which will provide further valuable analytical information is the use of diode array detectors. These, with the new generations of micro- and minicomputers, can collect spectral scans at several time points (as many as 10 per second) during the elution of a component from a column, making it possible to monitor purity and provide selective identification of peptides or proteins with unique absorbance features. Ultimately we would expect to find the combination of mass spectrometry and microbore HPLC giving direct identification and even sequence information for small peptides.

6.2. SIZE EXCLUSION

The development of non-crushable packings suitable for size exclusion (gel permeation) chromatography of proteins and peptides has made it possible to perform rapid size fractionations of protein or polypeptide mixtures by HPLC. Previously only slow size separations could be made using compressible gels which were usually based on polysaccharide or polyacrylamide matrices such as Sephadex or Biogel. The introduction of Ultra-gel and Sepharose has provided more stabilized packings which give better performance in open column chromatography. Various methods are used to obtain beaded supports with particle and pore sizes suitable for high pressure size exclusion chromatography. Etched glass (CPG), controlled pore silica, fused micro-beads (Zorbax) and a variety of packings of unknown proprietary compositions are available. Manufacturers seek to obtain packings having uniform small particles (5–10 μm) and pore sizes within these particles which do not vary in size by more than a few per cent. The behaviour of a protein or polypeptide during size exclusion chromatography depends on the physical and chemical properties of the sample, the eluent and, of course, the type of column packing used.

6.2.1. Physical parameters

A review of physical parameters can be found in Yau *et al.* (1979) and in Regnier and Gooding (1980).

A number of concepts are worth understanding in relation to size exclusion chromatography. The total mobile phase (V_t) is distributed both inside (V_i) and outside (V_o) the porous particles which pack the column. Movement of solvent into and out of the pores occurs by diffusion. The distribution of a particular solute between the inner and the outer solvent is defined by the distribution coefficient, K_D, where

$$K_D = V_e - V_o / V_i = (V_e - V_o)/(V_t - V_o).$$

V_e is the effluent volume of a given solute.
V_o is the void volume and represents the interstitial liquid volume between the particles.
V_t is the total bed volume of solvent, i.e. $V_i + V_o$.

Molecules which cannot enter the pores and hence equilibrate with the liquid in the pores ($K_D = 0$) are eluted with a V_e which equals V_o. Small molecules with total permeation ($K_D = 1$) elute as a single peak with a V_e equal to the total bed volume V_t. Thus the K_D can vary between zero and one. Large molecules move through the column more rapidly than smaller ones. In perfect size exclusion chromatography the minimum elution volume is V_o and the maximum $V_o + V_i$. The void volume V_o can be measured by chromatography of a large molecule which is unable to enter the pores (for example, ferritin, mol. wt. 467 000, or Dextran blue, mol. wt. $\sim 2 \times 10^6$) and the V_t by chromatographing a small inert compound (such as [^{14}C]-glucose or ^3H_2O) which will have no interaction with the column. It should be remembered that the use of small peptides or modified amino acids such as DNP-lysine to measure V_t can give false values for this parameter due to interaction with the column packing. To describe the resolution of two components usefully the band width of the peaks must be considered. This can be derived from the relationship shown.

$$\text{Resolution} = \frac{V_{e(2)} - V_{e(1)}}{2(pw_1 + pw_2)}$$

The peak width (pw) can be measured as the standard deviation of each peak (pw_1 and pw_2). The efficiency of a column is described by the height equivalent to a theoretical plate (HETP)—an expression derived from a separation process which can be carried out in discrete steps. Such a process is countercurrent extraction where phases fully equilibrate and can then be separated at each step or theoretical plate. In chromatography, the phases are in constant motion and it is only possible to calculate the column height which would be equivalent to a plate. The HETP or the total number of theoretical plates in a column can be derived from chromatography of a standard compound. Manufacturers often use different standards to those used by a particular investigator. The number of plates can vary with the standard used, the column characteristics and the operating conditions, thus the chromatographer should be somewhat wary of quoted figures in comparing columns from different manufacturers. As the ability of these manufacturers to produce and pack small high quality beads has increased so has the HETP decreased, that is the total number of plates has increased. Several physical factors contribute to band broadening (as well as chemical factors—see 6.2.2). These include the velocity of the mobile phase, eddy diffusion caused by flow perturbation of the phase as it passes the beads and longitudinal diffusion. The use of small beads (5–10 μm) which have the least variation in size and pore dimensions together with good packing techniques help to reduce band spreading. Band spreading is also caused by wall effects and the use of 7.5–8 mm i.d. columns helps to reduce such effects. A compromise must of course be reached between the cost of the packing and the column diameter and length, thus the column dimensions employed represent a cost-effective compromise. Since the diffusion coefficients for large proteins are significantly smaller than those for peptides, band widths and column efficiencies are much more affected by increasing the column mobile phase velocity for proteins than peptides. Increasing the column length increases the efficiency of a column only if the flow rate is optimal for the apparent size of the compounds being analysed. Different manufacturers' gel permeation columns have differing suggested optimal flow rates for resolution of mixtures of the same proteins. In the manufacturer's literature for the Toyo Soda columns sold as Blue columns by LKB, optimal flow rates of about 0.05 ml/min are suggested for maximum resolution. This flow rate is below that which can be generated by many pumps and of course leads to longer analysis times. The optimum flow rate on the Waters I-125 columns for the same protein is 0.25 ml/min (as determined in our laboratory). In each case a suitable compromise must be reached depending on the results needed.

In addition to the factors already discussed above, the resolution is also related to the ratio of the total pore volume (V_i) to the column void volume (V_o). The greater the value of V_i/V_o the higher will be the resolution. The simplest way to achieve higher values of V_i/V_o is to increase column length. A length of 600 mm is a good cost-effective solution to most size exclusion problems. It is worth comparing this ratio for different columns before purchase of a gel permeation column.

6.2.2. Chemical factors

The polysaccharide- and polyacrylamide-based gels used in conventional size exclusion chromatography show very little chemical interaction with proteins or peptides. However, this is frequently not the case with HPLC gel permeation supports. The surface of silica supports contains acidic silanol groups. These can interact with positively charged groups on the protein, causing adsorption or repulsion of negatively charged species and thus preventing efficient sieving of proteins in the pores of the gel. To overcome these effects the silanol groups are modified with neutral hydrophilic side chains. Usually modification is incomplete and thus the unreacted silanol groups or other charged groups can influence the elution of proteins to varying extents depending on the mobile phase and the pI of the individual protein.

The two types of column presently used by most chromatographers are the Waters I series (I-60, I-125 and I-250) protein analysis columns and the Toyo Soda TSKSW series (2000, 3000 and 4000) columns which are sold under licence by many manufacturers. Other columns available are Synchropak GPC (Synchrom, Linden, Ind. U.S.A.), Lichrosorb and Lichrosphere DIOL (Merck), glycophase GPG (Pierce), Aquapore-OH (Brownlee), Superose (Pharmacia) and Spheron (Lachema). The precise nature of some of these proprietary packings is not known.

The influence of chemical factors in evaluating columns has been discussed by Regnier and Gooding (1980) with particular relevance to Synchropak GPC 100, and a detailed evaluation of Lichrosphere DIOL columns was made by Schmidt *et al.* (1980). The observations made with these columns are applicable to varying extents to other types of columns. It must be borne in mind that the silica-based materials can only be used at pH values between 2 and 7.7 to avoid damage to the packings. The effect of ionic strength on the elution volume of four proteins with different pI values is shown in Fig. 6.1. In this analysis Schmidt *et al.* (1980) chromatographed pepsin (pI 1), lysozyme (pI 11.0), ovalbumin (pI 4.7) and chymotrypsinogen (pI 9.5) at pH 5 on Lichrosphere DIOL. As the ionic strength

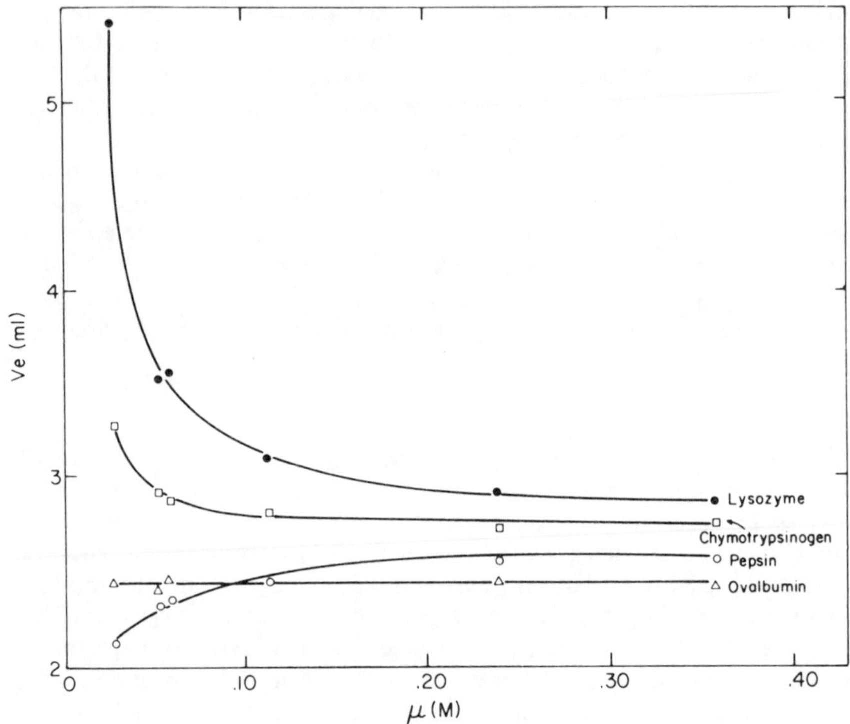

Fig. 6.1. The effect of ionic strength on the elution volume of proteins with different pI values from a Lichrosphere DIOL size exclusion column. Ionic strengths were created using sodium acetate and sulphate buffers at pH 5. (From Schmidt *et al.*, *Anal. Chem.*, **52,** 179, 1980)

increased the ionic exclusion effect on the acidic protein pepsin and the adsorption effects on the basic proteins chymotrypsinogen and lysozyme were reduced while the elution volume of ovalbumin (with a pI of 4.7) was unchanged. The suppression of these ionic effects is more effective at the same concentration for some salts since their ionic strengths are greater (e.g. phosphate versus acetate).

In the absence of interactions between the column packing and the protein there should be a linear relationship between the log of mol. wt. and elution volume. The elution characteristics for the Lichrosphere DIOL columns used by Schmidt *et al.* (1980) are shown in Fig. 6.2. In this case the analysis was made at an ionic strength of $\mu = 0.36$ M. As can be seen lysozyme and chymotrypsinogen still exhibit abnormally large V_e values at this ionic strength. The effect can be partly attributed to hydrophobic interactions since the addition of ethylene glycol to the mobile phase reduces the V_e. The observed anomalous elution of cytochrome C and haemoglobin was related to aggregation. Thus cytochrome C at pH 7.5

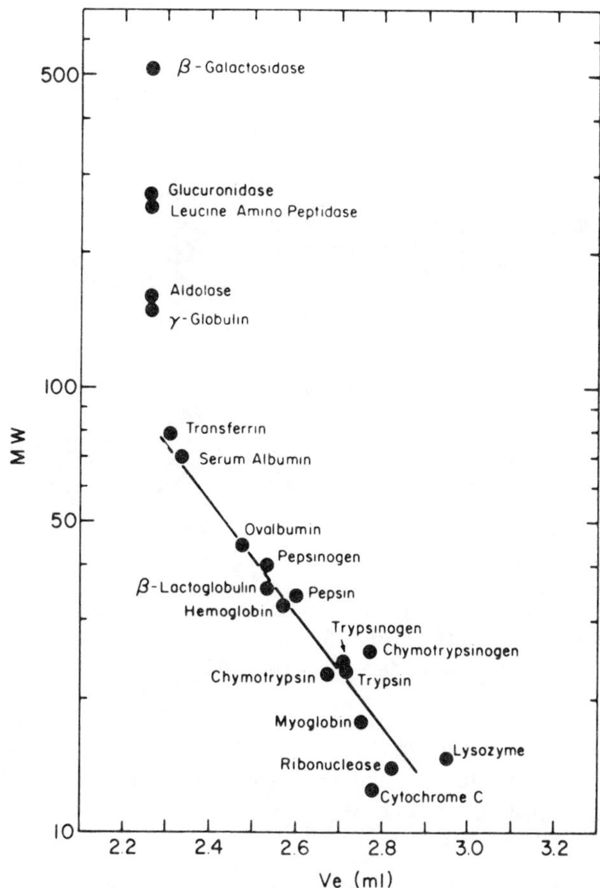

Fig. 6.2. Correlation of elution volume and apparent molecular weight for various proteins on size exclusion chromatography. Analytical conditions are as described in the legend to Fig. 6.1 at ionic strength 0.36. (From Schmidt *et al.*, *Anal. Chem.*, **52,** 180, 1980)

(0.05 M-sodium phosphate, 0.1 M-sodium chloride) was found to have a V_e corresponding to a mol. wt. of 12 500 while at pH 5 in 0.1 M-sodium acetate a mol. wt. of 20 000 was reported. For comparison, calibration curves and details of separation conditions for a variety of standard proteins on Synchropak GPC, Waters I and Toyo Soda series columns are shown in Figs. 6.3, 6.4 and 6.5.

Bearing in mind the factors which govern the apparent mol. wt. of a protein in solution and understanding the physical and chemical interactions of proteins with the column it is possible to use these columns for estimation of 'apparent'

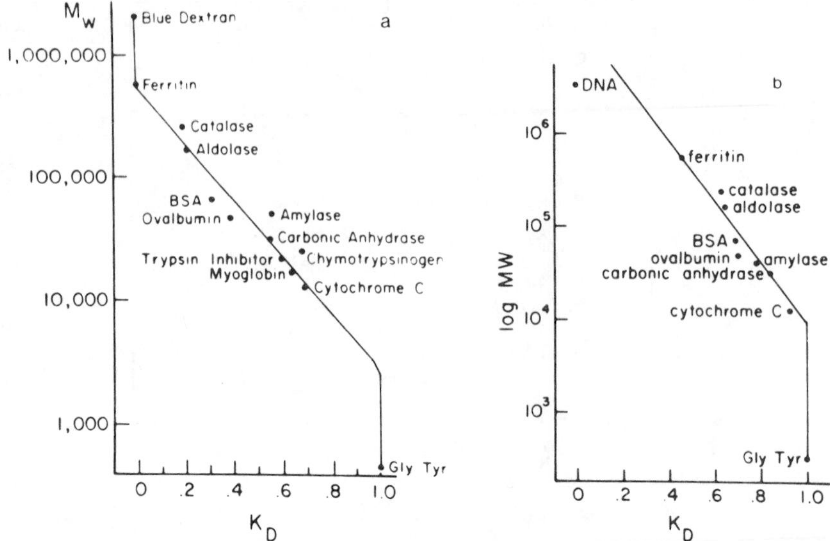

Fig. 6.3. A molecular weight calibration curve for proteins analysed by size exclusion chromatography on Synchropak TPC 100(a) and 500(b) column. The eluent was 0.1 M-KH_2PO_4 buffer at pH 7 with a flow rate of 0.5 ml/min. (From Regnier and Gooding, *Anal. Biochem.*, **103,** 4, 1980)

mol. wts. under non-denaturing conditions with an accuracy approaching 10%. The activity of the protein should be conserved under these conditions and recovery of most proteins is high.

6.2.3. Precise determination of molecular weight

The accurate determination of the mol. wt. of proteins cannot be carried out unless the proteins are chromatographed as random coiled polypeptide chains. Thus all disulphide bands must be broken either by full reduction, performic acid oxidation or sulfitolysis. Chromatography can be carried out in 0.1% aqueous SDS but the best results have been obtained with chromatography carried out in 6 M-guanidine HCl solutions. The effects of carrying out separation in SDS and guanidine HCl can be seen in Figs 6.6 and 6.7 (see Kato *et al.*, 1980a, b, c). Excellent correlation of mol. wt. with the distribution coefficient (K_D) can be achieved. A detailed study of the use of Toyo Soda TSK SW 3000 and 4000 columns using 6 M-guanidine HCl has been made by Ui (1979) who was able to produce data very similar to that obtained using agarose gels by Mann and Fish (1972). The use of guanidine HCl in HPLC instruments is not usually recommended since the chloride ions vigorously attack any weak points in the 'pacified' stainless steel. However, scrupulous attention to correct shut-down procedures using extensive washing of all steel parts will allow use of guanidine HCl. Studies in our own laboratory (Waterfield & Scrace, 1981) have employed 6 M-urea solutions containing 0.2 M-formic acid as a mobile phase in place of guanidine HCl as a denaturing agent. A good correlation between mol. wt. and K_D was observed but the exclusion limit of the column (I-125 in this case) was reduced to 65 000 due to the unfolding of the polypeptide chains. Urea solutions are useful for structural studies where column interactions need to be suppressed and solubility is a problem. The urea may be removed by adsorption of the peptides to Sep-Pak cartridges followed by a wash with 0.08 % TFA or 10 mM-ammonium acetate and elution of peptides with 50 % aqueous acetonitrile (containing 10 mM-ammonium acetate or 0.08 % TFA). If the peptides are large enough and do not adsorb to the tubing Spectropor dialysis tubing may be used. Tubing with a mol. wt. cut-off of 1 000–3 500 has proved satisfactory. These size exclusion columns can also be used for adsorption and partition chromatography by taking advantage of the adsorption and partition capabilities of the incompletely modified stationary phases. Fractionation of proteins by these techniques is discussed later.

6.3. ADSORPTION AND PARTITION OF POLYPEPTIDES AND PROTEINS

The use of reverse phase HPLC systems for peptide separation has become widespread since 1976 (for reviews see Hearn 1982; Regnier, 1983). The extension of techniques suitable for small peptides to the fractionation of larger polypeptides has, however, in most cases not yet produced such satisfactory methodologies. The resolution of protein or polypeptide mixtures on typical reverse phase packings is beset with problems which stem from the properties of these complex molecules. Proteins can display a diverse series of sites on their surfaces that can interact in different ways with the column packing. Many of these interactions can cause the loss of biological activity, as is well known to those who have tackled classical enzyme purifications. Thus interactions of proteins with surfaces and with organic solvents and the effects of extremes of pH and variations in salt concentrations are both employed and avoided in various separation techniques. Preservation of biological activity may severely limit the choice of chromatographic conditions while the solution of a structural problem may make it possible to exploit extreme conditions.

It is frequently found that a polypeptide or protein may exhibit several distinct forms which are separable, for often unknown reasons, during reverse phase HPLC. Such behaviour will obviously complicate fractionation and in the present state of the art it is still advisable to try several different strategies to obtain conditions where the sample behaves as a homogeneous species. As always, a review of the current literature will usually give clues to an initial strategy which may be successful. Almost always a series of chromatographic conditions will need to be evaluated and of course this requires sufficient material for these experiments. In many cases the analyst would be better advised to use FPLC for protein fractionation rather than experiment with reverse phase methods. Only a brief review of methods will be given here. Perhaps these will stimulate the interest of the biochemist. There are few systematic and original approaches to the problem of dealing with larger peptides, polypeptides and proteins. The studies of Rivier (1978) provided a stimulus for developing techniques using trialkyl ammonium and tetraethyl ammonium phosphate and formate buffers to examine the behaviour of some large peptides which could be chromatographed with varying efficiencies. Later a study of 32 polypeptides was reported by O'Hare and Nice (1979) (see also Nice and O'Hare, 1979) who obtained high resolution separations of such proteins as myoglobin, cytochrome C and lysozyme using 0.1 M-phosphate (pH 2.1, total molarity 0.2 M) and a gradient of acetonitrile. The use of both salt and low pH was necessary for optimum resolution and the concentration of organic modifier was very important. Working at low pH (approximately 2) leaves the side chains of aspartic and glutamic acids uncharged and hence the hydrophobicity of the polypeptide is increased, while the addition

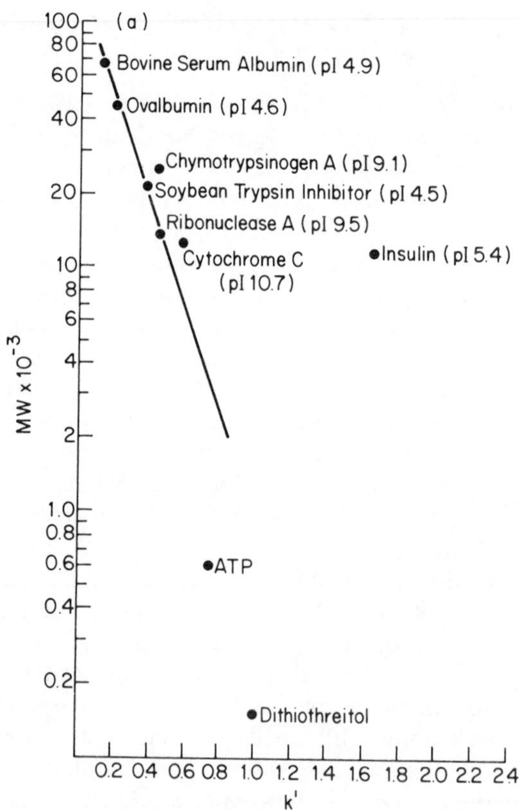

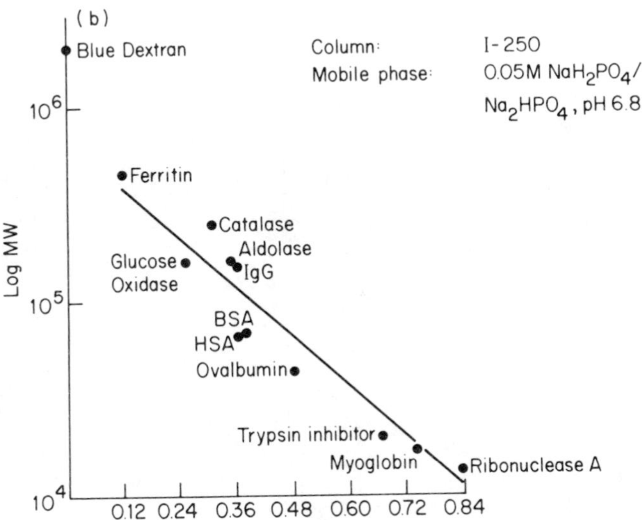

Fig. 6.4. Calibration curves for size exclusion chromatography of proteins on Waters I-125(a) and I-250(b) columns. Two Waters I-125 columns in series were used in (a) and the buffer was 0.1 M-KH_2PO_4 buffer pH 7 at a flow rate of 2ml/min. (From Jenik & Porter, *Anal. Biochem.*, **111,** 185, 1981. In (b) an I-250 column was used with 0.05 M-NaH_2PO_4 buffer pH 6.8. By permission of Waters Associates.)

of salt results in suppression of some of the ionic effects. The authors' choice of organic modifier and salt reflects the very useful ability to monitor columns at 215 nm where the peptide band absorbance may be detected. The disadvantage of this buffer system is the need to remove salt for most structural studies of polypeptides. Later studies by Nice *et al.* (1979) illustrate the application of acidic-salt buffers in the purification of polypeptide hormones and the use of mini columns for sample preparation and concentration. Bennett has used low pH systems with mobile phases containing TFA or HFBA and acetonitrile as organic modifier. TFA has been used by Bennett *et al.* (1980) for fractionation of peptides up to 10 000 mol. wt. and HFBA for separation of CNBr peptides of collagen which vary in mol. wt. from 3 000 to 60 800. Prior adsorption to reverse phase packings has been used as a sample preparation and concentration step (Van der Rest *et al.*, 1980). In this case HFBA proved more useful than TFA or phosphoric acid and a Vydac C18 column (Separation Group, Hespera, Ca., USA) was employed. This column has a 30 nm pore size (compared to the 6–10 nm of C18 μBondapak from Waters, Maynard, Mass. U.S.A.). This larger pore size should allow entry of larger polypeptides into the particle pores.

The separation of the CNBr fragments of human foetal globin was used to examine the optimum conditions for HPLC separation on a series of columns

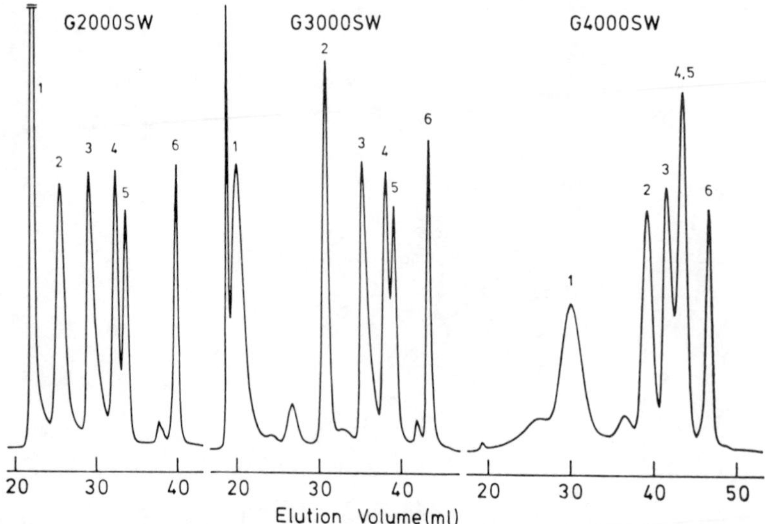

Fig. 6.5. Fractionation of proteins by size exclusion on TSK-SW columns with different pore sizes. Standard proteins are 1: thyroglobulin, 2: bovine serum albumin, 3: β-lactoglobulin, 4: myoglobin, 5: cytochrome C, 6: glycylglycylglycyl glycine. (From Kato *et al.*, *J. Chromatogr.*, **190,** 302, 1980.)

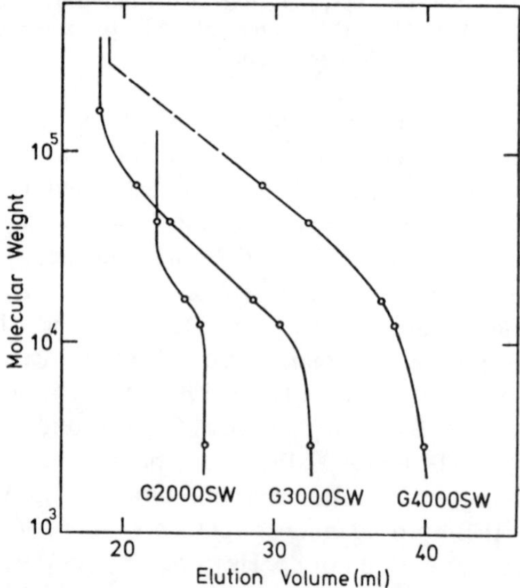

Fig. 6.6. Calibration curves for proteins in SDS solutions analysed on TSK-SW size exclusion columns. The buffer was 0.1 M-NaH_2PO_4 at pH 7 containing 0.1% SDS. (From Kato *et al.*, *J. Chromatogr.*, **193,** 34, 1980.)

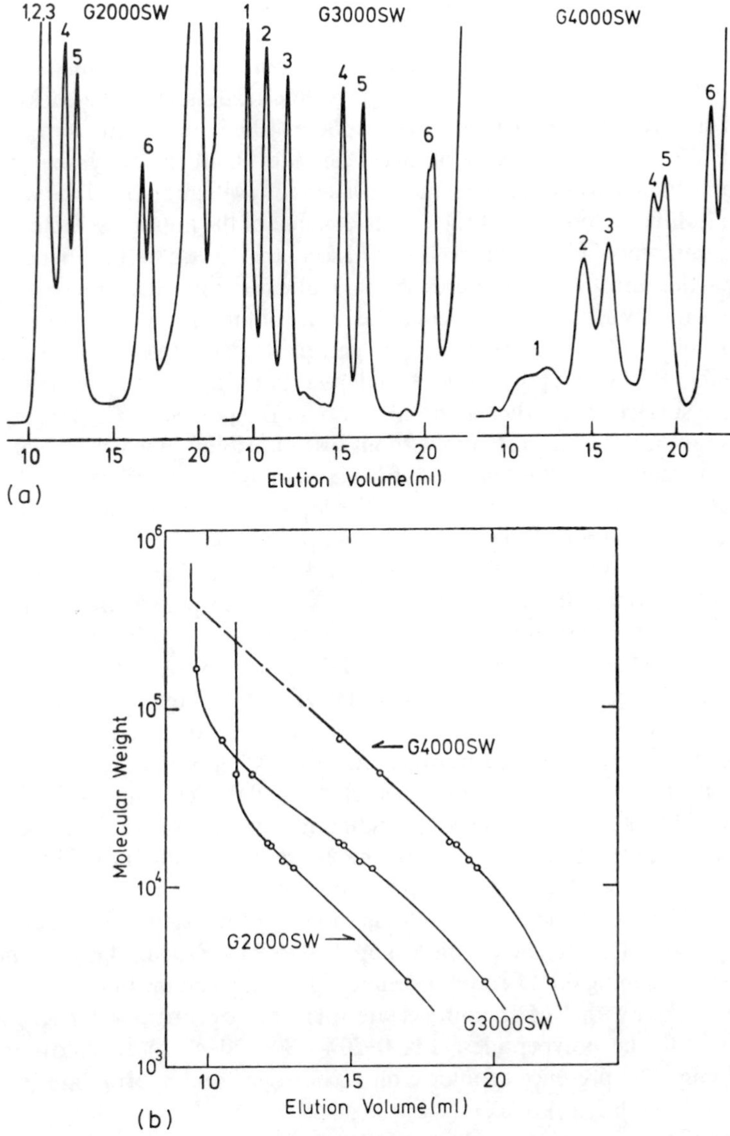

Fig. 6.7. Size exclusion chromatography of proteins on TSK-SW columns in guanidine solutions. Proteins were reduced and alkylated with iodoacetamide. (a) Separations on three different TSK-SW columns. The eluent was 6 M-guanidine HCl containing 0.1 M-NaH_2PO_4, pH 6. Standard proteins are 1: thyroglobulin, 2: bovine serum albumin, 3: ovalbumin, 4: myoglobin, 5: cytochrome C, 6: insulin. (b) Calibration curves for TSK-SW columns for proteins in 6 M-guanidine HCl. ((a) and (b) from Kato *et al.*, *J. Chromatogr.*, **193**, 459, 1980)

varying in pore size, particle shape and nature of the bonded phase (Pearson *et al.*, 1981). Using TFA-propan-1-ol buffers the authors show that recovery of the peptides is higher by a factor of 1.7–2 with pore sizes of 30 nm or 100 nm compared to 10 nm. They suggest that the geometrical shape of the packing may play a role, because the packings Lichrosphere (10 nm) and Lichrosorb (10 nm) showed differences in recovery of larger peptides. Unfortunately, the series of columns available was not sufficient for systematic evaluation of all the variables, making it difficult to interpret the effects of some of the important factors in the chromatography of polypeptides. It would seem that access to the inner channels of the particle through larger pores should make for more efficient reverse phase chromatography and this does seem to be borne out in the reports where this has been examined. As more experience is gained it seems that the major factor influencing HPLC of polypeptides in larger pore columns is related to the accessible surface area rather than the nature of the pore size. The length of the column can be reduced to 50 mm without loss of resolution which suggests that adsorption may have a role that needs further examination in separations of larger polypeptides. An elegant example of the development of reverse phase and normal phase techniques applied to polypeptide and protein purification has been made (see review by Rubinstein, 1979). Lichrosorb RP-18 packings have been used to purify β-lipotropin (Rubinstein *et al.*, 1977a), β-endorphin (Rubinstein *et al.*, 1977b) and Lichrosorb RP-8 in conjunction with Lichrosorb DIOL to purify leucocyte interferon (Rubinstein *et al.*, 1979). The average pore size of the Lichrosorb RP-8 and the Lichrosorb DIOL was 10 nm. Propan-1-ol was favoured as an organic modifier. Two-stage fractionations using pH 4 and 7.5 have been applied in several purifications, for example that of the leucocyte interferon. The ionic interactions exhibited by RP-8 column packings were suppressed by using the volatile salt pyridinium formate or with the non-volatile salt sodium acetate. Since it is not possible to monitor columns at 215 nm with these buffers, automated post-column fluorescence methods were used (Böhlen *et al.*, 1975). The use of Lichrosorb RP-8 is shown in Fig. 6.8 which is taken from Rubinstein *et al.* (1979). The sample was applied to RP-8 (10 μm, 4.6 × 250 mm) in 4 M-urea containing 0.1 M-sodium acetate pH 7.5 in a volume of 62.5 ml and the column washed with 1 M-sodium acetate (pH 7.5). A propan-1-ol gradient was used to elute the polypeptides (1 h, 0–20 %, 3 h, 20–40 %) at a flow rate of 0.25 ml/min. The presence of interferon was assayed and appropriate fractions pooled (see Fig. 6.8, (a)).

The application of a large volume of the sample in urea/acetate buffers to the RP-8 column shows the advantages in speed and concentration to be gained from exploiting the adsorptive properties of the reverse phase column. Peptides can also be desalted from urea solutions on Sep-Pak (Waters) cartridges (Waterfield & Scrace, 1981) and with mini-columns (Nice *et al.*, 1979) Sep-Pak packing material has also been used for sample concentration before HPLC. (Bennett *et al.*, 1980). Certain reverse phase columns can also be used for normal phase

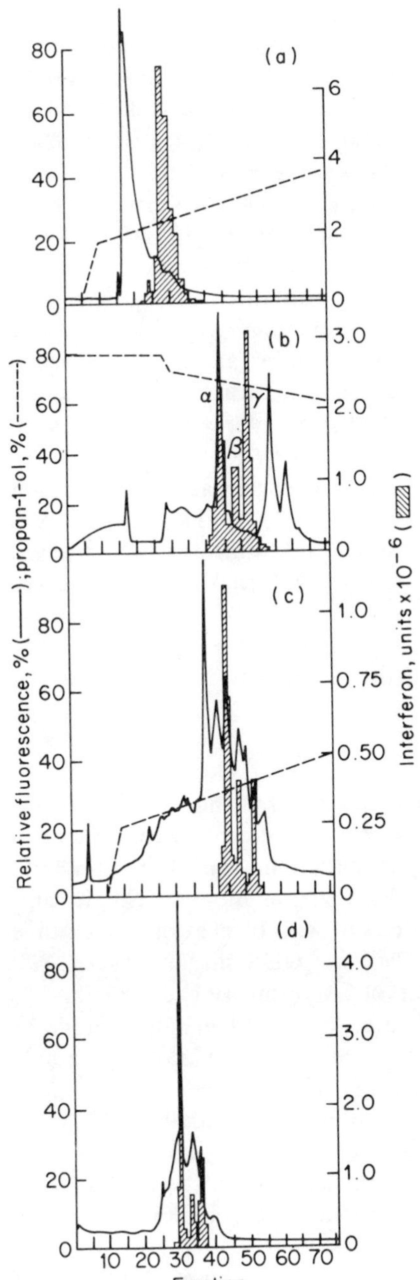

Fig. 6.8. Purification of interferon on Lichrosorb RP-8 using reverse phase and normal phase HPLC. (a) 63 ml of 4 M-urea/0.1 M-sodium acetate solution containing interferon was applied to the column which was then washed with 1 M-sodium acetate buffer (pH 7.5) and eluted with propan-1-ol (see text). (b) The eluate containing interferon from (a) was pooled and brought to 80 % propan-1-ol. The column was eluted with a gradient of 72.5–50 % propan-1-ol in 0.1 M-sodium acetate at 0.25ml/min. (c) Fractions from normal phase chromatography (γ) were pooled and separated by reverse phase HPLC in pyridine-formate buffers using a propan-1-ol gradient. (d) Re-chromatography of material containing interferon from separation (c). Similar chromatographic conditions to (c) were used.

Note: Columns were analysed by post-column fluorescence.

(From Rubinstein *et al.*, *Proc. Natl. Acad. Sci. U.S.A.*, **76,** 640, 1979.)

chromatography as is elegantly shown in the studies of Rubinstein (Rubinstein, 1979). Following elution from the RP-8 column used in the reverse phase mode, fractions containing interferon were pooled and propan-1-ol added to a concentration of 80 %. The material was then applied to a Lichrosorb DIOL column and eluted with a decreasing concentration of propan-1-ol, making it possible to purify the interferon (see Fig. 6.8(b)). Finally two reverse phase columns were run to achieve complete purity (see Fig. 6.8(c) and (d)).

Normal phase chromatography is particularly applicable to purification of hydrophobic proteins soluble in high levels of organic solvent. Rubinstein (1979) suggests that although hydrophobic proteins may co-precipitate with the bulk of proteins when relatively impure at high organic solvent concentrations, once they have been purified by prior reverse phase chromatography they would then be expected to stay in solution in high concentrations of organic solvents. The use of Lichrosorb DIOL for gel permeation was discussed earlier (p. 185). Recently LKB Toyo Soda introduced a phenyl-modified TSK column suitable for such normal phase chromatography.

The group at Roche have also examined Lichrosphere packing with 50 nm pore sizes. A comparison of C8 bonded packings with 10 nm and 50 nm pores showed that the capacity of the column and the peak width for large proteins (bovine serum albumin and collagen) is improved with packings having larger pores. However, they found that chromatography of proteins of mol. wt. 10 000–20 000 is possible using the packing with 10 nm pore size.

A number of important points arise from this selected review of the literature.

6.3.1. Sample preparation and loading

Reverse phase supports can be used for preliminary purification and concentration of polypeptides. Nice *et al.* (1979) describe the use of mini-columns filled with Partisil-10 ODS (10 μm, 5.5–6 nm pore size, Whatman, U.K.) for bulk fractionation of cell extracts and culture media. Using solutions at acid pH helps to inactivate proteases and elution is carried out with batches of acidic saline containing acetonitrile. Bennett *et al.* (1980) use ODS silica from Sep-Pak cartridges and extract tissue with a solution of 5 % formic (v/v) containing 1 % NaCl (w/v) and 1 % TFA (v/v). Tissues are delipidated first in acetone at −200°C, treated with hexane and acetone and dried. Following sample application, cartridges are washed with 0.1 % TFA and eluted with 80 % acetonitrile containing 0.1 %TFA. The eluate is diluted 1 : 4 with 0.1 % TFA and pumped into the analytical column. Rubinstein *et al.* (1979) applied 60 ml of partially purified interferon from a prior bulk fractionation on a Sephadex G-100 column in 4 M-urea solutions directly on to a reverse phase column (see p. 194). Waterfield and Scrace (1981) used Sep-Pak cartridges to recover peptides from urea–formic acid solutions collected from gel permeation columns. Elution from Sep-Pak was achieved with acetate buffers containing acetonitrile. These techniques allow the extraction of biological material and subsequent application of large volumes to reverse phase columns under conditions where the peptides or proteins are specifically adsorbed by the separation column. As discussed by Rubinstein (1979) the adjustment of the organic modifier can be used to allow even greater selectivity in the initial adsorption step.

The application of a sample to reverse phase columns may be one of the most important steps in fractionation. In our laboratory we usually inject samples dissolved in urea or guanidine solutions when fractionating peptides for structural study. This helps to dissolve samples, prevent aggregation and may allow even adsorption onto the column. Clearly adsorption as an even surface layer which allows optimum mobile phase-solute interactions is vital. The use of larger volumes for sample application can aid later chromatography.

6.3.2. Column selection for reverse phase separations

The use of columns with 10 nm pore sizes and 5–10 μm particles is satisfactory for many polypeptides but the loading capacity and band width of eluted polypeptides reported for larger proteins (20 000 mol.wt or greater) seems more efficient with pore sizes of 50 nm or greater. The use of buffers containing high salt at low pH helps prevent interactions with unmodified silanol groups. Some columns such as Ultrapore RPSC (Altex), Vydac, Zorbax C8 and Bioseries (Dupont) are highly ‘capped’ to render free silanols unreactive and may show advantages in recovery of peptides. The most satisfactory side-chain length is probably C3, C4, or C8 at present. A particularly good separation on an Ultrapore RPSC column (C3) is shown in Fig. 6.9.

6.3.3. Mobile phase

Most separations use low pH mobile phases (pH about 2) that contain 0.1 % TFA, 0.1 M-phosphate or acetate buffers. Mobile phases having pH values of 4 or 7 have proved useful for sequential sub-fractionations (Rubinstein, 1979). The choice of organic modifier at present seems to be acetonitrile or propan-1-ol. propan-1-ol and 1 M-sodium acetate (pH 7.5) or with a gradient of propan-1-ol conditions must be adapted to experimental situations and the need for preserving function is of paramount importance in the selection of conditions.

Three systems are worth initial study:

1. 0.1 M-NaH_2 PO_4-H_3PO_4 (pH 2.1, total phosphate 0.2 M) with acetonitrile gradients of 10–40 % (Nice *et al.*, 1979).
2. 1 M-pyridine, 2 M-formic acid (pH 4.0) with a linear gradient of 20–40 % propan-1-ol and 1 M-sodium acetate (pH 7.5) or with a gradient of propan-1-ol over the range 0–40 % (Rubinstein *et al.*, 1979).
3. 0.1 % TFA or HFBA with a 0–60 % acetonitrile gradient (Bennett *et al.*, 1980)

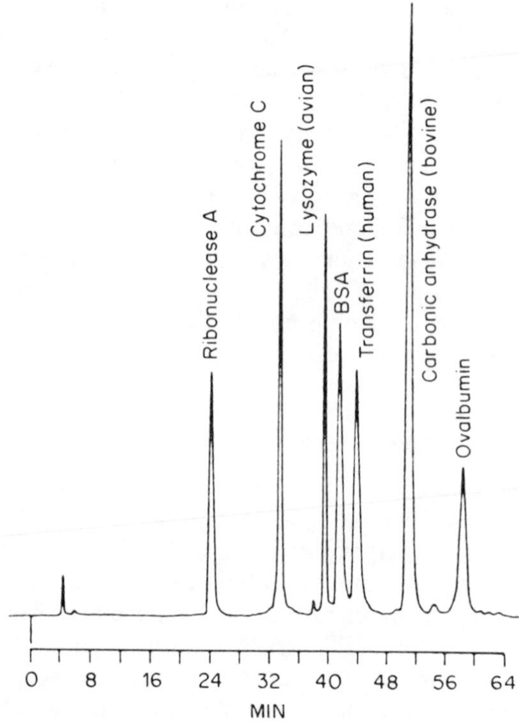

Fig. 6.9. Separation of native proteins by reverse phase HPLC using a wide pore short alkyl chain column. Proteins were fractionated in 0.01 M-TFA using an acetonitrile gradient from 20–70% at 0.2 ml min^{-1}. The column was an Ultrapore RPSC (4.6 × 75 mm Beckman) which was monitored at 280 nm. Approximately 20 μg of each protein was analysed.

(see Fig. 6.9). This phase is volatile, can be lyophilized and allows monitoring at 206–215 nm.

6.4. REVERSE PHASE SEPARATION OF PEPTIDES

The use of reverse phase HPLC for the resolution of mixtures of peptides was introduced by Rivier in 1976 and independently also by Hancock, Bishop and Hearn in the same year. Since these reports were published many researchers have explored the potential of various mobile phases and columns to obtain rapid high resolution methodologies which can be used for purification of biologically active peptides or fragments for structural study. A variety of reverse phase columns have been used and because of subtle differences in packings (often given the same name) obtained from the ever-increasing numbers of manufacturers, many variations of similar methodologies can be found in the literature. Separations can be influenced by the organic phase modifier, water content of the mobile phase, ionic strength, pH, temperature and the capacity for ion-pair and ion-exchange effects.

Rather than give a complete review of the literature a number of basic methods will be selected which give satisfactory resolution. The majority of these have been tried in the author's laboratory.

6.4.1. Mobile phase selection

The choice of mobile phase will depend on a number of factors.

1. Compatability with post-column assays. Selection will depend on the assay. Organic modifiers may have to be removed for many bioassays. This can usually be done by evaporation in a stream of nitrogen. Care should be taken that salt and pH conditions are compatible with assays.
2. The need for lyophilization. Recovery of peptides is usually required and thus the ability to lyophilize the mobile phase is important. Losses of peptides during desalting are usually large. Suitable phases are ammonium acetate and bicarbonate, triethylammonium formate, TFA, HFBA, formic acid and acetic acid.
3. The optical properties required. Since the peptide bond absorbance can be monitored at 206–215 nm the elution of all peptides can be quantitatively followed. This is one of the great advantages of HPLC and should be a major factor in selecting mobile phases. Organic acids are sufficiently transparent at 206–215 nm for use, although the background absorbance due to these will rise with high concentrations. The organic modifiers can be selected using tables available from most manufacturers or from text books on HPLC. Propan-1-ol, acetonitrile, methanol and ethanol are suitable. Purity will affect the shift in background absorbance which occurs when gradients are used at high absorbances. In some cases special solvent grades are available for working in the range 200–215 nm.
4. The size, charge and hydrophobicity of the peptides. Analysis of large peptides (greater than 30–40 residues) was dealt with earlier. Since reverse phase columns resolve peptides by hydrophobic interactions, an increase in hydrophobicity of the peptide will cause an increase in the retention time. Becuase one is restricted to pH 2–7.5 by the stability of the bonded silicon particles, the effects of pH mainly concern the carboxyl and α-amino groups of the peptides.

Thus a pH lower than the pI values of glutamic and aspartic acid side chains will increase the hydrophobicity since the side-chain carboxyls will be largely uncharged. Through ion pairing the polarity of the charged side chains can also be modified to change the hydrophobicity of the peptides. For example at pH 4, ammonium salts in the mobile phase will allow ion pairing and ionic suppression of the carboxyl groups. Ion pairing can be used to increase or decrease the polarity of the peptides through interaction with the anionic or cationic side chains. The retention time of the ion-paired peptide can be influenced by partition chromatography of the complex or by adsorption of the ion-pairing reagent by the bonded hydrophobic phase which will result in ion-exchange effects. In most cases for routine separations, mobile phase containing TFA, phosphoric acid or ammonium acetate will suffice. The use of more complex ion-pairing agents may be necessary for particular separations, e.g. mixtures of small predominantly acidic peptides. The papers of Hearn and his group should be consulted for an excellent series of investigations on ion pairing (see review by Hearn, 1982).

6.4.2. Peptide separation using phosphoric acid

Phosphoric acid has been used by a number of groups for structural studies. Fullmer and Wasserman (1979) used 0.1 % phosphoric acid at pH 2.2 together with an acetonitrile mobile phase for fractionation of peptides from calcium-binding proteins. These authors showed the use of HPLC to monitor the extent of enzyme digestion with time for the preparative separation of peptides and showed the use of monitoring the effluent at 210, 254 and 280 nm. In this study the phosphoric acid which remained after lyophilization of fractions did not interfere with amino acid analysis, end group determination using dansyl Cl or automated sequencing analysis. Direct injection of enzyme digests was used.

6.4.3. Peptide separation using trifluoroacetic and heptafluorobutyric acids

The use of TFA and HFBA for separation of peptides was described (Bennett *et al.*, 1980; Van der Rest *et al.*, 1980). We have found that these acids are ideal for peptide work and show a typical separation of the tryptic peptides of cytochrome C in Fig. 6.10. In this case a Vydac C18 column was used with an acetonitrile gradient. These buffers are volatile and transparent at 215 nm. TFA and HFBA are used in sequence analysis and can be obtained in highly purified form (Sequential grade, Pierce Chemical Co.).

6.4.4. Relationship of retention properties and structure

A relationship between hydrophobicity and retention time should exist for reverse phase separations. O'Hare and Nice (1979) chromatographed a series of peptides and calculated hydrophobicities using partition coefficients derived from octanol/water partition experiments. A relationship between the sum of the values for the most hydrophobic amino acids and the retention times was found. It should be noted, however, that significant discrepancies were reported. Meek and Rossetti (1981) used a chaotropic agent (0.1 M-sodium perchlorate with 5 mM-sodium phosphate or 0.1 % phosphoric acid to give pH values of 7.4 or 2.1 with acetonitrile gradients) to minimize conformational effects and improve peak shape. They obtained a good correlation between the retention coefficient and retention time. This retention coefficient was calculated using peptides of known structure (see also Meek, 1980). Hearn and Grego (1981) made a detailed study of the theory governing distribution equilibria and showed that a reasonable prediction of elution order could be made using summated hydrophobic fragmental constants for the amino acid side chains. The accumulation of more data will enable anomalous retentions to be rationalized.

Numerous convincing examples of the use of reverse phase HPLC for preparative and analytical fractionation of peptides have shown the value of these methods. A recent advance which is valuable in peptide analysis is the use of multiple wavelength monitoring to evaluate purity or selectively identify particular peptides. Several manufacturers now make computer-based spectral scanners which will record base, slope and apex of spectra for purity evaluation (see Fig. 6.10). In particular cases wavelength ratios may be useful.

6.5. FAST PROTEIN LIQUID CHROMATOGRAPHY

Because of the impact of FPLC on biopolymer fractionation an outline of this technique is worthy of special mention (for a review see Richey, 1982).

Pharmacia have introduced macroporous monobead columns which have a particle size of 9.8 (+2 %) μm, which are stable over the pH range 2–12. These have either strong anion or cation substituents for ion exchange, or surface modifications suitable for chromatofocusing (Mono Q, Mono S or Mono P columns). Very high capacities can be achieved with rapid flow rates (1 ml/min) at moderate pressures (10–20 bar) generated by inert piston pumps. Glass columns can be used because of the low back-pressures. A wide range of buffers which includes detergents and chaotropes such as guanidine HCl can be employed. Pharmacia have developed buffers that give optimum performance over particular pH ranges to avoid the pH variation inherent but not normally important in open column ion-exchange separations. Separation times can be extremely short (about 1–5 min) because the degree to which resolution depends on speed is much less than with other ion-exchange techniques. Knowledge of the elution behaviour of proteins on conventional ion exchangers is transferable to FPLC and usually results in increased resolution. If no information is known about the protein mixture to be fractionated, then because of the rapid speed of analysis, the conditions for optimum resolution can be found by screening with different buffers and plotting results to generate a family of elution curves. The literature produced by Pharmacia gives a very good outline of the potential application and operation of these columns. It is possible (but perhaps not advised by Pharmacia) to use these columns on many HPLC machines with

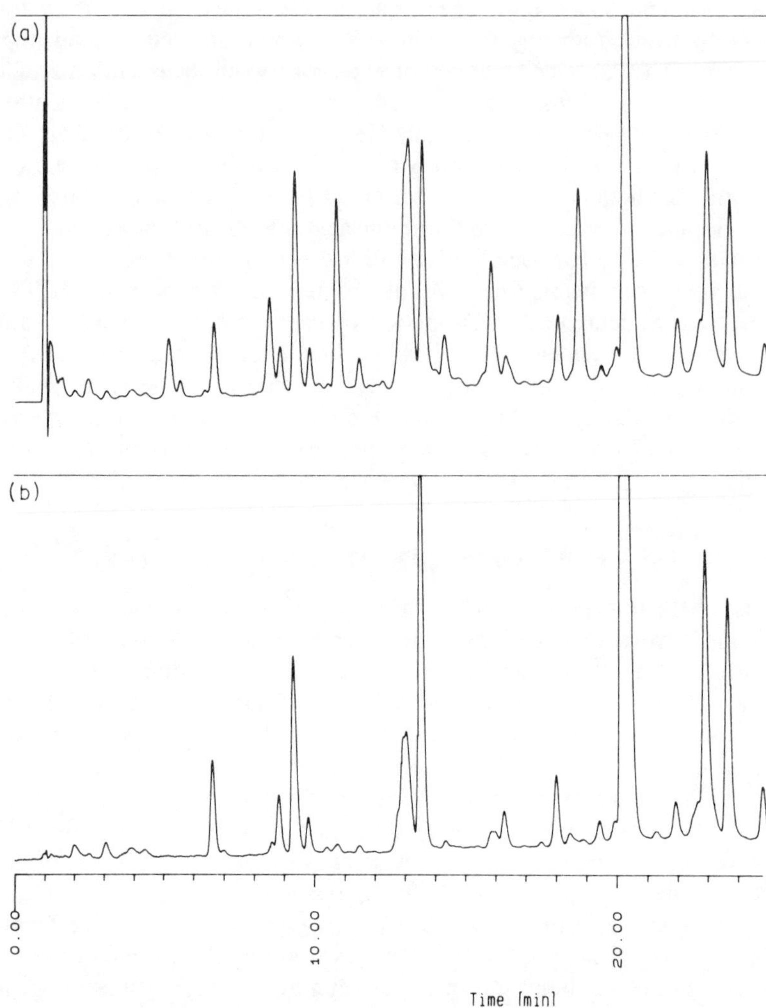

Fig. 6.10. Fractionation of tryptic peptides of cytochrome C by reverse phase HPLC. The separation was achieved on a Vydac RP (C18) column (4.5 × 75 mm) using 0.08 % TFA with a gradient from 1 to 40 % acetonitrile. The column was monitored with a Hewlett Packard 1040 diode assay detector. The absorbance of the effluent at 215 and 280 nm is plotted in (a) and (b), respectively. Spectral scans for peptides without aromatic residues (— — —) and for those containing tyrosine (————), tryptophan (· · · · · · · ·) and the haem group (— - —) are shown in (c). With this apparatus, selected peptides can be identified for subsequent analysis.

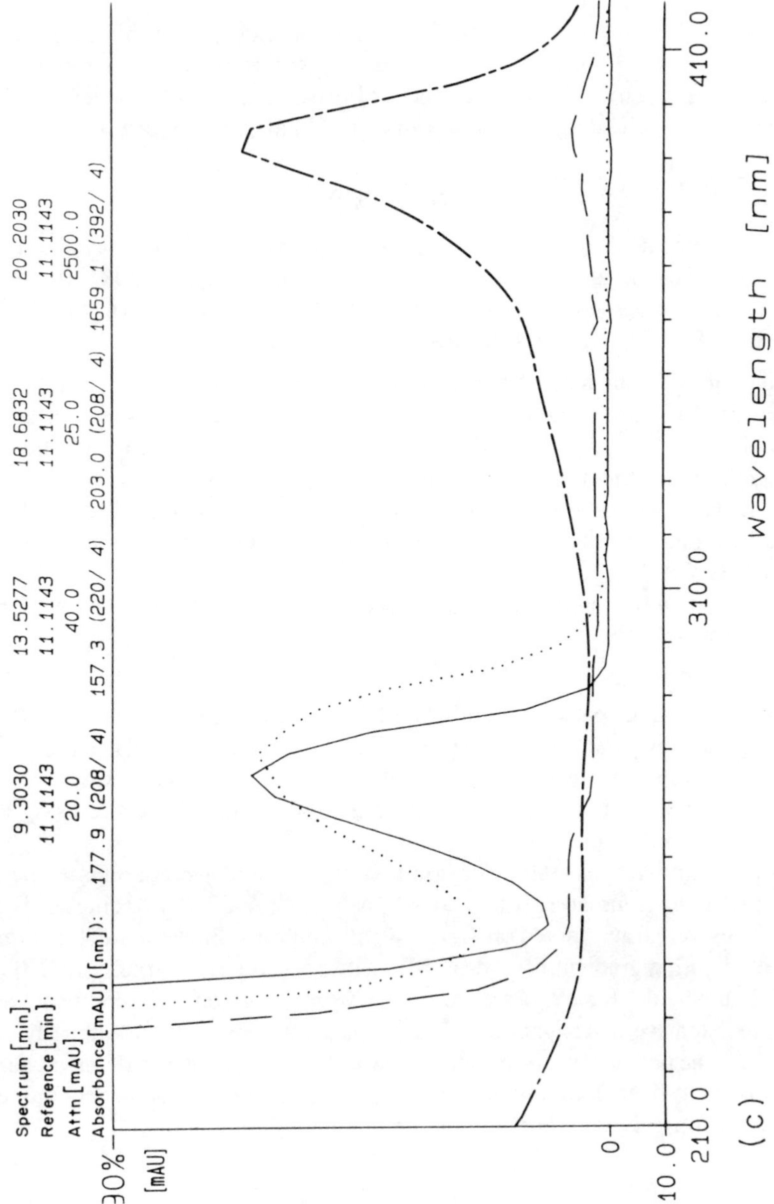

Spectrum [min]: 9.3030 13.5277 18.6832 20.2030
Reference [min]: 11.1143 11.1143 11.1143 11.1143
Attn [mAU]: 20.0 40.0 25.0 2500.0
Absorbance[mAU]([nm]): 177.9 (208/ 4) 157.3 (220/ 4) 203.0 (208/ 4) 1659.1 (392/ 4)
30%
[mAU]
0
10.0
210.0
310.0
410.0
Wavelength [nm]
(c)

appropriate fittings and pressure safety cut-offs to protect the glass columns. However, the use of stainless steel equipment with chloride-containing buffers at pH extremes should be avoided.

Chromatofocusing using Mono P columns provides a new high resolution technique which will have a major impact on resolutions of protein mixtures. Advantage is taken of charge differences of proteins at pH values close to their isoelectric points. Details can be found in the Pharmacia literature.

6.6. ION-EXCHANGE

The development of ion-exchange supports suitable for HPLC has come from the work of Regnier and his collaborators (reviewed by Regnier, 1982, 1983) and from several commercial companies—notably Pharmacia with the FPLC technique. Four major types of packings are available.

1. Glycophases, which are derived from a surface film of epoxy polymer linked covalently through glyceryl-propyl silane groups to glass particles. These are available with diethylaminoethyl, carboxymethyl, sulphonylpropyl and quaternary ammonium functional side chains as CPG glycophases.
2. Pellicular coatings containing the DEAE functional side-chains which are available commercially as Synchropak columns (Synchrom Inc., Linden, Indiana, U.S.A.).
3. Toyo Soda ion-exchange columns based on modified size exclusion packings (sold by several agents).
4. FPLC packings (Pharmacia).

These ion-exchange packings can be used for rapid separations similar to those made on conventional soft gel supports. These packings, particularly the FPLC ones, are being more widely used because of the good results which can be achieved. Regnier (1982) reviews some dramatic separations made using ion-exchange supports.

Vanecek and Regnier (1980) examined anion exchange columns in detail, particularly the commercially available Synchropak AX-300. Loadings of 10 mg of proteins were fractionated on 4.1 × 250 mm columns. Flow rates of 0.5 ml/min were used with a gradient of 0 to 0.5 M-sodium acetate in Tris buffer (pH 8) run over about 90 min. Recoveries of 90–100 % were reported for different enzymes.

More extensive investigations of this type of support will undoubtedly take place over the next few years and we can expect very versatile rapid ion-exchange separations to be in common use. At the present time FPLC is the technique of choice for ion-exchange HPLC fractions.

6.7. REFERENCES

Bennett, H. P. J., Browne, C. A. & Solomon, S. (1980). *J. Liquid Chromatogr.*, **3,** 1353–1365.

Böhlen, P., Stein, S., Stone, J. & Udenfriend, S. (1975). *Anal. Biochem.*, **67,** 438–445.
Fullmer, C. S. & Wasserman, R. H. (1979). *J. Biol. Chem.*, **254,** 7208–7212.
Hearn, M. T. W. (1982). *Advances in Chromatogr.*, **20,** 1–82.
Hearn, M. T. W. & Grego, B. (1981). *J. Chromatogr.*, **218,** 497–507.
Jenik, R. A. & Porter, J. W. (1981). *Anal. Biochem.*, **111,** 184–188.
Kato, Y., Komiya, K., Sasaki, H. & Hashimoto, T. (1980a). *J. Chromatogr.*, **190,** 297–303.
Kato, Y., Komiya, K., Sasaki, H. & Hashimoto, T. (1980b). *J. Chromatogr.*, **193,** 29–36.
Kato, Y., Komiya, K., Sasaki, H. & Hashimoto, T. (1980c). *J. Chromatogr.*, **193,** 458–463.
Mann, K. G. & Fish, W. W. (1972). *Methods Enzymol.*, **26,** 28–42.
Meek, J. L. (1980). *Proc. Natl. Acad. Sci. U.S.A.*, **77,** 1632–1636.
Meek, J. L. & Rosetti, Z. L. (1981). *J. Chromatogr.*, **211,** 15–28.
Nice, E. C. & O'Hare, M. J. (1979). *J. Chromatogr.*, **162,** 401–407.
Nice, E. C., Capp, M. & O'Hare, M. J. (1979). *J. Chromatogr.*, **185,** 413–427.
O'Hare, M. J. & Nice, E. C. (1979). *J. Chromatogr.*, **171,** 209–226.
Pearson, J. D., Mahoney, W. C., Hermodson, M. A. & Regnier, F. E. (1981). *J. Chromatogr.*, **207,** 325–332.
Regnier, F. E. (1982). *Anal. Biochem.*, **126,** 1–7.
Regnier, F. E. (1983). *Methods Enzymol.*, **91,** 137–190.
Regnier, F. E. & Gooding, K. M. (1980). *Anal. Biochem.*, **103,** 1–25.
Richey, J. (1982). *Technical Review Series*, Pharmacia Fine Chemicals, Uppsala, Sweden.
Rivier, J. E. (1978). *J. Liquid Chromatogr.*, **1,** 343–366.
Rivier, J. E., Spiess, J., Perrin, M. & Vale, W. (1979). In *Biological/Biomedical Applications of Liquid Chromatography*, Hawk, G. L. (Ed.), *Chromatographic Science Series*, vol. 12, Marcel Dekker, New York and Basel, pp. 223–241.
Rubinstein, M. (1979). *Anal. Biochem.*, **98,** 1–7.
Rubinstein, M., Stein, S., Gerber, L. D. & Udenfriend, S. (1977a). *Proc. Natl. Acad. Sci. U.S.A.*, **74,** 3052–3055.
Rubinstein, M., Stein, S. & Udenfriend, S. (1977b). *Proc. Natl. Acad. Sci. U.S.A.*, **74,** 4969–4972.
Rubinstein, M., Rubinstein, S., Familletti, P. C., Miller, R. S., Waldman, A. A. & Pestka, S. (1979). *Proc. Natl. Acad. Sci. U.S.A.*, **76,** 640–644.
Schmidt, D. E. Jr., Giese, R. W., Conron, D. & Karger, B. L. (1980). *Anal. Chem.*, **52,** 177–182.
Ui, N. (1979). *Anal. Biochem.*, **97,** 65–71.
Vanecek, G. & Regnier, F. E. (1980). *Anal. Biochem.*, **109,** 345–353.
Van der Rest, M., Bennett, H. P. J., Solomon, S. & Glorieux, F. H. (1980). *Biochem. J.*, **191,** 253–256.
Waterfield, M. D. & Scrace, G. T. (1981). In *Biological/Biomedical Applications of Liquid Chromatography*, Hawk, G. L. (Ed.), *Chromatographic Science Series*, vol. 18, Marcel Dekker, New York and Basel, pp. 135–158.
Yau, W. W., Kirkland, J. J. & Bly, D. D. (1979). In *Modern Size-Exclusion Chromatography*, Wiley, New York, p. 27.

Practical Protein Chemistry—A Handbook
Edited by A. Darbre
© 1986, John Wiley & Sons Ltd

WILLIAM JOHN GULLICK
Protein Chemistry Laboratory,
Imperial Cancer Research Fund,
Lincoln's Inn Fields,
London WC2 3PX, U. K.

7

Peptide Mapping of Proteins

CONTENTS

7.1. INTRODUCTION

Peptide maps on proteins are generally run for two purposes, either to see if molecules from different sources are structurally related or to identify changes in structure of an individual molecule which have arisen as a consequence of normal or aberrant metabolism. Examples of the former category include:

1. comparison of proteins with the same function isolated from different tissues or species;
2. comparison of products of translation produced *in vitro* with equivalent molecules isolated from cells;
3. comparison of subunits of multimeric proteins.

Peptide maps on individual proteins that may change structurally can be divided into two groups: firstly, proteins that mature structurally into stable cellular constituents by post-translational mechanisms such as limited proteolysis, glycosylation or prosthetic group attachment and, secondly, proteins that undergo reversible structural alterations such as phosphorylation or acetylation which may modulate their activity.

The principle of peptide mapping is to fragment proteins into several pieces by peptide bond cleavage and then to compare the physical behaviour of these polypeptides or peptide mixtures. The fragments may be generated by any reproducible method that can cleave peptide bonds such as by proteolytic enzymes or by chemical treatments. The subtleties of peptide mapping reside in the methods of separation and visualization of the fragment mixture and the interpretation of the maps produced.

Three general methods of peptide separation are described here. The first is by electrophoresis in polyacrylamide gels containing SDS, usually in one dimension. The polypeptide fragments are resolved by their relative mobilities which may or may not give a reliable estimate of their mol. wts. More importantly the pattern of fragments of one protein may be used as a paradigm for comparison with fragments derived from another sample. This method is colloquially referred to as Cleveland mapping after the name of the first author of the paper introducing the procedure (Cleveland *et al.*, 1977). SDS gels can separate polypeptides ranging in mol. wt. from 10^3 to 10^6 daltons but they are most reliable with proteins of apparent mol. wt. of 10^4 to 10^5 daltons. Consequently this method is not generally used for proteins smaller than 10^4 daltons.

The second method consists of two-dimensional peptide mapping. It is known as 'finger-printing' and was introduced by Ingram (1956, 1957) to locate and subsequently identify the single amino acid difference between normal adult and sickle cell haemoglobin. The protein is degraded usually by trypsin or α-chymotrypsin digestion (Ingram also used 12 M-HCl at 37°C for 2–3 days) to provide a mixture of many non-overlapping short peptides (10–20 amino acid residues). The peptides are then separated by electophoresis in the first dimension and liquid chromatography in the second dimension. Finger printing is a more powerful tool than Cleveland mapping, because more peptides are produced and they are separated to form a pattern of spots which is characteristic for any individual protein.

The third method that has been developed for peptide mapping employs a similar initial approach as two-dimensional mapping in that the fragments produced are terminal digestion products with low mol. wts. The fragments in solution are separated either by reverse phase HPLC (RPHPLC) using columns packed with silica coated with alkane chains of various lengths or by ion-exchange HPLC (IEHPLC) on silica coated with ionizable groups. The fragments are resolved by virtue of their different retention times on the column. The two major advantages of this method are speed (an individual run may take as little as thirty minutes), sensitivity and the easy recovery of the peptides for characterization, amino acid analysis and sequencing. Optical detectors are used to measure the absorbance of the peptides at one or several wavelengths and the presence of radioactivity determined either by flow-through detectors or by counting fractions individually. Peptides may also be derivatized with fluorescent compounds and detected with extreme sensitivity.

7.2. PEPTIDE MAPPING IN THEORY

The general classes of protein structure that one might reasonably expect to encounter in comparing two proteins are as follows:

1. Identical structures: entirely homologous amino acid sequences with identical post-translational modifications (if any).
2. Entirely homologous amino acid sequences but with differently modified side chains or terminii, for instance by acetylation, glycosylation, phosphorylation etc. (for a review of protein modifications see Wold, 1981).
3. Fully homologous but one protein being shorter, either at the *N*- or *C*-terminus, or both.
4. Identical through part of the sequence, unrelated or partially related through the rest.
5. Related with significant amino acid homology but with scattered amino acid changes, insertions or deletions throughout the primary structure.
6. Entirely unrelated in structure.

There are of course several permutations of these general categories.

The two simplest situations are type (1) and type (6) where either identical or completely different maps would be produced. The more fragments that are visualized the more reliable the comparison becomes since complicated patterns contain more information with which to examine relationships.

Proteins that have homologous amino acid sequences but differ in post-translational modification (type 2) will give identical maps unless the modification alters the mobility of the fragment upon which it resides. Cleveland maps which separate large polypeptides in one dimension are less likely to detect their differences than are two-dimensional maps where the hydropathy and charge to mass ratio of the smaller fragments will be more radically affected. Since post-translational modification is confined to limited numbers of amino acids in a sequence some fragments will be identical and some will differ. However, this will also be the pattern produced if a pair of proteins are identical throughout part of their sequences but differ completely through the rest (type 4).

If one protein of a pair is truncated at either the *N*- or *C*-terminus (type 3) the maps produced will be similar but the shorter protein will be missing one or several fragments present in the map of the other. There will also be one fragment in each of the maps which is not common, derived from the part of the sequence adjacent to or surrounding the point of cleavage. If a protein is shortened at both its *N*- and *C*- terminus there will be two fragments that differ in mobility in each map.

Perhaps the most difficult situation to interpret by peptide mapping is where two proteins have related amino acid sequences but contain scattered amino acid substitutions, deletions and insertions (type 5). This may arise where a gene is duplicated and over a period of time genetic drift has occurred. Peptide maps of two such molecules will be more similar the less that these changes have occurred. However, even quite extensive homology of this type may not become apparent. The subunits of the acetylcholine receptor from *Torpedo californica*, for example, have approximately 40% sequence homology (Noda *et al.*, 1983) yet give different peptide maps (Gullick *et al.*, 1981 and Fig. 7.1). Thus peptide mapping has considerable limitations under these circumstances and may be misleading if not interpreted carefully.

7.3. PEPTIDE MAPPING IN PRACTICE

7.3.1. Cleveland mapping

Labelled or unlabelled proteins of interest may be purified from a mixture by conventional biochemical methods such as ammonium sulphate precipitation and open column chromatography or by more sophisticated approaches such as immune precipitation (Cooper *et al.*, 1983) or affinity chromatography (Gullick & Lindstrom, 1982a; Einarson *et al.*, 1982) which give as a final product a protein in solution. Frequently, however, purification to homogeneity is difficult by these means and SDS polyacrylamide gel electrophoresis is used as a final and very powerful resolving system. The original Cleveland mapping method (Cleveland *et al.*, 1977) involved lightly staining the protein in the primary gel with Coomassie Blue (Diezel *et al.*, 1972) and then rapidly destaining and excising the relevant slice of wet gel.

An alternative approach that is useful with radioactively labelled proteins is to remove the gel from the electrophoresis apparatus and to rinse it in about twenty volumes of distilled water to reduce the amount of buffer salts in the gel and then to dry it down without fixation (if the gel is fixed in acetic acid and methanol the intact protein is often difficult to elute from the rehydrated gel). The dried gel is

Fig. 7.1. Peptide mapping on SDS polyacrylamide gels of the subunits of acetylcholine receptor. (a) Peptide map using V8 protease: (1–4) 5 μg of α, β, γ and δ subunits, (5) molecular weight marker proteins, (6–9) α, (10–13) β, (14–17) γ and (18–21) δ digested with 0.0005, 0.005, 0.05 and 0.25 μg of V8 protease, respectively. (b) Peptide maps using papain: (1–4) 5 μg of α, β, γ and δ subunits, (5) molecular weight markers, (6–9) α, (10–13) β, (14–17) γ and (18–21) δ digested with 0.0001, 0.001, 0.01 and 0.1 μg of papain, respectively. Arrows indicate those bands which were stained with periodate–Schiff reagent to detect carbohydrate. (From Gullick *et al.*, 1981, by permission of the American Chemical Society, Washington, D.C.)

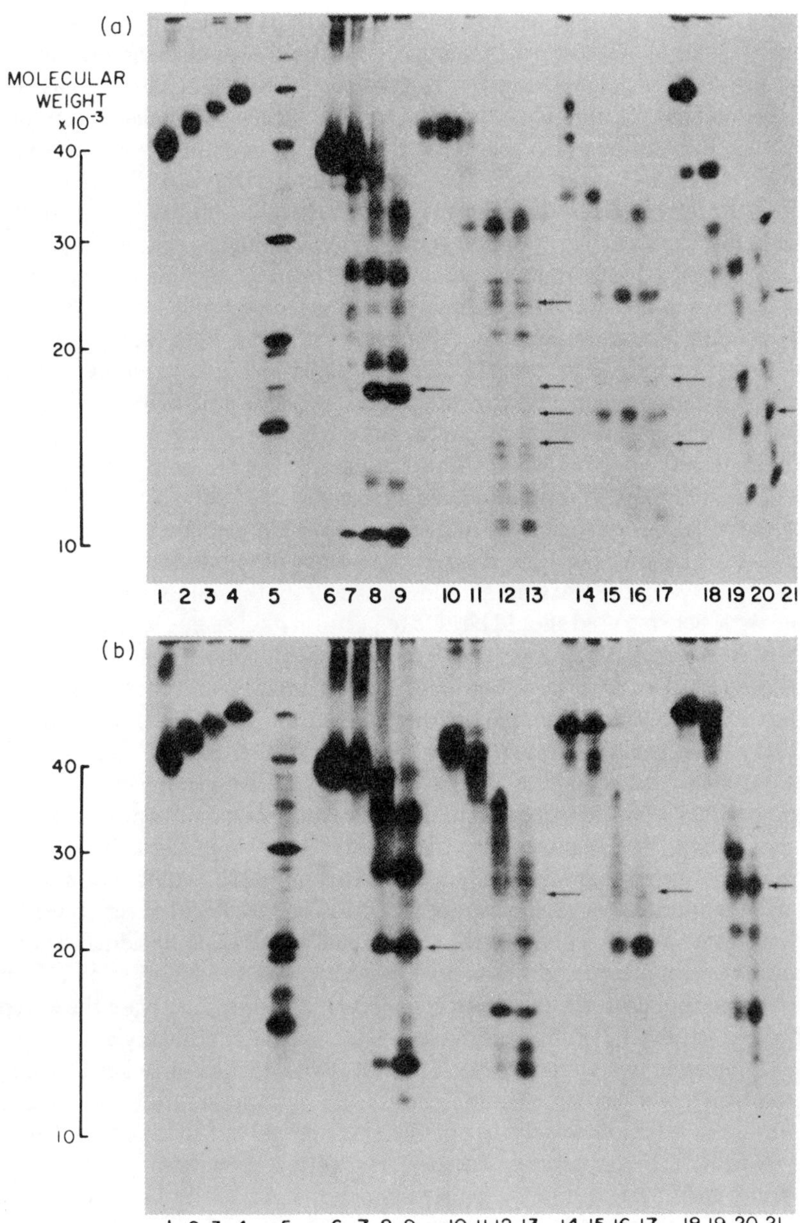
(a)
MOLECULAR WEIGHT x 10^{-3}
40
30
20
10
1 2 3 4 5 6 7 8 9 10 11 12 13 14 15 16 17 18 19 20 21
(b)
40
30
20
10
1 2 3 4 5 6 7 8 9 10 11 12 13 14 15 16 17 18 19 20 21

then autoradiographed to visualize the proteins and the desired band excised using the X-ray film as a template.

A second SDS gel is prepared with an acrylamide concentration that gives a mobility of the intact test protein of about 0.2 relative to the bromophenol blue dye front (Fig. 7.1.). Gels formed with gradients of acrylamide are also used and are particularly useful when the size of the protein fragments cannot be predicted. A stacking gel should be poured with a low enough concentration of acrylamide so that it does not limit the mobility of the intact protein or the proteolytic enzyme (if any) and long enough so that the samples stack into a tight band before they reach the resolving gel. The wells should be wide enough to fit in the rehydrated gel pieces and fairly deep so that both the enzyme solution and the gel piece(s) can be accommodated.

The choice of cleavage reaction depends on the nature of the substrate protein and the experience and prejudice of the experimenter. Certainly the most common types of reagents are proteolytic enzymes and probably the most popular is V8 protease isolated from the culture filtrate of *Staphylococcus aureus*, V8 strain (Drapeau *et al.*, 1972). This enzyme cleaves on the *C*-side of aspartic acid and glutamic acid residues under most conditions but may be restricted to cleaving only on the *C*-side of glutamic acid in the appropriate buffer (Houmard & Drapeau, 1972). In the normal Laemmli discontinuous gel solutions (Laemmli, 1970) the enzyme should cleave at both amino acids. Since the proteolytic enzyme used should be able to perform its cleavage in the stacking gel it must be active in 0.1% (w/v) SDS and preferably higher concentrations. V8 protease (Bordier & Crettol-Jarvinen, 1979; Gullick and Lindstrom, 1982b), papain (Gullick *et al.*, 1981), elastase and α-chymotrypsin (Gullick & Lindstrom, 1982a) and trypsin (Bordier & Crettol-Jarvinen, 1979) have all been used.

If the substrate protein is available in solution the enzyme may be added directly to it *in vitro* and incubated until the cleavage reaction has proceeded to the desired extent. The products are then loaded onto a gel and the fragments separated in the normal way (Gullick & Lindstrom, 1982b; Gullick & Lindstrom, 1983). It is not necessary to inactivate the protease since it will be separated from its substrate in the resolving gel. A reducing agent is often added after the digestion is completed and the sample heated to 100°C for 2–5 min. This may give sharper bands in the resolving gel and will certainly inactivate the enzyme.

More generally the protein substrate is within a zone of polyacrylamide. In this case the enzyme in solution (about 10 μl ideally) is loaded into the relevant gel lanes which already contain running buffer. The enzyme solution should ideally be similar in composition to that of the stacking gel but made more dense by including 10–20% glycerol or sucrose. The gel piece containing the substrate protein is then pushed into the gel lane above the protease solution taking care not to disturb the dense layer. Gel spacers slightly thicker than those used in the primary gel can be helpful but it is usually quite easy to squeeze the gel piece between the plates of a gel with the same dimensions as the primary gel. A snug fit is desirable since there is less danger of the gel piece falling into the enzyme solution and displacing it. Loading the protease underneath the gel piece is better because an even concentration of enzyme will be present across the track. If one overlays the enzyme it should be on a cushion of denser buffer which covers the gel piece to give an even depth of enzyme solution.

The voltage is then applied and the substrate protein stacked out of the gel piece together with the native protease. Once a thin stack has formed the current is turned off for a period of time, generally 30–60 min, to allow the enzyme to act. The active undenatured protease will then cleave some of the susceptible bonds of the denatured and unfolded substrate protein. The current is then turned back on and the fragments separated in the resolving gel.

Polypeptides resolved by SDS gels may be visualized by several methods. The protein may be labelled with radioactive isotopes either metabolically or *in vitro* and the fragments detected by autoradiography or fluorography (Bonner & Laskey, 1974). The labelling may be specific by an affinity label such as a natural ligand or synthetic ligand analogue (Gullick & Lindstrom, 1982a), or by enzymes or kinases that can introduce radioactive groups provided as substrates. The label may be more generally introduced by iodination of available tyrosine residues (Christopher *et al.*, 1978; Parsons & Kowal, 1979; Salacinski *et al.*, 1981; Markwell, 1982) or metabolically by cell (Cooper *et al.*, 1983) or organ (Stein & Sussman, 1983) culture in medium supplemented with one or more radioactive amino acids. Specific labelling will tend to label only some of the polypeptide fragments but may be useful for examining local homologies such as surrounding a ligand binding site (Gullick & Lindstrom, 1982a) or sites of phosphorylation.

Gels may also be stained with reagents that detect polypeptides generally such as Coomassie Blue (Diezel *et al.*, 1972) or with silver (Merril *et al.*, 1979; Oakley *et al.*, 1980; Merril *et al.*, 1981), or more specifically to detect glycopeptides (Green & Pastewka, 1975; Weber & Hof, 1975; Eckhardt *et al.*, 1976) with periodate Schiff reagent (Gullick *et al.*, 1981), modified silver staining (Dubray & Bezard, 1982) or with radioactive (Koch and Smith, 1982) or fluorescent lectins (Furlan *et al.*, 1979). (See also Chapter 8 for staining methods.)

Polypeptides may be transferred out of the gel to nitrocellulose sheets (Towbin *et al.*, 1979; Burnette, 1981; Gershoni & Palade, 1983) positively charged membranes (Gershoni & Palade, 1982) or diazotized paper (Christophe *et al.*, 1982; Seed, 1982) and visualized by the specific binding of one or more layers of enzymically or radioactively labelled antibodies (Gullick & Lindstrom, 1982a) or lectins (Glass *et al.*, 1981; Hawkes, 1982). Antisera may detect all of the fragments, whereas monoclonal antibodies or lectins will bind to one or a limited number of fragments. Since antibodies are capable of detecting an individual protein in a mixture it is possible using this method to compare two proteins one of which may never have been purified. The relatively large size of the polypeptides resolved by Cleveland mapping is advantageous for immunodetection since they are likely to retain some uncleaved sequence-type determinants.

In order to optimize the fragment pattern so that many pieces of a suitable size are produced, different amounts of enzyme should be tried with a fixed amount of substrate protein (Cleveland *et al.*, 1977 and Fig. 7.1). If one has no knowledge of the proteolytic susceptibility of the substrate protein (which may vary considerably) then a good starting enzyme amount is about 1 μg per gel lane and perhaps one tenth and ten times as much additionally if there is enough substrate protein available. On the basis of patterns obtained with these starting amounts of enzyme the system may be optimized.

Chemical cleavage can also be used. In fact when a very small amount of unlabelled substrate protein is available and silver staining is necessary to visualize the fragments, chemical cleavage is advantageous since there will be no contribution to the staining pattern by the enzyme (Lischwe & Ochs, 1982). Cleavage at methionine residues can be achieved with cyanogen bromide (Chen-Kiang *et al.*, 1979; Nikodem & Fresco, 1979; Lonsdale-Eccles *et al.*, 1981), at asparagine-proline bonds with formic acid (Lam & Kasper, 1980; Sonderegger *et al.*, 1982) and at tryptophan residues with *N*-chlorosuccinimide (Lischwe & Ochs, 1982). The gel piece is treated in a test tube to fragment the protein and then loaded onto a second gel to resolve the fragments.

Many variations on this theme of peptide mapping in gels have been published. Several proteins may be resolved in a primary gel and then the whole gel lane cut out and laid horizontally across a second gel. The acrylamide strip is then sealed into place with agarose containing a proteolytic enzyme. The system is stacked and the proteins digested as usual. In this manner peptide maps of several proteins in a mixture (of necessity with different relative mobilities in the primary gel) can be obtained with little effort (Bordier & Crettol-Jarvinen, 1979; Lam & Kasper, 1980; Tijssen & Kurstak, 1983). However, one problem with this approach is that the first gel may contain different amounts of each protein but only a single enzyme concentration can be used. The different enzyme to substrate ratios may result in the digestion of the proteins to different extents and this must be taken into consideration when examining the fragment patterns.

When proteins are available in very small quantities their fragments may be transferred from SDS gels to diazotized paper with which they react covalently or to nitrocellulose to which they bind with non-covalent interactions. If appropriate antisera or monoclonal antibodies are available, fragments can be detected with extreme sensitivity. Gullick and Lindstrom (1982a) digested a mixture of the subunits of the acetylcholine receptor from three different species of animals and resolved the fragments on a one-dimension gel. The polypeptides were transferred to diazophenylthioether-paper (Symington *et al.*, 1981; Seed, 1982) and antisera against each individual subunit type and monoclonal antibodies were used to visualize fragments of that subunit out of the complete mixture, thus considerably simplifying their comparison (Fig. 7.2, (a)–(g)). The peptide maps did not appear to be related but, in the case of (g) a monoclonal antibody visualized a large fragment in each of the lanes 4, 5 and 6 with the same mobility.

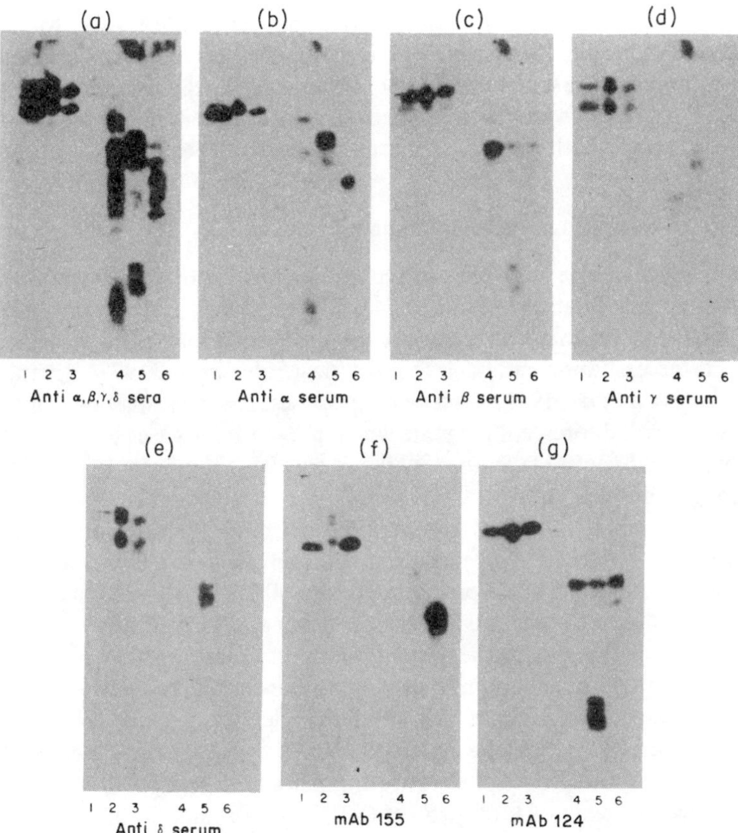

Fig. 7.2. Intact and V8 proteolysed acetylcholine receptors from Torpedo, Electrophorus and bovine muscle separated on a 15% SDS polyacrylamide gel and then transferred electrophoretically to diazophenylthioether paper. Lane 1: torpedo acetylcholine receptor (10 ng); lane 2: electrophorus acetylcholine receptor (20 ng); lane 3: bovine acetylcholine receptor (50 ng); lane 4: torpedo receptor (40 ng) digested with V8 protease (1:10 w/w); lane 5: electrophorus receptor (80 ng) treated as in lane 4; lane 6: bovine receptor (100 ng) treated as in lane 4 probed with (a) mixture of equal amounts of antisera to Torpedo α, β, γ and δ-subunits, (b) anti α-serum, (c) anti β-serum, (d) anti γ-serum, (e) anti δ-serum, (f) monoclonal antibody to α subunits, (g) monoclonal antibody to β subunits. (From Gullick & Lindstrom, 1982a, by permission of the American Chemical Society, Washington, D.C.)

This indicated that the receptors from each species possessed some degree of homology since in each case the enzyme generated a fragment of the same size which contained the same sequence type determinant.

There will no doubt be further developments in Cleveland mapping. For example, experiments in which rare mRNA molecules are reverse transcribed, cloned and sequenced to provide predicted amino acid sequences are becoming quite common. Antisera or monoclonal antibodies can be prepared against synthetic peptides from the predicted sequence and used to detect proteins transferred to diazotized paper without the protein ever having been purified.

7.3.2. Two-dimensional peptide mapping

Purified proteins are digested with trypsin or sometimes α-chymotrypsin in solution, as precipitates (Gibson, 1974) or in gel pieces (Elder *et al.*, 1977). The fragments may be unlabelled, but usually only small quantities of material are available necessitating radioactive labelling (Elder *et al.*, 1977; Bryant *et al.*, 1979; Feitelson *et al.*, 1981). Unlabelled peptides, however, may be reacted after separation with fluorescent reagents which may in some cases be almost equally sensitive as radioactive labelling (Schiltz *et al.*, 1977; Stephens, 1978; Fishbein *et al.*, 1980; Whittaker & Moss, 1981).

Protein samples produced for peptide mapping are often in rather large volumes. In order to concentrate them and to remove unwanted buffer components the proteins are precipitated with 50 % (w/v) trichloracetic acid (2 h at 4°C is sufficient to precipitate most proteins) and the precipitate collected by centrifugation. When using very small amounts of radioactive proteins a cold carrier protein such as human immunoglobulin should be added to encourage complete precipitation. Since neither the carrier protein nor the enzyme are radioactive the quantities added are only limited by the fact that the sample must ultimately be spotted onto the separating plate in a very small volume (0.25–5 μl). Large amounts of total protein (>100 μg) are undesirable since they will tend to give a streaky appearance to the map and leave a high proportion of the radioactive material at the origin.

The pelleted sample is dissolved or suspended in a volatile buffer (0.5 ml of 50 mM ammonium bicarbonate) to which the enzyme is added (1–50 μg for radioactive samples, 1:100 enzyme to substrate weight ratio for unlabelled samples). The mixture is left to digest for 6–24 h at 37°C. A second addition of trypsin may be necessary to obtain complete digestion since trypsin autodigestion leads to a loss of enzyme activity with time.

Radiolabelled proteins in gel slices should be equilibrated with the ammonium carbonate buffer and digested in the same way. The small fragments will be more soluble than the intact substrate protein and will passively elute out of the gel. The supernatant fluid is then concentrated by lyophilization for spotting onto the plate. This also has the effect of removing the volatile buffer components. Some procedures include an alkylation and oxidation step prior to enzyme digestion (Gibson, 1974; Stephens, 1981).

The dried mixture of small peptides is then dissolved in a small volume (10 μl) of electrophoresis buffer for spotting onto glass plates coated with 0.1–1.0 mm thickness of silica or cellulose. Unfortunately commercially available plates frequently have rather uneven coatings. This can be checked by holding the plate against a light box and looking for dark streaks or patches which indicate different thicknesses. Poor plates should be rejected or used for initial trial runs to establish the length of time and pH for electrophoresis or the composition of the chromatography solvent. The supporting medium may also consist of sheets of Whatman 3 MM paper (as used by Ingram, 1956, 1957), but losses are considerable due to adsorption of the peptides. However, this is useful for testing the homogeneity of peptides (see section 7.3.2.1).

The sample should be applied in very small aliquots (0.25 μl) so that the origin is as compact as possible. Individual applications can be dried rapidly using a hair drier or fan. Charged, coloured dye substances are often applied to the plate. After electrophoresis these should have migrated in a reproducible manner which gives a visual indication that the electrophoresis went satisfactorily. Electrophoresis is in buffers containing pyridine, acetic acid, water and sometimes butanol (Gibson, 1974; Stephens, 1978; see Table 7.1). The pH of the solution may be varied to give different separations of the charged peptides. The order of electrophoresis followed by chromatography is conventional but not necessary. The plate is dampened with electrophoresis buffer, taking care not to smudge the applied sample, and run at 1000 V for 40–90 min (for 200 × 200 mm plates). It should be pointed out that this is potentially a dangerous procedure and all the electrical apparatus should be periodically checked for faults by a qualified electrician. The electrophoresis plate is removed from the apparatus and dried overnight in a fume hood. The solvents used are quite volatile and when dry there should be no smell of acetic acid on the plate. If several samples are to be compared it is important that the individual electrophoretic runs should be repeated in exactly the same way with the same buffer solution to improve reproducibility. It is desirable for the same reasons to carry out chromatography of plates at the same time in the chromatography tank.

The composition of the solvent for chromatography is critical for the mobility of the peptides. More organic solvent in the mixture tends to increase the relative mobility of the hydrophobic peptides since the stationary phase of the

Table 7.1. Conditions for fingerprinting tryptic peptides on silica gel G thin layer plates. TLC plates are commonly 200 × 200 mm, 0.1–0.25 mm thick.

First dimension: electrophoresis
- pH 3.5, pyridine/acetic acid/water (2:20:978, by vol.) 1000 V, 45 min
- pH 6.5, pyridine/acetic acid/water (100:3:897, by vol.) 1000 V, 40 min
- pH 4.7, butan-1-ol/pyridine/acetic acid/water (2:1:1:18, by vol.)

Second dimension: chromatography at 25° C
- Chloroform/methanol/ammonium hydroxide (2:2:1, by vol.)
- Propan-1-ol/ammonium hydroxide (7:3, v/v)
- Butan-1-ol/pyridine/acetic acid/water (97:75:15:60, by vol.), pH 5.3

chromatography system is hydrophilic. Several solvent systems are in use (Gibson, 1974; Stephens, 1978) but modifications may be necessary to optimize a particular separation (Table 7.1).

The plates are placed in a glass chromatography tank for chromatography upwards in the second dimension. The origin of sample application should be about 1 cm above the solvent level. Development is allowed to continue until the solvent reaches the top or nearly the top of the plate. With 200 × 200 mm plates this takes about 5 h. The plates are then removed and air dried and then either autoradiographed directly or dipped into, or sprayed with, an enhancing solution for isotopes of low energy (Bonner & Stedman, 1978). Plates may be stained for peptides in several ways (Schiltz *et al.*, 1977) but most frequently the plates are sprayed with fluorescamine in acetone (Stephens, 1978) and the spots examined under u.v. light (366 nm). (See also Chapter 8). A permanent record may be obtained in the form of an autoradiograph on X-ray film or the plates should be photographed soon after fluorescent staining (within 24 h at least).

When there is some doubt from the maps obtained as to how related two samples are, it is usual for an additional map to be run, onto which equal amounts of each sample are spotted. When visualized, the coincidence or otherwise of spots is more apparent and more trustworthy.

Problems that occur in this technique include use of unevenly coated plates, excessive amounts of protein applied (or too small an amount of radioactivity) and lack of optimization of digestion and running conditions. It is worth noting also that although most peptides are positively charged in electrophoresis buffer there may still be some that have a net negative charge and it is useful to perform a trial run, spotting the sample in the lower centre of a plate and electrophoresing for only 30 min. The plate can then be exposed or stained to determine the mobilities of the peptides in the first dimension and the ideal duration of electrophoresis. Some particularly well-resolved two-dimensional maps are shown in Fig. 7.3 to show what may be achieved.

Fig. 7.3. Separation of tryptic peptides of scallop and rabbit actin on thin layer plates of silica gel G. Upper pair: tryptic peptides from scallop (a) and rabbit (b) actin showing unique peptides (arrows); electrophoresis, pH 3.5; chromatography, chloroform/methanol/ammonium hydroxide (2:2:1, by vol.). Spots for arginine and lysine are shown. Centre pair: tryptic peptides of scallop (c) and rabbit (d) actin showing unique (peptides (arrows), peptide positional differences (arrow with star) and numerous secondary cleavage products; electrophoresis, pH 6.5; chromatography, propan-1-ol/ammonium hydroxide (7 : 3, v/v). Lower pair: rabbit actin prepared electrophoretically (e), containing SDS during digestion and mapping, compared with actin prepared conventionally (f), in the absence of SDS; certain spots (circles) are less intense in the presence of SDS, while others (arrows) are more distinct; electrophoresis, pH 3.5; chromatography, propan-1-ol/ammonium hydroxide (7 : 3, v/v). All samples in this figure were loaded with 1 nmol in 2.5 μl. See Table 7.1 for details of electrophoresis and chromatography. (From Stephens, 1978, by permission of Academic Press, New York and London.)

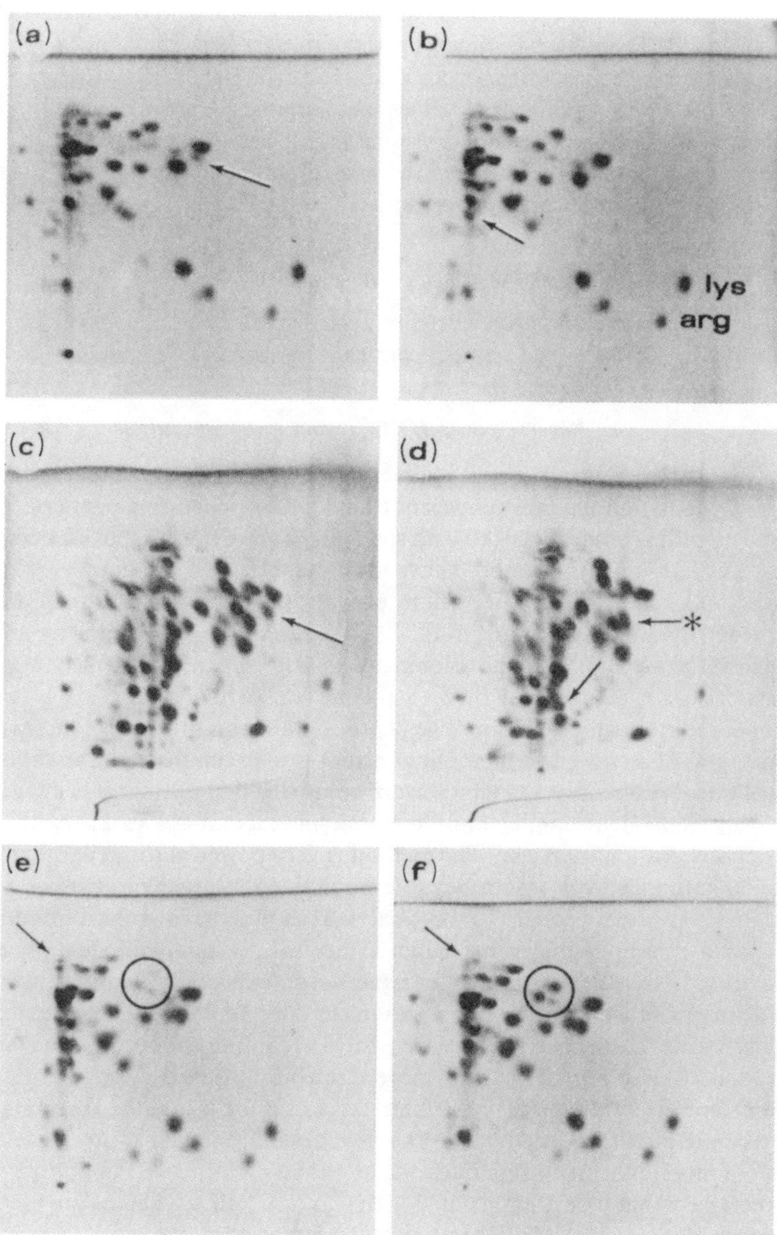
(a)
(b)
lys
arg
(c)
(d)
*
(e)
(f)

7.3.2.1. Test for purity of peptides

The homogeneity of peptides may be tested on large sheets of Whatman 3 MM paper (460 × 570 mm) by two-dimensional separation. In the first dimension use electrophoresis with pyrididine–acetic acid–water (10:100:2800, by vol.) pH 3.6, at 2.2 kV for 80 min. (Other pH values may be used.) Chromatograph in the second dimension with n-butanol–pyridine–acetic acid–water (150:100:30:120, by vol.). Localize spots with fluorescamine or ninhydrin-cadmium reagent. RPHPLC is both more rapid and sensitive.

7.3.3. Peptide mapping by HPLC

There are three different approaches to analytical peptide mapping by HPLC, i.e. reverse phase, ion-exchange and gel permeation separation techniques.

7.3.3.1. Reverse phase HPLC using gradient elution

See Snyder & Kirkland (1979) and Fullmer & Wasserman (1979). This is the most commonly used and the most powerful technique for separating peptides. This method employs columns packed with small (3–10 μm), porous particles coated with alkane or phenyl groups which act as a non-polar stationary phase. Mixtures of small peptides generated by enzymic (Fullmer & Wasserman, 1979; Nelson & Taylor, 1981; Chang *et al.*, 1982) or chemical (Van Der Rest *et al.*, 1980) cleavage are applied to the column in a solvent chosen so that all, or at least the majority, of the peptides bind to the surface of the column packing by hydrophobic interactions. The mobile phase containing a gradient of increasing volume ratio of any organic modifying solvent is then run through the column. The peptides are eluted as extremely sharp zones, the first to emerge being those which are most hydrophilic and therefore interact most weakly with the hydrophobic stationary phase. The method is very powerful for peptides up to about 50 amino acids in length but is less suitable for larger proteins.

In RPHPLC the protein to be analysed should be dissolved in dilute ammonium bicarbonate or other buffer at neutral or slightly alkaline pH and digested with trypsin or another enzyme as described for two-dimensional peptide mapping. When digestion is complete the mixture may be derivatized with chemicals to improve the sensitivity of detection of the components after separation (Chang *et al.*, 1982) or more frequently injected directly onto the column. One disadvantage of pre-column derivatization is that the absorbing or fluorescing group introduced makes the peptides more similar in structure and therefore more difficult to separate.

Generally, column packings coated with C_{18} or C_8 alkane chains are used as stationary phases although other types of coatings are available (Snyder & Kirkland, 1979). The choice of mobile solvent is critical and many different solvent mixtures and chromatographic conditions have been employed for particular separations (for a detailed discussion see Snyder & Kirkland, 1979). There have been attempts to predict the retention times of peptides in a particular solvent system and column packing (Meek, 1980; Browne *et al.*, 1982; Wilson *et al.*, 1981) but more often a qualitative comparison of elution profiles is sufficient to establish the relationship of the samples examined.

A few solvent systems have been widely used. One such system is to equilibrate the column with 0.1% (v/v, 0.013 M) trifluoroacetic acid (TFA) in water (Mahoney & Hermodson, 1980). The peptide mixture is then injected onto the column and a few column volumes of 0.1% TFA is passed through to determine that all or the majority of the sample components have been retained. A linear gradient of acetonitrile in 0.1% TFA is then run through the column to elute the individual peptides. Almost all the peptides which have been examined in this system will elute by the time the acetonitrile concentration has reached 60%. To avoid refractive index changes due to poor mixing of solvent the secondary solvent may be prepared as 60% acetonitrile, 40% water containing 0.1% TFA. The solvents should be as pure as possible since even minor contaminants may be concentrated on the column and then eluted by the gradient. A flow rate of 1 ml/min is generally optimal since slower elution does not increase resolution. At the end of the gradient the column should be gradually equilibrated back to the starting solvent to be ready for the next sample application. Other buffers such as ammonium acetate or sodium phosphate at neutral pH may give different separations.

A practical example of such an approach is the comparison of normal with mutant haemoglobin molecules (Boissel *et al.*, 1981; Sugihara *et al.*, 1981). In one study (Boissel *et al.*, 1981) the β chains of several human haemoglobins were fragmented by tryptic digestion and the resultant peptides separated on a C_{18} column equilibrated with ammonium acetate buffer employing a gradient of acetonitrile. One peptide in each pattern was displaced in relation to normal human haemoglobin (Fig. 7.4). The peptides with unusual retention times were collected for amino acid analysis and sequencing. In each case a single amino acid was found to differ from normal haemoglobin. The separation time by RPHPLC in this experiment was 80 min, thus combining speed and resolution with recovery of the peptide.

7.3.3.2. Ion-exchange HPLC

A second method quite frequently used analytically is IEHPLC where the column packing is coated with ionizable groups which retain small peptides by charge interactions. Elution is performed by increasing ionic strength or changes in pH of the mobile phase sometimes in tandem with increasing volume ratios of organic solvents (Takahashi *et al.*, 1981; Isobe *et al.*, 1982). This technique is best for relatively small molecules although it may be used under some circumstances for separating proteins (Snyder & Kirkland, 1979).

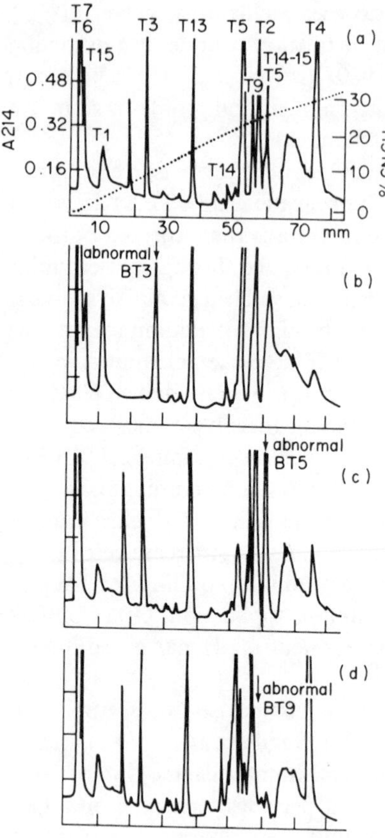

Fig. 7.4. Separation of peptides from tryptic digests (0.8 mg) of human haemoglobin β chains by reverse phase HPLC: (a) normal; (b) haemoglobin Cocody (pI = 7.205); (c) haemoglobin Avicenna (pI = 7.225); (d) haemoglobin Korle Bu (pI = 7.210). (From Boissel *et al.*, 1981, by permission of Elsevier/North Holland Biomedical Press, Netherlands.)

IEHPLC relies upon the same type of ionic interactions employed in classical, open column, ion-exchange chromatography. The sample is applied to a column in low ionic strength solvent with a pH at which the stationary phase and the sample are ionized but possess different charges. A gradient of increasing ionic strength or changing pH is then run through the column to elute the peptides. This exploits the differences in the p*K* values of their ionizable groups to effect a resolution. The advantages of this method over conventional open column procedures are that the components present in the sample are not greatly diluted during the separation and the detection methods are very sensitive. For both RPHPLC and IEHPLC the sample may if necessary be applied in a very large volume since the peptides should adhere to the column until eluted by the gradient.

7.3.3.3. Gel permeation HPLC

The third method of separation is gel permeation HPLC. This has rarely been used for analytical mapping (Tomono *et al.*, 1982) since columns that can resolve complex mixtures of small peptides of less than 5000 daltons are not currently available. There are, however, columns which can separate larger peptides and proteins and these are sometimes used as a purification step or to desalt protein solutions prior to digestion and analytical peptide mapping by RPHPLC or IEHPLC.

7.3.3.4. Detection systems

The detection methods employed for both RPHPLC and IEHPLC peptide mapping are absorbance of light, fluorescence emission and measurement of radioactivity. There are other methods such as change in refractive index and electrochemical detection but these are rarely used (see Snyder & Kirkland, 1979, for a detailed review).

The absorbance of light at a particular wavelength is by far the most common method of detection in HPLC. The only prerequisite for this approach is that the solvent (particularly the secondary gradient solvent) should not absorb significant amounts of light at the chosen wavelength. The TFA-acetonitrile gradient system described above is particularly good in this respect allowing detection down to about 210 nm (Mahoney & Hermodson, 1980). Several wavelengths may be monitored at once so that peptides containing particular groups such as tyrosine and tryptophan may be identified. Detectors are now available which can perform spectral scans of an individual peak at both its leading and trailing edges as it emerges from the column to establish its identity and purity.

If very small amounts of peptide fragments are available they may be derivatized with compounds such as *o*-phthalaldehyde or fluorescamine for fluorescence detection (Hughes *et al.*, 1979). The principle involved is that u.v. of a particular wavelength that excites the fluorogenic reagent is focused on the flow cell while another lens at 90° C to the light source collects the fluorescent energy emitted. The increase in sensitivity over absorption detection may be as much as one hundred times but the method requires an additional pump for post-column derivatization (see Chapter 8, section 8.16.3.1).

Radioactive proteins are commonly used and the radioactive peptides may be monitored by flow-through counters or more often by collection of fractions which are measured in conventional gamma or liquid scintillation counters (Nelson & Taylor, 1981).

7.4. CONCLUSION

The best method for separating peptides currently available is RPHPLC. It is extremely simple to perform, it is rapid, sensitive, particularly with non-radioactive samples, and has very good resolving power. Also, the separated sample components are in solution which facilitates further studies such as amino acid analysis and microsequencing.

7.5. REFERENCES

Boissel, J-P., Wajcman, H., Fabritius, H., Cabannes, R. & Dominique, L. (1981). *Biochem. Biophysica Acta*, **670,** 203–206.
Bonner, W. M. & Laskey, R. A. (1974). *Eur. J. Biochem.*, **46,** 83–88.
Bonner, W. M. & Stedman, J. D. (1978). *Anal. Biochem.*, **89,** 247–256.
Bordier, C. & Crettol-Jarvinen, A. (1979). *J. Biol. Chem.*, **254,** 2565–2567.
Browne, C. A., Bennett, H. P. J. & Solomon, S. (1982). *Anal. Biochem.*, **124,** 201–208.
Bryant, M. L., Nalewaik, R. P., Tibbs, V. L. & Todaro, G. J. (1979). *Anal. Biochem.*, **96,** 84–89.
Burnette, W. N. (1981). *Anal. Biochem.*, **112,** 195–203.
Chang, J-Y., Knecht, R., Ball, R., Alkan, S. S. & Braun, D. G. (1982). *Eur. J. Biochem.*, **127,** 625–629.
Chen-Kiang, S., Stein, S. & Udenfriend, S. (1979). *Anal. Biochem.*, **95,** 122–126.
Christophe, D., Brocas, H. & Vassart, G. (1982). *Anal. Biochem.*, **120,** 259–261.
Christopher, A. R., Nagpal, M. L., Carroll, A. R. & Brown, J. C. (1978). *Anal. Biochem.*, **85,** 404–412.
Cleveland, D. W., Fischer, S. G., Kirschner, M. W. & Laemmli, U. K. (1977). *J. Biol. Chem.*, **252,** 1102–1106.
Cooper, J. A., Reiss, N. A., Schwartz, R. J. & Hunter, T. (1983). *Nature (Lond.)*, **302,** 218–223.
Diezel, W., Kopperschlager, G. & Hoffmann, E. (1972). *Anal. Biochem.*, **48,** 617–620.
Drapeau, G. R., Boily, Y. & Houmard, J. (1972). *J. Biol. Chem.*, **247,** 6720–6726.
Dubray, G. & Bezard, G. (1982). *Anal. Biochem.*, **119,** 325–329.
Eckhardt, A. E., Hayes, C. E. & Goldstein, I. J. (1976). *Anal. Biochem.*, **73,** 192–197.
Einarson, B., Gullick, W., Conti-Tronconi, B., Ellisman, M. & Lindstrom, J. (1982). *Biochemistry*, **21,** 5295–5302.
Elder, J. H., Pickett, R. A., Hampton, J. & Lerner, R. A. (1977). *J. Biol. Chem.*, **252,** 6510–6515.
Feitelson, M. A., Wettstein, F. O. & Stevens, J. G. (1981). *Anal. Biochem.*, **116,** 473–479.
Fishbein, J. C., Place, A. R., Ropson, I. J., Powers, D. A. & Sofer, W. (1980). *Anal. Biochem.*, **108,** 193–201.
Fullmer, C. S. & Wasserman, R. H. (1979). *J. Biol. Chem.*, **254,** 7208–7212.
Furlan, M., Perret, B. A. & Beck, E. A. (1979). *Anal. Biochem.*, **96,** 208–214.
Gershoni, J. M. & Palade, G. E. (1982). *Anal. Biochem.*, **124,** 396–405.
Gershoni, J. M. & Palade, G. E. (1983). *Anal. Biochem.*, **131,** 1–15.
Gibson, W. (1974). *Virology*, **62,** 319–336.
Glass, W. F., Briggs, R. C. & Hnilica, L. S. (1981). *Anal. Biochem.*, **115,** 219–224.
Green, M. R. & Pastewka, J. V. (1975). *Anal. Biochem.*, **65,** 66–72.
Gullick, W. J. & Lindstrom, J. M. (1982a). *Biochemistry*, **21,** 4563–4569.
Gullick, W. J. & Lindstrom, J. M. (1982b). *J. Cell. Biochem.*, **19,** 223–230.
Gullick, W. J. & Lindstrom, J. M. (1983). *Biochemistry*, **22,** 3312–3320.
Gullick, W. J., Tzartos, S. & Lindstrom, J. (1981). *Biochemistry*, **20,** 2173–2180.
Hawkes, R. (1982). *Anal. Biochem.*, **123,** 143–146.
Houmard, J. & Drapeau, G. R. (1972). *Proc. Natl. Acad. Sci. U.S.A.*, **69,** 3506–3509.
Hughes, G. J., Winterhalter, K. H. & Wilson, K. J. (1979). *FEBS Letts.*, **108,** 81–86.
Ingram, V. M. (1956). *Nature (Lond.)*, **178,** 792–794.
Ingram, V. M. (1957). *Nature (Lond.)*, **180,** 326–328.
Isobe, T., Takayasu, T., Takai, N. & Okuyama, T. (1982). *Anal. Biochem.*, **122,** 417–425.
Koch, G. L. E. & Smith, M. J. (1982). *Eur. J. Biochem.*, **128,** 107–112.
Laemmli, U. K. (1970). *Nature (Lond.)*, **227,** 680–685.

Lam, K. S. & Kasper, C. B. (1980). *Anal. Biochem.*, **108,** 220–226.
Lischwe, M. A. & Ochs, D. (1982). *Anal. Biochem.*, **127,** 453–457.
Lonsdale-Eccles, J. D., Lynley, A. M. & Dale, B. A. (1981). *Biochem. J.*, **197,** 591–597.
Mahoney, W. C. & Hermodson, M. A. (1980). *J. Biol. Chem.*, **255,** 11199–11203.
Markwell, M. A. K. (1982). *Anal. Biochem.*, **125,** 427–432.
Meek, J. L. (1980). *Proc. Natl. Acad. Sci. U.S.A.*, **77,** 1632–1636.
Merril, C. R., Switzer, R. C. & Van Keuren, M. L. (1979). *Proc. Natl. Acad. Sci. U.S.A.*, **76,** 4335–4339.
Merril, C. R., Dunau, M. L. & Goldman, D. (1981). *Anal. Biochem.*, **110,** 201–207.
Nelson, N. C. & Taylor, S. S. (1981). *J. Biol. Chem.*, **256,** 3743–3750.
Nikodem, V. & Fresco, J. R. (1979). *Anal. Biochem.*, **97,** 382–386.
Noda, M., Takahashi, H., Tanabe, T., Toyosato, M., Kikyotani, S., Hirose, T., Asai, M., Takashima, H., Inayama, S., Miyata, T. & Numa, S. (1983). *Nature* (*Lond.*), **301,** 251–255.
Oakley, B. R., Kirsch, D. R. & Morris, N. R. (1980). *Anal. Biochem.*, **105,** 361–363.
Parsons, R. G. & Kowal, R. (1979). *Anal. Biochem.*, **95,** 568–574.
Salacinski, P. R. P., McLean, C., Sykes, J. E. C., Clement-Jones, V. V. & Lowry, P. R. (1981). *Anal. Biochem.*, **117,** 136–146.
Schiltz, E., Schnackerz, K. D. & Gracy, R. W. (1977). *Anal. Biochem.*, **79,** 33–41.
Seed, B. (1982). *Nucleic Acids Res.*, **10,** 1799–1810.
Snyder, L. R. & Kirkland, J. J. (1979). *Introduction to Modern Liquid Chromatography*, Wiley, New York.
Sonderegger, P., Jaussi, R., Gehring, H., Brunschweiler, K. & Christen, P. (1982). *Anal. Biochem.*, **122,** 298–301.
Stein, B. S. & Sussman, H. H. (1983). *J. Biol. Chem.*, **258,** 2668–2673.
Stephens, R. E. (1978). *Anal. Biochem.*, **84,** 116–126.
Stephens, R. E. (1981). *Biochemistry*, **20,** 4716–4723.
Sughihara, J., Imamura, T., Imoto, T. & Yanase, T. (1981). *Biochem. Biophys. Acta*, **669,** 105–108.
Symington, J., Green, M. & Brackmann, K. (1981). *Proc. Natl. Acad. Sci. U.S.A.*, **78,** 177–181.
Takahashi, N., Isobe, T., Kasai, H., Seta, K. & Okuyama, T. (1981). *Anal. Biochem.*, **115,** 181–187.
Tijssen, P. & Kurstak, E. (1983). *Anal. Biochem.*, **128,** 26–35.
Tomono, T., Suzuki, T. & Tokunaga, E. (1982). *Anal. Biochem.*, **123,** 394–401.
Towbin, H., Staehelin, T. & Gordon, J. (1979). *Proc. Natl. Acad. Sci. U.S.A.*, **76,** 4350–4354.
Van Der Rest, M., Bennett, H. P. J., Solomon, S. & Glorieux, F. H. (1980). *Biochem. J.*, **191,** 253–256.
Weber, P. & Hof, L. (1975). *Biochem. Biophys. Res. Comm.*, **65,** 1298–1302.
Whittaker, R. G. & Moss, B. A. (1981). *Anal. Biochem.*, **110,** 56–60.
Wilson, K. J., Honegger, A., Stotzel, R. P. & Hughes, G. J. (1981). *Biochem. J.*, **199,** 31–41.
Wold, F. (1981). *Ann. Rev. Biochem.*, **50,** 783–814.

Practical Protein Chemistry—A Handbook
Edited by A. Darbre
© 1986, John Wiley & Sons Ltd

ANDRÉ DARBRE
Department of Biochemistry,
King's College London, Strand,
London WC2R 2LS, U.K.

8

Analytical Methods

CONTENTS

8.1. INTRODUCTION

This chapter deals with analytical techniques which have not been discussed in other chapters in this volume, but which are of general use in protein chemistry. The casual reader may feel that too many alternative methods are given here, when a single method might have been considered sufficient. The author is aware of this possible criticism, but defends himself by pointing out that no method has proved to be perfect for all proteins. In addition, the research may be concerned with widely differing quantities of available protein and the resources in chemicals, apparatus and other laboratory facilities will be equally variable amongst laboratories interested in protein studies. The range of techniques under discussion has had to be limited and for that reason certain topics such as isotachophoresis and chromatofocusing have not been included.

The problems of contamination of samples from extraneous sources are becoming increasingly important as the sensitivity of analytical methods increases and the protein chemist is often constrained to work with very small samples. A survey of sources of contamination in the gas–liquid chromatography of amino acids was shown to include finger prints, dirty glassware, water, hydrochloric acid and latex gloves, as well as the more obvious saliva, skin fragments and cigarette smoke (Rash *et al.*, 1972). More than 150 peptides were detected from a single fingerprint (Marshall, 1984a; see section 8.19.2.2).

The future prospect for very high sensitivity amino acid analysis (< pmol) may lie in the preparation of chemicals and buffers and the manipulation of protein separation methods under aseptic conditions. However, for the present, precautions may be taken with regard to glassware and water, and reagents and chemicals should be of the purest grade.

8.2. GLASSWARE

Glassware (particularly tubes used for hydrolysis) should be cleaned with hot sulphuric acid/nitric acid mixture (3:1, v/v) overnight. Rinse and dry upside down in an air oven at 110° C. Chromic acid is not recommended for cleaning glassware because of residual oxidizing properties which remain with the glass. Prior to use, test tubes or ignition tubes should be heated in an oven at about 550–625° C in a glass vessel under aluminium foil for several hours to destroy any contaminating amino acids. They should be stored in a closed dust-free container.

When working at the μg level, losses of protein by adsorption to the walls of the containing vessel are very important. Such losses may be minimized by treating the surface or by modifying the solvent. Polypropylene or polycarbonate tubes may be better than glass. Thus, tubes may be soaked for 5 min with 2% (v/v) dimethyldichlorosilane in carbon tetrachloride (Hager & Burgess, 1980) or other silylating reagent. Trimethylsilyl chloride in toluene (5% v/v) is efficient. After reaction, drain off the solution, wash the glassware with acetone and dry upside down in the oven at 110°C.

The silanization of glass plates for polyacrylamide gel electrophoresis was described (Neuhoff, 1984; after Radolo, 1980). Glass plates are cleaned with detergents, water and finally acetone. Immerse or spray the plate with 0.2% solution of Polyfix D (Desaga, Heidelberg) in ethanol/water (1:1, v/v). Dry under an infrared lamp at a distance of approximately 20 cm (producing about 60°C). This drying process is important. The silanized glass plates can be stored for several weeks.

Glass may be reacted with 3-aminopropyltrimethoxysilane first and then glutaraldehyde, followed by the addition of BSA and subsequent reduction with sodium borohydride or the glass can be coated with 1% (w/v) solution of PEG 20 000 in deionized water and then dried at 110°C (Aplin & Hughes, 1981).

It is better to modify the solvent than to introduce impurities by pretreating glass and plastic surfaces. Suelter and DeLuca (1983) reduced losses of protein by adsorption to glass and plastic surfaces by adding glycerol (50% final concentration) or Triton X-100 (0.2 mM final concentration) to the solvent. For plastic containers, glycerol or Triton X-100 may be used at low or high ionic strength. For glass, use glycerol at low ionic strength and Triton X-100 at high ionic strength. However, it should be noted that not all proteins will exhibit the same adsorptive properties.

8.3. WATER

For many purposes deionized water should be adequate. The Milli-Q system which incorporates ion-exchange resins and carbon filters provides 18-megohm water which contains less than 50 ppb of organic material and is pyrogen-free (Millipore (U.K.) Limited, Harrow, Middx.).

Triple distillation of water in all-glass apparatus does not remove all contaminants. One method for removing amino acids and ammonia is to pretreat water by boiling with ninhydrin (2 – 4 g/1) before distillation. The method described below uses hypochlorite to destroy amino acids. Excess hypochlorite is then reduced and the water distilled (Hare, 1977).

Method

Add 2 500 ml of water to a 5–1 distilling flask. Add 0.5 g of NaOH (pellets) and

6.5 ml of NaOCl (4–6%, w/v). Reflux overnight. Add 4 ml of conc. H_2SO_4 (caution!) and 2 g of $SnCl_2 . 2H_2O$. Distil into a flask protected by a citric acid trap. Check the water for the presence of chlorine by adding 1 g of KI to 10 ml of water. If a yellow colour appears, add more $SnCl_2$. Collect 2 l of distilled water. It is better to use water as freshly made as possible. See section 8.16.1 for testing purity of citrate buffers (Hare, 1977).

8.4. HYDROCHLORIC ACID

Some brands of commercial HCl contain amino acid contaminants and these can be removed by triple distillation. It is better to store purified HCl in small quantities in sealed glass vessels rather than in a larger vessel which is continually being opened and exposed to contamination. Hydrochloric acid (12 M) may also be prepared by passing HCl gas through a sulphuric acid trap into water until saturated. Protect with a citric acid trap.

Constant boiling HCl (5.7 M) is prepared by distillation of conc. HCl and pure water (9:11, v/v). See section 8.6.1 for its use in protein hydrolysis.

Test for the purity of HCl

Evaporate 1 ml of HCl on a rotary evaporator or under nitrogen gas. Dissolve the residue in 100 μl of pH 2 solution (1 ml of 6 M-HCl in 100 ml of water) and inject 50 μl on the amino acid analysis column. Test for combined amino acid contaminants by heating 1 ml of HCl at 150°C for 25 min. Evaporate and analyse as before. Compare the results against a blank, consisting of 50 μl of pH 2 solution (see Hare, 1977).

8.5. CONCENTRATION OF PROTEIN SOLUTIONS

Methods used for concentrating proteins from dilute solution include precipitation by salts, acids and solvents such as acetone or methanol, ultrafiltration, dialysis, evaporation and lyophilization. The smaller the amount of protein the greater is the problem, e.g. trichloroacetic acid does not precipitate protein at the 1–25 μg level (Bensadoun & Weinstein, 1976).

Proteins are efficiently concentrated by electrophoresis using a steady-state packing gel (Kapadia & Chrambach, 1972; Chrambach & Nguyen, 1978; Wachslicht & Chrambach, 1978; Nguyen & Chrambach, 1979; Nguyen *et al.*, 1980) or by isotachophoresis (Öfverstedt *et al.*, 1981). The use of electrophoresis-sedimentation for the concentration of macromolecules (Posner, 1976) or electrophoresis to remove macromolecules from non-ionic solutions (Allington *et al.*, 1978) involves membrane barriers which may adsorb macromolecules irreversibly. This problem was met by using liquid membrane barriers (Wagner *et al.*, 1983, 1984; *vide infra*). Proteins in dilute solution (100–800 ng/ml) were determined after precipitation and radiolabelling with [^{3}H]-1-fluoro-2,4-

dinitrobenzene (Nishio & Kawakami, 1982). Carboxymethyldextran was used to displace several proteins from DEAE-cellulose in highly concentrated form (Torres & Peterson, 1979).

Lyophilization is the method most commonly used for concentrating protein solutions. Use a solid carbon dioxide-acetone (ethanol or isopropanol) bath to freeze the solution and an oil pump protected by a cold trap to remove water and volatile buffers. An alkali trap is required if acids are being removed. Strong formic acid solutions should be diluted with water to about 30% concentration before freezing.

The Speed-Vac Concentrator (Savant Instruments, Inc.) may also be used for the rapid removal of water, organic solvents and volatile buffers from the protein solution, under vacuum in a low speed centrifuge with variable temperature-controlled conditions. Different sized tubes may be used. The method is particularly useful for concentrating eluate fractions from chromatography columns or amino acid sequencing instruments. It is quicker and more convenient than evaporation of individual samples with a stream of nitrogen.

Protein residues may be difficult to dissolve, particularly after extraction with fat solvents or after drying at high temperatures. If the protein is present as a thin film, use 1 M-NaOH for 1 h at room temperature, or heat 1 M-NaOH at temperatures up to 100°C for 10 min or longer. Aqueous hexafluoroacetone (Burkhardt & Wilcox, 1967; Cromwell & Stark, 1969) and trifluoroacetic acid (Mahoney & Hermodson, 1980) are also good solvents. Insoluble structural proteins such as collagen and keratin were treated with H_2O_2 and they became easily soluble in 0.05% SDS-1 M-NaOH at 100°C for their determination by the biuret reaction (Goshev & Nedkov, 1979).

Care must be taken when freeze-drying under high vacuum solutions containing single amino acids or small peptides (as well as other types of compounds), particularly when micro-quantities are involved. There is a danger of serious losses which could invalidate quantitative studies. It was shown with freeze-drying for 90 min using a vacuum between 5 and 10 μm that 89% of 2-[^{14}C]-L-phenylanine and 86% of *N*-feruloyl glycyl-L-phenylanine were lost from the cold (EtOH-CO_2) glass vessel (Van Sumere *et al.*, 1983).

Dialysis may be used to remove non-volatile buffers, salts, performic acid, etc. with Visking tubing. There is also commercially available a wide range of cellulose filters with molecular weight cut-off values ranging from 1 000 to 50 000 (Spectrapor membrane from Spectrum Medical Industries Inc.). Protein solutions may be concentrated using these membranes in conjunction with ultrafiltration apparatus. Adsorption of proteins to dialysis tubing is a danger (see Chapter 6, section 6.2.3).

Peptides dissolved in large volumes of sodium acetate buffer, pH 7.5, containing 4 M-urea are quickly concentrated by passing them down a Lichrosorb C_8 reverse-phase column. Monitoring the column effluent with acetate or formate present is best with fluorescence detection. Peptides may also be desalted from urea and buffer constituents by using mini-columns (Nice *et al.*, 1979) or Sep-pak (Waters) cartridges (Bennett *et al.*, 1980; Waterfield & Scrace, 1981; see also Chapter 6, section 6.2.3).

Using a Centricon microconcentrator small samples (up to 2 ml) may be concentrated down to 25–40 μl in 30–60 min using low-adsorptive Centricon membranes in a centrifuge with a fixed-angle rotor. Desalting or buffer exchange may be carried out simultaneously with concentration (Amicon/Scientific Systems Division, Cherry Hill Drive, Danvers, MA 01923, U.S.A.).

Because SDS is a denaturing agent it may be necessary to remove it completely. Electrophoresis and anion-exchange resins often lead to low protein recoveries. The extraction of SDS with high protein recoveries in the μg and mg range (70 to 100 %) was achieved by means of ion-pair extraction with triethylammonium or tributylammonium cations (Henderson *et al.*, 1979).

8.5.2. Methods

The following examples were selected to show protein concentration by different methods.

8.5.2.1. Quantitative recovery of protein in dilute solution

This is the method of Wessel and Flügge (1984).

Add 0.4 ml of methanol to 0.1 ml of protein solution. Vortex mix and then centrifuge 10 s at 9 000 g. Add 0.1 ml of chloroform (0.2 ml where the sample contains a high concentration of phospholipid). Vortex mix and centrifuge 10 s at 9 000 g. Add 0.3 ml of water, vortex mix and centrifuge 1 min at 9 000 g. Remove the upper layer and discard. Add 0.3 ml of methanol to the lower chloroform phase. Mix and centrifuge 2 min at 9 000 g. Remove the supernatant and dry the protein pellet under a stream of air.

For protein determination dissolve in 50–100 μl of SDS (5 %, w/v) or dissolve directly in 50 μl of electrophoresis buffer containing 5 % SDS for electrophoresis.

The method can be used for both soluble and hydrophobic proteins and is not affected by the presence of detergents, lipids, salts, buffers and 2-mercaptoethanol.

Recoveries were of the order of 92–100 % for amounts of protein 40–120 μg in 0.1 ml of original solution.

8.5.2.2. Electrophoresis using liquid membrane barriers

The method of Wagner *et al.* (1983, 1984) is described. Electrophoresis with a non-linear pH gradient was used to concentrate proteins carrying different charges. It was faster than isoelectricfocusing. Two successive layers, one of an amino acid (histidine or valine) and one of carrier ampholytes (Servalyte), acted to provide a liquid membrane as a barrier to the proteins. Thus, 200 ml of ribonuclease solution was concentrated 48.3-fold in 4 h, with 96.6% recovery (Wagner *et al.*, 1984). A potential gradient (400 V, 5 h) was applied to a system consisting of Pharmalyte (pH 2.5–5.0), histidine (pI 7.64) and Ampholyte (pH 9–11). The acidic and basic isoenzymes of horseradish peroxidase (pI 4.0, 8.4) accumulated at the two interfaces (Wagner *et al.*, 1983).

8.5.2.3. Concentration of μg quantities of protein

This is the method of Mahuran *et al.* (1983). The concentration of μg quantities of protein from dilute solution by precipitation with trichloroacetic acid has been improved by the addition of deoxycholate in the method given here. Detergent, which is often used when isolating membrane proteins, is subsequently extracted.

(a) Precipitation of protein

1. Add 120 μl of sodium deoxycholate (2% w/v) to the protein solution to give a concentration of 80 μg/ml. Mix. Stand on ice for 30 min.
   The detergent concentration may vary from 50–500 μg/ml of sample without affecting the recovery.
2. Add 10 ml of cold trichloroacetic acid (24% w/v) to give a final concentration of 6% (w/v). Mix. Allow to stand on ice for 1 h for complete precipitation of the protein-detergent complex.
3. Centrifuge at 2400 g at 4°C in a swing-bucket rotor for 45 min.
4. Carefully remove the supernatant by suction.
5. Dissolve the pellet in 0.5 ml of 62.5 mM Tris-HCl, pH 6.8, containing 3% SDS made 0.5 M in $NaHCO_3$ (final pH 8.8).
6. Dialyse the solution overnight at room temperature against 21 of 6.25 mM Tris-HCl, pH 6.8, containing 0.3% SDS.
7. Lyophilize the solution in 7-ml screw-top vials or tubes.

(b) Extraction of detergent

1. Add to the dried sample 1 ml of extraction solution consisting of acetone/glacial acetic acid/triethylamine (90:5:5, by vol.).
2. Quickly transfer the solution to a tapering centrifuge tube.
3. Stand on ice for 1 h to precipitate the protein.
4. Centrifuge on a desk-top centrifuge at room temperature.
5. Remove the supernatant by suction.
6. Wash the pellet once with 1 ml of extraction solution and then with 1 ml of acetone.
7. Dry under nitrogen. Yields 40–80%.

8.6. TOTAL HYDROLYSIS OF PROTEINS FOR AMINO ACID ANALYSIS

The hydrolysis of proteins by acid or base leads to the loss of some amino acids and amino acid derivatives. The most general method for total hydrolysis is the use of 6 M-HCl for 18–24 h at 110°C (Moore & Stein, 1963). However, tryptophan, asparagine, glutamine, glycosides, carboxylate, sulphate and phosphate esters are destroyed and there are partial losses of serine, threonine and tyrosine. Also, cystine and methionine are either partially destroyed or oxidized to cysteic acid or methionine *S*,*S*-dioxide, respectively. Peptide bonds involving valine, isoleucine and leucine are often difficult to hydrolyse and it may be necessary to hydrolyse samples for 48, 72, 96 or even 120 h. The rate of liberation and destruction of individual amino acids depends largely on the individeal protein and also on the presence of salts or metals (as in metallo-proteins). It is usual to extrapolate graphically back to zero time for serine, threonine and other amino acids being destroyed with progressive hydrolysis and to infinite time for branched amino acids (Tristram & Smith, 1963). However, after a 24-h hydrolysis period an *ad hoc* assumption may be made that 10% serine and 5% threonine are destroyed.

Methionine is subject to oxidation to methionine *S*-oxide during hydrolysis (and also during amino acid analysis) and mercaptoethanol should be added to 6 M-HCl. Tyrosine may be converted to 3-chloro- or 3-bromo-tyrosine during HCl hydrolysis and this may be prevented by adding phenol.

The hydrolysis of proteins for amino acid analysis is discussed in more detail elsewhere (Light & Smith, 1963; Moore & Stein, 1963; Eveleigh & Winter, 1970; Roach & Gehrke, 1970; Savoy *et al.*, 1975; Blackburn, 1978a; Perham, 1978).

8.6.1. Methods of acid hydrolysis

8.6.1.1. Hydrochloric acid

Use constant boiling HCl (see section 8.4) or best quality commercial 6 M-HCl, and add 0.02% (v/v) mercaptoethanol and 0.25% (w/v) phenol. Bubble N_2 through the acid for about 5 min to remove any dissolved oxygen.

Preparation of sample for hydrolysis

Pyrex test tubes or ignition tubes (which may be made by sealing off short lengths of Pyrex tubing) should be pre-cleaned and heated as described (section 8.2).

A sample (1–5 mg) of air-dried or lyophilized protein is weighed into a tube. More usually, with smaller amounts, a known volume of protein solution is lyophilized directly in the tube. The sample should be salt-free and proteins should be in volatile buffer solutions.

Add 6 M-HCl to the protein (1 ml for 5 mg of protein: 10–200 μl for smaller quantities) in the tube. Rotate the tube at an angle of about 45° with the fingers while allowing a gas–oxygen flame to play on the glass at the upper end of the tube. Allow the glass to thicken and draw a constriction to about 1 mm bore. (If necessary a small glass rod may be attached to the lip of the tube by heating and the constriction may then be drawn with less danger of burning the fingers.) See Fig. 8.1.

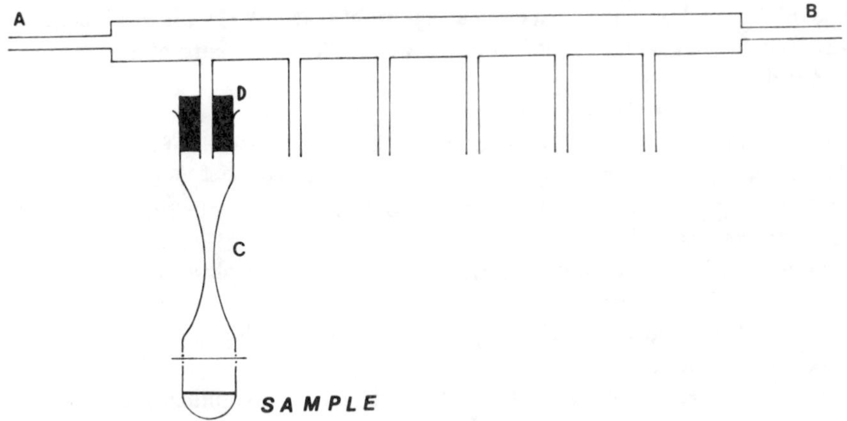

Figure 8.1. Manifold for evacuating tubes for hydrolysis. A steel manifold has a main body with 20 mm outer diameter with side arms 6 mm outer diameter set about 65 mm apart. Three-way taps can be attached by pressure tubing at positions A and B. The hydrolysis tubes with a constriction (C) are fitted onto silicone rubber sleeves (D) (see text).

Freeze the tube contents in a solid CO_2–acetone bath and then attach the tube to a vacuum line through a short length of Tygon or rubber tubing, with an oil pump evacuating to below 50 μm Hg. With a number of samples, it is convenient to use a steel or glass manifold, as shown in Fig. 8.1. The tubes fit on to the rubber sleeves (D) of the side arms. A three-way tap at A enables the vacuum to be applied or released.

It is often considered that 3–5 min of evacuation is sufficient. The samples should be kept under observation. If there is any rising of the mass in the tube, or frothing, apply heat briefly to the upper part of the tube, when it will be seen to subside. If the above procedure is not successful, momentarily open tap A to release the vacuum and then close the tap. This may have to be repeated several times. By having a second three-way tap at B, connected to a nitrogen cylinder, it is possible to alternate evacuation through A with nitrogen purging through B. This method rigorously excludes oxygen. Some workers allow the protein sample to melt in order to ensure that any dissolved oxygen may be removed.

When evacuation is complete, with the vacuum pump still operating, hold the tube near the bottom with one hand and with the other apply a small oxygen flame to the constriction. Allow the constriction to collapse and at the same time twist slightly to seal off and remove the tube.

A method was described for preparing in about 1 h up to 12 samples for acid or alkali hydrolysis in a vacuum desiccator (Phillips, 1981).

Hydrolysis conditions

The tube should be placed in a circulating-air oven at 110° C with good temperature control. This is important (Moore & Stein, 1963). A gas chromatograph oven is useful.

After hydrolysis for the appropriate period of time (24, 48, 72 h), gentle centrifugation may be used to spin down the liquid on the walls of the tube. Score the tube with a file. Crack open by moistening the file mark and applying the white-hot tip of a length of glass rod. Fire-polish the rim of the tube. Unopened tubes are best stored at −20° C.

Some proteins, e.g. elastin are extremely difficult to hydrolyse and may require longer periods of hydrolysis.

It is best to remove the HCl as quickly as possible, using a stream of N_2, the Speed-Vac Concentrator or the rotary evaporator.

When there are many hydrolysates it is convenient to lyophilize the samples in batches. Cover the tubes with parafilm and make pin holes over each tube. Freeze the hydrolysates in an acetone–solid CO_2 bath. Place the samples in a vacuum desiccator over NaOH or KOH (flake is better than pellet) and lyophilize with an oil pump protected by an alkali trap, usually overnight. Although this is common practice, Moore & Stein (1963) reported on further losses of threonine and serine and the appearance of artifact peaks on the amino acid chromatogram, as a result of slow removal of HCl.

A method was described for the rapid removal of HCl from large numbers of protein hydrolysates without frothing, involving the use of a centrifuge (Kaldy & Kereliuk, 1978). See section 8.5 for Speed-Vac Concentrator.

It is advisable to filter protein hydrolysates through a Millipore membrane (0.45 μm) before applying an aliquot to the amino acid analyser.

8.6.1.2. Rapid hydrolysis method

A rapid hydrolysis method for peptides and resin-attached peptides was reported (Westall & Hesser, 1974). They used propionic acid/12 M HCl (1 : 1, v/v) *in vacuo* at 160° C for 15 min. Where serine analysis was important the conditions were 150° C for 15 min.

8.6.1.3. Hydrolysis without gas-oxygen flame

For the occasional hydrolysate and where an oxygen flame is not available a Rotaflo stopcock with the lower limb sealed off may be used (Darbre, 1971). The Teflon plunger limits the temperature to 110° C. Vacuum reaction tubes of this type are available (Pierce).

Hydrolyses may also be carried out under nitrogen atmosphere. Weigh the sample in a small screw-cap vial fitted with a Teflon liner (Pierce). If in solution, the sample can be dried down in the vial. Add 6 M-HCl (100–500 μl). Flush with nitrogen and cap immediately. Hydrolysis can be carried out in a heated aluminium block at 155° C for 20 min. This short hydrolysis time is claimed to give similar results to those obtained at 110° C for 22 h. Longer hydrolysis times may also be used (Hare, 1977). Ultrasonication also removes dissolved air while the sample is flushed with nitrogen (Kaiser *et al.*, 1974).

It was reported that hydrolysis with 6 M-HCl in vials under nitrogen at 145° $\pm$ 2° C for 4 h gave equivalent results for ribonuclease to those obtained at 110° $\pm$ 1° C for 24 h, but it was emphasized that plots of yield versus hydrolysis time must be made for each amino acid to obtain the best values for the amino acid composition of a protein (Roach & Gehrke, 1970; see also Kaiser *et al.*, 1974).

8.7. TRYPTOPHAN

Tryptophan is mostly destroyed during the usual HCl hydrolysis conditions, particularly in the presence of carbohydrate. Attempts to overcome this problem have been made by adding thioglycollic acid to HCl, or by using *p*-toluene sulphonic acid or methanesulphonic acid or mercaptoethanesulphonic acid (see Methods in section 8.7.1).

Alkaline hydrolysis is the method of choice where tryptophan results are paramount but molar calculations must be based on the results of an independent acid hydrolysate (Hugli & Moore, 1972). Alkaline solutions used include barium hydroxide (Robel, 1967), NaOH (Hugli & Moore, 1972) and LiOH (Lucas & Sotelo, 1980).

5-Methyl tryptophan is a useful internal standard in alkaline hydrolysis and its recovery can be used to correct for Trp destruction (Wilkinson *et al.*, 1976), but it may coelute on some columns with lysinoalanine which is generated during alkaline hydrolysis (Ziegler *et al.*, 1967).

8.7.1. Tryptophan by alkaline hydrolysis

8.7.1.1. Methods

Method 1: NaOH (Hugli & Moore, 1972)

Reagents

Commercial partially hydrolysed starch (50 g) washed with acetone and dried is used as the anti-oxidant.

1 % octan-1-ol in toluene, to suppress foaming.

50 % NaOH (w/v). Dilutions are made freshly.

Method

The protein (1–5 mg) is placed in a polypropylene centrifuge tube (109 × 5.0 mm). Add 25 mg of starch, 0.5 ml of 4.25 M-NaOH (or 0.1 ml of protein solution and 0.5 ml of 5 M-NaOH) and 5 μl of 1 % octan-1-ol. Place in a Pyrex tube and make a constriction in the upper part of the tube without heating the plastic liner. Cool the lower part of the tube in an acetone–solid CO_2 bath to chill, but not to freeze, the contents. Evacuate the tube under vacuum below 50 μmHg and seal. Repeated tapping of the tube and momentarily breaking the vacuum and cooling of the tube may be needed to control the foaming. Heat at 110° C for 16 or 24 h normally, or up to 96 h. If Ile-Trp or Val-Trp bonds are present hydrolyse at 135° C for 48 h.

Open the cooled tube with minimum aperture. Add 0.5 ml of 0.20 N-sodium citrate buffer, pH 4.25 without Brij 35. Transfer the solution quantitatively by Pasteur pipette with washings to a 5.0 ml volumetric flask containing 420 μl of 6 M-HCl cooled in solid CO_2. Make up to volume with citrate buffer.

Comments

Tryptophan may be analysed on a short cation-exchange column (80 × 9 mm) (see Moore *et al.*, 1958) with Na^+ citrate buffer, pH 5.4. Neutral and acidic amino acids come off together in 20 min, tryptophan at 35 min, lysinoalanine (formed during alkaline hydrolysis) at 53 min, followed by Lys (65 min), His (80 min) and NH_3 (95 min). The reason for avoiding the usual buffer at pH 2.2 for the analysis is that 10 % of Trp is lost at 25° C. Integral values of 100 ± 3 % were reported.

Method 2: LiOH (Lucas & Sotelo, 1980)

Hydrolyse the protein in a Pyrex screw-capped tube with Teflon liner. Use 1 ml of 4 M-LiOH per 25 mg of sample. Flush the tube with nitrogen gas, freeze in acetone–solid CO_2 and tightly close. Hydrolysis at 110° C for 20 h or 145° C for 4 h gave no significant difference in the results ($p < 0.05$). Cool and neutralize the hydrolysate with orthophosphoric acid (85 %). Bring to a known volume. An aliquot is diluted with 0.2 M phosphate buffer, pH 7.0 for analysis.

Comments

Extrapolation to zero time indicated that about 5 % of Trp was destroyed with 4–8 h hydrolysis at 145° C. These workers compared alkaline hydrolysis methods and found LiOH to be best. However, this method is relatively new and untried.

8.7.2. Tryptophan by acid hydrolysis

8.7.2.1. Methods

Method 1: HCl and thioglycollic acid (Matsubara & Sasaki, 1969)

Hydrolyse the protein with freshly prepared 6 M-HCl containing 4 % thioglycollic acid at 108–110° C for 24–64 h. The yield for Trp is about 90 %. Results for other amino acids are unaffected, except proline. Carbohydrate is tolerated in small amount. When 50-molar excess of glucose was present in the protein, no Trp was recovered.

Method 2: *p*-Toluene sulphonic acid (Liu & Chang, 1971)

p-Toluene sulphonic acid monohydrate is recrystallized from HCl-H_2O and freed of all HCl by heating *in vacuo* at 45° C over NaOH pellets. (Test by means of 1 % $AgNO_3$ in 50 % HNO_3.)

Hydrolyse 2–4 mg of protein with 1 ml of 3 M-*p*-toluene sulphonic acid containing 0.2 % (w/v) of 3-(2-aminoethyl)-indole *in vacuo* at 20–30 μm at 110° C for 22, 48 or 72 h. Cool, add 2.0 ml of 1 M-NaOH and make up to 5 ml. Filter and take an aliquot for analysis. Trp and other amino acids may be determined. The limit of carbohydrate is 2 mg.

Method 3: Mercaptoethanesulphonic acid (Penke *et al.*, 1974)

Hydrolyse the protein with 1 ml of 3 M-mercaptoethanesulphonic acid at 110° ± 2° C for 24 and 72 h. Cool, add 2 ml of M-H_2SO_4 and transfer quantitatively to a 5.0 ml flask.

Comments
The method is better than method 1 and about equal to method 2. Recovery for Trp is 96 % after 22 h and 91 % after 72 h for chymotrypsin. The method is recommended for pure protein only.

Method 4: Methanesulphonic acid (Simpson *et al.*, 1976)

Reagent
4 M-Methanesulphonic acid is prepared by diluting conc. acid. Add solid 3-(2-aminoethyl)-indole to give 0.2 % w/v. Flush with nitrogen and store at 4° C. Also obtainable from Pierce.

The protein (0.2–2.0 mg) is hydrolysed *in vacuo* (20 μm) at 115° C for 22, 48 and 92 h with 1 ml of 4 M-methanesulphonic acid. Use 0.3 ml for samples less than 1.0 mg. Partially neutralize with 3.5 M-NaOH. Make up to a known volume. Centrifuge or filter for amino acid analysis.

Comments
Losses of Trp (3.2 %), Thr (5.9 %) and Ser (9.6 %) were less than those obtained in method 2. However, when the carbohydrate content is more than 20 %, Trp losses are great and cannot be accurately determined. See Inglis (1983) for the determination of tryptophan and other amino acids.

8.7.3. Tryptophan in the intact protein

Tryptophan in the intact protein may be determined spectrophotometrically (Goodwin & Morton, 1946; Edelhoch, 1967; Spande & Witkop, 1967), fluorometrically (Shelton & Rogers, 1971) or after conversion to a coloured derivative (Spies & Chambers, 1949; Barman & Koshland, 1967; Scoffone *et al.*, 1968; Basha & Roberts, 1977). Magnetic CD at 293 nm (Barth *et al.*, 1972) and a scintillation method using 3-diazonium-1,2,4-(5-^{14}C) triazole (De Traglia *et al.*, 1979) were used with a limited number of proteins. Tryptophan was determined in sub-pmol quantities by a modification of the Pictet-Spengler reaction by boiling it with potassium ferricyanide and formaldehyde to form 9-hydroxymethyl-β-carboline which was quantified by spectrofluorometry or reverse phase HPLC (Inoue *et al.*, 1983).

Amongst amino acids in proteins, tryptophan is one of the smallest in amount and therefore its accurate determination enables the precise mol. wt. of the protein to be obtained.

8.7.3.1. Determination with 2-hydroxy-5-nitrobenzyl bromide

The method of Barman and Koshland (1967) is described.

Reagents

Urea recrystallized from ethanol–water.
2-Hydroxy-5-nitrobenzyl bromide (HNB-Br).
Trichloroacetic acid (TCA), 50%.

Method

The protein to be analysed is incubated at 37° C for 16–20 h in 1 ml of 10 M-urea which had been adjusted to pH 2.7 with conc. HCl. Cool to room temperature. Add 5 mg of HNB-Br in 0.1 ml of dry acetone by gravity feed from a 0.1 ml pipette kept immersed in the protein while stirring vigorously by magnetic stirrer or Vortex mixer. Any ppt which forms (mainly HNB-hydroxide) is removed by centrifuging.

Pass the labelled protein down a coarse Sephadex G-25 column (230 × 11 mm) previously equilibrated with 0.18 M-acetic acid (pH 2.7) or with 10 M-urea (pH 2.7) to keep the protein in solution. Elute at about 100 ml/h. The protein fractions (1–4 ml) are pooled. Precipitate the protein with 50% TCA to give a final concentration of 5% TCA. If urea is used, dilute 5-fold with water before adding the TCA. Collect the precipitate by centrifugation after 30 min or after standing overnight at 4° C. Wash twice with 5 ml of ethanol–HCl solution (2 ml conc. HCl + 98 ml 95% ethanol). Finally dissolve in 1 ml of conc. HCl. Take 0.1 ml aliquot and adjust to pH > 12 with 2.5 M-NaOH. Make up to 2.5 ml total volume. Read at

410 nm. (ε 18000/M/cm). Further aliquots may be hydrolysed with 6 M-HCl for amino acid analysis.

8.8. AMIDE RESIDUES

Asparagine and glutamine are converted to aspartic and glutamic acids, respectively, under the usual protein hydrolysis conditions using 6 M-HCl.

The total of asparagine and glutamine may be determined by the ammonia released under selected acid-catalysed hydrolysis conditions, such as conc. HCl at 37° C for 10 days or 2 M-HCl at 100° C for 2–8 h. Under these conditions there is insignificant interference due to deamidation of serine, threonine, cystine and tryptophan. An aliquot of the hydrolysate is made alkaline in a Conway microdiffusion apparatus and after overnight diffusion the ammonia is determined by a ninhydrin photometric method (Wilcox, 1967). Glutamine is more easily deamidated than asparagine (Blombäck, 1967). Asparagine and glutamine may be determined separately by chemical (Wilcox, 1967) and enzymic (Tower, 1967) procedures. See review by Balis (1971).

Proteins are often contaminated with free ammonia even after dialysis. Purification may be achieved by passing a solution of protein down a column of cation-exchange resin in the H^+ form.

8.8.1. Methods for amide determination

8.8.1.1. Diffusion method

This is the method of Wilcox (1967).

Control value

Dissolve 10 mg of protein in 1.0 ml of 0.03 M-methanolic HCl and precipitate at once with 2 ml of ethyl ether. Centrifuge. Repeat the extraction twice more. Collect all the extracts and take to dryness *in vacuo*. Transfer the residue quantitatively to the outer chamber of a Conway diffusion vessel. Place 1 ml of 0.02 M-H_2SO_4 in the centre well. Carefully and quickly make the sample residue alkaline with 1.0 ml of a saturated solution of sodium tetraborate adjusted to pH 10.5 with 2 M-NaOH. The lid must be well sealed with a weight on top. Allow diffusion to take place overnight at room temperature. Take the contents of the centre well quantitatively, make up to a known volume with water and determine ammonia by a ninhydrin photometric method.

Amide nitrogen in protein

Incubate 2 mg protein/ml 2 M-HCl at 100° C for 2–8 h. Cool. Centrifuge if necessary. Transfer aliquots 100–250 μl to the outer chambers of Conway vessels, as for free ammonia, but use more concentrated borate buffer (15 g boric acid in 100 ml of 2 M-NaOH). Take approx 0.2 μmol NH_3 for each assay. Plot NH_3/ time and determine intercept at zero time.

8.8.1.2. Diffusion method with trinitrobenzenesulphonic acid

The method of Whitaker *et al.* (1980) is described.

Reagents

2,4,6-Trinitrobenzenesulphonic acid (TNBS) (Sigma or BDH) recrystallized once from 1 M-HCl/water (Fields, 1972).
TNBS solution (1.1 M) 100 mg/200 μl of water prepared freshly and kept frozen in the dark.
Ammonium sulphate (analytical grade, Merck) dried 3 h at 100° C and stored in a desiccator over silica gel.
Boric acid, 0.1 % (w/v)
$Na_2B_4O_7$ saturated aqueous solution

A. 0.1 M-Na_2SO_3 (made weekly)
B. 0.1 M-NaH_2PO_4
C. 0.1 M-$Na_2B_4O_7$ in 0.1 M-NaOH
D. 1.5 ml of A + 98.5 ml of B (prepare daily)

Method

Hydrolyse the protein with 10 M-HCl at 37°C for 10 days. In the outer well of a Conway diffusion dish place the hydrolysed protein sample with water to a total volume of 1.5 ml. In the centre well place 0.9 ml of 0.1 % boric acid. Grease the lid with Gliseal. Raise the lid slightly and add quickly 3 ml of saturated sodium tetraborate solution adjusted to pH 10.5 with 2 M-NaOH. Close the lid and put a weight on top. After 20 h at room temperature quantitatively transfer the contents of the centre well to a tared test tube and make up to a final weight of solution of 1.5 g. This allows three analyses by the end-point method of Fields (1972) described below. S = 6 nmol. Water and known amounts of ammonium sulphate are determined at the same time.

End-point method of Fields

Add the sample to 0.5 ml of solution C and make up to 1.0 ml. Add 0.02 ml of TNBS solution. Mix rapidly. After 5 min, stop the reaction with 2.0 ml of solution D. Read at 420 nm.

ε (TNP-α-amino groups)	22 000/M/cm
ε (TNP-ε-amino groups)	19 200/M/cm
ε (TNP-thiol groups)	2 250/M/cm

8.8.1.3. Amino acid analyser

The method of Inglis *et al.* (1974) is described. Hydrolyse the protein (2–3 mg) with 1 ml of hydriodic acid (constant boiling containing 0.03 % of hypophosphorous acid) (BDH-MAR) in sealed tubes *in vacuo* at 107°C in a forced air-draught oven for 1.5 h for soluble proteins or 2 h for insoluble proteins. Alternatively use 2 ml of 2 M-HCl at 110° C for 2 h and 14 h. Dilute with 5 ml of 0.1 M-HCl and load 1 ml onto an amino acid analyser column (Beckman column (75 × 9 mm i.d.) with Aminex A-5 resin). Determine the ammonia peak against a blank value without hydrolysis.

Using 1.5 mg protein for hydrolysis and 2.5 mg for a blank determination, a single amide residue/mol protein mol. wt. 7 000 was established. Hydriodic acid for hydrolysis and distillation by the micro-Kjeldahl method gave comparable results. The analytical column should have a Teflon disc with about 1 cm of resin at the top of the column to remove unhydrolyzed protein. This extra resin should be removed before adding NaOH to the column. Discarded resin can be cleaned by heating at 70°C with 12 M-sulphuric acid.

8.8.1.4. Bis(1,1-trifluoroacetoxy)iodobenzene

The method of Soby and Johnson (1981) is described.

Reagent

Bis(1,1-trifluoroacetoxy)iodobenzene (BTI) may be prepared (Parham & Loudon, 1978). Dissolve 5.0 g of phenyliodosyl acetate (Aldrich) in 25 ml of anhydrous TFA with slight heating and allow to stand at room temperature for 1 h. Remove the solvent *in vacuo*. Recrystallize from hexane-TFA. M.P. 123–126° C. Store in a dark bottle at −20°C.

Method

Carboxamide degradation. In a Pyrex test tube, mix 200 μg of protein in 2 ml of 0.01 M-TFA or 5 M-guanidine HCl −0.01 M-TFA with 2 ml of freshly prepared BTI in dimethylformamide (36 mg/ml). Seal *in vacuo* (1–2 Torr.). Heat at different temperatures (35°, 60°, 110° C) for increasing periods of time (4 and 18 h). Remove excess reagents by dialysis against water (Spectrapor 3 dialysis membrane). Extract the dialyzed sample three times with an equal volume of n-butyl acetate. Lyophilize. Hydrolyse with 6 M-HCl *in vacuo* at 110° C for 24 h.

Asparagine is converted quantitatively to 2,3-diaminopropionic acid and glutamine to 2,4-diaminobutyric acid. These acids are normally not resolved from lysine on the amino acid analyser, therefore the amide content of the protein is determined by difference analysis of aspartic and glutamic acid determined with and without prior treatment with BTI. Some amino acids (Cys, His, Lys, Met, Thr) are affected by treatment with BTI.

8.8.1.5. Enzymic hydrolysis and gas-liquid chromatography

The method of Hediger *et al.* (1973) is described.

Enzymic hydrolysis

Digest 0.2 μmol of peptide in 0.2 ml of 0.05 M-NH_4 HCO_3 buffer, pH 8.0 with 1 % (w/w) of Pronase for 24 h at 37° C. Follow with further digestion with 4 % (w/w) of aminopeptidase M for 17 h. Lyophilize.

Gas–liquid chromatography

Derivatize asparagine and glutamine by esterifying with 0.2 ml of 1.25 M-HCl in butan-1-ol at 100°C for the critical period of 7 min in a Teflon-lined screw-cap vial. Cool in ice and take to dryness with a stream of nitrogen. Acylate with 0.1 ml of trifluoroacetic anhydride and methylene dichloride (1 : 3, v/v) and heat for 20 min at 100° C. A suitable internal standard is n-butyl stearate, with 0.325 % (w/w) ethylene glycol adipate on heat-treated Chromosorb G on a column 1.5 m × 4 mm (Gehrke & Stalling, 1970). Other amino acids may be determined by GLC at the same time. Esterification under the above conditions is critical because of the partial conversion of the amides to aspartic acid and glutamic acid.

8.9. SULPHUR-CONTAINING AMINO ACIDS

8.9.1. Methods of determination

8.9.1.1. Cysteine and cystine

This method (Moore, 1963) determines cysteine plus cystine as cysteic acid. In addition, methionine may be determined as methionine *S,S*-dioxide. Carbohydrate-containing samples may be used. Some amino acids such as histidine, tryptophan and tyrosine are destroyed (see also Hirs, 1967).

Reagents

Performic acid: Add 1 ml of 30 % H_2O_2 to 9 ml of 88 % formic acid. Let the mixture stand for 1 h at room temperature. Cool to 0°C.

Standard solution

Dissolve 20 mg of cystine, 30 mg of methionine and 20 mg alanine in 2 ml of 1 M-NaOH and make up to 10 ml with water. Use 10 μl for analysis within 1–2 h.

Method

Add 2 ml of performic acid solution to the protein (containing about 0.1 mg cystine) in a Pyrex tube and allow to stand at 0°C for 4 h for soluble proteins or overnight for proteins which do not dissolve. Then add 0.30 ml of 48 % HBr with swirling of the tube. Take to dryness at 40°C with a rotary evaporator. Hydrolyse the protein with 3 ml of 6 M-HCl *in vacuo* at 110°C for 18 h. Take to dryness and analyse by ion-exchange chromatography. The yield of methionine *S,S*-dioxide is $100 \pm 2\%$ and cysteic acid $94 \pm 2\%$. The number of half-cystine + cysteine or methionine residues is best calculated by reference to the molar quantity of a stable amino acid (alanine, leucine, aspartic acid or glutamic acid).

8.9.1.2. Sulphydryl groups and half-cystine

The method of determination of both free sulphydryl groups and half-cystine residues in a protein (Simpson *et al.*, 1976) is shown in the equation 8.1.

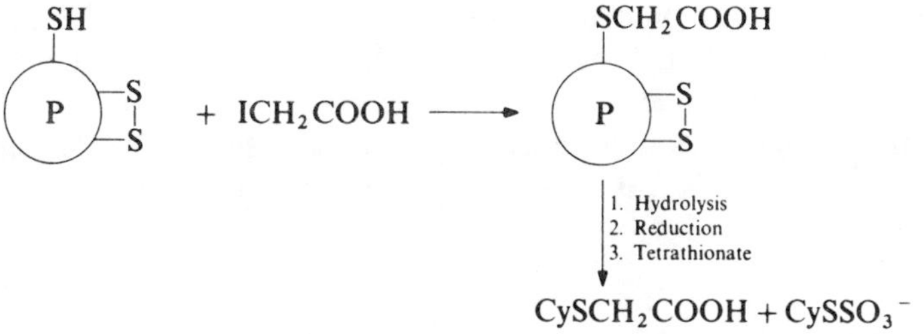

$$\text{P(SH)(S–S)} + ICH_2COOH \longrightarrow \text{P}(SCH_2COOH)\text{(S–S)} \xrightarrow{\text{1. Hydrolysis; 2. Reduction; 3. Tetrathionate}} CySCH_2COOH + CySSO_3^- \qquad (8.1)$$

Free–SH groups

Dissolve the protein (0.3 μmol) in 0.1 M-Tris-HCl buffer, pH 8.3 containing 1 mM EDTA at 4°C in the absence or presence of 6 M-guanidine HCl and in the presence or absence of DTT (0.10 ml, 10 μmol) and react with 10-fold molar excess of recrystallized ICH_2COOH. Potassium phosphate buffer, pH 6.7 may be used. Hydrolyse with 4 M-methanesulphonic acid (see 8.7.2). Determine carboxymethylcysteine with the amino acid analyser.

Half-cystine residues

Cool 1 ml of protein hydrolysate. Add 0.3 ml of pyridine (redistilled over ninhydrin) and 0.9 ml of 4 M-NaOH to give pH 6.8. Add DTT (4 μmol in 1 ml water). Flush with nitrogen, seal with parafilm and incubate at 37°C for 1 h.

After reduction, add solid Na tetrathionate (60 mg, 200 μmol) and allow to stand at 25°C for a minimum of 5 h, to convert all cysteine to *S*-sulphocysteine. Rotary evaporate *in vacuo* at 30°C. Redissolve in 0.5 ml of water and take to dryness (to remove pyridine). Dissolve the residue in pH 2.2 buffer, filter and determine *S*-sulphocysteine with the amino acid analyser.

8.9.1.3. Disulphide bonds

The method described (Thannhauser *et al.*, 1984) is claimed to be more sensitive, rapid and convenient than previous methods. S = 10^{-8} mol of disulphide bonds with an error of $\pm 3\%$.

The disulphide bond is cleaved with excess sodium sulphite (equation 8.2). The reagent used to follow this cleavage is 2-nitro-5-thiosulphobenzoate (NTSB) which reacts only with thiols to form 2-nitro-5-thiobenzoate (NTB) (equation 8.3). The formation of NTB is followed spectrophotometrically, $\varepsilon_{412\,\text{nm}}$ 13 600/M/cm

$$R-S-S-R' + SO_3^{2-} \rightleftharpoons R-S-SO_3^- + R'S^- \qquad (8.2)$$

$$R'S^- - + \text{NTSB} \rightleftharpoons R'SSO_3 + \text{NTB} \qquad (8.3)$$

Reagent preparation

NTSB stock solution (0.5 mM) is prepared as shown in equation 8.4.
Dissolve 0.1 g of DTNB (Ellman's reagent) (2.5×10^{-4} mol, Aldrich) in 10.0 ml of 1 M-Na_2SO_3 (1×10^{-2} mol) at 38°C and adjusted to pH 7.5. Bubble oxygen through the solution and follow the reaction to its maximum (99%), by measuring the concentration of NTB at 412 nm. This takes about 45 min. Store this stock solution of NTSB at −20°C: it is stable for 6 months. Use without purification.

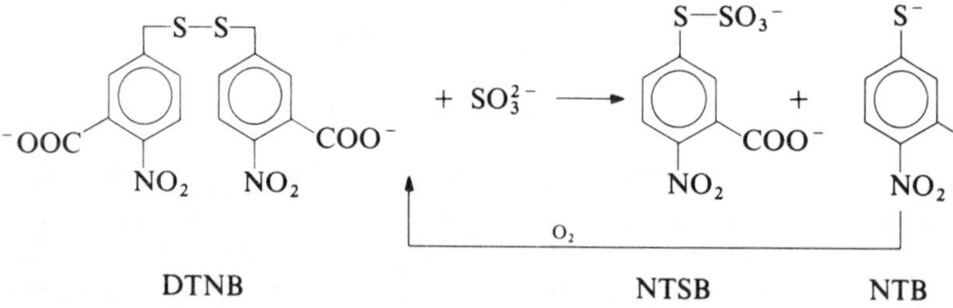

(8.4)

NTSB assay solution

Dilute the NTSB stock solution 100 times with a solution containing 0.2 M-Tris base, freshly prepared 0.1 M-$Na_2S_2O_3$, 3 mM-EDTA and 2–3 M-guanidine thiocyanate (Eastman) and adjust to pH 9.5 with HCl. The solution is stable for 2 weeks at room temperature.

Assay procedure for disulphide

Pipette 0.01–0.20 ml of peptide or protein solution (with -*S*-*S*- concentration 0.5–2 mM) into 3.0 ml of NTSB assay solution. Incubate (in the dark) for 5 min for peptides or 25 min for proteins. Measure absorbance at 412 nm against a blank of 3.0 ml of NTSB assay solution and the appropriate amount of water.

Comments
Guanidine is used to denature the protein and make the disulphide bonds accessible. Urea must not be used, because any ammonium cyanate present will react with thiols. The disulphide cleavage and thiol titration are carried out simultaneously in the same reaction vessel and there is no need to remove dissolved oxygen. The method is applicable to peptide mapping, with specific blocking of cystine and cysteine residues and subsequent recovery of unblocked coumpounds. Damodaran (1985) reported that the NTB anion undergoes a photochemical reaction with a consequent loss of absorbance at 412 nm and recommends that the incubation period of the assay should be carried out in the dark.

Ellman's reagent was introduced for the determination of free thiol groups in proteins (Ellman, 1959) and was subjected to a re-examination (Riddles *et al.*, 1979).

8.9.1.4. 3-Sulphinoalanine and cysteic acid

This method by Ida and Kuriyama (1983) makes use of HPLC.

3-Sulphinoalanine (L-cysteine sulphinic acid) (Csa) and cysteic acid (Cya) (as well as *O*-phosphoserine) coeluted within a few minutes when a cation-exchange column was used for the separation of a mixture of amino acids. However, they were well resolved when using a strong base anion-exchange resin (ISA-07/S2504, Shimadzu Seisakusho, Kyoto, Japan) with 0.05 M-KH_2PO_4 eluent. Retention times: Csa 22 min; Cya 35 min; Ser(P) 41 min.

Post-column derivatization with OPA in the presence of 2-mercaptoethanol (see section 8.16.3.1) gave a linear fluorescence peak area yield for Csa (20 pmol–5 nmol) and for Cya (10 pmol–5 nmol).

8.9.1.5. Sulphydryl group with p-chloromercuribenzoate

Organic mercurials, first introduced by Hellerman (1937) are still the most specific and sensitive reagents for sulphydryl groups on the intact protein. The method described is that of Boyer (1954).

Reagent

p-Chloromercuribenzoate. To purify, dissolve in 1 M-NaOH and centrifuge if necessary. Add 1 M-HCl to precipitate. Dissolve the precipitate in dil. NaOH, and reprecipitate twice more. Wash 3 times with distilled water. Dry in a thin layer *in vacuo* over P_2O_5. Dissolve in dilute acetate or phosphate buffer at pH 7–8.

Method

The protein solution is adjusted to pH 4.6 and the reagent is added in dilute neutral or slightly alkaline solution at room temperature. Plot a time course for increasing absorbance at 232 nm. Use a 10-fold molar excess of reagent over -SH group. Calculate the total increase in absorbancy.

Comments
p-Chloromercuribenzoate solutions deteriorate within a few days. Standardize solutions (after centrifuging if necessary) by reading at 232 nm, at pH 7 (ε_M1.69 $\times 10^4$), or 234 nm at pH 4.6 (ε_M1.74 $\times 10^4$). EDTA if present may give a spuriously high value.

8.9.1.6. Methionine and methionine S-oxide

Periodate is commonly used for oxidative cleavage of the carbohydrate moieties of glycoproteins (Clamp & Hough, 1965; De La Llosa *et al.*, 1980) and oxidative cleavage of disulphide bonds. Cysteine is determined after performic acid oxidation (Hirs, 1967). Methionine in these proteins is oxidized to methionine *S*-oxide to a greater extent than was realized formerly (Yamasaki *et al.*, 1982), even when using low concentrations of periodate (5 mM) and this can be easily overlooked. Methionine *S,S*-dioxide which may be formed is stable to normal acid hydrolysis conditions used for proteins and can be quantified by amino acid analysis (Means & Feeney, 1971; Friedman *et al.*, 1979). Methionine *S*-oxide under the same conditions is decomposed to methionine mainly, with some methionine methyl sulphonium salt, homocystine and homocysteic acid (Floyd al., 1963). Thus, methionine *S*-oxide is usually quantified by amino acid analysis after alkaline hydrolysis, or indirectly quantified as methionine *S,S*-dioxide (Neumann, 1972).

8.10. ARGININE

8.10.1. Methods of determination

8.10.1.1. Arginine in the intact protein

The method by Tomlinson and Viswanatha (1974) is described.

Reagents

A. 0.1% α-naphthol in 50% ethanol (prepared fresh daily).
B. 10% KOH.

C. 5% urea.
D. Potassium hypobromite solution (0.64 ml Br_2 in 100 ml of 5% KOH, prepared fresh daily).
E. Arginine stock solution: 1.0 μmol/ml.

Method

The arginine may be free or protein bound. Take 1 ml of test solution containing 0.1–1.0 μmol of arginine. Add 1 ml of reagent A then 1 ml of B. Mix carefully then add 1 ml of C. Mix and add very rapidly with continuous shaking, 2 ml of D (a critical step!). A reagent blank is prepared at the same time with 1 ml of water instead of test solution. Allow to stand for 20 min at room temperature. Read at 520 nm. Determine the concentration from a standard curve 0.1–1.0 μmol/ml.

8.10.1.2. Determination with p-nitrophenylglyoxal

This is the method of Yamasaki *et al.* (1981).

Reagents

A. 0.1 M-sodium pyrophosphate/0.15 M-sodium ascorbate, pH 9.0. Prepare fresh daily using nitrogen-purged water.
B. *p*-Nitrophenylglyoxal (10% w/v) in methanol. Prepare fresh daily.
C. Subtilisin (0.1% w/v) in 0.01 M-sodium phosphate, pH 7.9.
D. Bovine trypsin (0.1% w/v) in 1.0 mM-HCl/20 mM-$CaCl_2$.

p-Nitrophenylglyoxal is prepared from *p*-nitroacetophenone (Aldrich) by the method of Steinbach and Becker (1954), as follows:

Reflux for 1 h a solution of 39 g (0.30 mol) of selenous acid, 24 ml of water and 150 ml of glacial acetic acid and 49.8 g (0.30 mol) of *m*-nitroacetophenone. Cool. Filter off precipitated selenium. Distil. First remove water and acetic acid at 15 mm. The yellow product distils at 118–128°C (0.6 mm). On standing, the viscous liquid becomes a glassy solid with an indefinite m.p. 37–46°C. Yield 0.15 mol.

Method

Use 0.20 ml of protein sample (1–2 mg) containing 0.1–1.0 μmol of arginine in deionized water or 0.1 M-sodium pyrophosphate, pH 9.0 and add 40 μl of reagent C and 40 μl of D. Incubate at 37°C for 3 h. Cool. Make up to 3.0 ml with A.

Add with mixing 25 μl of reagent B and incubate at 30°C for 30 min. Cool below 20°C in an ice bath and read at 475 nm against a reagent blank. The arginine content is obtained from a standard curve prepared from a serial dilution of stock

arginine solution (0.03–0.33 mM L-arginine in 3 ml of reagent A) + 25 μl of B.
Arginine ε_{475} 0.715/M/cm

8.11. PHOSPHORYLATED AMINO ACIDS

Phosphoester bonds in *O*-phospho-serine, -threonine and -tyrosine are hydrolysed by strong acids and short hydrolysis times are essential to prevent degradation of phosphoamino acids in proteins (Rothberg *et al.*, 1978). Because of their acid lability there is no established method for determining the absolute amounts of these compounds in proteins. [^{32}P]-labelled amino acids may be detected by autoradiography or by scintillation counting.

Phosphorylated peptides may be separated from non-phosphorylated peptides and from each other by reverse phase HPLC under isocratic conditions at 22°C on a Spherisorb C_{18} column, using phosphate buffers with pH 3.2–4.5 and n-hexane sulphonic acid as counter ion (Fransson *et al.*, 1982). About 0.1 nmol of peptide can be detected at 210 nm but fluorimetric detection or the use of ^{32}P increases the sensitivity about 100-fold.

8.11.1. Hydrolysis of protein

Hydrolyse the protein with 6 M-HCl at 110°C for 1–4 h. Filter through a 0.22 μm Millipore filter. Remove HCl *in vacuo*. Dissolve the residue in 0.01 M-TFA with or without unlabelled standards (Capony & Demaille, 1983).

Phosphohistones (100 μg) were hydrolyzed with 6 M-HCl *in vacuo* at 100°C for 4 h or incubated with 100 μg of a protease from *S. griseus* in 50 mM Tris-HCl buffer, pH 7.5 at 20–24°C for 20 h (Swarup *et al.*, 1981).

Enzymic hydrolysis of [^{32}P]-phosphoprotein. Dissolve the protein in 100 μl of 0.1 M-ammonium bicarbonate made 2 mM with EDTA (neutralized) and containing 250 nmol each of unlabelled Ser(P), Thr(P) and Tyr(P). Add trypsin (200 μg/ml) and incubate for 12 h at 37°C. EDTA inhibits any phosphatase activity which may be introduced with the protease. Subsequently, add aminopeptidase M (200 mequiv. units) and incubate for a further 12 h at 37°C (Julian Downward, 1984, personal communication).

8.11.2. Methods of analysis

8.11.2.1. Thin layer chromatography

1. Use cellulose MN-300-coated plates with eluent butan-1-ol/isopropanol/formic acid/water (3:1:1:1, by vol.) (Swarup *et al.*, 1981, no details given).
2. Use polyamide plates (Cheng-Chin, Taiwan). Develop in one dimension with 5% propionic acid containing 0.013–0.025% SDS. Dry. Spray with 0.2% ninhydrin. Heat in the oven at 50°C with observation. Thr(P) and Ser(P) comigrate, $R_f = 0.70$. Tyr(P) = 0.54.

8.11.2.2. *Electrophoresis*

The method of Brautigan *et al.* (1981) after Hunter & Sefton (1980) is described. pH 1.9 buffer. Glacial acetic acid/formic acid (88 % by vol.)/water (78:25:897, by vol.). pH 3.5 buffer. Glacial acetic acid/pyridine/water (50:5:945, by vol.). Purify pyridine by refluxing 18 h over *p*-toluene sulphonyl chloride and then distil.

Use TLC plates (Brinkmann, Polygram CEL MN300, 200 × 200 mm, 0.1 mm). Dissolve the sample in pH 1.9 buffer and carry out electrophoresis at pH 1.9, 1.5 kV for 75 min, towards the anode. After drying, carry out electrophoresis in the second dimension with pH 3.5 buffer, 1 kV for 1 h. After drying the plate, spray with buffered ninhydrin (0.1 %) (section 8.14.2). Tyr(P) gives a grey colour; Ser(P) and Thr(P) give a blue colour. Prepare an autoradiogram if radioactive.

Standards may be prepared using 0.5 μl of solutions of neutralized phosphoamino acids (20 mg/ml). See Fig. 8.2 for the separation obtained.

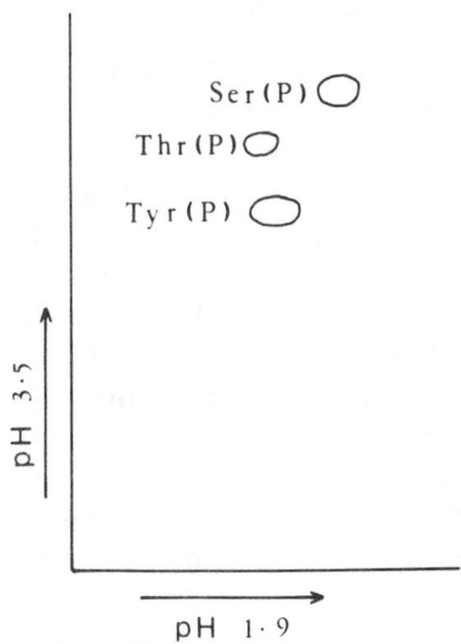

Figure 8.2. Two-dimensional separation of phosphoamino acids by electrophoresis. First dimension at pH 1.9. Second dimension at pH 3.5. (Brautigan *et al.*, 1981).

8.11.2.3. *Amino acid analyser*

This is the method of Capony & Demaille (1983). Use short column of Moore and Stein (80 × 6.0 mm I.D.) packed with Durrum DC6A resin. Dissolve the sample in 0.01 M-TFA, with or without unlabelled standard phosphoamino acids (1.5 nmol of each), and apply to column. Elute isocratically at 50°C with 0.01 M-TFA at 0.4 ml/min. Mix with OPA reagent (see section 8.16.3.2) (flow rate 0.2 ml/min) in a mixing tee for fluorometric determination. When working with [^{32}P]-labelled sample, the OPA reagent is replaced by water and 1 ml fractions are collected for radioactive counting.

Approximate retention times: Ser(P) 19 min; Thr(P) 28 min; Tyr(P) 64 min. Cysteic acid appears in the void volume: other amino acids are not eluted.

Regenerate the column with the following solutions:

A. 1.0 M-Na^+ (34.8 g NaCl, 39.2 g trisodium citrate/l, plus 1 ml of phenol to prevent bacterial growth)
B. 0.2 M-NaOH
C. 1 M-TFA
D. 0.01 M-TFA

Quantitative analysis: using standard compounds the following yields were obtained by fluorescence measurement:
With Thr(P) = 100, then Ser(P) = 88 % and Tyr(P) = 71 %.
Thr(P) vs. Thr = 89 %; Ser(P) vs. Ser = 71 %; Tyr(P) vs. Tyr = 71 %.

8.11.2.4. High pressure liquid chromatography

1. Use a Beckman Ultrasil Ax anion-exchange column (250 × 4.6 mm) equilibrated with 15 mM-potassium phosphate buffer, pH 3.8. Apply the sample and elute with the same buffer for 25 min at a flow rate of 2.0 ml/min. Non-radioactive standards are detected by absorbance at 206 nm. Approximate retention times: Thr(P) 6 min; Ser(P) 8.6 min; Tyr(P) 9 min and Pi 10.5 min. (Swarup *et al.*, 1981).
2. Load the sample on a Synchrom AX-300 anion-exchange column (250 × 4.6 mm). Elute with phosphate buffer (as in (1) above), 1 ml/min. Derivatize the eluate fractions with OPA (Lee & Drescher, 1978) and monitor at 338 nm. Fractions (1 ml) are counted on a liquid scintillation counter. Retention times are Thr(P) 12 min, Ser(P) 17 min and Tyr(P) 23 min (Julian Downward, 1984, personal communication).
3. Use a column (250 × 4.6 mm) packed with Partisil-10 SAX anion-exchange resin (Whatman). Apply the sample and elute with 10 mM-potassium phosphate buffer, pH 3.0, containing 12.5 % (v/v) methanol, at a flow rate of 1 ml/min at room temperature. Detect with OPA-ET. Approximate retention times: Thr(P) 26 min; Tyr(P) 32 min; Ser(P) 38 min and Pi 60 min. (Yang *et al.*, 1982).
4. Dansylation and HPLC of phosphoamino acids (Congote, 1982) Hydrolyse the protein for 2 h at 110°C with conc. HCl (Pierce). Evaporate under a stream of nitrogen gas. Add 100 μl of 0.2 M-Na_2CO_3 and evaporate again. Suspend the sample in 100 μl of water and 100 μl of freshly prepared dansyl chloride solution (2.5 mg DNSCl per ml of acetone). Incubate in the dark for 1 h at 37°C. Take to dryness with a stream of nitrogen. Dissolve the DNS derivative in 100 μl of 0.5 % TFA and use for HPLC.

HPLC. Use a Protesil 300 diphenyl column (Whatman). Wash the column with water/TFA (1 000:1, v/v) or water/acetonitrile/TFA (970:30:1, by vol.). Elute the sample with a linear gradient of 30 ml water/acetonitrile/TFA (970:30:1, by vol.) and 30 ml water/acetonitrile/TFA (900:100:1, by vol.). Monitor the eluate at

280 nm. Collect radioactive samples for scintillation counting. The approximate retention times are 12 min (Ser(P)), 16 min (Thr(P)) and 22 min (Tyr(P)).

HPLC with μ Bondapak C_{18} (Waters). Elute with water/acetonitrile/TFA (925:75:1, by vol.) isocratically. Approximate retention times are Ser(P) 15 min and Thr(P) 22 min. Tyr(P) is not seen.

5. Ion-pair reverse phase HPLC of phosphoamino acids (Aitken & Morrice, 1984; Morrice & Aitken, 1985).

Partisil 5 ODS2 column (150 × 4.6 mm) at room temperature with eluent 0.5 mM tetrabutyl amonium hydroxide/0.5 mM *o*-phthalate, pH 6.3, 1 ml/min. Read at 243 nm, 0.1 a.u.f.s., by indirect detection. S = 200 pmol. Approximate retention times: Asp & Glu 5.3 min; Pi 6.9 min; Thr(P) 11.6 min; Ser(P) 12.8 min; Tyr(P) 15.9 min.

8.12. γ-CARBOXYGLUTAMIC ACID

Since γ-carboxyglutamic acid (Gla) was discovered in prothrombin in 1974 and later in other proteins and in urine, quantitative methods suggested for its determination include the amino acid analyser (Hauschka, 1977; Tabor & Tabor, 1977; Gundberg *et al.*, 1979; James, 1979; Madar *et al.*, 1979; Hauschka *et al.*, 1980; Ishikawa & Wuthier, 1981), GLC (Matsu-ura *et al.*, 1981), specific colorimetric analysis, 0.5 to 5×10^{-4} M (Pecci & Cavallini, 1981) and HPLC (Kuwada & Katayama, 1981, 1983).

Gla is completely decarboxylated in 1 M-HCl at 100°C for 16 h *in vacuo* and alkaline hydrolysis is used for proteins (Kuwada & Katayama, 1981).

Peptides containing Gla residues may be detected by a 'thermal diagonal' technique. The peptide mixture is subjected to paper electrophoresis at pH 6.5. The paper is dried by heating at 80°C for about 1 h and then subjected to electrophoresis at pH 6.5 in the second dimension. Decarboxylated peptides move off the diagonal (Magnusson *et al.*, 1974; Morris, H., personal communication).

8.12.1. Hydrolysis of protein

1. Hydrolyse protein with 2.5 M-KOH at 100°C for 16 h *in vacuo* (or under N_2). Centrifuge to remove any precipitate. Cool the supernatant in ice-water and adjust to pH 4–5 by adding dropwise 60 % and then 6 % $HClO_4$. Allow to stand on ice for 30 min. Centrifuge to remove $KClO_4$ precipitate. Take 2–10 μl of the supernatant for analysis (Kuwada & Katayama, 1981).
2. Hydrolyse protein with 0.4–1.0 ml of 2.0 M-KOH at 110°C for 22 h. Add 0.2 ml of freshly prepared saturated $KHCO_3$. While vortexing, neutralize with 70 % $HClO_4$ to pH 7 (pH indicator paper). Centrifuge. Use the clear supernatant for analysis. Gla is stable at pH 7 at −20°C (Hauschka, 1977).

8.12.2. Methods of analysis

8.12.2.1. Thin layer chromatography

Aluminium cellulose plate 0.1 mm (Merck) eluted with n-butanol/formic acid/water (75:15:10, by vol.) Gla gives a violet spot with ninhydrin with $R_f = 0.24$. A colorimetric method of determination of Gla after reaction with acetaldehyde and nitroprusside was described (Pecci & Cavallini, 1981).

8.12.2.2. Amino acid analyser

Cation-exchange column

The elution time of Gla is very sensitive to the pH of the first buffer. With the Beckmann 121 M analyser and pH 2.68 citrate buffer, elution times were 13.0 min (Cya), 17.4 min (taurine) and 30.5 min (Gla). Other amino acids were eluted later. The ninhydrin colour factor for Gla is 39 % of that for Glu (Hauschka, 1977).

Pyroglutamic acid elutes closely to Gla from the column using pH 3.25 buffer and it may be necessary to confirm the identity of a peak in the Gla region of the chromatogram. The Gla eluent fraction is collected. An aliquot is hydrolysed with 1 ml of 2.5 M-NaOH at 105°C for 24 h. A second aliquot is similarly hydrolysed with 6 M-HCl. Pyroglutamic acid is completely hydrolysed to Glu in both cases. The concentration of Gla is calculated from the difference between acid-formed and alkali-formed Glu. Use [^{14}C]-cysteic acid as an internal standard for recovery. The overall recovery of Gla as Glu is 95–98 % (Ishikawa & Wuthier, 1981).

Anion-exchange column

With an anion-exchange column, Gla is eluted after other amino acids. With an Aminex A-27 or A-28 anion-exchange column (125 × 9.0 mm Bio-Rad) equilibrated with 0.3 M-sodium acetate buffer pH 4.6, and eluted with the same buffer, 70 ml/h at 30°C, Gla eluted at 55 min (Tabor & Tabor, 1977). Similar results were reported (James, 1979).

8.12.2.3. High pressure liquid chromatography

The following method is from Kuwada and Katayama (1983).

Pre-column derivatization

Mix 50 μl of protein hydrolysate (or urine) with 50 μl of OPA/ET reagent (section 8.16.3.2) and vortex vigorously. After 2 min add 100 μl of 0.1 M-KH_2PO_4 in 66 % acetonitrile. Inject 2–20 μl onto the column.

HPLC

Use a short column (50 × 4.6 mm) packed with Nucleosil 5SB 5 μm particles (Macherey-Nagel). This is a silica-based strong anion exchanger with $-N(CH_3)_3$ functional group. Elute isocratically at 47°C, 2 ml/min with the mobile phase 0.2 M-sodium citrate buffer, pH 5.28/acetonitrile (1:1, v/v). Repeated analyses may be carried out under these conditions without regeneration of the column. The excitation monochromator of the detector is set at 240 nm and the emission passed through a 418 nm cut-off. Most amino acids elute together in 2 min, followed by Glu, Asp, Cya, S-carboxymethylcysteine and finally Gla (7 min). The peak area/pmol Gla relationship is linear from 0.3 to 100 pmol.

Comment
More recently the OPA-Gla derivative was determined by HPLC using β-carboxyaspartic acid as internal standard (Haroon, 1984).

8.13. ACETYL AND FORMYL GROUPS

The report of an *N*-acetyl peptide isolated from the *N*-terminus of tobacco mosaic virus (Narita, 1958) soon led to the discovery of many proteins having a blocked N^α-terminus. Proteins with N^α-acetyl alanine, aspartic acid, glycine, serine and threonine from a variety of sources were listed (Brown & Roberts, 1976; Croft, 1980; see also reviews Allfrey *et al.*, 1984; Wold, 1984).

N-formyl methionine is well established in protein chain initiation on the ribosome, *N*-formyl valine in Gramicidin A (Sarges & Witkop, 1964) and *N*-formyl glycine in melittin (Kreil & Kreil-Kriss, 1967) and spontaneous *N*-formylation between L-lysine and formaldehyde has been demonstrated (Trezl *et al.*, 1983).

The acetyl group has been determined by several different enzymic methods (Korff, 1954; Rose, 1955; Soodak, 1957; Stegink, 1967; Kuo & Younathan, 1973; Guynn & Veech, 1975), colorimetrically as the hydroxamic–ferric complex (Ludowieg & Dorfman, 1960), by steam distillation followed by titration (Johansen *et al.*, 1960; Augustyniak & Martin, 1965) or as the acetohydrazide and 1-acetyl-2-DNP-hydrazide (Phillips, 1963) or as the 1-acetyl-2-dansyl hydrazide (Schmer & Kreil, 1969). The enzymic methods make use of purified enzymes and many of them require 100 nmol or more of acetate in the sample, although that of Guynn & Veech (1975) requires only 3–12 nmol of acetate/ml for fluorescence detection, equivalent to 0.015–0.05 μmol/g wet wt. of tissue.

Formate was determined by the Barker and Somers colorimetric method by heating with 2-thiobarbituric acid to form a chromophore, λ max. 450 nm. Phosphates and acetic acid did not interfere. However, the sensitivity was low, with a linear absorbance curve in the range 1–10 mM (Flynn & Whitmore, 1983).

Provided that a short *N*-terminal peptide can be isolated from the protein, then this can be hydrolysed and analysed for its *N*-acetyl amino acid content by using chromatography and electrophoresis (Yoshida, 1972). If the peptide is amenable to direct study by MS then this provides by far the most convenient method for identification of the *N*-blocked amino acid and the primary sequence of the peptide may be obtained at the same time. About 50 nmol of peptide is required (Auffret *et al.*, 1978; see also Chapter 19).

There have been several methods for determining acetate in proteins by GLC (Schroeder *et al.*, 1962; Ward & Coffey, 1964; Ward *et al.*, 1966; Hoenders & Bloemendal, 1967; Henkel, 1971; Bricknell & Finegold, 1973) and pyrolysis (Williams & Siggia, 1977). Problems included the tailing of acetic acid on the column, the appearance of 'ghost' peaks on the chromatogram and a low response factor with the detector.

Chauhan and Darbre (1981a) liberated acetyl and formyl groups from proteins and peptides by hydrolysis with alkali. The use of acid for hydrolysis led to the liberation of free acetic and formic acids and losses were always incurred. The liberated acetate and formate were converted to phenacyl esters by means of crown ether catalysis with dicyclohexyl-18-crown-6. About 20 nmol of acetyl or formyl group in the sample was necessary, because the phenacyl esters were determined with the flame ionization detector. (There is still a problem in making full use of the sensitivity of GC. Normally it is only possible to inject onto the column a small fraction of the final derivatized sample.) In this case 1–2 nmol of phenacyl esters were injected. The method described below is more sensitive because pentafluorobenzyl (PFB) esters were formed which could be determined with the electron capture detector.

8.13.1. Determination of acetyl and formyl groups

8.13.1.1. Gas-liquid chromatography method

The method of Chauhan and Darbre (1982c) is described.

Apparatus

GLC with electron capture detector (or flame ionization detector for lower sensitivity). Glass capillary column (30 m × 0.27 mm I.D.) coated with 5% Chromosorb R and 5% PP Seb by a single coating method (Chauhan & Darbre, 1981b). Direct injection of the sample onto the capillary column with sample volumes up to 2.0 μl without inlet heater (Chauhan & Darbre, 1981c).

Reagents

α-Bromo-2,3,4,5,6-pentafluorotoluene (PFB Br) (Aldrich); 15-crown-5 (Fluka); Acetonitrile (redistilled over P_2O_5).

Method

Hydrolyse 5 nmol or more of *N*-acetylated compound with 20 μl of 1 M-NaOH in a 0.5 ml polypropylene microcapped centrifuge tube in an autoclave at 15 p.s.i. (123°C) for 3 h. After hydrolysis add a known amount (about 5 nmol) of internal standards (n-butyric and isovaleric acids). Add 20 μl of 1 M-HBr and 2 μl of 1 M-$NaHCO_3$ solution. Transfer 5 μl aliquots to silanized glass tubes (50 × 3.0 mm i.d.) and take to dryness with a stream of nitrogen. Add acetonitrile solution (10 μl) containing PFB Br (50 nmol) and 15-crown-5 (50 nmol). Seal the tube and incubate at 80°C for 2 h with occasional shaking. Open the tube and dilute a 1 μl aliquot in 2 ml of toluene. Inject 1 μl of toluene solution onto the column isocratically at 110°C without using an inlet heater. The retention times are approximately 6 min (PFB-Br), 9 min (PFB-formate), 11 min (PFB-acetate), 15 min (PFB-propionate) and 18 min (nitrobenzene). Nitrobenzene may be used as an internal standard.

It is essential to carry out a blank experiment in the absence of the test material. Traces of acetate and formate arise from the chemicals used. Control values (usually about 0.5 mol or less of acetate/mol protein) are obtained by preparing the PFB esters without previous hydrolysis of the protein sample. The sensitivity of the method is limited by the control values and the purity of the protein sample and not the sensitivity of detection by the electron capture detector ($S \simeq 3$ fmol PFB esters).

8.14. DETECTION OF COMPOUNDS AFTER PAPER AND THIN LAYER CHROMATOGRAPHY

After drying the chromatogram it should be viewed under u.v. light for the detection of fluorescent compounds. Solvent fronts may be seen which indicate any irregularity of solvent flow.

The following procedures may be used as dip or spray reagents as appropriate.

8.14.1. Fluorescent methods of detection

1. Dip the paper in a solution of fluorescamine (25 mg/l) in acetone containing triethylamine (0.5 %, v/v). Examine under u.v. light. $S = 1$ nmol of peptide. If the peptides are to be recovered a weaker (4 mg/ml) solution may adequate for detection. After washing with acetone, other staining procedures may be applied.
2. Fluorescamine: 10 mg/100 ml of acetone. Triethylamine: 1 % (v/v) in acetone. The chromatogram is dried to remove solvents, washed with acetone and dried. Spray with triethylamine solution and dry in the hood (5 min). Spray with fluorescamine solution. Dry. Spray with acetone and dry. Visualize peptides by their fluorescence under a u.v. lamp (336 nm). Repeat the entire procedure for detection of weakly reacting peptides.

After marking the spots, heat the chromatogram in an oven at 110°C for 3 h. Peptides with a proline residue at the *N*-terminus can then be detected (Gracy, 1977).

3. *o*-Phthalaldehyde: 30 mg/100 ml of acetone. Triethylamine, 1% (v/v) in acetone containing 0.05% (v/v) 2-mercaptoethanol.

Carry out the procedure described in (2) with these two reagent solutions. The reagent is cheaper and as efficient as fluorescamine.

8.14.2. Ninhydrin detection

Ninhydrin staining may be used after fluorescence detection. Following an acidic solvent it is best to make the reagent slightly alkaline with a weak base, and after an alkaline solvent make the reagent slightly acid with a weak acid.

1. Ninhydrin (0.2% in 95% acetone) with the addition of a few drops of collidine or glacial acetic acid. After dipping or spraying allow to dry at 60°C for about 20 min or at 100°C for 5–10 min with continual observation. Excessive heat causes a dark background. Alternatively the plate may be developed for 24 h in the dark, without heating, when a non-coloured background is obtained. Most amino acids give a violet colour, but there are some differences (Asp: bluish-red, Pro and Hyp: Yellow). Sensitivity limits about 1 μg.

2. Ninhydrin (300 mg), 3 ml of glacial acetic acid and 100 ml of n-butanol. Data on detection limits, ranging from 0.001 μg alanine to 0.1 μg proline and aspartic acid are given by Fahmy *et al.* (1961).

3. Buffered ninhydrin solutions. Ninhydrin (300 mg), glacial acetic acid (20 ml) and collidine (2,4,6-trimethyl pyridine) (5 ml) made up to 100 ml with ethanol (Jones & Heathcote, 1966). Ninhydrin (0.1% w/v) in acetone/glacial acetic acid/collidine (100:30:4, v/v) (Brautigan *et al.*, 1978).

4. Ninhydrin-cadmium acetate (Heathcote and Washington, 1967).

Colours obtained with ninhydrin fade with time, but this reagent gives permanent colours (mainly red, but with proline, yellow).

Dissolve 0.5 g of cadmium acetate in a solution of 50 ml of water and 10 ml of glacial acetic acid. Make up to 500 ml with acetone. Portions of this solution are taken and solid ninhydrin added to give a final concentration of 0.2 g %.

Spray and then heat at 60°C for about 15 min. Note the results and note again after 24 h at room temperature. Alternatively, do not heat, but store at room temperature overnight in the dark to allow the colour to develop. The background is cleaner. Some compounds such as taurine and threonine develop the colour very slowly. S = 0.5 nmol.

8.14.3. Miscellaneous detection methods

Chlorine-starch-iodide (Perham, 1978)

This test may be applied after ninhydrin and depends on the conversion of peptide-NH to peptide-N-Cl derivatives.

If pyridine buffers have been used, wash the paper with acetone. Expose the paper to chlorine gas for a few minutes, during which time any ninhydrin stains will be bleached. Dry the paper in air (in the hood) to remove excess chlorine. Dip the paper in an aqueous solution of 2% (w/v) starch and 2% (w/v) KI (1 : 1, v/v). Peptides show up as blue-black spots (the larger the peptide the more intense the colour) on a pale blue background.

Arginine (Yamada & Itano, 1966)

Reagents
A. 0.02% phenanthrenequinone in anhydrous ethanol.
B. 10% NaOH in 60% ethanol.

Method
Mix equal volumes of reagents A and B shortly before use. Dip or spray the chromatogram. Dry in air for 20 min. A fluorescent spot is visible under u.v. $S = 10^{-4}\,\mu$mol. Ninhydrin and Pauly's diazo reagent may be subsequently used.

Arginine (Irreverre, 1965)

Reagents for modified Sakaguchi reaction
A. 0.0125% α-naphthol in absolute ethanol. Keeps for one week in the cold in a brown bottle.
B. 1.5 M-NaOCl. Keeps 3–4 months at 0–4°C.
C. Iodine solution. 22.4 g of iodine and 30 g of KI dissolved in 400 ml of distilled water. Keeps indefinitely at room temperature.
D. 10% NaOH.

Method
Dip the chromatogram in Reagent A and dry. Spray the chromatogram with three bursts of spray solution (1.5 ml of A + 23.5 ml of B) followed by one burst of spray (2.4 ml of C + 20 ml of D).

The spray solutions must be used within 10 min of mixing. Arginine, arginine-containing peptides and many guanidino compounds give red colours, stable in the dark for several days. ($S = < 1\,\mu$g). A ninhydrin reagent can also be applied over this spray, provided that it is made acidic with acetic acid to neutralize excess alkali.

Histidine and tyrosine (Brenner & Niederweiser, 1967)

Pauly's diazo reagent
A. 0.4 M-sodium sulphanilate
B. 0.4 M-sodium nitrate
C. 0.25 M-HCl
D. 1 M-sodium carbonate

Method
Mix the reagents A–D in the proportions 1 : 1 : 8 : 10, by vol. The reagent is specific for histidine, tyrosine and certain of their derivatives. Histidine and histidine-peptides give bright red colours and tyrosine a brown colour.

Tyrosine (Perham, 1978)

Dip in 0.1 % (w/v) 1-nitroso-2-naphthol in acetone. Allow to dry. Then dip in a mixture of acetone/conc. nitric acid (9 : 1, v/v). Dry the paper and heat carefully at 105°C for 2–3 min without charring. Tyrosine gives a deep red colour. This test may be used after ninhydrin.

Tryptophan and imidazoles (Smith, 1969)

Ehrlich reagent
p-Dimethylaminobenzaldehyde in conc. HCl (10 % w/v) diluted in acetone (1 : 4, v/v). Mix just before use.

Method
After spraying, allow to dry at room temperature. Tryptophan and tryptophan-containing peptides slowly develop a purple colour. This test may be used after ninhydrin.

Sulphur amino acids (Scoffone & Fontana, 1975)

1. *Sodium azide: iodine*
Spray the dry chromatogram with 0.05 M-iodine in 50 % ethanol containing 1.5 % of sodium azide. The spots are best seen under u.v. light. S = 0.5 μg methionine.

2. *Platinic iodide*
Mix in the following order 4 ml of 0.002 M-platinic chloride, 0.25 ml of 1 M-potassium iodide, 0.4 ml of 2 M-HCl and 76 ml of acetone. Dip or spray heavily. Cysteine, cystine and methionine and reducing agents give a white spot on a red-

purple background. Any phenols, or lutidines used as developing solvents must first be removed by ether, acetone etc.

3. *Sodium nitroprusside*

Reagent A. Dissolve sodium nitroprusside (1.5 g) in 5 ml of 1 M-sulphuric acid. Add 95 ml of methanol and 10 ml of 28 % of ammonium hydroxide. Filter and store at 0–4°C.

Reagent B. Dissolve 2 g of sodium cyanide in 5 ml of water and make up to 100 ml with methanol.

(a) To detect cysteine. Use reagent A.

(b) For cystine, dip or spray with reagent A. Allow to dry slightly and while still damp use reagent B.

(c) For both cysteine and cystine. Prepare the reagents at double strength and treat with an equal mixture of A and B.

Asparagine (Pasieka & Borowiecki, 1965)

After ninhydrin development of the chromatogram dip in 5 % aq. borate solution; wash in a water dip and dry. Asparagine shows as a blue spot.

Asparagine gives a ninhydrin reaction on paper chromatograms but its presence may be obscured by neighbouring arginine, aspartic acid, glycine, serine or other amino acids. The borate treatment results in a blue colour for positive identification. For the purposes of separation, at pH 5.8 with HVE, asparagine moves slightly towards the cathode, while aspartic acid moves to the anode.

Phosphate (Schmidt and Thannhauser, 1945)

Reagent

Ammonium molybdate, 1 g (finely ground) dissolved in 8 ml of water, 8 vol. Conc. HCl, 3 vol. Perchloric acid, 12 M (S.G. 1.72), 3 vol. Acetone, 86 vol.

Method

Prepare the reagent freshly by mixing the volumes shown.

Dip the paper and allow to dry. Expose to u.v. light for at least 30 min. Nucleoside mono-, di- and tri-phosphates appear as blue spots.

Sugars (Bounias, 1980)

Spray with a 6.5 mmol solution of *N*-(1-naphthyl) ethylenediamine dihydrochloride (Baker) in methanol containing 3 % (v/v) sulphuric acid. Heat for 5 min at 100°C. Coloured spots (varying from brownish-red for ascorbic acid to blue-grey for deoxyglucose) appear on a white ground. The method can be quantified.

Iodine vapour

Expose dried paper and thin layer chromatograms to iodine vapour in a glass tank. Sometimes amino acid derivatives, indoles and other compounds are located by their rapid staining to give dark areas on a paler ground.

See Barrett (1974) for other non-destructive detection methods.

Further reading

Details of many locating reagents may be obtained (Block *et al.*, 1958; Dawson *et al.*, 1969; Smith, 1969). Easley (1965) reported on the application of different reactions in sequence on the chromatogram for peptide mapping.

8.15. THIN-LAYER CHROMATOGRAPHY OF AMINO ACIDS

TLC is the simplest and most versatile method for amino acid separation but quantitation is only approximate. The identification of individual spots can only be done by careful comparison with standard amino acids separated under identical conditions at the same time. Brenner and Niederweiser (1967) described methods for the preparation of both Silica Gel G and cellulose plates and the separation of many amino acids by two-dimensional chromatography. Commercially available TLC plates give reproducible results.

8.15.1. Separation methods

1. Jones & Heathcote (1966). A separation of 24 amino acids, including leucine and isoleucine, was achieved on plates (200 × 200 mm) coated with cellulose MN 300 (Macherey, Nagel & Co.) 100 μm thick. The total time required was 6 h.

Solvent for first dimension: propan-2-ol/formic acid/water (40:2:10, by vol.); for second dimension: *tert*-butanol/methyl ethyl-ketone/0.88NH_3/water (50:30:10:10, by vol.).

2. Haworth & Heathcote (1969). An improved solvent system allowed the separation of 40 amino acids and related compounds on cellulose MN-300 plates without binding agent. Solvent for first dimension: propan-2-ol/2-butanone/1 M-HCl (60:15:25, by vol.); for second dimension: 2-methylpropanol/2-butanone/propanone/methanol/water/(0.88)NH_3 (40:20:20:1:14:5, by vol.). Wash the plates before use to remove impurities. The tank should be saturated with each solvent before use. After running the first solvent, dry the plate in cold air for 15 min then at 60°C for 15 min to remove HCl, before running in the second dimension. The running time is about 2.5 h for each dimension. See Fig. 8.3 for the separation of 24 amino acids.

The same solvent system was used for the identification of 76 *N*-containing metabolites found in biological fluids by means of selective staining reagents (Heathcote *et al.*, 1970). See section 8.14 for spray reagents.

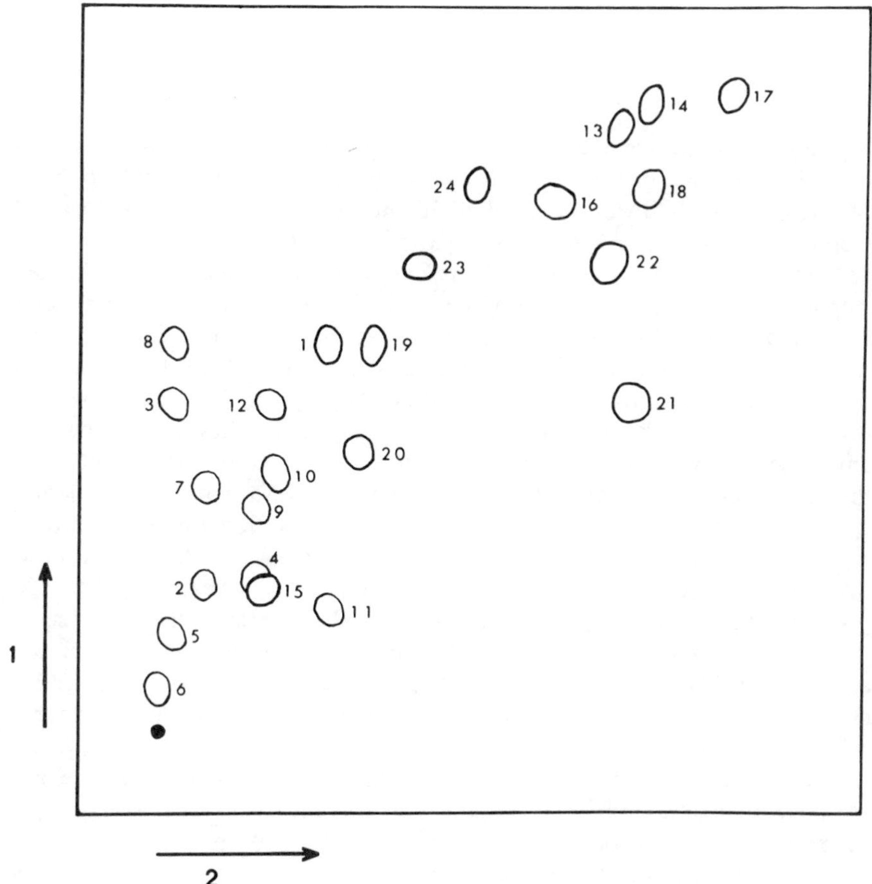

Figure 8.3. Two-dimensional thin layer chromatography of amino acids on cellulose plate. See text for details of solvent systems.
1 Ala, 2 Arg, 3 Asp, 4 Asn, 5 Cys, 6 Cys Cys, 7 Cya, 8 Glu, 9 Gln, 10 Gly, 11 His, 12 Hyp, 13 Ile, 14 Leu, 15 Lys, 16 Met, 17 Nle, 18 Phe, 19 Pro, 20 Ser, 21 Thr, 22 Trp, 23 Tyr, 24 Val.
(Redrawn with permission from Haworth & Heathcote, 1969.)

3. Dale & Court (1981). Avicel F TLC plates (Analtech, Luton, U.K.) were used and six solvent systems investigated for one- or two-dimensional chromatography. The hR_f values for 35 amino acids were reported. Loads of up to 0.05 M could be used for preparative work.
4. Sherma *et al.* (1983). The separation of 19 amino acids was studied on reverse phase TLC plates, including C_{18} chemically bonded silica gel.
5. Nabi *et al.* (1983). TLC separation of 24 amino acids on stannic tungstate is of interest, but offers no advantages.

8.16. COLUMN CHROMATOGRAPHY OF AMINO ACIDS

8.16.1. Columns and buffers

Sulphonated polystyrene cross-linked with divinylbenzene was used for ion-exchange chromatography of amino acids (Moore & Stein, 1951, 1954a, b, 1963; Moore *et al.*, 1958; Spackman *et al.*, 1958) and is still widely used. Accelerated methods of analysis were discussed (Spackman, 1967).

Amino acids in their cationic form at low pH bind to the negatively charged sulphonic acid groups. As the pH of the eluting buffer (usually sodium citrate) is raised, amino acids lose their positive charges and thus with decreasing affinity for the column, the amino acids are eluted successively. The main factor in determining the relative order of elution is the pH of the solution. However, cations (Na^+) in the buffer compete for the sulphonic acid sites and although the amino acids have a stronger affinity for the resin they are continually being displaced because of the much higher concentration of Na^+ ions. Also the nature of the amino acid side chain influences the order of elution, e.g. tyrosine ($pK_a = 9.1$) which has an aromatic side chain is eluted after leucine ($pK_a = 9.7$) instead of earlier, as expected. One column (0.9 × 120 cm) packed with Amberlite IR-120 resin was used to separate the acidic and neutral amino acids using as eluents 0.2 M-Na^+, 0.1 M-citrate buffer at pH 3.25 followed by 0.2 M-Na^+, 0.1 M-citrate at pH 4.25. A separate column (0.9 × 15cm) was used with 0.35 M–Na^+, 0.12 M-citrate at pH 5.28 for the analysis of basic amino acids. See Svasti (1980) for a critical discussion.

A dual column system is not entirely satisfactory. Two identical samples are required for loading onto the columns and this increases the possibilities of quantitative errors. Also, the instrumentation is more complex.

Difficulties arise when there are large differences in the ratios of the amino acids present, when some peaks are on-scale (and can be determined) and others are off-scale and cannot be determined. In such a case it is necessary to re-run the sample using smaller quantities and a full analysis is then obtained by correlating the quantities of the amino acids in the two runs. The problem may be avoided by using a simultaneous recording of the separated amino acids at higher and lower sensitivities of detection, or to use a linear/logarithmic transformation of the electrical signal going to the strip chart recorder or integrator when all peaks will then be determined.

Many different buffer mixtures have been described to separate all the protein amino acids on a single column in a single run. Sometimes three or four discrete buffer changes are involved. These are easily defined and controlled, but usually a rise in the baseline (a buffer step or buffer change peak) is seen with each change of buffer resulting from chromatographic displacement of retained impurity peaks and small changes in buffer refractive index (Barbarash & Quarles, 1982).

Continuous buffer gradients (Peterson & Sober, 1959) were less popular because they could not be reproduced accurately, but with modern HPLC technology it is possible to overcome the problems associated with both buffer gradients and the accurate pumping of small volumes. Benson (1972) described a system with three buffers having pH 3.25, 3.50 and 3.65 with sodium molarity increasing 0.020, 0.70 and 1.60 M, respectively. Under these conditions problems may occur due to the column packing expanding or contracting with the changing molarity of the eluent. In contrast, Hare (1972) used a constant molarity buffer system with four buffers having 0.20 M–Na^+ but with increasing pH 3.25, 4.15, 5.25 and 10.1.

In order to work at the highest sensitivity possible, background 'noise' shown on the strip chart recorder trace should be reduced as much as possible by using pure water (section 8.3) and buffers.

Test for purity of buffers

If sodium citrate buffers are used, dissolve 20 mg of $Na_3C_6H_5O_7.2H_2O$ in 1 ml of water (approx. 0.2 M-Na^+ concentration) and add 40 μl of 'clean' M-HCl to give pH 2.0. Inject 100 μl of this solution onto the ion-exchange column and compare the resultant chromatogram with those of the water blank and the HCl chromatogram (Hare, 1977).

The factors which may be varied to establish the required separation include pH, molarity and flow rate of the buffer and column temperature. Buffers should be made with highly purified water and the purest chemicals available. The pH should be adjusted to within 0.001 pH unit and solutions kept under nitrogen. It is usual to include in the first buffer (pH 3.25) an alcohol (e.g. 2% (v/v) isopropanol or ethanol) to improve the separation between threonine and serine. Commercial apparatus is usually sold with complete technical information for its use, but the individual worker will certainly have to make adjustments to the analysis programme by trial and error.

Figs. 8.4 and 8.5 show amino acids separated with Beckman instrumentation. Note in Fig. 8.4 the typical baseline rise obtained when a new buffer is introduced onto the column, as shown after 42 and 56 min. Fig. 8.5 shows a separation carried out at higher sensitivity with the hydrolysate of a glycoprotein where glucosamine and galactosamine are shown.

Reviews on amino acid analysis may be consulted (Hamilton, 1966, 1967; Benson, 1975, 1977; Hare, 1977; Blackburn, 1978b); Pfeifer & Hill, 1983).

The ion-exchange method for the determination of amino acids in plasma (Moore & Stein, 1954a) was modified by the introduction of lithium citrate buffers for the separation of the numerous ninhydrin-positive compounds present in this and other biological fluids. A single column was used to separate 65 compounds in about 20 h (Perry *et al.*, 1968). By using a two-column system, 41 compounds were separated in 7.5 h (Benson, 1972) and 55 compounds in 9.5 h (Murayama & Shindo, 1977).

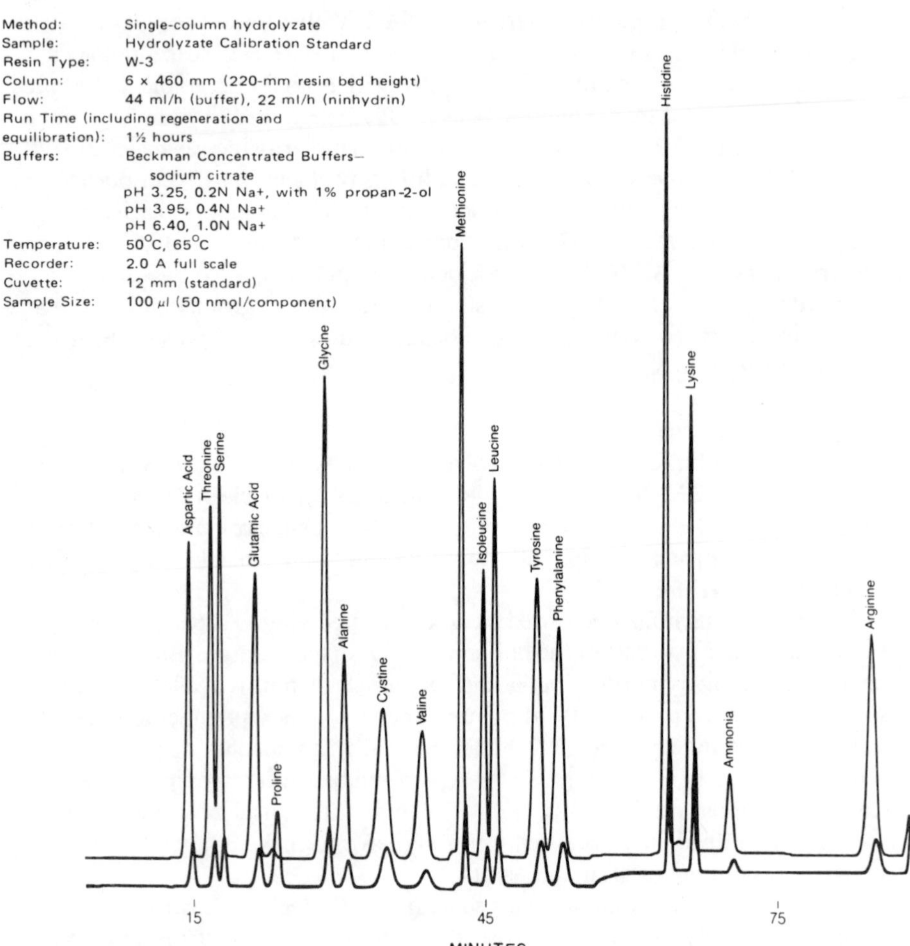

Figure 8.4. Single-column separation of amino acids by ion-exchange chromatography. Column (6 × 460 mm) packed with resin type W-3 to 220 mm resin bed height; Beckman sodium citrate buffers, pH 3.25, 0.2 N Na^+, with 1 % propan-2-ol; pH 3.95, 0.4 N Na^+; pH 6.40, 1.0 N NA^+; temperature 50° C, 60° C; flow rates 44 ml/h (buffer), 22 ml/h(ninhydrin); 12 mm cuvette; 2.0 A full scale; sample size, 100 µl (50 nmol/component) of standard calibration mixture. The separation was by courtesy of Beckman.

8.16.2. Internal standards

An accurately known amount of an internal standard can be added to a protein solution before hydrolysis. The recovery of the internal standard is used to determine the recoveries of the protein amino acids. If a two-column method is used it is almost certain that different standards will be required. Because of the

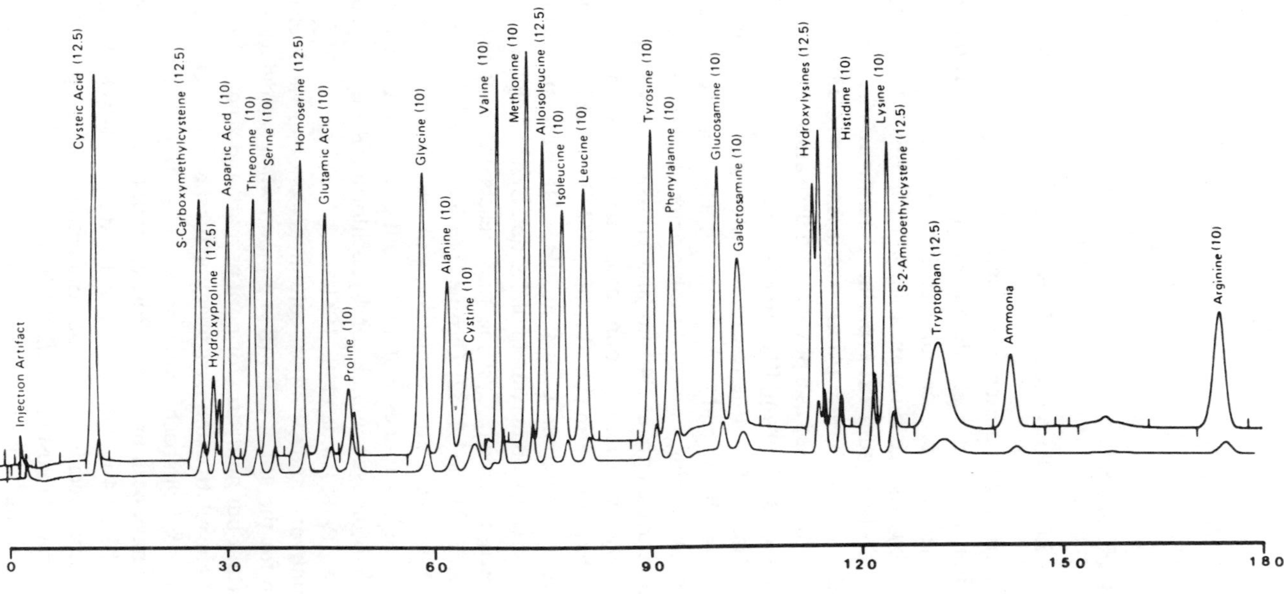

Figure 8.5. Single column separation of amino acids in a glycoprotein hydrolysate. Instrument, Beckman Model 119CL Analyser; column (6 × 460 mm) packed with resin type W-3H to 370 mm resin bed height; Beckman sodium citrate buffers, pH 3.33, 0.20 N Na^+; pH 4.25, 0.20 N Na^+; pH 6.04, 1.0 N Na^+; temperature 30°C, 60°C; flow rates 30 ml/h (buffer), 15 ml/h (ninhydrin); 12 mm cuvette; 0.5 A full scale; sample size, 100 μl (nmol concentrations shown in parentheses). The separation was by courtesy of Beckman.

individual nature of each column and variations existing between laboratories it may be necessary to experiment to find a suitable compound. Norleucine, norvaline or nitrotyrosine should be tried for the long column and diaminopimelic acid or homoarginine for the short basic column (see review, Riordan & Giese, 1977).

8.16.3. Detection of amino acids

8.16.3.1. Post-column derivatization

The eluted amino acids from the ion-exchange column (until the early 1970s) were always mixed with ninhydrin in a post-column reaction and passed through a heated mixing coil. The developed colour was first read at 570 nm in one colorimeter, followed by passage through a second colorimeter to obtain a reading at 440 nm for proline and hydroxyproline determination.

Fluorescamine reacts selectively with primary amines and was introduced by Udenfriend *et al.* (1972). It was used as a post-column derivatization method for the determination of amino acids giving 10 to 100-fold increase in sensitivity over ninhydrin (Benson & Hare, 1975). Fluorescamine is not only expensive, but is unstable in aqueous medium and has been largely superseded by *o*-phthalaldehyde (OPA). A fluorimeter is used for the determination of the amino acids with excitation wavelength at 340 nm and emission at 450 nm. OPA (introduced by Roth, 1971) was used in amino acid analysers by many workers and is more sensitive than ninhydrin (Lund *et al.*, 1977). It is stable in aqueous medium and fluorphor formation occurs at pH 9–11, with an excitation maximum of 360 nm and emission maximum of 455 nm. Lysine yields only a low fluorescence unless Brij is added. Cysteine only reacts in the presence of iodoacetic acid (Cooper & Turnell, 1982). Cystine gives a low fluorescence and is best determined as cysteic acid after oxidation with performic acid. Proline, hydroxyproline and secondary amines do not react, but can do so after oxidation with chloramine-T, *N*-bromosuccinimide or hypochlorite (Roth & Hampaï, 1973; Böhlen & Mellet, 1979; Ishida *et* al., 1981).

A comparison of ninhydrin and OPA post-column detection methods resulted in a recommendation for the use of OPA (S = 5 pmol) when secondary amino acids were not important but ninhydrin for both primary and secondary amino acids (S = 100 pmol). The relative standard deviation for the amino acids was 6 % (range 3–24 %) (Cunico & Schlabach, 1983).

Other new methods have been proposed. Post-column fluorimetric detection of amino acids with 7-chloro-4-nitrobenzo-2-oxa-1, 3-diazole (NBD-Cl) enables pmol quantities of amino acids to be determined, including proline and hydroxyproline (S = 1 pmol) (Yoshida *et al.*, 1982). Also, 4-fluoro-7-nitrobenzo-2,1,3-oxadiazole was first introduced (Watanabe & Imai, 1983), with detection limits of 5 pmol for proline and 20 pmol for other amino acids.

8.16.3.2. Pre-column derivatization

Post-column derivatization involves technical difficulties of mixing each separated amino acid in the column effluent with the reagent solution under the appropriate conditions for optimal reaction. Pre-column derivatization is simpler and the derivatives may be separated by HPLC on a reverse phase column. (Some workers have reported that reproducibility of results does not compare with post-column derivatization.) Phenylthiohydantoin amino acid derivatives have been widely studied as a result of the Edman degradation for protein sequencing (see Chapter 13) and several methods of analysis are available including HPLC (Zimmerman *et al.*, 1976; Johnson *et al.*, 1979). These derivatives have not been used for the determination of amino acids *per se*. All the 20 protein dimethylaminoazobenzenethiohydantoin (DABTH) amino acids were separated in 30 min by HPLC (Yang & Wakil, 1984). These derivatives are discussed in Chapter 14.

Pre-column derivatization of amino acids and peptides for HPLC was briefly reviewed (Szókán, 1982).

Dansyl amino acids

Dansyl amino acids have been widely used in protein sequencing using TLC for their separation (see Chapters 10 and 11). More recently they have been successfully separated on reverse phase HPLC columns (Bayer *et al.*, 1976; Hsu & Currie, 1978; Wilkinson, 1978; Schmidt *et al.*, 1979; Tapuhi *et al.*, 1981; Weiner & Tishbee, 1981; Koroleva *et al.*, 1982; Grego & Hearn, 1983; Simmons & Meisenberg, 1983).

They can be detected at the 100 pmol level by u.v. and at the fmol level with fluorescence (Bayer *et al.*, 1976; Koroleva *et al.*, 1982), or chemiluminescent methods (Koboyashi & Imai, 1980; Honda *et al.*, 1983).

Dansyl amino acids have not yet produced results to compare with those obtained by OPA derivatization. One disadvantage is the multiplicity of peaks which may be obtained from a single amino acid, e.g. *O*-dansyl tyrosine and *O,N*-di-dansyl-tyrosine. One advantage is that proline and hydroxyproline may be detected together with other amino acids. Wilkinson (1978) suggested that the dansyl amino acids might be useful for the analysis of peptides at high sensitivity and for the quantitation of *N*- and *C*-terminal amino acids. *C*-terminal amino acid amides cleaved enzymically from amidated peptides were detected at the 50 pmol level with u.v. (Simmons & Meisenberg, 1983). Dansyl amino acids, including their D- and L-isomers were also separated by HPLC (Lam *et al.*, 1980; Lam, 1982).

Ortho-phthalaldehyde derivatives

o-Phthalaldehyde reacts with amino acids in the presence of β-mercapto-ethanol at 25°C to give a substituted isoindole (Simons & Johnson, 1976). Compounds lacking an α-hydrogen atom give less than 1 % of the fluorescence response of an equivalent amount of glycine (Cronin & Hare, 1977) unless the reaction is carried out at 100°C (Cronin *et al.*, 1979). Cystine gives a fluorescence yield about 3 % that of glycine at 25° or 100°C.

In the presence of mercaptoethanol, the reaction of OPA with primary amino acids is complete in 0.1–1.0 min. But, the OPA derivatives of glycine, lysine, hydroxylysine and phosphoserine have half-lives of 1–4 min (Turnell & Cooper, 1982). The stability of OPA-2-mercaptoethanol derivatives was studied (Cooper *et al.*, 1984b) and their separation in 22 min by HPLC was reported (Cooper *et al.*, 1984a). The effect of fluorescence decay can be minimized by rapid injection onto the column after derivatization. Lysine and hydroxylysine give an improved fluorescence yield (40–50 %) if Brij 35 is added to the reagent solution. Cystine can be quantified as cysteic acid or carboxymethylcysteine.

Secondary amino acids (proline and hydroxyproline) do not react with OPA directly. They can be oxidized with hypochlorite and the liberated primary amino groups will then react. Alternatively, proline and hydroxyproline are determined in a separate analysis by derivatization with 4-chloro-7-nitrobenzofuran (NBD) (Umagat *et al.*, 1982, see Method).

OPA-amino acid adducts are prepared and then loaded onto the HPLC column for their determination with a fluorimeter. Excellent separations have been reported (Hill *et al.*, 1979, 1982; Hodgin, 1979; Lindroth & Mopper, 1979; Jones *et al.*, 1981; Larsen & West, 1981; Liebezeit & Dawson, 1981; Umagat *et al.*, 1982; Jones & Gilligan, 1983a, b; Cooper *et al.*, 1984a).

The electrochemical detector has been used for OPA-mercaptoethanol derivatives of amino acids. This detector offers scope not only for increasing the sensitivity of detection, but also for the selective detection of some derivatives, e.g. basic amino acids (Joseph & Davies, 1983).

Automatic analysis of the fluorogenic derivatives requires that they should be stable and for this reason the OPA-mercaptoethanol derivatives are not useful (Lindroth & Mopper, 1979; Turnell & Cooper, 1982). OPA-ethanethiol derivatives have a longer half-life, but they still show a decreasing fluorescence of 20–50 % over 6–8 h (Simons & Johnson, 1977). However, these were used for automatic on-line pre-column derivatization of plasma amino acids (Fleury & Ashley, 1983).

Cloete (1984) reported on an automated routine method for OPA derivatives with an Altex Ultrasphere ODS column (250 × 4.6 mm) particle size 5 μm, with a Brownlee guard column with 5 μm C_{18} cartridge. More than 2 000 samples were injected on a single column.

OPA was introduced into the programme for automated sequence analysis to block primary amines when proline was exposed at the *N*-terminus of the peptide. This resulted in decreased background interference and longer sequences being analysed (Brauer *et al.*, 1984).

Ortho-phthalaldehyde reagents

OPA reagent: Lee *et al.* (1979)
Dissolve 62 g of boric acid and 25 g of KOH in 900 ml of water. Adjust to pH 10.4 with 45 % KOH and make up to 1 l. Filter through a 0.22 μm Millipore membrane and add 2 ml of mercaptoethanol, 6 ml of 30 % (w/v) Brij 35 and 1.2 g of OPA dissolved in 15 ml of methanol. 'Age' overnight in the dark under nitrogen at room temperature. The reagent is stable for 1 month in the dark under nitrogen at 4°C.

The reagent was used for amino acids at the 10 pmol level (Lee *et al.*, 1979) and for phospho-amino acids (section 8.11.2.3).

OPA/ET Reagent: Kuwada & Katayama (1983)
Dissolve 100 mg of *o*-phthalaldehyde in 5 ml of methanol. Add 50 μl of ethanethiol (ET) and 10 ml of 0.15 M-sodium borate buffer pH 10.5, containing 0.2 % of Brij-35. Mix. Flush with nitrogen and stand in the dark for at least 16 h before use. To maintain reagent strength and 10 μl of ET every four days. (See section 8.12.2.3.)

Pre-column derivatization with OPA

The method of Jones and Gilligan (1983a, b) is described.

Reagents
Dissolve 50 mg of *o*-phthaldialdehyde in 1.25 ml of methanol and add 50 μl of 2-mercaptoethanol and 11.2 ml of 0.4 M-borate buffer (pH adjusted to 9.5 with 4 M-NaOH). Stable 1–2 weeks.

Or use fluoraldehyde-OPA reagent solution (Pierce). This contains OPA in borate buffer with added Brij 35 and 2-mercaptoethanol. Stable 6 months if stored at 4°C.

Derivatization
Standards or unknowns are dissolved in water. Proteins hydrolyzed with 6 M-HCl are lyophilized and the residue dissolved in water. If methanesulphonic acid is used for hydrolysis, neutralize the hydrolysate with 4 M-NaOH. Dilute if necessary with 0.4 M-borate buffer (pH 9.5) to give a final concentration of about 25 nmol maximum of any single amino acid per ml. Before hydrolysis the protein may be oxidized or carboxymethylated.

Take 5–10 μl aliquots of the sample and mix with 5 μl of the reagent solution. After 1 min, add 20–100 μl of 0.1 M-sodium acetate (pH 7.0). Mix. Analyse a 20 μl sample immediately. 2-Aminoethanolamine is the internal standard.

(To analyse serum, urine and cerebrospinal fluid: mix 25 μl vigorously with

75 μl of acetonitrile. Centrifuge at 10 000 g for 2 min. Take 10 μl aliquots for OPA derivatization.)

HPLC

Ultrasphere ODS column (lengths varying from 45 to 250 mm × 4.6 mm i.d.) with particle sizes 5 or 3 μm or Microsorb C_{18} column (100 × 4.6 mm i.d.).

Use a guard column (CO:PELL ODS sorbent, 40 μm particle size) particularly with 3 μm C_{18} resins.

The length of the column does not greatly affect the resolution of the amino acids, but shorter columns (45–100 mm) reduce the analysis time and are recommended. The resolution of glycine and threonine can be a problem. There is considerable variability between columns, even those from the same commercial source. In some cases glycine and threonine are resolved completely, whereas in other cases complete resolution cannot be obtained, despite changes in the elution system.

Fluorimeter with 9 μl cuvette; filters for excitation at 305–395 nm and emission at 420–650 nm.

A gradient is formed between two degassed solvents.

A. 0.1 M-sodium acetate, pH 7.2 (protein hydrolysates) or pH 6.4 (physiological fluids).

B. Absolute methanol (HPLC grade)

There is a linear relationship between peak area and pmol injected in the range 500 fmol to several hundred pmol. At levels below 5 pmol large deviations from linearity are observed with some amino acids such as serine, glycine and alanine, due to background contamination. At this level it is necessary to make corrections by subtracting values obtained by 'blank' runs. S = 25–50 fmol for most amino acids. The best working range is 500 fmol to 1 nmol.

Fig. 8.6 shows the separation of amino acids in 13.5 min on an Ultrasphere 3 μm ODS column. Fig. 8.7 shows the elution profiles obtained with hydrolysates of vasopressin and somatostatin under similar conditions.

Comment

More than 60 amino acids and amines have been separated by the OPA-pre-column technique. More recently analysis times of less than 7 min have been recorded with a Perkin Elmer 3 × 3, C_{18} column (4.6 × 32 mm, 3 μm particle size) (Jones, 1984, personal communication). Cysteine is determined as carboxymethylcysteine and methionine as methionine S-oxide. Proline and hydroxyproline are determined by the method of Umagat *et al.* (1982) below.

Pre-column derivatization with 4-chloro-7-nitrobenzofuran (NBD)

This method of Umagat *et al.* (1982) is used for proline (NBD) and hydroxyproline. An aliquot of the neutralized sample used for OPA derivatization is diluted if necessary to give a final concentration of *ca.* 20 nmol proline/ml.

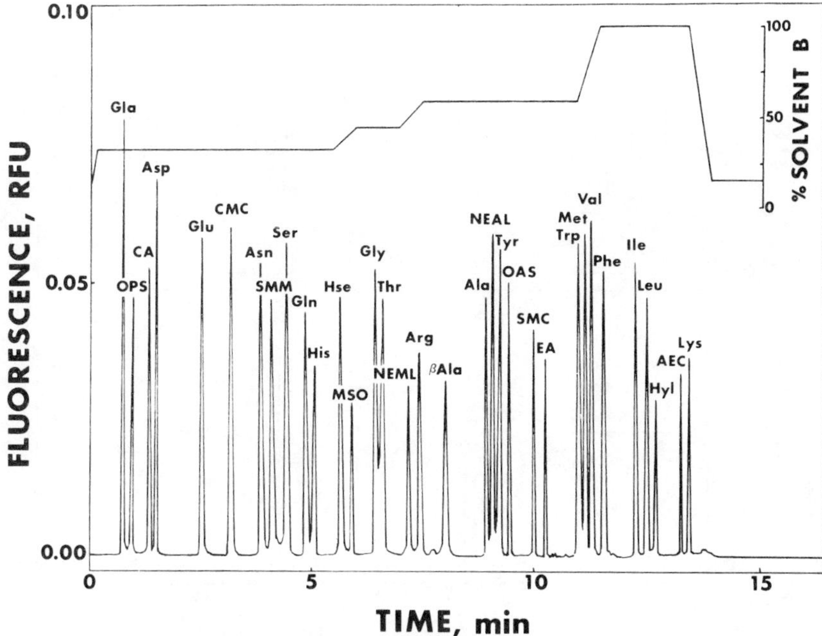

Figure 8.6. Elution profile of amino acid standards derivatized by the reaction with OPA. Each peak represents 40 pmol. The HPLC system consisted of two Beckman Model 100A pumps, a Beckman Model 421 microprocessor for generation of elution gradients and a Gilson Model 121 filter fluorimeter. The fluorimeter was equipped with a 9-μl cuvette and filters for excitation at 305–395 nm and emission at 420–650 nm. The emission filter was fitted with a 2-mm diameter aperature. The range setting of the fluorometer was 0.1 RFU and the time constant was 0.5 s. All sample injections were performed with a Beckman Model 210 injection valve equipped with a 20-μl sample loop. Chromatographic peaks were recorded and integrated by a Beckman C-R1A Integrator system. The reverse phase HPLC column used for the separation was an Ultrasphere ODS (75 × 4.6 mm i.d.; particle size 3 μm) fitted with a guard column (30 × 2.1 mm) packed with CO:PELL ODS sorbent (particle size 40 μm). Flow rate was 1.5 ml min^{-1} and initial column pressure was 3 100 psi. The separation was provided by courtesy of Dr. Barry N. Jones.

Mix equal volumes of the sample, 0.4 M-borate buffer (pH 9.5) and the NBD solution (2 mg/ml methanol). Heat at 60°C for 5 min in a closed screw-cap vial. Stop the reaction by cooling to 0°C. Inject aliquots onto the column.

HPLC

Use an Ultrasphere column (250 × 4.6 mm i.d.) with 5 μm ODS particles.
Solvent A. Tetrahydrofuran/0.05 M-sodium acetate pH 6.6 (1:99, v/v).
Solvent B. Methanol.

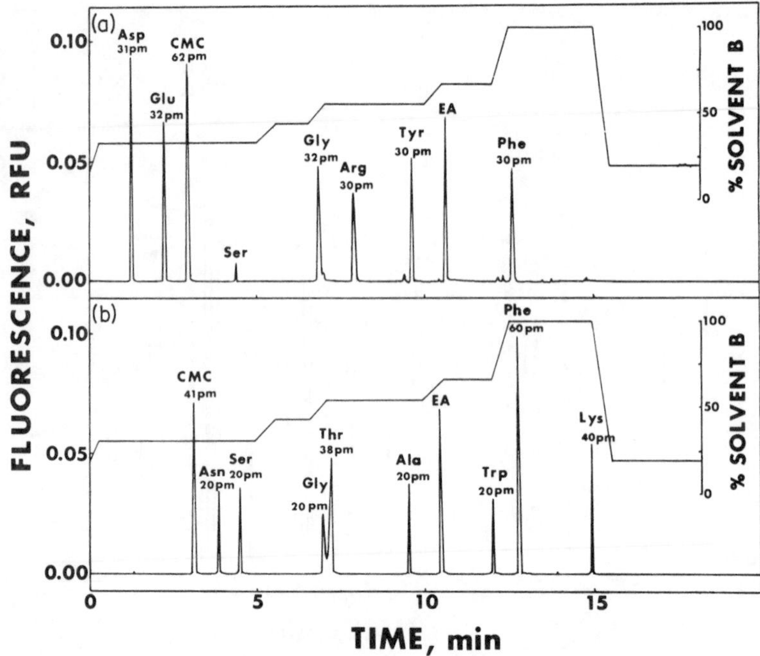

Figure 8.7. Elution profiles of hydrolysates of two carboxymethylated polypeptides. Quantitation of each peak in pmol is given. (a) Acid hydrolysate of vasopressin. The quantitation is in good agreement with the known composition of this polypeptide; Asp, Glu, CMC_2, Gly, Arg, Tyr, Phe. (b) Enzymic hydrolysate of somatostatin. Based on the amino acid sequence of somatostatin, the expected composition should be CMC_2, Asn, Ser, Gly, Thr_2, Ala, Trp, Phe_3, Lys_2. The separations were supplied by courtesy of Dr. Barry N. Jones.

Linear gradient from 30% B to 45% B at a rate of 1.5% B/min. Flow rate 1.0 ml/min.
Monitor at an excitation wavelength of 220 nm with a 370 nm cut-off filter and a sensitivity setting of 0.1 μA.

NBD-hydroxyproline and proline adducts are eluted within 6 min under these conditions.

Phenylthiocarbamyl derivatives

The chemistry of the reaction between PITC and amino acids is well known as a result of the Edman degradation procedure, where the phenylthiohydantoin derivatives are normally determined (Edman & Begg, 1967). The reaction between amino acids and PITC to form phenylthiocarbamyl (PTC) derivatives is the basis for determining amino acids (so-called PICO-TAG™ analysis, commercially-available from Waters Associates). These derivatives are stable, easily prepared and have been successfully determined after separation by reverse phase HPLC. Both primary and secondary amino acids react and are separated in one run. Detection is with a u.v. detector at 254 nm. This is cheaper and simpler than the fluorimeter used for OPA derivatives. Labelled PITC with ^{35}S can also be used. (Knoop *et al.*, 1982; Cohen *et al.*, 1984; Heinrikson & Meredith, 1984).

Pre-column derivatization with phenylisothiocyanate
The method of Heinrikson & Meredith (1984) is described.

Reagents
PITC (in 1 ml vacuum-sealed ampoules), triethylamine (Sequanal grade) and amino acids from Pierce. Pyridine and triethylamine distilled from ninhydrin and calcium hydride. Acetonitrile and methanol, HPLC grade from Burdick & Jackson. Coupling buffer: acetonitrile/pyridine/triethylamine/water (10:5:2:3, by vol.) Stable one month at −5°C. PTC-amino acids are stable if stored frozen at pH 6.8.

Derivatization
Dry down protein hydrolysates or solutions of amino acids in a small test tube. Dissolve the residue in 100 μl of coupling buffer and take down to dryness on the rotary evaporator (or Speed-Vac Concentrator). This preliminary drying procedure removes traces of HCl and avoids the appearance of an interfering peak which appears on the HPLC chromatogram near histidine.

Add 100 μl of coupling buffer and 5 μl of PITC. Allow to react for 5 min at room temperature. Take to dryness as before under high vacuum (50–100 μm). Dissolve the resulting PTC-amino acids in 250 μl of 0.05 M-ammonium acetate, water or water/acetonitrile (7:2, v/v). Inject 1–10 μl (100–1000 pmol of each amino acid) onto the reverse phase HPLC column. An aliquot can be injected onto a conventional amino acid analyser column to test for completion of the coupling reaction.

HPLC
Octyl and octadecyl columns (250 × 4.6 mm, 5 μm particles) are suitable. Those from Altex, Dupont and IBM were tested.
Solvent systems are:

A. 0.05 M-ammonium acetate, pH 6.8, adjusted with phosphoric acid.
B. 0.1 M-ammonium acetate, pH 6.8 in (1) acetonitrile/water (50:50, v/v)
(2) methanol/water (80:20, v/v)
(3) acetonitrile/methanol/water (44:10:46, v/v)
C. Acetonitrile/water (70:30, v/v)

Excellent separations were obtained using solvent gradients with the column temperature maintained at 52°C, in a total running time of 30–40 min. Absorption was measured in the u.v. at 254 nm. The details for one gradient system are given here.

Time (min)	%A	%B	%C
0	100	0	0
15	85	15	0
30	50	50	0
30.1	0	0	100
40	100	0	0

Table 8.1 gives the molar absorptivities of PTC-amino acids in the order of elution from the column.

Table 8.1. Molar absorptivities of PTC-amino acids in solvent system A, B(1), C (see text) (from Heinrikson and Meredith, *Anal. Biochem.*, **136,** 65–74, 1984).

Amino acid	Molar absorptivity 254 nm (1 mol/cm)
Ala	15 772
Glu	14 465
CMC	≏ 15 500
Hyp	15 420
Ser	14 824
Gly	15 326
His	12 645
Thr	15 152
Ala	16 316
Arg	13 874
Pro	15 235
Tyr	16 524
Val	16 295
Met	16 716
Ile	21 375
Leu	17 069
Phe	16 293
Lys	30 224

The method described was reported by Cohen *et al.* (1984) and details were supplied by courtesy of Waters Associates.

The procedure for amino acid analysis is called PICO. TAG™ analysis. Up to 12 samples can be derivatized at a time and after 10–20 min the entirely volatile mixture is removed by vacuum in specially designed apparatus. This minimizes the occurrence of interfering reagent peaks and allows the use of large excesses of PITC. The linearity of response is excellent in the range 5–1000 pmol. Automated analysis is possible. All protein amino acids, as well as asparagine, glutamine, cysteic acid, carboxymethylcysteine, homoserine, methionine-*S*-oxide, methionine-*S*,*S*-dioxide, hydroxyproline and hydroxylysine may be analysed on a specified PICO.TAG™ column. Separations achieved on this column in 12 min are shown in Fig. 8.8 with 250 pmol of Pierce H Standard and in Fig. 8.9 under the same conditions with 1 pmol of standard solution.

8.17. GAS–LIQUID CHROMATOGRAPHY OF AMINO ACIDS

Amino acids must be converted to volatile derivatives for their analysis by gas–liquid chromatography. There is only one single-step derivatization method, that of trimethylsilylation, which has been developed for the qualitative and quantitative analysis of all the protein amino acids (Gehrke *et al.*, 1969; Gehrke & Leimer, 1971). It has the serious disadvantage of giving multiple peaks for some amino acids, but it could be useful for the determination of aspartic and glutamic acids in the presence of asparagine and glutamine.

The most successful methods involve two steps of derivatization: (1) esterification of carboxyl groups and (2) acylation of other reactive groups (amino, imino, hydroxy, sulphydryl, guanidino). Derivatives which have been shown to give good separations for all or most of the protein amino acids are listed in Table 8.2.

Excellent reviews (Weinstein, 1966; Blau, 1968; Hušek & Macek, 1975; Blackburn, 1978b; MacKenzie, 1981) and a specialized volume (Zumwalt *et al.*, 1986) on amino acid analysis by GLC may be consulted for further detailed information.

There is no difficulty in obtaining satisfactory qualitative results. Quantitative work requires more persistence and there may be problems particularly with histidine and arginine. Cystine may be degraded on certain stationary phases (see Gehrke *et al.*, 1979; Darbre, 1986). The most extensive quantitative studies were carried out with the TFA amino acid n-butyl esters by Gehrke and his group and the HFBA amino acid isobutyl esters by MacKenzie and Tenaschuk. Details are given here for the preparation of the latter derivatives, with which this author has had some experience.

Most workers have used the Fischer method of alcohol esterification with various HCl concentrations. Modifications in the procedure were introduced in order to increase the yields. A direct single-step procedure is usual. An inter-esterification step was introduced (Gehrke *et al.*, 1965; Stalling *et al.*, 1967), whereby the amino acids were converted to methyl esters and these were then inter-esterified to n-butyl esters. This improved esterification and avoided the problems of low solubility of some amino acids in higher alcohols. However, these workers later reverted to direct esterification (Roach & Gehrke, 1969b).

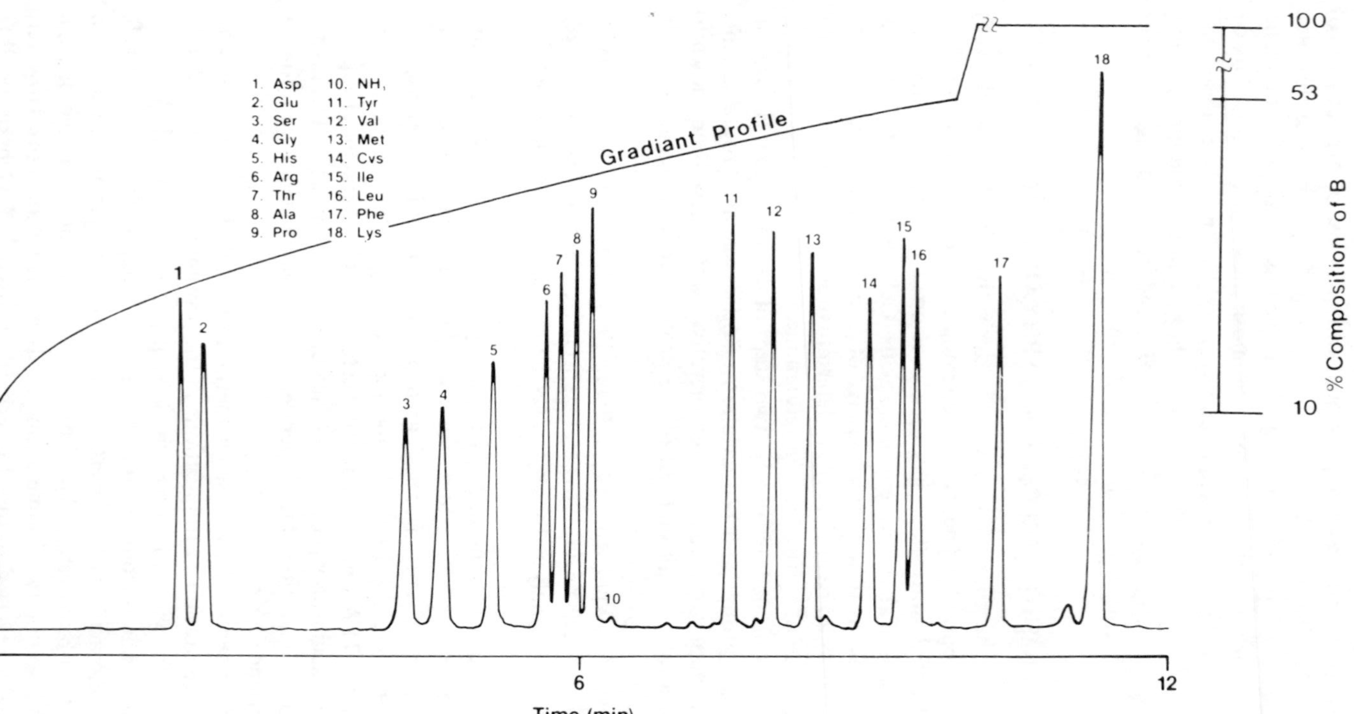

Figure 8.8. Pico-Tag™ method of amino acid analysis with 250 pmol of Pierce H standard solution. Conditions: Pico-Tag™ column (Waters) at 38° C; detector 254 nm at 0.1 a.u.f.s. Solvent A. 6 % acetonitrile, 94 % 0.14 M-sodium acetate, 3.6 mM triethylamine, pH 6.40. Solvent B. 60 % acetonitrile in water. Flow rate, 1 ml/min; Curvilinear gradient, 0–46 % B in 10 min, wash at 100 % B (Waters™ Curve 5). Supplied by courtesy of Waters Associates.

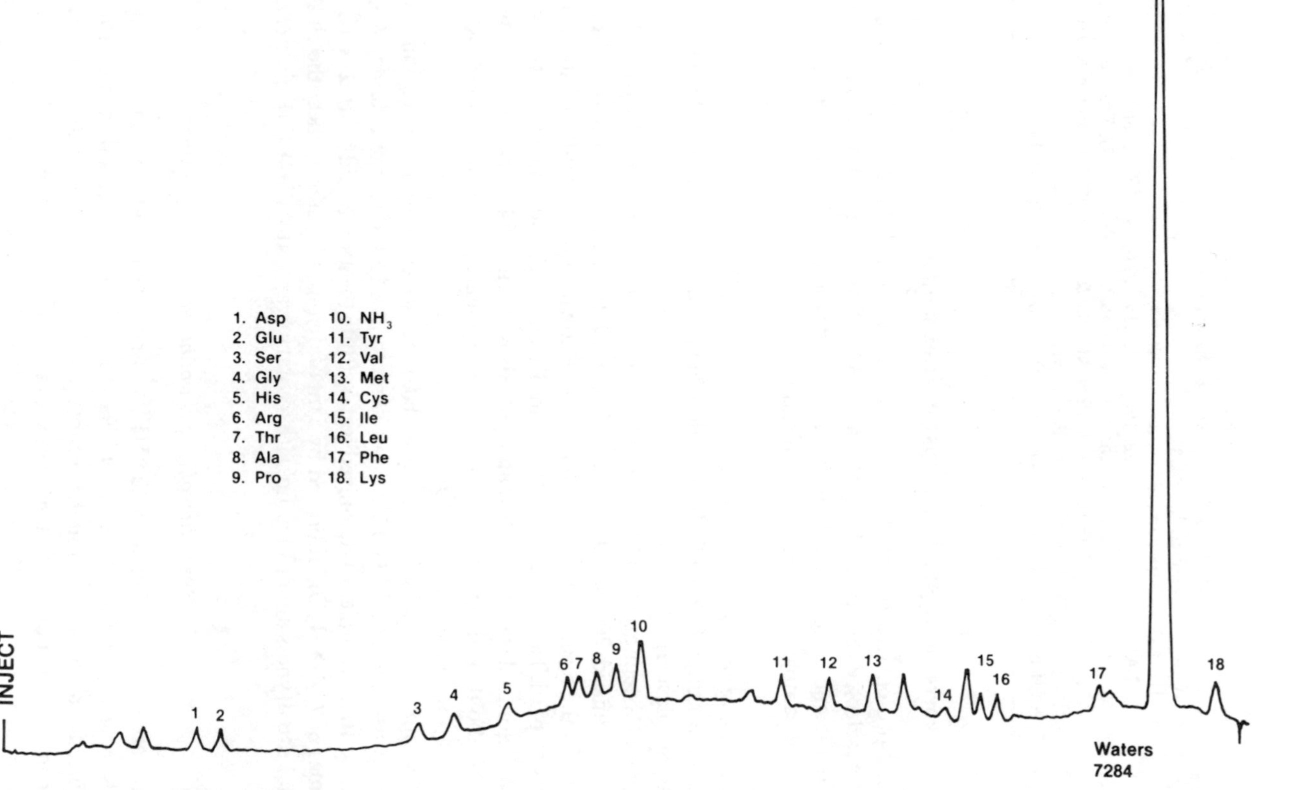

Figure 8.9. Pico-Tag™ analysis of 1 pmol of Pierce H standard solution. Conditions as in Fig. 8.8, except detector set at 0.005 a.u.f.s. Supplied by courtesy of Waters Associates.

Table 8.2. Some derivatives of amino acids for gas–liquid chromatography.

Ester	Acyl group	References
Methyl	TFA	Darbre & Islam, 1968; Cliffe *et al.*, 1973
n-Propyl	TFA	Gamerith, 1983a, b
	HFB	Kirkman, 1974; March, 1975
n-Butyl	TFA	Roach & Gehrke, 1969a; Gehrke *et al.*, 1971; Zumwalt *et al.*, 1985
Isobutyl	HFB	Mackenzie & Tenaschuk, 1974, 1979a, b; Pearce, 1977; Siezen & Mague, 1977; Desgres *et al.*, 1979; Moodie & Burger, 1981; Chauhan & Darbre, 1982a, b; Moodie *et al.*, 1983
Isoamyl	HFB	Zanetta & Vincendon, 1973; Felker & Bandurski, 1975

8.17.1. General procedure for preparation of sample

Add to a protein hydrolysate a known amount of internal standard. Norleucine is usually selected because this separates well from other amino acids, but pipecolinic and tranexamic acids and homoarginine have also been used. Some biological samples may require purification.

Purification of biological samples

Plasma samples (100–500 μl) are deproteinized with sulphosalicylic acid (50mg/ml plasma).

1. Pack a Pasteur pipette (40 × 3 mm I. D.) with a 10 mm column of Dowex-50W (8 % cross-linked, 200–400 dry mesh, H^+ form) with small glass wool plugs above and below the resin. Spike the sample with a known amount of internal standard. Adjust to pH 1.0 and layer onto the resin. Remove non-cationic impurities with 2 ml of deionized water. Elute the amino acids with 2 ml of 4 M-NH_4OH at a flow rate of about 1 drop every 5 to 10 s. The eluate is taken to dryness for derivatization (Desgres *et al.*, 1979).
2. The amino acids dissolved in 0.1 M-HCl are placed onto a 7-ml cation resin bed of Amberlite CG-120 (H^+), 100–120 mesh in a 150 × 15 mm column. Wash with 2 × 10 ml volumes of deionized water. Elute the amino acids with 2 × 10 ml volumes of 7 M-NH_4OH followed by 5 ml of deionized water. Collect the eluate and take to dryness under vacuum with a rotary evaporator at 60°C (Kaiser *et al.*, 1974).

8.17.1.1. Heptafluorobutyryl amino acid isobutyl esters

Details are given for preparing the HFB amino acid isobutyl esters (method of MacKenzie & Tenaschuk, 1974, 1979a, b). The flame ionization detector is ubiquitous and sensitive enough for most purposes. For special requirements the electron capture and alkali flame detectors (Chauhan & Darbre, 1982a, b) and mass spectrometry (MacKenzie & Hogge, 1977; Desgres *et al.*, 1979) were used.

After drying down samples both before and after esterification, the last traces of water may be removed azeotropically by adding methylene chloride and taking to dryness. Sonication or vortex mixing may also be used to ensure complete mixing and reaction of dried residues with the reagent solutions for esterification and acylation.

Preparation of alcohol-HCl solutions for esterification

The alcohol should be dried over granulated calcium sulphate (ca. 50 g/l) *or* refluxed 3–4 h over magnesium turnings (add a crystal of iodine to catalyze the reaction), decanted and refluxed for 3–4 h over calcium hydride and redistilled.

The desired molarity of HCl in the alcohol may be obtained by passing dry HCl gas into the alcohol to saturation or to the appropriate weight, gravimetrically, and then adjusting the volume after titration with alkali. Store in small bottles in the cold.

When small amounts are required for occasional samples, add the appropriate amount of acetyl chloride to the alcohol in the cold (cautiously, with several additions because of the exothermal reaction) and make up to volume.

Esterification

The amino acid sample (10–100 nmol of each amino acid) is taken to dryness at 50°C in a small Reactivial (Pierce) using a stream of dry nitrogen gas. Add isobutanol-3 M-HCl (50–100 μl) and heat at 120°C for 30 min. After about 5 min the vial is agitated for 20–30s with a vortex mixer and returned to the heater block for the remaining period. Cool to room temperature. Remove excess reagent with nitrogen gas. This is rapid if the vial is at 50°C. Methylene chloride may be added for azeotropic removal of water, but we have not found this to be necessary. (Gamerith (1983b) reported similarly with the propan-1-ol esters.)

Acylation

Add heptafluorobutyric anhydride (50 μl) and heat at 150°C for 10 min. After cooling to room temperature, the sample is taken just to dryness with nitrogen. Excessive exposure to a stream of nitrogen gas will lead to losses of the more volatile derivatives.

Dissolve the derivatives in dry distilled acetonitrile or ethyl acetate (50 or 100 μl). Inject 1–2 μl onto the GLC column.

If histidine is present, take up the sample in 50 μl of a mixture of ethyl acetate and acetic anhydride (1:1, v/v) and heat the vial at 150°C for 3 min. After cooling to room temperature, the sample may be diluted with ethyl acetate for GLC.

Derivatization is best carried out in heater blocks with the holes partly filled with silicone oil such that this is level with the liquid inside the Reactivial. This allows refluxing in the upper part of the vial.

Losses may be incurred when Reactivials are incompletely sealed and small Pyrex tubes (50 × 3 mm i.d.) are sometimes preferable. This involves some glass-sealing and cracking open of tubes, for each derivatization step.

Our recommendation is to carry out the analysis as soon as possible after derivatization (within 2–3h preferably), particularly with regard to cysteine, cystine, methionine, arginine and histidine.

Columns

Quantitative studies and good separations of the HFBA amino acid isobutyl esters were reported by several groups of workers. Different columns were used (given below) but the order of elution was the same. Variations in the temperature programme may be introduced to improve the separation desired. Carrier gas flow rates, injection port and detector temperatures may be adjusted for the instrument and column being used.

1. Capillary glass column (25m × 0.4 mm i.d.) coated with 5% Chromosorb R and 15% OV-101. Carrier gas: hydrogen, 3ml/min. Make-up gas: nitrogen, 30 ml/min. Hydrogen, 27 ml/min. Air, 350 ml/min. Column temperature 90°C with programme 4°C/min. Order of elution over 35 min: Ala, Gly, Val, Thr, Ser, Leu, Ile, Nle, Pro, Cys, pipecolinic acid, Hyp, Met, Asp, Phe, Glu, Lys, Tyr, Arg, CMC, His, Homoarg, Trp, Cys Cys (Chauhan & Darbre, 1982a, b).
2. Capillary column (25 m × 0.23 mm i.d.) coated with OV–101 stationary phase from LKB (Desgres *et al.*, 1979).
3. Glass SCOT capillary column (60 m × 0.5 mm i.d.) with SE-30 stationary phase from SGE (Melbourne) (Pearce, 1977).
4. Glass column (3.5 m × 2.5 mm i.d.) packed with Gas-Chrom Q, 100–120 mesh coated with 3% SE-30 (w/w) from Applied Science Laboratories (MacKenzie & Tenaschuk, 1974).
5. Glass column (3.1 m × 2 mm i.d.) packed with Chromosorb W HP, 100–120 mesh coated with 3% SE-30 (MacKenzie & Tenaschuk, 1979a, b).
6. Glass column (3.5 m × 2 mm i.d.) packed with Gas-Chrom Q, 80–100 mesh coated with 3% SE-30 from Supelco (Bellafonte, Pa., U.S.A.) (Pearce, 1977).
7. Glass column (6 m × 2 mm i.d.) packed with Gas-Chrom Q, 80–100 mesh coated with 3% OV-101 (Siezen & Mague, 1977).

8.18. PROTEIN DETERMINATION

8.18.1. Introduction

Proteins may be determined in a variety of ways and these are summarized below. Bovine serum albumin is often used as a standard. It should be dried at 60°C *in vacuo* over P_2O_5 to constant weight, and this may be conveniently carried out using a Fischer pistol apparatus. Alternatively, with BSA in solution use $A^{1\%}_{1\text{cm }280}$ 6.6 (Tanford & Roberts, 1952). When only small quantities of protein are available, it is not usually possible to determine the weight of the sample on an ash-free and moisture-free basis (Hirs, 1967). Amino acid analysis is commonly used today. Lewis (1979) studied a sample of 350 proteins and considered nitrogen-based assay procedures gave most consistent results.

8.18.1.1. Gravimetric method

This method requires mg quantities of material.

8.18.1.2. Nitrogen determination

This method also requires mg quantities of material and the methods are time-consuming and difficult. The Kjeldahl method involves complete destruction of the protein with sulphuric acid and added catalysts. The nitrogen of the protein is trapped as ammonium sulphate and subsequently determined as ammonia after steam distillation from the hydrolysate made alkaline by the addition of concentrated NaOH. The ammonia is usually determined by trapping in standard acid and back-titration, although other methods such as colorimetry or the ammonia gas-sensing electrode (Clark, 1979) may be used. Apparatus for the Kjeldahl method with on-line titration is available (Buchi Laboratory-Techniques Ltd., Flawil, Switzerland). The method is still widely used in the foodstuff and brewing industries, but in protein chemistry it has been outmoded by amino acid analysers, which are about 100 times more sensitive for ammonia determination and also provide a total amino acid analysis at the same time. (See Ballentine, 1957; Bradstreet, 1965).

8.18.1.3. Amino acid analysis

The protein is hydrolysed completely and the hydrolysate subjected to amino acid analysis. Single amino acids are easily determined at the 5–10 nmol level, although many specialist laboratories work routinely at the 50–100 pmol level.

8.18.1.4. Colorimetric methods

Many of these methods are easily used but may not all be applicable to the protein under study. Erroneous results are sometimes obtained because the test protein may have a different structure to that of the standard (usually BSA).

1. The biuret method is one with specific reaction between alkaline cupric ion and the peptide bond. It is simple but limited in sensitivity ($S \simeq 300\ \mu g$ protein/ml). There have been many variations of the reagent, introduced to improve colour

yield, stability of colour and elimination of interference from sulphydryl compounds, such as dithiothreitol.

2. The most widely used is that of Lowry *et al.* (1951). It is basically dependent upon the colour obtained on reacting the Folin-Ciocalteu phenol reagent with Tyr residues, although other chromogenic amino acids such as Trp, His and Cys and peptide linkages are involved. The standard curve is not linear. The method is generally applicable unless the protein does not contain Tyr. Detergents (often used to solubilize proteins from membranes) interfere and many modified methods have been proposed.

3. Non-specific dye-binding. The determination of protein by a simple colorimetric method by reacting protein with a dye is an attractive proposition. Many different dyes have been used. The methods are frequently rapid, sensitive and reproducible under given conditions, but the extent of dye-protein binding depends on the individual protein and this is subject to interference. Also, the absorbance-protein concentration is often a non-linear relationship and the slope of the curve may differ greatly between proteins.

Protein in leaf tissue was determined with xylene brilliant cyanin G and naphthalene blue black (Bramhall *et al.*, 1969). Sulphobromophthalein was reacted with protein under acid conditions and the precipitated acid dye-protein complex was collected by centrifugation, redissolved in alkaline solution and read at 580 nm (Bonting & Jones, 1957; McGuire *et al.*, 1977). A specific method for collagen (10 μg–100 μg) made use of Sirius Supra red F3BA (Junqueira *et al.*, 1979). Coomassie Brilliant Blue was successfully introduced (Bradford, 1976; Sedmak & Grossberg, 1977; Spector, 1978).

A comparative study of the protein content of samples from cultured cells and subcellular fractions was made using the Lowry and Bradford methods (Chiappelli *et al.*, 1979). The Lowry method gave higher values, but the ratio of the values obtained by the two methods was constant (mean $1.58 \pm$ S.D. 0.09). Thus the Bradford method was recommended because of its rapidity.

No single colorimetric method can be guaranteed to give accurate results with all proteins, because of the individual nature of proteins and also in some cases because of the presence of contaminating detergents or lipid material, resulting from the method of extraction or the source of the protein. Thus, the reaction of proteins with homobifunctional cross-linking reagents was shown to cause decreased dye response with Coomassie Blue or silver staining (Leftak, 1983). Several methods are given here which should meet most requirements.

8.18.1.5. Spectrophotometric methods

Spectrophotometric methods have much to commend them. They are simple, rapid and non-destructive. The earliest, well-established method of Warburg and Christian (1941/42) by making use of the 280/260 nm absorbance ratios allows protein to be determined in the presence of nucleic acid. Several methods have since been proposed involving readings at different wavelengths. Details are given later.

8.18.1.6. Fluorometric methods

Fluorimetric analysis offers a non-destructive and highly sensitive method of analysis, provided that a fluorogenic group can be introduced onto the protein or polypeptide. The analysis of amino acids was discussed in section 8.16.3 *et seq.*

8.18.2. Protein determination with the Folin–Ciocalteu reagent

8.18.2.1. Method of Lowry et al. (1951)

Reagents

A. 2% (w/v) Na_2CO_3 in 0.10 M-NaOH.
B. 0.5% $CuSO_4.5H_2O$ in 1% Na or K tartrate.
C. Alkaline copper solution. Mix 50 ml reagent A with 1 ml reagent B. Discard after one day.
D. Mix 50 ml of 2% (w/v) Na_2CO_3 with 1 ml of reagent B. Discard after one day.
E. Folin-Ciocalteu reagent. This can be prepared in about 1 day but it is now readily available commercially with a long shelf life if kept at 0°–4°C. Titrate an aliquot of the reagent to phenolphthalein end-point (pH 10.0). On the basis of this titration dilute the reagent about 2-fold to make it 1 M in acid (E).

Standard

Crystalline BSA or other pure protein. Prepare a solution with 500 μg/ml. Pipette into separate tubes for determination 10 μl–200 μl (5 μg–100 μg) and make up to 200 μl with water.

Method

In a 3 to 10 ml test tube pipette 0.2 ml of protein solution (5 μg to 100 μg). Add 1 ml of reagent C. Mix with a vortex mixer and allow to stand at least 10 min. Add rapidly 0.10 ml of reagent E and mix immediately (total volume 1.3 ml). After 30 min or longer read at 750 nm against a blank. Read concentration from standard curve. Any multiple of the volumes may be used. Many cuvettes require 3 ml of solution.

If the protein solution is dilute ($\simeq$25 μg/ml) use 0.5 ml + 0.5 ml of exactly double strength reagent C + 0.1 ml of reagent E, as described.

Sometimes proteins dissolve badly in 0.1 M-alkali reagent A. In such a case, add 0.1 ml of 0.1 M-NaOH to the dried protein film (5–100 μg). It may be necessary to heat to dissolve the protein. Add 1 ml of reagent D (no NaOH) to the dissolved protein and after a further 10 min, add 0.1 ml of reagent E.

Comments
The amount of colour varies with different proteins. The colour/protein relationship is only approximately linear between 15 μg and 40 μg of protein (30–80 μg, if reagent volumes are doubled). The standard curve is non-linear, but the method is simple, reproducible and sensitive (S $\simeq$ 0.2 μg protein). From studies of the calibration curve, a simple linear transform for protein concentrations upto 1.0 mg/ml was proposed (Coakley & James, 1978).

Several modifications have been recommended to avoid problems caused by interfering substances. In many cases the incorporation of these substances into the standard protein and reagent blank solutions will allow adequate compensation for an accurate determination.

The presence of SDS does not affect the reaction (Lees & Paxman, 1972). 2-Mercaptoethanol interferes and may be removed by *completely* drying the sample in vacuum before adding the reagents (Makkar *et al.*, 1980). Where interfering substances such as phenolic catecholamines are present (Pollard *et al.*, 1978) it is necessary to perform a preliminary precipitation of proteins with trichloroacetic acid (Hoffman *et al.*, 1976; Nikodijevic *et al.*, 1976; Zinder *et al.*, 1976). A modified procedure was described using chloramine-T for protein solutions containing purines and/or sulphydryl compounds (Higuchi & Yoshida, 1977, 1978, 1980).

An extensive review of the method listing many compounds which interfered with the determination (e.g. purines, glycine, hydrazine, ammonium sulphate (> 0.15 % final concentration) and phenols was published (Peterson, 1979).

8.18.2.2. Simplified method of Lowry et al.

The method of Peterson (1977) is described.

Stock reagents

1. Copper tartrate-carbonate (CTC). Add a solution of 20 % sodium carbonate slowly with stirring to a solution of copper sulphate-tartrate, to give a final conc. of 0.1 % $CuSO_4.5H_2O$, 0.2 % potassium tartrate, 10 % sodium carbonate. Stable at least 2 months at 20°C.
2. 0.8 M-NaOH.
3. 10 % SDS (Sigma, Pierce).
4. Folin-Ciocalteu reagent (2 M from Fisher Scientific).

Working solutions

1. 0.15 % Sodium deoxycholate (DOC).
2. 72 % TCA
3. BSA 0.5 mg/ml. Discard after 5 days in refrigerator.

4. Reagent A. Mix equal volumes of stock CTC, NaOH, SDS and water. (Keeps 2 weeks at 20°C). A small amount of precipitate does not interfere if well shaken before use.
5. Reagent B. Dilute Folin-Ciocalteu reagent with water (1 : 5, v/v).

Method

Protein (5–100 μg) is made up to a total volume of 1.0 ml with water. Add 0.1 ml of 0.15 % DOC. Allow to stand for 10 min at 20°C. Add 0.1 ml of 72 % TCA. Mix. Centrifuge at 3 000 g for 15 min. Decant off supernatant completely. (This precipitation step may be omitted if interfering substances are absent or appropriately controlled.)

Take up the protein pellet in 1 ml of water. Add 1 ml of reagent A. Mix to dissolve. Allow to stand for 10 min at 20°C. Add 0.5 ml of reagent B and mix immediately. After 30 min read at 750 nm. Colour loss is 1–2 %/h at 20°C. A linear transformation is obtained by plotting log absorbance/log protein concentration. The DOC-TCA precipitation method allows the rapid quantitative recovery of soluble and membrane proteins.

The protein (1–20 μg) may be determined by reducing all volumes by a factor of five.

8.18.2.3. Lowry method giving linear response

Method 1: Hartree (1972)

Reagents
A. 2 g of KNa tartrate + 100 g of Na_2CO_3 dissolved in 500 ml of 1.0 M-NaOH and made up to 1 l with water.
B. 2 g of KNa tartrate + 1 g of $CuSO_4.5H_2O$ dissolved in 90 ml of water and 10 ml of 1 M-NaOH added.

Reagents A and B will keep for 6 months if stored in polyethylene containers.
C. Folin–Ciocalteu reagent diluted with water (1:15, v/v). The solution should be prepared daily and when titrated with 1 M-NaOH to pH 10 (phenolphthalein) should be 0.15–0.18-M in H^+.

Procedure
Protein samples and standards are diluted to 1.0 ml in 13 mm diameter test tubes. Add 0.9 ml of reagent A. Mix and place in a water bath at 50°C for 10 min. Cool to room temperature and add 0.1 ml of reagent B. Leave at room temperature for at least 10 min, then add 3 ml of reagent C with extreme rapidity and mix within 1 second. Heat at 50°C for 10 min. Cool and read at 650 nm. (Protect the cuvettes from etching by washing with HCl at the end of the work.)

Comments

Absorbancies at 650 nm are 30% lower than those at 750 nm, but colour yields by this method are higher than those of the original Lowry method and give a linear relationship with the amount of protein (15–110 μg). Also, it is claimed that the insolubility of some proteins does not present a problem by this method. Sucrose and Triton X-100 affect the result as they do in the original method.

Method 2: Hess *et al.*, (1978)

Reagents

A. Dissolve 0.2 g of sodium tartrate and 10 g of Na_2CO_3 in 55 ml of 1 M-NaOH. Dilute to 100 ml with distilled water.
B. Dissolve 2 g of sodium tartrate and 1 g of $CuSO_4.5H_2O$ in 90 ml of water. Add 10 ml of 1.0 M-NaOH.
C. Dilute Folin-Ciocalteu reagent with distilled water (1 : 2, v/v). Prepare freshly each day.

Procedure

The protein is made up to 150 μl in an acid-washed tube. Add 100 μl of 2 M-NaOH. Vortex mix. Add 180 μl of reagent A. Mix. Allow to stand for 30–45 min at room temperature. Add 20 μl of reagent B. Mix. Allow to stand for 20 min or longer. Start vortex-mixing and run in quickly 300 μl of reagent C, followed by a second aliquot of 300 μl. Allow to stand for 30–60 min at room temperature. Read at 650 or 750 nm.

Prepare a standard curve 0–50 μg.

Comments

The standard curves were both linear up to absorbance 1.1, with that for 750 nm being slightly higher. The improved results are due to the high concentration of reagent C. SDS does not interfere.

8.18.2.4. Lowry method in the presence of detergents

Method 1: Wang and Smith (1975)

Protein may be determined in the presence of SDS and Triton X-100.

Reagents

A. Alkaline copper solution. Dissolve 250 mg of copper EDTA (ethylene diamine tetra-acetic acid) in 800 ml of distilled water. Add 100 ml of 20% Na_2CO_3 (w/v) and 10 ml of 10 M-NaOH. Mix and make up to 1 l with water.

B. 10% SDS solution.
C. Folin-Ciocalteu reagent (Harleco) diluted with water (1:1, v/v).

Standard
Dissolve 50 mg of BSA in 50 ml of 0.1% Triton X-100 (v/v). Aliquots are diluted to contain 40, 60, 80, 100 and 120 μg/0.2 ml.

Procedure
Place 0.2 ml of protein solution in a test tube and add 1 ml of reagent A. Mix on a vortex mixer and stand at room temperature for 15 min. Add 1 ml of reagent B. Mix on a vortex mixer. Add 0.1 ml of reagent C and mix. Allow to stand for 30 min then read at 500 nm.

Comments
The addition of 10% SDS is essential, because it prevents formation of a precipitate when the Folin reagent is added to samples containing Triton X-100. A linear relationship between colour and protein (40–120 μg) was established. Triton X-100 up to 1% concentration of the sample is permissible. If higher concentrations than 1% of Triton X-100 are encountered, then it is necessary to add equivalent amounts to the protein standard to correct for excess colour formation. Mather & Tamplin (1979) removed Triton X-100 by a single extraction with isoamyl alcohol before using the Lowry method.

Method 2: Cadman *et al.* (1979)

Procedure
Use the same reagents as in method 1.
The protein is made up to 0.7 ml solution. Use BSA standard 0–60 μg. Add 1 ml of reagent A. Mix and allow to stand at room temperature for 15 min. Add 0.5 ml of reagent B. Add 0.1 ml of reagent C and mix immediately. Allow to stand for 30 min at room temperature and read at 500 nm. Both standards and samples should contain the same amount of detergent.

Comments
In the presence of SDS (final concentration 4.35%) other detergents were permissible at a final concentration of 1% (Cutscum, Nonidet-P40, Tween 20) and 0.75% for cetyltrimethylammonium bromide.

8.18.3. Protein determination with the biuret reaction

8.18.3.1. Micro-biuret method

This is the method of Itzhaki & Gill (1964).

Reagents

A. 0.21 % $CuSO_4.5H_2O$ in 30 % NaOH. (To prevent precipitation of cupric hydroxide, add 21 ml of aq. $CuSO_4.5H_2O$ solution (1 % w/v) to 75 ml of 40 % NaOH and make up to 100 ml with water).

B. 30 %NaOH.

Both reagents in Pyrex bottles keep for at least 6 months.

Procedure

To 2 ml of protein solution (0.02 mg to 0.53 mg/ml) add 1 ml of reagent A. Mix thoroughly. Stand at room temperature for 5 min and read at 310 nm against a blank. Prepare tubes with the following mixtures:

2 ml protein solution + 1 ml reagent A	X_1 = absorbance reading
2 ml distilled water + 1 ml reagent A	X_2
2 ml protein solution + 1 ml reagent B	Y_1
2 ml distilled water + 1 ml reagent B	Y_2

Absorbance for protein = $(X_1 - X_2) - (Y_1 - Y_2)$
Read against standard curve.

Comments

The method was adapted from one previously published (Zamenhoff, 1957) and can be used in the presence of DNA. DNA concentrations up to 0.7 mg/ml in the final reaction mixture did not interfere, i.e. DNA/protein ratios of 10 : 1 or greater.

The colour was developed in less than 5 min and was stable for 2.5 h for 7 proteins studied, but in the case of lysozyme there was 11 % increase in colour over this period of time. It is about 1/16 as sensitive as the Lowry method, but it is less specific for type of protein and gives a linear absorption/protein relationship. The reaction is unaffected by the presence in the 2 ml protein solution of NaCl (1.5 M), Na acetate (0.1 M), Na formate (0.75 M), perchloric acid (0.67 M) or Na_2H-phosphate (0.01 M). Urea concentrations higher than 2.0 M in the protein solution required small corrections by subtracting the corresponding absorption for the urea alone.

An interesting method was proposed in which the protein is first reacted with biuret reagent and then with 2 M Folin-Ciocalteu reagent. It was claimed to give good reproducibility (± 2 % for 50–600 μg protein/ml). The non-linear absorbance/protein relationship was expressed as a straight line by a Hill plot (Ohnishi & Barr, 1978).

8.18.3.2. Biuret method for samples containing thiols

The method of Pelley *et al.* (1978) is described.

Reagent

Dissolve 3.9 g of $CuSO_4.5H_2O$ and 6.7 g of Na_2EDTA in 700 ml of distilled water. Add 200 ml of 20% NaOH with constant stirring to give a final volume of 900 ml.

Method

A solution containing 50–1000 μg of protein in a test tube is made up to total volume of 0.1 ml. Add 1.0 ml of biuret reagent. Incubate for 5 min at 60°C. Read at 545 nm.

Comments
The method is not sensitive but is simple and the absorbance curve is linear up to 1000 μg protein. S = 50 μg. The colour is stable for 1 h minimum. Iodoacetamide (12.5 μg) is added to eliminate interference due to large amounts of dithiothreitol. This EDTA-chelated reagent (Chromy *et al.*, 1974) as described is superior to tartrate-chelated reagent (Gornall *et al.*, 1949) for eliminating interference by sulphydryl compounds.

Johnson (1978) used a copper–diethyldithiocarbamate complex in the presence of deoxycholate to obtain a linear standard curve with BSA concentration up to 50 μg/ml.

8.18.3.3. Biuret method for samples containing detergents

The method of Watters (1978) is described.

Reagent

1.5 g of $CuSO_4.5H_2O$, 4.9 g of disodium tartrate dihydrate and 7.5 g of NaOH in 1 l of water. Stable 4 months in plastic bottle at room temperature.

Method

Add 0.5 ml of protein solution (containing 0.2–0.4 mg of protein and 2% detergent) to 2.5 ml of biuret reagent. Mix. Allow to stand at room temperature for 30 min. Read at 540 nm against a blank containing 0.5 ml of 2% detergent solution.

The modified reagent avoids turbidity with detergents which are used in extracting proteins. BSA gave a linear absorbance curve over the range 0.4–8 mg/ml.

8.18.3.4. Biuret method for samples of whole fatty tissues

The method of Beyer (1983) is described.

Procedure

Use the reagent above (Watters, 1978).
Extract the aqueous tissue extract with 3 ml of acetone-petroleum ether (35°–60°C) (1 : 1, v/v) by vortex mixing for 30 s. Centrifuge and decant. Add 0.1 ml of 10% sodium deoxycholate, pH 8.0 and 2.9 ml of reagent to the extracted residue. Seal the tube with Parafilm and mix by gentle inversion. Sonicate 2 × 15 s at 20 KC with a Branson Sonifier tapered micro-tip with No. 2 power output. Place the tube in a boiling water bath for 30 s for full colour development. Cool to room temperature by standing in ice. Read at 540 nm against a reagent blank.

8.18.4. Protein determination with Coomassie Brilliant Blue

8.18.4.1. Method of Bradford (1976)

Reagents

Dye solution. Dissolve 100 mg of Coomassie Brilliant Blue (CBB) G-250 (Sigma) in 50 ml of 95% ethanol. To this solution add 100 ml of phosphoric acid (85% w/v) and make up to 1 l with distilled water. Filter. Keeps for 2 weeks at 20°C. A commercial kit is available (Bio-Rad, Pierce).

Method

Protein solutions with 10 μg to 100 μg protein are pipetted into test tubes (12 × 100 mm) and made up to a total volume of 0.1 ml with 0.15 M-NaCl solution. Add 5 ml of dye solution and mix by inversion or vortex mixing. After 2 min and before 1 h read at 595 nm against a reagent blank (0.1 ml of the appropriate buffer and 5 ml of dye solution).

The method may be scaled down using 1–10 μg protein in a total volume 0.1 ml + 1 ml of dye solution. Read in a 1 ml cuvette at 595 nm.

Cuvettes should be washed with glassware detergent, followed by water and acetone to eliminate any protein-dye complex binding to the glass surface. Alternatively the cuvettes should be soaked in 0.1 M-HCl overnight.

Comments
There is some non-linearity of response, but this dye-binding assay is about four times more sensitive than the Lowry method. The colour is stable ±4% for 1 h. Phenolic catecholamines do not interfere and the method was applied to adrenal gland fractions (Pollard *et al.*, 1978). Sodium, potassium and magnesium chloride, ammonium sulphate and ethanol have no effect. Strongly alkaline buffering agents, Tris, acetic acid, 2-mercaptoethanol, sucrose, glycerol, EDTA and trace quantities of detergents have some effects but these can be eliminated by running appropriate controls. See also Spector (1978) for a list of reagents which influence the reaction.

Critiques of the Bradford method indicated that the method was indeed rapid and reproducible but that absorbance at 595 nm/μg protein plots were not linear and in addition the slopes of the curves between proteins varied considerably (Pierce & Suelter, 1977; Van Kley & Hale, 1977). Thus a standard curve for a specific protein should not be applied indiscriminately to an unknown protein. Because of these problems, it was suggested that a better method would be readings at 280 nm (protein > 50 μg/ml) or 205 nm (protein < 50 μg/ml) (Pierce & Suelter, 1977).

Spector (1978) re-examined the Bradford method and found the extinction coefficient of the dye-protein complex to remain constant over the protein concentration of 0.8 to 10 μg/ml of protein. He recommended that the volume of dye-reagent (0.5–5.0 ml) should be selected which dilutes the protein sample to give a final concentration of 0.8 to 10 μg/ml. This gives a linear Beer's Law relationship for accurate determination of $\leqslant 0.5$ to 50 μg of protein.

Proteins in concentrations as low as 2 μg/ml were determined using Coomassie Brilliant Blue R (Sigma). The protein was collected on a glass fibre, disc, washed with trichloroacetic acid, stained with CBB R, and the protein-dye complex removed with methanolic NaOH for colorimetric determination at 590 nm (McKnight, 1977). When protein determinations involve preliminary precipitation onto filter discs, NaOH may be used to improve the subsequent colorimetric or radioactive determination (Durham & Lopez-Solis, 1979).

The Bradford method can be used for proteins dissolved in SDS-mercaptanol-Tris electrophoresis buffer. BSA determined under these conditions gives a linear standard curve (10–90 μg of protein) similar to that given by the Lowry procedure, but when the methods were applied to whole cell homogenates there were reproducible differences in the slopes of the lines shown by the two methods (Rubin & Warren, 1977).

8.18.4.2. Micro-assay using Bradford method

This is the method of Brogdon & Dickinson (1983). Add 100 μl quantities of protein samples (e.g. HPLC eluate) or standards into microtitre plate wells. Add 200 μl of protein-dye reagent diluted with water (1 : 5, v/v). Shake the plates vigorously and after 10 min read at 595 mm. (Titertek Multiskan spectrophotometer; custom-prepared filter from Flow Labs, Inc.)

8.18.5. Protein determination with sulphobromophthalein

The method of McGuire *et al.* (1977) is described.

Reagent

Mix together 2.0 ml of 1 M-HCl, 0.5 ml of 1 M-citric acid, 10 μl of disodium sulphobromophthalein (Baker Chemical) 50 mg/ml water.

2M-NaOH containing 1 mg/ml of sodium deoxycholate (DOC).
Wash solution. Mix 1 M-HCl, 1 M-citric acid and 1 M-NaOH (4:1:2, by vol.)

Procedure

In a 1.5 ml polypropylene centrifuge tube add the protein sample (10 μl) in water or buffer. Add the same volume of 2 M-NaOH containing DOC. Mix. Allow to stand for 1 h at room temperature if necessary to solubilize the protein. Add 50 μl of reagent. Mix and centrifuge (Eppendorf Model 3200 at full speed, ca. 12 000 g) for 1 min. Remove the unbound dye solution with a Pasteur pipette. Add 0.5 ml of wash solution. Centrifuge as before. Remove wash solution and discard. Resuspend the pellet in 0.5 ml of 0.1 M-NaOH by vortex mixer. Read at 580 nm. A linear standard curve is given by 0.5 μg–10 μg protein. If the dye concentration is increased to 250 mg/ml water used in the reagent, a linear curve is obtained to 100 μg of protein.

Comments
The assay does not work in the presence of molecules which interfere with dye-protein binding, e.g. Triton X-100 (0.1 %) or SDS (0.1 %). Less reproducible results are given with a bench centrifuge generating about 2 000 g. The assay response is not necessarily constant from protein to protein. The method is recommended for protein in tissue samples. There is no critical time between steps.

8.18.6. Trinitrobenzenesulphonic acid method in the presence of lipid

The method of Mokrasch (1970) is described.

Reagents

SDS 5 % (w/v) in water.
Potassium tetraborate, 0.1 M.
2,4,6-Trinitrobenzenesulphonic acid (TNBS) (Eastman white label) (It may be recrystallized from 1 M-HCl). Prepare a 0.03 M solution in *tert.* butanol/*sec.* butanol (9 : 1, v/v). Keeps 2 months at 0–2°C.
Chloroform or diethyl ether. Formic acid, 98 %.

Procedure

The protein sample (0.04 to 0.5 mg) is treated with 20 μl of 5 % SDS solution and dried with a stream of gas (N_2, CO_2 or air). Add 1 ml of 0.1 M-K tetraborate and mix with a vortex mixer. Add 0.20 ml of the TNBS solution and shake again, and at 10 min intervals during a 30-min incubation period at 24–26°C (or 10 min at 40°C) or leave in a shaking incubator.

Add 3 ml chloroform or diethyl ether, mix vigorously and centrifuge about 1 min at 1 000 g. Discard the organic phase containing lipid material etc. Acidify the aqueous phase with 0.2 ml of 6M-HCl. Re-extract with 3 ml chloroform or ether and discard the organic phase. Add 2.0 ml of 98 % formic acid and heat at 100°C for 20 min. After cooling, the yellow solution is read at 345 nm.

Comments

Ovalbumin and BSA are suitable standards. The recovery was within 10 % of theoretical. Other methods of protein determination in the presence of lipid were not found to be reliable. A similar method was developed for the determination of amines, amino acids and proteins in mixture (Mokrasch, 1967) and free amino groups in proteins (Habeeb, 1966). Reacting amino acids and peptides with TNBS was used for their detection in HPLC (Vitt *et al.*, 1983).

8.18.7. Atomic absorption

This method allows the determination of proteins in coloured or turbid solutions (Davies & Holdsworth, 1979).

Copper reagent

Add 10 ml of 0.20 M-$CuSO_4$ to 590 ml of 0.167 M-NaOH containing 2 g of Na K tartrate $4H_2O$. AG2-X8 ion exchange resin. Atomic absorption spectrophotometer. Varian-Techtron model 1000 with air/acetylene flame.

Method

Add 1 ml of copper reagent to 1 ml of protein solution (0.05–1.0 mg/ml). Mix. After 2–5 min add 250 mg AG2-X8 ion exchange resin and shake at intervals for 5 min. Allow the resin to settle. Take 1 ml of supernatant and add to 3 ml of distilled water. Determine Cu by atomic absorption at 324.8 nm. Carry out a reagent blank and prepare a standard curve with BSA (0.05–1.0 mg/ml). Reduce instrumental noise and prevent blockages by aspirating 0.5 ml of glacial acetic acid between samples. Chlorophyll, haem, ruthenium red, etc. do not interfere. This is useful where other methods are not applicable. The protein solution should contain less than 0.4 M salts or amino acids. Insoluble proteins may be dissolved in 1 % (w/v) SDS.

8.18.8. Spectrophotometric methods

8.18.8.1. Absorbance A_{280}/A_{260}

Most proteins show maximal absorbance at 280 nm, due mainly to their content of tyrosine and tryptophan. Often, contaminating nucleic acids may be present which absorb strongly at 280 nm, although maximum absorption is at 260 nm. Warburg and Christian were the first to solve this problem. They determined the extinction coefficients of various proteins and nucleic acids at 280 nm and 260 nm, from which a factor (F) was calculated as shown in Table 8.3. Range 50–500 μg protein/ml.

Table 8.3. Protein determination by ratio of optical densities A 280/260 nm.

$A_{280/260}$	Nucleic acid %	F
1.75	0.0	1.116
1.52	0.5	1.054
1.36	1.0	0.994
1.16	2.0	0.899
1.03	3.0	0.814
0.939	4.0	0.743
0.874	5.0	0.682
0.822	6.0	0.632
0.784	7.0	0.585
0.753	8.0	0.545
0.730	9.0	0.508
0.705	10.0	0.478
0.64$	14.0	0.377
0.595	20.0	0.278

Method: Warburg and Christian (1941/1942)

Measure the optical densities of the protein sample at 280 nm and 260 nm. Calculate the ratio of these absorbance values and find the appropriate value for F in Table 8.3 and substitute in the equation:

Protein concentration (mg/ml) $= F \times 1/d \times A_{280}$, where d is the cuvette width in cm.

Alternatively, use the following calculation (Layne, 1957):

Protein concentration (mg/ml) $= 1.55\ A_{280} - 0.76\ A_{260}$.

8.18.8.2. *Absorbance* A_{230}/A_{260}

Method: Kalb & Bernlohr (1977)

Protein concentration (μg/ml) $= 183\ A_{230} - 75.8\ A_{260}$. These wavelengths were selected because they correspond approximately to the minimum and maximum, respectively, of the nucleic acid spectrum. Small errors in wavelength setting have little effect on correction for nucleic acid, but protein absorbances change by 1 %

/0.1 nm near 230 nm. High concentrations of ammonium sulphate, glycerol and sucrose (which interfere with the Lowry method) do not interfere if blanks are taken into account.

8.18.8.3. Absorbance A_{235}/A_{280}

Method: Whitaker & Granum (1980)

Protein concentration (mg/ml) = $(A_{235} - A_{280})/2.51$

Comments
Nucleic acids have approximately the same absorbance at 235 nm and 280 nm and therefore do not affect the result. Based on the amino acid composition of 208 proteins the combined absorbance values of Trp, Tyr, Phe, His, Met, CySSCy and CySH gave a value of 0.853 and for 8 proteins studied here the value was 0.849.

The protein concentration determined from A_{235} nm alone shows a 2.82-fold increase in sensitivity over that at 280 nm alone. This method has a sensitivity of 45% of the Lowry method.

Another method depends on measuring A_{233}/A_{240} (Groves *et al.*, 1968).

8.18.8.4. Absorbance A_{215}/A_{225}

Method: Waddell (1956)

Serially dilute the protein solution and take readings. Protein concentration

$$(\mu g/ml) = (A_{215} - A_{225}) \times 144$$

Comments
Murphy and Kies (1960) reported that NaCl and $(NH_4)_2SO_4$ did not interfere. Certain buffers absorb strongly at 215 nm but can be used at low concentration (0.005 M), e.g. acetate, citrate, barbiturate, succinate, phthalate, if proper blanks are taken into account. The method was critically reappraised (Wolf & Maguire, 1983) and was recommended for both pure and mixtures of proteins, within the linear range 1.5–45 μg/ml.

8.18.8.5. Absorbance A_{280}/A_{205}

Method: Scopes (1974)

Protein solutions should be diluted with 0.1 M-K_2SO_4 containing 5 mM-KH_2PO_4 adjusted to pH 7.0 with KOH. The absorbance of this buffer is approximately 0.08/cm at 205 nm, depending on the CO_2 content. Adsorption losses of protein are minimized by preparing dilutions in polyethylene containers.

The silica cells should be cleaned with conc. nitric acid and corrected for light scatter in the far u.v.

The side chains of Trp, Tyr, Phe and His absorb considerably at 205 nm. Arg, Met and Cys absorb to a lesser extent. A correction is made for Trp and Tyr (but not for other amino acids because for most purposes this is not important) by measuring (on a more concentrated sample) the absorption at 280 nm.

The recommended method of calculation is

$$\varepsilon_{205}{}^{\mathrm{mg/ml}} = 27.0 + 120 \times (A_{280}/A_{205})$$

Comments

Scopes listed 12 proteins with ε_{280} 0.50–2.65 and ε_{205} 29.0–36.4. Only abnormally high concentrations of Phe are likely to give answers outside the expected $\pm 2\%$ for determination of protein by spectrophotometry.

A high quality spectrophotometer is required for readings at 205 nm and all traces of interfering compounds having high absorbance at this wavelength, e.g. $HCOOH$, CH_3COOH, CH_3COCH_3 must be rigidly excluded.

8.19. STAINING METHODS FOR PROTEINS IN GELS

Proteins separated by gel electrophoresis were first detected by silver staining by Kerenyi and Gallyas (1972), although the full potential of the method was only established later (Switzer *et al.*, 1979).

Silver staining methods for proteins are reported to be about 100–200 times more sensitive than those based on Coomassie Brilliant Blue (CBB). Six silver staining methods were studied and the best ones could resolve 0.5 ng protein/mm^2 polyacrylamide gel (Ochs *et al.*, 1981).

The silver staining sensitivity varies with the protein, as shown in a comparative study of four different methods applied to human parotid saliva proteins and the choice of staining method is highly critical to the interpretation of results (Friedman, 1982). The patterns obtained after staining with CBB and silver often differ and it was reported that proteins such as calmodulin and troponin C can be detected by CBB but not by silver stain. It was found that such proteins required a glutaraldehyde treatment for silver staining and the washing procedure required careful control to avoid loss of soluble proteins from the gel (Schleicher & Watterson, 1983). Also, Irie *et al.*, (1982) found the silver stain to be less sensitive with basic proteins than neutral proteins and introduced a double staining method with CBB and silver.

Partial peptide maps were prepared from ng quantities of protein in SDS-polyacrylamide gels (Lischwe & Ochs, 1982). They visualized the peptides by silver staining and considered this to be quicker, less expensive and less hazardous than radiolabelling proteins and preparing autoradiograms or fluorograms. They used the commercially available kit (Gelcode) from Upjohn Diagnostics with the method of Sammons *et al.* (1981) for 1.5 mm thick gels, and recommended for thinner gels the silver-staining kit from Bio-Rad using the method of Wray *et al.* (1981).

The mechanisms of silver staining may involve phospho-amino acids (Satoh & Busch, 1981) and sulphydryl and carboxyl groups (Olert *et al.*, 1979). Higher sensitivity of detection of the protein is obtained after pre-treatment with glutaraldehyde (Oakley *et al.*, 1980) or formaldehyde (Wray *et al.*, 1981) and prior reduction with dithiothreitol (Morrissey, 1981). Dion and Pomenti (1983) concluded that glutaraldehyde (but not formaldehyde) enhanced the sensitivity of staining and that the protein staining densities were almost linear with the mol percentage of lysine present.

The original method (Merril *et al.*, 1979; Switzer *et al.*, 1979) was modified by many workers to make it both simpler and cheaper (Irie, 1980; Oakley *et al.*, 1980; Merril *et al.*, 1981a, b, 1982; Morrissey, 1981; Poehling & Neuhoff, 1981; Sammons *et al.*, 1981; Wray *et al.*, 1981). See Poehling & Neuhoff (1981) for a critical review.

A silver staining method for histones in Triton-acid-urea gels was claimed to be the first reliable method, giving nearly linear staining intensity 0–50 ng of protein (Mold *et al.*, 1983).

It was shown that the silver staining method with the repeated exposure of the polyacrylamide gels to hyper- and hypotonic solutions enabled 10 fg of protein to be detected (Ohsawa & Ebata, 1983).

Silver staining normally requires many transfer steps involving manual manipulation and incomplete removal of unreacted stain may result in considerable background staining. A continuous flow washing procedure was introduced to avoid these problems. This gave a linear response in the range 2–70 ng of muscle protein per band (Giulian *et al.*, 1983).

A new polyacrylamide gel drying technique using microwaves was described (Gersten *et al.*, 1983). With a 0.75 mm gel this took 3.5 min compared to 35–45 min with a conventional gel dryer (Bio-Rad): 3 mm gels were dried in 1 h compared to 4–5 h.

8.19.1. Dye staining methods for proteins

Dye staining methods for proteins in gels are not so sensitive as silver staining methods but they are simple to carry out. CBB is most commonly used, but many modifications of the method are in use.

Proteins stained with CBB may be photographed under white light illumination using Polaroid type 665 film at f32, with exposure time 0.5 s with a yellow filter (Steck *et al.*, 1980).

1. Merril *et al.* (1981a). Stain polyacrylamide gels with CBB R250 (0.1% w/v, Serva, Heidelberg) in 50% methanol/12% acetic acid overnight with gentle

stirring. Destain with several changes of 10% ethanol/5% acetic acid. Photograph through a 550 nm narrow band interference filter (Baird Atomic, Bedford, MA.).

SDS-polyacrylamide gel electrophoresis of proteins tightly bound to DNA showed two artifact bands (corresponding to 54 and 68 kdalton) which were stainable with CBB and silver, when mercaptoethanol was present in the buffer (Tasheva & Dessev, 1983).

2. Porro *et al.* (1982). Stain polyacrylamide gels overnight in CBB R250 (0.2% w/v) in 45% methanol/10% acetic acid and destain in 10% methanol/7% acetic acid.

After electrophoresis, agarose gels are moistened with distilled water, pressed with filter paper and dried at 37°C for 30 min. Stain for 15 min with CBB R250 (0.1% w/v) in 45% ethanol/10% acetic acid. Destain with 45% ethanol/10% acetic acid.

3. Coomassie Blue detection of proteins and polypeptides after isoelectric focusing (Trah & Schleyer, 1982).

Coomassie Brilliant Blue R250 cannot normally be used to detect proteins on polyacrylamide gels after isoelectric focusing because the carrier ampholytes used are themselves polyamino polycarboxylic acids and react with the dye. To overcome this problem colloidal suspensions of CBB R250 and CBB G250 were used in perchloric (Reisner *et al.*, 1975) or trichloroacetic (Bibring & Baxandall, 1978; Righetti & Chillemi, 1978) acid solution. The following method gives good results.

After isoelectric focusing, the gel slab 0.5 mm thick is bathed in an aqueous solution of 0.15% CBB G-250, containing 40% methanol and 4% formaldehyde for 3 h. Destain with 10% methanol overnight, after several changes of solution in the beginning. A faint blue background cannot be eliminated, but it does not interfere with detection.

Photograph under white light with Polaroid Type 665 Land pack film with orange filter, using f 8, $t = \frac{1}{8}$ s.

The method is applicable to native polyacrylamide gel systems and in the presence of SDS or 8 M-urea. Proteins and polypeptides (mol. wt. 1 665–68 000) with isoelectric points pH 2.2–10.2 were isoelectrically focused in a pH 3.5–pH 10.5 gradient. S = 20 μg for the basic tridecapeptide α-melanostimulating hormone.

4. Coomassie Blue staining during electrophoresis (Varghese & Diwan, 1983). This method permits simultaneous staining with electrophoresis in acidic gels and was used particularly for histones.

Add CBB R 250 (0.01%, w/v) to the cathode buffer and electrophorese in 0.9 M-acetic acid, pH 2.5 at 4 mA/gel tube and 100 V until the dye reaches the top of the gel. Destain for 2–3 h to visualize minor bands.

5. Mold *et al.* (1983). The following pre-silver staining methods were used for histones on polyacrylamide gels. Immerse the gel for at least 3 h in 500 ml of a

solution containing 1 % amido black, 40 % methanol and 7 % acetic acid. Destain overnight in 4 l of 32 % ethanol/8 % acetic acid and then wash with 4 × 1 l changes of water for 1 h.

An alternative method is to pre-stain with gentle stirring for no longer than 3 h in 500 ml of a solution containing 0.05 % 2,7-naphthalene sulphonic acid (Eastman), 50 % methanol and 10 % acetic acid. Destaining is unnecessary.

8.19.2. Silver staining of proteins

N.B. Diamine-silver solutions are potentially explosive and should not be stored. After use the discarded diamine should be destroyed by the addition of HCl and diluted with water for disposal.

In order to recover the silver add saturated NaCl solution. Remove the aqueous supernatant by aspiration and collect the precipitated AgCl for recycling (Wray *et al.*, 1981).

After silver staining, permanent records of the gels may be made using either a Polaroid print (3 × 5 in) or a negative that can be printed subsequently. A method was described using direct duplicating film (Kodak X-Omat) with u.v. light for 1–1.5 s followed by treatment with Kodak liquid X-ray developer and fixer. The technique does not work with CBB (Gibson, 1983).

8.19.2.1. Polyacrylamide gels with photodevelopment

Method A

This method takes less than 50 min to visualize the protein (Merril *et al.*, 1981a).

1. After electrophoresis, "fix" the protein by soaking the gel for 10 min in a solution containing 50 % methanol and 12 % acetic acid.
2. Wash the gel twice with a solution containing 10 % ethanol and 5 % acetic acid.
3. Soak the gel in 0.5 % potassium ferricyanide for 5 min with gentle shaking.
4. Pour off the ferricyanide solution and wash the gel three times with water, 20s for each wash.
5. Add a solution containing 0.2 g of silver nitrate, 0.2 g of ammonium nitrate, 0.5 ml of 37 % formaldehyde (Fisher Scientific) and 0.06 mg of benzotriazole in 100 ml of water.
6. Illuminate the gel for 20 min in a light box with a 160-W fluorescent grid lamp Model T-12 (Aristo Grid Lamp Products, Washington) with a light emission equivalent to a 1500-W tungsten light source.
7. Pour off the silver nitrate solution and without rinsing add 200 ml of a 3 % sodium carbonate solution containing 0.5 ml of 37 % formaldehyde and 0.6 mg of benzotriazole/l. Gently shake with continued illumination. A brown precipitate appears within about 1 min and the sodium carbonate solution should be poured

off and replaced with fresh solution. The illumination in sodium carbonate solution with shaking can be continued until the desired level of staining is achieved.
8. Staining may be terminated at any time by pouring off the sodium carbonate solution and soaking with 1 % acetic acid for 5 min.
9. Wash the gel with water before storing.

Method B

This method allows the visualization of protein and nucleic acid within 10 min. (Merril *et al.*, 1984).

1. Fix the gel for 5 min with 200 ml of a solution of methanol/acetic acid/water (100:20:80, by vol.) containing 4 g of citric acid and 0.4 g of NaCl.
2. Rinse briefly with 200 ml of deionized water to remove NaCl.
3. Place the gel in 200 ml of a solution of methanol/acetic acid/water (100:20:80, by vol.) containing 4 g of silver nitrate with illumination as in Method A. Take serial photographs during image development and use the photograph taken at 10 min for quantitative analysis.
4. Stop development by placing the gel in the dark. S = 0.5 ng protein. DNA produces a negative image.

8.19.2.2. Polyacrylamide gels

This is the method of Marshall (1984a).

Reagents for silver staining

Diamine solution. Mix commercial (30 %) methylamine solution with NaOH solution (0.36 %, w/v) (1:5, v/v) and add approximately 10 ml of this solution to 4.0 ml of 20 % (w/v) $AgNO_3$ solution, until the brown precipitate just clears. Adjust to 100 ml with water.
Take care: see section 8.19.2.
Developer. Add 15 ml of 1 % (w/v) citric acid and 1.5 ml of commercial (40 %) formaldehyde to 3 l of water.

Reagents for destaining

A. Dissolve 11.1 g of NaCl and 11.1 g of $CuSO_4$ in 285 ml of water and add ammonia solution (25 %) until the precipitate just clears to give a clear deep blue solution (final vol. ≃ 300 ml).
B. Dissolve 44 g of sodium thiosulphate pentahydrate in 85 ml of water (final vol. ≃ 100 ml).
C. Kodak hypo clearing agent, 20 g in 800 ml of water.

Method of staining

1. Soak the polyacrylamide gel (0.8–3.0 mm thick) after electrophoresis overnight in 50% methanol, 10% acetic acid (100 ml/gel).
2. Wash at 60°C for 10 and 20 min with change of water (200 ml/gel).
3. Incubate at 60°C for 30 min in 0.1% (w/v) formaldehyde solution (200 ml/gel) (prepared by adding 2.5 ml of 40% formaldehyde to 1 l of water).
4. Cool in water (200 ml/gel) for 10 min at room temperature.
5. Shake at room temperature for 10 min in diamine solution (100 ml/gel).
6. Discard the diamine and quickly rinse the gel with two changes of water and two changes of developer (100 ml/gel each time).
7. Continue the development with changes of the solution at 5-min intervals for about 30 min, until the background is black with 'silver mirror' deposit. Protein spots gradually intensify over 5 to 25 min.
8. Wash the gel for 10 and 20 min in two changes of water and soak overnight in fresh water.

Method of destaining

Mix reagents A and B in the proportions 3:1 (v/v) and dilute with an equal volume of water. Incubate the gels in the solution (100 ml/gel) for 1–4 min until the background becomes yellow. Quickly rinse in water and incubate in reagent C (100 ml/gel) for 30 min. Wash for 10, 20, and 30 min in three changes of water.

Comment
This method is based on earlier work (Marshall & Latner, 1981; Marshall, 1983). It gives high sensitivity (S = < 0.01 ng of BSA/mm^2) as a result of using low formaldehyde concentration and a procedure for over-staining followed by destaining. With 2-dimensional electrophoresis more than 200 peptides were detected from 30 μl of urine and more than 150 from a single fingerprint. See also Marshall (1984b) for silver staining of human salivary proteins.

8.19.2.3. Large-pore polyacrylamide gels

This is the method of Perret *et al.* (1983) after Marshall and Latner (1981). Large-pore gels are required for the electrophoretic fractionation of macromolecules with mol. wts. greater than 10^6. Agarose gels turn dark with silver staining and this technique cannot be applied (Boulikas & Hancock, 1981). The following silver staining method was used in the presence of SDS and the large-pore gel consisted of 2.55% polyacrylamide cross-linked with 2.75% methylene-bis-acrylamide.

Method

1. After electrophoresis of proteins, soak the gel in formaldehyde (37%) (Fluka)/ethanol/water (15:27:63, by vol.) for 1 h and then leave overnight in a different concentration bath (1:25:75, by vol.).
2. Shake the gel at 60°C in methanol/acetic acid/water (5:7:88, by vol.) for at least 4 h, with changes of solution every 20–30 min.
3. Shake the gel at 60°C for 15 min with a fresh solution of 4% (w/v) paraformaldehyde (Merck) in 1.43% (w/v) sodium cacodylate (Fluka) pH 7.3 for 15 min at 60°C.
4. Wash the gel at 60°C with distilled water for at least 3 h. Change the water every 20–30 min.
5. Place the gel at room temperature in silver solution (section 8.19.2.2) for 10 min.
6. Rinse with water for 2–3 min.
7. Develop with a freshly prepared aqueous solution of citric acid (0.005% w/v, Merck) and formaldehyde (0.019% w/v, Fluka). Change the solution frequently until the black protein bands appear.
8. Stop the development with methanol (50%, v/v).

8.19.2.4. Agarose gels after isoelectric focusing

This method of silver staining only takes 10 min and is about 20 times more sensitive than CBB R250 and it can be used after previous CBB staining (Lasne *et al.*, 1983).

Fixing and drying

After isoelectric focusing, the proteins in the gel are fixed by immersion in a solution of 10% trichloroacetic acid–1% sulphosalicylic acid for 10 min. Wash the gel in 95% ethanol for 10 min. Dry under warm air.

Autoradiography

For low energy isotopes use Ultrofilm ^{3}H LKB 2208-190 and develop with Kodak X-ray Developer LX24 and X-ray Fixer AL4. Coomassie Blue staining. Immerse the gel in a solution containing 35% ethanol, 10% acetic acid and 0.6% CBB R250 for 5 min. Destain with 35% ethanol–10% acetic acid for 10 min.

Silver staining

1. After fixing and drying, the gel is placed in a solution containing 50% methanol, 12% trichloroacetic acid, 2% $CuCl_2$ and 1% $ZnCl_2$ for 5 min.

2. Immerse the gel in a mixture of 0.01 % $K MnO_4$ and histological dye Harris' haematoxylin (Chroma 2 E 0565) (1:2, v/v). The composition of the dye is ethanol 10 ml, haematoxylin 1 g, aluminium potassium sulphate 20 g, yellow mercuric oxide 0.5 g).
3. Rinse the gel in distilled water for a few seconds, until the orange colour fades and turns to pale mauve.
4. Prepare solution A: 8 g of A. R. sodium carbonate (anhydrous) in 100 ml of distilled water (prepared daily); Solution B: Dissolve in the following order 0.28 g of ammonium nitrate, 0.2 g of silver nitrate, 1 g of silicotungstic acid (SiO_2, $12WO_3.26H_2O$) and 0.73 ml of 35 % formaldehyde in 80 ml of water. Make up to 100 ml (keeps for months at room temperature). Immediately before use, add an equal volume of solution B to solution A. Immerse the gel. Within 1–3 min the protein bands become reddish brown and finally black.
5. Stop the reaction by washing the gel in distilled water for a few seconds.

S = 0.045 μg of protein per band.
See also Vesterberg & Gramstrup-Christensen (1984) for silver staining in agarose gels.

8.19.2.5. Polyacrylamide and immuno-precipitates in agarose gels

This method (Porro *et al.*, 1982) uses a single-step silver ion reduction after treatment of the protein with glutaraldehyde. Washing steps are eliminated.

Reagents

Glutaraldehyde solution (Merck-Schuchardt). Diamino silver (0.2 % w/v). Mix together NaOH solution (0.36 g/21 ml water), 1.4 ml of NH_4OH (28 % w/v) and 1 ml of silver nitrate (20 % w/v) and make up to 100 ml with water. *Take care*, see section 8.19.2.

Polyacrylamide gels

After electrophoresis fix the protein in 45 % methanol–10 % acetic acid mixture for a minimum of 30 min. Wash with water to remove excess organic solvents. Bathe the gel in 200 ml of glutaraldehyde solution (1 % w/v) with efficient stirring for 15 h at room temperature.

The concentration of glutaraldehyde is critical, but 1 % w/v is suitable for gels 120 × 140 mm and 0.5 % w/v for gels 30 × 45 mm., with thicknesses of 0.75 to 1.5 mm.

Transfer the gel into 200 ml of a solution of diamino silver for about 30 min. Efficient stirring prevents silver deposits on the gel. Overstained backgrounds can be selectively destained with acetic acid (10 % v/v) in a few min.

Wash with water (or buffered saline at pH 7.2 if acetic acid was used).

Photograph. Store in plastic bag. S = 0.03 ng/mm^2 for bovine serum albumin, and 0.5 ng/mm^2 for ovalbumin. This is about 100 times more sensitive than that obtained with CBB.

Agarose gels

After rocket immunoelectrophoresis on agarose gel, moisten the gel with distilled water or saline; press with filter paper; dry at 37°C for 30 min. Treat the gel with glutaraldehyde solution for 15 min. Transfer to 100 ml of diamino silver for 5–10 min. Overstaining can be reduced as before. Use efficient stirring throughout.
Note glutaraldehyde concentration to be used:

Antiserum in agarose (% v/v)	Glutaraldehyde (% w/v)	Staining of 'rockets'
0.05–0.5	1	Negative form
>0.5–1	0.33	
0.05–0.25	3	Positive form
>0.25–1	5	

8.19.2.6. Silver staining followed by blue toning

The method of Berson (1983) is described.
Image intensification of silver-stained proteins was increased 3 to 7-fold by blue toning. Densitometric evaluation showed that the linearity of staining was maintained between 0.05 and 1.7 pmol/mm^2.

Reagents

A. 5% ferric chloride, w/v in water.
B. 3% oxalic acid, w/v.
C. 3.5% potassium hexacyanoferrate III, w/v.
These stock solutions are stable for several months at room temperature. Toning bath: prepare just before use. Mix 10 ml each of reagents A, B and C and 70 ml of water.
The solution should be brown: discard if blue-green.

Method

The polyacrylamide gels were silver-stained (Morrissey, 1981) and washed in tap water for at least 30 min before being soaked in the toning bath for between 0.5 and 3 min. The protein bands turn blue. Stop by washing for 10 min in tap water. Stabilize by soaking for 10 min in 20% methanol, 5% acetic acid (v/v) and store in a heat-sealable food-storage bag. Evaluate densitometrically at 550 nm for silver staining and 660 nm after blue toning, using a Gilford Model 250 spectrophotometer. Photograph gels with Panatomic Film (Kodak).

8.19.2.7. Double staining with Coomassie Blue and silver

This is the method of Irie *et al.* (1982).
Prestaining with CBB combined with formaldehyde fixation increases the intensity and discrimination of protein spots on gels.

Fixing and prestaining

1. After electrophoresis the polyacrylamide gels are immersed in formaldehyde/acetic acid/ethanol/water (5:10:25:60, by vol.) containing CBB R-250 (0.05 %, w/v) for 1 h and then with the same solvents (0.5:10:25:64:5, by vol.) containing CBB R-250 (0.05 % w/v) for more than 3 h or overnight.
2. Remove free dye by destaining. Wash the gel once with acetic acid/isopropanol/water solution (10:25:65, by vol.) followed by acetic acid/isopropanol/water (10:10:80, by vol.) with repeated changes until the background is clear.

Silver staining: modified from Merril *et al.* (1979)

1. Wash the fixed gel (prestained or not) in 10 % ethanol in water several times to remove the acetic acid.
2. Wash with 4 % paraformaldehyde/sodium cacodylate (1.43 %, w/v) adjusted to pH 7.3 with HCl for 30 minutes.
3. Wash with 10 % ethanol, 4 × 5 min each time.
4. Wash for 30 min in cupric nitrate/silver nitrate solution (made by dissolving 3.5 g silver nitrate in 100 ml of water followed by addition of 1.5 ml of cupric nitrate solution (0.5 %, w/v) and then the simultaneous addition of 4 ml of pyridine and 8 ml of reagent grade absolute ethanol).
5. Wash in fresh (within 5 min) diamine solution for 10–12 min prepared as follows: Mix 30 ml of 19.4 % (w/v) silver nitrate solution and 22.2 ml of a NaOH/NH_4OH solution (stock solution contains 100 ml of 0.36 % NaOH, 45 ml of fresh conc. NH_4OH and 55 ml of 20 % (v/v) ethanol).
6. Treat with reducer solution for 1 min. This contains 2.5 ml of 10 % formaldehyde (dilute commercial formaldehyde with water, 1:10 (v/v), 6 ml of 1 % citric acid and 100 ml of ethanol in 1 l of water. Repeat with the same wash for 1 min. Now rinse several times with a second reducer solution (5 ml of 10 % formaldehyde, 5 ml of 1 % citric acid and 100 ml of ethanol in 1 l of water). Observe continuously until the appearance of brown or black protein spots is optimal.

7. Stop development by immersion in 1 % acetic acid. The gel can be kept in this solution for at least 2 months; alternatively vacuum dry on a filter paper at 100°C. S = 0.02–0.2 ng/mm^2 for neutral proteins. S $\simeq$ 1.0 ng/mm^2 for basic proteins.

8.20. STAINING OF PROTEINS ON NITROCELLULOSE PAPER

The transfer of electrophoretically-resolved proteins to nitrocellulose sheets is a process known as 'Western' blotting (Burnette, 1981). Methods of visualization include amido black, India ink and silver staining. The specific immunodetection of reactive antigens on nitrocellulose was first described by Towbin *et al.* (1979; see also section 8.20.3).

8.20.1. Amido Black

This is the method of Wojtkowiak *et al.* (1983).
Stain the nitrocellulose sheet with immobilized protein with Amido Black (0.1 % solution in 45 % methanol and 10 % acetic acid). Destain in 2 % acetic acid in 90 % methanol. S = 1 μg of protein.

8.20.2. India ink

This is the method of Hancock and Tsang (1983).
The proteins are resolved by SDS-gel electrophoresis on a 3.3 % to 20 % gradient gel and electrophoresed onto nitrocellulose paper (Schleicher and Schuell) (Tsang *et al.*, 1983).

The nitrocellulose paper with the blotted proteins is washed four times for 10 min each with 250 ml of 0.15 M–NaCl in 0.01 M-Na_2HPO_4/NaH_2PO_4 buffer, pH 7.2 containing 0.3 % Tween 20 (PSB-TW) at 37°C in a plastic water-tight box in a shaking water bath. After each wash, rinse the paper with deionized water.

Use Pelikan Fount India drawing ink for fountain pens (Pelikan AG, D-3000 Hanover 1, Germany).

Stain with 1 μl ink/ml of PSB-TW for a minimum of 2 h, but preferably overnight with agitation. Use about 200 ml of stain solution for one 180 × 200 mm sheet. A small increase in sensitivity of the stain is obtained by adding separately 10 % glutaraldehyde. No special destaining is required; just rinse with deionized water for 5 min.

The sensitivity of staining varies with the protein, from 80 ng for β-galactosidase to 520 ng for ovalbumin. Protein bands are light grey on a yellow-grey background. The strips can be held in PSB-TW for one month, with no loss of sensitivity or resolution.

The ink used is more sensitive than other inks, Amido Black, Fast Green or CBB R250.

8.20.3. 2,4-Dinitrofluorobenzene and antibody detection

This is the method of Wojtkowiak *et al.* (1983).
The transferred proteins on nitrocellulose sheet are reacted with DNFB. The dinitrophenylated protein is detected by antibody to DNP and visualized by using the peroxidase-antiperoxidase reagent method. DNFB staining detects 10 ng of transferred protein. It is about 100 times more sensitive than CBB staining and avoids problems often encountered in silver staining.

Reagents

A. 50 mM-$NaHCO_3$ in 50 % dimethylsulphoxide.
B. 10 mM-sodium phosphate-buffered 0.14 M-NaCl, pH 7.2.
C. 3 % BSA–10 % heat inactivated calf serum solution in 10 mM-sodium phosphate-buffered 0.14 M-NaCl, pH 7.2.
D. 50 mM-Tris-HCl buffer, pH 7.5 containing 0.3 mg/ml of 3,3′-diaminobenzidine and 0.005 % H_2O_2.

Procedure

1. Incubate the nitrocellulose sheet in reagent A containing 0.00001 to 0.1 % DNFB at room temperature for 10 min. Longer incubation times or higher concentrations of DNFB make it difficult to remove excess. Wash several times with A.

Reactive antigens are studied as follows by the method of Towbin *et al.* (1979).

2. Wash the transfers with reagent B, 5 × 5 min each.
3. Rock in reagent C at 40°C for 1 h.
4. Incubate at 4°C overnight in rabbit anti-serum to DNP (Miles) diluted (1 : 200, v/v) with C.
5. Wash the transfers with reagent B, 4 × 5 min each.
6. Incubate for 30 min at room temperature with goat anti-rabbit IgG (Serasource) diluted (1 : 40, v/v) with reagent C.
7. Wash in reagent B, 4 × 5 min each.
8. Incubate for 30 min at room temperature with peroxidase-antiperoxidase complex (Miles) diluted (1 : 100, v/v) with C.
9. Wash in B, 3 × 5 min each.
10. Place in reagent D to detect peroxidase activity.

8.20.4. Silver and antibody blot method

Proteins resolved on SDS-polyacrylamide gels are transferred to nitrocellulose paper and silver-staining detects the proteins. The antibody blot method of Yuen *et al.* (1982) described here is applicable because the proteins retain their antigenic properties and can be detected by specific antibodies.

Electrotransfer of proteins to nitrocellulose

The proteins are separated by SDS-polyacrylamide gel electrophoresis (Laemmli, 1970). Electrotransfer the proteins to nitrocellulose paper as follows. Pre-soak nitrocellulose paper (Schleicher & Schuell) in transfer buffer (24 mM-Tris, 192 mM-glycine, 20 % methanol, pH 8.3) and then place over the gel. Sandwich the gel-paper unit between two sheets of Whatman 3 MM paper (also pre-soaked). Fix in Electroblot unit (E-C Apparatus) with gel towards cathode and paper towards anode and transfer at 5.0 W for 4 h with continuous buffer circulation.

Silver staining

The method of staining is after Oakley *et al.* (1980) and Merril *et al.* (1981a).
(a) After protein transfer, wash the nitrocellulose paper overnight in 0.01 M-Tris-HCl, pH 7.4.
(b) Wash with two changes of glass-distilled water of 5 min each.
(c) Soak in 100 ml of potassium ferricyanide (0.5 %, w/v) for 5 min with shaking on a light box.
(d) Briefly wash with glass-distilled water.
(e) Transfer to 100 ml of a solution containing 0.2 g silver nitrate, 0.2 g ammonium nitrate and 0.5 ml of 37 % formaldehyde. Shake the paper in this solution for 20 min on a light box.
(f) Discard the silver nitrate solution. Wash the paper twice with distilled water for 10 s each and with freshly distilled water for 3 min.
(g) Leave the paper overnight in 100 ml of an aqueous solution containing 3 g sodium carbonate and 0.25 ml of 37 % formaldehyde.
(h) Wash twice with glass-distilled water for a total of 15 min.
(i) Leave in 100 ml of a solution containing 1.4 ml of conc. NH_3 solution, 0.07 g of NaOH and 0.2 g of silver nitrate for 15 min.
(j) Wash briefly twice with glass-distilled water.
(k) Add 100ml of a solution containing 5.0 mg citric acid and 0.019 % formaldehyde. The paper can now be stored in water. Proteins appear as light bands on a yellowish-brown background. When proteins are transferred to nitrocellulose paper from gels that do not contain SDS, the proteins appear as yellowish-brown bands which are difficult to distinguish from the background.

Detection by specific antibodies

Buffer A: 10 mM-Tris-HCl, pH 7.5 containing 0.05 M-NaCl, 2 mM-EDTA, 4 % BSA fraction V and 0.1 % sodium azide.
Buffer B: 10 mM-Tris-HCl, pH 7.5 containing 0.05 M-NaCl and 2 mM-EDTA.

1. Equilibrate the silver-stained nitrocellulose sheet overnight in 0.01 M-Tris-HCl, pH 7.4.
2. Change to buffer A with continual shaking for 24 h.
3. Change to fresh buffer A containing appropriate amounts of antibodies, with shaking for 16 h.
4. Wash with buffer B, 5×10 min each.
5. Add fresh buffer A containing 2×10^5 cpm of ^{125}I-labelled protein A (Sigma) and leave for 3–4 h.
6. Wash with buffer B, 6×10 min each.
7. Dry in air.
8. Expose against Kodak X-Omat AR film with the help of an intensifying screen. Store at $-70°C$. Develop with Kodak KLX developer and Kodak fixer.

8.21. PROTEIN ATTACHED TO SOLID SUPPORTS

In affinity chromatography and solid-phase sequencing it is often necessary to determine the amount of protein bound to inert matrices, such as agarose, dextrans and cellulose. In principle, this could be done by the hydrolysis of a sample to determine total N, but often non-protein N is introduced into the matrix when CNBr activation is used or a N-containing spacer arm is introduced. Quantitative amino acid analysis of the protein-matrix sample is often unsatisfactory because of the interference by high concentrations of carbohydrates released. An alternative method is that of difference analysis, by determining the amount of protein added to the matrix and the amount recovered after extensive washing. However, the following methods enable a direct measurement to be made quickly. See also Chapter 12, section 12.3.3.

8.21.1. Matrix-bound protein

The method described is that of Marciani *et al.* (1983) after Westley and Lambeth (1960).
A known amount of Cu^{2+} is added to the matrix-bound protein. The amount of Cu^{2+} complexed is a measure of the total bound protein and this is determined by measuring the free Cu^{2+} by reaction with diethyldithiocarbamate.

Reagents

A. 0.167 M-NaOH containing KNa tartrate $4H_2O$, 3.4 g/l.
B. Add 10 ml of 0.2 M-$CuSO_4$ to 590 ml of reagent A.
C. Na diethyldithiocarbamate (0.1 g) dissolved in 100 ml of 0.025% solution of crystalline BSA.
D. Bio-Rad AGI-X8 resin, 200–400 mesh, chloride form suspended in deionized water (200 mg/ml).

Procedure

The protein-gel beads are thoroughly washed to remove salts and/or free protein. An aliquot is centrifuged for 3 min at 300 g to obtain a measure of the settled gel volume. Add deionized water to obtain a slurry of water/gel beads in the ratio 1 : 1 to 1 : 4, depending on the amount of bound protein. Take aliquots of the slurry to give a range of 0.01 to 0.10 ml settled gel and transfer into test tubes. Make up the samples to 1.5 ml with distilled water.

Add 1 ml of reagent mixture (3–6 ml of reagent B made up to 10 ml with reagent A). Mix. After 15 min at room temperature add 5 ml of reagent C. Mix. Centrifuge at low speed. Read the dark yellow supernatant at 446 or 486 nm against a water blank. Use a sample of untreated beads for the control.

Standard curve

Prepare this at the same time using 0.5 ml samples of standard solutions containing 50–800 μg of protein in separate test tubes. Add 1 ml of reagent B. Mix. After 15 min at room temperature add 1 ml of Bio-Rad beads (D) and 5 ml of reagent C. Centrifuge and read as before.

8.21.2. Sepharose-bound protein

The method of Golovina *et al.* (1977) is described.

Reagent

Dissolve 50 g of polyethylene glycol (mol. wt. 20 000) in 50 ml of water.

Method

Place a suspension of Sepharose-bound protein (0.2–1.0 ml) in a quartz cell with 1 cm light path. Add polyethylene glycol solution by syringe to give a total volume of 3 ml in the cell. Stir without introducing air bubbles. Read at 280 nm against a blank without protein.

A linear absorbance/protein curve was obtained. S = 20 μg protein/0.5 ml of gel. A calibration curve may be obtained by adding increasing amounts of protein to a gel suspension (apo-glyceraldehyde-3-phosphate dehydrogenase here).

8.21.3. Amino groups on solid supports

These two methods were described by Lee and Loudon (1979).

8.21.3.1. Picrate method

Reagents

Picric acid, 0.1 M in methylene chloride.
N,N-diisopropylethylamine, distilled from CaH_2.

Method

Wash the solid support (about 50 mg) in a small sintered glass funnel (1 cm i.d.) twice with 2 ml of picric acid solution followed by washing with 40 ml of pure methylene chloride. Add *N,N*-diisopropylethylamine (2 ml) then add MeOH (2 ml) and finally *N, N*-diisopropylethylamine (2 ml), removing the liquid each time by suction. The collected washings are diluted to 250 ml with 95 % ethanol. Read at 358 nm. ε 14 500 for the amine-picrate complex.

8.21.3.2. Chloride titration method

Reagents

1 M-HCl.
10 % triethylamine in dimethylformamide.
Mercuric nitrate solution, standardized against KCl.
Bromophenol blue, 1 % (w/v) in ethanol.
Diphenylcarbazone, 1 % (w/v) in ethanol.

Method

Wash the support (about 50 mg) with 20 ml of 1 M-HCl followed by 40 ml of deionized distilled water and finally 5 ml of methanol.

Successively wash with 4 ml of 10 % triethylamine in DMF, 2 ml of DMF and 4 ml of methanol. Dilute the combined washings with 20 ml of water and 80 ml of ethanol. Add 5 drops of bromophenol blue and adjust the mixture to a yellow colour by the careful addition of dilute nitric acid (0.5 M). Add 5 drops of 1 % diphenylcarbazone indicator and titrate with standard mercuric nitrate solution, until a sudden change to a pink colour is seen, indicating excess mercuric ion.

Comments
The authors found these two methods to be the most convenient of four investigated for determining amino groups on controlled pore glass beads.

The determination of amino groups by Schiff base formation with salicaldehyde and measurement at 315 nm was reported by Chou & Chien (1978) to be better than with 2-hydroxy-1-naphthaldehyde as used by Schmitt & Walker (1977).

8.21.4. Amino groups on insoluble matrix beads

This is the method of Antoni *et al.* (1983), after Mokrasch (1967) and is for hydrophilic matrices, because the reaction described only occurs in an aqueous environment.

Take a volume of packed gel containing not more than 4–5 μmol of amino groups. Dilute to 9 ml with 0.1 M-$K_2B_4O_7$. Add 1 ml of 0.01 M-trinitrobenzoic acid. At the same time prepare a reference sample with 9 ml of 0.1 M-$K_2B_4O_7$ but no gel. Stir at 37°C for 2 h. Allow the gel to precipitate, or centrifuge. Take 1 ml of supernatant, dilute with 5 ml of 0.1 M-$K_2B_4O_7$ and add 0.5 ml of 0.03 M-glycine. Run simultaneous blank samples with 1 ml of supernatant, 5 ml of 0.1 M-$K_2B_4O_7$ and 0.5 ml of water. After 25 min at 25°C, add 10 ml of cold methanol and read each sample against its own blank at 340 nm.

The concentration of amino groups on the gel is determined from the difference in absorbance between the reference and the sample readings, using ε 12 400. Repeated measurements on the same gel differ by less than 10%.

8.21.5. Chloranil colour test for amino groups

The method by Christensen (1979a, b) is described.
This rapid method may be used to detect free amines and primary or secondary amino groups on resins to check for completion of coupling in solid phase peptide synthesis, or in blocking amino groups.

Reagents

Saturated solution of chloranil, 2,3,5,6-tetrachloro-1,4-benzoquinone in toluene.

Method

1. Add 1 mg of peptide resin to a small test tube.
2. For the detection of primary amino groups add 200 μl of acetaldehyde *or* for secondary amino groups add 200 μl of acetone.
3. Add 50 μl of chloranil solution. Swirl 5 min at room temperature.

In the presence of free amino groups a green or blue colour is formed on the beads. S = 5 μmol/g resin, which is about the same as for the ninhydrin test.

8.21.6. Colour test for derivatized gels

The following test is by Cuatrecasas (1970).
Add 1 ml of saturated sodium borate to an aqueous slurry (0.2–0.5 ml) of agarose or polyacrylamide gel. Add 3 drops of 3% (w/v) of aqueous solution of 2,4,6-trinitrobenzenesulphonate and allow to stand for 2 h at room temperature for

complete reaction. Unsubstituted polymer, carboxylic acid and bromoacetyl derivatives give a yellow colour; derivatives containing primary aliphatic amines, orange, aromatic amines, red-orange and unsubstituted hydrazide derivatives, deep red.

8.21.7. Colorimetric assay for carbodiimides

This assay method was described by Jacobson and Fairman (1980).

Reagents

2 M-pyridine: HCl and 1 M-1,2-diaminoethane: HCl at pH 7.0 for water soluble carbodiimides (read at 400 nm).
2 M-pyridine: HCl and 1 M-piperazine: HCl at pH 7.0 for water insoluble carbodiimides (read at 420 nm).

Method

Add 0.1 ml of water-soluble carbodiimide to 1.0 ml of the reagent solution. After exactly 20 min at 22°C read at 400 nm. Use dimethylformamide solution for water-insoluble carbodiimides and read at 420 nm.

Comments
The assay is specific for the carbodiimide and not products formed during coupling reactions. The absorbance increases linearly for 6 h and a standard curve is prepared under the same conditions. The absorbance is linearly dependent on the concentration of carbodiimide. Concentrations from 50 μM to 50 mM may be determined. For low concentrations incubate at 45°C to increase rate of colour development.

8.21.8. Fluorimetric determination of carbodiimide

The method of Chen (1983) is described.
Vortex mix 100 μl of test solution in a glass-stoppered test tube with 100 μl of 20 mM-*trans*-aconitic acid in anhydrous dioxan (distilled and dried over metallic sodium). Warm immediately at 50°C for 10 min. Then mix in 50 μl of 0.2 M-pyridine in anhydrous dioxan and dilute with 2.5 ml of methanol. Measure immediately with excitation wavelength at 400 nm and emission wavelength at 490 nm against a reagent blank. A linear fluorescence/carbodiimide relationship up to 50 nmol was shown. S = 50 pmol.

Trans-aconitic acid may be used as a spray reagent on TLC plates to detect carbodiimides.

8.22. CARBOHYDRATE-CONTAINING PROTEINS

8.22.1. Introduction: *a contribution by J. R. Clamp.*

Glycoproteins occur widely in all living organisms. They may be intracellular or extracellular, present as membrane components or in solution, existing as fibrillary structures or as typical globular proteins. Probably more than half of the proteins that have been examined in detail contain covalently attached carbohydrate. Thus whenever a new protein is investigated, the possibility that it contains carbohydrate should always be borne in mind.

The effect that the presence of carbohydrate confers on the protein depends very much on the amount. Glycoproteins vary greatly in their content of carbohydrate from 1% or less to over 80%. One or two small oligosaccharide units attached to a reasonably sized protein will scarcely affect the typical protein properties. Glycoproteins with 80% or more carbohydrate on the other hand behave more like polysaccharides. Most glycoproteins, however, have levels of carbohydrate that lie between these two extremes. Any protein should therefore be suspected of being a glycoprotein particularly if it shows anomalous behaviour on gel-permeation chromatography, ultraviolet absorption, ultracentrifugation, protein staining, charge heterogeneity and so on.

If a protein is suspected of containing carbohydrate, the simplest way to confirm this for most workers in the protein field is to look for hexosamines during amino acid analyses. The commonest hexosamines in glycoproteins are glucosamine and galactosamine. Most methods for amino acid analysis include data on the behaviour of these two hexosamines in the analytical system. If not, it is relatively simple to run the appropriate standards. It is important to realise that not all glycoproteins contain hexosamine and that substantial amounts of carbohydrate will be destroyed during the vigorous hydrolysis required for amino acid analysis.

The monosaccharides that one might expect to find in a typical mammalian glycoprotein include L-fucose (6-deoxy-L-galactose); D-mannose; D-galactose; *N*-acetyl-D-glucosamine (2-acetamido-2-deoxy-D-glucose) and sialic acid (various neuraminic acid derivatives). In addition, some glycoproteins contain D-glucose or *N*-acetyl-D-galactosamine (2-acetamido-2-deoxy-D-galactose). The most satisfactory methods for separating, identifying and estimating these seven monosaccharides are based upon gas chromatography. There are two main methods using this technique, namely the alditol acetate method and the methyl glycoside/TMS method.The first involves aqueous acid hydrolysis followed by reduction of the liberated aldoses to the corresponding glycitols which are then acetylated. The second method depends upon methanolysis, yielding the methyl glycosides together with the methyl ester of sialic acid, which are then converted to the trimethylsilyl derivatives.

It should be noted that proteins which have been passed down Sephadex columns are likely to be contaminated with glucose residues.

8.22.2. Methyl glycoside/trimethylsilyl method *by J. R. Clamp*

This section was contributed by J. R. Clamp.
The methyl glycoside/TMS method was developed in our laboratory and this is the method that will be described in detail.

Methanolysis

This is carried out using 1.0 M-HCl in methanol. Dry methanol is obtained by distillation from the prepared Grignard reagent. Dry HCl gas from a cylinder (BDH Ltd.) is then bubbled into the dry methanol which is kept cool on an ice bath. Fifteen min at a reasonably fast bubbling rate into 250 ml of methanol will usually give a molarity in excess of 1.0 and the molarity can then be adjusted to 1.0 by the addition of further dry methanol. The molarity is determined in the usual way by titration of an aliquot sample with sodium hydroxide solution. The methanolic-HCl is transferred to 10 ml ampoules which are flushed with nitrogen, sealed and stored at −20°C until required.

The amount of glycoprotein that is required for analysis depends upon its carbohydrate content. The ideal amount would yield between 25 and 125 nmol of each monosaccharide. However, with an unknown glycoprotein one must make an inspired guess and between 1.0 and 2.0 mg of glycoprotein gives a reasonable analysis for material with an average content of carbohydrate.

This amount of glycoprotein, together with the internal standards, mannitol (100 nmol) and perseitol (100 nmol), are dried in a 2 ml ampoule. Methanolic-HCl (0.5 ml) is added, the ampoule is flushed with nitrogen, sealed and heated at 85°C for 24 h.

After methanolysis, the mixture is neutralized with silver carbonate. The powder is added in small amounts with thorough trituration and the pH is assessed by testing the film of liquid adhering to the spatula with narrow range pH paper.

re-*N*-acetylation

Methanolysis causes some de-*N*-acetylation of the *N*-acetylhexosamines. Re-*N*-acetylation is therefore performed by adding acetic anhydride (0.1 ml) to the neutralized mixture. This has the additional advantage of preventing the adsorption of sugars on to the silver chloride–silver carbonate particles. The mixture is left for 6 h and then centrifuged. The supernatant is transferred to a 5 ml pear-shaped flask. Methanol (0.5 ml) is added to the sediment remaining in the ampoule, which is thoroughly triturated, centrifuged and the supernatant transferred as before. This process is repeated once more and the combined supernatant solutions are flash-evaporated at 35°C. It is important that all traces of acetic anhydride are removed and if the flask has the characteristic odour of acetic anhydride, methanol should be added and flash evaporation repeated. The third internal standard is added, namely arabinitol (50 nmol in methanol) and the flash-evaporation step repeated.

The 5 ml pear-shaped flask now contains the methyl glycoside-methyl ester derivatives of the glycoprotein monosaccharides, the internal standards arabinitol, mannitol and perseitol, together with any amino acids or peptides that have arisen from the polypeptide chain. These latter contaminants do not interfere with the analysis in any way.

At this stage, the dried sample may be kept stored in a vacuum desiccator over phosphorus pentoxide, until required for analysis.

Trimethylsilylation

The free hydroxyl groups in the sugars must be substituted before gas chromatography can be carried out and trimethylsilylation is ideal for this purpose. The reaction is rapid and quantitative at room temperature and the introduction of a large number of methyl groups increases the sensitivity with the flame ionization detector. A satisfactory reagent for carrying out trimethylsilylation is a mixture of trimethylchlorosilane/hexamethyldisilazane/pyridine (1:1:5, v/v). This mixture is made up weekly and kept in a stoppered flask. There is a slow formation of ammonium chloride on standing and the mixture is centrifuged each time and the supernatant used for the reaction.

The dried sample is triturated with 0.05 ml of the trimethylsilylation mixture and left for 30 min at room temperature in the stoppered flask, preferably in a dry-box. At the end of this time, the flask is centrifuged and 1 to 5 μl of the supernatant is injected into the gas chromatograph.

Gas-liquid chromatography

The gas chromatographic details are important for it is our experience that the majority of the problems encountered during setting up this method are associated with poorly conditioned columns. The columns are glass with dimensions of about 2 m × 2 to 3 mm (i.d.). The packing material is a high quality diatomaceous earth, acid-washed and silanized, coated with 3 to 4% SE-30. Suitable material is available from most gas chromatographic suppliers. After injection, the chromatograph is temperature-programmed from 120°C to 210°C at 1°/minute.

Suitable monosaccharide mixtures should be taken through the whole procedure for calculation of the response factors relative to the internal standards. The neutral hexoses usually present no problems. However, the polar sugars, namely the *N*-acetyl hexosamines and sialic acid are sensitive indicators of the condition of the column. The sugar peaks should be perfectly symmetrical so that the appearance of small or tailing peaks indicates a poorly conditioned column. To bring the column into condition, inject the trimethylsilylation mixture into the column every few h at 120°C with a slow nitrogen flow. Gradually the column will be brought into condition although the process may take several days or even weeks depending on the state of the column. The improvement can be assessed by injecting a standard sugar mixture at intervals.

Further details of the method may be found (Bhatti *et al.*, 1970; Chambers & Clamp, 1971; Clamp *et al.*, 1971, 1972; Clamp, 1974a,b).

8.22.3. Uronic acids, hexosamines and galactose of glycosaminoglycans by methylglycoside/trimethylsilyl method

The method of Dierckxsens *et al.* (1983) is described.

The carbohydrate derivatives are prepared by a modified procedure of Clamp *et al.* (1971, see also previous section).

Gas-liquid chromatography

Fused silica column (25 m × 0.32 mm i.d.) with liquid phase CP ™ Sil 5 (Chrompak, U.K.); flame ionization detector; carrier gas (N_2) flow rate 0.7 ml/min with make-up gas to detector 30 ml/min; inlet temperature 220°C; detector temperature 280°C; temperature programme 135 to 215°C at 1°C/min. A separation of pertrimethylsilyl methyl glycosides, including arabinitol, mannitol, mannose, glucose (2 peaks) and galactose (4 peaks) was obtained in less than 60 min. Reproducibility was good in the range 3–30 μg. The yields of monosaccharide derivatives were more than 87 % based on hexosamine determination, for all glycosaminoglycans studied.

8.22.4. Sensitive methyl glycoside/trimethylsilyl method

The method of Chaplin (1982) is described.

Dry down the glycoprotein sample (0.5–10 μg) or standard carbohydrates (0.1–1 nmol) over P_2O_5 under vacuum in a 100 μl Reactivial. Mesoinositol may be used as the internal standard. Add dry methanolic HCl (40 μl, 0.625 M) and methyl acetate (10 μl). Seal the vial with a Teflon-lined septum. Vortex mix and heat at 70°C for 16 h. Cool. Add *tert.* butyl alcohol (10 μl) and take to dryness with a stream of oxygen-free nitrogen.

Re-*N*-acetylation

Add dry methanol (50 μl), pyridine (5 μl) and acetic anhydride (5 μl) successively and with intermediary mixing. Allow to stand for 15 min. Thoroughly dry with nitrogen and then over P_2O_5 under vacuum. Silylation. Add 20 μl of silylating reagent (Sylon HTP, Supelco, Pa.). Vortex mix. Leave for 1 h at room temperature. Take to dryness. Immediately dissolve the residue in 14 μl of hexane and apply all this sample to the GLC column.

Gas-liquid chromatography

Fused silica WCOT capillary column (25 m × 0.32 mm i.d.) with CP-Sil 5 liquid phase (Chrompak, U.K.); temperature programme, 140°C for 2 min, then × 8°C/min up to 260°C; flame ionization detector.

A separation showing 29 TMS derivatives of simple and amino sugars and uronic acids was achieved in less than 16 min. S < 100 pmol for each peak detected.

8.22.5. Gas chromatography–mass spectrometry method

The method of Lowe and Nilsson (1984) is described.

The intact glycoproteins are methylated, hydrolysed, reduced and then acetylated. After separating out the products by column chromatography the methylated hexitol acetates are analysed by mass spectrometry. See review on GC, HPLC and MS of complex carbohydrates (McNeil *et al.*, 1982).

8.22.6. Neutral and amino sugar alditol acetates by gas-liquid chromatography

This is the method of Henry *et al.* (1983).

This method is rapid and simple. The amino sugars are first deaminated with nitrous acid, to give as main products 2,5-anhydromannose (from glucosamine), 2,5-anhydrotalose (from galactose) and glucose (from mannosamine). The alditol acetates prepared from the amino sugars directly have very long retention times relative to those of the neutral sugars on the column used here.

Deamination

Add 100 mg of solid sodium nitrite to an aqueous solution of sample (0.1 ml) containing up to 2 mg of amino sugar. Cool in ice and add 0.1 ml of 9 M-sulphuric acid. After 5 min allow to warm to room temperature. After a further 25 min make alkaline with 0.2 ml of 15 M-ammonia solution then reduce and alkylate, by the method of Blakeney *et al.* (1983).

Reduction

Add 1 ml of sodium borohydride solution (2% w/v, in anhydrous dimethylsulphoxide) to 0.1 ml of 1 M-ammonia sugar solution. Reduce at 40°C for 90 min. Destroy excess borohydride by adding 0.1 ml of glacial acetic acid.

Acetylation

Add the catalyst 1-methylimidazole (0.2 ml) followed by acetic anhydride (2 ml) to the reduced monosaccharide. Mix. After 10 min at room temperature, add water

(5 ml) to destroy excess acetic anhydride. When cool, add dichloromethane (1 ml) and vortex mix. After separation of the phases, remove the lower phase for GLC. Inject 2 μl onto the column and store the remainder in a Reactivial at −20°C.

Gas–liquid chromatography

Use a SCOT glass capillary column (28.5 m × 0.5 mm i.d.) with Silar 10C liquid phase. Temperature programme, 190°C for 4 min followed by 4°C/min to 250°C.

Fifteen alditol acetates of deaminated amino sugars were separated in less than 20 min. The method was applied to chitin and glycoproteins.

8.22.7. Alditols by thin layer and gas-liquid chromatography

The method of Wang *et al.* (1983) is described.

Reduced and permethylated oligosaccharides (1–2 mg containing up to 15 glycose units) were separated on pre-coated silica gel 60 TLC plates (250 μm, Merck) without activation of the plates using benzene/methanol (16:1 or 10:1, v/v) as developing solvents. Apply the sample as a band and develop the plates twice in tanks lined with solvent-saturated filter paper.

Visualize by spraying with 0.5% orcinol in 2 M-sulphuric acid. S = 0.5 μg of sugar.

Disaccharides are separated on the basis of glycosidic linkage position, anomeric configuration and sugar content.

Gas–liquid chromatography

Fused silica capillary column (12 m × 0.25 mm) chemically bonded with DB-1 (0.1 μm thickness, J. & W. Scientific); carrier gas, helium at a linear velocity of 37 cm/s; oven temperature programmed from 150° to 330°C at 5°C/min. S = 30 ng for sophorose (Glc β1-2Glc).

8.22.8. Glucosamine, galactosamine, tryptophan and lysinoalanine with the amino acid analyser

The method of Johnson (1983) is described.

Hydrolysis for amino sugars

Hydrolyse *in vacuo* with 4 M-HCl at 100°C for 5 h. Take to dryness. Dissolve in 3 ml of 2% *N*-ethylmorpholine acetate, pH 9 and pass through a 50 × 9 mm column of Dowex 2- × 4 resin equilibrated with 1% *N*-ethylmorpholine acetate pH 9 to remove peptides and hydrolysis products which might interfere with the analysis of the amino sugars (Fanger & Smyth, 1970).

Collect the first 20 ml of eluate from the Dowex column. Add 1 ml of glacial acetic acid and 100 μl of 50% glycerol in ethanol (Dawson & Mopper, 1978) and dry. (The glycerol reduces losses due to reaction with glass surfaces). Dissolve the residue in water. Dry. Redissolve in water. Dry. Dissolve in buffer B (*vide infra*) containing 0.4% thiodiglycol for loading onto the column.

Hydrolysis for tryptophan

Alkaline hydrolysis is carried out with 4.8 M-NaOH with 25 mg of starch and 10 μl of 5% octan-1-ol in toluene (Hugli & Moore, 1972; see section 8.7.1.1) with 5-methyl tryptophan as internal standard. Polypropylene cryotubes with loosely fitting caps (the sealing rings are removed) (Vanguard International) are used as liners for 150 × 18 mm glass tubes. The tubes are sealed *in vacuo* and heated at 120°C for 24 h in an oven (not a heating block, because losses occur during refluxing). Cool. Neutralize with 420 μl of ice-cold 6 M-HCl diluted to 5.0 ml with buffer B containing 0.4% thioglycollate. Centrifuge for 4 min at 20000 g to remove any precipitate before loading onto the column.

The amino sugars may be separated using short columns and special buffers (Plummer, 1976) or on longer columns with slow flow rates (Fauconnet & Rochemont, 1978). This is often inconvenient and here a Beckman 119CL amino acid analyser is used with a 6 × 260 mm column of W-3H resin and ninhydrin detection with standard Beckman buffers as listed.

Buffers

A. (prepared from 10 × concentrate, pH 3.25)
   0.2 N-sodium citrate, 1% propan-2-ol, 0.25% thiodiglycol and octanoic acid, adjusted to pH 3.33 with 10 M-NaOH.
B. (prepared from 5 × concentrate, pH 4.12)
   0.2 N-sodium citrate, 0.2 N-sodium chloride, 0.25% thiodiglycol, adjusted to pH 3.95 with conc. HCl.
C. (prepared from 5 × concentrate, pH 6.4)
   0.2 N-sodium citrate, 0.8 N-sodium chloride, 0.05% octanoic acid, pH 6.4.

Column conditions

For analysis of protein amino acids:

0–38 min	buffer A
38–62 min	buffer B
62–115 min	buffer C
115–120 min	0.2 M-NaOH
120–140 min	buffer A to re-equilibrate

Change of temperature from 47° to 65°C at 28 min.
Buffer flow rate 35 ml/h.
Ninhydrin flow rate 17.5 ml/h.

For analysis of Trp, 5-Me Trp, Glc NH_2 and Gal NH_2:

0–20 min	buffer B
20–75 min	buffer C
75–80 min	0.2 M-NaOH
80–100 min	buffer B to re-equilibrate

Column temperature 70°C, using the low-temperature dial on the instrument.

Retention times (approx.)
Acidic and neutral amino acids elute within 25 min. Tyr 29 min, Phe 34 min, Glc NH_2 40 min, Gal NH_2 42 min, Lysinoalanine 46 min, His 48 min, Lys 52 min, Trp 59 min, NH_3 65 min, Arg 78 min, 5-Me Trp 79 min.

8.22.9. Neutral and amino sugars with the amino acid analyser

This is the method of Perini and Peters (1982).
Neutral and amino sugars from glycoproteins and oligosaccharides were separated on an amino acid analyser column with Durrum DC 4A resin. The eluted sugars were determined fluorometrically by post-column OPA derivatization. S = 8 pmol.

8.23. ACKNOWLEDGEMENTS

The author is grateful for financial support over the years from the United Kingdom Medical Research and Science Research Councils, University of London Central Research Fund, G. D. Searle & Co. Ltd., the Wellcome Trust and the Ministry of Defence (The Physiological Laboratory, AMTE.).

8.24. REFERENCES

Aitken, A & Morrice, N. (1984). *Methods in Protein Sequence Analysis, Fifth Internat. Conf.*, Walker, J. (Ed.), abst. p.30, Churchill College, Cambridge.
Allfrey, V. G., Di Paolo, E. A. & Sterner, R. (1984). *Methods Enzymol.*, **107,** 224–240.
Allington, W. B., Cordry, A. L., McCullough, G. A., Mitchell, D. E. & Nelson, J. W. (1978). *Anal. Biochem.*, **85,** 188–196.
Antoni, G., Presentini, R. & Neri, P. (1983). *Anal. Biochem.*, **129,** 60–63.
Aplin, J. D. & Hughes, R. C. (1981). *Anal. Biochem.*, **113,** 144–148.
Auffret, A. D., Williams, D. H. & Thatcher, D. R. (1978). *FEBS Lett.*, **90,** 324–326.
Augustyniak, J. Z. & Martin, W. G. (1965). *Can. J. Biochem.*, **43,** 291–292.
Balis, M. E. (1971). *Methods Biochem. Anal.*, **20,** 103–133.
Ballentine, R. (1957). *Methods Enzymol.*, **3,** 984–995.
Barbarash, G. R. & Quarles, R. H. (1982). *Anal. Biochem.*, **119,** 177–184.
Barman, T. E. & Koshland, D. E., Jr. (1967). *J. Biol. Chem.*, **242,** 5771–5776.
Barrett, G. C. (1974). *Advances in Chromatogr.*, **11,** 145–179.
Barth, G., Voelter, W., Bunnenberg, E. & Djerassi, C. (1972). *J. Am. Chem. Soc.*, **94,** 1293–1298.
Basha, S. M. M. & Roberts, R. M. (1977). *Anal. Biochem.*, **77,** 378–386.

Bayer, E., Grom, E., Kaltenegger, B. & Uhmann, R. (1976). *Anal. Chem.*, **48,** 1106–1109.
Bennett, H. P. J., Browne, C. A. & Solomon, S. (1980). *J. Liquid Chromatogr.*, **3,** 1353–1365.
Bensadoun, A. & Weinstein, D. (1976). *Anal. Biochem.*, **70,** 241–250.
Benson, J. R. (1972). U.S. Patent No. 3 686 118.
Benson, J. R. (1975). In *Instrumentation in Amino Acid Sequence Analysis*, Perham, R. N. (Ed.), Academic Press, London, New York, San Francisco, pp. 1–39.
Benson, J. R. (1977), *Methods Enzymol.*, **47,** 19–31.
Benson, J. R. & Hare, P. E. (1975). *Proc. Natl. Acad. Sci. U.S.A.*, **72,** 619–622.
Benson, J. V. Jr. (1972). *Anal Biochem.*, **50,** 477–493.
Berson, G. (1983). *Anal. Biochem.*, **134,** 230–234.
Beyer, R. E. (1983). *Anal. Biochem.*, **129,** 483–485.
Bhatti, T., Chambers, R. & Clamp, J. R. (1970). *Biochim. Biophys. Acta*, **222,** 339–347.
Bibring, T. & Baxandall, J. (1978). *Anal. Biochem.*, **85,** 1–14.
Blackburn, S.(1978a). In *Amino Acid Determination*, Blackburn, S. (Ed.), 2nd edn, Marcel Dekker, New York and Basel, pp. 7–37.
Blackburn, S. (1978b) In *Amino Acid Determination*, Blackburn, S. (Ed.), 2nd edn. Marcel Dekker, New York and Basel, pp.109–187.
Blakeney, A. B., Harris, P. J., Henry, R. J. & Stone, B. A. (1983). *Carbohydr. Res.*, **113,** 291–299.
Blau, K. (1968). In *Biomedical Applications of Gas Chromatography*, Szymanski, H. A. (Ed.), vol. 2, Plenum Press, New York, pp. 1–52.
Block, R. J., Durrum, E. L. & Zweig, G. (1958). *Paper Chromatography and Paper Electrophoresis*, 2nd edn, Academic Press, New York.
Blombäck, B. (1967), *Methods Enzymol.*, **11,** 398–413.
Böhlen, P. & Mellet, M. (1979), *Anal. Biochem.*, **94,** 313–321.
Bonting, S. L. & Jones, M. (1957). *Arch. Biochem. Biophys.*, **66,** 340–353.
Boulikas, T. & Hancock, R. (1981). *J. Biochem. Biophys. Methods,* **5,** 219–228.
Bounias, M. (1980). *Anal. Biochem.*, **106,** 291–295.
Boyer, P. D. (1954), *J. Am. Chem. Soc.*, **76,** 4331–4337.
Bradford, M. M. (1976). *Anal. Biochem.*, **72,** 248–254.
Bradstreet, R. B. (1965). *The Kjeldahl Method for Organic Nitrogen*, Academic Press, New York & London.
Bramhall, S., Noack, N., Wu, M. & Loewenberg, J. R. (1969). *Anal. Biochem.*, **31,** 146–148.
Brauer, A. W., Oman, C. L. & Margolies, M. N. (1984). *Anal. Biochem.*, **137,** 134–142.
Brautigan; D. L., Ferguson-Miller, S., Tarr, G. E. & Margoliash, E. (1978). *J. Biol. Chem.*, **253,** 140–148.
Brautigan, D. L., Bornstein, P. & Gallis, B. (1981). *J. Biol. Chem.,* **256,** 6519–6522.
Brenner, M. & Niederwieser, A. (1967). *Methods Enzymol.*, **11,** 27–31.
Bricknell, K. S. & Finegold, S. M. (1973). *Anal. Biochem.*, **51,** 23–31.
Brogdon, W. G. & Dickinson, C. M. (1983). *Anal. Biochem.*, **131,** 499–503.
Brown, J. L. & Roberts, W. K. (1976). *J. Biol. Chem.*, **251,** 1009–1014.
Burkhardt, W. A. III & Wilcox, P. E. (1967). *Biochem. Biophys. Res. Commun.*, **28,** 803–808.
Burnette, W. N. (1981). *Anal. Biochem.*, **112,** 195–203.
Cadman, E. Bostwick, J. R. & Eichberg, J. (1979). *Anal. Biochem.*, **96,** 21–23.
Capony, J.-P. & Demaille, J. G. (1983). *Anal. Biochem.*, **128,** 206–212.
Chambers, R. E. & Clamp, J. R. (1971). *Biochem. J.*, **125,** 1009–1018.
Chaplin, M. F. (1982). *Anal. Biochem.*, **123,** 336–341.
Chauhan, J. & Darbre, A. (1981a). *J. Chromatogr.*, **211,** 347–359.

Chauhan, J. & Darbre, A. (1981b). *J. High Resol. Chromatogr. Chromatogr. Commun.*, **4,** 11–16.
Chauhan, J. & Darbre, A. (1981c). *J. High Resol. Chromatogr. Chromatogr. Commun.*, **4,** 260–265.
Chauhan, J. & Darbre, A. (1982a). *J. Chromatogr.*, **227,** 305–321.
Chauhan, J. & Darbre, A. (1982b). *J. Chromatogr.*, **236,** 151–156.
Chauhan, J. & Darbre, A. (1982c). *J. Chromatogr.* **240,** 107–115.
Chen, S.-C. (1983). *Anal. Biochem.*, **132,** 272–275.
Chiapelli, F., Vasil, A. & Haggerty, D. F. (1979). *Anal. Biochem.*, **94,** 160–165.
Chou, Y.-S. & Chien, H.-C. (1978). Sheng Wu Hua Hsueh Yu Sheng Wu Wu Li Li Chin Chan, **22,** 12–14.
Chrambach, A. & Nguyen, N. Y. (1978). In *Electrokinetic Separation Methods*, Righetti, P. J., van Oss, C. I. & Varderhoff, J. W. (Eds.), Elsevier/North Holland, Amsterdam, pp. 337–367.
Christensen, T. (1979a). *Acta Chem. Scand.*, **B33,** 763–766.
Christensen, T. (1979b). *Pept. Struct. Biol. Funct., Proc. Sixth Am. Pept. Symp.*, Gross, E. & Meienhofer, J. (Eds.) Pierce, Rockford, Illinois, pp. 385–388.
Chromy, V., Fischer, J. & Kulhanek, V. (1974). *Clin. Chem.*, **20,** 1362–1363.
Clamp, J. R. (1974a). *Biochem. Soc. Trans.*, **2,** 64–66.
Clamp, J. R. (1974b). *Biochem. Soc. Symp.*, **40,** 3–16.
Clamp, J. R. & Hough, L. (1965). *Biochem. J.*, **94,** 17–24.
Clamp, J. R., Bhatti, T. & Chambers, R. E. (1971). *Methods Biochem. Anal.*, **19,** 229–344.
Clamp, J. R., Bhatti, T. & Chambers, R. E. (1972). In *Glycoproteins*, Gottschalk, A. (Ed.) Elsevier, Amsterdam, pp. 612–652.
Clark, A. J. (1979). *Laboratory Practice*, **28,** 1076–1078.
Cliffe, A. J., Berridge, N. J. & Westgarth, D. R. (1973). *J. Chromatogr.*, **78,** 333–341.
Cloete, C. (1984). *J. Liquid Chromatogr.*, **7,** 1979–1990.
Coakley, W. T. & James, C. J. (1978). *Anal. Biochem.*, **85,** 90–97.
Cohen, S. A., Tarvin, T. L., Bidlingmeyer, B. L. & Tarr, G. E. (1984). *Methods in Protein Sequence Analysis*, Fifth Internat. Conf., Walker, J. (Ed.), abst. p. 31, Churchill College, Cambridge.
Congote, L. F. (1982). *J. Chromatogr.*, **253,** 276–282.
Cooper, J. D. H. & Turnell, D. C. (1982). *J. Chromatogr.*, **227,** 158–161.
Cooper, J. D. H., Lewis, M. T. & Turnell, D. C. (1984a). *J. Chromatogr.*, **285,** 490–494.
Cooper, J. D. H., Ogden, G., McIntosh, J. & Turnell, D. C. (1984b). *Anal. Biochem.*, **142,** 98–102.
Croft, L. R. (1980). *Handbook of Protein Sequence Analysis*, 2nd edn, Wiley, Chichester, New York, Brisbane, Toronto.
Cromwell, L. D. & Stark, G. R. (1969). *Biochemistry,* **8,** 4735–4740.
Cronin, J. R., & Hare, P. E. (1977). *Anal. Biochem.*, **81,** 151–156.
Cronin, J. R., Pizzarello, S. & Gandy, W. E. (1979). *Anal. Biochem.*, **93,** 174–179.
Cuatrecasas, P. (1970). *J. Biol. Chem.*, **245,** 3059–3065.
Cunico, R. L. & Schlabach, T. (1983). *J. Chromatogr.*, **266,** 461–470.
Dale, T. & Court, W. E. (1981). *Chromatographia,* **14,** 617–620.
Damodaran, S. (1985). *Anal. Biochem.*, **145,** 200–204.
Darbre, A. (1971). *Laboratory Practice*, **9,** 726.
Darbre, A. (1986). In *Amino Acid Analysis by Gas Chromatography,* Zumwalt, R. W., Kuo, K. C. T. & Gehrke, C. W. (Eds.), CRC Press, Cleveland, Ohio.
Darbre, A. & Islam, A. (1968). *Biochem. J.*, **106,** 923–925.
Davies, D. & Holdsworth, E. S. (1979). *Anal. Biochem.*, **100,** 92–94.
Dawson, R. & Mopper, K. (1978). *Anal. Biochem.*, **84,** 186–190.

Dawson, R. M. C., Elliott, D. C., Elliott, W. H. & Jones, K. M. (1969). Data for Biochemical Research, 2nd edn, Oxford University Press, London, pp. 509–557.
De La Llosa, P., Abed, A. E. & Roy, M. (1980). *Canad. J. Biochem.*, **58,** 745–748.
Desgres, J., Boisson, D. & Padieu, P. (1979). *J. Chromatogr.*, **162,** 133–152.
De Traglia, M. C., Brand, J. S. & Tometsko, A. M. (1979). *Anal. Biochem.*, **99,** 464–473.
Dierckxsens, G. C., De Meyer, L. & Tonino, G. J. (1983). *Anal. Biochem.*, **130,** 120–127.
Dion, A. S. & Pomenti, A. A. (1983). *Anal. Biochem.*, **129,** 490–496.
Duggan, E. L. (1957). *Methods Enzymol.*, **3,** 501–504.
Durham, J. P. & Lopez-Solis, R. O. (1979). *Anal. Biochem.*, **100,** 98–99.
Easley, C. W. (1965). *Biochem. Biophys. Acta*, **107,** 386–388.
Edelhoch, H. (1967). *Biochemistry*, **6,** 1948–1954.
Edman, P. & Begg, G. (1967). *Eur. J. Biochem.*, **1,** 80–91.
Ellman, G. L. (1959). *Arch. Biochem. Biophys.*, **82,** 70–77.
Eveleigh, J. W. & Winter, G. D. (1970). In *Protein Sequence Determination*, Needleman, S. B. (Ed.), Springer-Verlag, Berlin, Heidelberg, New York, pp. 92–95.
Fahmy, A. R., Niederwieser, A., Pataki, G. & Brenner, M. (1961). *Helv. Chim. Acta,* **44,** 2022–2026.
Fanger, M. W. & Smyth, D. G. (1970). *Anal. Biochem.*, **34,** 494–499.
Fauconnet, M. & Rochemont, J. A. (1978). *Anal. Biochem.*, **91,** 403–409.
Felker, P. & Bandurski, R. S. (1975). *Anal. Biochem.*, **67,** 245–262.
Fields, R. (1972). *Methods Enzymol.*, **25,** 464–468.
Fleury, M. O. & Ashley, D. V. (1983). *Anal. Biochem.*, **133,** 330–335.
Floyd, N. F., Cammaroti, M. S. & Lavine, T. F. (1963). *Arch. Biochem. Biophys.*, **102,** 343–345.
Flynn, G. H. & Whitmore, T. N. (1983). *Anal. Biochem.*, **131,** 42–45.
Fransson, B. Ragnarsson, U. & Zetterquist, O. (1982). *J. Chromatogr.*, **240,** 165–171.
Friedman, R. D. (1982). *Anal. Biochem.*, **126,** 346–349.
Friedman, M., Noma, A. T. & Wagner, J. R. (1979). *Anal. Biochem.*, **98,** 293–304.
Gamerith, G. (1983a). *J. Chromatogr.*, **256,** 267–281.
Gamerith, G. (1983b). *J. Chromatogr.*, **256,** 326–330.
Gehrke, C. W. & Leimer, K. (1971). *J. Chromatogr.*, **57,** 219–238.
Gehrke, C. W., Lamkin, W. M., Stalling, D. L. & Shahroki, F. (1965). *Biochem. Biophys. Res. Commun.*, **19,** 328–334.
Gehrke, C. W., Nakomoto, H. & Zumwalt, R. W. (1969). *J. Chromatogr.*, **45,** 24–51.
Gehrke, C. W., Kuo, K. & Zumwalt, R. W. (1971). *J. Chromatogr.*, **57,** 209–217.
Gehrke, C. W., Younker, D. R., Gerhardt, K. O. & Kuo, K. C. (1979). *J. Chromatogr. Sci.*, **17,** 301–307.
Gersten, D. M., Zapolski, E. J. & Ledley, R. S. (1983). *Anal. Biochem.*, **129,** 57–59.
Gibson, W. (1983). *Anal. Biochem.*, **132,** 171–173.
Giulian, G. G., Moss, R. L. & Greaser, M. (1983). *Anal. Biochem.*, **129,** 277–287.
Golovina, T. O., Cherednikova, T. V., Mevkh, A. T. & Nagradova, N. K. (1977). *Anal. Biochem.*, **83,** 778–781.
Goodwin, T. W. & Morton, R. A. (1946). *Biochem. J.*, **40,** 628–632.
Gornall, A. G., Bardawill, C. J. & David, M. M. (1949). *J. Biol. Chem.*, **177,** 751–766.
Goshev, I. & Nedkov, P. (1979). *Anal. Biochem.*, **95,** 340–343.
Gracy, R. W. (1977). *Methods Enzymol.*, **47,** 195–243.
Grego, B. & Hearn, M. T. W. (1983). *J. Chromatogr.*, **255,** 67–77.
Groves, W. E., Davis, F. C., Jr. & Sells, B. H. (1968). *Anal. Biochem.*, **22,** 195–210.
Gundberg, C. M., Lian, J. B. & Gallop, P. M. (1979). *Anal. Biochem.*, **98,** 219–225.
Guynn, R. W. & Veech, R. L. (1975). *Methods Enzymol.*, **35,** 302–307.
Habeeb, A. F. S. A. (1966). *Anal. Biochem.*, **14,** 328–336.
Hager, D. A. & Burgess, R. R. (1980). *Anal. Biochem.*, **109,** 76–86.

Hamilton, P. B. (1966). *Adv. Chromatogr.*, **2,** 3–62.
Hamilton, P. B. (1967). *Methods Enzymol.*, **11,** 15–27.
Hancock, K. & Tsang, V. C. W. (1983). *Anal. Biochem.*, **133,** 157–162.
Hare, P. E. (1972). *Space Life Sciences*, **3,** 354–359.
Hare, P. E. (1977). *Methods Enzymol.*, **47,** 3–18.
Haroon, Y. (1984). *Anal. Biochem.*, **140,** 343–348.
Hartree, E. F. (1972). *Anal. Biochem.*, **48,** 422–427.
Hauschka, P. V. (1977). *Anal. Biochem.*, **80,** 212–223.
Hauschka, P. V., Henson, E. B. & Gallop, P. M. (1980). *Anal. Biochem.*, **108,** 57–63.
Haworth, C. & Heathcote, J. G. (1969). *J. Chromatogr.*, **41,** 380–385.
Heathcote, J. G. & Washington, R. J. (1967). *Analyst*, **92,** 627–633.
Heathcote, J. G., Washington, R. J., Haworth, C. & Bell, S. (1970). *J. Chromatogr.*, **51,** 267–275.
Hediger, H., Stevens, R. L., Brandenberger, H. & Schmid, K. (1973). *Biochem. J.*, **133,** 551–561.
Heinrikson, R. L. & Meredith, S. C. (1984). *Anal. Biochem.*, **136,** 65–74.
Hellerman, L. (1937). *Physiol. Rev.*, **17,** 454–484.
Henderson, L. E., Oroszlan, S. & Konigsberg, W. (1979). *Anal. Biochem.*, **93,** 153–157.
Henkel, H. G. (1971). *J. Chromatogr.*, **58,** 201–207.
Henry, R. J., Blakeney, A. B., Harris, P. J. & Stone, B. A. (1983). *J. Chromatogr.*, **256,** 419–427.
Hess, H. H., Lees, M. B. & Derr, J. E. (1978). *Anal. Biochem.*, **85,** 295–300.
Higuchi, M. & Yoshida, F. (1977). *Anal. Biochem.*, **77,** 542–547.
Higuchi, M. & Yoshida, F. (1978). *Agr. Biol. Chem.*, **42,** 1669–1674.
Higuchi, M. & Yoshida, F. (1980). *Anal. Biochem.*, **105,** 90–96.
Hill, D. W., Walters, F. H., Wilson, T. D. & Stuart, J. D. (1979). *Anal. Chem.*, **51,** 1338–1341.
Hill, D., Burnworth, L., Skea, W. & Pfeifer, R. (1982). *J. Liquid Chromatogr.*, **5,** 2369–2393.
Hirs, C. H. W. (1967). *Methods Enzymol.*, **11,** 59–62.
Hodgin, J. C. (1979). *J. Liquid Chromatogr.*, **2,** 1047–1059.
Hoenders, H. J. & Bloemendal, H. (1967). *Biochim. Biophys. Acta*, **147,** 183–185.
Hoffman, P. G., Zinder, O., Bonner, W. M. & Pollard, H. B. (1976). *Arch. Biochem. Biophys.*, **176,** 375–388.
Honda, K., Sekino, J. & Imal, K. (1983). *Anal. Chem.*, **55,** 940–943.
Hsu, K.-T. & Currie, B. L. (1978). *J. Chromatogr.*, **166,** 555–561.
Hugli, T. E. & Moore, S. (1972). *J. Biol. Chem.*, **247,** 2828–2834.
Hunter, T. & Sefton, B. M. (1980). *Proc. Natl. Acad. Sci. U.S.A.*, **77,** 1311–1315.
Hušek, P. & Macek, K. (1975). *J. Chromatogr.*, **113,** 139–230.
Ida, S. & Kuriyama, K. (1983). *Anal. Biochem.*, **130,** 95–101.
Inglis, A. S. (1983). *Methods Enzymol.*, **91,** 26–36.
Inglis, A. S., Roxburgh, C. M. & Takayangi, H. (1974). *Anal. Biochem.*, **61,** 25–31.
Inoue, S., Tokuyama, T. & Takai, K. (1983). *Anal. Biochem.*, **132,** 468–480.
Irie, S. (1980). Seigaku, **52,** 411–413.
Irie, S., Sezaki, M. & Kato, Y. (1982). *Anal. Biochem.*, **126,** 350–354.
Irreverre, F. (1965). *Biochim. Biophys. Acta*, **111,** 551–552.
Ishida, Y., Fujita, T. & Asai, K. (1981). *J. Chromatogr.*, **204,** 143–148.
Ishikawa, Y. & Wuthier, R. E. (1981). *Anal. Biochem.*, **114,** 388–395.
Itzhaki, R. F. & Gill, D. M. (1964). *Anal. Biochem.*, **9,** 401–410.
Jacobson, B. S. & Fairman, K. R. (1980). *Anal. Biochem.*, **106,** 114–117.
James, L. B. (1979). *J. Chromatogr.*, **175,** 211–215.
Johansen, P. G., Marshall, R. D. & Neuberger, A. (1960). *Biochem. J.*, **77,** 239–247.

Johnson, D. A. (1983). *Anal. Biochem.*, **130,** 475–480.
Johnson, M. K. (1978). *Anal. Biochem.*, **86,** 320–323.
Johnson, N. D., Hunkapiller, M. W. & Hood, L. E. (1979). *Anal. Biochem.*, **100,** 335–338.
Jones, B. N. & Gilligan, J. P. (1983a). *J. Chromatogr.*, **266,** 471–482.
Jones, B. N. & Gilligan, J. P. (1983b). *Am. Biotechnology Lab.*, **46,** 47–51.
Jones, B. N., Pååbo, S. & Stein, S. (1981). *J. Liquid Chromatogr.*, **4,** 565–586.
Jones, K. & Heathcote, J. G. (1966). *J. Chromatogr.*, **24,** 106–111.
Joseph, M. H. & Davies, P. (1983). *J. Chromatogr.*, **277,** 125–136.
Junqueira, L. C. U., Bignolas, G. & Brentani, R. R. (1979). *Anal. Biochem.*, **94,** 96–99.
Kaiser, F. E., Gehrke, C. W., Zumwalt, R. W. & Kuo, K. C. (1974). *J. Chromatogr.*, **94,** 113–133.
Kalb, V. F., Jr. & Bernlohr, R. W. (1977). *Anal. Biochem.*, **82,** 362–371.
Kaldy, M. S. & Kereliuk, G. R. (1978). *Laboratory*, **27,** 868.
Kapadia, G. & Chrambach, A. (1972). *Anal. Biochem.*, **48,** 90–102.
Kerenyi, L. & Gallyas, F. (1972). *Clin. Chim. Acta*, **38,** 465–467.
Kirkman, M. A. (1974). *J. Chromatogr.*, **97,** 175–191.
Knoop, D. R., Morgan, E. T., Tarr, G. E. & Coon, M. J. (1982). *J. Biol. Chem.*, **257,** 8472–8480.
Koboyashi, S. & Imai, K. (1980). *Anal. Chem.*, **52,** 424–427.
Korff, R. W. von (1954). *J. Biol. Chem.*, **210,** 539–544.
Koroleva, E. M., Maltsev, V. G., Belenkii, B. G. & Viska, M. (1982). *J. Chromatogr.*, **242,** 145–152.
Kreil, G. & Kreil-Kriss, G. (1967). *Biochem. Biophys. Res. Commun.*, **27,** 275–280.
Kuo, S. C. & Younathan, E. S. (1973). *Anal. Biochem.*, **55,** 1–8.
Kuwada, M. & Katayama, K. (1981). *Anal. Biochem.*, **117,** 259–265.
Kuwada, M. & Katayama, K. (1983). *Anal. Biochem.*, **131,** 173–179.
Laemmli, U. K. (1970). *Nature* (*Lond.*), **277,** 680–685.
Lam, S. K. (1982). *J. Chromatogr.*, **234,** 485–488.
Lam, S., Chow, F. & Karmen, A. (1980). *J. Chromatogr.*, **199,** 295–303.
Larsen, B. R. & West, F. G. (1981). *J. Chromatogr. Sci.*, **19,** 259–265.
Lasne, F., Benzerara, O. & Lasne, Y. (1983). *Anal. Biochem.*, **132,** 338–341.
Layne, E. (1957). *Methods Enzymol.*, **3,** 447–454.
Lee, K. S. & Drescher, D. G. (1978). *Int. J. Biochem.*, **9,** 457–467.
Lee, C. C. Y. & Loudon, G. M. (1979). *Anal. Biochem.*, **94,** 60–64.
Lee, H.-M., Forde, M. D., Lee, M. C. & Bucher, D. J. (1979). *Anal. Biochem.*, **96,** 298–307.
Lees, M. B. & Paxman, S. (1972). *Anal Biochem.*, **47,** 184–192.
Leftak, I. M. (1983). *Anal. Biochem.*, **135,** 95–101.
Lewis, R. N. A. H. (1979). *Anal. Biochem.*, **99,** 136–141.
Liebezeit, G. & Dawson, R. (1981). *J. High Resol. Chromatogr. Chromatogr. Commun.*, **4,** 354–356.
Light, A. & Smith, E. L. (1963). In *The Proteins*, Neurath, H. (Ed.), vol. 1, 2nd edn, Academic Press, New York, London, pp. 1–44.
Lindroth, P. & Mopper, K. (1979). *Anal. Chem.*, **51,** 1667–1674.
Lischwe, M. A. & Ochs, D. (1982). *Anal. Biochem.*, **127,** 453–457.
Liu, T.-Y. & Chang, Y. H. (1971). *J. Biol. Chem.*, **246,** 2842–2848.
Lowe, M. E. & Nilsson, B. (1984). *Anal. Biochem.*, **136,** 187–191.
Lowry, O. H., Rosebrough, N. J., Farr, A. L. & Randall, R. J. (1951). *J. Biol. Chem.*, **193,** 265–275.
Lucas, B. & Sotelo, A. (1980). *Anal. Biochem.*, **109,** 192–197.
Ludowieg, J. & Dorfman, A. (1960). *Biochim. Biophys. Acta*, **38,** 212–218.
Lund, E., Thomsen, J. & Brunfeldt, K. (1977). *J. Chromatogr.*, **130,** 51–54.

MacKenzie, S. L. (1981). *Methods Biochem. Anal.*, **27,** 1–88.
MacKenzie, S. L. & Hogge, L. R. (1977). *J. Chromatogr.*, **132,** 485–493.
MacKenzie, S. L. & Tenaschuk, D. (1974). *J. Chromatogr.*, **97,** 19–24.
MacKenzie, S. L. & Tenaschuk, D. (1979a). *J. Chromatogr.*, **171,** 195–208.
MacKenzie, S. L. & Tenaschuk, D. (1979b). *J. Chromatogr.*, **173,** 53–63.
Madar, D. A., Willis, R. A., Koehler, K. A. & Hiskey, R. G. (1979). *Anal. Biochem.*, **92,** 466–472.
Magnusson, S., Sottrup-Jensen, L., Petersen, T. E., Morris, H. R. & Dell, A. (1974). *FEBS Lett.*, **44,** 189–193.
Mahoney, W. C. & Hermodson, M. A. (1980). *J. Biol. Chem.*, **255,** 11199–11203.
Mahuran, D., Clements, P., Carrella, M. & Strasberg, P. M. (1983). *Anal. Biochem.*, **129,** 513–516.
Makkar, H. P. S., Sharma, O. P. & Negi, S. S. (1980). *Anal. Biochem.*, **104,** 124–126.
March, J. F. (1975). *Anal. Biochem.*, **69,** 420–442.
Marciani, D. J., Wilkie, S. D. & Schwartz, C. L. (1983). *Anal. Biochem.*, **128,** 130–137.
Marshall, T. (1983). *Electrophoresis*, **4,** 269–272.
Marshall, T. (1984a). *Anal. Biochem.*, **136,** 340–346.
Marshall, T. (1984b). *Electrophoresis*, **5,** 245–250.
Marshall, T. & Latner, A. L. (1981). *Electrophoresis*, **2,** 228–235.
Mather, I. H. & Tamplin, C. B. (1979). *Anal. Biochem.*, **93,** 139–142.
Matsubara, H. & Sasaki, R. M. (1969). *Biochem. Biophys. Res. Commun.*, **35,** 175–181.
Matsu-ura, S., Yamamoto, S. & Makita, M. (1981). *Anal. Biochem.*, **114,** 371–376.
McGuire, J., Taylor, P. & Greene, L. A. (1977). *Anal. Biochem.*, **83,** 75–81.
McKnight, G. S. (1977). *Anal. Biochem.*, **78,** 86–92.
McNeil, M., Darvill, A. G., Åman, P., Franzén, L. -E. & Albersheim, P. (1982). *Methods Enzymol.*, **83,** 3–45.
Means, G. E. & Feeney, R. E. (1971). *Chemical Modification of Proteins*, Holden-Day, San Francisco.
Merril, C. R., Switzer, R. C. & Van Keuren, M. L. (1979). *Proc. Natl. Acad. Sci. U.S.A.*, **76,** 4335–4339.
Merril, C. R., Dunau, M. L. & Goldman, D. (1981a). *Anal. Biochem.*, **110,** 201–207.
Merril, C. R., Goldman, D., Sedman, S. A. & Ebert, M. H. (1981b). *Science*, **211,** 1437–1438.
Merril, C. R., Goldman, D. & Van Keuren, M. L. (1982). *Electrophoresis*, **3,** 17–23.
Merril, C. R., Harrington, M. & Alley, V. (1984). *Electrophoresis*, **5,** 289–297.
Mokrasch, L. C. (1967). *Anal. Biochem.*, **18,** 64–71.
Mokrasch, L. C. (1970). *Anal. Biochem.*, **36,** 273–277.
Mold, D. E., Weingart, J., Assaraf, J., Lubahn, D. B., Kelner, D. N., Shaw, B. R. & McCarty, K. S., Sr. (1983). *Anal. Biochem.*, **135,** 44–47.
Moodie, I. M. & Burger, J. (1981). *J. High Resol. Chromatogr. Chromatogr. Commun.*, **4,** 218–223.
Moodie, I. M., Walsh, D. L. & Burger, J. A. (1983). *J. Chromatogr.*, **261,** 146–152.
Moore, S. (1963). *J. Biol. Chem.*, **238,** 235–237.
Moore, S. & Stein, W. H. (1951). *J. Biol. Chem.*, **192,** 663–681.
Moore, S. & Stein, W. H. (1954a). *J. Biol. Chem.*, **211,** 893–906.
Moore, S. & Stein W. H. (1954b). *J. Biol. Chem.*, **211,** 907–913.
Moore, S. & Stein, W. H. (1963). *Methods Enzymol.*, **6,** 819–831.
Moore, S., Spackman, D. H. & Stein, W. H. (1958). *Anal. Chem.*, **30,** 1185–1190.
Morrice, N. & Aitken, A. (1985). *Anal. Biochem.* **148,** 207–212.
Morrissey, J. H. (1981). *Anal. Biochem.*, **117,** 307–310.
Murayama, K. & Shindo, N. (1977). *J. Chromatogr.*, **143,** 137–152.
Murphy, J. B. & Kies, M. W. (1960). *Biochim. Biophys. Acta*, **45,** 382–384.

Nabi, S. A., Farooqui, W. U., Ziddiqui, Z. M. & Rao, R. A. K. (1983). *J. Liquid Chromatogr.*, **6,** 109–122.
Narita, K. (1958). *Biochem. Biophys. Acta*, **28,** 184–191.
Neuhoff, V. (1984). *Electrophoresis*, **5,** 251.
Neumann, N. P. (1972). *Methods Enzymol.*, **25,** 393–400.
Nguyen, N. Y. & Chrambach, A. (1979). *J. Biochem. Biophys. Methods*, **1,** 171–187.
Nguyen, N. Y., Di Fonzo, J. & Chrambach, A. (1980). *Anal. Biochem.*, **106,** 78–91.
Nice, E. C., Capp, M. & O' Hare, M. J. (1979). *J. Chromatogr.*, **185,** 413–427.
Nikodijevic, O., Nikodijevic, B., Zinder, O., Yu, M.-Y. W., Guroff, G. & Pollard, H. B. (1976). *Proc. Natl. Acad. Sci. U.S.A.*, **73,** 771–774.
Nishio, K. & Kawakami, M. (1982). *Anal. Biochem.*, **126,** 239–241.
Oakley, B. R., Kirsch, D. R. & Morris, N. R. (1980). *Anal. Biochem.*, **105,** 361–363.
Ochs, D. C., McConkey, E. H. & Sammons, D. W. (1981). *Electrophoresis*, **2,** 304–307.
Öfverstedt, L. G., Johansson, G., Fröman, G. & Hjerten, S. (1981). *Electrophoresis*, **2,** 168–173.
Ohnishi, S. T. & Barr, J. K. (1978). *Anal. Biochem.*, **86,** 193–200.
Ohsawa, K. & Ebata, N. (1983). *Anal. Biochem.*, **135,** 409–415.
Olert, J., Sawatzki, G., Kling, H. & Gebauer, J. (1979). *Histochemistry*, **60,** 91–99.
Parham, M. E. & Loudon, G. M. (1978). *Biochem. Biophys. Res. Commun.*, **80,** 1–6.
Pasieka, A. E. & Borowiecki, M. T. (1965). *Biochim. Biophys. Acta*, **111,** 553–556.
Pearce, R. J. (1977). *J. Chromatogr.*, **136,** 113–126.
Pecci, L. & Cavallini, D. (1981). *Anal. Biochem.*, **118,** 70–75.
Pelley, J. W., Garner, C. W. & Little, G. H. (1978). *Anal. Biochem.*, **86,** 341–343.
Penke, B., Ferenczi, R. & Kovács, K. (1974). *Anal. Biochem.*, **60,** 45–50.
Perham, R. N. (1978). Techniques for determining the amino acid composition and sequence in proteins, in *Techniques in Protein & Enzyme Biochemistry*, B110, Elsevier/North Holland, Shannon, pp. 1–39.
Perini, F. & Peters, B. P. (1982). *Anal. Biochem.*, **123,** 357–363.
Perret, B. A., Felix, R., Furlan, M. & Beck, E. A. (1983). *Anal. Biochem.*, **131,** 46–50.
Perry, T. L., Stedman, D. & Hansen, S. (1968). *J. Chromatogr.*, **38,** 460–466.
Peterson, E. A. & Sober, H. A. (1959). *Anal. Chem.*, **31,** 857–862.
Peterson, G. L. (1977). *Anal. Biochem.*, **83,** 346–356.
Peterson, G. L. (1979). *Anal. Biochem.*, **100,** 201–220.
Pfeifer, R. F. & Hill, D. W. (1983). *Adv. Chromatogr.*, **22,** 37–69.
Phillips, D. M. P. (1963). *Biochem. J.*, **86,** 397.
Phillips, R. D. (1981). *Anal. Biochem.*, **113,** 102–107.
Pierce, J. & Suelter, C. H. (1977). *Anal. Biochem.*, **81,** 478–480.
Plummer, T. H. Jr. (1976). *Anal. Biochem.*,**73,** 532–534.
Poehling, H.-M. & Neuhoff, V. (1981). *Electrophoresis,* **2,** 141–147.
Pollard, H. B., Menard, R., Brandt, H. A., Pazoles, C. J., Creutz, C. E. & Ramu, A. (1978). *Anal. Biochem.*, **86,** 761–763.
Porro, M., Viti, S., Antoni, G. & Saletti, M. (1982). *Anal. Biochem.*, **127,** 316–321.
Posner, I. (1976). *Anal. Biochem.*, **70,** 187–194.
Radolo, B. J. (1980). *Electrophoresis*, **1,** 43–56.
Rash, J. J., Gehrke, C. W., Zumwalt, R. W., Kuo, K. C., Kvenvolden, K. A. & Stalling, D. L. (1972). *J. Chromatogr. Sci.*, **10,** 444–450.
Reisner, A. H., Nemes, P. & Bucholtz, C. (1975). *Anal. Biochem.*, **64,** 509–516.
Riddles, P. W., Blakeley, R. L. & Zermer, B. (1979). *Anal. Biochem.*, **94,** 75–81.
Righetti, P. G. & Chillemi, F. (1978). *J. Chromatogr.*, **157,** 243–251.
Riordan, J. F. & Giese, R. W. (1977). *Methods Enzymol.*, **47,** 31–40.
Roach, D. & Gehrke, C. W. (1969a). *J. Chromatogr.*, **43,** 303–310.

Roach, D. & Gehrke, C. W. (1969b). *J. Chromatogr.*, **44,** 269–278.
Roach, D. & Gehrke, C. W. (1970). *J. Chromatogr.*, **52,** 393–404.
Robel, E. J. (1967). *Anal. Biochem.*, **18,** 406–413.
Rose, I. A. (1955). *Methods Enzymol.*, **1,** 591–595.
Roth, M. (1971). *Anal. Chem.*, **43,** 880–882.
Roth, M. & Hampaï, A. (1973). *J. Chromatogr.*, **83,** 353–356.
Rothberg, P. G., Harris, T. J. R., Nomoto, A. & Wimmer, E. (1978). *Proc. Natl. Acad. Sci. U.S.A.*, **75,** 4868–4872.
Rubin, R. W. & Warren, R. W. (1977). *Anal. Biochem.*, **83,** 773–777.
Sammons, D. W., Adams, I. D. & Nishizawa, E. E. (1981). *Electrophoresis*, **2,** 135–141.
Sarges, R. & Witkop, B. (1964). *J. Am. Chem. Soc.*, **86,** 1861–1862.
Satoh, K. & Busch, H. (1981). *Cell Biol. Int. Rep.*, **5,** 857–866.
Savoy, C. F., Heinis, J. L. & Seals, R. G. (1975). *Anal. Biochem.*, **68,** 562–571.
Schleicher, M. & Watterson, D. M. (1983). *Anal. Biochem.*, **131,** 312–317.
Schmer, G. & Kreil, G. (1969). *Anal. Biochem.*, **29,** 186–192.
Schmidt, G. J., Olson, D. C. & Slavin, W. (1979). *J. Liquid Chromatogr.*, **2,** 1031–1045.
Schmitt, H. W. & Walker, J. E. (1977). *FEBS Lett.*, **81,** 403–405.
Schroeder, W. A., Cua, J. T., Matsuda, G. & Fenninger, W. D. (1962). *Biochim. Biophys. Acta*, **63,** 532–534.
Scoffone, E. & Fontana, A. (1975). *Mol. Biol. Biochem. Biophys.*, **8,** 162–203.
Scoffone, E., Fontana, A. & Rocchi, R. (1968). *Biochemistry*, **7,** 971–979.
Scopes, R. K. (1974). *Anal. Biochem.*, **59,** 277–282.
Sedmak, J. J. & Grossberg, S. E. (1977). *Anal. Biochem.*, **79,** 544–552.
Shelton, J. R. & Rogers, K. S. (1971). *Anal. Biochem.*, **44,** 134–142.
Sherma, J., Sleckman, B. P. & Armstrong, D. W. (1983). *J. Liquid Chromatogr.*, **6,** 95–108.
Siezen, R. J. & Mague, T. H. (1977). *J. Chromatogr.*, **130,** 151–160.
Simmons, W. H. & Meisenberg, G. (1983). *J. Chromatogr.*, **266,** 483–489.
Simons, S. S., Jr. & Johnson, D. F. (1976). *J. Am. Chem. Soc.*, **98,** 7098–7099.
Simons, S. S., Jr. & Johnson, D. F. (1977). *Anal. Biochem.*, **82,** 250–254.
Simpson, R. J., Neuberger, M. R. & Liu, T. -Y. (1976). *J. Biol. Chem.*, **251,** 1936–1940.
Smith, I. (1969). In *Chromatographic and Electrophoretic Techniques*, Smith, I. (Ed.), vol. 1, 3rd edn, Heinemann Medical Books, Bath, pp. 104–169.
Soby, L. M. & Johnson, P. (1981). *Anal. Biochem.*, **113,** 149–153.
Soodak, M. (1957). *Methods Enzymol.*, **3,** 266–269.
Spackman, D. H. (1967). *Methods Enzymol.*, **11,** 3–15.
Spackman, D. H., Stein, W. H. & Moore, S. (1958). *Anal. Chem.*, **30,** 1190–1206.
Spande, T. F. & Witkop, B. (1967). *Methods Enzymol.*, **11,** 498–506.
Spector, T. (1978). *Andl. Biochem.*, **86,** 142–146.
Spies, J. R. & Chambers, D. C. (1949). *Anal. Chem.*, **21,** 1249–1266.
Stalling, D. L., Gille, G. & Gehrke, C. W. (1967). *Anal. Biochem.*, **18,** 118–125.
Steck, G., Leuthard, P. & Bürk, R. R. (1980). *Anal. Biochem.*, **107,** 21–24.
Stegink, L. D. (1967). *Anal. Biochem.*, **20,** 502–516.
Steinbach, L. & Becker, E. I. (1954). *J. Am. Chem. Soc.*, **76,** 5808–5810.
Suelter, C. H. & DeLuca, M. (1983). *Anal. Biochem.*, **135,** 112–119.
Svasti, J. (1980). *Trends Biochem. Sci.* (Pers. Edn) **5** (1) VIII–IX.
Swarup, G., Cohen, S. & Garbers, D. L. (1981). *J. Biol. Chem.*, **256,** 8197–8201.
Switzer, R. C. III, Merril, C. R. & Shifrin, S. (1979). *Anal. Biochem.*, **98,** 231–237.
Szókán, G. (1982). *J. Liquid Chromatogr.*, **5,** 1493–1498.
Tabor, H. & Tabor, C. W. (1977). *Anal. Biochem.*, **78,** 554–556.
Tanford, C. & Roberts, G. L. (1952). *J. Am. Chem. Soc.*, **74,** 2509–2515.

Tapuhi, Y., Schmidt, D. E., Lindner,W. & Karger, B. L. (1981). *Anal. Biochem.*, **115,** 123–129.
Tasheva, B. & Dessev, G. (1983). *Anal. Biochem.*, **129,** 98–102.
Thannhauser, T. W., Konishi, Y. & Scheraga, H. A. (1984). *Anal. Biochem.*, **138,** 181–188.
Tomlinson, G. & Viswanatha, T. (1974). *Anal. Biochem.*, **60,** 15–24.
Torres, A. R. & Peterson, E. A. (1979). *Anal. Biochem.*, **98,** 353–357.
Towbin, H., Staehelin, T. & Gordon, J. (1979). *Proc. Natl. Acad. Sci. U.S.A.*, **76,** 4350–4355.
Tower, D. B. (1967). *Methods Enzymol.*, **11,** 76–93.
Trah, T. J. & Schleyer, M. (1982). *Anal. Biochem.*, **127,** 326–329.
Trézl, L., Rusznák, I., Tyihák, E., Szarvas, T. & Szende, B. (1983). *Biochem. J.*, **214,** 289–292.
Tristram, G. R. & Smith, R. H. (1963). *Adv. Protein Chem.*, **18,** 227–318.
Tsang, V. C. W., Peralta, J. M. & Simons, A. R. (1983). *Methods Enzymol.*, **92,** Part E, 377–391.
Turnell, D. C. & Cooper, J. D. H. (1982). *Clin. Chem.*, **28,** 527–531.
Udenfriend, S., Stein, S., Böhlen, P., Dairman, W., Leimgruber, W. & Weigele, M. (1972). *Science*, **178,** 871–872.
Umagat, H., Kucera, P. & Wen, L.-F. (1982). *J. Chromatogr.*, **239,** 463–474.
Van Kley, H. & Hale, S. M. (1977). *Anal. Biochem.*, **81,** 485–487.
Van Sumere, C., Geiger, H., Bral, D., Folckenier, G., Vande Casteele, K., Martens, M., Hanselaer, R. & Gevaert, L. (1983). *Anal. Biochem.*, **131,** 530–532.
Varghese, G. & Diwan, A. M. (1983). *Anal. Biochem.*, **132,** 481–483.
Vesterberg, O. & Gramstrup-Christensen, B. (1984). *Electrophoresis*, **5,** 282–285.
Vitt, S. V., Vorob'ev, M. M., Paskonova, E. A. & Saporovskaya, M. B. (1983). *J. High Resol. Chromatogr. Chromatogr. Commun.*, **6,** 158–160.
Wachslicht, H. & Chrambach, A. (1978). *Anal. Biochem.*, **84,** 533–538.
Waddell, W. J. (1956). *J. Lab. Clin. Med.*, **48,** 311–314.
Wagner, A. P., Psarrou, E. & Wagner, L. P. (1983). *Anal. Biochem.*, **129,** 326–328.
Wagner, A. P., Psarrou, E. & Wagner, L. P. (1984). *Anal. Biochem.*, **137,** 248–255.
Wang, C.-S. & Smith, R. L. (1975). *Anal. Biochem.*, **63,** 414–417.
Wang, W.-T., Matsuura, F. & Sweeley, C. C. (1983). *Anal. Biochem.*, **134,** 398–405.
Warburg, O. & Christian, W. (1941/42). *Biochemische Z.*, **310,** 384–421.
Ward, D. N. & Coffey, J. A. (1964). *Biochemistry*, **3,** 1575–1577.
Ward, D. N., Coffey, J. A., Ray, D. B. & Lamkin, W. M. (1966). *Anal. Biochem.*, **14,** 243–252.
Watanabe, Y. & Imai, K. (1983). *Anal. Chem.*, **55,** 1786–1791.
Waterfield, M. D. & Scrace, G. T. (1981). In *Biological/Biomedical Applications of Liquid Chromatography*, Hawk, G. L. (Ed.), *Chromatographic Science Series*, vol. 18, Marcel Dekker, New York & Basel, pp. 135–158.
Watters, C. (1978). *Anal. Biochem.*, **88,** 699–701.
Weiner, S. & Tishbee, A. (1981). *J. Chromatogr.*, **213,** 501–506.
Weinstein, B. (1966). *Methods Biochem. Anal.*, **14,** 203–323.
Wessel, D. & Flügge, U. I. (1984). *Anal. Biochem.*, **138,** 141–143.
Westall, F. & Hesser, H. (1974). *Anal. Biochem.*, **61,** 610–613.
Westley, J. & Lambeth, J. (1960). *Biochem. Biophys. Acta*, **40,** 364–366.
Whitaker, J. R. & Granum, P. E. (1980). *Anal. Biochem.*, **109,** 156–159.
Whitaker, J. R., Granum, P. E. & Aasen, G. (1980). *Anal. Biochem.*, **108,** 72–75.
Wilcox, P. E. (1967). *Methods Enzymol.*, **11,** 63–76.
Williams, R. J. & Siggia, S. (1977). *Anal. Chem.*, **49,** 2337–2342.

Wilkinson, J. M. (1978). *J. Chrom. Sci.*, **16,** 574–552.
Wilkinson, M., Iaccolucci, G. A. & Myers, D. V. (1976). *Anal. Biochem.*, **70,** 470–478.
Wojtkowiak, Z., Briggs, R. C. & Hnilica, L. S. (1983). *Anal. Biochem.*, **129,** 486–489.
Wold, F. (1984). *Trends Biochem. Sci.*, **9,** 256–257.
Wolf, P. & Maguire, M. (1983). *Anal. Biochem.*, **129,** 145–155.
Wray, D., Boulikas, T., Wray, V. P. & Hancock, R. (1981). *Anal. Biochem.*, **118,** 197–203.
Yamada, S. & Itano, H. A. (1966). *Biochem. Biophys. Acta*, **130,** 538–540.
Yamasaki, R. B., Shimer, D. A. & Feeney, R. E. (1981). *Anal. Biochem.*, **111,** 220–226.
Yamasaki, R. B., Osuga, D. T. & Feeney, R. E. (1982). *Anal. Biochem.*, **126,** 183–189.
Yang, C.-Y. & Wakil, S. J. (1984). *Anal. Biochem.*, **137,** 54–57.
Yang, J. C., Fujitaki, J. M. & Smith, R. A. (1982). *Anal. Biochem.*, **122,** 360–363.
Yoshida, A. (1972). *Anal. Biochem.*, **49,** 320–325.
Yoshida, H., Sumida, T., Masujima, T. & Imai, H. (1982). *J. High Resol. Chromatgr. Chromatogr. Commun.*, **5,** 509–511.
Yuen, K. C. L., Johnson, T. K., Denell, R. E. & Consigli, R. A. (1982). *Anal. Biochem.*, **126,** 398–402.
Zamenhoff, S. (1957). *Methods Enzymol.*, **3,** 696–704.
Zanetta, J. P. & Vincendon, G. (1973). *J. Chromatogr.*, **76,** 91–99.
Ziegler, Kl., Melchert, I. & Lürken, C. (1967). *Nature* (*Lond.*), **214,** 404–405.
Zimmerman, C. L., Appella, E. & Pisano, J. J. (1976). *Anal. Biochem.*, **75,** 77–85.
Zinder, O., Nikodijevic, O., Hoffman, P. G. & Pollard, H. B. (1976). *J. Biol. Chem.*, **251,** 2179–2181.
Zumwalt, R. W., Kuo, K. C. T. & Gehrke, C. W. (Eds.) (1986). *Amino Acid Analysis by Gas Chromatography*, CRC Press, Cleveland, Ohio.

Practical Protection Chemistry—A Handbook
Edited by A. Darbre
© 1986, John Wiley & Sons Ltd

SHULAMITH WEINSTEIN*, M. H. ENGEL§ and P. E. HARE†
* *Department of Organic Chemistry, The Weizmann Institute of Science, Rehovet, Israel*
§ *School of Geology and Geophysics, The University of Oklahoma, Norman, OK 73019, U.S.A.*
† *Geophysical Laboratory, Carnegie Institution of Washington, Washington, D.C. U.S.A.*

9 Enantiomeric Analysis of a Mixture of the Protein Amino Acids by High Performance Liquid Chromatography

CONTENTS

9.1. INTRODUCTION

High performance chromatographic methods for the resolution of amino acids into their enantiomers are of value to many areas of research. Indeed, following the development of HPLC, resolution methods employing this technique were studied by several groups. Amino acids were resolved on supports such as polystyrene (Davankov & Semechkin, 1971) and polyacrylamide (Lefebvre *et al.*, 1977) to which chiral ligands for copper II complexation were covalently bound. Metal ions in conjunction with chiral ligands in the mobile phase were used in the reversed phase mode (Gilon *et al.*, 1979). A method was reported for the efficient and highly sensitive resolution of individual amino acids into their enantiomers, employing an aqueous mobile phase which contained as chiral additive a copper II-L-proline complex. The solid support used was either a cation-exchange resin or reverse phase silica. However, not all the common underivatized, protein amino acid enantiomers were resolved, nor were all the amino acids separated from one another (Hare & Gil-Av, 1979; Gil-Av *et al.*, 1980). More recently, the separation of enantioners of dansylated amino acids was achieved using a coupled-column (Tapuhi *et al.*, 1981).

Based on the work of Hare and Gil-Av, a series of chiral additives, *N,N*-dialkyl-amino acids coordinated to copper II, was used to effect resolution of the amino acids (Weinstein, 1982; Weinstein *et al.*, 1982). The present procedure exploits the excellent resolutions obtained by reverse phase chromatography with the chiral mobile phase consisting of *N,N*-di-n-propyl-L-alanine and cupric acetate (Cu-DPA) in aqueous solution or in water-acetonitrile mixtures. This chiral complex resolves *all* the protein amino acids into enantiomers; however, adjacent peaks of a few amino acids still overlap. This difficulty is overcome by using a two-stage method. First, the mixture of the protein amino acids is separated and collected in groups by a cation-exchange column, in a manner very similar to routine amino acid analysis, but employing volatile buffers. After evaporaion of the solvent, each group is submitted to enantiomeric analysis on a reverse phase column, using the chiral mobile phase. Subnanomole sensitivity is achieved by post-column derivatization with OPA and 2-mercaptoethanol with subsequent fluorometric detection. Proline, although resolved, does not react with OPA but can be detected by post-column derivatization with ninhydrin (Spackman *et al.*, 1958). Analysis of cystine and cysteine can be effected after oxidation to cysteic acid, which is resolved efficiently by the present procedure.

9.2. MATERIALS

To avoid contamination, the water is double glass distilled. All other chemicals and distilled solvents used are commercially available, with the exception of *N,N*-di-n-propyl-L-alanine, which was synthesized following a modified procedure of Bowman and Stroud (1950).

9.2.1. Preparation of *N,N*-di-n-propyl-L-alanine

Suspend 0.2 mol (17.8 g) of L-alanine (Sigma) in 200 ml of ethanol. Add 3 g of 10% palladium on activated coal catalyst (Fluka) and 43 ml of propionaldehyde (Fluka). The mixture is hydrogenated for 48 h in a Parr apparatus at 40–50°C and an initial hydrogen pressure of 50 psi. The catalyst is removed using a sintered glass filter and the filtrate evaporated to dryness. The reaction product (N,N-di-n-propyl-L-alanine) is crystallized from chloroform. The purity of the product is confirmed by TLC, NMR and C, H, N analysis.

9.2.2. Solutions for HPLC

Buffer I: 0.2 M pyridine with added glacial acetic acid to give pH 3.4.
Buffer II: 0.4 M pyridine with added glacial acetic acid to give pH 5.3.
OPA reagent: 30 g boric acid is dissolved in 1 l of water and KOH added to pH 9.5–10; 2.5 g EDTA is then added and dissolved. Dissolve 1.0 g of o-phthaldialdehyde and 0.5 ml mercaptoethanol in 10 ml of methanol and add slowly with stirring. The resulting solution is filtered through a 0.22 μm Millipore filter.

Chiral mobile phase: 8 mM (1.384 g) of N,N,-di-n-propyl-L-alanine and 4 mM (0.8 g) of cupric acetate are dissolved in 1 l of water (pH 5.3–5.5) and the solution is filtered through a 0.22 μm Millipore filter and degassed under vacuum.

9.3. HPLC PROCEDURE

A mixture of approximately 2 mg each of the racemic common protein amino acids (Table 9.1) is dissolved in 10.0 ml of water. Instead of cysteine and cystine use cysteic acid only. Omit proline and hydroxyproline.

9.3.1. Group separation

The amino acids are separated into four groups:

1. Acidic: Cya, Asp, Thr, Ser, Glu
2. Neutral: Gly, Ala, Val, Met, Leu, Ile
3. Aromatic: Phe, Tyr, Trp
4. Basic: Lys, His, Arg

Inject 20 μl of the amino acid solution onto a stainless steel column (50 × 4.6 mm) packed with 5 μm cation-exchange resin of the polystyrene-divinyl benzene-sulphonic acid type equilibrated with buffer I. The first three groups with the exception of Trp are eluted with buffer I. The chromatogram is shown in Fig. 9.1 and the chromatographic conditions are described therein. Buffer II is used to elute Trp, Lys, His and Arg. The OPA reagent, at a flow rate of 0.6 ml/min, is mixed with the column eluate and to allow sufficient time for the reaction to take place, this is run through a reaction loop and into the fluorescence detector (Waters, 420 AC). The first run establishes the exact retention times of the amino acids. Now the procedure is repeated under exactly identical conditons, omitting the post-column derivatization step and detection. The amino acids are collected directly from the outlet of the column at the appropriate time intervals, e.g. the acidic group between 1 and 9 min after injection, the neutral group between 10 and 15 min and the other amino acids are collected separately to avoid large elution volumes (Phe, 20–24; Tyr, 34–38; Trp, 62–66; Lys, 75–81; His, 84–89;

Table 9.1. Resolution of the protein amino acids using HPLC. Column: Spherisorb/LC-18, 5 μm, 150 × 4.6 mm; chiral additive: Cu (N,N-di-*n*-propyl-L-alanine)$_2$

Amino acid	Retention time* (min)	Amino acid	Retention time* (min)
Mobile phase: water; flow rate: 0.2 ml/min†		Mobile phase: 1% acetonitrile in water; flow rate: 0.25 ml/min†	
D-Lys	3.4	Gly	3.8
L-Lys	4.4	D-Ala	4.2
D-Arg	5.2	L-Ala	6.2
L-Arg	8.4	D-Val	15.4
Gly	6.4	L-Val	38.6
D-Ser	6.8	D-Met	34.6
L-Ser	8.4	L-Met	62.6
D-Ala	7.0	D-Ile	45.4
L-Ala	10.0	L-Ile	121.4
D-Thr	7.8	D-Leu	49.0
L-Thr	9.6	L-Leu	112.6
D-His	10.4		
L-His	13.8	Mobile phase: 14% acetonitrile in water; flow rate; 0.5 ml/min‡	
D-Cysteic acid	12.3		
L-Cysteic acid	13.6	D-Tyr	2.7
D-Asp	21.0	L-Tyr	3.5
L-Asp	24.6	D-Phe	14.7
D-Val	22.6	L-Phe	20.3
L-Val	54.2	D-Trp	29.1
D-Glu	40.0	L-Trp	32.7
L-Glu	70.2		
D-Met	49.0	Mobile phase: 5% acetonitrile in water; flow rate: 0.25 ml/min§	
L-Met	86.6		
		D-Ala	2.5
		L-Ala	3.1
		D-Pro	4.5
		L-Pro	10.5

* Retention times from column void volume (V_o) were determined separately for each amino acid.
† V_o was 5 min, estimated from the sample solvent signals.
‡ V_o was 3.7 min.
§ Ala and Pro were each resolved separately on a 150 × 4.6 mm Supelco LC-18 column, monitored by ninhydrin derivatization and light absorption, Ala at 570 and Pro at 440 nm. Vo was 5 min.

Arg, 175–188 min). Trp is added to Phe and Tyr. Lys, His and Arg are combined to form the basic group.

Alternatively, to avoid the long retention times for Trp and the basic amino acids, the original amino acid mixture can also be separated on a column (50 × 2.0 mm) using only buffer II (from the start of the run). This results in a more rapid elution of Trp, Lys, His and Arg. All the other amino acids elute very early without separation and have to be collected as in the above procedure, with buffer I.

Each group is evaporated to dryness using a stream of nitrogen and then dried further for 20 min under vacuum. The residues are each redissolved in 200 μl of the chiral mobile phases to be used in the next stage.

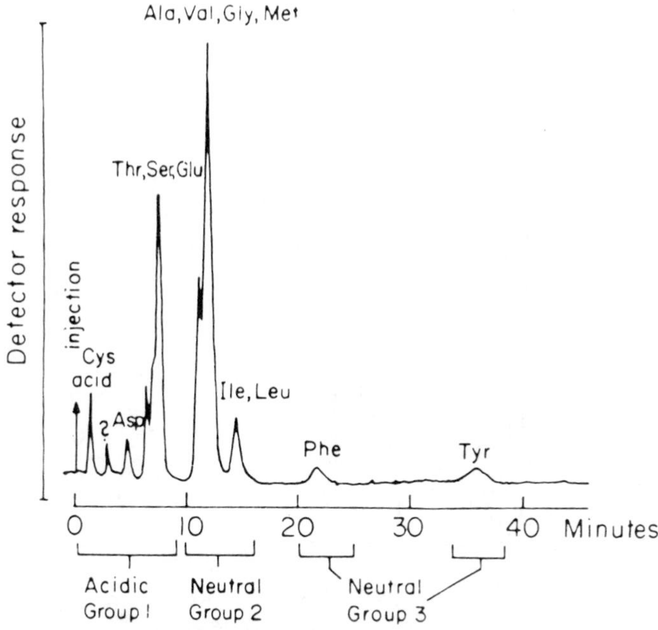

Fig. 9.1. Separation of the acidic and neutral amino acids in three groups on the cation exchange column with 0.2 M-pyridine/acetic acid buffer, pH 3.4. Temperature 50°C; flow rate, 0.17 ml/min, followed after 45 min with 0.4 M buffer, pH 5.3 to collect Trp, Lys, His and Arg. (Figs. 9.1–9.5 are reproduced from Weinstein *et al.*, 1982, by courtesy of Academic Press.)

9.3.2. Enantiomer analysis

Twenty microliter aliquots of each fraction are injected onto a stainless steel reverse-phase column (150 × 4.6 mm) packed with 5 μm Spherisorb LC-18 resin. The D- and L-enantiomers of the individual amino acids are resolved using the chiral mobile phase. Where necessary, acetonitrile is added to the eluent to decrease retention times. The amino acids eluting are monitored as in the cation-exchange resin stage. Retention times are given in Table 9.1.

Chromatograms showing resolution of the amino acid enantiomers in the four fractions and chromatographic conditions are given in Figs. 9.2–9.5. (Appropriate standards enriched in the L- and D-enantiomers confirmed that the order of emergence of *all* of the enantiomers is D- before L-.)

The acidic (Ser, Thr, Cya, Asp and Glu) and basic (Lys, Arg, His) groups are eluted with the chiral additive dissolved in water. For the group consisting of most of the neutral amino acids (Gly, Ala, Val, Met, Leu and Ile) acetonitrile is added (1 %, v/v) to the mobile phase to decrease the retention times of these

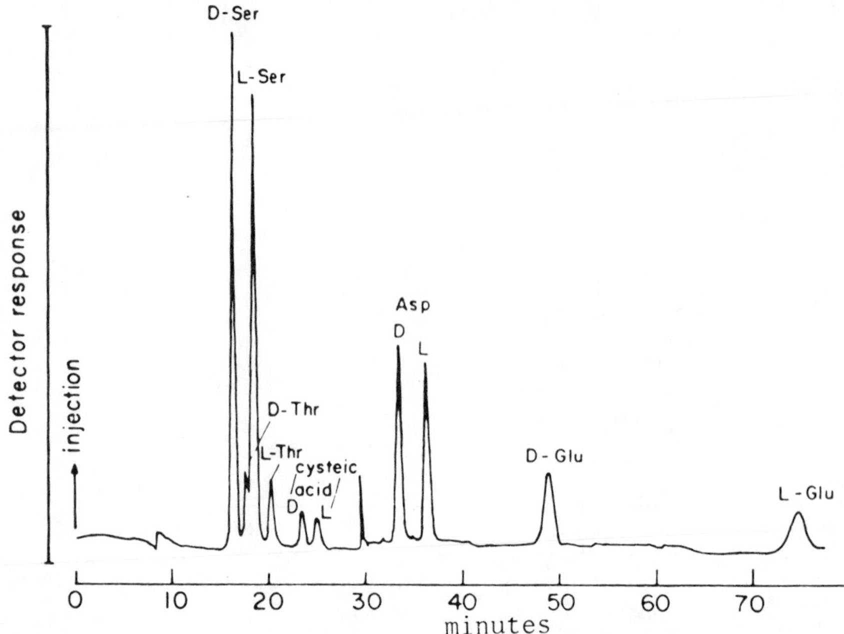

Figure 9.2. Separation of the acidic amino acids (group 1 from the cation exchange column) into enantiomers on the reverse phase column. Mobile phase: the chiral additive (Cu-DPA) in water. Temperature ambient; flow rate 0.1 ml/min, increased to 0.2 ml/min after 30 min.

amino acids. Group 3 (Phe, Tyr, Trp) is eluted with acetonitrile (14 %, v/v) in the mobile phase.

Within each fraction, complete resolution is achieved for all the amino acid enantiomers.

9.4. CONCLUSION

The two-stage chromatographic scheme can be easily applied as a routine procedure in any laboratory in which HPLC apparatus equipped with an appropriate detector is available. Commercial columns not requiring any modification are employed. The method can be improved by using gradient elution, adding acetonitrile gradually to the mobile phase to analyze groups 2 and 3 in one run. Another improvement would be a splitter valve at the outlet of the cation-exchange column diverting most of the eluate for collection and directing only a small part to the detector.

An alternative approach to enantiomer analysis is the collection of the amino acids from the cation-exchange resin one by one with a volatile buffer (0.1 M-

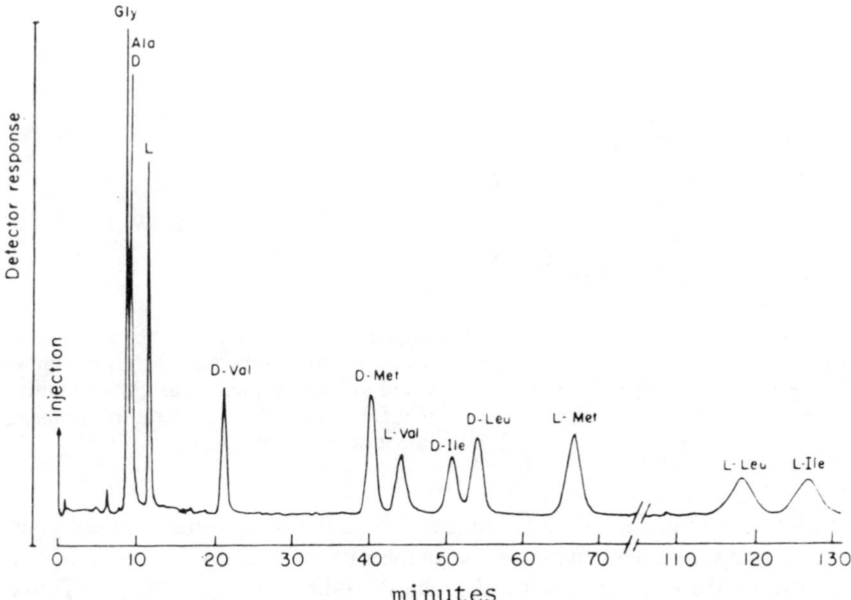

Figure 9.3. Separation of the neutral amino acids (group 2 from the cation exchange column) into enantiomers on the reverse phase column. Mobile phase: the chiral additive (Cu-DPA) in water and acetonitrile (1%, v/v). Temperature, ambient; flow rate 0.25 ml/min.

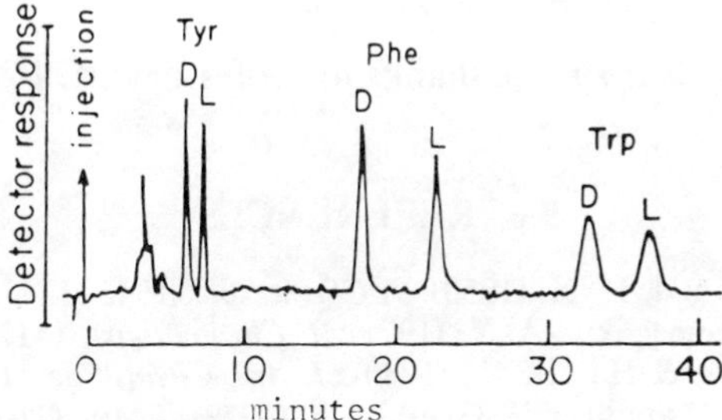

Figure 9.4. Separation of the aromatic amino acids (Tyr, Phe and Trp) into enantiomers on the reverse phase column. Mobile phase: the chiral additive (Cu-DPA) in water and acetonitrile (14%, v/v). Temperature, ambient; flow rate 0.5 ml/min.

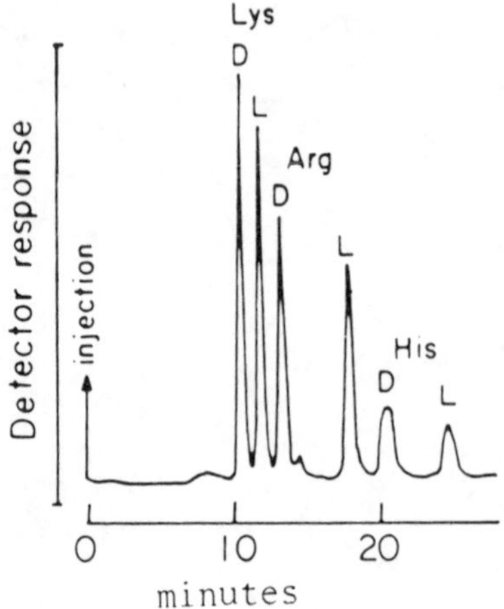

Figure 9.5. Separation of the basic amino acids into enantiomers on the reverse phase column. Mobile phase: the chiral additive (Cu-DPA) in water. Temperature, ambient; flow rate 0.2 ml/min.

pyridine/acetate buffer, pH 3.10, followed by a 0.4 M-pyridine/acetate buffer, pH 5.3), and subsequent enantiomeric analysis of each amino acid separately on the reverse-phase column with the chiral additive (Engel, 1981: personal communication). In special cases, in which not all of the protein amino acids are present, the ion-exchange separation may be eliminated.

Operating with a chiral mobile phase offers the option of reversal of the configuration of the chiral additive which would reverse the order of emergence of enantiomers and afford confirmation of the ratio of D- and L-isomers. A difference in the D-/L-value as determined with chiral additives with opposite configuration will reveal the presence of impurities.

9.5. ACKNOWLEDGEMENT

The authors express their sincere thanks to Professor E. Gil-Av for his support and advice.

9.6. REFERENCES

Bowman, R. E. & Stroud, H. H. (1950). *J. Chem. Soc.*, 1342–1345.
Davankov, V. A. & Semechkin, A. V. (1971). *J. Chromatogr.*, **141,** 313–353.
Gil-Av, E., Tishbee, A. & Hare, P. E. (1980). *J. Am. Chem. Soc.*, **102,** 5115–5117.
Gilon, C., Leshem, R., Tapuhi, Y. & Grushka, E. (1979). *J. Am. Chem. Soc.*, **101,** 7612–7613.
Hare, P. E. & Gil-Av, E. (1979). *Science*, **204,** 1226–1228.
Lefebvre, B. L., Audebert, R. & Quivoron, C. (1977). *Isr. J. Chem.*, **15,** 69–73.
Spackman, D. H., Stein, W. H. & Moore, S. (1958). *Anal. Chem.*, **30,** 1190–1206.
Tapuhi, Y., Miller, N. & Karger, B. L. (1981). *J. Chromatogr.*, **205,** 325–337.
Weinstein, S. (1982). *Angew. Chem.*, **94,** 221–222.
Weinstein, S., Engel, M. H. & Hare, P. E. (1982). *Anal. Biochem.*, **121,** 370–377.

Practical Protein Chemistry—A Handbook
Edited by A. Darbre
©1986, John Wiley & Sons Ltd

GREG WINTER

MRC Laboratory of Molecular Biology, Hills Road, Cambridge CB2 2QH, U.K.

10

Manual Sequence Strategy—a Personal View

CONTENTS

10.1. INTRODUCTION

At the core of virtually all modern sequencing strategies lies the Edman reaction (Edman, 1949; Edman & Begg, 1967). The *N*-terminal amino acid of the polypeptide is first coupled to phenylisothiocyanate to form the phenylthiocarbamyl peptide; this derivative is then cleaved with anhydrous acid to expose a new *N*-terminus and to release the original *N*-terminal amino acid as a 5′-thiazolinone (Figure 10.1). The excess reagents and by-products are extracted by an organic solvent wash. Repetition of this process with identification of the released thiazolinones or of the new *N*-termini enables the sequence of amino acids to be established from the *N*-terminal end. The length of sequence so obtained is limited by coupling, cleavage and extraction yields at each step, by blocking of the *N*-terminus, by 'premature' cleavage at arginine and histidine (Schroeder, 1967a, 1972b) and by random internal cleavage of the polypeptide chain (Kopple & Bächli, 1959). For example, insoluble peptides may only partly couple with phenylisothiocyanate; phenylthiocarbamyl peptides desulphurized by traces of oxygen (Ilse & Edman, 1963) are not cleaved by anhydrous acid and hydrophobic peptides may be extracted into the organic solvent and lost. Peptides with *N*-terminal glutamine (Blombäck, 1967), tryptophan (Uphaus *et al.*, 1959) and various cysteine derivatives (Smyth & Utsumi, 1967; Press, 1967, Doolittle, 1972; Wade *et al.*, 1975) can cyclize and block the degradation as can the rearrangement of Asn-Gly sequences (Swallow & Abraham, 1958; Edman, 1970; Jörnvall, 1973) and O → N migration of acyl-serine and acyl-threonine (Smyth *et al.*, 1963; Edman, 1970; Jörnvall, 1970). The sequence determination therefore ceases as the signal dies away or becomes swamped by a rising background signal.

The repetitive yields of manual sequencing are usually lower than those of automated methods, due probably to greater difficulties in excluding oxygen (with resultant oxidative desulphuration) and to the less efficient solvent wash. Since the trifluoroacetic acid cleavage is a *reversible* reaction (Edman & Begg, 1967), as shown in Fig. 10.1, unextracted thiazolinones from previous steps will inhibit the cleavage. In the dansyl-Edman technique (Hartley, 1970), although not in the DABITC-PITC technique (Chang *et al.*, 1978), removing an aliquot of peptide for dansylation and/or electrophoresis additionally reduces the repetitive yield. Manual sequencing suffers the further disadvantage of qualitative identification which can become difficult against a rising background. Thus manual sequencing is limited to relatively short runs of sequence ($<$ 15 residues) and the *ultimate goal of a manual sequence strategy therefore involves breaking a protein into short peptides.* Invariably this relies on enzymic cleavage since those residues amenable to chemical cleavage, such as methionine, tryptophan or cysteine are usually few and far between.

10.2. CHOICE OF STRATEGY

The flowsheet (Fig. 10.2) summarizes the choice of strategy to be considered. The protein should be at least 95% pure with non-identical sub-units separated and purified (see Chapter 1). Lipid, carbohydrate or cofactors should be removed. Digestion strategies will be suggested by the molecular weight of the polypeptide

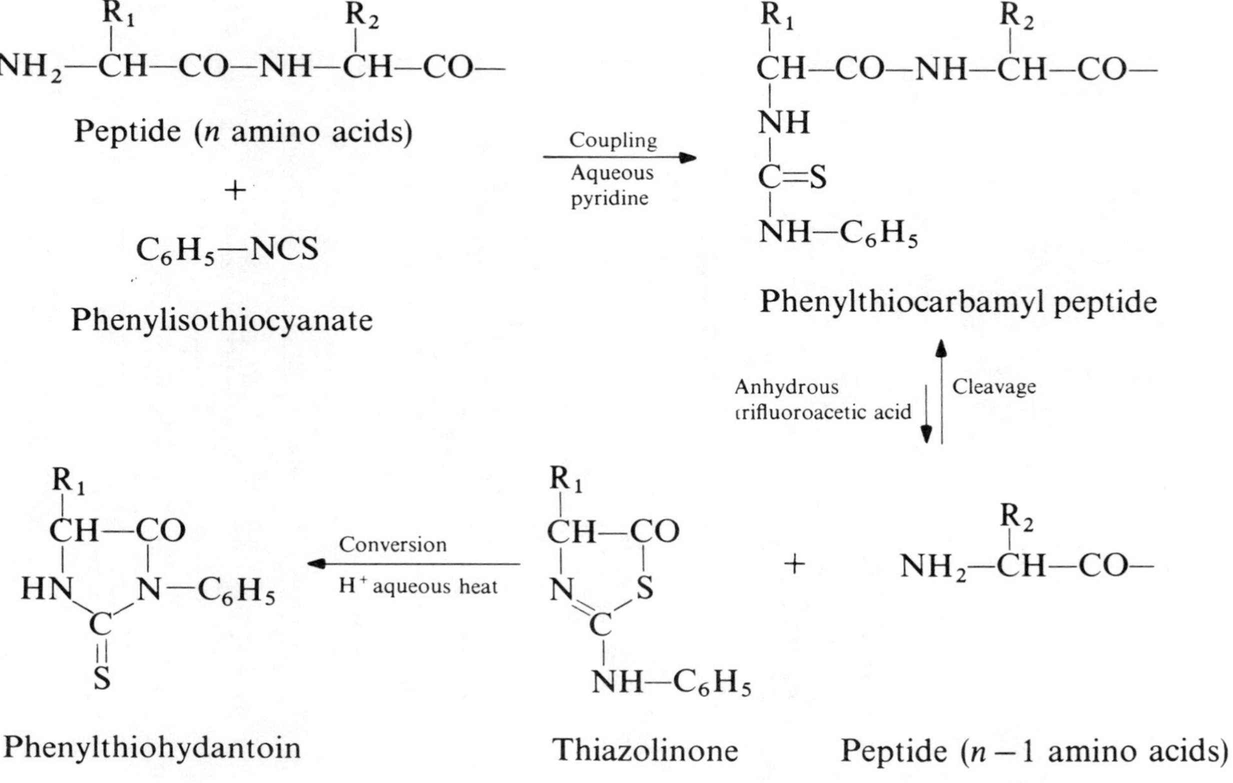

Figure 10.1. Scheme 1.

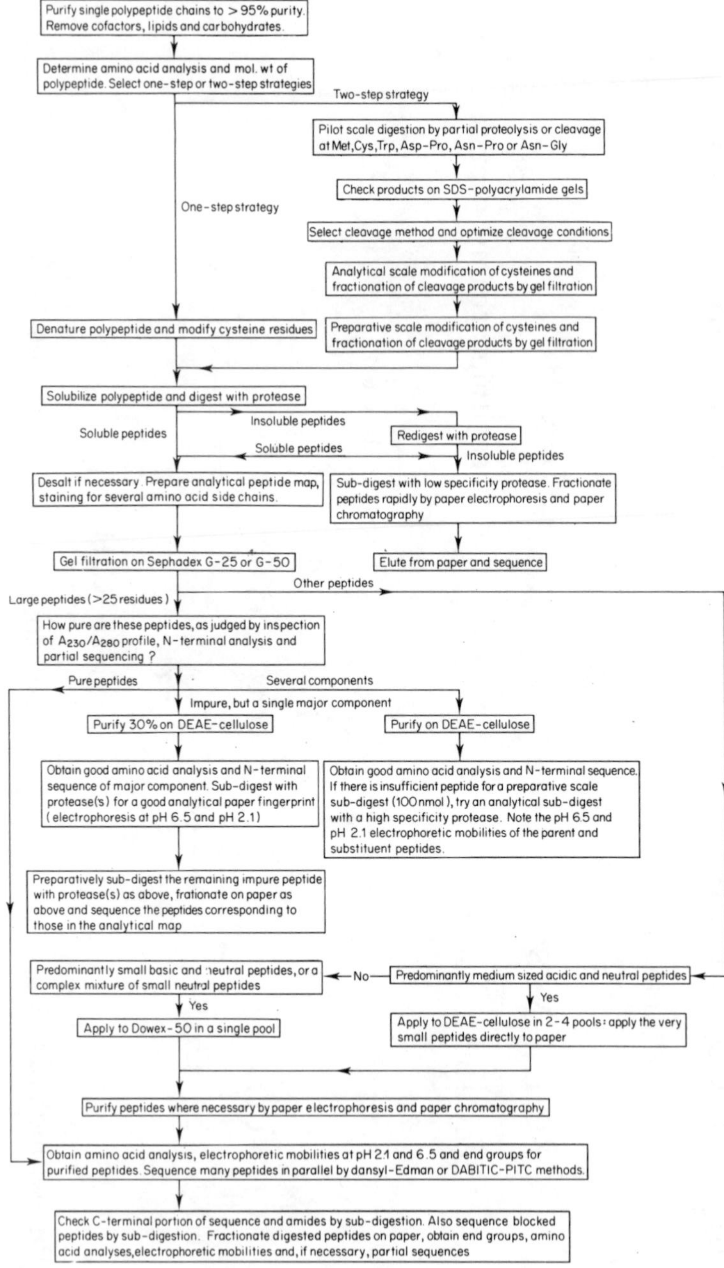

Fig. 10.2. Flow sheet: a strategy for manual sequencing

and its amino acid analysis, since the average size of peptides expected from digestion with trypsin (cuts at lysine and arginine) or other enzymes can then be deduced (Table 10.1). Usually the more specific enzymes, such as trypsin, chymotrypsin, staphyloccocal protease and *A. mellea* protease are used. However, with a large polypeptide chain difficulties in separating numerous (> 40) peptides from a single digest may necessitate either blocking some potential points of cleavage or using a two-step strategy. Thus tryptic cleavage at lysine residues can be prevented by derivatizing with maleic anhydride (Butler & Hartley, 1972). The maleylated lysine derivatives carry a negative charge and are usually deblocked with acid after tryptic digestion and before fractionation of the peptides by ion-exchange or electrophoresis. Arginine residues can also be blocked to tryptic cleavage by cyclohexanedione in the presence of borate (Patthy & Smith, 1975). For a blocking strategy to succeed, it is essential to ensure that the protein is completely modified and, later, that the peptides are completely deblocked.

Table 10.1. Protease specificities

Enzyme	Main specificities at X	Comments
Trypsin	*At basic residues*: *C*-terminal to Lys, Arg and amino-ethyl-Cys (Smyth, 1967)	No hydrolysis at X-Pro bonds, reduced hydrolysis if X is preceded or followed by an acidic residue. Digests selectively at Arg at alkaline pH (Green & Bartelt, 1977) and active in 2 M-urea.
Chymotrypsin	*At aromatic residues*: *C*-terminal to Trp, Tyr, Phe and Leu (Smyth, 1967).	No hydrolysis at X-Pro bonds or oxidized Trp, reduced hydrolysis if X is preceded or followed by an acidic residue. Active in 2 M-urea.
Elastase	*At aliphatic residues*: *C*-terminal to Ala, Val, Gly and Ser (Hartley & Shotton, 1971).	Under partial digestion conditions the site of cleavage is unpredictable. Active in 2 M-urea.
Staphyloccocal protease	*At acidic residues*: *C*-terminal to Glu and Asp (Drapeau *et al.*, 1972; Houmard and Drapeau, 1972; Drapeau, 1977).	Cleavage conditions may be adjusted for either Glu or Glu and Asp. Also cleaves at carboxymethyl-Cys: reduced hydrolysis if X is followed by a bulky residue or surrounded by acidic residues. Active in 4 M-urea (Drapeau, 1977).
Thermolysin	*N*-terminal to Ile, Leu, Val and Phe (Matsubara, 1970; Matsubara & Feder, 1971; Heinrikson, 1977).	No hydrolysis at X-Pro bonds, digestion at 60–80°C may help solubilize the polypeptide.
Pepsin	*At aromatic residues*: *N*-terminal to Trp, Tyr, Phe and Leu (Smyth, 1967).	Cleaves also at a wide range of other amino acids: the acidic digestion conditions may help to solubilize the polypeptide.
Armillaria mellea protease	*N*-terminal to Lys and amino-ethyl-Cys (Shipolini *et al.*, 1974).	Cleaves occasionally at other residues: for example, *N*-terminal to Arg, Phe and Trp (G. Winter, unpublished) and *C*-terminal to Arg (Doonan *et al.*, 1975).

Although diagonal methods (Perham, 1967; Butler & Hartley, 1972; Patthy & Smith, 1975) can be used in conjunction with a blocking strategy to selectively isolate tryptic peptides containing internal lysine or arginine residues, the direct digestion of an unblocked protein by clostripain (Mitchell, 1977) or *A. mellea* protease (Shipolini *et al.*, 1974) is probably a superior alternative to either of these blocking strategies for a primary digest.

In the two-step strategy, cleavage with chemical methods (Table 10.2) at infrequent amino acids, such as tryptophan, methionine or cysteine, or at surface

Table 10.2. Chemical cleavage methods possibly suitable for a two-step stragegy

Reagent	Main specificity	Partial cleavage due to:	Additional cleavages due to:
Cyanogen bromide (Gross, 1967).	At Met to give *C*-terminal homoserine lactone. *C*-terminus blocked to carboxypeptidases.	Met *S*-oxide not cleaved: Met-Ser, Met-Thr (Schroeder *et al.*, 1964, Met (Cys-Cys) (Doyen & Lapresle, 1979) and Met-Glu (Corradin & Harbury, 1970) only partially	Acid cleavage at Asp-Pro and Asn-Pro (Piszkiewiez *et al.*, 1970) and oxidative cleavage at Trp (Artavanis, 1974).
2-Nitro-5-thiocyanobenzoic acid followed by alkaline cleavage (Degani & Patchnornik, 1974).	At Cys to give an *N*-terminal iminothiazolidine-4-carboxyl residue. *N*-terminus blocked to Edman degradation or aminopeptidases but can be deblocked by catalytic reduction (Schaffer & Stark, 1976).	Mixed disulphide formation (Price, 1976) and β-elimination of *S*-cyano derivative (Stark,	Phe-Thr and Phe-Ser if catalytic reduction is used for deblocking *N*-terminus (Schaffer & Stark, 1976).
Dimethylsulphoxide-HBr (Savige & Fontana, 1977) or *N*-chlorosuccinimide (Shechter *et al.*, 1976).	At Trp to give a *C*-terminal lactone. Cys and Met oxidized.		
CNBr in formic and heptafluorobutyric acids (Ozols & Gerard, 1977).	At Trp, provided Met is first photoxidized.		
Hydroxylamine (Bornstein, 1970).	At Asn-Gly.	Extent of cleavage may depend on prior exposure of polypeptide to acid or alkali (Konigsberg & Steinman, 1977).	Asn-Leu (Bornstein & Balian, 1970) and Asn-Ala (Steinman *et al.*, 1974).
Dilute acid (Piszkiewicz *et al.*, 1970).	At Asp-Pro and Asn-Pro.		

loops in the native protein by limited proteolysis (Porter, 1958; Karlsson *et al.*, 1969; Lowey *et al.*, 1969; Platt *et al.*, 1973), may generate a few large peptides. Subdigestion of each of the large peptides in turn dramatically reduces the number of small peptides in any one fractionation. The difficulties with the two-step strategy lie in achieving complete and specific cleavage in the first step and in separating and ultimately overlapping the large polypeptides. For example, cyanogen bromide cleaves the polypeptide chain at methionine to give *C*-terminal homoserine; it does not cleave methionine sulphoxide and only partially cleaves Met-Ser and Met-Thr bonds. Tryptophan may undergo an oxidative cleavage while the acid labile Asp-Pro bond and some amides may be cleaved under the acidic digestion conditions. The large denatured polypeptides produced are often insoluble and elute poorly from ion-exchange columns. Multiple charge forms of each peptide resulting from partial deamidation, or a mixture of *C*-terminal homoserine and homoserine lactone peptides also complicate separations based on charge. Overlapping the CNBr peptides requires the isolation of all methionine peptides from a separate digest of the entire protein. However, with foresight most of these problems may be overcome. Before the digest, any methionine sulphoxide can be reduced to methionine with 2-mercaptoethanol or dithiothreitol under denaturing conditions. (A direct amino acid analysis of the protein does not reveal the presence of methionine sulphoxide since under the acidic hydrolysis conditions the sulphoxide is reduced to methionine: however, methionine sulphoxide can be measured indirectly (Shechter *et al.*, 1975)). A mixture of *C*-terminal homoserine and homoserine lactone can be converted to either one form or the other (Ambler, 1965; Mutt & Said, 1974) and the large insoluble peptides may be solubilized in 8 M-urea and run on ion-exchange columns in 8 M-urea (Hartley, 1964). Alternatively these peptides may be solubilized in 100 % formic acid and separated on a Biogel gel filtration column in 20–70 % formic acid (Adelstein & Kuehl, 1970) or solubilized with citraconic anhydride, fractionated on a Sephadex or Biogel column at neutral pH and finally deblocked with acid. The overlap peptides may be selectively isolated from a digest of the intact protein using the methionine diagonal method (Tang & Hartley, 1967). Cysteine peptides are perhaps best identified by labelling these residues with [^{14}C]iodoacetic acid and purifying the radioactive peptides (Weeds & Hartley, 1967, 1968). Tryptophan may be detected by noting the main A_{280} peaks on column fractionation of peptides and the intrinsic fluorescence or staining of tryptophan peptides on paper fractionation.

Since the separation of large polypeptides by gel filtration invariably gives better yields than separation by ion-exchange, the cleavage method employed in a two-step strategy is perhaps best decided on the basis of analytical SDS-polyacrylamide gel electrophoresis. A pattern of a few well-separated bands on polyacrylamide gels augers well for a gel filtration strategy. SDS-polyacrylamide gel electrophoresis is additionally useful in identifying partial and incomplete cleavages as in, for example, a CNBr digest which reveals more bands than methionine residues per mol polypeptide.

10.3. ENZYMIC DIGESTION

Before digesting the protein with enzymes, it must be denatured and the cysteines converted to a hydrophilic derivative. Disulphide bonds are first reduced under denaturing conditions with dithiothreitol (Konigsberg, 1972) and the cysteines modified with iodoacetic acid, iodoacetamide (Hirs, 1967b) or ethyleneimine (Shotton & Hartley, 1973). Alternatively the cysteines may be oxidized with performic acid (Hirs, 1967a). The choice of technique depends very much on the strategy. A detailed discussion of these modification reactions will be found in Chapter 2.

Aminoethylation can be used to introduce bonds susceptible to cleavage by trypsin or *A. mellea* protease and carboxymethylation with [^{14}C]iodoacetic acid to radioactively label the cysteines. Performic acid treatment results in the oxidation of methionine and tryptophan as well as cysteine and usually gives a more soluble product than reductive alkylation.

Ideally, after protease digestion a given portion of the polypeptide chain will be represented in only one peptide, thus maximizing the yield of this single species and reducing the complexity of peptide fractionation. Since a protease shows a range of affinities and cleavage rates at different points in the polypeptide chain, it is usually impossible to achieve this ideal. However, it is possible to limit digestion to the main sites of protease affinity by careful choice of the conditions. In general, by decreasing the protease/substrate ratio or by using a larger volume of buffer, the stronger binding modes are promoted at the expense of the weaker; by decreasing the time of digestion the ratio of those fragments produced by fast cleavage to those produced by slow cleavage is enhanced. Suggested 'mild' digestion conditions for several proteases are listed in Table 10.3. Alternatively peptide bond cleavage may be followed by pH-stat (Edelman *et al.*, 1968) and the digest terminated as the rate of uptake of acid or base levels out. Another type of 'unspecific' cleavage may arise from contaminating proteases; chymotrypsin

Table 10.3. Suggested protease digestion conditions for a primary digest

Enzyme	Ratio of protease to substrate (w/w)	Temperature (°C)	Time (h)
Trypsin	1/100	37	2
Chymotrypsin	1/50 plus 1/100 soya bean trypsin inhibitor	37	2
Elastase	1/100	37	2
Staphyloccocal protease	1/30	37	4
Armillaria mellea Protease	1/100	37	4

Protein substrate present at about 1 mg/ml in 0.5% (w/w) ammonium bicarbonate, pH 8.0. Similar conditions may be used for sub-digests of purified peptides present at about 1 μmol/ml in 0.5% (w/w) ammonium bicarbonate, pH 8.0. For sub-digests, thermolysin may be used at 1/100 (w/w) at 60°C for 2 h and pepsin at 1/100 (w/w) at 37°C for 4 h in 5% formic acid.

contamination can be removed from trypsin by purifying the trypsin on an affinity column and trypsin contamination in chymotrypsin preferentially titrated out by addition of trypsin inhibitor to all chymotryptic digests.

Usually the denatured protein is insoluble in the ammonium bicarbonate digestion buffer and must first be solubilized, or at least converted to a fine suspension. The protein is therefore lyophilized and dissolved in a minimal volume of denaturant, such as 6 M-guanidinium chloride, 8 M-urea, 100 % formic acid or strong ammonia solution. This is squirted into a much larger volume of buffer, although very insoluble proteins can be digested in 2 M-urea with trypsin, chymotrypsin and elastase (Harris, 1956) or in 4 M-urea with staphyloccocal protease (Drapeau, 1977). If the protein has been solubilized in formic acid or ammonia, the digestion buffer is adjusted, respectively, with ammonia or formic acid to ensure that the final pH on addition of protein is pH 8.0. Since ammonium formate is volatile, no desalting is necessary later. However the use of urea and especially guanidinium chloride as denaturants entails a desalting step at some later stage.

The digestion of a fine suspension can be followed by removing aliquots, spinning down the precipitate and measuring release of A_{280} peptides or [^{14}C]carboxymethylcysteine into the supernatant. Complete digestion can be checked by running a small aliquot on an SDS-polyacrylamide gel when the band due to intact protein should be absent. The main sample should meanwhile be frozen and kept at −20°C; should the digest be incomplete, the sample can be thawed and digestion allowed to continue. Finally, insoluble polypeptide can be centrifuged away from the soluble peptides and both portions freeze-dried. It may be worthwhile redigesting the insoluble polypeptide with higher concentrations of protease and combining the soluble peptides thereby released with the major pool of soluble peptides. See also Chapter 3 for enzymic digestion of proteins.

10.4. FRACTIONATION STRATEGY FOR SOLUBLE PEPTIDES

Design of the optimum strategy for peptide fractionation requires knowledge of the total numbers, the sizes, the charges and the solubility characteristics of the digested peptides. As a general rule, large peptides are less soluble than small peptides, while the charge and size of peptides are often related to the protease used. For example, the large peptides produced by trypsin tend to be acidic, since the chances that a tryptic peptide contains at least two acidic residues will be greater than the chances that it contains either an internal lysine or arginine, or two histidines. However, in designing a fractionation strategy, an analytical paper fingerprint of the soluble peptides, stained with ninhydrin and stains selective for various side chains, will prove more useful than a catalogue of general rules. A typical fingerprint would involve electrophoresis at pH 6.5 followed by chromatography in butanol/acetic acid/water/pyridine (BAWP) at right angles to the

Table 10.4. High voltage paper electrophoresis and paper chromatography

	Electrophoresis			
	pH 6.5	pH 3.5	pH 2.1	BAWP Chromatography
Buffer (v/v)				
Butanol	–	–	–	15
Acetic acid	0.3	5	4	3
Formic acid	–	–	–	–
Pyridine	10	0.5	–	10
Water	89.7	94.5	45	12
Coolant	Toluene	White spirit	White spirit	–
Position of origin	30 cm from anode	20 cm from anode	10 cm from anode	10 cm from top. Also mark positions of bends at 2 cm and 7 cm.
Loading capacity for a mixture of peptides	< 50 nmol/cm of any one peptide of Whatman No. 1 <200 nmol/cm of any one peptide on Whatman 3 MM			
Time	45 min, 3 kV	90 min, 3 kV	45 min, 3 kV	Overnight

electrophoretic dimension (see Table 10.4). Since there may be a large number of neutral peptides, the pH 6.5 neutral band is cut out and electrophoresed at pH 2.1. A map which revealed a preponderance of large acidic peptides would suggest a strategy involving a resin such as DEAE-cellulose, whereas a preponderance of small neutral and basic peptides would suggest a resin such as Dowex-50. A good peptide fingerprint at this stage is also a useful tally for all the peptides which appear in the digest. In the event of missing sequences, the purified peptides can be checked against the map to ensure that none of the peptides on the map have been lost during the purification.

The simple purification strategy described below—gel filtration, DEAE-cellulose or Dowex-50, and high voltage paper electrophoresis/paper chromatography—would be suitable for an average polypeptide of up to 40 000 mol. wt. The large peptides which fractionate and elute poorly from paper are first removed by gel filtration and the majority of peptides are purified rapidly on paper after partial fractionation on ion-exchange columns. Not all digests are amenable to this type of fractionation strategy, or indeed to manual sequencing. For example, membrane proteins are poorly digested by enzymic cleavage with high specificity proteases such as trypsin or staphyloccocal protease; hydrophobic peptides form mixed micelles on size fractionation, precipitate on ion-exchange columns and extract into the coolant of high voltage paper electrophoresis and into the solvent wash of the Edman reaction. The best strategy for such proteins may require cleavage by chemical methods to give a few large peptides, fractionation in denaturing solvents or polyacrylamide gels and sequencing by automated methods after coupling the peptide to a solid phase. Alternatively, a strategy based on digestion with a low specificity protease such as pepsin or elastase might also be appropriate. This would give a large number of small peptides which could be fractionated on Dowex-50 and the impure peptides sequenced by mass spectrometry mixture analysis (Morris *et al.*, 1971). In recent years HPLC techniques have been developed for rapid peptide separations (see Chapter 6).

10.4.1. Gel filtration

Gel filtration is an ideal first step since yields are usually excellent and the large peptides often require no further purification. The soluble peptides derived from digestion with most proteases should be soluble in ammonium bicarbonate and can therefore be loaded on a Sephadex G-50 or G-25 column and run in 0.5% ammonium bicarbonate, with monitoring of the eluate spectrophotometrically at 280 nm (tryptophan and tyrosine absorption) and 230 nm (peptide bond absorption). However, if the peptides are insoluble in the small loading volume, solid urea should be added until they become soluble. The peptides from a pepsin digest may be insoluble at neutral pH and should therefore be dissolved in a small volume of 100% formic acid, diluted to 5% formic acid, loaded on a Sephadex G-50, Biogel P-60 or P-30 column, run in 5% formic or acetic acid and the eluate monitored spectrophotometrically.

Small aliquots from each fraction are also analysed by high voltage paper electrophoresis and apparently pure peptides checked by end-group analysis and perhaps rapid dansyl–Edman sequencing (see Chapter 11). Aliquots are taken for amino acid analysis from the peak tubes corresponding to the pure peptides and lyophilized. Large polypeptides (20–50 residues), although pure, will often appear as a smear on paper electrophoresis. However, the smaller peptides produced by a sub-digestion of the large peptide will usually give clear-cut bands. Thus, sub-digesting a small portion of each tube across a 'pure' optical density peak will sometimes reveal minor components otherwise unsuspected. Large impure peptides can often be successfully purified on DEAE-cellulose, although the yields may be low. Since sub-digestions will probably be required to complete or check the sequence it may be wise to retain two-thirds of the peptide from the DEAE-cellulose column. The yield from DEAE-cellulose, even if low, should be sufficient for amino acid analysis, a partial *N*-terminal sequence and an analytical scale sub-digest. The sub-digest can be analysed by paper electrophoresis/chromatography: preparative digests can then be made on the impure peptide, and the 'true' constituent peptides allocated by reference to the analytical fingerprint.

The bulk of the remaining peptides are either pooled *en masse* and applied to Dowex-50, or divided into a small number of pools and applied to DEAE-cellulose.

10.4.2. Dowex-50 column chromatography

From Dowex-50 (sulphonated polystyrene) peptides can be eluted with a gradient of increasing pH and pyridinium ion concentration (Schroeder, 1967a, 1972a). See Table 10.5 for details. At pH 2, the *C*-terminal carboxyl groups and side chain

Table 10.5. Ion-exchange columns

	Dowex-50	DEAE-Cellulose
Preparation of matrix	Wash *in situ* with 1 M-NaOH, 1 M-HCl and pH 5.0 buffer (2 M in pyridine: 645 ml pyridine and 573 ml glacial HAc diluted to 4 l). Check all sodium ion has eluted by a flame test, then wash with pH 3.1 buffer (0.2 M in pyridine: 64.5 ml pyridine and 1114 ml glacial HAc and diluted to 4 l) until column is equilibrated.	Recycle in a Büchner funnel according to manufacturer's instructions (wash with 0.2 M-NaOH, distilled water, 0.2 M-HCl and finally distilled water). Leave for several h in 10% ammonium bicarbonate. Wash with 0.1% ammonium bicarbonate pH 8.3 and pack into column. Pack down and equilibrate the exchanger overnight by pumping at ≈ 100 ml/h.
Column type and dimensions	Water-jacketed high pressure column (9.0 mm diam. × 500 mm).	25 ml glass pippette plugged with glass wool and with a syringe needle passed through a rubber bung for the top seal.
Column temperature	60°C	Room temperature
Flow rate	15 ml/h (high pressure pump, e.g. Milton Roy chromatographic minipump).	5 ml/h
Fraction size	2 ml	2 ml
Sample loading	Sample dissolved in 1 ml water with a few drops of conc. HCl added to bring to pH 2. Spin off any precipitate before loading, apply to top of column, blow in with nitrogen and fill dead space with pH 3.1 buffer.	Salt-free sample dissolved in 2 ml of 0.1% ammonium bicarbonate (pH 8.3). If necessary add solid urea to dissolve precipitate.
Wash 1	Wash with pH 3.1 buffer until acidic peptides have eluted (≈ one column volume). Check by pH 6.5 high voltage paper electrophoresis.	Wash with 0.1% ammonium bicarbonate (pH 8.3) until basic peptides have eluted. Check by optical density at 230 nm—several peaks may be expected.
Gradient	300 ml pH 3.1 buffer, 600 ml pH 5.0 buffer (1:2 gradient vessel). This gives a gradient linear in pyridinium ion concentration.	200 ml 0.1% ammonium bicarbonate (pH 8.3), 200 ml 1% ammonium bicarbonate (pH 8.3). This gradient elutes mainly the basic and neutral peptides. A second gradient of 200 ml 1% ammonium bicarbonate (pH 8.3), 200 ml 3% ammonium bicarbonate (pH 8.3), elutes mainly acidic peptides.
Wash 2	pH 5.6 buffer (8.5 M in pyridine, 684 ml pyridine and 180 ml glacial HAc diluted to 1 l).	50 ml 5% ammonium bicarbonate followed by 10 ml 6 M-guanidine HCl. Column must now be recycled.

carboxyl groups of aspartic acid and glutamic acid are protonated, as are *N*-terminal amino groups and the side chains of arginine, lysine and histidine. The peptides therefore adsorb to the sulphonic acid groups of the resin. As the pH is increased the carboxyl groups ionize and the peptides are released. Peptides are thereby fractionated according to their net charge and p*K* of carboxyl groups. Acidic peptides are eluted first, followed by neutral then basic peptides. The neutral peptides tend to elute in two groups: first the 'true' neutrals containing no internal basic or acidic groups, then 'other' neutrals with compensating basic and acidic groups. This reflects the higher pK_a of side chain compared with *C*-terminal carboxyl groups. Large peptides, hydrophobic peptides and very basic peptides (net positive charge of several units) are usually eluted in low yield. Pyridinium ion is not transparent to u.v. light at 230 nm or 280 nm and the peptide profile is therefore monitored by analytical high voltage paper electrophoresis at pH 6.5. From the analytical results, suitable poolings are made for preparative paper electrophoresis and the pools are concentrated by rotary evaporation.

10.4.3. DEAE-cellulose column chromatography

From DEAE-cellulose (diethylaminoethylcellulose) peptides can be eluted with a gradient of increasing bicarbonate ion (Roy & Konigsberg, 1972) (Table 10.5). At pH 8.0 the *C*-terminal carboxyl groups and side chain carboxyl groups of aspartic acid and glutamic acid are ionized, while the side chains of arginine and lysine are protonated. The imidazole ring of histidine is completely deprotonated and the *N*-terminal amino group is 50% deprotonated. Peptides are fractionated according to their net charge at pH 8.0.

Basic peptides are first eluted and followed by neutral then acidic peptides: by loading the sample in low ionic strength and eluting with a low ionic strength wash, basic peptides can be readily resolved from each other. Such 'hydrophobic' interactions with the matrix also allow the separation of related peptides differing only in neutral amino acids.

The elution of peptides can be followed by u.v. absorption at 280 nm and 230 nm and also by analytical high voltage paper electrophoresis. Suitable poolings are made for preparative high voltage paper electrophoresis, where necessary, and the fractions freeze-dried or rotary evaporated.

10.4.4. High voltage paper electrophoresis and paper chromatography

Paper electrophoresis and paper chromatography are rapid purification steps for most medium and small peptides (see 10.8.2). Paper electrophoresis is usually used at pH 6.5, pH 3.5 or pH 2.1. At these pH values the side chains of lysine and arginine are fully protonated, as are the *N*-terminal amino groups. At pH 6.5, the imidazole ring of histidine is partially protonated and carboxyl groups are fully ionized. At pH 3.5 the histidine is fully protonated and carboxyl groups partially protonated. At pH 2.1 the carboxyl groups are fully protonated. Only the sulphonic acid side chain of cysteic acid is deprotonated at pH 2.1. Peptides electrophorese with a mobility strictly dependent on their net charge and size (Offord, 1966) and are located on the paper by staining a side-strip with ninhydrin (suitable for one-dimensional maps) or by overstaining with low levels of fluorescamine (suitable for both one- and two-dimensional maps). Paper electrophoresis is, however, sensitive to salt and overloading and the detection of peptides by ninhydrin or fluorescamine staining relies on an unblocked *N*-terminal amino acid. Paper chromatography using BAWP (Waley & Watson, 1953; see Table 10.4) fractionates peptides according to their partition coefficient between water and an organic phase: hydrophobic peptides move faster than hydrophilic peptides. The resolution obtained with BAWP chromatography is not as great as with paper electrophoresis, but it is invaluable for separating peptides of similar size and charge.

10.5. FRACTIONATION STRATEGY FOR INSOLUBLE PEPTIDES

The insoluble peptides can often be solubilized and fractionated in denaturing solvents, but it is invariably faster to use sub-digestion. An 'aliphatic' core can be sub-digested with elastase or thermolysin and an 'aromatic' core with chymotrypsin, or with pepsin at acid pH. The peptides can then be fractionated rapidly on paper, or if the core represents a substantial portion of the molecule, by Dowex-50 and paper.

10.6. PEPTIDE PURITY AND SEQUENCING

'Pure' peptides should be checked for purity by analytical electrophoresis at pH 6.5 and pH 2.1, by amino acid analysis and end-group analysis. The electrophoretic mobilities at pH 2.1 and pH 6.5 (Offord, 1966), combined with the amino acid analysis, should allow the size of the peptide and the number of amides to be estimated. Since over 20 peptides can easily be sequenced in parallel each day it is important to gather a large pool (> 40 if possible) of pure peptides *before* starting to sequence (details of the dansyl–Edman and DABITC–PITC method are given in Chapters 11 & 14). To assign amides in the dansyl–Edman technique a small aliquot of peptide may be analysed by pH 6.5 electrophoresis before and after Edman cleavage of Asx or Glx. Alternatively, after sequencing, the peptide may be proteolytically sub-digested and the constituent peptides fractionated by pH 6.5 electrophoresis and analysed. The protease is selected so as to prise apart the amide ambiguities, a purpose for which staphyloccocal protease (splits at glutamic acid) is exceptionally useful. Furthermore, since the reliability of sequencing diminishes toward the *C*-terminus of the peptide, it is advisable to isolate the *C*-terminal peptide(s) and check the amino acid analysis and pH 6.5 mobility against that expected: if the *C*-terminus is incorrect, this could lead to incorrect overlapping and major errors of sequence transposition. Ideally the check on the *C*-terminal area should be combined with sub-digestion for amide assignment.

Errors in sequencing may derive from a multitude of sources. Histidine is less consistent in behaviour than other types of residue and 'premature' cleavage at histidine may cause histidine to be lost from the sequence, or more usually cause sequence heterogeneity beyond this position (Schroeder, 1967b, 1972b). Incomplete cleavage by anhydrous acid, due to the insufficient extraction and build up of thiazolinones will cause new residues to be 'ghosted' by (the) preceding residue(s): this becomes a source of error if a newly exposed residue, such as dansyl-Trp or dansyl-Pro is poorly visualized. This residue may then incorrectly be identified as the 'ghost' residue. Similarly 'ghosting' combined with extraction of the remaining non-polar portion of the polypeptide by solvent washes may also lead to errors of identification. For these reasons, repetition of sequence should be viewed with suspicion. In the dansyl–Edman technique, insufficient hydrolysis of the dansyl-peptide, especially of peptides with *N*-terminal dansyl-Ile or dansyl-Val, will generate dansyl dipeptides which may be incorrectly identified as dansyl-amino acids: for example, dansyl-Ile-Glu is occasionally confused with dansyl-Glu on polyamide sheets. Finally, errors may arise from *sequence-crossover* while sequencing an impure peptide. Assuming that a major sequence and a minor (contaminant) sequence are being read, the blocking of the major sequence at Gln, Trp or Asn-Gly or its preferential extraction in the solvent wash may cause the sequence of the minor component to be inadvertently read off from this position.

10.7. OVERLAPPING STRATEGY

Peptides produced by one protease can be overlapped by the peptides produced in a separate digest with a second protease. Some combinations of proteases, such as trypsin/chymotrypsin or trypsin/staphyloccocal protease are complementary, whereas others such as trypsin/*A. mellea* protease or chymotrypsin/pepsin are less appropriate since they will produce a large number of one-residue overlaps (Table 10.1). The second batch of peptides need not be fully sequenced, but at least a good amino acid analysis, end group analysis and electrophoretic mobility should be obtained. In addition, these peptides should preferably be sequenced through the overlap region or sub-digested and the amino acid analysis of the substituent peptides checked. Overlaps should extend at least two residues back from the *C*-terminus of one peptide and two residues into the *N*-terminus of the next. While the sequence is substantially incomplete, smaller overlaps than this can only be justified by additional supporting evidence. As the sequence nears completion, the chances of rivals being found for a given small overlap diminish, and these overlaps can therefore be tentatively allocated. At this stage, further strategy depends on the nature of the *lacunae*; if there are substantial missing sequences apparently scattered throughout the whole molecule, it may be advisable to undertake another total enzymic digestion. If the missing sequence is confined to one area, it may be possible to isolate that area in a large fragment. If the sequence is essentially complete, although with a few overlaps missing, it may be possible to resolve them by preparing paper fingerprints of large overlap polypeptides. It is at this stage that a blocking strategy may be most productively employed. For example, a protein, already digested with trypsin and chymotrypsin, but missing several lysine overlaps, could be modified with maleic anhydride, digested with trypsin and deblocked. The peptides could be fractionated on G-50 or G-75 Sephadex, a portion of each fraction redigested with trypsin and analysed by high voltage paper electrophoresis. Using suitable stains for key amino acids and noting carefully the mobilities of the substituent peptides, the missing overlaps might be tentatively identified. This could be checked by partial sequencing and amino acid analysis of the purified overlap peptide.

Assuming that the predicted and observed amino acid analyses and mol. wts. agree fairly well, the best proof of sequence completion is a rigorous and complete overlapping. Not only must all peptides sequenced be allocated to the final sequence, but also all the peptides noted in the analytical paper fingerprint should be identified. If peptides have been found in low yield which cannot be fitted to the final sequence, it may be worth while checking the published sequences of the proteases used and inspecting the sequences of peptides of similar mobility for evidence of sequence cross-over.

10.8. METHODS

10.8.1. Analytical paper electrophoresis and chromatography

Two-dimensional analytical fingerprints of proteins, stained with fluorescamine or ninhydrin and reagents specific for particular amino acid side chains are exceptionally useful in checking how well a proteolytic digest has worked and in designing a fractionation strategy. For example, suppose that the amino acid analysis of the protein revealed 15 Arg/mol subunit and yet on fractionating a tryptic digest on a two-dimensional map only five fluorescent spots could be detected after staining with phenanthrene quinone (Yamada & Itano, 1966). This would suggest that either 10 arginine residues remained unsplit in an insoluble core, or that 10 arginine tryptic peptides were large, insoluble and stuck at the origin. A single map can be stained successively (Bruton, 1975) with fluorescamine (Udenfriend *et al.*, 1972), phenanthrene quinone for arginine residues (Yamada & Itano, 1966) and finally the Pauly stain for histidine and tyrosine residues (Easley *et al.*, 1969). If the protein has been oxidized by performic acid, the tryptophan peptides can be detected by their intrinsic fluorescence (inspect map under u.v. light before staining with fluorescamine). See Chapter 8, section 8.14 for staining techniques.

The best display of the soluble peptides can only be determined empirically. However, pH 6.5 electrophoresis/BAWP chromatography, with the pH 6.5

neutral peptides re-run at pH 2.1 or pH 3.5, is often an effective strategy (Bruton, 1975). Digest 10–20 nmol of protein and freeze-dry to remove all traces of salt. Dissolve in a small volume of 98 % formic acid, dilute to 50 % formic and apply as a 1 cm band on Whatman 3 MM paper for pH 6.5 electrophoresis (Table 10.4). After electrophoresis cut out the neutral band and stitch onto the origin of Whatman 3 MM paper for BAWP chromatography: similarly, cut out the strip of basic peptides and the strip of acidic peptides and treat to BAWP chromatography. Now cut out the entire strip of neutral peptides which have been run on BAWP and stitch onto the origin of Whatman 3 MM paper for final pH 2.1 or pH 3.5 electrophoresis.

Analytical fingerprints are also useful for monitoring the elution of peptides from columns and making pooling decisions. Dry down a small aliquot ($\simeq$ 5 nmol of peptide) of every alternate fraction, redissolve in 50 μl of pH 6.5 electrophoresis buffer and apply as contiguous 1 cm bands to Whatman No. 1 paper. Run electrophoresis at pH 6.5, cut out the neutral band and stitch onto Whatman No. 1 paper for pH 2.1 electrophoresis. Stain the acidic, basic and refractionated neutral peptides with fluorescamine, phenanthrene quinone and the Pauly stain.

10.8.2. Preparative paper electrophoresis and chromatography

A simple mixture of peptides, such as a pool of acidic peptides eluted from a DEAE-cellulose column, may require only a single fractionation on paper whereas a complex mixture, such as that from the tail-end of a G-50 Sephadex column, may require several fractionations. The strategy is best chosen after a trial electrophoresis (for example, pH 6.5 electrophoresis with the neutral peptides re-run at pH 2.1). A complex mixture of neutral peptides will probably be best resolved by BAWP chromatography, otherwise electrophoresis is preferred since the bands are much sharper. In addition, impurities eluted after BAWP chromatography may catalyse the destruction of tyrosine during acid hydrolysis (overcome by adding 0.1 % phenol to 6 M-HCl) and prevent efficient dansylation (overcome by extracting the peptide with butyl acetate). If BAWP chromatography is used in conjunction with paper electrophoresis these problems can be avoided by using BAWP before electrophoresis. Good strategy also dictates that the final electrophoretic or chromatographic dimension should be on Whatman No. 1 paper since the endogenous levels of free amino acids are much lower than on Whatman 3 MM paper.

Peptides are dissolved in pH 6.5 buffer (see Table 10.4) and applied as a band along the origin of the paper. Insoluble peptides should be dried and dissolved in a small amount of 98 % formic acid or trifluoroacetic acid. This is then diluted to 50 % acid with water and the peptide loaded. The peptides are dried with a warm hair dryer although this may result in the partial oxidation of Met peptides which will then have different R_F values on BAWP chromatography. For electrophoresis, fluorescent markers (see Table 10.6) are loaded between the samples of

Table 10.6. Electrophoretic markers

Markers	Recipe
Xylene cyanol FF and Orange G (dyes)	Dissolve 500 mg Xylene cyanol FF and 500 mg Orange G in 10 ml water.
Dansyl-Arg-Arg, dansyl-Arg and dansic acid (fluorescent)	Dissolve 17.4 mg of arginine and 47.7 mg of arginyl-arginine in 10 ml of 0.5 M $NaHCO_3$. Add 80 mg of dansyl-chloride in 10 ml of acetone and incubate in a sealed tube for 2 h at 37°C. Store at 4°C.
Amino acid mixtures	One mg portions of the following amino acids are combined. 1 marker: Cya, Leu, Phe, Pro, Tyr, Val, Arg, Lys 2 marker: Asp, Glu, Ala, Gly, Ile, Ser, Thr, His Each set of amino acids is made up to 1 ml with 1 mM-HCl and the solution stored at 4°C.

peptide and dye markers are loaded at the edges of the paper. Amino acid markers (Table 10.6) are also useful if ninhydrin staining is routinely employed. In BAWP chromatography the position of the solvent front is a sufficient marker.

10.8.2.1. Single dimension

Less than 1 μmol each of a few peptides. Dissolve peptides in 0.5 ml of pH 6.5 buffer and load as a 15 cm band on Whatman No. 1 paper. Electrophorese then overstain with fluorescamine or stain a side-strip with ninhydrin.

10.8.2.2. Two dimensions

Less than 1 μmol of many peptides. Dissolve peptide in 1 ml of pH 6.5 buffer and load as a 15 cm band on Whatman 3 MM paper. Use electrophoresis or chromatography in first dimension, overstain with fluorescamine *and* stain side-strips with ninhydrin and also phenanthrene quinone. Cut out each peptide band *and interbands* and stitch each onto Whatman No. 1 paper: the interbands are not discarded since they may contain low-yield unique peptides which have been overshadowed by the high-yield peptides in the main bands.

Electrophorese in the second dimension and overstain with fluorescamine, or stain side-strips with ninhydrin. If BAWP chromatography is used in the first dimension the broad strips may be sharpened by careful wetting-up prior to the electrophoretic dimension.

Less than 500 nmol of several peptides. Dissolve peptides in 0.5 ml of pH 6.5 buffer and load as a 5 cm band on Whatman 3 MM paper. Use electrophoresis or chromatography in first dimension: cut out entire strip, stitch onto Whatman No. 1 paper and electrophorese in second dimension. Overstain with fluorescamine.

10.8.2.3. Two dimensions with three dimensions on the neutral peptides

Less than 500 nmol of several peptides. Dissolve peptides in 0.5 ml of pH 6.5 buffer and load as a 5 cm band on Whatman 3 MM paper. Electrophorese at pH 6.5, cut out the neutral band and stitch onto Whatman 3 MM paper for BAWP chromatography. After chromatography cut out the entire strip and stitch onto Whatman No. 1 for a final dimension of electrophoresis. From the first (pH 6.5) dimension, cut out the strip of acidic peptides and the strip of basic peptides and stitch onto Whatman No. 1 for the second dimension of electrophoresis.

10.8.3. Peptide staining

After electrophoresis or chromatography the dry paper should be inspected for intrinsic fluorescence (oxidised tryptophan) under u.v. light. Peptides are then visualized by overstaining with fluorescamine (Udenfriend *et al.*, 1972) and/or by staining side-strips with ninhydrin-cadmium acetate (Heilmann *et al.*, 1957). Fluorescamine overstaining opens up the possibility of detecting and preparatively eluting peptide from two-dimensional paper maps and in addition ensures that bands which have run askew on one-dimensional maps (likely after BAWP chromatography) are correctly cut out. Although the ninhydrin staining of side-strips on one-dimensional maps is much less sensitive and may not detect peptides with *N*-terminal Ile or Val, it does have some advantages over fluorescamine overstaining. With fluorescamine staining, bands are often found on elution to contain virtually no peptide, whereas with ninhydrin staining the intensity of colour is roughly proportional to the amount of peptide. Unlike fluorescamine, ninhydrin stains peptides with *N*-terminal proline and may differentiate peptides with different *N*-terminal amino acids. Although most *N*-terminal amino acids give a red colour, Gly and Ser give a yellow colour and Asn and Cya initially give an orange colour which turns slowly to red. The use of *both* fluorescamine and ninhydrin (and also perhaps stains specific for amino acid side chains) may help decide exactly where to cut in a complex one-dimensional fractionation.

10.8.3.1. Fluorescamine staining

Fluorescamine is stored as a 0.1 mg/ml solution in dry acetone and 1 ml added to 100 ml of 1% pyridine (v/v) in acetone. The paper is dipped and dried in an air-flow (warm or room temperature). After inspection under u.v. light, if no fluorescent bands have appeared, the paper is dipped in a *fresh* solution of 2 ml of fluorescamine stock solution plus 100 ml of 1% pyridine (v/v) in acetone. Thus gradually the amount of fluorescamine is increased until bands appear.

10.8.3.2. Ninhydrin–cadmium acetate staining

Ninhydrin is stored as a 0.25% (w/v) solution in acetone and cadmium acetate in aqueous acetic acid (1 g cadmium acetate $2H_2O$, 50 ml of glacial acetic acid and

100 ml of water). The paper is dipped in a mixture of 100 ml of ninhydrin solution plus 15 ml of cadmium acetate solution. The paper is dried at room temperature and then heated in an oven at 100°C for 3 min. After noting the colour of the bands, the paper is left overnight in the fume hood to bring out changes in colour or the appearance of slow-reacting peptides.

10.8.4. Peptide elution

Peptide bands are cut out and each strip fashioned with one pointed end and one blunt end. The blunt end is hooked over a trough containing the elution buffer and trapped there by a microscope slide. The pointed end is led into a test tube and the elution buffer allowed to move down the strip by capillary action. The whole apparatus is covered with a Perspex lid to keep the atmosphere saturated with buffer, and about 200 μl collected from each strip. Acetic acid (25–50 %) is used for the elution of basic peptides and 10–20 % pyridine (distilled) for others. The peptides are dried down with an oil pump over NaOH (for acetic acid) or P_2O_5 (for aqueous pyridine). The process is rapid if the tubes are placed in a metal heating block which has been warmed to 40°C.

10.9. REFERENCES

Adelstein, R. S. & Kuehl, W. M. (1970). *Biochemistry*, **9,** 1355–1364.

Ambler, R. P. (1965). *Biochem. J.*, **96,** 32P.

Artavanis, S. (1974). TIM from *B. stearothermophilus*. Ph.D. Thesis, University of Cambridge.

Blombäck, B. (1967). *Methods Enzymol.*, **11,** 398–411.

Bornstein, P. (1970). *Biochemistry*, **9,** 2408–2421.

Bornstein, P. & Balian, G. (1970). *J. Biol. Chem.*, **245,** 4854–4856.

Bruton, C. J. (1975). *Biochem. J.*, **147,** 191–192.

Butler, P. J. G. & Hartley, B. S. (1972). *Methods Enzymol.*, **25,** 191–199.

Chang, J. Y., Brauer, D. & Wittmann-Liebold, B. (1978). *FEBS Lett.*, **93,** 205–214.

Corradin, G. & Harbury, H. A. (1970). *Biochim. Biophys. Acta*, **221,** 489–496.

Degani, Y. & Patchornik, A. (1974). *Biochemistry*, **13,** 1–11.

Doolittle, R. F. (1972). *Methods Enzymol.*, **25,** 231–244.

Doonan, S., Doonan, H. J., Hanford, R., Vernon, C. A., Walker, J. M., Airoldi, L. P. da S., Bossa, F., Barra, D., Carloni, M., Fasella, P. & Riva, F. (1975). *Biochem. J.*, **149,** 497–506.

Doyen, N. & Lapresle, C. (1979). *Biochem. J.*, **177,** 251–254.

Drapeau, G. R. (1977). *Methods Enzymol.*, **47,** 189–191.

Drapeau, G. R., Boily, Y. & Houmard, J. (1972). *J. Biol. Chem.*, **247,** 6720–6726.

Easley, C. W., Zegers, B. J. M. & Vijlder, M. D. de (1969). *Biochim. Biophys. Acta*, **175,** 211–213.

Edelman, G. M., Gall, W. E., Waxdal, M. J. & Konigsberg, W. H. (1968). *Biochemistry*, **7,** 1950–1958.

Edman, P. (1949). *Arch. Biochem. Biophys.*, **22,** 475–476.

Edman, P. (1970). In *Protein Sequence Determination*, Needleman, S. B. (Ed.), Springer, Berlin, Heidelberg, New York, pp. 211–255.

Edman, P. & Begg, G. (1967). *Eur. J. Biochem.*, **1,** 80–91.
Greene, L. J. & Bartelt, D. C. (1977). *Methods Enzymol.*, **47,** 170–174.
Gross, E. (1967). *Methods Enzymol.*, **11,** 238–255.
Harris, J. I. (1956). *Nature* (*Lond.*), **177,** 471–473.
Hartley, B. S. (1964). In *Structure and Activity of Enzymes*, Goodwin, T. W., Harris, J. I. & Hartley, B. S. (Eds.), Academic Press, New York, p. 47.
Hartley, B. S. (1970). *Biochem. J.*, **119,** 805–822.
Hartley, B. S. & Shotton, D. M. (1971). In *The Enzymes*, Boyer, P. D. (Ed.), 3rd edn, vol. 3, Academic Press, New York, p. 323.
Heilmann, J., Barrollier, J. & Watzke, E. (1957). *Hoppe-Seyl. Z., Physiol. Chem.*, **309,** 219–220.
Heinrikson, R. L. (1977). *Methods Enzymol.*, **47,** 175–189.
Hirs, C. H. W. (1967a). *Methods Enzymol.*, **11,** 197–199.
Hirs, C. H. W. (1967b). *Methods Enzymol.*, **11,** 199–203.
Houmard, J. & Drapeau, G. R. (1972). *Proc. Natl. Acad. Sci. U.S.A.*, **69,** 3506–3509.
Ilse, D. & Edman, P. (1963). *Aust. J. Chem.*, **16,** 411–416.
Jörnvall, H. (1970). *Eur. J. Biochem.*, **16,** 25–40.
Jörnvall, H. (1973). *Abstr. Int. Cong. Biochem. 9th*, Colloquium D, Abstract Db 13, p. 454.
Karlsson, F. A., Peterson, P. A., Berggard, I. (1969). *Proc. Natl. Acad. Sci. U.S.A.*, **64,** 1257–1263.
Konigsberg, W. (1972). *Methods Enzymol.*, **25,** 185–188.
Konigsberg, W. H. & Steinman, H. M. (1977). In *The Proteins*, Neurath, H. & Hill, R. L. (Eds.), 3rd edn, vol. 3, Academic Press, New York and London, p. 54.
Kopple, K. D. & Bächli, E. (1959). *J. Org. Chem.*, **24,** 2053–2055.
Lowey, S., Slayter, H. S., Weeds, A. G. & Baker, H. (1969). *J. Mol. Biol.*, **42,** 1–29.
Matsubara, H. (1970). *Methods Enzymol.*, **19,** 642–651.
Matsubara, H. & Feder, J. (1971). In *The Enzymes*, Boyer, P. D. (Ed.), 3rd edn, vol. 3, Academic Press, London, p. 721.
Mitchell, W. M. (1977). *Methods Enzymol.*, **47,** 165–170.
Morris, H. R., Williams, D. H. & Ambler, R. P. (1971). *Biochem. J.*, **125,** 189–201.
Mutt, V. & Said, S. I. (1974). *Eur. J. Biochem.*, **42,** 581–589.
Offord, R. E. (1966). *Nature* (*Lond.*), **211,** 591–593.
Ozols, J. & Gerard, C. (1977). *J. Biol. Chem.*, **252,** 5986–5989.
Patthy, L. & Smith, E. L. (1975). *J. Biol. Chem.*, **250,** 565–569.
Perham, R. N. (1967). *Biochem. J.*, **105,** 1203–1207.
Piszkiewicz, D., Landon, M. & Smith, E. L. (1970). *Biochem. Biophys. Res. Commun.*, **40,** 1173–1178.
Platt, T., Files, J. G. & Weber, K. (1973). *J. Biol. Chem.*, **248,** 110–121.
Porter, R. R. (1958). *Nature* (*Lond.*), **182,** 670–671.
Press, E. M. (1967). *Biochem. J.*, **104,** 30c–33c.
Price, N. C. (1976). *Biochem. J.*, **159,** 177–180.
Roy, D. & Konigsberg, W. (1972). *Methods Enzymol.*, **25,** 221–231.
Savige, W. E. & Fontana, A. (1977). *Methods Enzymol.*, **47,** 459–469.
Schaffer, M. H. & Stark, G. R. (1976). *Biochem. Biophys. Res. Commun.*, **71,** 1040–1047.
Schroeder, W. A. (1967a). *Methods Enzymol.*, **11,** 351–361.
Schroeder, W. A. (1967b). *Methods Enzymol.*, **11,** 445–461.
Schroeder, W. A. (1972a). *Methods Enzymol.*, **25,** 203–213.
Schroeder, W. A. (1972b). *Methods Enzymol.*, **25,** 298–313.
Schroeder, W. A., Shelton, J. B. & Shelton, J. R. (1964). *Arch. Biochem. Biophys.*, **130,** 551–556.

Shechter, Y., Burstein, Y. & Patchornik, A. (1975). *Biochemistry*, **14,** 4497–4503.
Shechter, Y., Patchornik, A. & Burstein, Y. (1976). *Biochemistry*, **15,** 5071–5075.
Shipolini, R. A., Callewaert, G. L., Cottrell, R. C. & Vernon, C. A. (1974). *Eur. J. Biochem.*, **48,** 465–476.
Shotton, D. M. & Hartley, B. S. (1973). *Biochem. J.*, **131,** 643–675.
Smyth, D. G. (1967). *Methods Enzymol.*, **11,** 214–231.
Smyth, D. S. & Utsumi, S. (1967). *Nature* (*Lond.*), **216,** 332–335.
Smyth, D. G., Stein, W. H. & Moore, S. (1963). *J. Biol. Chem.*, **238,** 227–234.
Stark, G. R. (1977). *Methods Enzymol.*, **47,** 129–132.
Steinman, H. M., Naik, V. R., Abernethy, J. L. & Hill, R. L. (1974). *J. Biol. Chem.*, **249,** 7326–7338.
Swallow, D. L. & Abraham, E. P. (1958). *Biochem. J.*, **70,** 364–373.
Tang, J. & Hartley, B. S. (1967). *Biochem. J.*, **102,** 593–599.
Udenfriend, S., Stein, S., Böhlen, P., Dairman, W., Leimgruber, W. & Weigele, M. (1972). *Science*, **178,** 871–872.
Uphaus, R. A., Grossweiner, L. I. & Katz, J. J. (1959). *Science*, **129,** 641–643.
Wade, M., Laursen, R. A. & Miller, D. L. (1975). *FEBS Lett.*, **53,** 37–39.
Waley, S. G. & Watson, J. (1953). *Biochem. J.*, **55,** 328–337.
Weeds, A. G. & Hartley, B. S. (1967). *J. Mol. Biol.*, **24,** 307–311.
Weeds, A. G. & Hartley, B. S. (1968). *Biochem. J.*, **107,** 531–548.
Yamada, S. & Itano, H. A. (1966). *Biochim. Biophys. Acta*, **130,** 538–540.

Practical Protein Chemistry—A Handbook
Edited by A. Darbre
© 1986, John Wiley & Sons Ltd

GREG WINTER
MRC Laboratory of Molecular Biology,
Hills Road, Cambridge CB2 2QH, U.K.

11

Manual Sequencing by the Dansyl–Edman Reaction

CONTENTS

11.1. INTRODUCTION

Dansyl chloride reacts with the *N*-terminal amino acid of a polypeptide (and with the side chains of cysteine, tyrosine, lysine and histidine) to give a fluorescent dansyl peptide (Gray, 1964). The *N*-terminal amino acid is released on acid hydrolysis as its fluorescent dansyl derivative, and can be readily identified by two-dimensional chromatography on polyamide thin layer sheets (Gray, 1972a). Under the acid hydrolysis conditions, dansyl tryptophan and amides are completely destroyed, while dansyl proline and dansyl serine are partly destroyed. Of the side chain derivatives only ε-dansyl-lysine and *O*-dansyl tyrosine are stable. In the dansyl-Edman technique (Hartley, 1970; Gray, 1972b) the peptide is degraded sequentially by repeated Edman reaction and the new *N*-terminus released is identified by dansylating a small aliquot after each Edman cycle. Although the repetitive yield does suffer by continually removing aliquots of the peptide, the dansyl-Edman method avoids the solvent wash of the phenylthiocarbamyl (PTC) peptide necessary in the direct Edman degradation (Edman & Henschen, 1975) or the DABITC-PITC (Chang *et al.*, 1978) methods. Non-polar PTC peptides are readily extracted into the organic phase of this wash and may therefore be lost.

11.2. DANSYLATION

The efficient labelling of peptide with dansyl chloride (Fig. 11.1) requires an alkaline pH to keep the reactive α-amino group of the peptide unprotonated, and a mixed organic-aqueous system to dissolve the insoluble dansyl chloride. Dansylation of amino groups competes with hydrolysis of the dansyl chloride and the final extent of peptide labelling is determined by these two pathways. Pyridine catalyses the hydrolysis of dansyl chloride at the expense of coupling (Gray, 1964) while triethylamine catalyses both coupling and hydrolysis. It is therefore advisable to remove traces of pyridine salts (left over from the Edman reaction or from the elution of peptides from paper) by drying down the peptide in the presence of triethylamine. Traces of ammonium salts (left over from ammonium bicarbonate buffers) which otherwise compete for dansyl chloride to give dansic acid amide are also displaced by triethylamine. The use of triethylamine will result in additional dansyl derivatives, presumably due to primary and secondary amine contaminants, but on polyamide sheets these spots migrate above dansyl proline in solvent 2 (see Fig. 11.2).

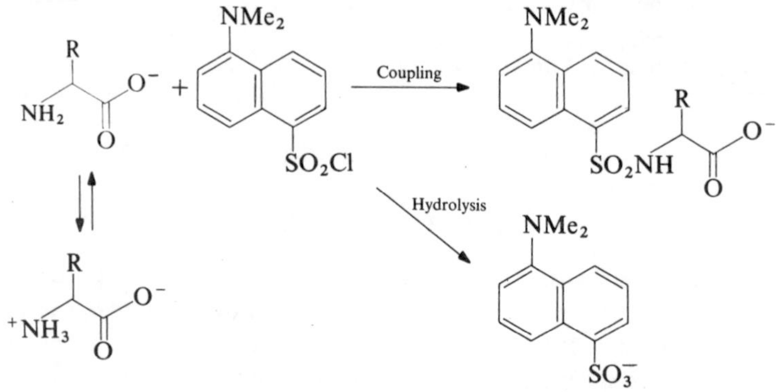

Figure 11.1. Dansylation of amino acid with competing hydrolysis of dansyl chloride to dansic acid.

Peptides may either be hydrolysed by 6 M-HCl at 100° C for 5–15 h or by a mixture of equal volumes of conc. HCl and propionic acids, 165° C for 12 min (G. Winter, unpublished). Hydrolysis in 6 M-HCl for 5 h minimizes the loss of dansyl proline and dansyl serine but leads to incomplete hydrolysis of the peptide bond linking *N*-terminal dansyl valine or dansyl isoleucine to other residues (especially valine or isoleucine). The resulting dansyl dipeptides are sometimes misidentified as dansyl amino acids, and therefore the approximate positions of some common dansyl dipeptides are given in Fig. 11.2. To a first approximation, dansyl dipeptides appear on the straight line joining the positions of the two

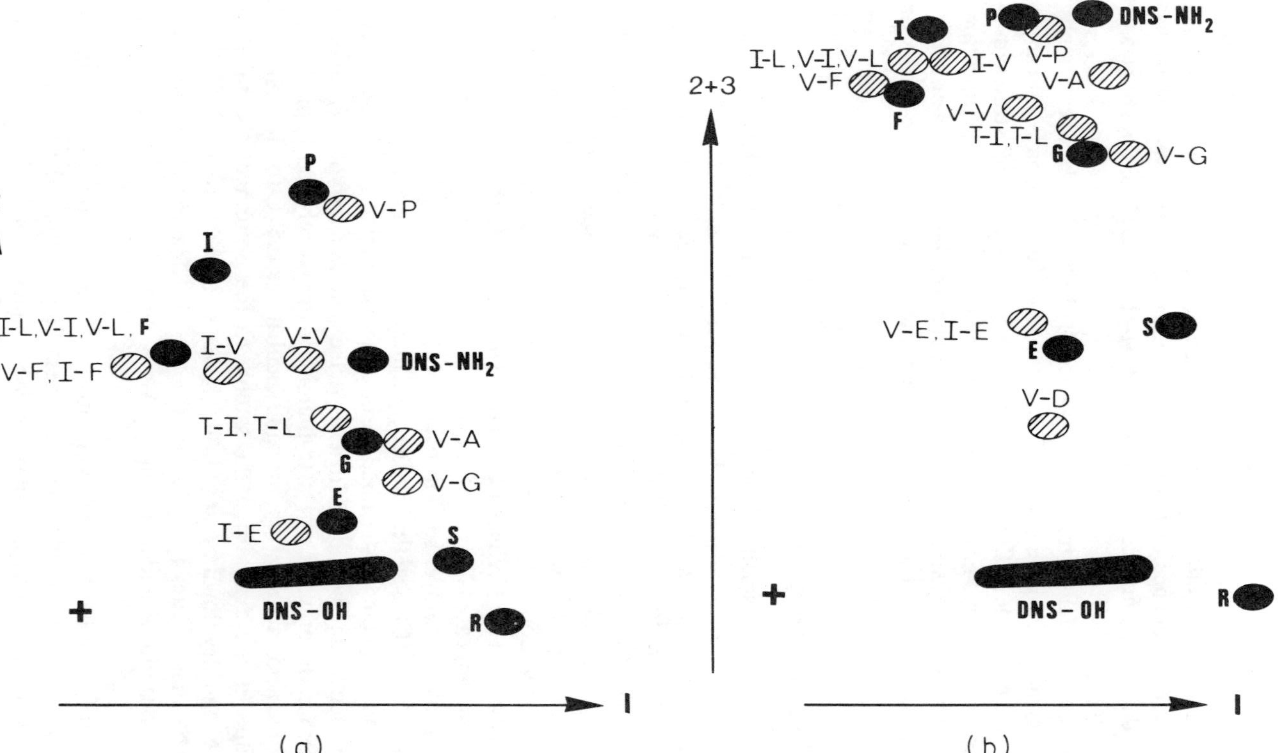

Fig. 11.2. Approximate positions of some dansyl dipeptides (hatched) relative to dansyl amino acid markers (solid) (a) after solvents 1 and 2 and (b) after solvents 1, 2, and 3. The dipeptides I–L, V–I and V–L run with F after solvent 2. DNS-OH: dansic acid, DNS-NH_2: dansic acid amide; DNS standards are R: arginine, E: glutamic acid, G: glycine, I: isoleucine, F: phenylalanine, P: proline, S: serine. The one-letter code is used for other amino acids, L: leucine; V: valine.

constituent dansyl amino acids on the thin layer sheets. Clearly the identification of an anomalous spot as a dansyl dipeptide can be checked by extending the time of hydrolysis (to over 24 h). Hydrolysis of dansyl derivatives with the HCl-propionic acid mixture results in greater degradation of dansyl proline and dansyl serine and a greater chance of finding dansyl dipeptides. However, the partial oxidation of dansyl methionine observed with 6 M-HCl hydrolysis does not occur (G. Winter, unpublished).

The destruction of dansyl tryptophan on acid hydrolysis can be overcome by adding 1 % tryptamine to the 6 M-HCl: alternatively, dansyl tryptophan can be cleaved from the peptide with chymotrypsin. The success of either strategem is variable, requiring up to 5 nmol of dansyl peptide for a clear-cut identification. Side chain amides can be distinguished from side chain carboxyl groups by pH 6.5 electrophoresis of the peptide both before and after Edman cleavage of Asx or Glx.

11.2.1. Method for dansylation

1. Dissolve peptide in a small volume of 1 % (v/v) aqueous triethylamine (Pierce Sequenal Grade). Transfer a small aliquot (~ 1 μl, 0.5 nmol) to the bottom of a dansyl tube using a 1–5 μl calibrated micropette (centrifuge if necessary to bring the sample to the bottom of the tube). Dansyl tubes (4 mm × 50 mm long) are made by drawing out glass tubing in a flame so that the bottom of the tube is conical. The tubes are pre-heated at 500–600° C overnight (see Chapter 8, p. 230).
2. Dry down sample and add 3 μl sodium bicarbonate solution (0.2 M) and 3 μl dansyl chloride solution (5 mg/ml in dry acetone) and seal the tube with parafilm.
3. Incubate 30 min at 50° C, checking that the yellow colour (unhydrolysed dansyl chloride) has disappeared.
4. Dry on an oil pump over sodium hydroxide.
5. Add 5 μl of 6 M-HCl. Seal the tube with a flame and incubate at 100° C for 6 h. Peptides with a suspected *N*-terminal carboxymethylcysteine should be hydrolysed under vacuum and peptides with a suspected *N*-terminal tryptophan should be hydrolysed in the presence of 1 % tryptamine. Alternatively add 10 μl of conc. HCl-propionic acid mixture (1 : 1, v/v) and incubate at 165° C for 12 min.
6. Open tube (if necessary bring the sample to the bottom of the tube by centrifuging) and dry down on an oil pump over sodium hydroxide.
7. Check polyamide plates (75 × 75 mm, Cheng Chin Trading Company, Ltd., Taiwan) for cleanliness under u.v. light. (Many workers now use 50 × 50 mm plates.)
8. Add 3 μl of 95 % ethanol to dansyl hydrolysate and spot out (max. diameter 1 mm) with a drawn out capillary on both sides of a polyamide sheet. The origin spots are back to back 6 mm in from two adjacent sides. Spot out 0.2 μl of dansyl markers (0.1 mg/ml each of DNS-Phe, DNS-Ile, DNS-Pro, DNS-Gly, DNS-Glu, DNS-Ser and DNS-Arg) on one side of the sheet only.

9. Run in each direction until the solvent covers 70% of the plate (10–30 min). In the first direction run with solvent 1 (1.5% formic acid), then dry *completely* with a warm hair dryer. Re-run at right angles to the first solvent in solvent 2 (benzene/acetic acid, 9:1, v/v). Dry and inspect both sides under u.v. light. These solvents resolve everything except DNS-glutamic acid/DNS-aspartic acid, DNS-serine/DNS-threonine and α-DNS-lysine/ε-DNS-lysine/DNS-arginine/DNS-histidine. DNS-alanine is sometimes obscured by dansic acid amide. Note down the identifications possible. Solvent 3 (ethylacetate/methanol/acetic acid, 20:1:1, by vol.) is run in the same direction as solvent 2 and will resolve DNS glutamic acid/DNS aspartic acid, DNS serine/DNS threonine and improve the resolution of DNS alanine/dansic acid amide (Fig. 11.3). Solvent 4 (0.05 M-tri-sodium phosphate/ethanol, 3:1, v/v) run in the same direction as solvents 2 and 3 is supposed to resolve the monosubstituted basic DNS amino acids. However, a clear cut identification is not always possible and usually requires comparison with an appropriate dansyl marker spotted adjacent to the unresolved basic amino acid derivatives.

10. The polyamide sheets may be re-used if washed with ammonia/acetone (75 ml of conc. ammonia solution, 925 ml of water and 1 l of acetone) for about 1 h. Metal holders for the polyamide sheets should not be left indefinitely in the wash solution, as this may lead to corrosion products being deposited on the polyamide sheets. These products quench the fluorescence of the dansyl derivatives. Examine the washed plates under u.v. light before use. Plates with strong fluorescent spots should be discarded.

11.3. EDMAN COUPLING AND CLEAVAGE

Critical practical features of the Edman coupling/cleavage reaction (Fig. 10.1, Chapter 10) are the use of pure reagents, the exclusion of oxygen from the coupling step and the complete drying of peptides after both coupling and cleavage. Reagents are therefore redistilled (Edman & Begg, 1967) and the coupling mixture is thoroughly flushed with nitrogen. Peptides, preferably in solid metal heating blocks which retain heat, are dried under vacuum in a heated vacuum desiccator. Although the dansyl-Edman method is slow for sequencing a single peptide (~ two cycles/day) and the repetitive yield is rather lower than with automated methods, the technique is ideally suited to rapidly sequencing many small peptides in parallel. However, the individual flushing of many peptides with nitrogen before the coupling step is time-consuming and can be automated by deploying a nitrogen manifold with multiple outlets. A simple manifold designed by the author incorporates 20 outlets (syringe needles coated with Teflon tubing) soldered into a hollow brass block and spaced so as to fit directly into the tubes within a heating block. Cling-film is then used in place of screw caps for sealing the tubes and is clipped tightly into place on all tubes simultaneously by a 20-holed

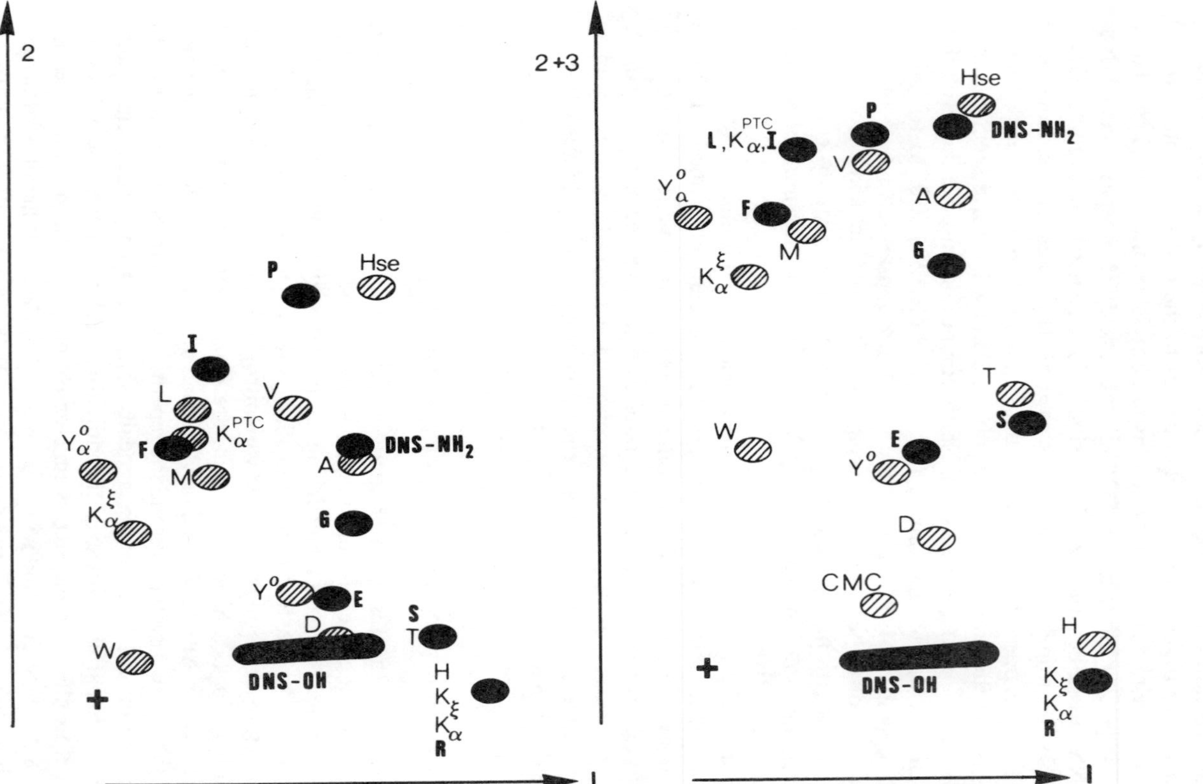

Fig. 11.3. Approximate positions of dansyl amino acids (hatched) relative to dansyl amino acid markers (solid). Hse: DNS-homoserine, Y°: *O*-DNS-tyrosine, Y_α°: *bis*-DNS-tyrosine, K_ε: ε-DNS-lysine, K_α: α-DNS-lysine, K_α^ε: *bis*-DNS-lysine, K_α^{PTC}: α-DNS-ε-PTC-lysine, CMC: DNS-carboxymethylcysteine, DNS-OH: dansic acid, DNS-NH$_2$: dansic acid amide. The one letter code is used for other amino acids.

Perspex sheet. A manifold incorporating variable temperature control of the tubes is commercially available (N-Evap from Organomation).

The following reagents (redistilled or of sequencing grade) and materials are required: phenylisothiocyanate (10 % v/v in pyridine, store in the dark at −20° C for no more than one week), aqueous pyridine (1:1 v/v, store in dark), trifluoroacetic acid, ethanol (95 % v/v), *n*-butyl acetate, nitrogen gas, Pyrex tubes (heat cleaned at 500–600° C) with screw caps fitted with Teflon seals, a thermostatically controlled heating block set at 50° C, vacuum desiccators and a vacuum line to oil pump. Phosphorus pentoxide and sodium hydroxide are used in the desiccator for drying PTC peptides after the coupling step. A separate desiccator containing sodium hydroxide is kept for drying peptides after the cleavage step.

11.3.1. Method for Edman degradation

1. Phenylisothiocyanate, pyridine and trifluoroacetic acid should be kept tightly sealed and flushed with nitrogen both initially and after use.
2. Dry down 20–50 nmol of polypeptide which must be salt free.
3. Add 100 μl of 50 % pyridine and flush with nitrogen for 1 min.
4. Add 50 μl of 10 % PITC in pyridine and flush with nitrogen for 1 min.
5. Seal tubes and incubate in heating block at 50° C for 30 min.
6. Transfer vacuum desiccator containing phosphorus pentoxide and sodium hydroxide to a 60° C oven.
7. Remove heated desiccator from oven and dry down peptides.
8. After 10 min check that solid residues (mainly PTC peptide and diphenylthiourea) are dry, pinkish and crystalline. If they appear oily, add 50 μl of 95 % ethanol and dry again.
9. Dissolve in 50 μl of anhydrous trifluoracetic acid, seal tubes and incubate at 50° C for 15 min. If the cleavage step is expected to expose *N*-terminal glutamine, reduce the incubation to 5 min. This will minimize the cyclization of glutamine to pyrrolidone carboxylic acid.
10. Dry down peptides over sodium hydroxide: once again add 50 μl of ethanol if the peptides do not appear to be completely dry.
11. Add 100 μl of deionized water. Use Vortex mixer thoroughly with 1 ml of butyl acetate and separate the phases on a bench centrifuge (5 min). Extract and discard the upper organic phase: be careful not to discard any material at the interface since it may consist of insoluble peptide. Repeat the butyl acetate wash two more times.
12. Dry down the peptide and take up in 1 % triethylamine solution. Remove a sample for dansylation: dry the main sample of the peptide and repeat the Edman cycle. If it is necessary to stop the cycle of reaction do so when the peptide has been dried after the cleavage step.

Note Samples may be rapidly taken to dryness with a vacuum centrifuge (Speed Vac Concentrator from Savant Instruments, Hicksville, N.Y.).

11.4. REFERENCES

Chang, J. Y., Brauer, D. & Wittmann-Liebold, B. (1978). *FEBS Lett.*, **93,** 205–214.

Edman, P. & Begg, G. (1967). *Eur. J. Biochem.*, **1,** 80–91.

Edman, P. & Henschen, A. (1975). In *Protein Sequence Determination*, Needleman, S. B. (Ed.), 2nd ed. Springer, Berlin, Heidelberg, New York, pp. 232–279.

Gray, W. R. (1964). Ph.D. Thesis, Cambridge University.

Gray, W. R. (1972a). *Methods Enzymol.*, **25,** 121–138.

Gray, W. R. (1972b). *Methods Enzymol.*, **25,** 333–344.

Hartley, B. S. (1970). *Biochem. J.*, **119,** 805–822.

Practical Protein Chemistry—A Handbook
Edited by A. Darbre
© 1986, John Wiley & Sons Ltd

BRIGITTE WITTMANN-LIEBOLD
Max-Planck-Institut für Molekulare Genetik,
Abteilung Wittmann,
Ihnestr. 63–73,
D-1000 Berlin 33 (Dahlem), Germany

12

Practical Methods for Solid-phase Sequence Analysis

CONTENTS

12.1. INTRODUCTION

The impressive progress in sequencing of proteins over the last few decades has been made possible by the introduction of the stepwise degradation of proteins and peptides by phenylisothiocyanate (Edman, 1950) and the automation of this reaction (Edman & Begg, 1967). The principles involved are still being applied to protein sequence analysis. However, in many laboratories modifications of the degradation are still performed manually, e.g. the combined dansyl–Edman method (Gray & Hartley, 1963), a semi-automatic version of the PITC degradation (Tarr, 1975), or the newly developed double coupling method with *N,N*-dimethylaminoazobenzene-4′-isothiocyanate and phenylisothiocyanate (Chang *et al.*, 1978). Reasons for this are:

1. small groups performing sequence analysis or those who are not specialized in protein chemistry can rarely afford the machines and the necessary servicing for automatic degradations;
2. the manual methods are sensitive and can be successfully performed by skilled workers;
3. simultaneous sequencing of 8–16 peptides per person can be made manually.

Although automatic sequencing requires labour, effort and money to keep the machines in optimum condition, they do have several advantages, e.g. reproducibility of conditions for all degradation cycles, easier exclusion of oxygen, improved efficiency of the degradations with higher repetitive yields and full 24-h operation. Furthermore, improved degradations are possible with insoluble peptides, or those which become highly hydrophobic during the degradation cycles. Because of these advantages, versions of the automatic Edman degradation by the liquid-phase film technique (Edman & Begg, 1967; Waterfield & Bridgen, 1975), the solid-phase method (Laursen, 1971; Laursen *et al.*, 1975) and the recent gas-liquid phase cartridge version (Hewick *et al.*, 1981) have been developed for protein and peptide sequencing (see Chapter 17 for a discussion of the latest development, the gas phase method). In this chapter, only solid-phase sequencing is considered.

The basic principle of solid-phase sequencing is to bind the peptide covalently to a solid support and to degrade it step-wise in bound form with phenylisothiocyanate. A variety of methods has been developed to couple peptides and proteins to different supports derived from polystyrene and glass beads provided with functional groups (Laursen, 1977b). The covalently bound peptide support can be packed into columns, thoroughly rinsed with the reagents of the degradation and washed extensively without problems of wash-out as in the liquid phase technique. After the cleavage reaction, the thiazolinone liberated from the amino terminus is washed off the column, leaving the residual peptide covalently bound and available for another cycle of reactions.

Operating and maintaining the solid-phase sequencer is easier than the liquid-phase sequencer, but new investigators in the field have difficulties in selecting the best method of attachment for a particular peptide. Therefore, this chapter deals mainly with the synthesis of resins and the attachment methods which we have found most useful. Currently, sequencing is performed over a wide range of sensitivity. Therefore, the attachment methods described in detail consider starting with amounts of 500 pico- to 200 nanomol of peptide material. Only those methods which in our hands have given satisfactory sequencing results with peptides derived from biological sources are reported here. For further information see recent articles on microsequencing (e.g. Hughes *et al.*, 1979; Salnikow *et al.*, 1981; Walker *et al.*, 1982) and reviews on sequencing techniques (Wittmann-Liebold & Kimura, 1984; Wittmann-Liebold & Ashman, 1985).

12.2. MATERIALS

12.2.1. Solvents

It is not imperative that the solvents used for solid-phase sequencing be of high purity. In most cases, solvents of analytical grade are adequate. Sequencing has been performed by different laboratories using solvents from various suppliers, with no noticeable difference in results.

However, microsequencing demands solvents free of contaminants which can give rise to by-products that affect the identification procedure(s) employed. Therefore, purification procedures are described below. Routine tests of their purity should be performed, as their quality varies from batch to batch, even when purchased from the same supplier. Their quality also depends on storage time.

Tests for aldehyde and peroxide should be performed and it is advisable to carry out attachment experiments without peptide, using the resulting resin in a sequencer run. If the gas chromatogram or the HPLC trace shows no extra peaks and no spots are visible on thin-layer sheets when the same quantities and sensitivity as with peptides are used, then the solvents (and reagents) are of satisfactory quality. Further, test runs with polypeptides of known structure are recommended.

12.2.1.1. Purification procedures

The following procedures are used (Wittmann-Liebold, 1981; Wittmann-Liebold *et al.*, 1975).

Acetone: pro analysis grade (Merck):

2.5 l passed through a column (i.d. 35 mm) of silica (height 80 mm) and neutral Al_2O_3 (height 100 mm) (grade Silica 100–200 active; Alumina N, activity I, Woelm Pharma, Eschwege) and redistilled over glass, b. p. 56°C. The solvent is kept over molecular sieve, type 13 x, 1/16″ pellets (Packard) or type 3 Å pellets, 2 mm (Merck).

Acetonitrile (CH_3CN): pro analysis grade (Merck):

purified as described for acetone, b. p. 81–82°C. Stored under nitrogen at 10°C.

1,2-Dichloroethane: pro analysis grade (Merck):

purified and stored as described for acetonitrile, b. p. 84°C.

Dimethylformamide (DMF): pro analysis grade (Merck):

redistilled as follows:

Add to 263 ml of DMF, 34 ml of benzene and 12 ml of H_2O. Distil and discard the first 60 ml of distillate up to 150°C. The product over 150°C is collected and redistilled over P_2O_5 under reduced pressure (water pump). DMF is stored over molecular sieve (see acetone) under nitrogen.

Ethanolamine: synthesis grade (Merck):

redistilled at 15 mm Hg at 77°C, mol. wt. 61.08.

Stored in ampoules under nitrogen at −20°C.

N-ethylmorpholine (NEM): synthesis grade (Merck):

distilled twice, over sodium borohydride and ninhydrin, and redistilled over a 30 cm column filled with glass rings, b. p. 138°–139°C. Stored in ampoules under nitrogen at −20°C.

Methanol: Uvasol grade (Merck):
purified and stored as described for acetonitrile, b.p. 64°C.
N-methylmorpholine (NMM): synthesis grade (Merck):
purified as described for N-ethylmorpholine, b.p. 115–116°C.
Stored in ampoules under nitrogen at −20°C.
Pyridine: pro analysis grade (Merck):
distilled successively over KOH, ninhydrin and KOH, b.p. 114°–116°C.
Stored in ampoules under nitrogen at −20°C.
Toluene: pro analysis grade (Merck):
purified as described for acetone, distilled from $CaSO_4$. 0.5 H_2O (dried at 500°C immediately before use); b.p. 110–111°C.
Stored over molecular sieve (see acetone).
Triethylamine (TEA): synthesis grade (Merck):
distilled over phthalic anhydride, b.p. 89°C.
Stored in ampoules under nitrogen at −20°C.
Triethylenetetramine: practical grade (Fluka):
purified as described for triethylamine, distilled under reduced pressure (water pump).
Stored in ampoules under nitrogen at −20°C.
Trifluoroacetic acid (TFA): purum grade (Fluka):
redistilled from $CaSO_4$. O.5 H_2O (dried at 500°C immediately before use) over a column filled with glass rings, b.p. 72–73°C.
Stored in 50 ml aliquots in sealed glass stoppered flasks under nitrogen at −20°C.
Water: twice glass distilled, use freshly prepared.

12.2.2. Reagents

Methylisothiocyanate (MITC): purum grade (Fluka).
p-Phenylisothiocyanate (PITC): purissimum grade (Fluka):
redistilled under reduced pressure (oil pump) under nitrogen, while keeping the distillate at 4°C. Stored under nitrogen in small ampoules at −20°C.
p-Phenylene diisothiocyanate (DITC): (Eastman):
If the preparation has a yellow colour, it is recrystallized from boiling acetone: 2 g of DITC is dissolved in 10 ml of acetone and allowed to cool slowly, yield 1.55 g of needles which give a clear, colourless solution in DMF, m. p. 129–131°C.
4-*N*, *N*-dimethylaminoazobenzene-4′-isothiocyanate (DABITC):
(Fluka or Pierce), recrystallized from boiling acetone: 1 g dissolved in 70 ml, filtered through a paper filter and allowed to cool slowly; yield about 0.7 g of orange-red or brown needles which should have m. p. 169–170°C.
1-ethyl-3-(3-dimethylaminopropyl)-carbodiimide/HCl (EDC): (Serva),
m. p. 112–115°C.

3-aminopropyl triethoxysilane: (Pierce)
N-β-aminoethyl-(3-aminopropyl) trimethoxysilane: (Pierce).

12.2.3. Glass beads for dilution of resins

The following glass beads are available:

Reflex beads (Kristall Dragon Werk, Georg Wild, Bayreuth, Germany no. 31/18).
Glass beads, 200–235 mesh (Microbeads Division of Cataphote Corporation, Jackson, Missippi).

The untreated beads are washed with petroleum ether and then heated in conc. HCl for about 2 h to remove any contaminating metal ions. The beads are washed thoroughly with water and methanol on a sintered glass filter and dried at 100°C.

Acid-washed, trimethylsilyl-treated glass beads (Pierce) are available and ready for use. It is claimed that they give improved column flow rates. However, unsilanized beads are widely used.

12.2.4. Supports

Polystyrene beads (styrene/1 % divinylbenzene copolymer):

Biobeads S-X1, minus 400 mesh (BioRad Laboratories). Immediately before use, the beads are washed several times with benzene, chloroform, dioxane and methanol. They are stored at 4°C.

Biobeads S-X1, 200–400 mesh, chloromethylated (BioRad Laboratories).

Aminopolystyrene (APS) and triethylenetetramine polystyrene (TETA) resins (Pierce):

As the quality of the available supports varies and their stability in air is limited, it is recommended that these resins from polystyrene beads be prepared (see below) and stored under nitrogen in ampoules at −20°C.

Controlled pore glass; Corning CPG-10, 200–400 mesh, (Electronucleonics, Fairfield, N. J.; Serva, Heidelberg; Pierce, Rockford, Ill., Macherey & Nagel, Düren).

CPG is available with different particle and pore sizes. The most commonly used beads have 40–80 μm diameter (mesh size 200–400) with a nominal pore diameter of 75 Å. However, more recently, beads with larger pore diameters have been used (85–240 Å). Obviously, depending on the chain length of the polypeptide to be attached, larger pore sizes may be more suitable for the attachment of proteins (Machleidt & Wachter, 1977). However, our own observations do not support this.

Aminopropyl glass:

APG 10/75, 200–400 mesh (Pierce)

β-N-aminoethyl-(3-aminopropyl) glass:
β-APG, 200–400 mesh (Pierce).
Only colourless APG or *β*-APG should be used; our own preparations (see below) give more reproducible results and better attachment yields.

12.2.5. Attachment buffers

Attachment buffer 1: *N*-methylmorpholine/TFA buffer, pH 9.5:
Mix 5 ml of NMM and 5 ml of H_2O giving pH 11.4.
Add about 4 drops of TFA and adjust to pH 9.5.
Use freshly prepared and keep under nitrogen.

Attachment buffer 2: Pyridine/HCl, pH 5.0:
Add 8 ml of pyridine to 80 ml of H_2O.
Add 5.3 ml of 32% HCl and H_2O to a final volume of 100 ml at pH 5.0.
Keep under nitrogen.

Attachment buffer 3: Anhydrous buffer, DMF/triethylamine:
Add 2 ml of triethylamine to 200 ml of DMF.
The attachment under anhydrous conditions in DMF without any addition of TEA is also in use (see carboxyl-attachment).

Attachment buffer 4: Bicarbonate buffer, pH 9.0:
0.1 M $NaHCO_3$ is adjusted to pH 9.0 with 1 M-NaOH and contains 10% 2-propanolamine (v/v).

12.2.6. Reagent solutions and buffers in the sequencer

The purified reagents and solvents are as previously described.

Coupling buffer 1: Pyridine/*N*-methylmorpholine/trifluoroacetic acid:
28 ml of NMM are dissolved in 190 ml of H_2O.
Add 3.75 ml of TFA and water to a final volume of 200 ml at pH 8.1.
Add pyridine to the NMM-buffer, pH 8.1 in the ratio 3:2, (v/v) final pH is about 7.9.
The buffer should be freshly prepared and kept under nitrogen.

Coupling buffer 2: Anhydrous buffer, DMF/triethylamine:
Add 2.5 ml of TEA to 97.5 ml of DMF.

Coupling buffer 3: This buffer is used for micro sequencing with DABITC/PITC.
Pyridine/*N*-ethylmorpholine/trifluoroacetic acid/DMF:
4.5 ml of NEM are dissolved in 22.5 ml of H_2O under nitrogen.
Add a few drops of TFA and water to a final volume of 30 ml at pH 9.1.
Add 20 ml of DMF and 20 ml of pyridine to 10 ml of the NEM buffer, degas and keep under nitrogen.

MITC-solution: Methylisothiocyanate in acetonitrile:
Dissolve 1 g of MITC in 1 ml of acetonitrile.

PITC-solution 1: 10% PITC in acetonitrile:
Add 5 ml of PITC to 50 ml of acetonitrile under nitrogen.

PITC-solution 2 (for microsequencing with DABITC/PITC): 5% PITC in DMF:
Add 1 ml of PITC to 19 ml of DMF under nitrogen.
DABITC-solution (for microsequencing): 0.5% DABITC in DMF:
Add 100 mg of DABITC to 20 ml of DMF under nitrogen.
Prepare freshly every 24 h.
Solvent for microsequencing
Methanol/0.2% triethylamine:
Add 1 ml of TEA to 500 ml of MeOH under nitrogen.

12.3. PREPARATION OF RESINS

12.3.1. Polystyrene resins

Starting material for the synthesis of derivatized polystyrene resins is a styrene/1% divinylbenzene copolymer, Bio-Beads S-X1 (BioRad). Fig. 12.1 shows the steps of synthesis for some resins. The sequencing resins must display two essential properties: the functional reactive groups must be accessible to the peptides and reagents and the resin must be chemically and mechanically stable to the conditions of the Edman degradation.

12.3.1.1. Swelling properties

The swelling properties of the resin are determined in a small flat-bottomed tube and the height of the resin is measured before and 1 h after the addition of the solvent (Laursen, 1971). The swelling ratio V_S/V_D, where V_D is the dry volume and V_S the swollen volume of the resin, should be at least 3 to 5, to ensure that the reagent can penetrate well into the hydrophobic resin. After synthesis of polystyrene resins the swelling properties should be checked against the values in Table 12.1. The beads of the ideal sequencing support should be small and uniform in size for packing in a chromatography column and unaffected in swelling volume and stability by the different solvents and reagents used. However, resins which swell appreciably in pyridine, dimethylformamide, 1,2-dichloroethane, methanol and trifluoroacetic acid have the best properties for the degradation. These requirements are only partially met by cross-linked polystyrene resins and changes in volume with these resins during the degradation cycle lead to column blockage or unequal penetration of the reagents and solvents. Therefore, these resins are diluted with glass beads with a 30-fold excess by weight in the reaction column (section 12.5.2.).

12.3.1.2. Preparation of aminopolystyrene

The method is mainly as described (Laursen *et al.*, 1972; Laursen, 1975b) with some alterations.

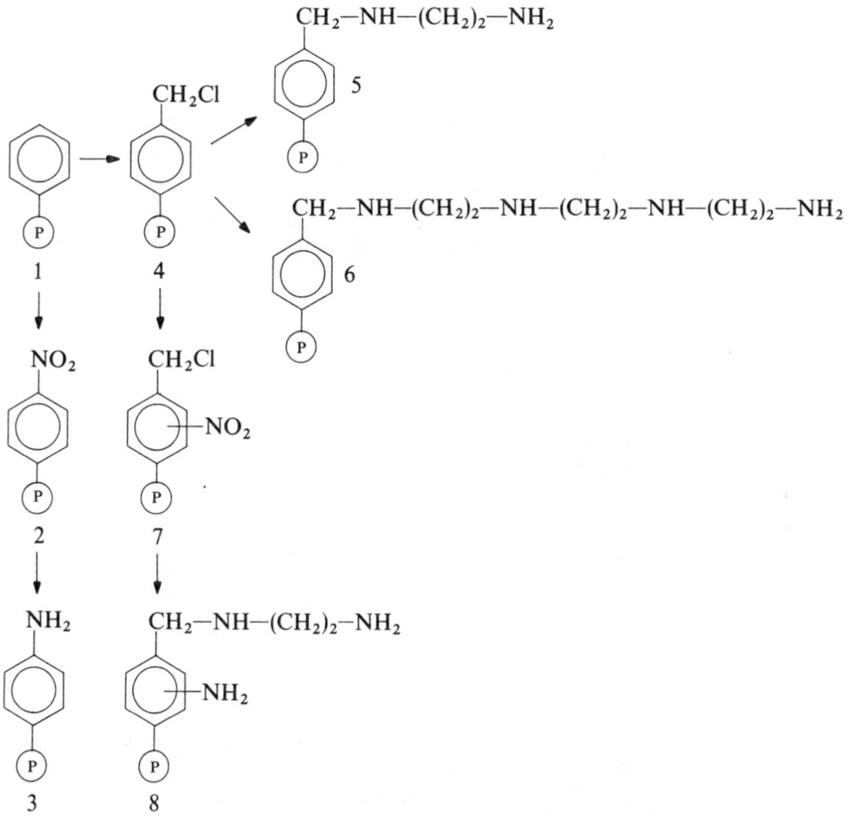

Figure 12.1. Stages for preparing resins from cross-linked polystyrene: 1. cross-linked polystyrene; 2. nitro-polystyrene; 3. amino-polystyrene; 4. chloromethyl-polystyrene; 5. ethylenediamine polystyrene; 6. triethylenetetramine polystyrene; 7. nitrochloromethyl polystyrene; 8. aminoethylenediamine polystyrene. The preparation of 3 and 6 is described in the text. For the synthesis of the other resins see Laursen (1975a).

Synthesis of nitropolystyrene

Wash 5 g of polystyrene (Biobeads S-X1, minus 400 mesh) thoroughly with benzene, chloroform, dioxane and MeOH in that order (with 200 ml of each for 30 min). Wash through a 60-micron metal sieve with MeOH to remove and discard larger particles and dry the beads *in vacuo* (yield approximately 4 g). Cool 55 ml of fuming nitric acid (90 %) in a cooling bath to -3°C and add the resin in small portions over a period of 15 min under stirring. The temperature must be kept below -3°C. Stir the mixture for 1 h at 0°C. Pour the mixture onto 500 ml of ice and collect the resin on a sintered glass filter. Wash alternately with dioxane

Table 12.1. Swelling properties of polystyrene resins. The dry volume (V_D) and the swollen volume (V_S) of the resin and the swelling ratio V_S/V_D are determined as follows (after Laursen, 1971): measure the volume of 100 mg of each resin after centrifugation, treat with 2 ml of each solvent, stir, keep at ambient temperature for 60 min and centrifuge to measure the swollen volume. The V_S/V_D ratios are listed below.

Solvent		MeOH		PYR		DMF		TFA	
	V_D	V_S	V_S/V_D	V_S	V_S/V_D	V_S	V_S/V_D	V_S	V_S/V_D
Biobeads S-X1 minus 400 mesh	140 μl	200 μl	1.4	750 μl	5.4	400 μl	2.3	200 μl	1.4
Biobeads S-X1 chloromethylated 200–400 mesh	150 μl	170 μl	1.1	600 μl	4.0	550 μl	3.7	300 μl	2.0
Aminopolystyrene*	155 μl	200 μl	1.3	250 μl	1.6	700 μl	4.5	850 μl	5.5
TETA polystyrene†	180 μl	500 μl	2.8	600 μl	3.3	500 μl	2.8	650 μl	3.6

* Preparation described section 12.3.1.2.
† Preparation described section 12.3.1.3.

(4 × 200 ml) and water (4 × 200 ml). Wash with MeOH (2 × 200 ml). Pass through a 60-micron sieve to disperse clumped resin particles. Dry the nitropolystyrene resin *in vacuo*. Yield 6–7 g.

Synthesis of aminopolystyrene

Stir suspension of 6.3 g of nitropolystyrene with 115 ml of DMF at 140°C (oil bath). Add slowly 48 g of $SnCl_2 \cdot 2H_2O$ in 40 ml of DMF (exothermic reaction) and finally maintain the temperature at 140°C for 30 min. Cool the mixture to 100°C and add 41 ml of conc. HCl. Heat at 95°C for 1 h and collect the resin on a sintered glass filter. Wash with 2 M-HCl (2 × 200 ml) and with H_2O (2 × 200 ml) and pass through a 100-micron sieve to disperse clumped resin particles. Repeat the washing with 2 M-HCl and water and then DMF/triethylamine (3 : 1, v/v; 2 × 200 ml) until no chloride ions are detected in the washings. The black-green coloured resin is washed with water and again passed through a 100-micron sieve. Finally, wash with MeOH (2 × 200 ml) and pass through a 60-micron sieve. Dry the yellow-brown resin at 60°C *in vacuo*. Yield: 5.5 g.

Storage of the resin

Aminopolystyrene resin is only stable for about one month at ambient temp. Store small amounts in sealed ampoules under nitrogen at −20°C. The ampoules may be stored for up to a year.

Swelling properties

The resin swells easily in DMF and TFA, but not so well in pyridine and MeOH (Table 12.1).

12.3.1.3. Preparation of triethylenetetramine polystyrene

Method

The preparation of TETA resin is mainly as described (Horn & Laursen, 1973; Laursen, 1975b).

Wash chloromethylated Biobeads S-X1 (BioRad) with benzene, chloroform, dioxane and methanol and dry *in vacuo*. Stir 1 g of freshly washed resin with 15 ml of triethylenetetramine for 30 min at room temperature and then heat at 115°C (oil bath) for 90 min. Collect the resin on a sintered glass filter and wash thoroughly with 20 ml of triethylamine for 30 min at room temperature. Filter. Wash alternately with MeOH and water (3 × 200 ml each), wash with H_2O and centrifuge. Boil the resin in 50 ml of 1 M-HCl for 2 h to remove traces of unbound TETA. Filter. Wash with water, triethylamine, water and MeOH (2 × 1 ml each) and dry *in vacuo* overnight. Yield ≃ 1 g.

Storage

As for aminopolystyrene resin. The resin is not as stable as aminopolystyrene. After storage for some months, background contamination rises during Edman degradations with this resin. Swelling properties are good (Table 12.1).

Washing of resin directly before use

Wash the resin with DMF (2 × 1 ml), then with MeOH (2 × 1 ml). Stir the resin with 15 ml of triethylamine for 20 min at room temp. Wash alternately with H_2O and MeOH. Finally wash with MeOH and dry *in vacuo*.

12.3.2. Glass supports

Porous glass beads as a support for solid-phase sequencing were introduced in 1973 (Machleidt *et al.*, 1973; Wachter *et al.*, 1973, 1975).

12.3.2.1. Properties of Corning controlled pore glass

Corning controlled pore glass (CPG) is a 96% silica glass, containing 3–4% B_2O_3 and 0.5–1% Na_2O, plus trace amounts of several metal oxides. CPG is available untreated or with several different surface treatments, which prepare it for specific procedures.

Corrosion rate of CPG

CPG is very stable but the corrosion rate is a function of temperature, time, pH, composition, volume of the solution and the surface area of the glass (the smaller the pore size the greater the surface area). Glass beads are sensitive to hydrogen fluoride. A decreasing stability of glass supports is observed over a period of time due to TFA.

Advantages of glass supports for sequencing

Porous glass beads retain a virtually constant volume throughout the degradation cycle and they can be employed undiluted in homogenous columns, without swelling or shrinking problems. They are not sensitive to pressure in columns and they allow high flow rates. They are chemically stable except in the presence of strong bases, especially at high temperature. Therefore, the attachment conditions for peptides to a glass support and the coupling buffers should be kept below pH 10.

A novel use of glass supports was described by Laursen (1977a). Protein was coupled by its amino terminus to a glass support and then partially degraded with

CNBr. The peptide mixture released from the support was then analyzed by a double label dansyl method. The technique allowed the alignment of CNBr peptides in proteins. Other cleavage methods could be applied.

12.3.2.2. Preparation of amino-glass supports

For the preparation of aminopropyl glass (APG) and β-*N*-aminoethyl-(3-aminopropyl) glass (β-APG) controlled pore glass beads (CPG 10/75, 200–400 mesh) were reacted with 3-aminopropyl-triethoxysilane (APTS) or with β-aminoethyl-(3-aminopropyl)-trimethoxysilane (AAPTS) (Robinson *et al.*, 1971; Wachter *et al.*, 1973; Bridgen, 1975; Frank, 1979). These glass beads with nominal pore diameter 75 Å, led to constant binding capacities of 150–170 nmol/mg CPG (Wachter *et al.*, 1973). See Fig. 12.2.

Method

Degas 4 g CPG beads (CPG 10/75, 200–400 mesh, Serva) 2 h at 180°C *in vacuo* (water pump), cool *in vacuo*. Add 30 ml of dry toluene and 3 ml of APTS or AAPTS. Degas and heat at 75°C for 24 h (in stoppered flask) with gentle stirring, preferably under nitrogen. Filter the resin on a sintered glass filter and wash alternately with toluene, acetone and MeOH (2 × 50 ml each). Dry *in vacuo* over P_2O_5 at room temperature. Store at 4°C under nitrogen.

12.3.2.3. Preparation of isothiocyanate glass

To synthesize an isothiocyanate aminopropyl glass support (DITC-APG) to which lysine peptides and peptides containing aminoethylcysteine can be bound, the amino groups of the aminopropyl glass are activated by reacting them with *p*-phenylene diisothiocyanate (Fig. 12.2) (Wachter *et al.*, 1973; Machleidt *et al.*, 1975).

Method

25 mol of DITC/mol of amino group and 2–3 vol of DMF per vol of glass are recommended for the reaction. 1 g of DITC is dissolved in 13 ml of DMF. Add 2 g of APG in portions with gentle stirring over 1 h. The mixture is then kept at room temperature for 2 h. The beads are washed on a sintered glass filter with acetone and DMF (2 × 50 ml), and MeOH (2 × 10 ml). The resin is dried *in vacuo*.

Storage

As for aminopolystyrene resin.

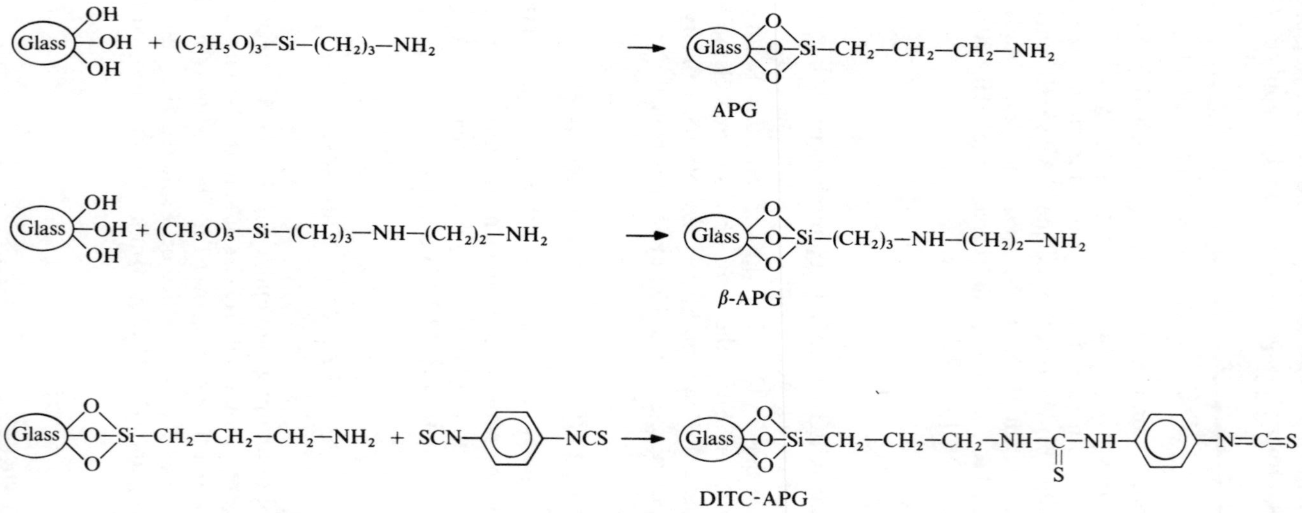

$$\text{Glass}(\text{OH})_3 + (C_2H_5O)_3\text{–Si–}(CH_2)_3\text{–}NH_2 \longrightarrow \text{Glass}(\text{O})_3\text{Si–}CH_2\text{–}CH_2\text{–}CH_2\text{–}NH_2$$

APG

$$\text{Glass}(\text{OH})_3 + (CH_3O)_3\text{–Si–}(CH_2)_3\text{–NH–}(CH_2)_2\text{–}NH_2 \longrightarrow \text{Glass}(\text{O})_3\text{Si–}(CH_2)_3\text{–NH–}(CH_2)_2\text{–}NH_2$$

β-APG

$$\text{Glass}(\text{O})_3\text{Si–}CH_2\text{–}CH_2\text{–}CH_2\text{–}NH_2 + \text{SCN–}C_6H_4\text{–NCS} \longrightarrow \text{Glass}(\text{O})_3\text{Si–}CH_2\text{–}CH_2\text{–}CH_2\text{–NH–C(=S)–NH–}C_6H_4\text{–N=C=S}$$

DITC-APG

Figure 12.2. Preparation of aminopropyl glass (APG), β-N-aminoethyl-(3-aminopropyl) glass (β-APG) and isothiocyanate glass (DITC-activated glass).

12.3.3. Determination of amino groups on resins or glass beads

Determination of the amino content will allow the stability of the support to be measured over a period of time and comparisons to be made with different batches of the same resin or other supports. See also Chapter 8, section 8.21 for other methods.

12.3.3.1. Methods

Method 1: Laursen, 1975b

The amino groups of the support are alkylated with trinitrobenzene sulphonic acid. Hydrolysis with KOH cleaves the Si–O bonds with the glass and also converts the trinitrophenyl groups to picric acid, which may be measured. The method is reproducible to within a few percent.

> Mix 50 mg of amino glass with 15 mg of trinitrobenzene sulphonic acid in 1 ml of 0.1 M-$NaBO_4$, pH 9.0. Mix by horizontal rotation in a small test tube for 1.5 h. Wash with H_2O (3 × 8 ml) by centrifugation. Add 1 ml of 2 M-KOH to the moist resin and keep at 45°C for 60 min with stirring. Dilute with water and measure at 358 nm.

The number of amino groups is calculated using $\varepsilon_{358} = 14\,100$ for picric acid at alkaline pH. Typical loadings are cited as 20–150 μmol amino group per g of amino glass (Laursen, 1975a).

Method 2: Gisin, 1972

The swollen resin is reacted with picric acid, which forms a salt with amino groups. The resin is treated with an excess of strong base to release the picrate from the bound form into solution where it may be estimated.

> Swell 50 mg of resin in 1,2-dichloroethane for 5 min. Neutralize with 5 % (v/v) diisopropylethylamine in dichloromethane for 3 min. Treat with 0.1 M-picric acid in dichloromethane for 3 min. Wash with dichloromethane (5 × 2 ml). Elute the picrate with 5% diisopropylethylamine in dichloromethane (2 × 3 ml). Wash with dichloromethane (3 × 2 ml). Dilute with 95 % ethanol and measure spectrophotometrically.

The molar absorption of diisopropylethylamine picrate is constant ($\varepsilon_{358} = 14\,500$) in the concentration range 1–20 × 10^{-5} M if the ethanolic solution contains less than 20 % dichloromethane.

Method 3; Schmitt & Walker, 1977

The amino support is converted to an immobilized Schiff's base by means of

2-hydroxy-1-naphthaldehyde which is then made soluble by benzylamine and measured spectrophotometrically.

> Mix 1 to 5 mg of resin with 2 ml of 0.2 M-2-hydroxy-1-naphthaldehyde in DMF and react 15 h at room temp. Wash the resin with DMF (4 × 3 ml) and then with ethanol until the u.v. absorption at 270 nm of the washes is zero. Add 2 ml of 0.4 M-benzylamine in ethanol and keep 45 min at room temperature. Centrifuge. Measure the absorption at 420 mn; the released Schiff's base in ethanol has $\varepsilon_{420} = 10\,900$.

The 'coupling capacity' of solid supports containing amino groups was defined as the number of amino groups per unit weight of resin available under the conditions of coupling.

12.4. ATTACHMENT METHODS

12.4.1. DITC-method for attachment of lysine peptides

Laursen *et al.* (1972); Wachter *et al.* (1973). Peptides containing lysines positioned either internally or at the carboxyl-terminal end, as well as peptides containing aminoethylcysteine, can be effectively coupled *via* their free amino groups to aminopolystyrene resins or to aminopropyl glass (APG and β-APG) with DITC (Fig. 12.3). A variation of the DITC-method utilizes *N*-(p-isothiocyanato-benzoyl)-DL-homserine lactone as bifunctional reagent for the attachment of lysine peptides to amino glass supports and this was reported to result in high attachment yields (Herbrink *et al.*, 1975).

12.4.1.1. Advantages, limitations and attachment procedure

Advantages

Attachment yields are usually excellent, about 80–90 %. In comparison with some other polystyrene resins, when aminopolystyrene-bound peptides are degraded by solid-phase techniques there is little background contamination due to side reactions. In the past, small peptides up to 25–30 residues were best linked to the aminopolystyrene resin, whereas bigger fragments and proteins were preferably bound to glass supports (APG and β-APG). More recently, the attachment of smaller sized peptides to glass beads has been improved (see 12.4.2–12.4.4).

Limitations

1. The attachment conditions may cause solubility problems with some peptides with resulting low attachment yields. To prevent losses, control tests should be performed during and after the attachment procedure (see 12.4.5). Some of the

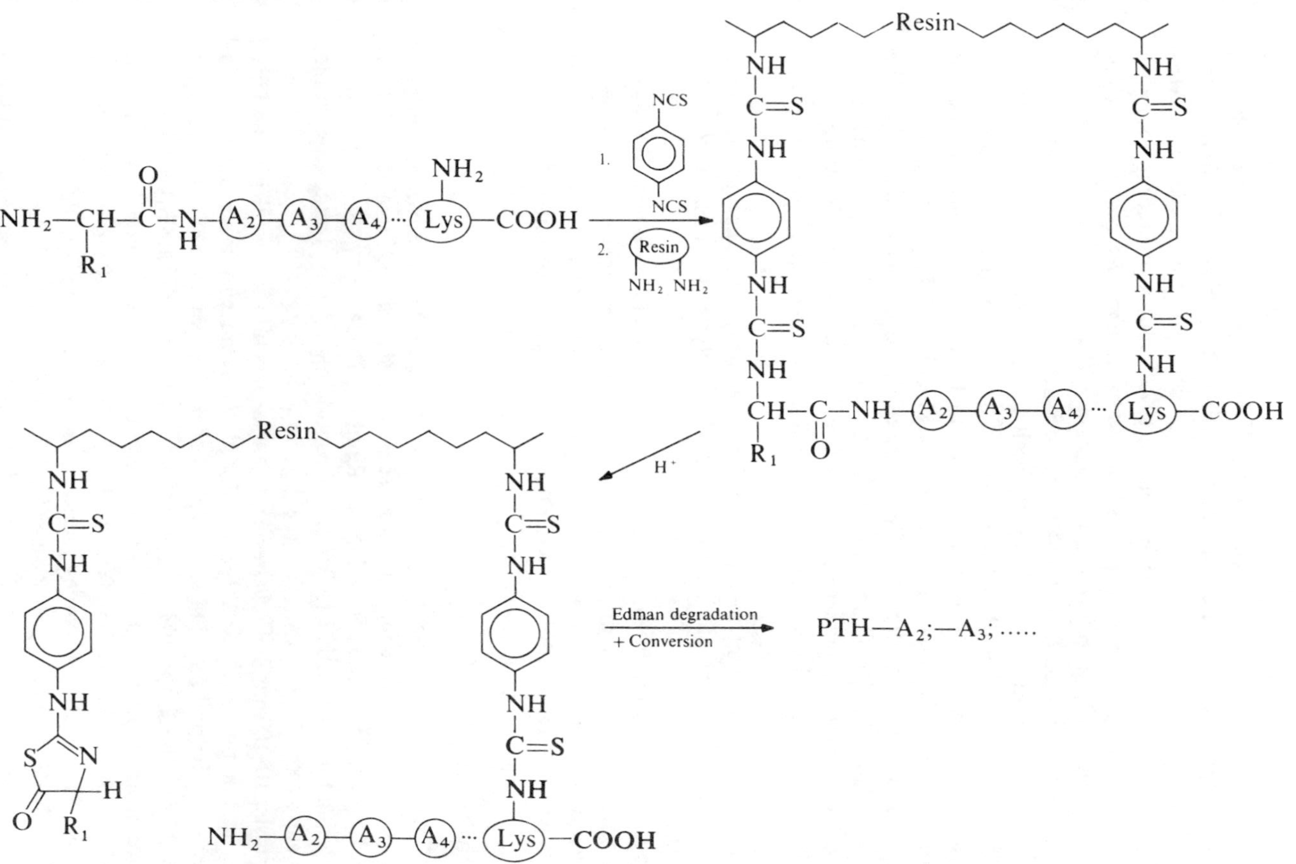

Figure 12.3. Attachment of peptides to amino supports by activation of their α- and ε-amino groups with p-phenylene diisothiocyanate.

peptides, which are insoluble in the attachment buffer described below, can be solubilized by the addition of pyridine or DMF. If this fails, a change in the pH of the buffer might increase the solubility (pH 7.5 to pH 10 may be employed). Sometimes peptides become soluble after they have been treated with trifluoroacetic acid and taken to dryness. Attachment with the anhydrous DMF-triethylamine buffer (Wachter & Werhahn, 1979) helps to overcome solubility problems. Lysine peptides can also be attached to aminopolystyrene by means of activation with EDC at pH 3 to 5 (see section 12.4.4).

2. The *N*-terminal amino acid and lysine residues are bound covalently to the resin and cannot be identified. However, small quantities of the *N*-terminal amino acid can often be detected, if the attachment of the *N*-terminal amino group is not complete. In this case, the methylthiohydantoin derivative of the first amino acid is formed if methylisothiocyanate is used for blocking of the excess resin amino groups, as done usually after the attachment of the peptide (Laursen, 1972). We replaced MITC by PITC so that the first residue could be identified as its phenylthiohydantoin-amino acid derivative. In microsequencing, the degradation yields are not affected by this treatment, provided that redistilled PITC has been used.

3. The DITC-attachment method is difficult with arginine peptides. Arginine residues are converted to ornithine by hydrazinolysis (Laursen *et al.*, 1972; Morris *et al.*, 1973).

However, during treatment with hydrazine, although different conditions were tried, internal cleavages of peptide bonds (such as asparagine bonds) were frequently observed, which gave misleading results during the degradation. This attachment method cannot be recommended for arginine peptides.

Attachment procedure

Coupling at pH 9.5 to aminopolystyrene or amino glass beads, according to Laursen *et al.*, 1972 (for modifications see Terhorst *et al.*, 1973; Wachter *et al.*, 1973; Laursen, 1975b, 1975c; Schiltz & Reinbolt, 1975).

Loading the resin with 1 to 3 nmol of peptide/mg resin or 1 nmol/mg amino glass support.

Transfer 10–100 nmol of *salt-free* peptide to a glass tube (5.0 × 130 mm) and dry *in vacuo* (oil pump) preferably in a Speed Vac Concentrator (Savant) or in desiccator or lyophilize. Redissolve in 50 μl of attachment buffer 1 under nitrogen (see 12.2.5). Check the pH, it should be > 9.0. Check solubility of peptide. Withdraw an aliquot (2–5 μl) for TLC control-test (see section 12.4.5.1). Add 1.0 mg of DITC (5 μmol) in 100 μl of DMF and stir for 30 min at 50°C under nitrogen. The solution becomes yellowish and no sediment should appear.

Wash 30–50 mg aminopolystyrene (or 100 mg aminopropyl glass) with DMF and then swell in 200 μl of DMF for 15 min at room temperature. Add this pretreated resin to the DITC-activated peptide in 2 portions. Stir the mixture for

60 min at room temperature under nitrogen. To react excess amino groups of the resin add 100 μl of PITC-solution 1 and 100 μl of coupling buffer 1 (see 12.2.6). Stir for 60 min at 30°C under nitrogen. Centrifuge. Keep the supernatant to test for remaining free peptide (see 12.4.5.2). Wash the resin with MeOH (3 × 8 ml). Centrifuge. Dry the resin *in vacuo*. In the case of polystyrene resin mix with 900 mg of glass beads (see 12.2.3.) and pack into the column.

12.4.2. DITC-method using isothiocyanate glass

12.4.2.1. Advantages, limitations and attachment procedure

Advantages

1. Proteins and large peptides give higher attachment yields if glass supports are used instead of polystyrene resins.
2. The solubility of the peptide in the attachment buffer is a problem which can be overcome. In cases where the peptide is not soluble in the normal buffer, addition of pyridine and/or DMF may be tried before the addition of the resin; therefore the peptide is not lost. When using amino supports and activation of the peptide by DITC (see section 12.4.1) the reagent tends to precipitate (as a urea derivative) depending on the amount of water needed to dissolve the peptide.
3. The method saves time.

Limitations

The method cannot be applied to peptides without lysines or (aminoethyl) cysteines. Peptides having these amino acids as internal residues can only be sequenced up to this position. The method cannot be recommended for arginine peptides.

Attachment procedure

According to Machleidt *et al.*, 1973; Wachter *et al.*, 1973; Machleidt & Machleidt, 1977; Machleidt & Wachter, 1977; Wittmann-Liebold & Lehmann, 1980.

Load the support with 1 nmol of peptide/mg of glass. Dissolve 10–200 nmol of peptide or protein in 30–100 μl of attachment buffer 1 under nitrogen. Check pH value, perform solubility test (see section 12.4.5). Add 10–100 mg of DITC-APG in dried form and stir for 60 min at 30°C under nitrogen. Add 50–100 μl of ethanolamine and stir for 60 min at 30°C under nitrogen. Wash extensively with DMF, MeOH and ether (2 × 2 ml each). Check supernatant for remaining peptide. Dry the resin *in vacuo*.

Modification

In cases where the peptide is not soluble in NMM-buffers the attachment can be performed in one of the following: 50% pyridine/water; DMF after the addition of triethylamine (attachment buffer 3) or *N*-methylmorpholine; sodium bicarbonate buffer, pH 9.0 (attachment buffer 4) with 1% SDS if necessary. The attachment yield of insulin B-chain in NMM–TFA buffer, pH 9.0 was found to be between 45 and 50%, whereas in sodium bicarbonate at pH 9.0 it was 66–70% (Chang, 1979).

12.4.3. Homoserine lactone method

Homoserine lactone-activated peptides are attached covalently to TETA-resin (Horn & Laursen, 1973; Laursen, 1977b). They can also be attached by this method to glass supports, APG or β-APG.

12.4.3.1. Advantages, limitations and attachment procedure

Advantages

1. Homoserine peptides are attached with high yields of 80 to 90%.
2. They can be degraded up to their carboxyl-terminal end.
3. The released amino acid derivatives have low levels of contamination, provided that the resin is stored under nitrogen in ampoules at −20°C and is washed thoroughly immediately before use.
4. The homoserine-containing peptides may be separated from the homoserine-free *C*-terminal peptide after cyanogen bromide cleavage by this method.

Limitations

1. The TETA resin is rather unstable and cannot be stored for long periods.
2. It has to be washed carefully before use.
3. TETA resin is less effective with lysine peptides than is aminopolystyrene and aminopropyl glass; more background is raised during repeated degradations. *N*-acetylimidazole treatment of peptides freshly coupled to TETA resins was reported to minimize the development of spurious ninhydrin positive spots during identification of the PTH-amino acid derivatives.

Attachment procedure

Prewashing the resin for loading
1 nmol peptide/mg of resin. Wash 50 mg of TETA-resin or 100 mg of aminopropyl glass with DMF (4 × 2 ml). In the case of TETA-resin wash also with triethylamine (2 × 1 ml). Wash with MeOH (2 × 2 ml) and with DMF (2 × 1 ml).

Lactonisation of peptide
Dry the peptide *in vacuo* over P_2O_5. Add anhydrous TFA (distilled) (50–100 μl/5–20 nmol; 0.5 ml/50–200 nmol) and allow to stand for 1 h at room temperature under nitrogen. Dry in a rotary evaporator or in Speed Vac Concentrator (Savant) and then *in vacuo* over KOH. Repeat the treatment with TFA for 15 min.

Attachment
Dissolve the peptide lactone in 175 μl of DMF and 25 μl of triethylamine; if the peptide is not soluble make the DMF-peptide 20 % with respect to triethylamine. Withdraw aliquot, 2–5 μl for test of solubility. Add the peptide solution to the pre-washed TETA-resin (or 20 mg aminopropyl glass) under nitrogen. Rinse the peptide tube with DMF (2 × 100 μl) and add to the resin mixture. Add 50 μl of triethylamine, sonicate and keep for 90 min at 45°C with gentle stirring under nitrogen. Wash in DMF and MeOH and centrifuge. Use the supernatant for peptide analysis.

Saturation
To the resin add 200 μl of PITC in acetonitrile (1 : 5, v/v) and 200 μl of coupling buffer 1 (see section 12.2.6). Stir for 30 min at 30°C under nitrogen. Centrifuge. Wash the resin with DMF (1 × 2 ml) and MeOH (2 × 2 ml). Dry the resin *in vacuo.*

12.4.4. Carboxyl attachment with carbodiimide

The carboxyl-terminal group of the peptide is activated by means of a water-soluble carbodiimide and covalently bound to amino-polystyrene, aminopropyl glass or β-APG. Originally, *C*-terminal activation was performed on *N*-protected peptides (Previero *et al.*, 1973) using TETA-resin and water-free alkaline conditions for the attachment at 40° to 50°C. Instead, the unprotected peptides are used and are activated under mild conditions to reduce the reaction of the side-chain carboxyl groups (Wittmann-Liebold & Lehmann, 1975; Wittmann-Liebold *et al.*, 1977; Beyreuther, 1977; Reinbolt *et al.*, 1977, 1979; Laursen *et al.*, 1980). Attachment at pH 3 to 5 was found to be superior to alkaline conditions when using aminopolystyrene resin. Alternatively, anhydrous conditions are used for the reaction with glass support (Wachter & Werhahn, 1979; Salnikow *et al.*, 1981). Further, a combined lysine- and carboxyl-attachment method has been employed (Schiltz & Reinbolt, 1975; Schiltz, 1977).

12.4.4.1. Advantages, limitations and attachment procedure

Advantages

1. The carboxyl method can be applied to almost all types of peptides including arginine- and lysine-containing tryptic peptides.

2. The acidic conditions for attachment are suitable for many peptides and offer an alternative for those tryptic lysine peptides which are not soluble under the alkaline attachment conditions used in the DITC-method.
3. In cases where a peptide contains an internal lysine whose position is unknown the *C*-terminal attachment method described here is more reliable.
4. This method can be applied with small samples of peptides for microsequencing.

Limitations

1. The attachment yield of small peptides varies between 40 and 70% (see Salnikow *et al.*, 1981). Peptides with *C*-terminal lysine couple to a low extent.
2. Larger peptides or proteins with more than 30 residues are bound to the resin in sufficient amounts, often 80 to 90%. Studies using bradykinin and insulin B-chain as standard test peptides show the former with 70–75% attachment and the latter about 60% (Wittmann-Liebold & Lehmann, 1980). With proteins the attachment yield can rise to 90%. However, the better they are bound covalently the worse their degradation yields become (probably because of steric hindrance). Therefore, care has to be taken not to quantitatively attach all of the carboxylic groups. We find it best if the protein is attached *via* these groups to the extent of about 10–30% on a theoretical basis with predominant binding of the *C*-terminal carboxyl group to the resin or glass. However, for proteins better degradation results are obtained with the DITC method (Lehmann and Wittmann-Liebold, 1984).

Procedure I: Attachment to aminopolystyrene:

Prewashing the resin
Add 1 ml of attachment buffer 2 (see section 12.2.5) to 30–50 mg of aminopolystyrene resin and stir at room temperature until a violet colour appears. Wash the resin with H_2O (2 × 8 ml) and DMF (2 × 1 ml) and centrifuge.

Attachment
Dry the peptide solution *in vacuo*. The peptide should not contain salts and must be free of organic acids, e.g. acetate and formate. Dissolve 5–50 nmol of peptide in 50 μl of attachment buffer 2 under nitrogen. Withdraw an aliquot for solubility test. Add the peptide solution to the prewashed resin. Rinse the peptide tube with 200 μl of DMF and add to the resin mixture. Repeat with 100 μl of DMF. Add 2 mg of EDC (dissolved in 20 μl of H_2O and 80 μl of DMF, freshly prepared) under nitrogen. Stir for 60 to 120 min at 30°C (in case of Glx- and Asx-containing peptides attach for 30–60 min only). Centrifuge. Use the supernatant to test for free unbound peptide. Wash the resin with DMF, methanol and coupling buffer 1 (2 × 200 μl).

Saturation
Add 200 μl of coupling buffer 1 and then 200 μl of PITC-solution 1 (see section 12.2.6) under nitrogen. Stir gently for 30 min at 30°C. Wash the resin with DMF (2 × 200 μl) and then with MeOH (2 × 200μl). Centrifuge. Dry the resin *in vacuo.*

Procedure II: Attachment to aminopropyl glass

Pretreatment of peptide
Dry 5–50 nmol of peptide *in vacuo.* Dissolve in 50–100 μl of anhydrous TFA and keep at room temperature for 15 min under nitrogen. Dry immediately *in vacuo* over KOH pellets for 20–30 min (prolonged drying might cause insolubility!).

Prewashing of APG
Wash APG (1mg/1 nmol peptide) with DMF (4 × 2 ml/50 mg).

Attachment
Add 2 mg of EDC freshly dissolved in 200 μl of DMF to the dried peptide. Sonicate mixture for 1 min. Add 50 mg of APG and degas. Incubate at 40°C for 60 min with gently stirring under nitrogen. Centrifuge. Analyse the supernatant for unbound peptide.
Wash the peptidyl glass with DMF and methanol (2 × 200 μl). Centrifuge and dry *in vacuo.*

12.4.5. Control of attachment

12.4.5.1. Solubility test

The peptides are routinely tested for solubility in the buffer employed for the attachment. Dissolve the peptide in the buffer. Withdraw an aliquot corresponding to 2 nmol. Dry *in vacuo* or apply directly onto a TLC plate (cellulose-coated Cel 300, Macherey and Nagel, Düren, Germany). Develop in pyridine/n–butanol/acetic acid/water (10 : 15 : 3 : 12, by vol.).

Spray with ninhydrin in acetone (3 g/100 ml). Only one spot should be obtained at an intensity corresponding to 2 nmol of the tested peptide. If aliquots of more than 2.5 μl are withdrawn from the peptide solution the salt must be removed by lyophilization. When microsequencing this test should be made by a more sensitive technique in order to save sample, e.g by HPLC or sensitive analysers.

12.4.5.2. Attachment control

After attachment of the peptide, the supernatant is tested for remaining unbound peptide by withdrawing an aliquot, hydrolysing it and performing amino acid analysis. In cases where peptide is found in the supernatant, the attachment procedure has to be repeated under different conditions. This can be performed after desalting the peptide if enough of the starting peptide material is recovered.

Attachment yield determination

A portion of the peptidyl resin or glass (10 mg) is washed with TFA (2 × 0.5 ml) to remove any non-covalently bound peptide and dried *in vacuo*. The resin is hydrolysed *in vacuo* with 250 μl of 12 M-HCl/propionic acid (1:1, v/v) for 24 to 48 h at 110°C (Scotchler *et al.*, 1970) or for 2 h at 130°C (Westall *et al.*, 1972) in ampoules under nitrogen. Thereafter the resin particles are removed by filtration, the solution evaporated to dryness and the residue redissolved in amino acid analyser citrate buffer, pH 2.2. The attachment yield is calculated from the amino acid analysis. Hydrolysis of resin-bound peptide can also be performed *in vacuo* with 12 M-HCl/acetic acid/phenol (2:1:1, by vol.) for 24 h at 110°C (Gutte & Merrifield, 1971).

Test runs in the sequencer

The performance of the machine and the purity of the solvent and reagents employed may be checked by using insulin B-chain and bradykinin as standard peptides, in the range 10 to 100 nmol per attachment (Wittmann-Liebold & Lehmann, 1980).

12.5. DEGRADATION IN THE SOLID-PHASE SEQUENCER

The construction of the solid phase sequencer is described elsewhere (Laursen, 1971, 1972, 1975b; Laursen *et al.*, 1975).

12.5.1. Column

A thick glass tube (230 mm × 2.5 mm internal diameter) is polished at both ends to ensure a good seal. The fittings are made from Kel F and Teflon (Laursen, 1971). The column is placed in a heated aluminium block and a constant temperature at 45°C is maintained by a thermostat.

12.5.1.1. Micro-column

For sequencing less than 20 nmol of peptide, miniaturized versions of this column are in use, e.g. a glass column of 15 mm × 2 mm internal diameter filled with peptidyl glass beads mixed with filling beads. Alternatively, the micro-column is made from Teflon tubing (Fig. 12.4) the internal volume of which corresponds to the amount of peptide-bound glass beads used (Salnikow *et al.*, 1982). The Teflon tubing is constricted at the bottom end and closed by the aid of an adaptor (made from KELF) with a 2 μm filter (porous Teflon, see Fig. 12.4). The dead-volume at

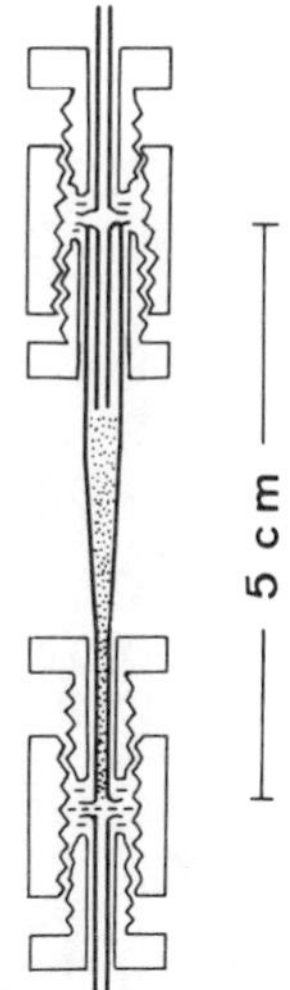

Fig. 12.4. Column device for micro-sequencing. The column is made from Teflon tubing, the size of which depends on the amount of support. Narrow-bore Teflon tubing is inserted at the top end and a porous Teflon filter, 2 μm (Zitex, Chemplast Inc., Wayne, N.J.) closes off the bottom end of the column. The micro-column is placed in a thermostated glass tube (see Fig. 12.6). (Salnikow *et al.*, 1981; Wittmann-Liebold, 1982).

the column inlet is reduced by the insertion of a short length of narrow bore Teflon tubing. The sequencing tubing is accommodated in a thermostatically controlled glass barrel (Fig. 12.6).

12.5.2. Packing of the column

Depending on the size of column and amount of peptide loaded resin it is mixed with up to 900 mg of glass beads (see section 12.2.3) and packed into the column (dry packing) (Laursen, 1975a). Micro-columns are filled without further dilution or the resin is sandwiched between layers of inert glass beads.

12.5.3. Sequencer programme

Sequencing programmes vary, depending on the apparatus and the solvents and coupling buffers chosen. Before starting the sequencer programme, the filled reaction column and the connecting tubes are manually washed with all solvents employed, e.g. methanol and 1,2-dichloroethane. In the two-column version of the sequencer the first column starts with a methanol-wash followed by the coupling reaction with the delivery of coupling buffer and PITC-solution. Simultaneously, the second column is started with a 1,2-dichloroethane wash, followed by rinsing with TFA. Alternately, the second column can be used for automatic conversion. The thiazolinone released into the second column after the first cycle is converted simultaneously with the coupling stage of the second degradation in the first column (see below). A programme for micro-sequencing is shown in Fig. 12.5 and its adaptation to a LKB-sequencer is given in Table 12.2.

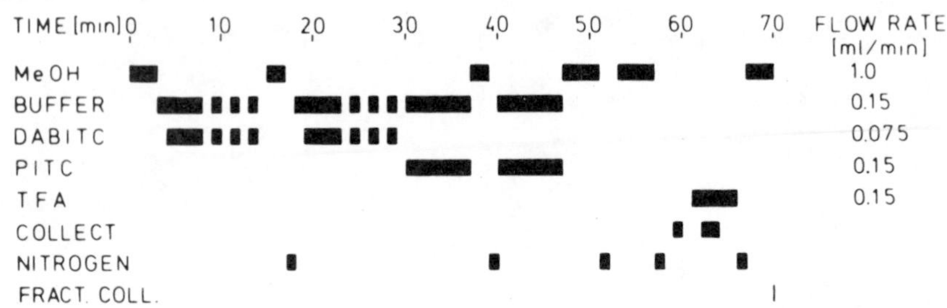

Fig. 12.5 Improved programme for solid-phase micro-sequencing with DABITC/PITC. MeOH, methanol with 0.2% propylamine; buffer, coupling buffer 3; DABITC, 0.5% in DMF; PITC, 5% in DMF; TFA, anhydrous; (Salnikow *et al.*, 1981).

Table 12.2. Modifications in a LKB solid-phase sequencer for DABITC/PITC method. A LKB 4020 solid-phase sequencer with a two-column system was adapted for microsequencing (Salnikow *et al.*, 1981). The programme is shown in Fig. 12.5. The modifications were as follows:

(a) A nitrogen purge line was inserted as shown in Fig. 12.6 for immediate nitrogen purge (1 bar) and drying of the sequencing column at coupling, cleavage and wash stages (shown in the programme).
(b) The original column was replaced by a micro Teflon tubing (see Fig. 12.4(a)) which is thermostated within a glass barrel (Fig. 12.4(b)) at 50°C.
(c) Column 2 is used for automatic conversion (30% TFA, 65 min at 50°C) by introduction of a conversion Teflon coil (40 cm × 1.5 mm) into which the thiazolinone derivative/TFA effluent of column 1 is pumped together with water (according to Birr, 1975).
(d) All connecting lines were shortened and miniaturized where possible.

12.5.4. Coupling

The most widely used buffer in the degradation is coupling buffer 1 (Laursen, 1971). Recently, this buffer has been replaced by an anhydrous buffer made from DMF with 2.5% of triethylamine or *N*-methylmorpholine (Machleidt *et al.*, 1975; Wachter & Wehrhahn, 1979). For sequencing with DABITC/PITC coupling buffer 3 (12.2.6) gave satisfactory results (see Fig. 12.5).

The sequencer buffers which are anhydrous or contain DMF are superior to the original solutions, probably because of the hydrophobic environment of the peptide bound to the resin. These buffers may be used with both glass and polystyrene supports. In general, the PITC reagent is dissolved in acetonitrile at a concentration of 5%. In combination with the DABITC/PITC degradation both reagents are dissolved in DMF (see Fig. 12.5), because DABITC is not so readily hydrolysed in this solvent.

12.5.5. Solvents

Usually 1,2-dichloroethane and methanol are used after coupling. Dichloroethane can be replaced by benzene. A recent programme employs only methanol for microsequencing (Salnikow *et al.*, 1981).

Improved yields were obtained with 0.2% propylamine in methanol and repeated nitrogen purges (see Fig. 12.5).

12.5.6. Cyclization

The solid-phase technique uses TFA for cyclization. The amount of acid is critical: short delivery causes incomplete cyclization and long treatment tends to increase unspecific cleavages of peptide bonds and results in increasing contamination of the released thiazolinones. A time of 8 min at 45°C is recommended. The optimum delivery time of the acid may be determined by using bradykinin which has proline residues in positions 2, 3 and 7. If there is less than 10% proline 'overlap' after the fourth degradation, then the time is satisfactory. Further, the yield of the third proline in position 7 should not be drastically decreased in comparison to the earlier ones.

The rinsing of the TFA becomes more effective if a nitrogen purge is provided prior to the TFA delivery (Machleidt *et al.*, 1975).

12.5.7. Conversion

12.5.7.1. Conversion medium

The released amino acid anilino-thiazolinones are converted to the amino acid phenylthiohydantoins (PTHs) in aqueous acids or in anhydrous methanolic HCl. The following methods are in use:

1. 1 M-HCl for 10–15 min at 80°C (Edman, 1950);
2. 20% TFA for 10 min at 80°C (Laursen, 1971);
3. 20% TFA for 20–30 min at 55°C (Wittmann-Liebold *et al.*, 1975);
4. 30–40% TFA for 45–65 min at 49–50°C (Salnikow *et al.*, 1981);
5. 1 M-methanolic HCl (made by adding 1 to 2 M-acetyl chloride to methanol with ice cooling) for 5–10 min at 55°C (Tarr, 1975; Horn & Bonner, 1977).
6. 2.4 M-methylamine in propan-2-ol (or methylamine gas) is proposed for the conversion to phenyl thiocarbamyl amino methylamides (Apella *et al.*, 1977).

The optimum conditions for the quantitative conversion of the individual amino acid thiazolinones to the PTH derivatives vary considerably. Glycine and lysine are converted at the lowest rate, whereas serine and threonine derivatives are converted quickly and undergo partial destruction under harsher conditions. Hence, a generalized procedure for the conversion reaction has to compromise between these extremes with attendant disadvantages. When sequencing costly peptides special treatment of individual cycles at the conversion stage is desirable.

The conversion in HCl is superior to TFA for the conversion of serine and threonine derivatives. Aqueous TFA is preferable for the conversion of glutamine and asparagine thiazolinones because desamidation of the amides to the free acids can be minimized, if the dilute acid is removed as quickly as possible after the conversion. On the other hand, conversion in dilute HCl has the disadvantage that glutamine and asparagine are almost entirely converted to the corresponding acids and therefore does not allow a safe differentiation.

The reaction in anhydrous methanolic HCl offers some advantages over aqueous conversion media, e.g. milder conditions and greater volatility of the reagent. A disadvantage, however, is that esterification of acid derivatives occurs which gives rise to two different derivatives in the identification procedure if this reaction is incomplete. This in turn depends on the stability of the conversion medium which is low at higher temperatures and should always be prepared freshly.

12.5.7.2. Manual conversion

In the solid-phase sequencers the released amino acid thiazolinone derivatives are collected in the tubes of the fraction collector in volumes of about 1.5 ml TFA and 3 ml of MeOH. In the manually-performed conversion the thiazolinones are stored overnight at room temperature in this mixture. However, the thiazolinones are sensitive to oxygen and hence partial destruction of several amino acid derivatives takes place, e.g. serine, threonine, arginine and histidine. Further, glutamic acid and aspartic acid derivatives are partly converted to their methyl esters if methanol is used as solvent after TFA. To avoid this problem 1,2-dichloroethane or DMF may be used as first solvent after cyclization to collect the thiazolinone. Further, a dilution procedure was recommended to improve the recovery of PTH-serine and -threonine (Wittmann-Liebold & Lehmann, 1975), as described below.

Add 3–7 ml of water to the tubes prior to the delivery of the amino acid thiazolinone. Next day, the contents of the tubes are evaporated in a rotary evaporator or in the Speed Vac Concentrator and subjected to the conversion.

Conversion procedure

Add 0.3 ml of 20 % TFA in water under a stream of nitrogen and keep for 30 min at 55°C under nitrogen. Dry *in vacuo* and keep under reduced pressure (oil pump) for 5 h to remove volatile by-products. Dissolve in MeOH and identify the PTH amino acids by gas-, thin layer- or high pressure liquid-chromatography.

Wachter and Werhahn (1979) used a heated desiccator for both the drying and conversion which simplified the process in their experience.

12.5.7.3. *Automatic conversion*

Devices were described which allow the quick conversion of the labile amino acid thiazolinones to their relatively stable PTHs in a nitrogen atmosphere. The advantages are that less destruction is observed with correspondingly cleaner background, less desamidation of glutamine and asparagine occurs (depending on the conversion medium employed), a higher yield of serine is obtained and time is saved. The first conversion device was designed as a double-walled, thermostated flask which allowed the temperature to be kept at 55°C and the quick removal of the acidic effluent by a stream of nitrogen (Wittmann-Liebold, 1973; Wittmann-Liebold *et al.*, 1976). A converter was described, utilizing a jacketed reaction cell (Bridgen, 1977a). Another device made use of a reaction coil, through which the sequencer effluent was passed with the addition of appropriate amounts of water to perform the conversion at 80°C (Birr, 1975). In this way the PTH derivatives were stored in acidic solution after their formation. An automatic converter is commercially available which in principle works similarly to the flask-type (Sequemat p-6 Auto Converter, for details see Horn & Bonner, 1977) and can be connected to either liquid- or solid-phase sequencer. It was claimed that PTH-glutamic acid and PTH-aspartic acid methyl esters were formed quantitatively without appearance of the free acids and asparagine and glutamine were not desamidated. However, using the commercial converter with this medium more than one derivative is formed and quantitation is complicated. More recently, adaptation of the second sequencer column provided in the solid-phase machine purchased from LKB has enabled automatic conversion without much cost or difficult modification (Salnikow *et al.*, 1981). Substantially, the conversion is made according to Birr (1975) using 30% TFA in water at 50°C for about 60 min in a programme adapted for micro-sequencing with DABITC/PITC (see Fig. 12.6 and Table 12.2).

12.5.7.4. *On-line HPLC identification*

Automation of the sequential degradation of polypeptides ideally requires automatic conversion of the amino acid thiazolinones and this becomes a necessary prerequisite for sequencing minute amounts of peptide. We have had wide experience with different types of sequencers and a further development includes the automated recording of the degradation by performing the identification of the released PTH amino acid on-line with the degradation in the machine, e.g. by HPLC techniques. The first version of an on-line HPLC identification was realized by Machleidt & Hofner (1980, 1981) as applied to their solid phase sequencer. This on-line system was constructed for a sequencer which releases the thiohydantoin derivatives dissolved in solvent with TFA and water. Firstly, the acid is removed and, secondly, the sample is injected onto the HPLC column.

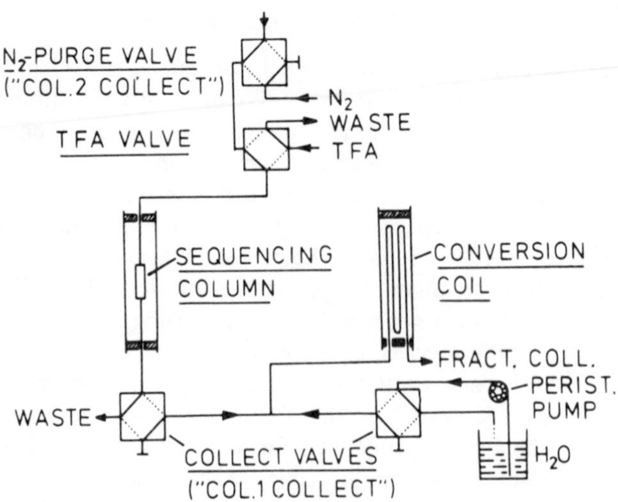

Fig. 12.6 Scheme of the modified LKB 4020 solid-phase sequencer for nitrogen purge and automatic conversion (Salnikow *et al.*, 1981).

With conversion devices which provide for drying of the acid (e.g. the flask-type) the installation of the automatic HPLC-identification is simple: an aliquot of the thiohydantoin/solvent solution (the effluent from the converter into the fraction collector) is injected onto the HPLC column by means of one automatic HPLC injection valve (Wittmann-Liebold, 1982). We use isocratic elution of the thiohydantoins on a RP_8, 3 μm column (4 × 250 mm) at 61°C with 19.5% propan-2-ol, 1% tetrahydrofuran and 1.5% of 10 mM sodium acetate buffer, pH 5.3, in the recycling mode and inject about 1/4 to 1/2 of the sample volume (Wittmann-Liebold & Ashman, 1985). The remainder of the sample is dried in the fraction collector and used for additional identification with a HPLC gradient system. Isocratic elution in the on-line system is easier and allows the control of the degradation and the performance of the machine, and complete resolution of all amino acid derivatives can be achieved. Our experience shows that in automatic on-line identification, storage of part of the released thiohydantoins is necessary in case of failure of the HPLC unit, or for further identification of modified amino acids.

12.6. DISCUSSION

12.6.1. Solid-phase sequencing of small peptides

Solid-phase sequencing methods are available for almost all types of polypeptides. They were developed for peptides containing approximately 30 residues or less, but can be applied to larger peptides and proteins as well. Two methods of attachment are in general use; the attachment of the α- and ε-amino groups of the peptides to polystyrene and glass supports by the activation of *p*-phenylene diisothiocyanate and the attachment of the *C*-terminal carboxyl group by activation with water-soluble carbodiimides. The most useful supports for both methods are aminopolystyrene resin and aminopropyl glass. Glass supports tend to release the amino acid derivatives of better quality. Therefore, they have been preferred for peptide attachment in recent years. The attachment yields are between 60 and 90 %, depending on the method adopted, the quality of the resins and the provision that the peptide is suitably soluble under the conditions used. Besides these general methods, homoserine lactone peptides can be coupled to TETA polystyrene, APG or β-APG with excellent attachment yields. Care must be taken to obtain good quality synthesized TETA resin, combined with prewashing before use and storage under nitrogen. Other types of supports, such as poly-*N*-alkyl-acrylamide-based resins (Atherton *et al.*, 1975; Cavadore *et al.*, 1976) or macroporous polystyrene derivatives (Inman *et al.*, 1977) have been proposed. But these resins have not found wide application yet for the practical structural analysis of polypeptides derived from biological sources. See Chapter 16 for recent approaches in designing new resins for solid-phase sequencing.

12.6.2. Solid-phase sequencing of large peptides and proteins

Glass supports are preferable when sequencing larger peptides and proteins. The process uses either the attachment of activated lysine residues, while fragments containing no such groups or mainly arginine residues only attach less readily. A generally applicable method for coupling larger lysine-deficient fragments to supports would be by the activation of the *C*-terminal amino acid by carbodiimides. Aminopolystyrene has attachment yields for proteins which are too low, but with aminopropyl glass or β-*N*-aminoethyl-(3-aminopropyl) glass the yields are increased to between 20 and 60 %. But not only has the attachment yield to be recorded but the sequencing ability of the protein-bound support has to be controlled. Experience has proved that insolubilized proteins might have poor release of *N*-terminal residues during degradation although binding to the support was excellent. More experience shows that solid-phase sequencing of proteins on the micro-scale can best be done by attachment *via* the lysine groups.

12.6.3. Comparison of liquid- and solid-phase techniques

The superiority of one technique over the other in the determination of primary structures of polypeptides depends on the peptide and protein characteristics, such as their solubility, amino acid composition and chain length. Also, the sequencing methods selected depend on the amount of material available and the sensitivity of the methods used for the identification of the released derivatives.

The advantages of the solid phase method are as follows:

1. the relatively low cost of the sequencer and the ease of mechanical maintenance:
2. the low cost of reagents and solvents;
3. the covalently bound peptide cannot be lost in the wash procedures of each successive degradative step;
4. the presence of SDS does not interfere with attachment of the protein to the resin (Walker *et al.*, 1982).

The disadvantages are due mainly to incomplete attachment of the polypeptides to the support material and sometimes to the partial destruction of the resins, which is progressive with the number of degradative cycles. A more detailed comparison of automatic sequencing techniques is presented elsewhere (Wittmann-Liebold, 1980, 1981).

12.6.4. Microsequencing

The development of highly sensitive identification methods using HPLC which avoids the derivatization procedures necessary for GLC has meant that microsequencing is now possible with amino acid derivatives being identified in the pmol range. Microsequencing is limited by the purity of the derivative, its stability during and after the completion of the cycle and the presence of contaminants. Clean solvents are vital and contamination-free degradation can only be obtained from sequencers which have a constant performance, minimal dead-volumes and corrosion-free valves and connections. The quality of the resin in solid-phase and the carriers in liquid-phase work are limiting factors. Another complication is that the yields of individual derivatives vary considerably. Whereas some are detectable when starting with minute amounts of peptides, others give very low yields or are not detectable against an increasing background of side-products. The performance of the machine should be carefully controlled to ensure runs in which the risk of 'swamping' the derivative with side-products is kept to a minimum. The performance of a sequencer can be judged by 'blank' degradation runs performed without peptide or protein. The fractions obtained from these degradation cycles should be tested in the identification procedure employed using relatively large quantities. Control runs with standard test proteins or peptides should be carried out to calibrate the apparatus.

It should also be noted that the isolation methods used for the purification of peptides and proteins greatly influence their solubilities and hence their suitability for a given degradation technique. For instance, traces of residual buffer salts or amines decrease the efficiency of the degradation and may cause partial *N*-terminal blocking of the peptide (e.g. by ammonium acetate). This is critical for sequencing small amounts of material. Isolation of minute polypeptide amounts have to be made under reducing conditions to avoid irreversible blocking of the *N*-terminal residue by contaminating aldehydes.

12.6.4.1. Degradation with labelled Edman reagent

Several methods are described in the literature, in which the quantity of the sample can be decreased for sequencing. One such method utilizes radioactively labelled PITC for the degradation (Laursen, 1971; Bridgen, 1975, 1977b; Bridgen & Waxdal, 1977). These methods depend on the purity of the labelled reagent since its impurities and the side-reactions of the Edman degradations result in labelled by-products and subsequent ambiguities of interpretation. The strength of the radioactive background also depends on the tightness of valves and connections in the sequencer. An easier approach is to degrade peptides and proteins labelled *in vivo*. Difficulties encountered with intrinsic or extrinsic labelling in microsequence analysis are discussed elsewhere (McKean, 1980).

12.6.4.2. Degradation with DABITC

An efficient approach for manual microsequencing analysis was described in which the normal PITC reagent is replaced by the homologue 4-*N*,*N*-dimethylaminoazo-benzene-4′-isothiocyanate (Chang *et al.*, 1978). The method is very sensitive and these derivatives can be quickly identified in pmol quantities by TLC or quantified using HPLC (Chang *et al.*, 1980; Lehmann & Wittmann-Liebold, 1984). For further details see Chapter 14. The DABITC/PITC method was adapted for microsequencing in the solid phase (Hughes *et al.*, 1979; Wittmann-Liebold & Lehmann, 1980) and more recently an improved programme was published which gives unambiguous results starting with minute amounts of sample (Salnikow *et al.*, 1981).

12.7. ACKNOWLEDGEMENTS

I thank Mr. Arnold Lehmann for his help in preparing this article and Mr. Keith Ashman and Mr. Neil Rawlings for reading the English version of the manuscript.

12.8. REFERENCES

Apella, E., Inman, J. K. & Dubois, G. C. (1977). See Previero & Coletti-Previero (1977), pp. 121–133.

Atherton, E., Clive, D. L. J. & Sheppard, R. C. (1975). *J. Am. Chem. Soc.*, **97,** 6584–6585.

Beyreuther, K. (1977). See Previero & Coletti-Previero (1977), pp. 107–114.

Birr, C. (1975). See Laursen (1975a), pp. 115–129.

Birr, C. (Ed.) (1980). *Methods in Peptide and Protein Sequence Analysis*, Proc. Third Internat. Conf., Elsevier/North Holland, Heidelberg, Amsterdam, New York, Oxford.

Bridgen, J. (1975). *FEBS Lett.*, **50,** 159–162.

Bridgen, J. (1977a). *Methods Enzymol.*, **47,** pp. 385–391.

Bridgen, J. (1977b). *Methods Enzymol.*, **47,** pp. 321–335.

Bridgen, J. & Waxdal, M. J. (1977). See Previero & Coletti-Previero (1977), pp. 153–162.

Cavadore, J. C., Derancourt, J. and Previero, A. (1976). *FEBS Lett.*, **66,** 155–157.

Chang, J. Y. (1979). *Biochem. Biophys. Acta*, **578,** 188–195.

Chang, J. Y., Brauer, D. & Wittmann-Liebold, B. (1978). *FEBS Lett.*, **93,** 205–214.

Chang, J. Y., Lehmann, A. & Wittmann-Liebold, B. (1980). *Anal. Biochem.*, **102,** 380–383.

Edman, P. (1950). *Acta Chem. Scand.*, **4,** 238–293.

Edman, P. & Begg, G. (1967). *Eur. J. Biochem.*, **1,** 80–91.

Elzinga, M. (Ed.) (1982). *Methods in Protein Sequence Analysis*, Proc. Fourth Internat. Conf., Humana Press, Clifton, N. J.,

Frank, R. (1979). Doctoral Thesis, Ruprecht-Karl-Universität, Heidelberg.

Gisin, B. F. (1972). *Analyt. Chim. Acta*, **58,** 248–249.

Gray, W. R. & Hartley, B. S. (1963). *Biochem. J.*, **89,** 379–380.

Gutte, B. & Merrifield, R. B. (1971). *J. Biol. Chem.*, **246,** 1922–1941.

Herbrink, P., Tesser, G. I. & Lamberts, J. J. M. (1975). *FEBS Lett.*, **60,** 313–316.

Hewick, R. M., Hunkapiller, M. W., Hood, L. E. & Dreyer, W. J. (1981). *J. Biol. Chem.*, **15,** 7990–8005.

Horn, M. J. & Bonner, A. G. (1977). See Previero & Coletti-Previero (1977), pp. 163–176.

Horn, M. J. & Laursen, R. A. (1973). *FEBS Lett.*, **36,** 285–288.

Hughes, G. J., Winterhalter, H., Lutz, H. & Wilson, K. H. (1979). *FEBS Lett.*, **108,** 92–97.

Inman, J. K., Dubois, G. C. & Apella, E. (1977). See Previero & Coletti-Previero (1977), pp. 81–94.

Laursen, R. A. (1971). *Eur. J. Biochem.*, **20,** 89–102.

Laursen, R. A. (1972). *Methods Enzymol.*, **25,** 344–359.

Laursen, R. A. (Ed.) (1975a). *Solid Phase Methods in Protein Sequence Analysis*, Proc. First Internat. Conf., Pierce, Rockford, Illinois.

Laursen, R. A. (1975b). In *Immobilized Enzymes, Antigens, Antibodies and Peptides*, Weetall, H. H. (Ed.), Marcel Dekker, New York, pp. 567–634.

Laursen, R. A. (1975c). See Laursen (1975a), pp. 3–9.

Laursen, R. A. (1977a). See Previero & Coletti-Previero (1977), pp. 95–106.

Laursen, R. A. (1977b). *Methods Enzymol.*, **47,** pp. 277–288.

Laursen, R. A., Horn, M. J. & Bonner, A. G. (1972). *FEBS Lett.*, **21,** 67–70.

Laursen, R. A., Bonner, A. G. & Horn, M. J. (1975). See Perham (1975), pp. 73–110.

Laursen, R. A., Obar, R., Chin, F. & Whitrock, K. (1980). See Birr (1980), pp. 9–20.

Lehmann, A. & Wittmann-Liebold, B. (1984). *FEBS Lett.*, **176,** 360–364.

Machleidt, W. & Hofner, H. (1980). See Birr (1980), pp, 35–47.

Machleidt, W. & Hofner, H. (1981). In *High Performance Liquid Chromatography in Protein and Peptide Chemistry*, Lottspeich, F., Henschen, A. & Hupe, K. P. (Eds.), Walter de Gruyter Verlag, Berlin, pp. 245–258.

Machleidt, W. & Machleidt, I. (1977). See Previero & Coletti-Previero (1977), pp. 233–246.

Machleidt, W. & Wachter, E. (1977). *Methods Enzymol.*, **47,** 263–277.

Machleidt, W., Wachter, E., Scheulen, M. & Otto, J. (1973). *FEBS Lett.*, **37,** 217–220.

Machleidt, W., Hofner, H. & Wachter, E. (1975). See Laursen (1975a), pp. 17–30.

McKean, D. J. (1980). See Birr (1980), pp. 143–152.

Morris, H. R., Dickinson, R. J. & Williams, D. H. (1973). *Biochem. Biophys. Res. Commun.*, **51,** 247–255.

Perham, R. N. (Ed.) (1975). *Instrumentation in Amino Acid Sequence Analysis*, Academic Press, London, New York, San Francisco.

Previero, A. & Coletti-Previero, M.-A. (Eds.) (1977). *Solid Phase Methods in Protein Sequence Analysis*, Proc. Second Internat. Conf., North-Holland, Amsterdam, New York, Oxford.

Previero, A., Derancourt, J., Coletti-Previero, M.-A. & Laursen, R. A. (1973). *FEBS Lett.*, **33,** 135–138.

Reinbolt, J., Tritsch, D. & Wittmann-Liebold, B. (1977). See Previero & Coletti-Previero (1977), pp. 209–218.

Reinbolt, J., Tritsch, D. & Wittmann-Liebold, B. (1979). *Biochimie*, **61,** 501–522.

Robinson, P. J., Dunnill, P. & Lilly, M. D. (1971). *Biochem. Biophys. Acta*, **242,** 659–661.

Salnikow, J., Lehmann, A. & Wittmann-Liebold, B. (1981). *Anal. Biochem.*, **117,** 433–442.
Salnikow, J., Lehmann, A. & Wittmann-Liebold, B. (1982). See Elzinga (1982), pp. 181–188.
Schiltz, E. (1977). See Previero & Coletti-Previero (1977), pp. 247–256.
Schiltz, E. & Reinbolt, J. (1975). *Eur. J. Biochem.*, **56,** 467–481.
Schmitt, H. W. & Walker, J. E. (1977). *FEBS Lett.*, **81,** 403–405.
Scotchler, J., Lozier, R. & Robinson, A. B. (1970). *J. Org. Chem.*, **35,** 3151–3152.
Tarr, G. E. (1975). *Anal. Biochem.*, **63,** 361–370.
Terhorst, C., Möller, W., Laursen, R. A. & Wittmann-Liebold, B. (1973). *Eur. J. Biochem.*, **34,** 138–152.
Wachter, E. & Werhahn, R. (1979). *Anal. Biochem.*, **97,** 56–64.
Wachter, E., Machleidt, W., Hofner, H. & Otto, J. (1973). *FEBS Lett.*, **35,** 97–102.
Wachter, E., Hofner, H. & Machleidt, W. (1975). See Laursen (1975a), pp. 31–46.
Walker, J. E., Auffret, A. D., Carne, A., Gurnett, A., Hanisch, P., Hill, D. & Saraste, M. (1982). *Eur. J. Biochem.*, **123,** 253–260.
Waterfield, M. D. and Bridgen, J. (1975). See Perham (1975), pp. 41–71.
Westall, F. C., Scotchler, J. & Robinson, A. B. (1972). *J. Org. Chem.*, **37,** 3363–3365.
Wittmann-Liebold, B. (1973). *Hoppe-Seyler's Z. Physiol. Chem.*, **354,** 1415–1431.
Wittmann-Liebold, B. (1980). In *Polypeptide Hormones*, Beers, R. F. J. & Bassett, E. G. (Eds.), Raven Press, New York, pp. 87–120.
Wittmann-Liebold, B. (1981). In *Chemical Synthesis and Sequencing of Peptides and Proteins*, Liu, T.-Y., Schechter, A. N., Heinrikson, R. L. & Condliffe, P. G. (Eds.), Elsevier/North Holland, New York, pp. 75–110.
Wittmann-Liebold, B. (1982). See Elzinga (1982), pp. 27–63.
Wittmann-Liebold, B. & Ashman, K. (1985) In *Modern Methods in Protein Chemistry*, **2,** Tschesche, H. (Ed.), Walter de Gruyter, Berlin, New York, pp. 303–327.
Wittmann-Liebold, B. & Kimura, M. (1984). In *Methods in Molecular Biology*, **1,** Walker, J. M. (Ed.), Humana Press, Clifton, New Jersey, pp. 221–242.
Wittmann-Liebold, B. & Lehmann, A. (1975). See Laursen (1975a), pp. 81–90.
Wittmann-Liebold, B. & Lehmann, A. (1980). See Birr (1980), pp. 49–72.
Wittmann-Liebold, B., Geissler, A.-W. & Marzinzig, E. (1975). *J. Supramol. Struct.*, **3,** 426–447.
Wittmann-Liebold, B., Graffunder, H. & Kohls, H. (1976). *Anal. Biochem.*, **75,** 621–633.
Wittmann-Liebold, B., Brauer, D., & Dognin, J. M. (1977). See Previero & Coletti-Previero (1977), pp. 219–232.

Practical Protein Chemistry—A Handbook
Edited by A. Darbre
© 1986, John Wiley & Sons Ltd

MICHAEL D. WATERFIELD, GEOFFREY SCRACE
and NICHOLAS TOTTY,
Protein Chemistry Laboratory,
Imperial Cancer Research Fund,
Lincoln's Inn Fields,
London WC2A 3PX, U.K.

13

Analysis of Phenylthiohydantoin Amino Acids

CONTENTS

13.1. INTRODUCTION

The most widely used and probably the most reliable method of determining the amino acid sequence of peptides and proteins employs the Edman degradation with the end-group reagent phenylisothiocyanate (PITC). For smaller peptides PITC may be used to remove the amino terminal amino acid while a chromophore or fluorophore such as dansyl chloride or DABITC which reacts with the newly exposed amino terminus is used to identify the new amino terminus as described in Chapters 11 and 14. Both manual and automated methodologies are currently used for small and large polypeptides which rely on identification of the amino terminal amino acid as the phenylthiohydantoin (PTH) derivative. A large number of papers have been and continue to be published on the analysis of the PTH derivatives of amino acids.

In this chapter five different techniques will be referred to which reflect the evolution and impact of instrumentation development on methods of PTH analysis. The simplest, cheapest and least sensitive method is undoubtedly TLC which was used by Edman and Begg (1967) in their classical publication describing the automated sequencer. This method is suitable for both manual and automated techniques but suffers from the disadvantage that the analyses are qualitative rather than quantitative and analysis is limited to the nmol range. However, the method has the advantage that several samples can be analysed at the same time. It is possible to semi-quantify TLC analysis by elution of spots from the chromatographic plates, but this is not recommended. TLC analysis needs to be supplemented with a method such as back hydrolysis (or HPLC) for the PTH derivatives of arginine and histidine. It is advisable to restrict its application to analysis of short peptides or highly efficient automated runs on medium size peptides in nmol quantities. The third method involves GLC which was introduced into the field of sequence analysis in the mid-1960s. Rapid and quantitative determination of most but not all the PTH amino acids can be made by GLC, and it is this failure to give a complete analysis which has proved its major drawback. The fourth method uses HPLC which was introduced in the mid-1970s and is currently the technique of choice since it is rapid, sensitive, quantitative and capable of identifying all the PTH amino acids at the pmol level. The fifth analytical technique which is used by only a few laboratories involves mass spectrometry. This technique probably offers the ultimate in sensitivity but requires complex instrumentation that is available in only a few laboratories.

The choice of analytical technique is clearly related to cost, the availability of instrumentation and to the sensitivity needed for the analysis. HPLC is the method of choice at present, particularly for high sensitivity applications and the major part of this chapter will deal with this aspect.

13.2. ANALYTICAL METHODS

13.2.1. High pressure liquid chromatography

A veritable explosion in the publication of HPLC techniques for PTH amino acid analysis occurred in the late 1970s and early 1980s. A variety of instruments, columns and buffers have been evaluated in many different laboratories and it is hard for the novice to pinpoint which is the best method to use. This section describes some systems with which we have had direct experience and it is hoped that the cautionary notes regarding these and other systems will help the reader to select a chromatographic method.

13.2.1.1. Instrumentation

A large number of different instruments are available commercially in modular form or as complete units. It is often possible to combine modules from different manufacturers which represent a compromise in cost versus performance, but this choice should not be taken by the inexperienced analyst.

The basic requirements for an analytical system are considered briefly.

1. The solvent delivery system should produce linear gradients at flow rates between 0.1 and 2.75 ml/min at pressures up to 4–5000 p.s.i. As column efficiency is improved by reducing the particle size of packings the operating pressures increase (unless column length is decreased) and it is wise to anticipate the possible need to operate at the upper pressure limits. The solvents used to generate gradients can be premixed with a valve and pumped by a single (often multiple head) pump. Alternatively, multiple pumps which deliver different solvents to a mixing chamber can be used. In both cases electronic programmers are required to generate the gradients and frequently the gradient can be selected from a range of preset gradient shapes. It is important to realize that the reproducible resolution of the PTH amino acids is a fairly sophisticated application of HPLC and consequently the accurate generation of the gradients is very important. A system can be easily evaluated by examining the reproducibility of the separations over several identical gradient runs. The ability to hold a gradient isocratically at the end of the analysis is important and it is also an advantage to be able to introduce isocratic steps during an analysis. The electronic programmer should be capable of returning the column to the initial chromatographic conditions at the end of each run and to prolong column life the gradient should be taken up to 100% of buffer B (see separation on C_8 columns in Section 13.2.1.5) briefly between each run and the column left in buffer B when not in use. As with all chromatographic analytical techniques the column plays a crucial part in the resolution and with an efficient column the operator may not have to resort to flow programming or different gradient shapes with intermediate isocratic steps. However, it is useful to have these capabilities available in case they are required.
2. An injection system allowing sample introduction without loss of operating pressures and hence solvent flow is essential. Many operators will quickly find the advantages of automatic injectors outweigh the cost of such modules. These injectors are available in versions which do not waste any sample during loading, and volumes as small as 1 μl or up to several hundred microliters can be injected with accuracies approaching $\pm 1\%$.
3. The column jacket should be temperature controlled over a range of 20° to 65°C to allow optimization of performance.
4. The optical detection system should be able to monitor effluent at 254 nm at a back-pressure of up to 50 p.s.i. using small volume flow cells (about 10 μl or less). The provision of a second channel at 313 nm is useful to monitor the modified

PTH derivatives of serine and threonine formed during some types of automatic sequencing. The refractive index changes which occur during gradient programming have been minimized in the design of flow cells by some manufacturers. As column performance improves, the speed of analysis increases and with the introduction of microbore columns the residence time of any eluted component in the flow cell is reduced and newer detectors have faster time constants to deal with such analyses. The ultimate usable sensitivity of different detectors depends on the equipment used and the operating range on some instruments has been reduced to 0.001 OD full scale with reasonable signal to noise ratios.

To analyse 1–0.01 nmol quantities a sensitivity range down to 0.005 O.D. is adequate but below this quantity the operator will need a more sensitive detector. At these low detection levels optimum performance of all aspects of the instrument and high levels of solvent purity are needed because baseline drift during gradient elution becomes a significant problem. Further purification of solvents, buffers and water may be required.

5. A system is needed for recording data, identifying the PTH amino acids and determining their peak areas. Several integrators with different capabilities are available. Most machines use pre-calibration with a standard to define peak areas per nmol and time 'windows' used for peak identification. Some machines can deal with complex baseline changes and with fused peaks. Different sets of peak widths and noise parameters can often be changed during a run to optimize analysis in relation to peak shape changes. The instrument capabilities are more taxed at high sensitivity and more versatile instruments are needed to deal with pmol analysis. Single-channel integrators which include direct recording on thermal-sensitive paper are now available at modest prices. Some of these integrators can be interfaced to computers for storage of raw or processed data. The interface to the computer to collect raw data can be made directly from the detector through a 12 or 16 bit analog to digital converter and a suitable interface such as the IEEE, RS232 or 16 bit general purpose interfaces. In this case relatively slow data rates of 10–25 points/second are needed to handle the chromatographic analysis, and software for integration must be purchased or written for the particular computer used. It is possible to use signal smoothing and background subtraction to improve signal to noise levels once raw data has been gathered and stored in the computer. The HPLC system used in some of the studies described in this chapter is commercially available from Waters (Milford, Massachusetts, USA) and consists of a WISP autoinjector model 710B, two M 6000A pumps, a model 440 detector (254 nm and 313 nm) with a water-jacketed column. For operation at 0.002 and 0.001 OD full scale an Altex model 160 detector with autozero capability has been used. The 10 mV output may be used directly for integration on the Waters 730 Data Module. The pumps are controlled by a model 720 system controller and data is recorded and processed by a 730 data module. For direct collection of raw data the detector is connected to an analog to digital converter which is interfaced through the HPIB buss to a

Hewlett-Packard model 9845T (204) computer. Recently a Nelson Analytical multichannel system which employs a Hewlett-Packard 9816 computer has been used.

13.2.1.2. Chromatographic solvents

The optical purity of the water, salt solutions and organic solvents is most important. Buffers should be sterile and particle free. A suitable system for water purification is the cartridge system manufactured by Millipore (see Chapter 8, section 8.3). Organic solvents of acceptable grade can be obtained from Rathburn Chemicals (Walkerburn, Scotland) or Burdick and Jackson (USA). Other manufacturers also produce suitable solvents (e.g. BDH, Fisons & Merck). Background peaks and baseline drift can cause problems as the sensitivity of the analytical method is increased. If necessary, water can be pumped through a coarse grade of octadecylsilane packing, such as that used in Sep-Pak cartridges (Waters) before buffers are made up. The organic solvents can be treated with the best grade of charcoal and redistilled through Widmer columns or on a spinning band still. However, in the hands of a novice this procedure can often contaminate rather than purify solvents and is usually not cost-effective. Without any further purification the acetonitrile (Rathburn S grade) used in our laboratory causes a shift in baseline of 0.0005–0.001 OD units during the gradients.

The presence of contaminants in solvents and reagents used in sequencing are a separate problem also related to the sensitivity of the analysis. If necessary, reagents can be purified as described by Hunkapiller *et al.* (1983).

The quality of acetic acid and sodium acetate used for buffers is not a problem if the best grades are purchased. Solvents are degassed by bubbling with a fast stream of helium for 2–3 minutes and during use a slow stream of gas is passed through the buffer. Other methods of degassing produce less reliable results (in our hands) although any methods that the manufacturers recommend give reproducible results if followed meticulously each time the buffer is prepared. Our stock buffer is prepared as a 1 M solution which is sterilized by filtration. Running buffers are made up approximately every two days by dilutions of stock buffer. It is impossible to avoid the occasional growth of microorganisms in pure acetate buffers. To prevent any growth of bacteria or fungi a low concentration of acetonitrile (10%) is included in buffer A (see Separation on C_8 columns in section 13.2.1.5).

13.2.1.3. Chromatographic columns

The chromatographic capabilities of HPLC columns continue to improve through greater control of the manufacturing processes and through the development of higher performance and smaller particle-size packings. Three types of reverse phase packings are currently used for PTH analysis—the octyl, octadecyl and the cyanopropyl packings. Several manufacturers produce columns with these packings varying in particle type and particle size from 3–10 μm. These packings have varying degrees of substitution by the hydrocarbon chains and in the degree to which unmodified silanol groups on the packing are 'capped' or blocked. Over recent years we have tried many columns from different manufacturers and have noticed a distinct improvement in reproducibility of separations. However, significant differences in the resolution of some pairs of PTH amino acids continue to occur with different columns having apparently identical packings. In many cases only a single pair of PTH amino acids in a standard mixture of 18 cannot be separated. Sometimes, significant tailing is observed for the PTH amino acids of arginine and histidine. We have found the only solution to these separation problems is to try a different column. In our experience the Zorbax C_8 columns manufactured by Dupont have proved the most reliable in terms of reproducibility of PTH amino acid resolution and column life. These columns have been used for up to 2 000 injections with reproducibility from day to day of 0.01 min in elution times of the PTH amino acids. The only other column (in our hands) which gives satisfactory chromatography of the basic PTH amino acids is the CN column manufactured by IBM.

13.2.1.4. Preparation of standard solutions of PTH amino acids

Standard PTH amino acids (Pierce, Rockford, Illinois) are dissolved in acetonitrile or methanol at a concentration of approximately 7.5 nmol/μl and aliquots (duplicate) diluted to read the OD at 269 nm. Reference to the molar extinction coefficients in Table 13.1 allows the precise concentration to be determined and each standard is then diluted to 5 pmol/μl. Special conditions must be used for Arg-PTH and His-PTH. The hydrochloride salts of these standards can be converted to the soluble trifluoroacetyl salts by dissolving a known amount in 20 % trifluoroacetic acid (TFA) which is then evaporated in a stream of nitrogen. These PTH amino acids can then be dissolved in acetonitrile and diluted by reference to the extinction coefficients as described above.

A mixture is prepared with equal amounts of each PTH standard (including Norleucine-PTH for analysis on C_8 columns), and the volume adjusted to give a suitable volume for injection of 50 pmol (e.g. in 10 μl). Standards can be stored frozen individually at −20°C for months and as a mixture at −20°C for several weeks. The standardization of Ser-PTH and Thr-PTH is problematic and is discussed below.

13.2.1.5. Separation of PTH amino acids by reverse phase HPLC

Numerous HPLC methods for PTH analysis have been published and in many cases these offer insignificant advantages over that originally described by Zimmerman *et al.* (1977) who used an octadecyl (C_{18} ODS) column and acetate buffers with acetonitrile as the organic modifier. The only real problem

Table 13.1. Molar absorbances and molecular weights of PTH amino acids (from Edman & Henschen, 1975).

PTH amino acid	Molecular weight	ε_{269}
Alanine	206	16000
Asparagine	249	17200
Aspartic acid	250	16100
Arginine HCl	328	15900
Glutamic acid	264	15900
Glutamine	263	17000
Glycine	192	14900
Histidine HCl	309	15500
Isoleucine	248	17000
Leucine	248	16700
N-ε-PTC-lysine	398	29000
Methionine	266	17100
Phenylalanine	282	15500
Proline	232	14300
5-Carboxymethylcysteine	297	16600
Tryptophan	321	19700
Tyrosine	298	15600
Valine	234	16500

encountered in their method concerns the elution properties of Arg-PTH and His-PTH which chromatograph as broader peaks than the other PTH amino acids. Improvements in resolution can be achieved with the latest ODS columns which have smaller particle sizes and high theoretical plate counts but Arg-PTH and His-PTH remain a problem on most ODS columns. This problem may be related to the extent to which free silanol (acidic) groups remain after derivatization of the packings. Significant improvements can be obtained by using C_8 (Dupont) and CN (IBM) columns. The Zorbax C_8 column (Dupont) is particularly good for the basic PTH amino acids probably because the packing is highly 'capped' to block free silanol groups. Examples based on various methods derived from the literature and modified in our own laboratory will be described.

Separation on C_8 columns

Buffers:

1M-Sodium acetate stock solution adjusted to pH 4.1 with acetic acid. Buffers contain 0.008–0.02 M acetate buffer and buffer A has 10% and buffer B 80% acetonitrile. Buffers are degassed with helium and the column is jacketed at 43°C and run at a flow rate of 2 ml/min.

Column:

Zorbax C_8 4.6 × 250 mm (Dupont). At least 600 mm of tubing should be immersed in the heating block between the injector and top of the column to preheat the buffers.

Analytical conditions:
A linear gradient from 20% B to 40% B over 8 min can be used for the initial chromatogram and an example of such a separation is shown in Fig. 13.1(a). The elution positions of Arg-PTH and His-PTH depend on the salt concentration and, as can be seen from Fig. 13.1(b) and (c) these two compounds can be moved relative to the other PTH amino acids by varying the salt concentration of the buffer. The separation of the hydrophobic PTH amino acids is influenced by the percentage of buffer B at the top of the gradient. By carrying out 2 or 3 runs with different final concentrations of buffer B the separation can be optimized. A difference of 1% can be critical.

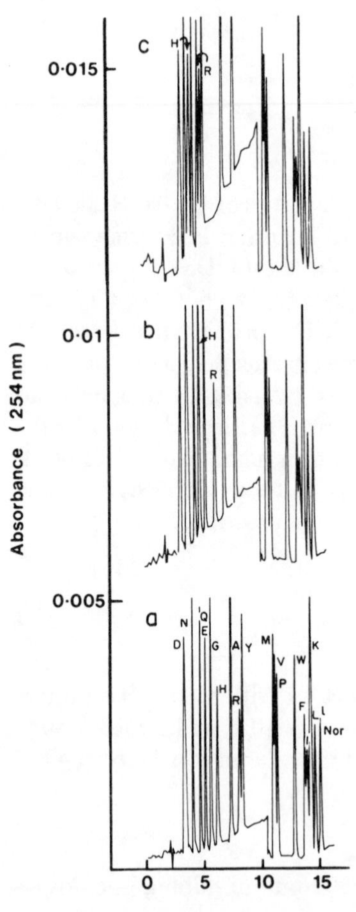

Fig. 13.1. HPLC separation of PTH amino acids using a Zorbax C_8 column – the effect of salt concentration in the buffers on resolution of arginine-PTH and histidine-PTH. Conditions for separation are given in section 13.2.1.5. The sodium acetate concentrations used are 0.005 M (a) 0.01 M (b) and 0.02 M (c). The single letter code is used to indicate amino acids. Norleucine-PTH is indicated as Nor.

Having optimized the chromatographic conditions it should not be necessary to change them for some weeks, depending on the reproducibility with which the operator can prepare buffers. As the column ages the salt concentration of the buffer will need to be increased to maintain the elution position of Arg-PTH and His-PTH. The precise elution position of these two relative to the other PTH amino acids and to background peaks from sequencer samples will need to be optimized. An example of the use of these columns in analyses of samples from a gas-phase sequencer is shown in Fig. 13.2.

Separations on C_{18} columns

The most commonly used columns for PTH amino acid analysis are probably those using C_{18} bonded phases made by Dupont (Zorbax ODS), Waters (C_{18} μBondapak) and Altex (Ultrasphere). Details of the use of these columns can be obtained from the manufacturers and publications in such journals as *Analytical Biochemistry* and *Journal of Chromatography*. Indeed a vast number of papers have been published employing innumerable variations of the operating parameters. Some of the problems encountered since 1977 which have led to this vast literature on a single method are perhaps due to variations in column packings which occur amongst columns from different manufacturers. Recently improved techniques for both synthesis of packings and for the packing of columns have been made, and it should now be possible to repeat separations more easily from laboratory to laboratory. The methodology described originally by Zimmerman *et al.* (1977) remains the basis for most useful separations. Sodium acetate buffers at pH 4 to 4.5 and concentrations 0.001 to 0.04 M, with acetonitrile or methanol as the organic modifier, employing a gradient of 20 to 40% organic modifier at temperatures of 40–45°C will give a very good separation apart from problems of tailing of Arg-PTH and His-PTH. Flow rates of up to 2.5 ml/min can be used and separations accomplished in 20 min or less. Acetonitrile is superior to methanol or ethanol because the background absorbance and viscosity at high concentrations are lower. It is clear that the chromatographic behaviour of Arg-PTH and His-PTH differs from that of the other PTH amino acids and by varying the buffer concentration their elution positions may be altered.

The separation of the PTH amino acids is influenced by temperature, flow rate, organic modifier concentration and the gradient as would be expected. In evaluating a new column to optimize performance it is useful to run analyses at four different temperatures (say 25, 35, 45 and 55°C) and to vary the final concentration of acetonitrile and the gradient rate to optimize the separation of the hydrophobic PTH amino acids. It is important to use a sufficient length of fine bore tubing between the injector and the column to pre-heat the buffers. The effect of pH can also be examined if necessary.

Examples of the use of C_{18} columns under various conditions can be found in the following literature.

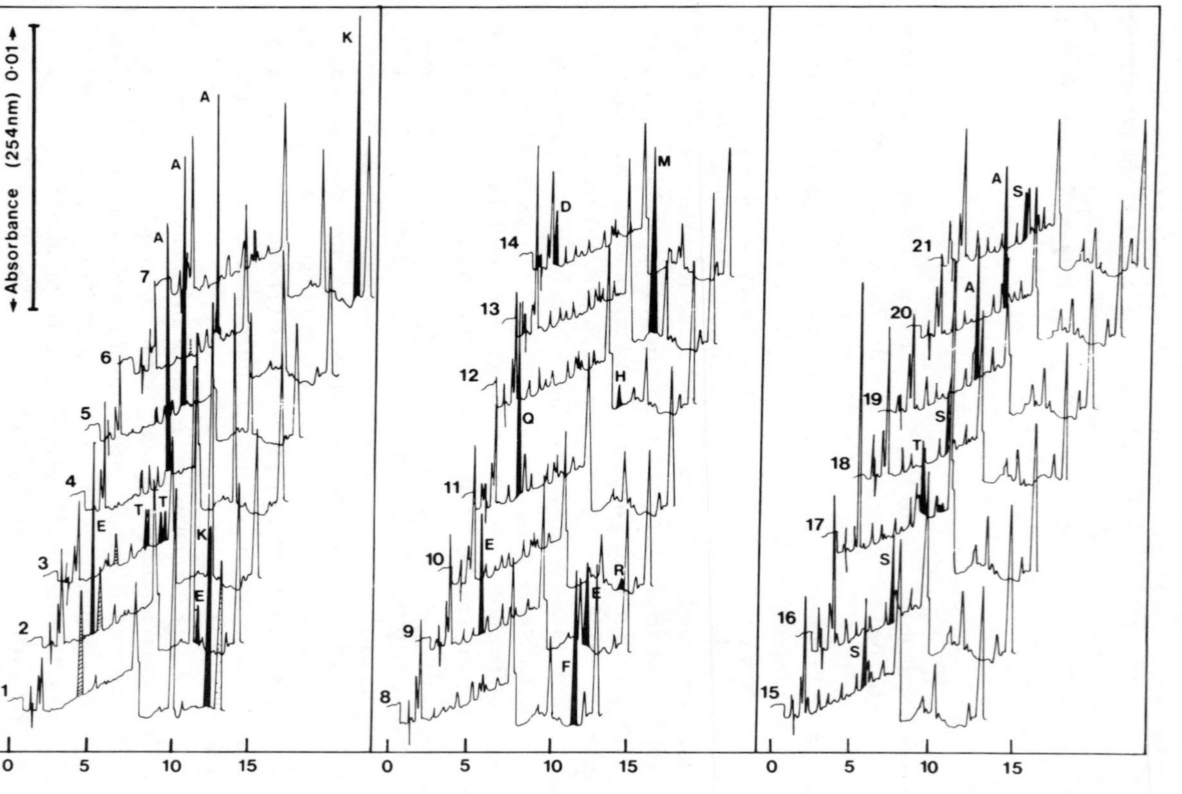

Fig. 13.2. HPLC analysis of the first 21 steps of Edman degradation of ribonuclease carried out in a gas-phase sequencer. Analytical conditions are as described in section 13.2.1.5 using 0.0025 M-sodium acetate. Details of background peaks are noted in the legend to Fig. 13.3.

1. Dupont Zorbax ODS (Dupont) with sodium acetate and acetonitrile—Zimmerman *et al.* (1977), Margolies and Brauer (1978) and Ultrasphere—Hawke *et al.* (1982).
2. C_{18} μBondapak (Waters) with sodium acetate and methanol—Zeeuws and Strosberg (1978), or Spherisorb ODS and ethanol—Sottrup-Jensen *et al.* (1980).
3. Ultrasphere ODS (Altex) with sodium acetate and acetonitrile containing tetrahydrofuran—Somack (1980).

Separation on CN columns

Buffers: 1 M-sodium acetate stock brought to pH 5.7 with acetic acid. Buffer A is 15% in buffer B and contains 0.015–0.04 M-sodium acetate buffer. Buffer B is 50% methanol and 50% acetonitrile. Buffers are degassed with helium and the column run at 1 ml/min and 32°C. The composition of these buffers is modified as described below to optimize separations.

Column: Cyanopropyl (IBM) 46 × 250 mm. Complete resolution of the common PTH amino acids on Zorbax CN (Dupont) columns was reported by Johnson *et al.* (1979), and subsequently on an IBM CN column by Hunkapiller *et al.* (1983). The major advantages for the CN column compared to that of the ODS are improved peak shape for Arg-PTH and His-PTH, increased column life (number of injections) and greater sensitivity (peak height/nmol of PTH). Since 1982 we have compared CN, C_8 and ODS columns and found that both CN and C_8 have advantages over ODS when the elution characteristics such as peak shape for Arg-PTH and His-PTH are compared. Our experience with C_8 columns suggests that the life of this column (2 000 injections/1 year) is comparable to the CN. We do not have sufficient experience yet with the Zorbax ODS (5 μm) to justify a conclusion regarding column life.

The optimization of separation on the CN columns is made by varying the salt concentration to position Arg-PTH and His-PTH at suitable points between the other PTH amino acids. Increasing the salt concentration will result in more rapid elution of these PTH amino acids. It should be borne in mind that the elution of background peaks from sequencer samples may dictate the placement of the basic PTH amino acids. As columns age it is necessary to increase the salt concentration to maintain the elution position of the basic PTH amino acids. To optimize the separation of the PTH derivatives of tyrosine, valine, proline and methionine the ratio of acetonitrile to methanol in buffer B should be changed. An increase in the methanol concentration will improve the proline/methionine separation and a decrease will improve the tyrosine/valine separation. To optimize the resolution of the PTH amino acids of asparagine, serine, threonine, glutamine and glycine the initial concentration of buffer B and rate of increase of buffer B in the gradient should be varied. If a solvent programmer is used which is capable of generating gradient shapes similar to the multiple step gradients described by Hunkapiller and Hood (1983), optimum separation can be achieved. The hold-up volume in any pre-column mixing device needs to be considered in reproducing these gradients.

Separation of PTH amino acids on phenylalkyl columns

Henderson *et al.* (1980) have reported a useful but slow separation using a μBondapak (Waters) phenylalkyl support.

13.2.1.6. Analysis of PTH amino acids from Edman degradations

The analysis of the majority of the common PTH amino acids liberated during manual or automatic Edman degradation is straightforward when sequencing amounts greater than 1 nmol. Above this level the contribution of impurities derived from the sequencer reagents does not usually interfere with HPLC.

Arginine-PTH and histidine-PTH

Since the analysis of all the PTH amino acids can be made using a single HPLC injection it is convenient to use TFA rather than HCl to convert the anilinothiazolinone amino acids to the PTH amino acids. This can be accomplished using 20% TFA for 20 min at 55°C. The acid is then removed with a stream of nitrogen and the sample redissolved in acetonitrile for HPLC injection. The TFA may need treatment with chromic acid and redistillation from dithiothreitol to ensure adequate purity and to remove oxidants (see Hunkapiller *et al.*, 1983). Arg-PTH and His-PTH are thus analysed as TFA rather than HCl salts. This procedure necessitates the preparation of standards as trifluoroacetate salts (see section 13.2.1.4). The elution position of these PTH amino acids is optimized by varying the salt concentration in buffers.

Aspartic acid-PTH and glutamic acid-PTH

These PTH amino acids are usually eluted with short retention times with HPLC and difficulties of quantitation can arise from the formation of multiple peaks (Harris *et al.*, 1980) or as a result of coelution with the dithiothreitol which is used as an antioxidant in sequencer reagents and solvents. These problems are not significant when using C_8 columns, as shown in analyses with a gas-phase sequencer (see Fig. 13.2). To increase the retention times of these acidic PTH amino acids they can be esterified (Hunkapiller & Hood, 1983).

Method: Add 50 μl of 1 M-acetyl chloride in methanol for 20 min at 55°C. Excess reagent is removed by evaporation under nitrogen and the samples redissolved in acetonitrile. Cool the methanol before carefully adding acetyl chloride (see Chapter 8, p. 283). The elution positions of these methyl ester derivatives are shown in Fig. 13.3. The accuracy of quantitation in sequencer samples is improved because the esters chromatograph in a region of low background noise (Hewick *et al.*, 1981; Hunkapiller & Hood, 1983). If necessary the ethyl esters can be prepared to obtain different elution positions relative to background peaks.

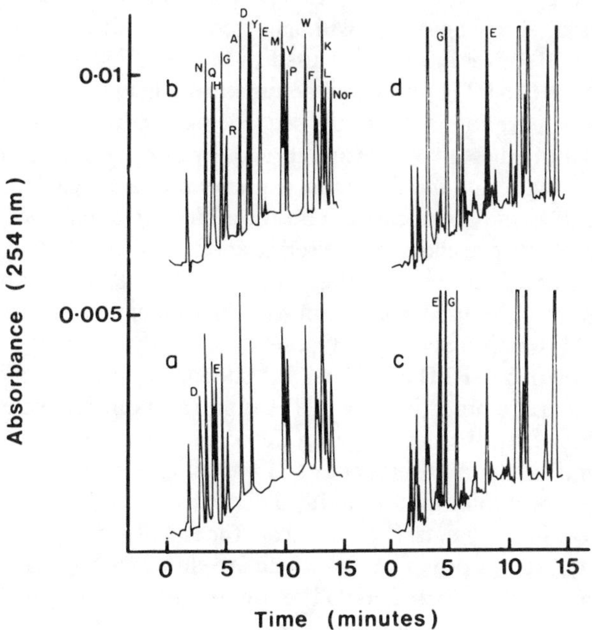

Fig. 13.3. The resolution of the methyl ester derivatives of aspartic acid-PTH and glutamic acid-PTH on HPLC analysis. Conditions for HPLC separation are given in section 13.2.1.5 and for esterification in section 13.2.1.6. Panel a shows separation of a standard mixture of PTH amino acids (40 pmol at OD 0.005) before and panel b after esterification. Panel c and d show analysis at high sensitivity (0.005 OD full scale) of step 2 of the gas-phase Edman degradation of ribonuclease. In panel c the glutamic acid-PTH is analysed directly and in panel d it is analysed as the methyl ester. Norleucine (Nor) was added to each sample before analysis. Typical background peaks are shown which are derived from dithiothreitol, the reaction productions of PITC and dimethylamine and hydrolysis products of PITC. Glycyl-glycine added to remove compounds reactive with amino groups is eluted as glycine-PTH.

Serine-PTH and threonine-PTH

The analysis of these two PTH amino acids is best carried out after prior precautions have been taken to add antioxidants such as dithiothreitol to all solvents and HFBA and TFA have been redistilled from dithiothreitol prior to use. It is also important to add the aqueous acid used for conversion to the anilinothiazolinone as soon as possible after collecting the liberated terminal amino acid derivative. It is possible that the addition of the butyl chloride extracts containing the PTH amino acids to the TFA solution in the conversion vessel followed by a brief drying period will help recoveries of these two PTH amino acids. Automatic converters are best used to perform this operation rapidly.

Ser-PTH and Thr-PTH purchased commercially may not be identical to the derivatives which are found under the sequencing conditions used. It is best to locate and quantify these two PTH amino acids by using an analysis of a standard protein containing serine and threonine (e.g. sperm whale myoglobin at position 3 for serine and bovine ribonuclease at position 3 for threonine (see Fig. 13.2)). By plotting the yield before and after the serine and threonine residues it is possible to calculate a correction factor for the recovery of these PTH amino acids. These proteins may also be used for optimizing instrument performance and the recovery of PTH amino acids (see Fig. 13.2).

The detection of Ser-PTH and Thr-PTH may be improved by monitoring the HPLC effluent at 314 nm if these two PTH amino acids are recovered as dehydro derivatives. However, the derivative produced by the sequencing system in use may vary significantly in absorbance at 314 nm since either authentic Ser-PTH and Thr-PTH, the dithiothreitol-modified derivatives or the dehydro derivatives may be formed. In our hands Thr-PTH from the gas-phase sequencer is found as a doublet or quartet of peaks, two of which are eluted each side of Tyr-PTH on the C_8 column (see Fig. 13.2). Ser-PTH elutes immediately prior to Ala-PTH in samples from the gas-phase sequencer (see Fig. 13.2). The derivatives which elute at these positions do not give a high absorption at 314 nm. Different results will be obtained with samples from solid-phase, spinning-cup and gas-phase sequencers. The best solution is to evaluate performance using standard proteins as mentioned above.

Cysteine-PTH

The method for sequence analysis of cysteine residues will depend on the nature of the prior derivatization of the sulphydryl group used during preparation of the polypeptide. The easiest PTH amino acid derivative of cysteine to quantify is that formed from the ^{14}C-carboxamidomethyl derivative obtained following reduction and subsequent alkylation of the sample with ^{14}C-iodoacetamide. A sequence run can thus be monitored by measuring radioactivity in the liberated PTH amino acids. Alternatively, the elution position of the standard cysteine-PTH derivative can be established and quantified. Cysteic acid-PTH can be analysed but elutes with a very short retention time and may be difficult to quantify.

13.2.1.7. High sensitivity analysis

When working with a sample size less than 1 nmol of polypeptide the appearance of background compounds, which are derived from reagents and solvents, and even the sample itself, may become a problem. Reagents may be purified or those from a different supplier may be tested. Examples of purification steps which may be necessary are discussed in the papers of Hewick *et al.* (1981) and Hunkapiller & Hood (1983). Common interfering contaminants or reaction products are mono- and diphenylthiourea, the product of reaction of dimethylamine and PITC (for gas-phase), Polybrene (for spinning-cup) and impurities in organic solvents and in the coupling buffer.

13.2.2. Gas–liquid chromatography

Several publications deal with the GLC analysis of PTH amino acids. Of particular interest are two practical accounts by Niall (1973) and Bridgen *et al.* (1975). Practical handbooks from Beckman Instruments (Palo Alto, California, USA) give a detailed background to GLC analysis.

13.2.3. Thin layer chromatography

A practical guide to one-dimensional analysis of PTH amino acids by TLC can be found in Edman and Henschen (1975). The big advantage of this method is that several steps can be analysed at the same time on one plate. More than 1 nmol of peptide must be available and preferably one should operate with 50–70 nmol. Overlap will make the interpretation difficult and it is advisable to use a second method for confirmation. The analysis of Arg-PTH and His-PTH should be made by back-hydrolysis or HPLC.

The two-dimensional method developed by Kulbe (1971, 1974; Summers *et al.*, 1973) is extremely good for analysis, since the standards can be applied on one side of the plate and the sample on the other side. The addition of a fluorescent indicator to the chromatographic solvent gives greater sensitivity and 0.1 nmol can be visualized easily. Analysis can be made on sheets of polyamide (50 × 50 mm) chromatographed in small covered beakers. The technique fails to analyse Arg-PTH and His-PTH but it is used in some laboratories as a valuable adjunct to other methods.

13.2.4. Back hydrolysis of PTH amino acids

Several different back-hydrolysis reagents have been described including alkali, hydriodic acid, methane sulphonic acid and HCl. The best procedure is that described by Mendez and Lai (1975) in which stannous chloride (0.1 %) in 6 M-HCl is used. Samples are dried in glass hydrolysis tubes and sealed under vacuum with the reagent after flushing with nitrogen to help reduce oxygen levels. Hydrolysis is at 150°C for 4 h. Samples are then dried and subjected to amino acid analysis.

Good recovery of most amino acids from thiazolinones can be obtained. Threonine is converted to γ-amino butyric acid. Serine is converted to alanine and tryptophan to glycine. Thus a second identification method is needed for serine and tryptophan. TLC, GLC or HPLC can fill this gap.

13.3. REFERENCES

Bridgen, J., Graffeo, A. P., Karger, B. L. & Waterfield, M. D. (1975). In *Instrumentation in Protein Sequence Analysis*, Perham, R. N. (Ed.), Academic Press, London, New York, San Francisco, pp. 111–145.

Edman, P. & Begg, G. (1967). *Europ. J. Biochem.*, **1,** 80–91.

Edman, P. and Henschen, A. (1975). In *Protein Sequence Determination*, Needleman, S. B. (Ed.), 2nd edn, Springer-Verlag, Berlin, Heidelberg, New York, pp. 232–279.

Harris, J. U., Robinson, D. & Johnson, A. J. (1980). *Anal. Biochem.*, **105,** 239–245.

Hawke, D., Pau-Miau, Y. & Shively, J. E. (1982). *Anal. Biochem.*, **120,** 302–311.

Henderson, L. E., Copeland, T. D. & Oroszlan, S. (1980). *Anal. Biochem.*, **102,** 1–7.

Hewick, R. M., Hunkapiller, M. W., Hood, L. E. & Dreyer, W. J. (1981). *J. Biol. Chem.*, **256,** 7990–7997.

Hunkapiller, M. W. & Hood, L. E. (1983). *Methods Enzymol.*, **91,** 486–493.

Hunkapiller, M. W., Hewick, R. M., Dreyer, W. J. & Hood, L. E. (1983). *Methods Enzymol.*, **91,** 399–413.

Johnson, N. D., Hunkapiller, M. W. & Hood, L. E. (1979). *Anal. Biochem.*, **100,** 335–338.

Kulbe, K. D. (1971). *Anal. Biochem.*, **44,** 548–558.

Kulbe, K. D. (1974). *Anal. Biochem.*, **59,** 564–573.

Margolies, M. N. & Brauer, A. (1978). *J. Chromatogr.*, **148,** 429–439.

Mendez, E. & Lai, C. Y. (1975). *Anal. Biochem.*, **68,** 47–53.

Niall, H. D. (1973). *Methods Enzymol.*, **27,** 942–1010.

Somack, R. (1980). *Anal. Biochem.*, **104,** 464–468.

Sottrup-Jensen, L., Petersen, T. E. & Magnusson, S. (1980). *Anal. Biochem.*, **107,** 456–460.

Summers, M. R., Smythers, G. W. & Oroszlan, S. (1973). *Anal. Biochem.*, **53,** 624–628.

Zeeuws, R. & Strosberg, A. D. (1978). *FEBS Lett.*, **85,** 68–72.

Zimmerman, C. L., Apella, E. & Pisano, J. J. (1977). *Anal. Biochem.*, **77,** 569–573.

Practical Protein Chemistry—A Handbook
Edited by A. Darbre
© 1986, John Wiley & Sons Ltd

J.-Y. CHANG
Pharmaceuticals Research Laboratories,
CIBA-GEIGY Limited,
4002 Basle, Switzerland

14

Manual Sequence Analysis of Polypeptides using 4-Dimethylaminoazobenzene-4′-isothiocyanate

CONTENTS

14.1. INTRODUCTION

Sequence determination of polypeptide by the conventional Edman (phenylisothiocyanate, PITC) degradation requires the identification of the released phenylthiohydantoin (PTH) amino acids (Edman & Henschen, 1975). PTH amino acids are colourless compounds, the majority having their absorption maxima at 269 nm with $\varepsilon = 16\,000$. They are detected by means of u.v. absorbance. The reagent 4-dimethylaminoazobenzene-4′-isothiocyanate (DABITC) results in the release of 4-dimethylaminoazobenzene-4′-thiohydantoin (DABTH) amino acids. These are coloured compounds, having their absorption maxima at 520 nm (in acidic media) with $\varepsilon = 47\,000$. It is thus apparent that identifiication of DABTH amino acids in the visible region is both

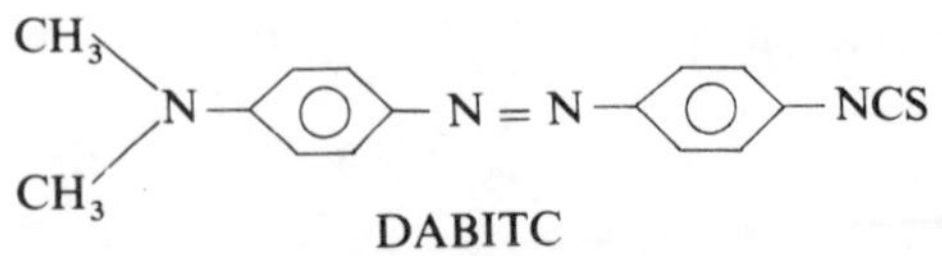

$$(CH_3)_2N-C_6H_4-N=N-C_6H_4-NCS$$

DABITC

more sensitive and more convenient. These derivatives were used for amino-terminal analysis (Chang, 1983a).

The manual DABITC-PITC double coupling method can be applied to both liquid-phase degradation (Chang *et al.*, 1978; Chang, 1981, 1983b) and solid-phase degradation (Chang, 1979b, 1983b).

14.2. MATERIALS

DABITC is prepared as described (Chang *et al.*, 1976; Chang, 1977) or purchased from Fluka (recrystallized before use) or Pierce. PITC (Merck) is redistilled under nitrogen at reduced pressure (b.p. ~ 92°C/12 mm Hg). Pyridine (Merck) is redistilled three times over (1) KOH (10 g/l), (2) ninhydrin (1 g/l) and (3) KOH (10 g/l). Trifluoroacetic acid (Fluka) is redistilled over $CaSO_4 . 0.5\ H_2O$ (10 g/l). Sequencer grade pyridine and TFA can also be obtained from Pierce. Other solvents are of commercial analytical grade and are used without further purification.

Polyamide sheets are purchased from Schleicher & Schüll (FRG) or Cheng-Chin Co. (Taiwan).

Amination of glass beads (CPG-75, Serva) and their activation with *p*-phenylene diisothiocyanate are described in Chapter 12, p. 387 *et seq.* (Bridgen, 1976).

14.3. LIQUID-PHASE DABITC-PITC METHOD

This is the method described by Chang *et al.* (1978) and Chang (1981).
The degradation cycle of the liquid phase DABITC-PITC method is outlined in scheme 1 (Figure 14.1). Several points should be observed during the manual operations.

1. DABITC is not very stable in pyridine and is better prepared fresh daily or at every degradation cycle.
2. Some hydrophobic peptides tend to precipitate as fine particles or thin films during the extractions. These precipitates are suspended at the interface of the aqueous organic phases after centrifugation and could be inadvertently removed by pipetting.

14.4. SOLID-PHASE DABITC-PITC METHOD

See Chang (1979b). Attachment of peptide and protein to DITC-activated amino glass and the solid-phase DABITC-PITC degradation cycle are performed as outlined in scheme 2 (Figure 14.2). Coupling of polypeptide to the solid support

Figure 14.1. Scheme 1: sequence of steps of liquid-phase DABITC-PITC degradation cycle.

Polypeptide 1–10 nmol. Add 80 μl of 50% aqueous pyridine and 40 μl of DABITC in pyridine (10 nmol/μl). Incubate under nitrogen at 52°C for 50 min. Add 10 μl of PITC. Incubate at 52°C for 30 min.

Extract with 500 μl (×4) of *n*-heptane/ethyl acetate (2:1, v/v). Dry the extracts *in vacuo*, or by nitrogen evaporation.

Add 50 μl of anhydrous TFA. Incubate (under nitrogen) at 52°C for 15 min. Remove solvent *in vacuo*, or by nitrogen evaporation.

Add 50 μl of H_2O. Extract the thiazolinone derivatives with 200 μl of butyl acetate and dry the extract *in vacuo* for conversion to DABTH amino acids.

Subject the residue to the next degradation cycle.

Figure 14.2. Scheme 2: sequence of steps for preparation of (a) bound polypeptide and (b) solid-phase DABITC-PITC degradation cycle (Chang, 1979b)

(a)

Polypeptide 1–10 nmol
↓+ 0.8 ml of 0.2 M-$NaHCO_3$-NaOH, pH 8.9
↓+ 20 to 25 mg DITC glass
↓N_2, 52°C, 1 h
↓+ 20 μl of ethanolamine
↓N_2, 52°C, 15 min
↓H_2O and CH_3OH wash
↓Dry *in vacuo*
↓+ 600 μl of 67% aqueous pyridine
↓+ 50 μl of PITC
↓N_2, 52°C, 30 min
↓CH_3OH wash
↓Dry *in vacuo*
↓+ 200 μl of trifluoroacetic acid
↓N_2, 52°C, 15 min
↓CH_3OH wash
↓Dry *in vacuo*
↓Subject to DABITC-PITC degradation

(b)

DABITC-PITC cycle
↓+ 400 μl of 50% aqueous pyridine
↓+ 200 μl of DABITC (15 nmol/μl in pyridine)
↓N_2, 52°C, 45 min
↓+20 μl of PITC
↓N_2, 52°C, 30 min
↓Remove supernatant after gentle centrifugation
↓Wash with 2 ml of pyridine and 2 × 2 ml of CH_3OH
↓Dry *in vacuo*
↓+ 200 μl of trifluoroacetic acid
↓N_2, 52°C, 15 min
↓Remove trifluoroacetic acid *in vacuo*
↓*Extract thiazolinones with 300 μl of CH_3OH
↓CH_3OH wash
↓Dry *in vacuo*
↓Subject to next cycle

*Collected for DABTH-amino acid identification, see text.

and their subsequent sequence analysis are carried out in the same glass-stoppered tube (10 × 50 mm) with a magnetic stirring bar (2 × 6 mm). Reaction is performed by placing the tube in a heating block (52°C) over a magnetic stirrer. A C-10 glass stopper fitted with a D2 sinter is used at each drying step to prevent the loss of glass beads. Glass beads must not be stirred vigorously to avoid pulverization.

14.5. CONVERSION OF THIAZOLINONE AMINO ACID TO THIOHYDANTOIN AMINO ACID

The extract of thiazolinone amino acid (obtained either from liquid-phase or solid-phase degradation) is evaporated and conversion to the thiohydantoin derivative is effected by 5 N-HCl/acetic acid (1:2, v/v) at 52°C for 50 min. The sample is dried and redissolved in a suitable volume of ethanol for TLC identification.

14.6. IDENTIFICATION OF DABTH AMINO ACIDS

14.6.1. Thin layer chromatography

Polyamide sheets (25 × 25 mm) are used to identify all DABTH amino acids except for DABTH-Ile and DABTH-Leu. The samples are applied at the origin (about 6 mm from the edges of two adjacent sides). The diameter of the spot is restricted to 1.0 mm maximum by using a hair drier with cold air. Two-dimensional development in a covered jar (30 × 80 × 80 mm) is by ascending solvent flow. No phase equilibrium is necessary. Solvent 1, water/acetic acid (2 : 1, v/v), is used for the first dimension and solvent 2, toluene/n-hexane/acetic acid (2 : 1 : 1, by vol.), is used for the second dimension.

To discriminate between DABTH-Leu and DABTH-Ile, use one-dimensional separation on silica gel plates with the solvent chloroform/ethanol (100 : 3, v/v) (Chang *et al.*, 1977). After drying, the developed plates are exposed to HCl vapour when all the yellow spots turn to a red or blue colour (Fig. 14.3).

Note that the successful identification of DABTH-amino acids relies on the skilful running of the small polyamide sheet and interpretation of the pattern of spots (Chang, 1977; Chang *et al.*, 1978).

Use a 1 μl micropipette (H. E. Pederson) with a fine, round tip (ground down on emery paper) to apply the samples to polyamide sheets.

Experience is required to identify the amino acid residues appearing as multiple derivatives, such as serine, threonine and lysine (Chang *et al.*, 1978; Chang, 1979a).

14.6.2. High pressure liquid chromatography

DABTH amino acids can also be analysed by reverse phase HPLC (Fig. 14.4) employing the conditions described in the legend of Fig. 14.2. This technique

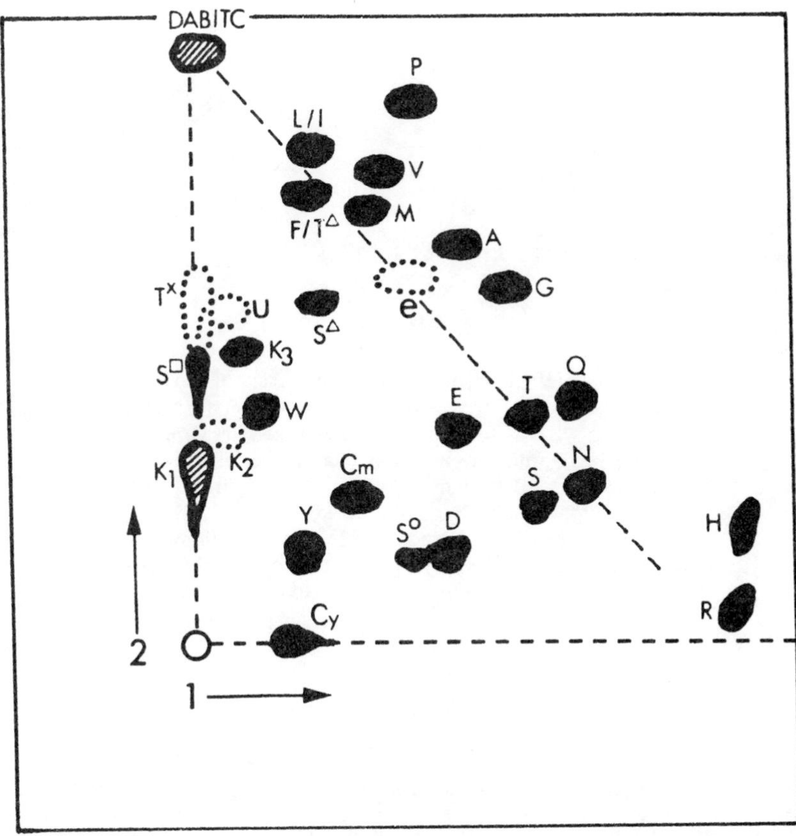

Figure 14.3. TLC (polyamide) separation of DABTH amino acids and their by-products.

Solvent 1: water/acetic acid (2:1, v/v); solvent 2: toluene/*n*-hexane/acetic acid, (2:1:1, by vol.). DABTH amino acids are identified by the one letter code for the corresponding amino acids. The colours of the derivatives are represented by solid areas (red), dotted areas (blue) and hatched areas (purple); 'e' is the marker DABTC-diethylamine; 'U' is a blue colour by-product of the excess reagent. For details of the by-products of serine, threonine and lysine refer to Chang (1979a).

gives quantitative results and is particularly useful when combined with the solid-phase DABITC-PITC degradation. The HPLC method is, however, unable to distinguish the colour differences between the blue 4,-*N*,*N*-dimethylaminoazobenzene 4′-thiocarbamyl and the red thiohydantoin derivatives which are seen with the TLC technique.

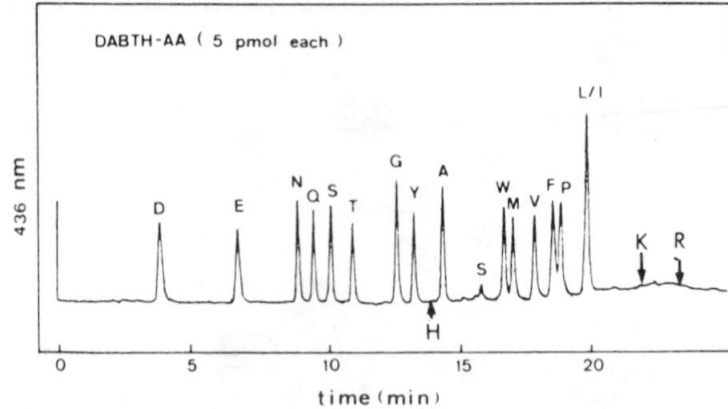

Figure 14.4. HPLC separation of 5 pmol each of DABTH-amino acids on a Zorbax ODS column. Chromatographic conditions were as follows. Solvent A was 35 mM acetate buffer, pH 5.0, and solvent B was acetonitrile. The gradient was 45%-70% B/0–10 min, kept at 70% B from 10 to 12 min, 70% B-80% B/12-14 min, kept at 80% B from 14–22 min, then 80% B-45% B/22–25 min. Column temperature was 22°C. The chart speed was 40 cm h^{-1}. Detector, 436 nm, at 0.005 absorbancy unit full scale.

14.7. CONCLUCIONS

The major advantage of the DABITC-PITC double coupling method over the conventional manual sequencing method is as follows:

1. Sensitivity: an amount of 1–10 nmol of peptide and protein is sufficient for sequence determination up to 20–30 residues. This marks a 10 to 20-fold increase of sensitivity over the conventional manual technique.
2. Efficiency: sequences up to 20–30 residues can be obtained routinely by this new method without major difficulty using either liquid- or solid-phase. This new method is therefore not significantly less efficient than the automatic sequencer.
3. Simplicity: only simple facilities are required. The use of TLC although not quantitative, is adequate to identify all the DABTH amino acids with reasonable confidence.

14.8. REFERENCES

Bridgen, J. (1976). *Biochemistry*, **15,** 3600–3604.
Chang, J.-Y. (1977). *Biochem. J.*, **163,** 517–520.
Chang, J.-Y. (1979a). *Biochem. Biophys. Acta*, **578,** 175–187.
Chang, J.-Y. (1979b). *Biochem. Biophys. Acta*, **578,** 188–195.
Chang, J.-Y. (1981). *Biochem. J.*, **199,** 557–564.
Chang, J.-Y. (1983a). *Methods Enzymol.*, **91,** 79–84.

Chang, J.-Y. (1983b). *Methods Enzymol.*, **91,** 455–466.
Chang, J.-Y., Creaser, E. H. & Beatley, K. W. (1976). *Biochem. J.*, **153,** 607–611.
Chang, J.-Y., Creaser, E. H. & Hughes, G. J. (1977). *J. Chromatogr.*, **140,** 125–128.
Chang, J.-Y., Brauer, D. & Wittmann-Liebold, B. (1978). *FEBS Lett.*, **93,** 205–214.
Edman, P. & Henschen, A. (1975). In *Protein Sequence Determination*, Needleman, S. B. (Ed.), 2nd ed., Springer, Berlin, Heidelberg, New York, pp. 232–279.

Practical Protein Chemistry—A Handbook
Edited by A. Darbre
© 1986, John Wiley & Sons Ltd

MARK SIEGELMAN and J. DONALD CAPRA
Department of Microbiology,
The University of Texas Health Science Center at Dallas,
5323 Harry Hines Blvd.,
Dallas, Texas 75235, U.S.A.

15

Current Techniques for Automated Liquid-phase Sequence Analysis

CONTENTS

15.1. INTRODUCTION

The development of the automated protein peptide sequencer (Edman & Begg, 1967) and its subsequent modifications and commercialization by the Beckman Corporation has clearly forged a major revolution in the approach to protein structural determination in the last 15 years. The impact of automated instrumentation on the field of protein sequencing can be appreciated by the rapid increase of compiled data (Dayhoff, 1972, 1973, 1976, 1978; Croft, 1980). There is a clear need for computer-assisted data storage and retrieval facilities in order to make full use of published protein sequences (see Editorial Note at the end of the chapter.)

Perhaps, however, the major impact has been on the biological questions that can be approached by protein structural determinations by contemporary scientists compared to those performing primary structural analyses 15 to 20 years ago. This, of course, reflects the extraordinary efficiency of the instrumentation and the sensitivity of detection systems that have allowed scientists to determine total primary structures on proteins with far less than 100 nmol of material whereas previously millimoles of material were required. This chapter will review some of the techniques currently in use in automated liquid phase sequence analysis. We will primarily focus on the use of the instrument, how samples are prepared for the instrument and how to best monitor machine performance. A discussion of sample preparation, carriers utilized and various major problems with automated sequencing instrumentation as it presently exists will be discussed. There are extensive reviews of each of these various areas which are available to the reader (Hermodson *et al.*, 1972; Niall, 1973; Waterfield & Bridgen, 1975).

15.2. SAMPLE PREPARATION

It is almost impossible to discuss sample size and suitability without regard to the specific instrument, program and systems available because so much of the discussion will hinge on whether a 1 M- or a 0.1-M-Quadrol program (Brauer *et al.*, 1975) is being utilized and whether polybrene with or without the use of a cold trap is being employed. In most instances these differences will vary the limits of analysis by an order of magnitude but not more. In addition, whether Quadrol or DMAA is being utilized as the coupling buffer also bears on these issues.

Thirteen years ago, when our laboratory was engaged in the amino acid sequence analysis of immunoglobulin heavy and light polypeptide chains, it was not unusual to use sample loads of approximately 10 to 20 mg in order to perform amino-terminal amino acid sequence analysis for 30–40 steps. Presently, utilizing a 0.1 M-Quadrol program, along with polybrene in a cold-trap modified instrument, we routinely sequence 1–2 mg to 60–80 steps (two-fold increase in efficiency). It is difficult to pin down precisely the reasons for these better results and it is likely that a combination of many factors are involved.

We routinely prepare our samples for the sequencer by lyophilizing them from either dilute acid, volatile aqueous buffers or water. Samples are typically stored in capped scintillation vials as 'white fluffy powder' until applied to the sequencer. It has been our experience that samples which are not 'white fluffy powders' are not as amenable to sequence analyses. The reasons for this are not clear but probably relate to solubility and contamination problems.

We almost always reduce and radiocarboxymethylate our samples prior to sequence analysis because of the difficulty in identifying cysteine residues by most of the conventional techniques of PTH identification and, therefore, a small portion of each phenylthiohydantoin derivative is subjected to scintillation counting.

Samples should not be applied to the sequencer cup if they contain salt as this may lead to washout of the entire sample. This is easily observed in radiosequencing procedures where the first two or three steps of the sequencer will show a progressive drop in radioactive counts reaching background level. Thus, the presence of salt in the sequencer cup must be avoided. In recent years, many investigators have applied samples in the presence of SDS. While this has not been routine practice in our laboratory, the presence of up to 1 % SDS has been useful in solubilizing samples, in particular, membrane proteins isolated originally on SDS gels, without detriment to subsequent sequencing.

The solvents used to introduce samples into the sequencer cup vary from laboratory to laboratory and there is no unanimity as to the best system. To our knowledge, no-one has studied this systematically. In our laboratory, samples are dissolved in 25 % acetic acid. If a sample is not soluble in this solvent, it is generally heated to 80°C for 10 to 30 min along with vigorous Vortex mixing. Should this prove to be unsuccessful, we routinely lyophilize the sample immediately and generally delay applying the sample for 24 h. For the next day, there are two alternatives that we have taken that have been successful. First, we try 1.0 M-ammonia. Many proteins are obviously soluble at higher pH values and by simply changing to this solvent about half of the proteins that are not soluble in 25 % acetic acid will be solubilized. If the protein is not soluble in ammonia, we generally lyophilize the protein again and delay its application into the sequencer an additional 24 h. The second alternative that we have used on many occasions is to directly apply the sample in a concentrated volatile organic acid such as formic acid or even HFBA. Indeed, many laboratories routinely apply samples in HFBA. Our own reasons for not doing this primarily revolve around the difficulty of handling and storing HFBA as a common laboratory solvent. It can, however, be 'delivered' from the sequencer manually which obviates some of the hazards of its use.

By and large, proteins which are known to be 'unblocked' but which do not sequence in the sequencer are problematic for a variety of reasons. After reduction and alkylation (in guanidine or urea), unless the pH is carefully maintained above 7.0 and the alkylating reagent concentration carefully controlled, there is the distinct probability of it reacting with the *N*-terminal amino acid. When this occurs, the protein appears 'blocked' to the sequencer. This particular problem is disastrous as there is no convenient way of removing the alkylating group. Another problem is when the protein is not soluble in one or more of the sequencer solvents or reagents. We have typically found that most of these problems occur with the Quadrol buffer. Often, by using DMAA, we have been able to achieve excellent results while in the presence of Quadrol buffer sequencing was impossible.

There are a number of procedures that are used prior to the execution of the cleavage cycle. Our own laboratory applies two maneuvers. The first is the so-called 'sample dry-down sub-routine' which effectively provides vacuum and then nitrogen flush to the cup in a repetitive fashion to provide an evenly distributed fine, dry sample film for subsequent chemical reactions. This sub-routine begins every run but is automatically by-passed on all subsequent cycles. In addition, we routinely run the first cycle (step 0) omitting PITC delivery (with the Rl valve in the off position). This does not delay sequencing by more than 10 min after the sample is dry, because immediately after 'Rl delivery' in the program, one can turn the Rl valve on and sequencing can proceed automatically. Microprocessor controlled programming makes this even simpler. The reasons for adequate sample 'dry-down' are obvious. Prior to the introduction of PITC and Quadrol, it is important that the sample be appropriately dried and evenly dispersed in the sequencer cup. We have routinely run samples without PITC in the first step to wash from the sample any artifacts that may show up on our usual PTH detection systems. This serves as an additional step in ensuring the uniform application of the sample to the sequencer cup.

15.3. QUADROL VERSUS VOLATILE BUFFERS

Since the introduction of the automated sequencer, Quadrol has remained the mainstay of buffering systems. It has virtually all the qualities essential for automated sequencing with the single drawback that it is necessary to extract the Quadrol with solvents like ethyl acetate in which some peptides are soluble. The volatile buffer systems of DMAA, DMBA and variations thereof gained widespread acceptance in the 1970s and it was only with the introduction of 0.1 M-Quadrol programming (Brauer *et al.*, 1975), along with the addition of the cold-trap, and the non-protein carrier polybrene, that led many investigators in recent years to return to Quadrol. DMAA has two clear advantages:

1. It has always been our impression that DMAA was a better 'solvent' than Quadrol and many proteins which are soluble in DMAA cannot be sequenced in Quadrol. Fortunately, this is rare and has occurred in our experience less than 1 % of the time.
2. Since DMAA is volatile, no ethyl acetate is required for extraction and only under extreme conditions is a benzene wash required to continue sequence analysis. This formerly intractable problem of extraction was largely overcome in Quadrol systems, firstly, by the use of polybrene and other carriers which have allowed the use of Quadrol in the sequencing of small peptides all the way to the *C*-terminus and, secondly, by decreasing the concentration of the Quadrol buffer.

Table 15.1 lists five peptides that were studied utilizing DMAA or Quadrol as the coupling buffer, with and without polybrene. The values indicate the yield (nmol) of the *C*-terminal amino acid in each of these four systems. Column A

Table 15.1. Comparison of 0.1 M-Quadrol and DMAA with and without polybrene. A: DMAA only; B: DMAA with polybrene; C: 0.1 M-Quadrol; D: 0.1 M-Quadrol with polybrene. The figures represent the number of nmol recovered of the *C*-terminal amino acid from 10 nmol of peptide (From McCumber *et al.*, 1980).

	A	B	C	D
Lysine peptides				
TISK	2.7	7.1	0	4.5
GQLLPQEK	2.0	5.1	0	4.2
TLFAGLCVIK	2.6	5.1	.25	5.0
Arginine peptide				
EQLNLR	2.9	5.8	2.0	4.7
Tryptophan peptide				
AGEEDCITSW	2.0	6.1	.20	5.7

represents the use of DMAA alone with a standard peptide program and column B with the addition of polybrene. The results in columns A and B were obtained using the same model 890C sequencer. Column C represents experiments done in a second sequencer equipped with a 0.1 M-Quadrol program and column D with the addition of polybrene as the peptide was loaded into the cup. In each experiment, 10 nm of peptide was used. All analyses were done by HPLC. The incorporation of polybrene gave improved results with different peptides as shown in Table 15.1.

The significant advantage of Quadrol versus DMAA in terms of machine performance (clear increased repetitive yields), the virtual absence of artifacts in most PTH identification systems with Quadrol programming and the clear advantage of Quadrol over DMAA in terms of the valving and vacuum systems, have made Quadrol much more popular in recent years. Many investigators have their instruments set such that they can easily switch from one system to the other. The design of the Beckman 890C protein peptide sequencer makes this easy.

(*Editorial Note* The spinning cup sequencer originally used 250 nmol of protein. Microsequencing in the 100 pmol range is now possible if Quadrol buffer is purified before use. This reduces background artifact peaks and allows longer sequences to be determined (Begg *et al.*, 1984)).

Quadrol buffer, as purchased, is diluted to 0.1 M-Quadrol with propan-1-ol/water (3:4, v/v). Add β-*N*-(2-aminoethyl-3-aminopropyl glass (LKB 4320–207) (10 mg/ml) and gently shake for 24 h at 20°C to remove traces of aldehyde. Allow the glass beads to settle and remove the buffer by pipette. Add DITC-β-*N*-(2-aminoethyl)-3-aminopropyl glass (LKB 4320–211) (2 mg/ml) to the 0.1 M-Quadrol and shake gently for 4 days at 20°C. Store the buffer over DITC glass at 20°C and pipette off buffer as required.

15.4. POLYBRENE

'No-one knows why it works, but it does': the introduction of polybrene (Klapper *et al.*, 1978; Tarr *et al.*, 1978) had a profound impact on the field of automated sequence analysis. Initially used in DMAA programs, polybrene is now extensively used with both 1 M-and 0.1 M-Quadrol programs and variations thereof. Polybrene can be purchased from Aldrich or in a 'purified state' as so-called 'Sequenol grade' from Pierce. There are a number of procedures described for the further purification of polybrene. Our own laboratory has found the Aldrich product to be satisfactory. When we initially studied the use of polybrene with DMAA and later with Quadrol, we studied the advisability of using 0.3 mg, 3 mg and 30 mg in the sequencer cup and found 3 mg to be best. More recent studies have indicated that approximately 1 mg is probably a more ideal amount to use although several laboratories use 10 mg. Since we routinely run samples in the sequencer 'one step without PITC', we have not been much concerned with polybrene artifacts detected by HPLC, although they do exist (Klapper *et al.*, 1978).

The precise location of these artifacts varies with the source of solvents and reagents, source of polybrene, different batches of polybrene, and different detection systems. Indeed, subtle variations in the solvent conditions for HPLC, various manufacturers' HPLC chromatographic columns or different columns by the same manufacturer lead to different artifact positions. These are generally not difficult to contend with, particularly if some 'wash step' is introduced into the sequencing procedure. Problems with artifacts can be overcome in one of several ways.

1. Polybrene can be further purified prior to use.
2. Polybrene can be applied to the sequencer cup without peptide or protein, or with glycyl-glycine and pre-cycled several cycles of automated degradation before adding the protein or peptide to the sequencer cup.
3. Peptide and polybrene can be added simultaneously and one or a series of 'wash steps' can be utilized. It is this latter procedure which we find most convenient and with rare exceptions this suffices to exclude most polybrene artifacts from interference. Occasionally artifacts occur in the first two or three cycles but they are easily distinguished from the amino acid background by the fact that they diminish over the first three cycles.

Polybrene should be and probably is quite stable, although our own experience is that polybrene left at room temperature in solution for more than one week inevitably leads to results that are mistrusted. Although we have never systematically studied this problem, both we and other laboratories have either made up fresh solutions of polybrene each day, or kept solutions of polybrene at 4°C for one week only. We dissolve polybrene in water (30 mg/ml) and store 1 ml in the freezer. Furthermore, each aliquot is refrozen after every use. Using this particular approach, we have not had any difficulties. Polybrene is inexpensive. The 100 g bottle of polybrene that we had five years ago is still being used. Replacements cost about $50, one of the least expensive items in automated sequencing. Finally, it should be noted that polybrene has not been found detrimental to any of the valving or detection systems within the Beckman sequencer. Our laboratory has processed over 20 000 samples both by HPLC and/or by back-hydrolysis with HI and analysis on a Durrum D-500 amino acid analyzer without noting any adverse effects.

15.5. SOLVENTS AND REAGENTS

Sequencer chemicals are the most expensive components of automated sequence analysis. A single instrument used continuously will require about $15 000 worth of chemicals per year. The source of these reagents is a subject of significant controversy in the field and attitudes range from the extremes of final purification of all chemicals (Begg *et al.*, 1978; Hunkapiller & Hood, 1978) to the purchase of the most inexpensive reagents and solvents available, and testing by individual investigators. Most investigators fall somewhere in between and our own laboratory has used several different sources of chemicals over the years and have developed the following strategy. We purchase all our reagents from Beckman. These are the most expensive sequencer chemicals on the market, but they are the only ones that are guaranteed sequencer-tested. They show the least variability from lot to lot and since they are each individually sequencer-tested, there is less likelihood of any major disasters when utilizing these products. However, their high cost has led several investigators including ourselves to test many other sources of reagents. By and large, our own experience has been that the reagents sold by Pierce are good although variable. The key reagent in our experience is HFBA and the vast majority of reported variations in repetitive yields is in our judgement due to variations in the quality of the HFBA. Begg has argued that commercially available HFBA contains 'water' and has described procedures for making it more anhydrous (Begg *et al.*, 1978). Our own experience has been that this is not generally useful and we have not found it particularly advantageous to reduce the water content from 30 ppm water down to the 4–6 ppm range.

Significant cost savings can be made by purchasing the solvents for the sequencer from sources other than Beckman. We have never experienced difficulty with benzene purchased from several sources, although our standard supplier is Burdick and Jackson. Ethyl acetate can be a problem, because many preparations of ethyl acetate contain materials which will destroy PTH amino acids.

We recommend a simple test with ethyl acetate. Take a standard amount of PTH amino acids (we normally use our gas chromatography or HPLC standard), mix it with a new batch of ethyl acetate (about 1.5 ml) and let is stand at room temperature for 30 min, and then dry it down under a nitrogen stream. The residue is then taken up and injected for GC or HPLC and the peak heights are measured. Our experience has been that either the ethyl acetate has no effect on the peak height or the PTH amino acid is destroyed and virtually no peak is detectable. We have rarely found ethyl acetate that gave an intermediate result. Thus, with this simple test, relatively inexpensive (Burdick & Jackson) or moderately expensive ethyl acetate (Pierce) can give good results. We have not tested chemicals from other firms in any systematic way, although many investigators have had extensive experience with other sources.

Additives to the various chemicals and solvents have been used less frequently in recent years than initially. The use of HCl and ascorbic acid in the ethyl acetate as originally recommended by Edman is now less frequently used and the use of reducing agents in the butyl chloride, while still in vogue, is used less frequently with more sensitive and alternative identification systems for HPLC analysis of serine and threonine.

Stills are required to purify chemicals and cost about $1 000. A spinning band still costs about $4 000–10 000 (B/R Instrument Corpn., Pasadena).

15.6. MAJOR PROBLEM AREAS WITHIN THE AUTOMATED SEQUENCER

15.6.1. Vacuum system

Most difficulties with automated sequencing are traceable to the vacuum system. Vacuum leaks and vacuum maintenance involve the most servicing time, both from the point of view of the field service engineer as well as the sequencer operator. The key element in correcting a vacuum leak is to isolate the source and this involves a series of straightforward maneuvers that allow the operator to determine whether the leak is at the cup level, between the cup and the low vacuum manifold or between the low vacuum manifold and the pump.

In most instances, the operator will ask whether the vacuum leak exists with the low vacuum to the cell turned on, or not. This typically will isolate the problem either at the cup level or elsewhere and then one normally traces back the vacuum leak rather than approaching it from the pump to the cup.

15.6.1.1. Leaks at the cup

If the problem is at the cup level there are two major sources of leaks. One concerns the delivery lines themselves and here both cracked diaphragms and improper placements of the tubes can often lead to significant vacuum leaks. Secondly, although much less commonly, leaks can occur around the scoop and nitrogen flush lines. Finally, the O-rings themselves need to be scrupulously checked for distortion, breaks or excessive swelling.

15.6.1.2. Leaks at the low vacuum manifold

There are three exit lines from the low vacuum manifold: one goes to the nitrogen ballast, the second goes to the valves and the third to the low vacuum solenoid. Each of these outlets should be blocked in turn and the vacuum retested. We normally use a greased rubber stopper. When the difficulty is in the line going to the nitrogen ballast, problems generally are in the nitrogen pilot valve. This three-way valve that comes off the nitrogen manifold is often a problem, particularly with DMAA machines. A second common problem with this valve is that there may be too much pressure from the nitrogen source and the valve will remain continually open. When the leak is traceable to a delivery valve it often points to one of the diaphragms being split. This can generally only be fixed by an experienced engineer. However, the most common leak will be traceable to the low vacuum solenoid and in machines equipped with DMAA this is a common source of difficulty. The low vacuum solenoid needs to be replaced frequently and is a source of frequent leaks.

15.6.1.3. Leaks between the pump and the low vacuum manifold

These are most easily diagnosed by disconnecting the line to the low vacuum manifold and plugging it with a greased rubber stopper. At this point, the low vacuum pump should be pulling below 50 millitorr and if it is not, either there are leaks between the low vacuum manifold and the pump or the pump itself needs to be replaced or repaired. The latter can be excluded by testing the vacuum of the isolated pump itself. Changing the oil at this point, swapping the two vacuum sensors and sensor lines and checking the fittings between the low vacuum manifold and the pump are obvious approaches.

15.6.2. Vacuum leaks in the single pump configuration

In the single pump configuration, the low vacuum pump is utilized only to actuate the delivery valves and generate vacuum in the fraction collector and is generally not a problem. Our low vacuum pump has been in operation for over two years without even a change of oil! The 'high' vacuum system, however, is occasionally a source of leaks and, in our experience, these leaks have been more difficult to trace, although less frequent than in the two-pump system. With the shut-off valve prior to the entry of the cold trap, one is immediately able to assess whether the leak is before the cold trap or not. This is particularly useful since many things that appear to be 'vacuum leaks' are, in fact, problems that are more related to an iced-up cold trap. This is easily assessed with the aid of this shut-off valve. If the vacuum is satisfactory here, the valve can be turned on and the back of the cold trap plugged with a large vacuum-greased rubber stopper. True leaks at the cold trap are due either to leaks through the drain line or poor sealing at either end of the cold trap. Complete defrosting of the unit and manipulation of that exit line usually solves the former problem, although generally only temporarily. A better design for draining is obviously needed. Sealing problems can usually be rectified by replacing or regreasing O-rings at those joints.

In our experience, vacuum leaks between the cold trap and the cup have been very unusual and neither the solenoids nor the air cylinder valve have been a serious source of trouble.

15.6.3. Programmer

Programmer problems are difficult to trace and generally require a specialist for repair. A common problem, however, which is difficult to diagnose but easy to repair, concerns the failure of the various counters (Veeder-Rooder) in the sequencer. These are typically diagnosed by a reagent delivery overflow shut-off which generally means that during the reagent delivery step, one of the counters has stuck. When this happens and reagent delivery exceeds a certain predetermined time, the machine goes into a standby state. In our experience, the vast majority of reagent overflow delivery time limit problems are traceable to the Veeder-Rooder counters. If this can be isolated, it can be easily reparied by replacing the counter. Other programmer problems which involve the brushes themselves are much more complicated and require an expert. Appropriate maintenance and cleanliness of the head of the programmer is also essential, although often neglected in many laboratories.

15.6.4. Relays

The electronics of the instrument are extremely reliable. Important weak points, however, are the relays. There are some simple indications to the source of problems caused by relay failures. The first is that the reagent overflow delivery limit will be tripped. The most common cause of this is a failure of the Veeder-Rooder counter; relay failure may also be apparent here. Another relay is involved in stepping the fraction collector and we have traced several failures of fraction collector advancement to relay problems. The only simple practical solution to this is to have several spare relays at hand.

15.6.5. Valves

The delivery valves utilized by the model 890C sequencer are the subject of considerable controversy. Many investigators have introduced the Wittmann-Liebold modifications to the 890C sequencer and have used difficulties with the delivery valves as one of the major reasons for their modification. One must be continually alert to detect inefficiency in the sequencer which is traceable to the delivery valves in order to spot problems early. This generally manifests itself as a decreased repetitive yield. A common problem involves cracked diaphragms. The failure of the diaphragms within the delivery valves is well known to anyone who owns a sequencer and simple tests for checking the diaphragms are not known to most people. In our laboratory, diaphragm problems are generally detected by pulling a vacuum into the cup and watching the delivery line with the valves in the 'off' position. The simple test involves delivering each reagent and solvent to the cup so that each delivery line is filled. At this point, applying vacuum to the cup should clear the line and then delivery should stop. It is important to appreciate that delivery should *completely* stop from *every* line. There must not be any leakage through these lines because during the long vacuum steps after coupling, cleavage and extraction, any delivery into the cup can be detrimental especially if it is one of the reagents. Nicked or eroded valve surfaces also occur and produce similar symptoms.

The waste collect valve is rarely a source of difficulty unless an automatic converter is being used.

The many solenoid valves throughout the machine also have a tendency to fail, which often causes loss of delivery of solvent or reagents, due to swollen or stuck valves. An ample supply of Skinner valve replacement kits are kept expressly for such occasions.

15.7. MONITORING MACHINE PERFORMANCE

In general, the successful operation of a sequencer involves careful attention to detail and prevention or early detection of problems. Be certain prior to the commencement of every run that the vacuum at the cup level is appropriate, that there is no backflow through the delivery valves, that there are sufficient solvents and reagents in the instrument and that reagent and solvent bottles are adequately sealed to air leaks to ensure a successful operation. It is surprising how many laboratories do not take the trouble to make these simple routine checks. Although we have utilized three automated sequencers in our laboratory for almost 13 years, we routinely sequence myoglobin on one instrument each week. While this can be considered a luxury, we have found it to be invaluable in the early detection of a number of difficulties with each instrument. In addition, virtually every run that involves a radioactive sample is monitored with an internal standard (a ^{35}S-labeled immunoglobulin produced by the MOPC21 plasmacytoma). Preparation is quite simple, as one simply grows the MOPC21 line in ^{35}S-methionine. When dialyzed, the supernatant from such a culture can be shown to contain on a reduced SDS gel two very clean radioactive peaks, corresponding to the mol. wts. of heavy and light chains. The MOPC21 light chain has methionine residues in positions 4, 11 and 13, while the heavy chain has no methionine up to position 34. Therefore, for most purposes, an entire dialyzed supernatant can serve as an internal standard. For more quantitative assessment of machine function, heavy and light chains can be separated and isolated on SDS gels or Sephadex columns and a single chain applied to the sequencer. This standard is ideal from a number of points of view. One can accurately calculate the repetitive yield on the basis of positions 4, 11 and 13. Therefore, this is useful for an overnight assessment of the performance of the sequencer. In our hands, the repetitive yield will vary between 92 and 97 % and is virtually always an accurate reflection of what we would find were we to do the same thing with myoglobin. Indeed, myoglobin can be included into this cocktail and sequenced for 12 steps (Val-Leu-Ser-Glu-Gly-Glu-Trp-Gln-Leu-Val-Leu-His-Val-Trp→) and steps 1, 2, 10 and 11 assessed by GC and HPLC analysis while the radioactive standard is assessed at positions 4, 11 and 13. If there are discrepancies between the two (more typically the myoglobin repetitive yield being lower than the radioactive standard), the problem is generally with the conversion system, e.g. the ethyl acetate, or with the heating bath, contaminants in the nitrogen blow-down system, overheating in either the reaction bath or in the dry-down bath, etc.

Two other means of monitoring machine performance are often used in our laboratory. Both concern the use of temperature profiles during the run. One can easily connect an inexpensive recorder to the temperature gauge of the reaction chamber and after a short period of time, the profile becomes akin to an electrocardiogram to a cardiologist. The opening and closing of various valves, the entry and exit of solvents and reagents and the efficiency of vacuum can be quite accurately assessed. Recently, we have taken to measuring the temperature in the cold bath by a similar procedure and have a single temperature gauge attached to probes that are installed in our cold traps. The probes are connected to a central receiving unit and the temperature is transmitted onto a linear chart recorder. The temperature in the cold bath reflects the effectiveness of the cold trap and therefore sequencer operation, although the major reason for its use is to detect failures of the cryocool unit, leaks in the vacuum systems that are around the cold trap and to act as a predictor of the most appropriate time to defrost the unit. The latter is also indicated by a rapid decrement in vacuum efficiency even after draining the trap. One serious problem particularly in humid climates with the cold trap modification involves condensation on the top of the unit and during every defrosting cycle. The ethanol bath becomes diluted with water. This effectively raises the temperature within the cold finger and thereby leads to inefficient freezing of solvents and reagents by the cold finger. The use of the temperature monitor in this instance is an excellent means of determining when and if this becomes a problem.

Editorial Note Sequence data can be obtained from the following addresses.
For nucleic acids: Genbank, 10, Moulton Street, Cambridge, MA 02238, U.S.A.
EMBL Nucleotide Sequence Data Library, EMBL, Postfach 10 22 09, D-6900 Heidelberg, West Germany.
For proteins:
Protein Sequence Database, National Biomedical Research Foundation,

Georgetown University Medical Center, 3900 Reservoir Road, N.W., Washington, DC 20007, U.S.A.
Professor R. F. Doolittle, Department of Chemistry, University of California, San Diego, La Jolla, California, U.S.A. 92093.
Addresses kindly supplied by Dr. Rodger Staden.

15.8. REFERENCES

Begg, G. S., Pepper, D. S., Chesterman, C. N. & Morgan, F. J. (1978). *Biochemistry*, **17,** 1739–1744.

Begg, G. S., Leslie, B. H. & Morgan, F. J. (1984). *Anal. Biochem.*, **138,** 30–33.

Brauer, A. W., Margolies, M. N. & Haber, E. (1975). *Biochemistry*, **14,** 3029–3035.

Croft, L. R. (1980). *Handbook of Protein Sequence Analysis*, 2nd edn, Wiley, Chichester, New York, Brisbane, Toronto.

Dayhoff, M. O. (1972). *Atlas of Protein Sequence and Structure*, vol. 5, National Biomedical Research Foundation, Washington.

Dayhoff, M. O. (1973). *Atlas of Protein Sequence and Structure*, vol. 5, suppl. 1, National Biomedical Research Foundation, Washington.

Dayhoff, M. O. (1976). *Atlas of Protein Sequence and Structure*, vol. 5, suppl. 2, National Biomedical Research Foundation, Washington.

Dayhoff, M. O. (1978). *Atlas of Protein Sequence and Structure*, vol. 5, suppl. 3, National Biomedical Research Foundation, Washington.

Edman, P. & Begg, G. (1967). *Eur. J. Biochem.*, **1,** 80–91.

Hermodson, M. A., Ericsson, L. H., Titani, K., Neurath, H. & Walsh, K. A. (1972). *Biochemistry*, **11,** 4493–4502.

Hunkapiller, M. W. & Hood, L. W. (1978). *Biochemistry*, **17,** 2124–2133.

Klapper, D. G., Wilde, C. E. & Capra, J. D. (1978). *Anal. Biochem.*, **85,** 126–131.

McCumber, L. J., Qadeer, M. & Capra, J. D. (1980). In *Methods in Peptide and Protein Sequence Analysis*, Proc. Third Internat. Conf., Birr, C. (Ed.), Elsevier/North Holland, Amsterdam, New York, Oxford, pp. 165–172.

Niall, H. D. (1973). *Methods Enzymol.*, **27,** 942–1010.

Tarr, G. E., Beecher, J. F., Bell, M. & McKean, D. (1978). *Anal. Biochem.*, **84,** 622–627.

Waterfield, M. & Bridgen, J. (1975). In *Instrumentation in Amino Acid Sequence Analysis*, Perham, R. N. (Ed.), Academic Press, London, New York, San Francisco, pp. 41–47.

Practical Protein Chemistry—A Handbook
Edited by A Darbre
© 1986, John Wiley & Sons Ltd

JOHN K. INMAN and ETTORE APPELLA
Laboratory of Immunology, National Institute of Allergy and Infectious Diseases and the Laboratory of Cell Biology, National Cancer Institute, National Institutes of Health, Bethesda, Maryland 20205, USA

16

Newer Methods of Solid- and Liquid-phase Sequence Determination—Personal Views

CONTENTS

16.1. INTRODUCTION

Over the years there has been a steady flow of reports on new approaches to amino acid sequence determination and on suggested improvements of existing methodology. It is usually difficult for the inexperienced investigator and often the busy specialist to ascertain whether a new procedure is a 'sure thing' or simply a good start that needs considerable development before becoming a reliable method. Most original reports are understandably enthusiastic and perhaps biased in a positive way. In this discussion, we present our thoughts on newer techniques and ideas which have interested us in the hope that this presentation will help new investigators to avoid some frustration and will guide more experienced workers who have a particular need for less proven methods. In addition, we have pointed out specific needs for yet undeveloped approaches and unborn ideas with the view of stimulating more pioneering research in methodology.

16.2. SOLID-PHASE SEQUENCING: NEWER APPROACHES

16.2.1. General comments

Laursen (1966) first reported the sequencing of peptides that had been attached covalently to an insoluble support and subsequently developed this technique into an automated method (Laursen, 1971). The impetus for this work came, certainly in part, from the promise and early success of solid-phase synthesis (Merrifield, 1963). Both in peptide synthesis and degradation, a repetitive sequence of reactions is carried out. Within each cycle the individual steps must have few side reactions and be completed in high yield in order to ensure a satisfactory result over many cycles. Reactions are driven to completion by using excess of reagents, and the task remains of repeatedly having to separate an intermediate product from by-products and excess reactants. This job, which is tedious and often wasteful in solution chemistry, is swiftly handled by filtration, or an equivalent process, in solid–liquid phase reaction systems. The promise of running highly efficient reactions in these systems with great saving of labor has been largely realized. However, new problems arise that compromise these gains; they originate from the accumulation of side-reaction effects involving the immobilized component and support structures.

Various developments have improved the techniques for performing Edman degradations on covalently immobilized peptides. At present, solid-phase sequencing is a practical approach that can be considered complementary to the spinning cup method of Edman and Begg (1967) provided one uses amounts of peptide in the range of about 10 to 200 nanomoles. Established solid-phase methodology will not be discussed here; instead the reader is referred to the proceedings of the first four international conferences devoted to protein sequence analysis (Laursen, 1975; Previero & Coletti-Previero, 1977a; Birr, 1980; Elzinga, 1982) and discussions by Laursen (1977a), Machleidt and Wachter (1977) and Wittmann-Liebold (Chapter 12, this volume).

In experienced hands solid-phase sequencing can sometimes be accomplished with less than 10 nmol of peptide when chemical methods of detection are used. We believe that significant improvements in the methodology are needed, and are indeed possible, before this approach can be carried out confidently at the subnanomole level. The potential for improvement of the solid-phase method is very large; suitable development of this approach should meet a growing need for sequence analyses on extremely limited samples of peptide. Very sensitive methods for chemical detection and identification of cleaved amino acid residues are presently available. In particular, the latest mass spectrometers promise great sensitivity. Even though very sensitive detection methods are at hand, the limiting level for obtaining reliable sequence information will be determined by the signal-to-noise ratio, that is, by the background. In principle, the solid-phase approach offers the possibility of obtaining extremely low background, but, for this potential to be realized, there must be considerable further development in all chemical and mechanical aspects of the methodology. The solid-phase approach specifically allows the following:

1. thorough washings with an almost universal choice of solvents without loss of peptide;
2. a wide choice of coupling reagents and solvents without constraints on volatility or extractability (large excesses of reagents are no major problem);
3. continuous removal of mobile by-products during the course of the reactions in order to minimize their possible side reactions with the peptide or support;
4. useful strategies involving specific internal cleavages that can be performed on the peptide before or after attachment, or between sequencing routines (Horn, 1975; Laursen, 1977a,b).

Background in subnanomole sequencing obviously can be reduced by employing a very small mass of support (a few mg or less) and miniaturizing the entire apparatus. Thus, all potentially contaminating surfaces are kept at a minimum. Existing hardware may prove inadequate as components and designs for automated microsequencers may have to evolve from semi-manual beginnings. Hydraulic rather than mechanical principles may serve better for control of delivery functions.

The greatest challenge lies with chemical innovation and development; the great flexibility of the solid-phase approach has barely been tapped. In addition to lowering background, sequencing efficiency (initial step yields and repetitive yields) will be improved only through applying a more detailed understanding of all chemical reactions involved (wanted and unwanted). Systematic re-investigation of the Edman degradation and further experimentation with other degradative methods is indicated. In the discussion which follows, we express what we feel are some interesting directions for the chemical development of the solid-phase approach. No attempt has been made to review or cover all new ideas which may have been suggested at one time or another.

16.2.2. Supports

Solid-phase sequencing has been performed almost exclusively with peptides covalently bound to functionalized matrices of three types: (a) polystyrene (1 % cross-linked with divinylbenzene), (b) porous glass or (c) polyacrylamide derivatives. The preparation and properties of such supports have been presented and discussed recently (Machleidt & Wachter, 1977).

An ideal support for immobilizing peptides should have the following characteristics:

1. low resistance to flow (in columns), entailing both low particle fragility and minimal volume change throughout each degradation cycle;
2. complete resistance of the matrix to breakdown;
3. a high specific surface area readily accessible to surrounding fluids;
4. adequate functional group density for efficient attachment;
5. convenience in functionalization;
6. chemical stability and inertness;
7. properties allowing favourable reaction kinetics (approaching free solution kinetics).

No presently known support can meet all the above criteria.

Low cross-linked polystyrene and polyacrylamide particles must be highly swollen with solvent in order that functional groups or attached peptides will be accessible to reagents; thus, the choice of solvents is important. Solvent changes often result in shrinkage or expansion of these gel-type polymers causing channeling or blockage in the reaction column. To minimize these problems, the peptide-polymer particles have been carefully mixed with many volumes of small glass beads. This additional material, however, increases the required volumes of reagent solutions, washes and eluents and can adsorb reagents or by-products. Consequently, there will be an elevation of background material in the analyzed samples which can be a serious limitation in microsequencing. In contrast, porous glass is rigid and does not present these difficulties. Thus, a wide variety of solvents can be used with no back-pressure problems, provided that attachment has been performed with due regard for the fragility of the particles. Unfortunately, glass surfaces are somewhat soluble in alkaline media, and care must be taken to avoid exposure to pH values above 8.5 or 9. Despite this problem, successful sequencing has been carried out with glass supports.

Polystyrene (low cross-linked) provides an unsatisfactory matrix for supporting proteins and very large peptides. Mutual incompatibility between solvents and the peptide or matrix causes structural changes that are unfavourable for reactive approach of reagents to the *N*-terminal groups. This type of problem was overcome in the case of a polyacrylic acid matrix by amidating it with sulfonated *p*-phenylenediamine (Cavadore & Vallet, 1978); the protein was attached to subsequently produced isothiocyanate groups, and the neighboring sulfonate ions provided a local hydrophilic environment needed for solvent compatibility.

All commonly used supports leave something to be desired in regard to their contribution to background; their involvement in non-ideal reaction kinetics is an important question since cycle-to-cycle repetitive yields seldom exceed 95%.

Further progress in developing new supports will depend on an understanding of the nature of chemical reactions in multi-phase systems. It will be necessary to abandon the notion that a solid support is equivalent to an inert wall to which the peptide is merely anchored. Investigators in the field of solid-phase peptide synthesis now follow the concept of a dynamic support–peptide–solvent system ruled by various kinetic and thermodynamic principles; for example, see the discussion by Fankhauser & Brenner (1973). Non-ideal reaction kinetics may not depend so much on impeded diffusion or mass transport inside the particles, but rather on impaired solvation of the peptide and support matrix (Hancock *et al.*, 1973). Solvation can be affected markedly by local internal phase separations. In certain micro-environments peptide and matrix can interact strongly in a way that is equivalent to precipitation which results in a closing off in a mechanical sense. In other situations local liquid–liquid partitioning may occur whereby reagent concentration is especially low in the micro-phase surrounding the peptide terminus. This problem could be avoided by use of a single-component solvent (Previero, 1977). Local repartitioning by either of the foregoing mechanisms may occur throughout a sequential degradation process and cause intermittent or permanent non-reactivity of certain chains. Pre-existing inhomogeneities within the matrix structure (uneven distribution of cross-links, etc.) could augment this problem. Functionalization of porous glass probably leaves the silica surface coated with a cross-linked siloxane polymer film rather than with a lawn of single-point attached groups. Thus, glass supports should exhibit similar local barriers to reaction as occur in gel polymers.

Several groups have reported on the preparation, properties, and use in solid-phase peptide synthesis of acrylic copolymers composed mainly of poly(*N*,*N*-dimethylacrylamide) (Arshady *et al.*, 1979; Atherton *et al.*, 1979; Stahl *et al.*, 1978; Stahl *et al.*, 1979a) or poly(*N*-acrylylpyrrolidine) (Stahl *et al.*, 1979b; Smith *et al.*, 1979). These polymers show a wider range of compatibility with polar and moderately non-polar solvents (and presumably also with attached peptide) than do polyacrylamide or polystyrene supports. Thus, the above functionalized copolymers offer attractive possibilities as supports for solid-phase degradations of peptides.

We have approached the design of supports by starting, first of all, with a rigid matrix in order to avoid the serious swelling-blockage problem. Macroporous polystyrene was chosen instead of porous glass so that very alkaline coupling reactions could be used. Also, polystyrene allows very versatile, stable and reproducible functionalization. Nonetheless, it was realized that this rigid, highly cross-linked matrix would still possess some flexibility at the molecular level and would present a great variety of semi-permanent (including unfavourable) micro-environments in the hydrophobic surface structures.

Our first goal was to place attachment functions (amino groups) on the ends of spacer arms. If the latter were polymers of appropriate size and composition, they would serve also as a co-solvent component and promote the formation of a phase separated from the matrix wherein the peptide and a variety of solvents would be compatible. In principle, conditions could then be found that would allow all peptide chains to be appropriately solvated and in good communication with reagents. Polyethylene glycol (polyoxyethylene) was selected as the 'space-modifying' polymer as suggested by the experimental approach to solid-phase co-solvents reported by Regen and Dulak (1977) and Regen (1977). Furthermore, stepwise peptide syntheses have been carried out successfully with the growing peptide attached to the ends of polyethylene glycol chains (Mutter & Bayer, 1974) with coupling kinetics similar to that observed in free solution (Bayer *et al.*, 1975). Derivatives of macroporous polystyrene were accordingly prepared as described in detail (Inman *et al.*, 1977). Preliminary sequencing trials were reported (Appella *et al.*, 1977). Excellent flow characteristics and very low background in analytical samples were found. However, subsequent experience with the Edman degradation of peptides attached to short spacers, compared with the same ones fixed to small polyethylene glycol chains (up to 13 oxyethylene residues), showed no consistent differences in initial or repetitive yields of PTH amino acids. We have concluded that either the total mass or the molecular size of the grafted polymer was not sufficient to produce the co-solvent effect or to prevent strong polystyrene-peptide interactions.

In conclusion, a possible approach to an ideal sequencing support would be the preparation of grafted copolymers in which one component acts as a rigid scaffold supporting the other polymer to which the peptide is covalently linked. The latter polymer would be chosen for its compatibility with the various solvents and reagents required for attachment and degradation. By itself, this co-solvent polymer, because of its solvent compatibility and openness of structure, would be mechanically unsuitable in a reaction column. The scaffold polymer, because of its surface complexities would itself be an unsatisfactory anchor. In brief, the above concept is the pellicular design extended to macroporous scaffolds. The right choice of the co-solvent polymer would be the most critical factor for success.

16.2.3. Attachment of peptides to supports and capping

The solid-phase approach entails special problems arising from the need to fix peptides to supports, usually through covalent bonds. Although peptides may be bound by adsorptive forces alone, such an immobilization would be far less secure; a liquid-phase elution must occur at least once during each reaction cycle, which could result in loss of peptide. Successful adsorption would require careful matching of individual peptides and adsorptive surfaces. Furthermore, strong adsorption may rule out satisfactory solvation and mobility of the peptide and hence optimal reaction kinetics (see section 16.2.2, discussion on supports). In spite of these potential drawbacks, Hewick *et al.* (1981) reported the successful sequencing of many peptides and proteins (>30 residues) with a miniaturized, solid-phase sequencer wherein the material to be studied is adsorbed to a polybrene-treated glass fiber disc (see Chapter 17). Loss of peptide is minimized by the high surface area of the support and by employing the most polar reagents (trimethylamine and trifluoroacetic acid) in a gaseous phase.

Covalent attachments are also beset with problems; however, the possibilities for new approaches or improved techniques seem open-ended. Nonetheless, a satisfactory, *general* means for attaching peptides solely through their *C*-terminal residues has yet to be found. Several excellent *C*-terminal attachment methods can be applied to special cases and are widely used. These procedures and a few general but non-*C*-terminal specific methods have been discussed by Wachter *et al.* (1975), Laursen (1977a) and Wittmann-Liebold (Chapter 12).

Direct coupling of peptides to supports involves a bimolecular mechanism. In order for the reaction to proceed to high yield with very small amounts of peptide, a substantial, effective concentration of nucleophilic (usually amino) groups must be provided by the carrier. Then, pseudo-first-order kinetics are possible since the amino groups are in large molar excess. Peptide amino groups are at low concentration and therefore need not be protected during attachment except in special circumstances (Laursen, 1977a). They should be blocked in some cases where the peptide is activated in a preliminary step before adding the support; protection of *N*-termini is seldom required for single-step attachments. Where protection is needed, we have introduced the *t*-butyloxycarbonyl (Boc) group with one of the newer reagents, 2-*t*-butyloxycarbonyloxyimino-2-phenylacetonitrile (Appella *et al.*, 1977). Amino protection (as with Boc) is necessary before reacting carboxyl groups with a carbodiimide in the interesting *C*-terminal attachment approach of Previero *et al.* (1975a) wherein subsequent incubation under alkaline conditions converts acidic side-chain functions into stable *N*-acylureas and a fraction of the *C*-termini into reactive oxazolinones.

Under the usual circumstances, activation of a peptide with reagents such as carbodiimides will lead to attachment through some of the side chain carboxyl groups. Complications in sequencing can then occur, such as blank results if all chains are attached at given internal positions in the sequence. Sequencing can proceed beyond these positions if additional attachment has occurred further on toward the *C*-terminus. If attachment reactions are not driven too hard, some chains will have free carboxyl groups at any given Asp or Glu position. Mech *et al.* (1976) observed that attachment (*via* carbodiimide) of small peptides having an internal Asp or Glu residue proceeded preferentially through the *C*-terminal carboxyl group. There were sufficient peptides having free acidic side chains to permit identification of Asp and Glu when sequencing. L'Italien and Laursen (1981) in fact successfully sequenced a large number of small peptides by solid-phase Edman degradations following their direct attachment to aminopolystyrene resin at pH 5.0 by means of a water-soluble carbodiimide. Their procedure was a modification of the method of Wittmann-Liebold and Lehmann (1975). The side-chain attachment problem thus can be handled in a general way through this thermodynamic bias and/or in a simply statistical manner by carefully limiting the yield of the attachment reaction.

Chang *et al.* (1977b) described improved conditions for coupling proteins to diisothiocyanate-activated porous glass beads. They reported on the immobilization of large peptides through histidine and tyrosine residues to diazotized arylamine supports or through cysteine side chains to an iodoacetamido derivative of glass beads. Birr and Garoff (1977) described the use of phenacyl ester bonds for the attachment of peptides or intermediate amine-bearing compounds. A number of investigators (e.g. Appella *et al.*, 1977; Di Bello *et al.*, 1977; Laursen, 1977a) have used carbodiimides in conjunction with active ester-forming additives such as *N*-hydroxysuccinimide (HOSu) or 1-hydroxybenzotriazole (HOBt) for the activation of peptide carboxyl groups. The rapid formation of active esters with these substances serves to suppress the formation of inactive *N*-acylurea derivatives. Carbodiimides and additives are necessarily present in large excess in order to promote reasonable reaction rates, but they may then react with each other to give undesirable by-products. Such a situation is known to occur for dicyclohexylcarbodiimide (DCC) and HOSu (Low & Kisfaludy, 1965; Gross & Bilk, 1968) or HOBt (Jakubke & Klessen, 1977). Another additive that has been used more recently in peptide synthesis, *N*-hydroxy-5-norbornene-2,3-dicarboximide, was shown to be quite stable in the presence of DCC (Fujino *et al.*, 1974). This additive, therefore, might serve more satisfactorily in solid-phase attachments; in addition, it is soluble both in water and organic solvents.

Lee and Riordan (1978) recently reported a simple and convenient means for activating peptide carboxyl groups prior to immobilization. When peptides are dissolved and allowed to stand in a mixture of trifluoroacetic acid (TFA) and trifluoroacetic anhydride, carboxylic mixed anhydrides are apparently formed. The volatile components are then removed, the activated peptide is dissolved in DMF and the amino resin is added. Temporary dissolution in TFA has the added advantage of promoting solubility of peptides in DMF (Machleidt & Wachter, 1977). Attachment yields of small to medium-sized peptides ranged from 55 to 98 %, reflecting the fact that trifluoroacetate is the better leaving group (Lee & Riordan, 1978). Partial deamidation of glutamine may have occurred in one example given. Cyclic imide formation is possible when peptides containing aspartic acid residues are treated in this way, but this question was not investigated. The above problems may be less serious if conditions can be found for activating only a portion of the carboxyl groups. Possibly *C*-terminal activation could then be selectively facilitated through intermediate oxazolinone formation.

An improvement over the diisothiocyanate method for attaching lysine-containing tryptic peptides involves the use of a hetero-bifunctional reagent, *p*-isothiocyanatobenzoyl-DL-homoserine lactone (Herbrink *et al.*, 1975; Tesser & Lamberts, 1976). Peptide amino groups are first combined with the reagent through its -SCN function. The lactone group and free amino termini (minus one residue) are regenerated by treatment with TFA; the peptide is then joined to an amine-bearing support by reaction with the lactone ring which remains attached to the *C*-terminal lysine side chain. The lactone function (Horn & Laursen, 1973) has been shown by many workers to be highly efficient in promoting attachment. A large excess of the above reagent is not required in order to prevent cross-linking of peptide since its two groups have very different reactivities.

If peptides can be initially linked to a support by an easily established, temporary bond, then a desired permanent attachment may be effected by an efficient intramolecular reaction. Roughly, this strategy was accomplished by Schiltz (1975) who attached non-lysine-containing peptides with the use of *p*-phenylene diisothiocyanate through their α-amino groups, then effectively joined the carboxyl group(s) to the support by means of an ordinarily less efficient carbodiimide coupling. The effective concentration of peptide at the surface of the support was greatly increased for the second reaction. Subsequent treatment with TFA released the *N*-terminus. A truly intramolecular attachment mechanism was explored by Wachter and Werhahn (1977) who employed a 'four-component condensation' approach originally introduced by Ugi (1962). The above two approaches are more advanced in concept and have an intriguing potential.

Intramolecular assistance to attachment reactions can also involve the nucleophilic partner held by the support. Such a mechanism may explain the increased reactivity of 1,2-diamines over monoamines as discussed by Horn and Laursen (1973). For example, matrices substituted with ethylenediamine or triethylenetetramine (TETA) are more reactive than ω-alkylamines toward the homoserine lactone ring. Therefore, more attention should be directed to the character of nucleophilic groups on supports so that they will be optimally efficient in conjunction with the chosen method of carboxyl activation.

Excess unused amino groups on the support can give rise to problems. If left alone, they will be 'capped off' with phenylthiocarbamyl groups during the first coupling step of an Edman degradation. Poor degradation yields (in our hands) and high background in analyses (Laursen, 1977a) can result. The use of methyl in place of phenyl isothiocyanate for the first degradative cycle gives considerably improved results in both respects. The finding of still better capping agents may be important for the success of very low level sequencing methods. It is essential that the capping reaction does not irreversibly block the peptide amino terminus or alter the amino acid side chains. Acetylation of excess amino support groups with mildly active esters of acetic acid could serve for capping if peptide amino termini were first protected with a Boc, or other removable group. Recently we have capped supports with methyl acetimidate after attaching Boc-peptides to macroporous polystyrene supports (G. C. DuBois, V. Alvarez & E. Appella, unpublished results); subsequently, these peptides were successfully sequenced. The resulting positively charged amidine groups on the support might usefully increase its hydrophilic character and solvation. In fact, the capping step provides a general opportunity to improve the carrier surface structure. However, capping should be designed especially with the view of avoiding subsequent side reactions between capped functions and bound peptides which can lead to poor degradation yields and a high background.

16.2.4. Edman degradations

Sequential degradations of peptides on insoluble supports have been carried out almost exclusively by the Edman method using reagents and conditions very similar to those employed in liquid-phase sequencing. For discussions of these procedures, the reader is again referred to sources cited in section 16.2.1. Future success in using the solid-phase approach for very low-level sequence analysis will depend upon important improvements in degradation chemistry as well as progress in support design and attachment methods. A major advantage of solid-phase chemistry is the flexibility in choice of solvents, washes, buffers, additives, reagents and temperatures. Extraction losses and volatility or solubility considerations impose very few constraints on operating parameters except where peptides are attached by adsorption or where reactions are carried out in a vapor–solid-phase mode.

High purity of reagents and solvents should be given the same consideration in solid- as in liquid-phase work. Deleterious impurities and side reactions can not be entirely eliminated, but their effects can be reduced by employing the smallest feasible volumes and concentrations of reagents. Solid-phase reactions permit a continuous flow-through of fresh reagent and removal of by-products, an advantage that can be easily realized in miniaturized systems where heat transfer (in preheating of reagents) is less of a problem.

16.2.4.1. Coupling reactions

In the coupling reaction (the formation of phenylthiocarbamyl peptides or their analogs) buffers do not play a role in film formation and retention of peptide; the reaction produces or consumes very little hydrogen ion and may be carried out with quite low concentrations of buffer. Additives such as protein denaturants, reaction catalysts and scavengers offer interesting possibilities for specific improvements in the coupling step; however, these advantages must be weighed against possible further side reactions. Scavengers such as primary amines, in molar concentration exceeding that of the peptide but not of the isothiocyanate, could aid in microsequencing. Small amounts of aldehydes or other highly reactive impurities could thus be removed. The free amine should still be low in concentration so that side reactions similar to those occurring after phenylthiocarbamyl capping of support groups would not be appreciable.

The choice of coupling solvent is critical. The issue of aqueous versus non-aqueous media rests in part on the importance of adequate solvation of bound peptides of various compositions. Water promotes solvation but also causes hydrolysis of the isothiocyanate giving rise to subsequent side reactions. An open mind on this question is recommended.

16.2.4.2. Cleavage reactions

Trifluoroacetic acid appears to be one of the best cleavage acids found so far. It is more easily purified than higher homologs of the perfluoro acids, and its volatility can be advantageous for the recovery and conversion of cleavage products (thiazolinones). In principle, other strong acids, such as BF_3, may be substituted. A critical test of various cleavage methods is the recovery of serine PTH or some other derivative of this amino acid suitable for identification.

In solid-phase approaches the thiazolinones are usually eluted with the cleavage acid. Since TFA, for example, is a strongly eluting solvent, various adsorbed substances tend to be stripped from the support at this stage and find their way into the analyzed sample. However, an advantage of this procedure is the option of continuous elution of acid during the cleavage step with collection of eluate at a lower temperature. If reaction columns are small, thiazolinones are cooled and preserved soon after they are formed. Cleavage reactions are carried out at elevated temperatures for periods sufficiently prolonged to cleave the most resistant residue, usually proline (Hermodson *et al.*, 1972; Brandt *et al.*, 1976). Another approach is possible. It is feasible to cyclize and cleave with acid vapor acting on the dry, bound peptide (Jentsch, 1975; Hewick *et al.*, 1981). After expelling the acid vapor with a stream of inert gas, the thiazolinone may be recovered with a less eluotropic solvent than TFA and thereby be associated with smaller amounts of background material. Acid cleavage would have to be completed before elution (or done in several stages with intermittent drying). The highly reactive thiazolinones then would be concentrated on the support surfaces at an elevated temperature and be prone to destruction. The proper choice of cleavage mode again must rest upon carefully controlled experiments, and much work still remains to be done.

The effects of water content of TFA need to be better understood and controlled. If the acid is too anhydrous, it will be in equilibrium with small levels of trifluoroacetic anhydride. The latter component can react with the phenylthiocarbamyl peptide to yield a non-cleavable by-product (Barrett, *et al.*, 1977). The anhydride may linger on the support by adsorption or through potentially reactive trifluoroacetyl derivatives of diverse support structures (e.g. -OH and -NH- groups; Schuttenberg & Schulz, 1976) which then could cause blocking of peptide *N*-termini following treatment with coupling buffer. On the other hand, too much water in the TFA may bring about an unacceptable degree of internal peptide bond splitting ('nicking') and reduce cleavage rates. It is not at all clear where the best compromise may lie. Convenient and reliable methods for measuring the water content of TFA are required.

16.2.4.3. Conversion reactions

Identification of thiazolinones usually follows their conversion into PTH-amino acid derivatives as in liquid-phase sequencing. Some recent advances in their detection and quantitation are discussed in section 16.3.3. A new conversion reaction, which has been employed in solid-phase and manual sequencing, is the rapid aminolysis of thiazolinones, in the dry state or in solution, with primary alkylamines to yield phenylthiocarbamyl-amino acid alkylamides (Inman & Appella, 1975, 1977; Appella *et al.*, 1977; Jörnvall *et al.*, 1978). These reactions require no heating or extracting; products from clean thiazolinones can be separated chromatographically directly after the removal of excess amine with a stream of nitrogen. The alkylamides are similar to PTH amino acids in chromatographic properties and u.v. absorptivity. Procedural details are given by Inman and Appella (1977). Several interesting conversion reactions have been described briefly by Previero and Cavadore (1975). 5-Methylthiazolinones may be *O*-acetylated with acetic anhydride and subsequently identified by gas chromatography, or they may be cleaved with aqueous ammonia to form the parent amino acid amides.

16.2.4.4. Identification and quantitation

The sensitivity of Edman sequence analysis can be increased by the use of modified coupling reagents. These methods, which employ radioactive or highly colored isothiocyanates, were originally developed in connection with manual and automated liquid-phase sequencing and will not be reviewed here. The use of [^{35}S]-phenylisothiocyanate (PITC) in automated solid-phase sequencing was described (Bridgen, 1976; Bridgen & Waxdal, 1977). Although the detection sensitivity claimed in the best experiments was in the picomol range, in usual practice, determinations below 100 pmol have not been practical (M. J. Waxdal, personal communication). Major problems are, firstly, the relatively low specific radioactivity of the available reagent and, secondly, the dilution of the reagent due to mechanical problems in delivery and to the large volume of the reaction bed. Horn and Bonner (1977) have suggested an improved system for delivery of [^{35}S]-PITC. Improved supports should permit a drastic reduction in reaction column volume. These would also help to reduce radioactive background which becomes a serious problem when pmol amounts of peptide are studied. We believe that the potential for very low-level analysis by means of radioactive isothiocyanates is likely to be in the future more promising for solid-phase than for liquid-phase sequencing.

A significant increase in sensitivity of solid-phase analysis can be realized without the expense and hazards of using radioactive reagents. Chang *et al.* (1977a) described a manual solid-phase method that employs *N*,*N*-dimethylaminoazobenzene-4′-isothiocyanate as coupling agent (see Chapter 14). This reagent yields highly colored thiohydantoin derivatives. More recently, the 4-*N*-*N*-dimethylaminoazobenzene-4′-thiohydantoin derivatives were identified and quantified by HPLC employing detection with visible light (436 nm) to give a threshold sensitivity of 5 to 10 pmol (Chang *et al.*, 1980; Wilson *et al.*, 1979). The use of this approach in automated solid-phase sequencing was reported (Hughes *et al.*, 1979). The coupling reaction at each cycle can be completed in a second stage with the more efficient phenylisothiocyanate; the resulting PTH derivatives after cleavage are colorless and do not interfere with identification of the colored thiohydantoins (Chang *et al.*, 1978). In our opinion, reactions carried out in a column either manually or automatically would offer considerable advantages over the tube method reported by Chang *et al.* (1977a). We have successfully performed manual solid-phase sequence analyses (using PITC) with small amounts of peptide-support packed above a tiny wad of glass wool in the taper of a Pasteur pipette. The latter was shortened at both ends; this small column was heated inside a test tube immersed in a propylene glycol bath, and washes were sometimes facilitated with a small positive pressure of nitrogen.

16.2.5. Other sequential degradations

The solid-phase approach opens the way for practical applications of sequential degradation methods other than those of the Edman type. Many reagents required for alternative chemical approaches are simply not suitable in liquid-phase systems because by-products and excess reagent cannot be cleanly separated from the peptide by extraction. Several promising avenues for further development of alternative methods are described below.

16.2.5.1. N-terminal degradation by thioacylation and acid cleavage

Barrett (1967), Previero and Pechere (1970) and Mross and Doolittle (1971) proposed methods of sequential degradation based on the coupling of a thioacyl group, RC(:S)-, to the *N*-terminal amino function of a peptide followed by acid cyclization and cleavage of the terminal residue as a thiazolinone derivative. This approach is different from the Edman method with regard to the mechanism and reagents involved in the coupling reaction; also, 2-alkyl- or aryl-thiazolinones instead of 2-anilino derivatives are formed as cleavage products. Several important advantages accrue. The weaker electron-withdrawing character of alkyl or aryl groups renders the thiocarbonyl sulfur atom more nucleophilic.

Thus, cyclization (and cleavage) occurs with less drastic acid treatment. Cleavage with TFA may be accomplished at room temperature within 20 min (Barrett 1967; Previero & Pechere, 1970). Weaker acids or diluted TFA might also be employed at slightly elevated temperatures. Perhaps most of the more serious problems arising during acid cleavages in the Edman degradation, such as 'nicking' the peptide chain, Ser and Thr dehydration, succinimide formation at Asn residues, Trp oxidation and cyclization of *N*-terminal Gln to pyroglutamyl, can thereby be reduced. Finally, the thiazolinones produced are generally more stable than the corresponding Edman intermediates, and they cannot isomerize to thiohydantoins. These circumstances should appreciably increase the yields of alternative conversion reactions used for identification.

Progress in development of this method of sequencing has been limited by the need for thioacylating reagents and conditions which approach the efficiency of phenylisothiocyanate in the coupling step. Reagents which have been explored are of the general type, RC(:S)-X, where R is an alkyl or aryl (aromatic) substituent and X is an activating leaving group, such as –SR′, –OR′ or –N < R′. Thioacyl chlorides (X = Cl) should be quite reactive, but they are too unstable to prepare or use. Thioacylating power is augmented as the electron-withdrawing character of the acyl substituent, R, increases. But this trend is accompanied by a slowing of the cyclization reaction, so that a compromise must be sought in selecting a suitable coupling reagent. A large number of compounds have been explored, and the ones most seriously considered for the study of sequential degradations are listed in Table 16.1. Charged leaving groups (items 1, 6 and 7) are extremely useful for increasing reagent solubility in semi-aqueous media, although non-aqueous solvents may be readily employed with uncharged

Table 16.1. Thioacylating agents with general structure RC (:S)-X used for studying sequential degradations of peptides

Item	Acyl substituent R	Leaving group X	References
1	phenyl	$-SCH_2COO^-$	Barrett (1967)
2	phenyl	$-SCH_2CN$	Previero & Pechere (1970)
3	phenyl	$-CH_2OCH_3$	Previero & Pechere (1970)
4	phenyl	4-nitrophenolate	Inglis & Maclaren (1971)
5	phenyl	succinimidyl	Cavadore (1978)
6	*m*-nitrophenyl	$-OCH_2CH_2N^+(CH_3)_3$	Barrett & Leigh (1975)
7	methyl	$-SCH_2COO^-$	Mross & Doolittle (1971, 1977) Doolittle *et al.* (1977)
8	methyl	$-SCH_2CH_3$	Doolittle *et al.* (1977)
9	methyl	$-SCH_3$	Previero *et al.* (1975b) Previero & Cavadore (1975) Previero (1977)
10	ethyl	$-SCH_3$	Previero *et al.* (1975b)
11	propyl	$-SCH_3$	Previero *et al.* (1975b)
12	isopropyl	$-SCH_3$	Previero *et al.* (1975b)

reagents in solid-phase work. The reactivities of simple alkyl carbodithioates (items 8–12) are impracticably low except in the presence of a general base catalyst such as triethylammonium acetate in highly polar solvents (Previero *et al.*, 1975b; Previero & Cavadore, 1975; Previero, 1977). Catalysis of the thioacylation reaction can be intramolecular as in the aminolysis of the choline ester of *m*-nitrothiobenzoic acid (Barrett & Leigh, 1975; Table 16.1, item 6). This reagent is also unusually resistant to hydrolysis. Cyanomethyl dithioesters (e.g. item 2) are fairly reactive (Previero & Pechere, 1970), but the cleaved leaving group polymerizes to produce very troublesome precipitates. An easily synthesized reagent (item 5) was recently described by Cavadore (1978).

Automated, solid-phase, sequence analysis, based on the thioacylation approach, was reported (Doolittle *et al.*, 1977; Mross & Doolittle, 1977; Previero, 1977). The amino acid residues were successfully identified through 26 cycles of degradation of insulin B chain and through 40 cycles with several proteins using the reagent thioacetylthioglycolic acid (Table 16.1, item 7). Repetitive yields averaged over 95% per cycle. Identification of the thiazolinones was accomplished after their conversion to thioacylamino acids or back hydrolysis to the parent amino acids. Inman and Appella (1975) rapidly converted a 2-phenylthiazolinone to a thiobenzoylamino acid methylamide with excess methylamine. This facile conversion reaction could possibly play an important future role in thioacylation degradations. Barrett and Chapman (1968) identified 2-phenylthiazolinones directly by mass spectrometry. These thiazolinones also have been converted to *N*-thiobenzoylamino acid anilides and identified as such by TLC (Barrett & Khokhar, 1969); however, the required aminolysis is apparently somewhat slower than the analogous reaction with methylamine.

16.2.5.2. *C*-terminal degradation

A number of methods for the stepwise, chemical degradation of peptides at their carboxyl termini have been proposed (see Chapter 18). These efforts, in fact, began before the Edman degradation was first reported, but they have not matched the latter's success. One promising approach, extensively studied by Stark (1972), is based on the reaction of a peptide *C*-terminal carboxyl group with ammonium thiocyanate in acetic anhydride to give a cyclic peptidyl thiohydantoin; this step is followed by an acid (or acetohydroxamate) cleavage to yield a thiohydantoin derivative of the terminal residue and the *n*-1 peptide. In the original procedures, each cycle involved separation and drying operations which were time consuming and tedious; only 2 to 6 cycles could be usefully completed. However, new hope was given to this approach by the use of solid-phase techniques (Darbre & Rangarajan, 1975; Darbre, 1977; Kassell *et al.*, 1977). In these references, *C*-terminal degradations in general were reviewed and discussed.

Solid-phase techniques have markedly facilitated the thiocyanate degradation, yet sequencing has not been feasible beyond about 6 residues. Special problems arise at, or next to, Asn, Asp, Glu and Pro residues. The latter amino acid splits off at the first step of its cycle, so that Pro-thiohydantoin appears along with the next amino acid in the sequence (Kassell *et al.*, 1977). The other amino acids mentioned cause low yields. Identification of thiohydantoins are carried out by various means without the benefit of any high optical absorptivity. All reagents must be very carefully purified. It is difficult to assess whether or not this degradation can be developed significantly further with additional thoughtful and systematic work. We hope that such research will be pursued because *C*-terminal degradations can very usefully complement *N*-terminal sequence methods. Furthermore, attachment of peptides to an insoluble support through their α-amino groups is generally easier to accomplish than *C*-terminal attachment.

Other *C*-terminal degradation methods have been proposed in conjunction with solid-phase techniques. Tarr (1975) coupled the peptide with an *S*-alkylthiuronium salt by means of a carbodiimide and then cleaved the *C*-terminal residue as an iminohydantoin derivative with aqueous base at pH 10 to 11.5. Up to 5 residues of small peptides were sequenced. This method has the advantage over the thiocyanate degradation of having milder reaction conditions but suffers from the limitation that *C*-terminal proline fails to cleave; Asp and Glu, in the terminal position, tend to form cyclic anhydrides when treated with a carbodiimide.

Additional approaches to *C*-terminal sequencing on insoluble supports were described recently. Previero and Coletti-Previero (1977b) demonstrated that a peptide ester could be cyclized to a *C*-terminal alkoxyoxazole with strong dehydrating agents; the *C*-terminal residue was then cleaved as an amino acid ester by alcohols in acidic media. Loudon and Parham (1978) presented quite a different degradative scheme that employed several new synthetic reagents. Both approaches, although exploratory, are interesting in that they suggest that new and challenging avenues to this worthwhile goal are always open to the creative chemist.

16.3. AUTOMATED LIQUID-PHASE SEQUENCING: IMPROVED TECHNIQUES

16.3.1. Normal range sequencing of proteins and large peptides

Two major problems still limit the extent to which proteins and large polypeptides may be sequenced by the automated, liquid-phase method (Edman & Begg, 1967). The first problem is carry-over (overlap) of amino acid residues between successive cycles due to incomplete coupling or cleavage reactions; the second one is the splitting of labile, internal peptide bonds during acid cleavage steps which allows new starting points for degradation. Both processes cause accumulation of background until unequivocal identification is no longer possible. Double coupling and double cleavage at each cycle has been suggested as a means for reducing overlap (Edman, 1975). However, this technique results in excessive losses of polypeptide from the spinning cup. Carrier proteins or polymers have been added to diminish such losses (Niall *et al.*, 1974; Silver & Hood, 1974), but unfortunately they undergo acidolysis to generate spurious amino termini. Increasing the temperature of the reaction vessel to 55°C also has been employed to reduce overlap resulting from incomplete coupling, particularly with steps involving proline residues (Smithies *et al.*, 1971).

Thomsen *et al.* (1976) suggested that reversing the order of delivery of PITC and buffer to the reaction cup would reduce internal peptide bond cleavages. They speculated that peptide bond cleavages are caused mainly by an acid catalyzed $N \rightarrow O$ acyl shift in which serine hydroxyl groups attack the peptide bonds involving serine amino groups. This shift can be reversed in alkaline media provided that competitive thiocarbamylation is prevented by initially excluding PITC. This change in procedure has not been entirely satisfactory since PITC in heptane is not miscible with Quadrol buffer. Another approach to the problem of internal peptide bond splitting is the measurement and control of water content of the heptafluorobutyric acid used for cleavage steps. A high content of water facilitates peptide bond cleavage through hydrolysis, but a low level of water, about 0.01 %, seems to reduce the undesirable bond splitting (Begg *et al.*, 1978). When the above approach was combined with the use of a different coupling buffer, such as *N*,*N*,*N*′,*N*′-tetrakis(2-hydroxyethyl)ethylenediamine (THEED) in place of Quadrol (Begg & Morgan, 1976), automatic sequencing with repetitive yields exceeding 98 % was possible.

Improved performance (up to 60 or 70 residues) in the primary structure determination of proteins was achieved with a modified sequencer equipped with a larger reaction cup, spinning at higher speed (1800/3600 rev/min) and with a cold trap attached to the high vacuum pump (Fairwell & Brewer, 1979).

16.3.2. Sequencing of small or hydrophobic peptides

It is generally known that automatic sequencing of small or hydrophobic peptides often fails because the extraction solvents, in combination with residual sequencing reagents, cause these peptides to be washed from the cup. Niall *et al.* (1969) used a volatile buffer prepared from dimethylallylamine in conjunction with a restricted vapor space to facilitate removal of reagents by evaporation. This minimized the need for solvent extraction. For the complete sequence determination of non-polar peptides, additional modifications both in sequencing procedures and programming are required.

Braunitzer *et al.* (1970) and Inman *et al.* (1972) modified peptides containing a *C*-terminal lysine or *S*-aminoethylcysteine residue by reaction with 4-sulfophenylisothiocyanate to produce a more hydrophilic derivative. The above reagent combines with both the *N*-terminal and side-chain amino groups. Treatment with acid causes cleavage of the *N*-terminal residue and exposes the second residue; the sulfophenylthiocarbamyl group attached to the side chain remains unaffected. One drawback of this modification is a tendency toward incomplete cleavage during the initial acid treatment which lowers the initial yield and introduces overlap at the beginning of a run. Also, identification of the first amino acid may be difficult. More recently, the isomer, 3-sulfophenylisothiocyanate, was reported to possess certain advantages in the above modification (Dwulet & Gurd, 1976), although limited use has been reported so far. Another hydrophilic modification is the amidation of side-chain carboxyl groups with 2-amino-1,5-naphthalenedisulfonate (Foster *et al.*, 1973). This derivatization procedure requires a carboxyl activation that can bring about blocking of an unprotected *N*-terminal amino group and may cause internal cyclization of aspartic acid residues.

Crewther and Inglis (1975) suggested an approach for minimizing extractive losses of peptide. Rather than attempting to remove all of the coupling reagents by extraction, the Quadrol is allowed to remain to help create a hydrophilic phase which better retains the peptide. However, leaving Quadrol in the cup is unsatisfactory because of prohibitive salt accumulation after a few cycles. It was recommended that a protein additive could serve as an anchoring substance. For example, parvalbumin (Rochat *et al.*, 1976) was used because it has a naturally blocked *N*-terminus and is not initially susceptible to Edman degradation. But, it does suffer random cleavages giving rise to extra degradations and an increase in background PTH amino acids. Retention in the cup of small or hydrophobic peptides is significantly increased by the presence of polybrene, a synthetic polymer that bears positively charged quaternary amino groups and is inert to sequencing conditions (Tarr *et al.*, 1978; Klapper *et al.*, 1978). We used polybrene in the sequencing of unusually long peptides having a high content of non-polar residues. For some peptides we obtained good results. On the other hand, we did not observe significant improvement using this additive in the analysis of other large fragments or whole proteins. Another problem we are observing with the use of polybrene is a lowering of initial yields. The mechanism for the usually diminished initial yield, an unsolved problem of Edman degradation chemistry, may be facilitated by polybrene (see also Chapter 15).

16.3.3. Microsequencing

Sequence analysis must be performed on many proteins which cannot be obtained easily in large quantity. Thus, many efforts have been made to modify existing methods so that smaller amounts of peptide (1 to 10 nmol) can be sequenced. Wittmann-Liebold (1973) and Wittmann-Liebold *et al.* (1975, 1977) modified the spinning cup sequencer, the main features being a superior vacuum system combined with new extraction procedures, the use of special leak-proof, zero dead-volume delivery valves and an automatic conversion device.

Hunkapiller and Hood (1978) combined these improvements with a more stringent purification of reagents and solvents and the use of polybrene. They sequenced 200 pmol of myoglobin through 47 residues with a repetitive yield of 95.5%. In order to improve performance, a new spinning cup sequenator with several novel design features was constructed (Hunkapiller & Hood, 1980; see Chapter 17). With this instrument, extended amino terminal sequences with 20 pmol of proteins or 200 pmol of peptides have been determined. Attempts to further increase the sensitivity of these techniques have been met with limited success. The remaining problems are still associated with losses encountered by wash-out during each cycle, and with contaminants of the PTH amino acids introduced by the sequencer process.

The use of sensitive, radioactive, analytical techniques has overcome some of the above problems. These procedures introduce a radioisotope label either biosynthetically into the amino acid residues (intrinsic labeling) or extrinsically through a radiolabeled sequencing reagent. Proteins labeled intrinsically *in vivo* or *in vitro* must be purified on a microscale; in many cases immunochemical purification can provide a product suitable for analysis. In different laboratories, intrinsically labeled proteins have beeen sequenced at the 0.1 to 1.0 pmol level (Palmiter *et al.*, 1977; Uehara *et al.*, 1980). One major drawback of this approach is that not all proteins can be labeled *in vivo* or in tissue culture because of their low abundance and/or low rate of synthesis.

The extrinsic labeling approach is exemplified by the work of Jacobs and Niall (1975). In each degradative cycle the protein labeled first with [^{35}S]-PITC, and the coupling completed with unlabeled reagent. These workers were able to identify about 20 residues on 5 to 15 nmol of protein. As discussed in section 16.2.4.4, Bridgen (1976) and Bridgen and Waxdal (1977) extended this technique to solid-phase sequencing. Both approaches have been limited by the fact that the spinning cup and the usual solid-phase reaction column have large surface areas and large quantities of reagents are required. These circumstances lead to high radioactive background in the analysis of very small samples and make analyses both hazardous and expensive.

Recently, HPLC employing reverse phase systems has been increasingly used in microsequencing work (see Chapter 6). This approach allows the isolation of μg quantities of pure peptides that are in suitable form for direct application to sequencing procedures (Yaun *et al.*, 1982). Another important adjunct to microsequencing is the use of mass spectrometry for the sequencing of small peptides (Morris, 1980; Biemann, 1981; Krutzsch, 1981). The extremely high sensitivity of the mass spectrometer may soon make such approaches the ones of choice.

16.4. CONCLUDING REMARKS

We believe that the need for rapid, reliable and highly sensitive sequence analyses of proteins and polypeptides will markedly increase in the years ahead. An understanding of the nature and control of cellular differentiation will require the unambiguous identification and characterization of many cell membrane receptors and effectors and of numerous peptides involved in transmitting signals between cells (hormones, releasing factors, etc.). Studies of the structure and organization of genes is being carried out by various techniques which give DNA sequences that may be translated into protein sequences. The checking of these sequences and the interpretation of the mRNA sequence will, in part, depend upon an independent primary structure determination of the peptide products. DNA sequencing alone will never suffice where any post-translational modification of the protein takes place.

When one approaches present sequencing methods with these present and future demands, it becomes apparent that new breakthroughs in methodology are required. Great progress has been made in developing automated sequencing equipment and the attendant techniques. But, there appears to be a limitation to further development of methodology that can be surmounted only through a vigorous reopening of research into the chemical aspects of sequence analysis. We feel that the greatest gains in improving sequencing at high sensitivity will be possible through applying new chemical knowledge to solid-phase methods where emphasis should be placed first on reaction kinetics, on the support and on miniaturization of the apparatus. Eventually, too, chemical and enzymic approaches, carried out in conjunction with mass spectrometric analyses, may play an important role in new methodology. We hope that many investigators will recognize the importance of this challenge.

16.5. REFERENCES

Appella, E., Inman, J. K. & DuBois, G. C. (1977). See Previero, A. & Coletti-Previero, M.-A. (Eds.), (1977a), pp. 121–133.

Arshady, R., Atherton, E., Gait, M. J., Lee, K. & Sheppard, R. C. (1979). *J. Chem. Soc., Chem. Commun.*, **1979**, 423–425.

Atherton, E., Gait, M. J., Sheppard, R. C. & Williams, B. J. (1979). *Bioorganic Chemistry*, **8,** 351–370.

Barrett, G. C. (1967). *J. Chem. Soc., Chem. Commun.*, **1967,** 487–488.

Barrett, G. C. & Chapman, J. R. (1968). *J. Chem. Soc., Chem. Commun.*, **1968,** 335–336.

Barrett, G. C. & Khokhar, A. R. (1969). *J. Chromatogr.*, **39,** 47–52.

Barrett, G. C. & Leigh, P. H. (1975). *FEBS Lett.*, **57,** 19–21.

Barrett, G. C., Hume, J. & Usmani, A. A. (1977). See Previero, A. & Coletti-Previero, M.-A. (Eds.), (1977a), pp. 57–68.

Bayer, E., Mutter, M., Polster, J. & Uhmann, R. (1975). In *Peptides 1974, Proc. Thirteenth European Peptide Symposium*, Wolman, Y. (Ed.), Wiley, New York, pp. 129–136.

Begg, G. S. & Morgan, F. J. (1976). *FEBS Lett.*, **66,** 243–245.

Begg, G. S., Pepper, D. S., Chesterman, C. N. & Morgan, F. J. (1978). *Biochemistry*, **17,** 1739–1744.

Biemann, K. (1981). In *Chemical Synthesis and Sequencing of Peptides and Proteins*, Liu, T.-Y., Schechter, A. N., Heinrikson, R. L. & Condliffe, P. G. (Eds.), Elsevier/North-Holland, New York, pp. 131–148.

Birr, C. (Ed.) (1980). *Methods in Peptide and Protein Sequence Analysis, Proc. Third Internat. Conf.*, Elsevier/North Holland, Amsterdam, New York, Oxford, pp. 1–531.

Birr, C. & Garoff, H. (1977). See Previero, A. & Coletti-Previero, M.-A., (Eds.), (1977a), pp. 177–183.

Brandt, W. F., Edman, P., Henschen, A. & Von Holt, C. (1976). *Hoppe-Seyler's Z. Physiol. Chem.*, **357,** 1505–1508.

Braunitzer, G., Schrank, B. & Ruhfus, A. (1970). *Hoppe-Seyler's Z. Physiol. Chem.*, **351,** 1589–1590.

Bridgen, J. (1976). *Biochemistry*, **15,** 3600–3604.

Bridgen, J. & Waxdal, M. J. (1977). See Previero, A. & Coletti-Previero, M.-A., (Eds.), (1977a), pp. 153–162.

Cavadore, J.-C. (1978). *Anal. Biochem.*, **91,** 236–240.

Cavadore, J.-C. & Vallet, B. (1978). *Anal. Biochem.*, **84,** 402–405.

Chang, J. Y., Creaser, E. H. & Hughes, G. J. (1977a). *FEBS Lett.*, **78,** 147–150.

Chang, J. Y., Creaser, E. H. & Hughes, G. J. (1977b), *FEBS Lett.*, **84,** 187–190.

Chang, J. Y., Brauer, D. & Wittmann-Liebold, B. (1978). *FEBS Lett.*, **93,** 205–214.

Chang, J. Y., Lehmann, A. & Wittmann-Liebold, B. (1980). *Anal. Biochem.*, **102,** 380–383.

Crewther, W. G. & Inglis, A. S. (1975). *Anal. Biochem.*, **68,** 572–585.

Darbre, A. (1977). *Methods Enzymol.*, **47,** 357–369.

Darbre, A. & Rangarajan, M. (1975). See Laursen, R. A., (Ed.), (1975), pp. 131–137.

DiBello, C., Marigo, A., Buso, O. & Lucchiari, A. (1977). *Tetrahedron Lett.*, no. 13, 1135–1136.

Doolittle, L. R., Mross, G. A., Fothergill, L. A. & Doolittle, R. F. (1977). *Anal. Biochem.*, **78,** 491–505.

Dwulet, F. E. & Gurd, F. R. N. (1976). *Anal. Biochem.*, **76,** 530–538.

Edman, P. (1975). In *Protein Sequence Determination*, Needleman, S. B., (Ed.), Springer, Berlin, Heidelberg, New York, pp. 232–279.

Edman, P. & Begg, G. (1967). *Eur. J. Biochem.*, **1,** 80–91.

Elzinga, M. (Ed.) (1982). *Methods in Protein Sequence Analysis, Proc. Fourth Internat. Conf.*, Humana Press, Clifton, New Jersey, pp. 1–589.

Fairwell, T. & Brewer, H. B., Jr. (1979). *Anal. Biochem.*, **99,** 242–248.

Fankhauser, P. & Brenner, M. (1973). In *The Chemistry of Polypeptides*, Katsoyannis, P. G. (Ed.), Plenum Press, New York, pp. 389–411.

Foster, J. A., Bruenger, E., Hu, C. L., Albertson, K. & Franzblau, C. (1973). *Biochem. Biophys. Res. Commun.*, **53,** 70–74.

Fujino, M., Kobayashi, S., Obayashi, M., Fukuda, T., Shinagawa, S. & Nishimura, O. (1974). *Chem. Pharm. Bull.*, **22,** 1857–1863.

Gross, H. & Bilk, L. (1968). *Tetrahedron*, **24,** 6935–6939.

Hancock, W. S., Prescott, D. J., Vagelos, P. R. & Marshall, G. R. (1973). *J. Org. Chem.*, **38,** 774–781.

Herbrink, P., Tesser, G. I. & Lamberts, J. J. M. (1975). *FEBS Lett.*, **60,** 313–316.

Hermodson, M. A., Ericsson, L. H., Titani, K., Neurath, H. & Walsh, K. A. (1972). *Biochemistry*, **11,** 4493–4502.

Hewick, R. M., Hunkapiller, M. W., Hood, L. E. & Dreyer, W. J. (1981). *J. Biol. Chem.*, **256,** 7990–7997.

Horn, M. J. (1975). See Laursen, R. A. (Ed.), (1975), pp. 51–60.

Horn, M. J. & Bonner, A. G. (1977). See Previero, A. & Coletti-Previero, M.-A., (Eds.), (1977a), pp. 163–176.

Horn, M. J. & Laursen, R. A. (1973). *FEBS Lett.*, **36**, 285–288.

Hughes, G. J., Winterhalter, K. H., Lutz, H. & Wilson, K. J. (1979). *FEBS Lett.*, **108,** 92–97.

Hunkapiller, M. W. & Hood, L. E. (1978). *Biochemistry*, **17,** 2124–2133.
Hunkapiller, M. W. & Hood, L. E. (1980). *Science*, **207,** 523–525.
Inglis, A. S. & Maclaren, J. A. (1971). *Proc. Australian Biochem. Soc.*, **4,** 31 Abs.
Inman, J. K. & Appella, E. (1975). See Laursen, R. A. (Ed.), (1975), pp. 241–253.
Inman, J. K. & Appella, E. (1977). *Methods Enzymol.*, **47,** 374–385.
Inman, J. K., Hannon, J. E. & Appella, E. (1972). *Biochem. Biophys. Res. Commun.*, **46,** 2075–2081.
Inman, J. K., DuBois, G. C. & Appella, E. (1977). See Previero, A. & Coletti-Previero, M.-A. (Eds.), (1977a), pp. 81–94.
Jacobs, J. W. & Niall, H. D. (1975). *J. Biol. Chem.*, **250,** 3629–3636.
Jakubke, H.-D. & Klessen, Ch. (1977). *J. f. Prakt. Chem.*, **319,** 159–162.
Jentsch, J. (1975). See Laursen, R. A. (Ed.), (1975), pp. 193–202.
Jörnvall, H., Inman, J. K. & Appella, E. (1978). *Anal. Biochem.*, **90,** 651–661.
Kassell, B., Krishnamurti, C. & Friedman, H. L. (1977). See Previero, A. & Coletti-Previero, M.-A. (Eds.), (1977a), pp. 39–48.
Klapper, D. G., Wilde, C. E., III & Capra, J. D. (1978). *Anal. Biochem.*, **85,** 126–131.
Krutzsch, H. C. (1981). In *Chemical Synthesis and Sequencing of Peptides and Proteins*, Liu, T.-Y., Schechter, A. N., Heinrikson, R. L. & Condliffe, P. G. (Eds.), Elsevier/North Holland, New York, pp. 149–159.
Laursen, R. A. (1966). *J. Am. Chem. Soc.*, **88,** 5344–5346.
Laursen, R. A. (1971). *Eur. J. Biochem.*, **20,** 89–102.
Laursen, R. A. (Ed.) (1975). *Solid Phase Methods in Protein Sequence Analysis, Proc. First Internat. Conf.*, Pierce, Rockford, Illinois, USA, pp. 1–282.
Laursen, R. A. (1977a). *Methods Enzymol.*, **47,** 277–299.
Laursen, R. A. (1977b). See Previero, A. & Coletti-Previero, M.-A. (Eds.), (1977a), pp. 95–106.
Lee, H.-M. & Riordan, J. F. (1978). *Anal. Biochem.*, **89,** 136–142.
L'Italien, J. J. & Laursen, R. A. (1981). *J. Biol. Chem.*, **256,** 8092–8101.
Loudon, G. M. & Parham, M. E. (1978). *Tetrahedron Lett.*, no. 5, 437–440.
Low, M. & Kisfaludy, L. (1965). *Acta Chim. Acad. Sci. Hung.*, **44,** 61–66.
Machleidt, W. & Wachter, E. (1977). *Methods Enzymol.*, **47,** 263–277.
Mech, C., Jeschkeit, H. & Schellenberger, A. (1976). *Eur. J. Biochem.*, **66,** 133–138.
Merrifield, R. B. (1963). *J. Am. Chem. Soc.*, **85,** 2149–2154.
Morris, H. R. (1980). *Nature (Lond.)*, **286,** 447–452.
Mross, G. A. & Doolittle, R. F. (1971). *Fed. Proc.*, **30,** 1241 Abs.
Mross, G. A. & Doolittle, R. F. (1977). In *Advanced Methods in Protein Sequence Determination*, Needleman, S. B. (Ed.), Springer, Berlin, Heidelberg, New York, pp. 1–20.
Mutter, M. & Bayer, E. (1974). *Angew Chem. Int. Ed. Engl.*, **13,** 88–89.
Niall, H. D., Penhasi, H., Gilbert, P., Myers, R. C., Williams, F. G. & Potts, J. T., Jr. (1969). *Fed. Proc.*, **28,** 661 Abs.
Niall, H. D., Jacobs, J. W., Van Rietschoten, J. & Tregear, G. W. (1974). *FEBS Lett.*, **41,** 62–64.
Palmiter, R. D., Gagnon, J., Ericsson, L. H. & Walsh, K. A. (1977). *J. Biol. Chem.*, **252,** 6386–6393.
Previero, A. (1977). *Methods Enzymol.*, **47,** 289–299.
Previero, A. & Cavadore, J.-C. (1975). See Laursen, R. A. (Ed.), (1975), pp. 63–72.
Previero, A. & Coletti-Previero, M.-A. (Eds.) (1977a). *Solid Phase Methods in Protein Sequence Analysis, Proc. Second Internat. Conf.,* North-Holland, Amsterdam, pp. 1–296.
Previero, A. & Coletti-Previero, M.-A. (1977b). See Previero, A. & Coletti-Previero, M.-A. (Eds.), (1977a), pp. 49–56.

Previero, A. & Pechere, J.-F. (1970). *Biochem. Biophys. Res. Commun.*, **40,** 549–556.
Previero, A., Derancourt, J., Coletti-Previero, M.-A. & Laursen, R. A. (1975a). *FEBS Lett.*, **33,** 135–138.
Previero, A., Gourdol, A., Derancourt, J. & Coletti-Previero, M.-A. (1975b). *FEBS Lett.*, **51,** 68–72.
Regen, S. L. (1977). *J. Am. Chem. Soc.*, **99,** 3838–3840.
Regen, S. L. & Dulak, L. (1977). *J. Am. Chem. Soc.*, **99,** 623–625.
Rochat, H., Bechis, G., Kopeyan, C., Gregoire, J. & Van Rietschoten, J. (1976). *FEBS Lett.*, **64,** 404–408.
Schiltz, E. (1975). See Laursen, R. A. (Ed.), (1975), pp. 47–50.
Schuttenberg, H. & Schulz, R. C. (1976). *Angew Chem. Int. Ed. Engl.*, **15,** 777–778.
Silver, J. & Hood, L. E. (1974). *Anal. Biochem.*, **60,** 285–292.
Smith, C. W., Stahl, G. L. & Walter, R. (1979). *Int. J. Peptide Protein Res.*, **13,** 109–112.
Smithies, O., Gibson, D., Fanning, E. M., Goodfliesh, R. M., Gilman, J. G. & Ballantyne, D. L. (1971). *Biochemistry*, **10,** 4912–4921.
Stahl, G. L., Walter, R. & Smith, C. W. (1978). *J. Org. Chem.*, **43,** 2285–2286.
Stahl, G. L., Smith, C. W. & Walter, R. (1979a). *J. Org. Chem.*, **44,** 3424–3425.
Stahl, G. L., Walter, R. & Smith, C. W. (1979b). *J. Am. Chem. Soc.*, **101,** 5383–5394.
Stark, G. R. (1972). *Methods Enzymol.*, **25,** 369–384.
Tarr, G. E. (1975). See Laursen, R. A. (Ed.), (1975), pp. 139–147.
Tarr, G. E., Beecher, J. F., Bell, M. & McKean, D. J. (1978). *Anal. Biochem.*, **84,** 622–627.
Tesser, G. I. & Lamberts, J. J. M. (1976). *Int. J. Peptide Protein Res.*, **8,** 559–563.
Thomsen, J., Bucher, D., Brunfeldt, K., Nexo, E. & Olesen, H. (1976). *Eur. J. Biochem.*, **69,** 87–96.
Uehara, H., Ewenstein, B. M., Martinko, J. M., Nathenson, S. G., Coligan, J. E. & Kindt, T. J. (1980). *Biochemistry*, **19,** 306–315.
Ugi, I. (1962). *Angew Chem.*, **74,** 9–22.
Wachter, E. & Werhahn, R. (1977). See Previero, A. & Coletti-Previero, M.-A. (Eds.), (1977a), pp. 185–192.
Wachter, E., Hofner, H. & Machleidt, W. (1975). See Laursen, R. A. (Ed.), (1975), pp. 31–46.
Wilson, K. J., Rodger, K. & Hughes, G. J. (1979). *FEBS Lett.*, **108,** 87–91.
Wittmann-Liebold, B. (1973). *Hoppe-Seyler's Z. Physiol. Chem.*, **354,** 1415–1431.
Wittmann-Liebold, B. & Lehmann, A. (1975). See Laursen, R. A. (Ed.), (1975), pp. 81–90.
Wittmann-Liebold, B., Geissler, A. W. & Marzinzig, E. (1975). *J. Supramol. Struct.*, **3,** 426–447.
Wittmann-Liebold, B., Brauer, D. & Dognin, J. (1977). See Previero, A. & Coletti-Previero, M.-A. (Eds.), (1977a), pp. 219–232.
Yaun, P.-M., Pande, H., Clark, B. R. & Shively, J. E. (1982). *Anal. Biochem.*, **120,** 289–301.

Practical Protein Chemistry—A Handbook
Edited by A. Darbre
© 1986, John Wiley & Sons Ltd

RODNEY M. HEWICK* and MICHAEL W. HUNKAPILLER†

*Genetics Institute, 225 Longwood Avenue, Boston, MA 02115, U.S.A.
†Applied Biosystems, Inc., 850 Lincoln Center Drive, Foster City, California 94404, U.S.A.

17

Microsequence Analysis using a Gas–Liquid Solid-phase Peptide and Protein Sequenator

CONTENTS

17.1. INTRODUCTION

The automated Edman degradation in the spinning cup sequenator (Edman & Begg, 1967) is the most generally used procedure for determining the amino terminal amino acid sequence of peptides and proteins. In recent years, the spinning cup sequenator and associated methodology has been extensively modified and refined (Wittmann-Liebold *et al.*, 1976; Hunkapiller & Hood, 1978; Klapper *et al.*, 1978; Tarr *et al.*, 1978; Wittmann-Liebold, 1980). With emphasis on the use of high purity reagents and solvents both in the sequenator and in the HPLC system used to identify the amino acid derivatives, it is currently possible to determine extended amino acid sequences (> 30 residues) on sub-nmol quantities of peptides and proteins (Hunkapiller & Hood, 1980).

The solid-phase sequencing system (Laursen, 1971) offered some initial advantages over the spinning cup sequenator, in that the sample is covalently attached to a derivatized glass or polystyrene support matrix. The covalent attachment eliminates the possibility of extractive losses of sample in the organic solvents used during the Edman degradation. Moreover, the solid-phase system has a simple reaction column packed with support matrix and lends itself to miniaturization. However, with the introduction of polybrene in spinning cup sequencing technology to help physically retain the sample within the reaction cell (Tarr *et al.*, 1978; Hunkapiller & Hood, 1978), sample wash-out has not proved to be a major problem even with hydrophobic peptides. On the other hand, covalent attachment of proteins and peptides to the support matrix in the solid phase system has proved difficult, with 20 to 50% attachment being fairly typical. In addition, the exact strategy of covalent attachment of protein is to some extent dependent on prior knowledge of the amino-acid content of the polypeptide chain. Moreover, the attachment chemistry is performed externally to the sequenator and introduces the use of extra reagents and time-consuming manipulations. A further disadvantage of the solid-phase sequencing procedure is that the information is often incomplete, with gaps in the sequence appearing where attachment to the support matrix has occurred and sometimes a complete loss of sample after the last attachment site is reached.

Recently a new type of miniaturized sequenator, called a gas–liquid solid-phase sequenator, has been developed which uses gas phase reagents at critical points in the Edman degradation (Hewick *et al.*, 1981; Hunkapiller *et al.*, 1983). The only liquids which interact with the sample film are phenylisothiocyanate and organic extraction solvents in which the protein is very poorly soluble. Embedding the sample in a matrix of polybrene further ensures resistance to washout of hydrophobic peptides and proteins associated with SDS. Thus in the gas–liquid solid-phase system, once the sample is applied and dried as a thin film in the reaction chamber, it is effectively immobile even though it is not covalently attached to a support matrix. Since the sample is stationary throughout the Edman degradation, a miniature flow-through reaction cartridge can be used to replace the elaborate and bulky spinning cup system. A direct consequence of the miniaturized cartridge configuration is a further increase in sensitivity over the spinning cup sequenator of Hunkapiller and Hood (1980). Also, the instrument consumes far less reagent, performs the Edman degradation more rapidly and is simpler in design.

This chapter reviews the more important aspects of the design of the gas–liquid solid-phase sequenator and describes in some detail procedures we have used successfully for sequencing < 100 pmol of polypeptide eluted from SDS polyacrylamide gels.

17.2. SEQUENATOR

The design of the gas–liquid solid-phase sequenator has been described in great detail previously (Hewick *et al.*, 1981), and only certain aspects of the instrument will be emphasized here. A schematic illustration of the instrument is shown in Fig. 17.1. A commercial version of the sequenator is available (see Editorial Note at the end of the chapter). The most significant modifications to the original prototype include:

1. electrical activation instead of pneumatic operation of the diaphragm valves used for reagent/solvent delivery and evacuation;
2. replacement of the original solid state programmer by a more flexible microprocessor-controlled version with the capacity for programming cycle-specific modifications to the degradation, e.g. extending cleavage times at cycles where proline is suspected.

17.2.1. Reaction cartridge

The cartridge is constructed from two pieces of Pyrex glass rod (25.4 × 25.4 mm diam.). Both ends of each rod are ground vacuum-flat and each piece of glass is ultrasonically machined so that it has a central hole 0.508 mm diam.). At one end of each piece of glass the capillary is flared out to form a shallow conical depression. These depressions form a small chamber when clamped together inside a metal cylinder, as shown in Fig. 17.2. The conical depression in the upper piece of glass is recessed to house the reaction surface—a Whatman GF/C glass fiber filter disc 12 mm diam. Clamped between the two mating glass surfaces is a fibrous porous Teflon disc, which provides a vacuum seal when crushed by the abutting glass surfaces at the periphery of the central chamber. The Teflon disc also provides a mechanical support for the GF/C disc. This arrangement is shown enlarged in Fig. 17.3.

A full description of the miniaturized delivery valves, reagent and solvent reservoirs, fraction collector and conversion flask which are used in the sequenator have been given previously (Hewick *et al.*, 1981).

17.2.2. Reagents and solvents

The reagents and solvents in the gas–liquid solid-phase sequenator and the approximate volumes used are listed in Table 17.1 and are compared with those used in the spinning cup sequenator (Hunkapiller & Hood, 1978).

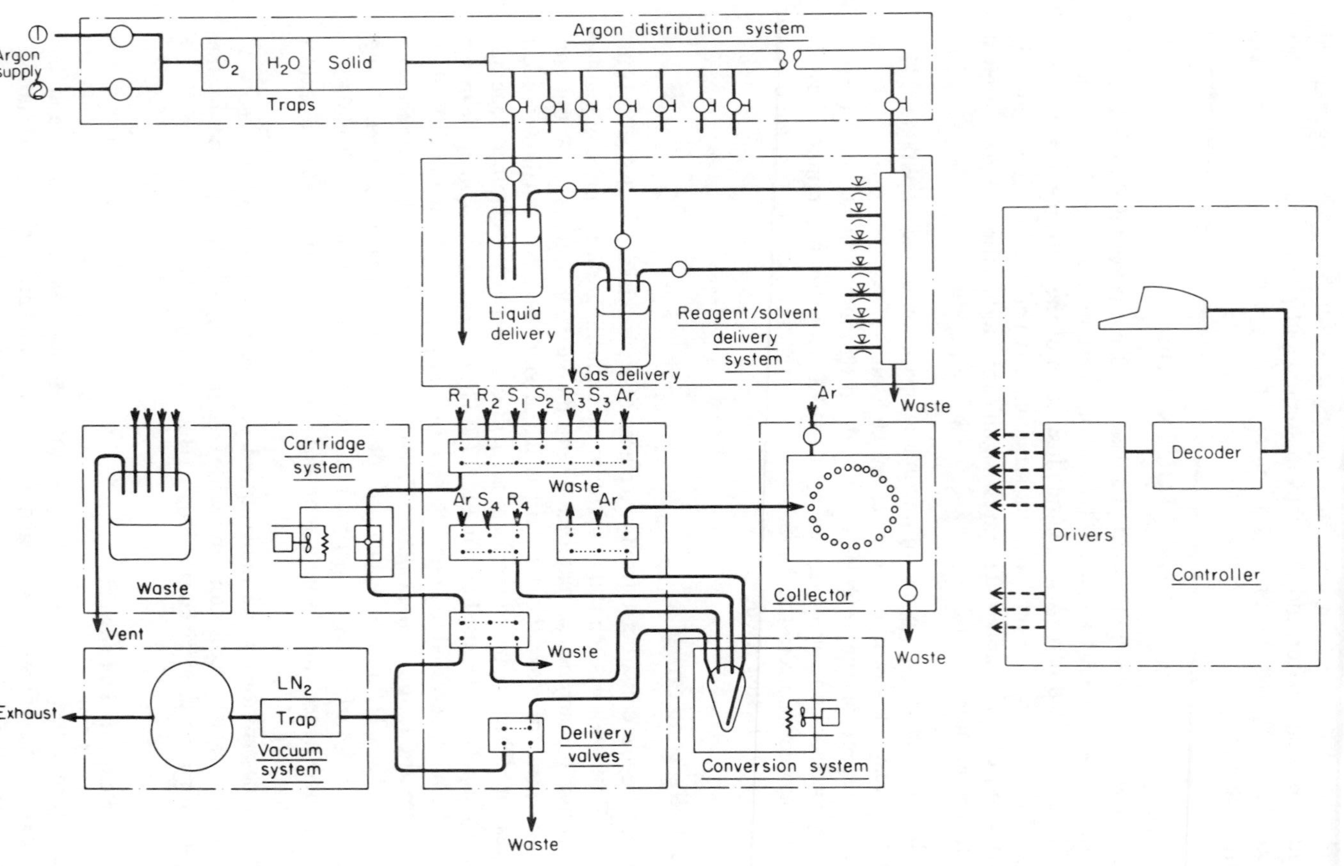

Figure 17.1. Schematic diagram of gas–liquid solid-phase sequenator.

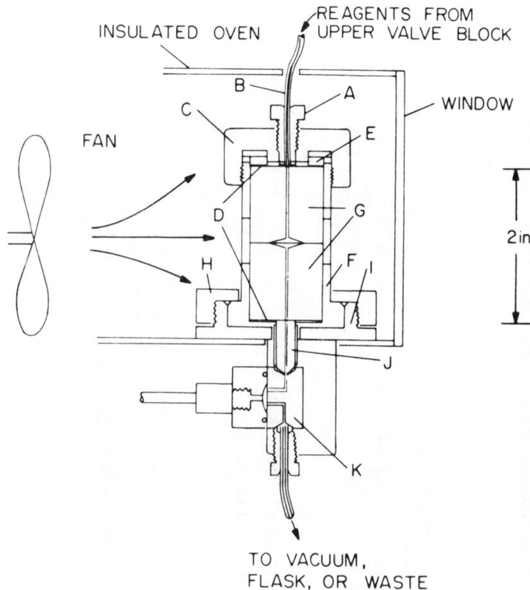

Figure 17.2. Cartridge assembly. A: aluminium top fittings; B: Teflon tubing; C: aluminium cap; D: Teflon washer; E: keyed aluminium washer; F: 304 stainless steel cartridge body; G: Pyrex glass rod; H: aluminium locking ring; I: aluminium mounting base; J: Kel-F bottom fitting; K: Kel-F valve block.

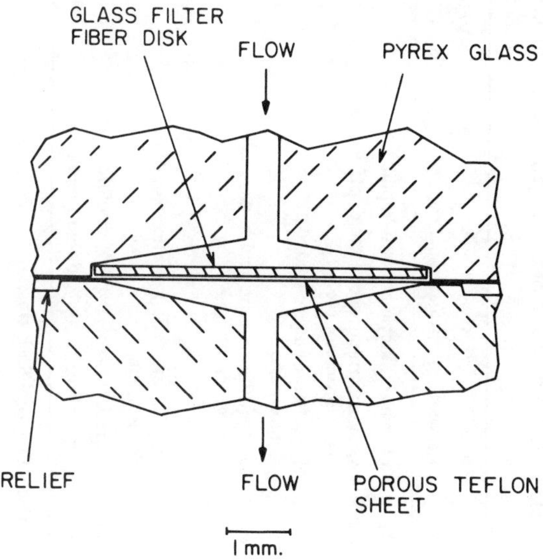

Figure 17.3. Enlarged detail of reaction chamber.

Table 17.1. Comparison of the reagent and solvent consumption per cycle of Edman degradation in a spinning cup and gas–liquid solid-phase sequenator

	Reagent/solvent (spinning cup)	Volume (cm^3)	Reagent/solvent (gas–liquid)	Volume (cm^3)
R1:	phenylisothiocyanate in n-heptane	0.4	15 % phenylisothiocyanate in n-heptane	0.05
R2:	Quadrol/trifluoroacetic acid, pH 9.0, in water/propanol (4:3)	0.7	25 % trimethylamine in water	5 cm^3/min argon flow
R3:	heptafluorobutyric acid, 0.01 % dithiothreitol	0.5	trifluoroacetic acid, 0.01 % dithiothreitol	5 cm^3/min argon flow
R4:	25 % trifluoroacetic acid in water, 0.01 % dithiothreitol		25 % trifluoroacetic acid in water, 0.01 % dithiothreitol	0.05
S1:	benzene	7	benzene	1.1
S2:	ethylacetate, 0.05 % acetic acid, 0.002 % dithiothreitol	16	ethylacetate, 0.05 % acetic acid, 0.002 % dithiothreitol	1.2
S3:	1-chlorobutane, 0.001 % dithiothreitol	7	1-chlorobutane, 0.001 % dithiothreitol	1.2
S4:	acetonitrile, 0.001 % dithiothreitol	4	acetonitrile, 0.001 % dithiothreitol	0.3*

* Excluding 1-ml flask rinse.

17.2.3. Operation of the sequenator

17.2.3.1. Sample loading

This is described with reference to Fig. 17.2. Fitting A is disconnected. The locking ring (H) is unscrewed and the cartridge lifted from the mounting base (I). The cartridge cap (C) is then unscrewed and the two glass pieces (G) removed from the cartridge body (F) for cleaning. The used GF/C disc and Teflon disc are discarded. The glass blocks are cleaned by squirting the following liquids over the ground glass surfaces and through the central capillary: (1) water; (2) 1 M-acetic acid; (3) water; (4) 1 M-NaOH; (5) water; (6) HPLC-grade methanol (to dry the glass). A new GF/C disc is inserted in the upper glass piece and an aqueous solution (0.025 ml) of polybrene (60 mg/ml) and glycylglycine (1 μmol/ml) is spotted onto the disc and dried under vacuum. The cartridge is reassembled with a new porous Teflon disc sandwiched between the two glass surfaces, and the sequenator program is run for at least four cycles. Following this precycling procedure, the cartridge is once again disassembled and the upper glass piece holding the GF/C disc is removed. Without dislodging the GF/C disc from its recess in the glass block, the sample is spotted and dried onto the disc in 25 μl aliquots. The cartridge is then reassembled and the sequence run is commenced.

17.2.3.2. Program

The sequenator program is essentially as described by Hewick *et al.* (1981), and is outlined in Table 17.2. Recent modifications include:

1. a reduction in the flow rate of ethyl acetate so that about 1.2 ml is delivered in a total extraction time of 250 s. As judged by examination of the sequenator fractions using HPLC (absorbance meter setting, 0.005 a.u.f.s.), this modification virtually eliminates 254 nm-absorbing by-products of the Edman degradation from the fractions;
2. an extension of the total cleavage time from 650 s to 800 s at 42°C.

17.2.3.3. Performance

The repetitive cycle yield of the sequenator in performing the Edman degradation was determined by sequencing various amounts of sperm whale apomyoglobin and analyzing the PTH amino acid derivatives by HPLC as described by Hewick *et al.*, (1981). The repetitive cycle yields for runs of myoglobin ranging from 10 nmol to 5 pmol varied from 98% (10 nmol) to 96% (500 pmol) to 94% (50 pmol) to 92% (5 pmol) (Fig. 17.4). The decrease in repetitive yield with reduced sample loading most likely reflects the effects of trace levels of oxidants still present in the sequenator system. At the 10 nmol level it was possible to identify the first 90 residues of myoglobin; at the 5 pmol level partial sequence

Table 17.2. Outline of gas-liquid solid-phase sequenator program

	Cartridge functions		Flask functions
	Vacuum		25% TFA
Waste ←	Cleavage (2) (TFA vapour)	450s	Conversion of ATZ* → PTH
	Flush Vacuum		
Waste ←	Extraction 1-chlorobutane		
	Flush Vacuum		
Waste ←	PITC flush Coupling (1) (TMA vapour)	500s	1800s
	Flush Vacuum		
Waste ←	PITC flush Coupling (2) (TMA vapour)	450s	
	Flush Vacuum		
Waste ←	Extraction 1. benzene 2. ethyl acetate		Vacuum
	Flush Vacuum		PTH extraction (acetonitrile) → Collector
Waste ←	Cleavage (1) (TFA vapour)	350s	Vacuum Flush dry
	Flush Vacuum		
	Extraction 1-chlorobutane	→ Transfer to flask →	Vacuum
	Flush Vacuum		

*Anilinothiazolinone.

data to residue 22 was obtained (Fig. 17.5). Those residues not identified at the 5 pmol level included the derivatives least soluble in 1-chlorobutane (e.g. histidine and arginine), and the most labile, i.e. serine, threonine, tyrosine and tryptophan.

The ability of the sequenator to sequence small hydrophobic peptides was evaluated using the octapeptide, human angiotensin II (Fig. 17.6). All eight derivatives were identified at the 5 nmol (5 μg) and 500 pmol (0.5 μg) level of sample loading. At the 50 pmol (0.05 μg) level, only the last two residues—proline and phenylalanine—could not be detected. The HPLC traces from this sequenator run are shown in Fig. 17.7.

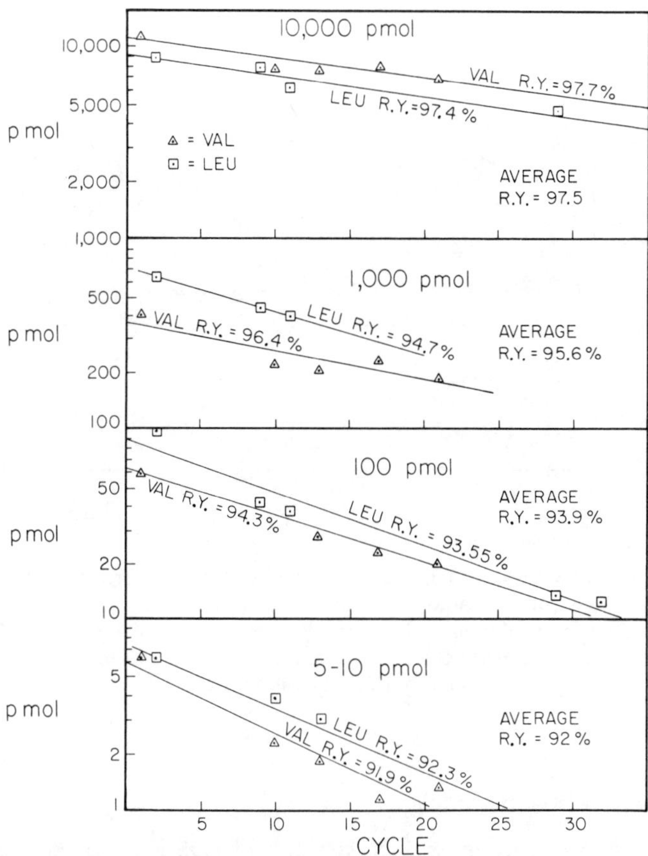

Figure 17.4. Sequenator repetitive cycle yield with various amounts of sperm whale apomyoglobin. Semilogarithmic plots of PTH valine (Δ) yields (cycles 1, 10, 13, 17 and 21) and PTH leucine (□) yields (cycles 2, 9, 11, 29 and 32) versus sequenator cycle are shown for analysis of 10 nmol, 1 nmol, 100 pmol and 10 pmol of protein. The repetitive yield (R.Y.) at each cycle was calculated from the slopes of the lines (least-squares-fit) for the plots of PTH yields.

17.3. PURIFICATION OF POLYPEPTIDE FOR MICROSEQUENCING

An important aspect of the gas–liquid solid-phase sequenator is its ability to handle small amounts of protein loaded in the presence of SDS. The high resolving power of SDS-PAGE makes it a powerful tool for purifying scarce proteins from complex mixtures of more abundant proteins. This section describes in some detail procedures used by the authors to sequence polypeptides present in SDS gels in only pmol quantities.

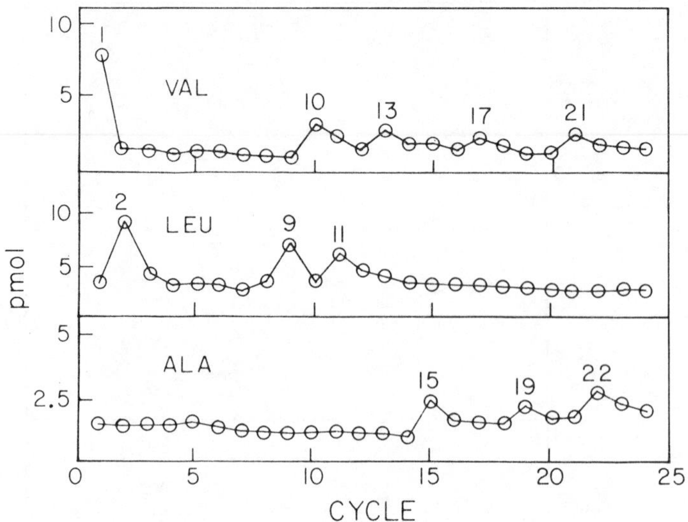

Figure 17.5. Yields of PTH valine, leucine and alanine derivatives from an NH_2-terminal amino acid sequence analysis of 10 pmol of sperm whale apomyoglobin. Yields were determined by converting HPLC peak heights for each derivative to pmol using values from a standard mixture of PTH amino acids.

17.3.1. Reagents and materials

Acrylamide, *N,N'*-methylene-bis-acrylamide and SDS is obtained from Biorad Laboratories (Richmond, CA.). Coomassie Brilliant Blue is obtained from Serva (West Germany). For elution and electrophoretic concentration of protein, Biorad SDS is recrystallized from ethanol after treatment with activated charcoal at 50°C. Other chemicals used should be of the best reagent grade available. Dialysis membrane (Spectrapor) is obtained from Spectrum Medical Industries Inc., Los Angeles, CA., and cleaned further by heating at 60°C for 30 min in (1) 1% sodium bicarbonate solution, (2) distilled water, (3) 0.1% SDS solution. The membrane is stored at room temperature in fresh 0.1% SDS solution + 0.02% sodium azide.

17.3.2. Apparatus and procedure

The sample is disrupted by heating for 5 min at 80°C in 0.08 M-Tris buffer pH 6.8, containing 2% SDS, 75 mM dithiothreitol (Calbiochem), 10% v/v glycerol and 0.005% bromophenol blue. After cooling to room temperature, the sample is made 5 mM in sodium thioglycollate and subjected to SDS-PAGE using a

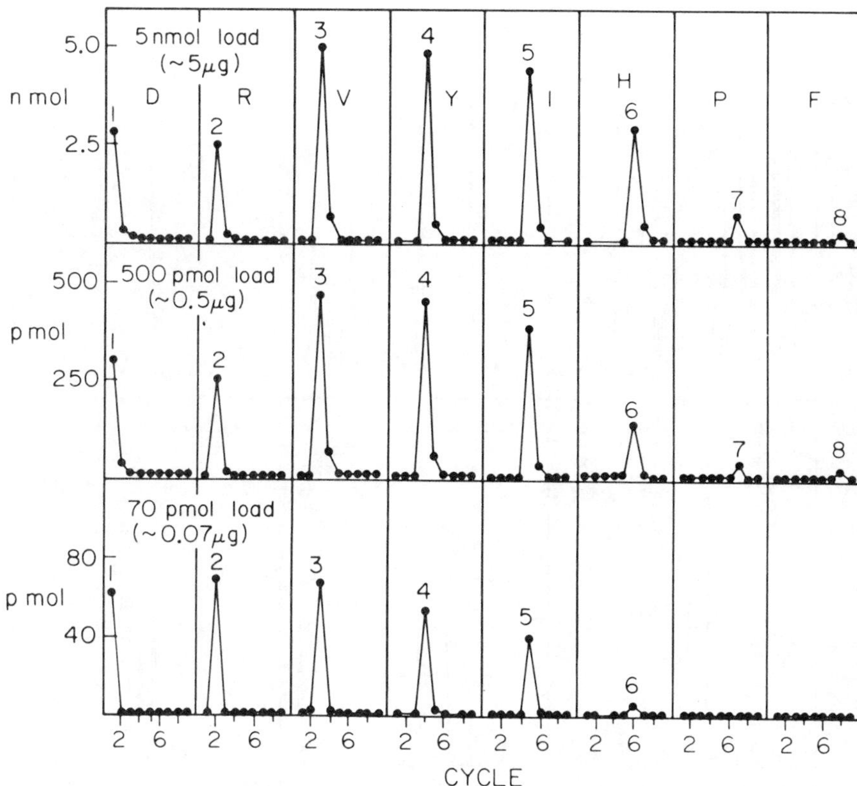

Figure 17.6. Yields of PTH derivatives from NH_2-terminal amino acid sequence analyses of 5 nmol, 500 pmol and 70 pmol angiotensin II (sequence: H-Asp-Arg-Val-Tyr-Ile-His-Pro-Phe-OH). Yields were calculated as described in the legend to Fig. 17.5.

modified version of the Laemmli (1970) discontinuous buffer/gel system with slab gels. After the electrophoresis is completed, the polyacrylamide gel is stained for 15 min in a solution of 25% isopropanol, 10% acetic acid and 0.5% Coomassie Blue and then destained in a solution of 25% methanol and 10% acetic acid at 4°C. The Coomassie Blue-stained protein band is excised with a razor blade, chopped into small cubes and allowed to soak in distilled water for 1 h to dilute out the acetic acid and methanol. Excess fluid is removed from the gel pieces and then they are soaked for 2 to 16 h at room temperature in 0.5–1.0 ml of a solution of 0.2 M-NH_4HCO_3, containing 1% SDS and 0.1% dithiothreitol. This procedure neutralizes residual acid and re-saturates the protein with SDS.

The electrophoretic elutor/concentrator is made of Plexiglas and comprises two chambers capped at the bottom by dialysis membrane of the appropriate molecular weight cut-off and linked to one another by an electrolyte bridge. The

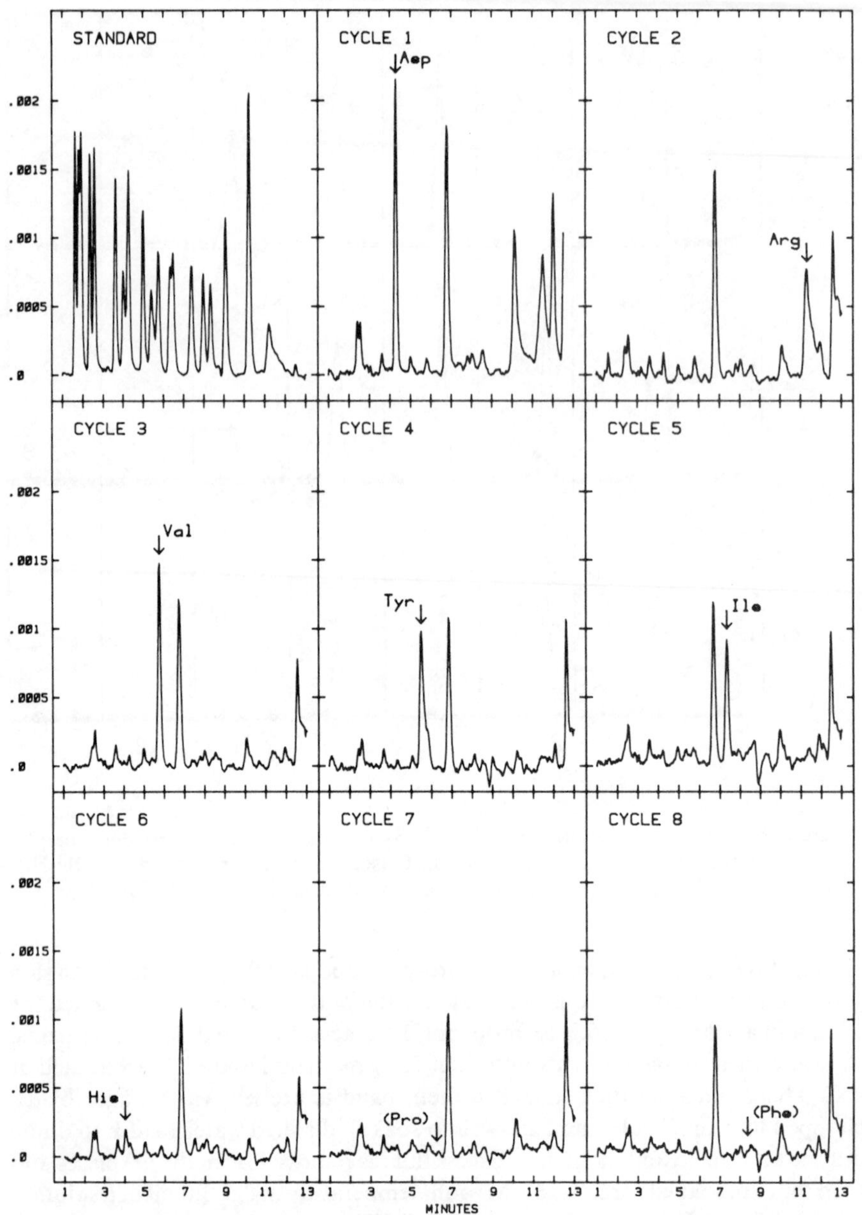

Figure 17.7. HPLC traces from an NH_2-terminal amino acid sequence analysis of 50 pmol of human angiotensin II. The absorbance meter of the HPLC detector was set at 0.005 absorbance units full scale. The chromatograms were recorded on a Hewlett-Packard series 3354 laboratory automation computer and the chromatographic solvent

apparatus is similar in design to the commercial model made by Isco Inc. (Lincoln, Nevada) except that the smaller chamber is flared out at the bottom to match the diameter of the larger sample-loading chamber. This design facilitates concentration of protein against the membrane of the smaller chamber but reduces the possibility of concentrating SDS there, since the membrane surface area allowing the migration of ions in and out of the elutor is the same.

The gel pieces and equilibration solution containing 1 % SDS are transferred to the loading chamber of the elutor and the apparatus completely filled with 0.05 M-NH_4HCO_3 containing 0.1 % SDS and 0.5 mM sodium thioglycollate. The electrode reservoirs are filled with the same solution. Following electrophoresis at 50 V for 16 h with electrode buffer recirculation, the electrode buffer solutions are replaced with a solution of 10 mM-NH_4HCO_3 containing 0.02 % SDS. Electrophoresis is continued for a further 20 h at 80 V. Using this procedure, the eluted protein may be quantitatively recovered in less than 100 μl of this solution by removing the tightly packed zone of Coomassie Blue and protein from the dialysis membrane using a Hamilton syringe. The sample is now loaded directly onto the GF/C disc reaction surface of the sequenator.

17.4. DISCUSSION

17.4.1. Sample immobilization and sensitivity of instrument

The gas–liquid solid-phase sequenator provides a further increase in amino acid sequencing sensitivity over that of previous instruments (Hewick *et al.*, 1981). The success of this new instrument stems from complete sample immobilization during the Edman degradation even in the absence of covalent attachment to a solid support phase. The peptide or protein is stationary for these reasons: (1) base (trimethylamine/water) is provided as a gas during the coupling reaction and acid (trifluoroacetic acid) is provided as a gas during the cleavage reaction; (2) the sample is anchored in a film of polybrene to resist washout in organic extraction solvents (benzene, ethyl acetate and 1-chlorobutane). Since the peptide or protein sample is simply dried onto the glass fiber disc which forms the reaction surface, initial sample losses associated with inefficient covalent attachment chemistry are

---

background from a blank injection (10 μl of acetonitrile) was subtracted by the computer to give the traces shown in the figure. The order of elution of PTH derivatives (12.5 pmol of each) in the standard mixture (upper left corner) is Asn, Ser, Thr, Gln, Gly, Ala, His, AspOMe, GluOMe, Tyr, Val, Pro, Met, Ile, Leu, Phe, Trp, Lys and Arg. Ten-microliter aliquots were injected, representing 40 % of each sample. The positions of the PTH derivatives assigned in the traces for cycles 1 through 6 are indicated by the arrows and the three-letter amino acid designations. The positions of the expected PTH derivatives for cycles 7 and 8 are indicated in a similar manner, although they are not assignable in this experiment.

avoided. The fact that the sample is immobile throughout repetitive Edman degradations makes it possible to miniaturize drastically the reaction chamber in the form of a small glass cartridge. No rotation of the reaction cell is required to maintain the sample as a thin film as in the case of the spinning cup sequenator, since the protein is initially applied and dried as a thin film and is never redissolved. Nor is there a requirement for precise metering of reagents and solvents because of the flow-through design of the reaction cartridge. The coupling and cleavage reactions are effected by bleeding vapor through the cartridge to the waste bottle. Solvents are slowly delivered through the cartridge to the waste bottle, or to the conversion flask in the case of anilinothiazolinone-amino acid extraction. The only initial metering requirement is that sufficient PITC is delivered to wet completely the glass fiber filter disc.

The high sensitivity and low background of chemical 'noise' which is characteristic of the gas–liquid solid-phase sequenator results from the way in which the sample is presented on the GF/C disc to the stream of reagents and solvents. The disc comprises a mesh of overlapping fibers and provides a very high total surface area onto which the gas- and liquid-permeable protein–polybrene matrix is dried. Yet the glass fiber disc presents a minimum overall dimension in the direction of fluid flow. This feature maximizes reagent and solvent exchanges with the sample. Using slow flow rates it is possible to achieve very efficient chemistry and very effective extractions with extremely small total reagent and solvent volumes (Table 17.1). It is because these delivered volumes are so small that the transfer of any accompanying 254 nm-absorbing impurity (still present even in ultra-pure reagents and solvents) into the fraction collector tube is minimized. For high sensitivity sequencing (< 100 pmol of material), the interpretation of the HPLC traces of each sequenator fraction is highly dependent on achieving low and superimposable baselines at 254 nm (Fig. 17.7).

17.4.2. Aspects of the chemistry

Recently it has been shown that anhydrous triethylamine vapor can be used instead of aqueous trimethylamine vapor during the coupling reaction of the Edman degradation in the gas–liquid solid-phase system without any noticeable effect on repetitive cycle yield or on sensitivity and background 'noise' (R. M. Hewick, C. Paul & M. Hunkapiller, unpublished data). This alternative reagent was tested on both small peptides and proteins. Amino acid background generated during sequence runs of myoglobin appeared to be no less when the Edman degradation was performed under completely anhydrous conditions using triethylamine in the system than when performed with the usual trimethylamine-water. This result indicates that the generation of amino acid background in the gas–liquid solid-phase system is not primarily due to hydrolysis of the peptide bond, but is caused by other chemical reactions, as has been suggested previously in the spinning cup system (Brandt *et al.*, 1980).

17.4.3. Range of samples sequenced

A representative listing of samples that have been analyzed is shown in Table 17.3. The present minimum amount of material required to obtain useful sequence data is 5–10 pmol (0.1–0.1 μg) for an average sized protein, and 50 pmol (0.05 μg) for a short peptide. Heavily glycosylated proteins such as erythropoietin (60% carbohydrate by weight) have as yet presented no special problems, although it is necessary to extend cleavage times before and after the glycosylated residue to minimize overlap due to incomplete cleavage. Sequences containing proline followed by amino acids with bulky side chains such as lysine, arginine and leucine (*Aplysia* neuropeptide B) are treated in a similar way.

Table 17.3. Polypeptides analyzed with the gas–liquid solid-phase sequenator.

Sample	Residues identified/ total residues	Amount pmol	μg
Angiotensin	8/8	500	0.5
Angiotensin	6/8	50	0.05
Somatostatin	14/14	1 400	2.0
Insulin, B chain	30/30	300	1.0
Neuropeptide B from *Aplysia*	31/34	500	2.0
Dynorphin	14/17	20	0.04
Myoglobin	90/153	10 000	165
Myoglobin	22/153	5	0.08
Larval cuticle protein 1 from *Drosophila**	55/166	850	15
Larval cuticle protein 3 from *Drosophila**	36/96	900	9
22 000-dalton membrane phosphoprotein from *Aplysia*†	23/200	15	0.3
Human T-cell growth factor†	15/160	25	1
Human histocompatibility antigen HLA-DR, a chain	49/300	700	23
Human histocompatibility antigen HLA-DR, β chain	39/240	500	13
Human erythropoietin‡	28/150	100	1.6
Human melanoma cell surface antigen†	13/850	60	5.5

* Purified by isoelectric focusing in polyacrylamide gels containing urea.
† Electrophoretically eluted from Coomassie Blue-stained, SDS-polyacrylamide gel.
‡ 60% carbohydrate by weight.

Perhaps of most significance is the ability of the present system to handle trace quantities of polypeptides electrophoretically eluted in the presence of SDS from stained SDS polyacrylamide gels (melanoma cell surface antigen, membrane phosphoprotein from *Aplysia*, Interleukin II or T-cell growth factor).

17.4.4. Future developments

Where a sample of protein can only be purified in trace amounts, the major current limitation to obtaining amino and sequence data is the quality of the sample that has been isolated. In many situations where the total yield of protein after purification is in the 10–100 pmol range, a large discrepancy exists between the quantity of starting material as determined by amino acid analysis and the amount of protein available for sequeneing as assessed by the yield of the initial PTH amino acid derivative from the Edman degradation. This difference in initial yield is not observed when dealing with trace quantities of protein taken directly from a larger source of pure material and most likely reflects partial blockage at the amino terminus incurred during sample preparation. In the case of polypeptides purified by SDS-PAGE, the situation can be improved by including suitable reducing agents in the loaded sample to protect minority proteins from oxidative modifications.

The drop in repetitive yield associated with decreased sample loading (Fig. 17.4) strongly suggests the presence of trace levels of oxidants in the sequenator system even though the reagents and solvents used are extensively purified. Initial studies indicate that further purification and scavenging of oxidants by including elevated levels of reducing agent, such as dithiothreitol, in the solvents may improve repetitive cycle yield at the very lowest levels of sample loading.

One of the most exciting aspects of the gas–liquid solid-phase sequenator is coupling the high resolving power of SDS-PAGE to high sensitivity sequencing. It is now possible to obtain partial sequence characterization of protein which is barely visible as a Coomassie Blue-stained band on a gel and which may represent a rarely expressed component of the cell. Partial amino acid sequence data of such a protein can be used to predict and synthesize oligonucleotide primer/probes for CDNA cloning of the mRNA from which the protein was translated (Noyes *et al.*, 1979). This procedure has already been used to clone rare message genes for interferons (Houghton *et al.*, 1980) and human histocompatability antigens (Sood *et al.*, 1981).

Editor's Note. The commercial instrument Model 470A (Applied Biosystems Inc., Foster City, CA, U.S.A.) was evaluated for microsequence analysis. Routine testing with 1 nmol of myoglobin after 22 cycles of degradation resulted in overall average initial and repetitive yields of $24.7 \pm 4.9\%$ and $96.1 \pm 1.9\%$ ($n = 5$), respectively (Esch, 1984). Applied Biosystems also provide at regular intervals a Users' Bulletin, which contains very useful information on all aspects of sequencing.

17.5. REFERENCES

Brandt, W. E., Henschen, A. & Holt, C. von (1980). *Hoppe-Seyler's Z. Physiol. Chem.*, **361,** 943–952.

Edman, P. & Begg, G. (1967). *Eur. J. Biochem.*, **1,** 80–91.

Esch, F. S. (1984). *Anal. Biochem.*, **136,** 39–47.

Hewick, R. M., Hunkapiller, M. W., Hood, L. E. & Dreyer, W. J. (1981). *J. Biol. Chem.*, **256,** 7990–7997.

Houghton, M., Eaton, M. A. W., Stewart, A. G., Smith, J. C., Doel, S. M., Catlin, G. H., Lewis, H. M., Patel, J. P., Entage, J. S., Carey, N. H. & Porter, A. G. (1980). *Nucleic Acids Res.*, **8,** 2885–2890.

Hunkapiller, M. W. & Hood, L. E. (1978). *Biochemistry*, **17,** 2124–2133.

Hunkapiller, M. W. & Hood, L. E. (1980). *Science*, **207,** 523–525.

Hunkapiller, M. W., Hewick, R. M., Dreyer, W. J. & Hood, L. E. (1983). *Methods Enzymol.*, **91,** 399–413.

Klapper, D. G., Wilde, C. E., III & Capra, J. D. (1978). *Anal. Biochem.*, **85,** 126–131.

Laemmli, U. K. (1970). *Nature* (*Lond.*), **227,** 680–684.

Laursen, R. A. (1971). *Eur. J. Biochem.*, **20,** 89–102.

Noyes, B. E., Mevarech, M., Stein, R. & Agarwal, K. L. (1979). *Proc. Natl. Acad. Sci. U.S.A.*, **76,** 1770–1774.

Sood, A. K., Pereira, D. & Weissman, S. M. (1981). *Proc. Natl. Acad. Sci. U.S.A.*, **78,** 616–620.

Tarr, G. E., Beecher, J. F., Bell, M. & McKean, D. J. (1978). *Anal. Biochem.*, **84,** 622–627.

Wittmann-Liebold, B. (1980). In *Polypeptide Hormones*, Beers, R. F., Jr. & Basset, E. G. (Eds.), Raven Press, New York, pp. 87–120.

Wittmann-Liebold, B., Graffunder, H. & Kohls, H. (1976). *Anal. Biochem.*, **75,** 621–633.

Practical Protein Chemistry—A Handbook
Edited by A. Darbre
© 1986, John Wiley & Sons Ltd

C. W. WARD
Division of Protein Chemistry, C.S.I.R.O., Parkville, 3052, Victoria, Australia.

18

Carboxyl Terminal Sequence Analysis

CONTENTS

18.1. INTRODUCTION

While major advances have been made in the structural analysis of proteins from the *N*-terminal end using Edman's phenylisothiocyanate degradation procedure, and extended degradations of 40 to 70 residues can now be attained on automated sequenators, the structural analysis of proteins from the *C*-terminal end is still unsatisfactory. This is in spite of the considerable effort that has gone into the development of some of these procedures and the very real need for *C*-terminal sequencing methods.

C-terminal sequencing procedures are required to aid structural determinations on peptides or proteins with naturally occurring *N*-terminal blocking groups such as formyl, acetyl and pyroglutamyl groups, or peptides which have become blocked through chemical side reactions during cleavage or fractionation operations. Such blocking reactions include carbamylation (Stark *et al.*, 1960), cyclization of *N*-terminal glutamine to pyroglutamic acid (Blombäch, 1967), iminothiazolinone formation following cyanocysteine cleavage (Jacobson *et al.*, 1973; Degani & Patchornik, 1974), cyclic amide formation by *N*-terminal *S*-carboxamidomethyl or *S*-carboxymethylcysteine (Smyth & Utsumi, 1967), (O → N) migration of acyl groups from the hydroxyl functions of amino terminal serine and threonine (Smyth *et al.*, 1962), or rearrangement of asparaginyl-glycine sequences (Bornstein, 1970; Jörnvall, 1973).

In addition to the demands of the protein chemist for adequate *C*-terminal sequencing methods, the rapid techniques of nucleic acid sequencing are going to place increasing demands on the availability of amino and carboxyl terminal sequence data. Such data are required to assist the correct placement of initiation codons and reading frames, to unravel the problems of overlapping expression of different gene products read from the same nucleic acid sequence in either the same or different reading frames (Barrell *et al.*, 1976; Fiers *et al.*, 1978); and to indicate whether proteolytic modification of larger precursor proteins has occurred. The amount of protein sequence data required depends on whether mRNA or genomic DNA is being sequenced. If mRNA is the original starting material for sequence analysis only limited amino terminal and carboxyl terminal amino acid sequences are required since the coding sequence of the translated region of mRNA exactly specifies the amino acid sequence of the protein when read in register (McReynolds *et al.*, 1978). However, when the nucleic acid sequences are derived from genomic DNA considerably more protein sequence data may be required since the sequence of bases coding for the final protein may not be continuous but may include large segments of non-translated sequence (introns), which are subsequently excised at the level of mRNA (Berget *et al.*, 1977; Leder *et al.*, 1977; Fiers *et al.*, 1978; Catterall *et al.*, 1978; Sakano *et al.*, 1979).

This article reviews the techniques used for the preferential isolation of *C*-terminal peptides, which can then be studied by normal *N*-terminal sequencing procedures, by *C*-terminal sequencing methods and/or methods for identification only of the *C*-terminal end-group. While the techniques for preferentially isolating *C*-terminal peptides and identifying *C*-terminal residues are satisfactory, the methods for direct *C*-terminal sequencing are poor and no currently available established procedure can provide the high repetitive yields and extended sequence analyses obtainable by the *N*-terminal Edman degradation.

18.2. IDENTIFICATION AND ISOLATION OF *C*-TERMINAL PEPTIDES

Several procedures have been proposed for the identification and isolation of the *C*-terminal region of proteins and peptides. In these approaches the *C*-terminal peptide is selectively resolved from the other peptides by HPLC, ion-exchange chromatography, two-dimensional high voltage electrophoresis or specific linking to solid resin supports.

18.2.1. Ion-exchange Procedures

These procedures are unlikely to be generally useful but may have special applications in specific instances. Zabin and Fowler (1972) attempted to locate the *C*-terminal peptide of β-galactosidase by tryptic digestion of maleylated protein (to restrict enzyme cleavage to arginine residues) followed by maleylation of the resulting peptide mixture and passage through Dowex-50 at pH 3 to 4. They reasoned that all peptides, except the *C*-terminal peptide, would be positively charged (due to arginine content) and should bind to the ion-exchange resin, allowing only the uncharged *C*-terminal peptide to pass through. In practice the approach was not successful since the *C*-terminal peptide of β-galactosidase was insoluble at acid pH. The method would not work for proteins that contain arginine as their *C*-terminal residue or whose *C*-terminal peptides contain Arg-Pro sequences or histidine residues.

Hargrave and Wold (1973) isolated the *C*-terminal peptides from insulin A-chain, lyosyme, cytochrome *C* and trypsin by first blocking all free carboxyl groups with glycinamide via the water-soluble 1-ethyl-3-(3-dimethylamino-propyl) carbodiimide (EDC) and then digesting with trypsin. All tryptic peptides, except the blocked *C*-terminus contain free carboxyl groups and will bind to the anion-exchange resin AG-1-X2 (Biorad). However, before chromatography the peptide mixture was pretreated with carboxypeptidase B because the positively charged arginine-containing peptides did not bind to this resin and coeluted with the *C*-terminal peptide. The free arginine was separated from the *C*-terminal peptide by gel filtration or cation-exchange chromatography. Although this multi-step procedure appears to be lengthy, the authors claimed that it was simple and relatively rapid.

18.2.2. Solid-phase resins

The introduction of solid-phase sequencing resins (see Machleidt & Wachter, 1977) has provided a powerful tool for the selective isolation of *C*-terminal peptides. In principle it should be possible to attach the protein selectively to the resin via its *C*-terminal carboxyl group, enzymically or chemically cleave this immobilized protein and selectively retain only the *C*-terminal peptide which is still attached to the resin. In practice, the problems of simultaneous activation of, and binding by, side chain carboxyl groups (Laursen, 1977) has limited the use of such resins as a general method for selective retention of the *C*-terminal portion of proteins.

Following CNBr cleavage solid-phase sequencing resins can be used to remove homoserine-containing peptides selectively from peptide mixtures. The method involves lactonization of the homoserine residues by treatment with anhydrous trifluoroacetic acid and subsequent aminolysis of the homoserine lactone with the amino resin (Horn & Laursen, 1973) as shown in Fig. 18.1.

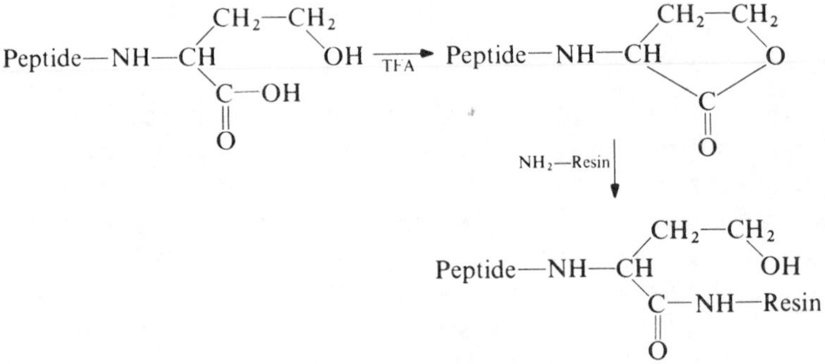

Peptide—NH—CH(CH_2—CH_2—OH)—C(=O)—OH $\xrightarrow{\text{TFA}}$ Peptide—NH—CH(CH_2—CH_2—O—C(=O)—, lactone ring)

$\downarrow$ NH_2—Resin

Peptide—NH—CH(CH_2—CH_2—OH)—C(=O)—NH—Resin

Figure 18.1. Procedure for selectively attaching peptides obtained by cyanogen bromide cleavage to amino resins. The homoserine residue is converted to homoserine lactone by treatment with trifluoroacetic acid and coupled to the solid phase resin by aminolysis of the lactone ring.

The homoserine-containing peptides are attached in high yield by their carboxyl terminus without the need for amino or carboxyl protection. This method can be used to isolate *C*-terminal homoserine-containing peptides from enzymic digests of large CNBr peptides (Laursen, 1977) as well as to remove the homoserine-containing peptides from the homoserine-free, *C*-terminal peptide (Horn & Laursen, 1973). Very large CNBr peptides may not couple efficiently because of their poor solubility.

18.2.3. Two-dimensional peptide mapping

A variety of peptide mapping approaches have been tried. Naughton and Hagopian (1962) identified the *C*-terminal peptide of foetal hemoglobin γ-chain by comparing the electrophoretograms of tryptic digests re-run in the second dimension at pH 6.5, with and without removal of the *C*-terminal arginine or lysine residues by carboxypeptidase B *in situ* on the paper. Theoretically, only the *C*-terminal peptide should remain on the diagonal following carboxypeptidase treatment. However, in human γ-hemoglobin chain three spots remained on the diagonal, viz. free lysine, the dipeptide Val-Lys, because dipeptides are very poorly attacked by carboxypeptidase (Neurath, 1960), and the true *C*-terminal peptide Tyr-His. The method should be suitable for proteins whose *C*-terminal residues are not arginine or lysine and for proteins whose tryptic peptides are readily digested by carboxypeptidase B.

Carlton and Yanofsky (1963) compared tryptic maps of undegraded and carboxypeptidase A + B-degraded protein to detect the *C*-terminal region of bacterial tryptophan synthetase A protein. The *C*-terminal sequence of this protein is Ala-Ala-Thr-Arg-Ser and partial tryptic cleavage resulted in the production on the peptide map of three spots due to Ala-Ala-Thr-Arg-Ser, Ala-Ala-Thr-Arg and free Ser. Only the spot for Ala-Ala-Thr-Arg disappeared after carboxypeptidase treatment because the other peptide and Ser co-migrated with other spots on the chromatogram. Such an approach is highly dependent on the resolution of peptides on the chromatogram. The larger the protein the greater the chance that modification of the *C*-terminal peptide will be missed.

The most elegant peptide mapping employed uses diagonal procedures based on the absence of a free carboxyl group in the *C*-terminal peptide of amidated proteins. The side chain and *C*-terminal carboxyl groups are first amidated before enzymic digestion. The *C*-terminal peptide is unique since it is the only peptide without a free carboxyl group. It can be selectively identified since its ionic charge, and therefore its electrophoretic mobility is almost the same at pH values on either side of pH 3.5, the pK_a of the α-carboxyl group. The mobilities of peptides which contain a free α-carboxyl group increase with decreasing pH below pH 3.5. After two-dimensional electrophoresis only the *C*-terminal amidated peptide will lie on the diagonal.

In the original procedure (Furka *et al.*, 1970) the protein was amidated with methylamine and *N*-cyclohexyl-*N'*-2-(4-β-morpholinyl)-ethyl carbodiimide methyl-*p*-toluenesulphonate and the enzymic digest subjected to high voltage electrophoresis at pH 6.5 in the first dimension and pH 1.8 in the second. Under these conditions *C*-terminal peptides containing histidine are not detected as they still move off the diagonal. In a modified procedure (Duggleby & Kaplan, 1975) this problem was avoided by carrying out electrophoresis in the first dimension at pH 4.4 so that the carboxyl groups were the only functional groups with altered ionization state at the pH used for electrophoresis in the second dimension. Ethanolamine was used to amidate the protein because it is readily soluble in water, it imparts good solubility properties to the modified protein and subsequent peptides, it is not found in normal proteins and it can be readily quantified on the amino acid analyser. These workers also used alanine amide and taurine as markers to define the exact position of the diagonal regardless of slight changes in absolute mobility or the effects of electro-osmosis.

18.2.3.1. Isolation of C-terminal peptides

The method of Duggleby and Kaplan (1975) for isolating the carboxyl terminal peptides from α- and β-chains of human hemoglobin is described. The amount of protein taken depends on the sensitivity of the detection system available.

Amidation

The protein (1.6 μmol, 50 mg) is dissolved in 3 ml of water containing a few drops of HCl. Add 2.6 g of urea. After stirring for 30 min, KCl (0.3 g) and ethanolamine (0.2 ml) are added and the pH immediately adjusted to pH 4.75 with HCl. Add 125 mg of EDC and maintain pH 4.75. Stir at room temperature for 4–6 h and then add 1 ml of 4 M-sodium acetate buffer pH 4.75. The solution is then dialysed against several changes of water.

Alternatively the amidation can be carried out in saturated guanidine HCl (1 ml) containing ethanolamine (20 μl, 300 μmol) adjusted to pH 4.7 with 1 M-HCl from a microsyringe (Dopheide & Ward 1981): EDC (15 mg, 80 μmol) is added with stirring and pH 4.7 is maintained for 6 h.

Enzyme digestion

Thermolysin digestion is performed in 1 % (w/v) ammonium bicarbonate at 37°C for 8 h using an enzyme: protein ratio of 1 : 35 (w/w). Digestion is terminated by lyophilization.

Peptide isolation by electrophoresis

The peptide mixture is dissolved in pH 4.4 buffer (pyridine/acetic acid/water, 6 : 10 : 1200, by vol.), applied to a 330 mm band on Whatman 3 MM chromatography paper together with the markers taurine and alanine amide (5 nmol/mm) and subjected to electrophoresis at pH 4.4 for 60 min at 6 V/mm. After drying, a guide strip is removed, stitched onto a full sheet of Whatman 3 MM paper and subjected to electrophoresis at pH 2.1 (formic acid/acetic acid/water, 1 : 4 : 45, by vol.) for 40 min at 6 V/mm. The electrophoretogram is stained with cadmium-ninhydrin (see p. 260). As shown in Fig. 18.2 two peptides are located on the diagonal delineated by the markers taurine and alanine amide. In preparative work the region of the pH 4.4 electrophoretogram corresponding to the two on-diagonal peptides can be cut out and electrophoresed at pH 2.1. The slowest moving peptides, which correspond to the two on-diagonal peptides, are then cut out and, if necessary, purified by further electrophoresis at pH 6.5 or 2.1.

The procedure has recently been used to locate the position at which the proteolytic enzyme bromelain cleaves the influenza virus hemagglutinin

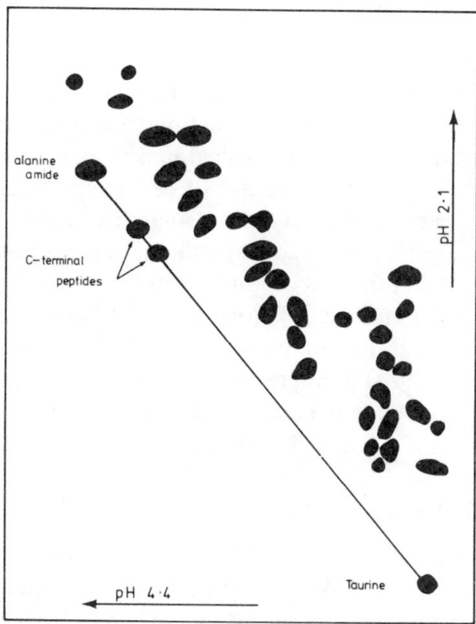

Figure 18.2. Selective isolation of C-terminal thermolytic peptides from the α- and β-chains of human hemoglobin. The thermolysin digest was subjected to electrophoresis at pH 4.4 in the first dimension and pH 2.1 in the second (Duggelby & Kaplan, 1975).

molecule when it is enzymically released from the virus surface (Dopheide & Ward 1981).

Duggleby and Kaplan (1975) point out that the method should have general applicability. All peptides derived from the carboxyl terminus of a protein will be on the diagonal irrespective of their amino acid sequence. Any enzyme can be used for the digestion provided that it gives peptides of sufficient length, and the sequence of the isolated peptide can be determined using standard N-terminal procedures. The presence of ethanolamine at the carboxyl terminus of the isolated peptide provides a built-in check that the peptide is derived from the carboxyl terminus of the protein, and although high coupling yields are desirable this is not an essential feature of the method since any fraction not amidated with ethanolamine will simply move off the diagonal and not interfere. Finally, the method should work with proteins that naturally contain an amidated C-terminus since such proteins contain the type of blocking group that is being introduced artificially.

18.3. *C*-TERMINAL END-GROUP DETERMINATION

The *C*-terminal amino acid in peptides or proteins can be detected by either chemical or enzymic procedures. The chemical methods include hydrazinolysis, selective tritiation, selective reduction, aldehyde formation, oxazole alcoholysis in acidic media, thiocyanate degradation or cyanamide degradation. The enzymic approach involves the use of carboxypeptidases, and has been enhanced in recent years by the increasing availability of enzymes from plant and microbial sources, which have broader specificities than the pancreatic enzymes. Carboxypeptidase digestions and chemical sequencing methods generally yield sequence data as well as end-group information. The novice is recommended to use enzymic methods if an analytical system with sufficient sensitivity is available. Jones *et al.* (1981) used pre-column OPA derivatization methods and HPLC separation to give the maximum quantitative sensitivity for this type of *C*-terminal analysis (see Chapter 8, section 8.16.3.2).

18.3.1. Selective tritiation

The simplest chemical method available is that of selective tritiation. It is highly sensitive, uses readily available reagents and is conducted under very mild conditions. Following tritium incorporation, exchangeable tritium is removed by repeated addition of water and evaporation, the sample is hydrolysed with 6 M-HCl (or methanesulphonic acid if tryptophan is suspected, Simpson *et al.*, 1976), and the amino acids resolved by normal peptide mapping techniques. The radioactivity in the separated residues is determined by liquid scintillation counting. Although originally only a qualitative procedure (Matsuo & Narita, 1975) it can now be used quantitatively.

Selective tritiation involves a sequence of three different reactions as shown in Fig. 18.3. It depends on the difference in chemical reactivity of the *C*-terminal and side chain carboxyl groups. The *C*-terminal carboxyl is unique because it can be activated through an intramolecular dehydration with the adjacent peptide function to yield an oxazolinone (Previero *et al.*, 1973). This reaction can be mediated by acetic anhydride. The oxazolinone so formed contains an active hydrogen which can be readily replaced by tritium when treated with pyridine and ^3H_2O. The oxazolinone undergoes base-catalysed racemization followed by hydrolytic ring-opening and regeneration of the *C*-terminal amino acid with the incorporated tritium attached to the α-carbon atom.

Matsuo *et al.* (1964) found that hydrogen–deuterium exchange occurred on the asymmetric carbon atom of amino acids under racemization conditions. Since the racemization during peptide synthesis was generally believed to involve the oxazolinone intermediate of the *C*-terminal amino acid they reasoned that production of such oxazolinones by chemical treatment should lead to a highly specific method for incorporating deuterium or tritium into the *C*-terminal amino acid residue in polypeptides.

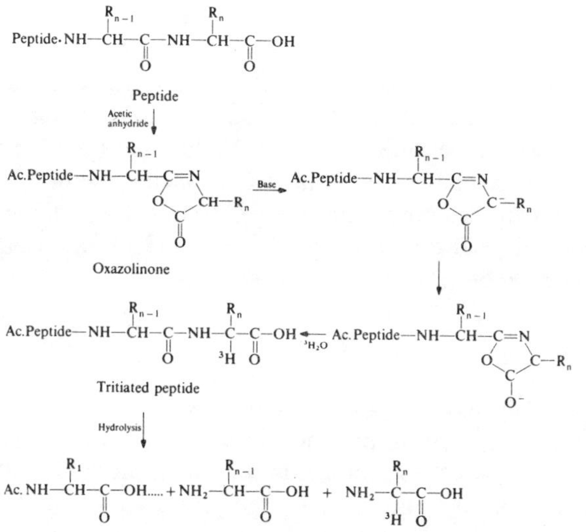

$$\text{Peptide·NH—}\underset{}{\overset{R_{n-1}}{CH}}\text{—}\underset{O}{\overset{}{C}}\text{—NH—}\overset{R_n}{CH}\text{—}\underset{O}{C}\text{—OH}$$

Peptide

Acetic anhydride ↓

Ac.Peptide—NH—CH(R_{n-1})—C=N (oxazolinone ring: O, CH—R_n, C=O) —Base→ Ac.Peptide—NH—CH(R_{n-1})—C=N (ring: O, C^-—R_n, C=O)

Oxazolinone

↓

Ac.Peptide—NH—CH(R_{n-1})—C(=O)—NH—C(R_n)(^3H)—C(=O)—OH ←^3H_2O— Ac.Peptide—NH—CH(R_{n-1})—C=N (ring: O, C—R_n, C—O^-)

Tritiated peptide

Hydrolysis ↓

Ac.NH—CH(R_1)—C(=O)—OH..... + NH_2—CH(R_{n-1})—C(=O)—OH + NH_2—C(R_n)(^3H)—C(=O)—OH

Figure 18.3. Reactions involved in the selective tritiation of *C*-terminal amino acids (after Matsuo & Narita, 1975).

Three variations of the selective tritiation method have been investigated. The original procedure (Matsuo *et al.*, 1965) involved a two-stage process: oxazolinone formation by heating with dicyclohexylcarbodiimide (DCC) or acetic anhydride in anhydrous dioxane, followed by the addition of 1 drop of pyridine and 0.5 ml D_2O to effect the base-catalysed racemization and hydrolytic ring-opening. They found that amino acids underwent complete deuteration exclusively on their α-carbon atom. Because deuterium exchange involved monitoring by NMR the ^2H_2O was replaced by the radioactive ^3H_2O, thus providing a simple and sensitive procedure for *C*-terminal amino acid detection.

The second variation (Matsuo *et al.*, 1966) was developed to allow *C*-terminal proline and aspartic acid residues to be selectively labelled. Under the conditions of the original procedure these residues did not form oxazolinones but were converted to either a mixed anhydride (proline) or a cyclic anhydride (aspartic acid). These anhydrides did not incorporate label when treated with pyridine and ^2H_2O but could be labelled when treated with deuterated acetic acid. This acid-catalysed labelling procedure for *C*-terminal proline and aspartic acid was then simplified to a single step process involving treatment of the peptide with a mixture of acetic anhydride and ^3H_2O.

The third variation (Matsuo *et al.*, 1966) was for application to proteins. Since most proteins were insoluble in the anhydrous dioxane used in peptide labelling, this solvent was replaced by an aqueous solution of pyridine, ^3H_2O and acetic anhydride. In this system the three reactions shown in Fig. 18.3 take place in one step. It is assumed that tritiation occurs via the transient formation of an oxazolinone which is in equilibrium with the uncyclized *C*-terminal peptide.

Holcomb *et al.* (1968) evaluated these three approaches to selective tritiation and concluded that the single-step base-catalysed procedure developed for proteins also gave the best results with peptides. They further found that, in contrast to the original two-step anhydrous solvent method, this single-step procedure led to tritium incorporation into *C*-terminal aspartic acid, presumably because the aspartic acid cannot form a cyclic anhydride in the aqueous solvent. Details are given below.

18.3.1.1. Tritiation

The method of Matsuo *et al.* (1966) is described.

Use small test tubes (13 × 100 mm). The sample (3–20 nmol) is dissolved in 0.05 ml (50 mCi) of ^3H_2O. Pyridine (0.1 ml) and acetic anhydride (0.025 ml) are added and the solution allowed to stand at room temperature for 3 h. The solvents are removed *in vacuo* at 40°C. Exchangeable tritium is removed by the addition and evaporation of 100 μl of water repeated six times. The tritiated sample is then hydrolysed directly in the same tube with 6 M-HCl–0.004 M-thioglycollic acid at 108°C *in vacuo* for 24 h.

Paper, thin-layer, gas-liquid, ion-exchange or high pressure liquid chromatography or electrophoresis can be used to identify the tritiated *C*-terminal amino acid. Two-dimensional peptide mapping on Whatman 3 MM paper is convenient as several samples can be analysed at the same time. The dried hydrolysate is dissolved in 50 μl of water and 2 μl of standard amino acid mixture (approx. 5 nmol of each amino acid) added. The sample is then subjected to HVE at pH 1.8 (3 000 V, 45 min). The portion of the chromatogram containing the partially resolved amino acids (approx. 100 mm strip) is sewn onto a second sheet of paper (200 × 200 mm) and subjected to ascending chromatography (approx. 3.5 h) in n-butanol/acetic acid/water/pyridine, (15 : 3 : 12 : 10, by vol.). After detection with ninhydrin/collidine reagent (see p. 260) the amino acid spots are cut out and their radioactivity determined in a liquid scintillation counter.

The application of this procedure to several chymotryptic peptides from influenza virus hemagglutinin light chain is shown in Fig. 18.4 and Table 18.1. The determination of the *C*-terminal amino acid is clear-cut for peptides C4a, C5 and C6. Background counts are low and the incorporation of tritium into the *C*-terminal amino acids is high. As shown in Fig. 18.4 all amino acids are effectively resolved with the exception of Leu/Ile. Although these two spots partially overlap, identification is easily made from the distribution of counts in the two fractions, as shown for peptide C3 in Table 18.1. Different solvent systems would also improve the separation (Holcomb *et al.*, 1968).

The final point to note is the virtual absence of labelling in internal glutamic acid and the variable labelling in internal aspartic acid. As reported by Baba *et al.*

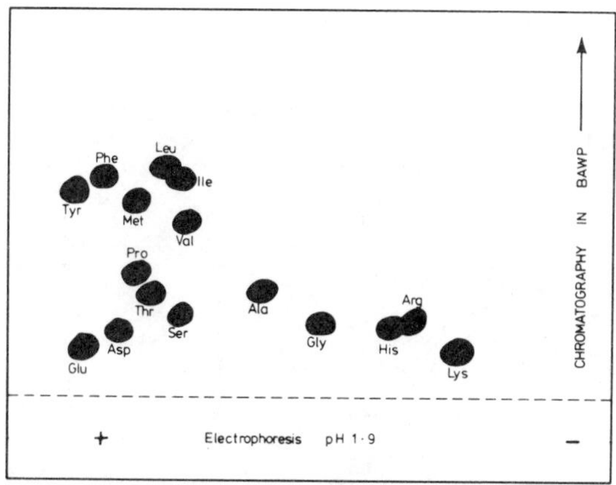

Figure 18.4. Two-dimensional paper separation of amino acid mixture after selective tritiation and hydrolysis. The hydrolysate and amino acid standard mixture is subjected to electrophoresis at pH 1.9 in the first dimension and chromatography in the second dimension with n-butanol/acetic acid/water/pyridine (15:3:12:10, by vol.).

(1972) aspartic acid residues involved in Asp-Gly sequences (peptide C3) undergo much higher tritium incorporation than other internal aspartic acid residues (see peptide C6).

Under the conditions described previously with pyridine/^3H_2O/acetic anhydride (4:2:1, by vol.), the base-catalysed hydrolysis of acetic anhydride is relatively fast, resulting in the consumption of acetic anhydride before oxazolinone formation is complete and consequently decreased reaction yields. Using a large excess of acetic anhydride does not give better results as reaction temperatures rise considerably, protein solubility is affected and unfavourable side reactions occur (Hsieh *et al.*, 1971, Cappugi *et al.*, 1971). When pyridine/^3H_2O/acetic anhydride (3:1:3, by vol.) was used in the cold, tritium incorporation into the *C*-terminal leucine residue of egg white lysozyme increased approximately 20-fold (Matsuo & Narita, 1975).

18.3.1.2. Tritiation of proteins

The method of Matsuo and Narita (1975) is described.

Protein (50–100 nmol) is placed in a small test-tube and 5 μl (0.5 mCi) of ^3H_2O and 10 μl of pyridine are added. The solution must remain clear, otherwise tritium

Table 18.1. Selective tritiation of some chymotryptic peptides from influenza virus hemagglutinin light chain

Amino acid	Radioactivity (counts/min) C3	C4a	C5	C6
Lys	218	83	35	72
His	50	26	55	91
Arg	50	24	44	146
Asp	1 860	170	58	248
Thr	33	32	22	26
Ser	37	31	30	48
Glu	121	45	41	126
Pro	38	12	5	23
Gly	45	34	36	63
Ala	39	20	22	29
Met	68	117	47	23
Ile	425	56	50	50
Leu	1 782	99	85	47
Tyr	53	73	120	4 523
Phe	63	2 451	3 221	68
nmol peptide used	7	4	6	11

Peptides (4–11 nmol) were tritiated and hydrolysed by the procedure described in section 18.3.1.1 and the amino acids resolved as shown in Fig. 18.4. The radioactivity associated with each amino acid is presented. The peptides examined had the following structures (Ward & Dopheide, 1979):

C3 Lys-Ser-Thr-Gln-Ala-Ala-Ile-Asp-Gln-Asp-Gly-Lys-Leu
C4a Asn-Arg-Val-Ile-Glu-Lys-Thr-Asn-Glu-Lys-Phe
C5 His-Gln-Ile-Glu-Lys-Glu-Phe
C6 Ser-Glu-Val-Glu-Gly-Arg-Ile-Gln-Asp-Leu-Glu-Lys-Tyr

labelling is unsatisfactory. If more solution must be added, all volumes of subsequent reagents must be adjusted to keep the same relative ratios described. Acetic anhydride (10 μl) is then added at 0°C and the mixture (in a Parafilm-sealed tube) is kept at 0°C for 5 min, then 20°C for a further 15 min. A second addition of pyridine (20 μl) and acetic anhydride (20 μl) is made at 0°C and the reaction mixture kept at 0°C for 5 min and then at 20°C for an additional 1 h. Excess acetic anhydride is decomposed by addition of tritiated water (5 μl) and by standing at 20°C for 1 h. The solution is evaporated to dryness and exchangeable tritium removed by the repeated addition and evaporation of 100 μl of 10% acetic acid. The tritiated protein is then directly hydrolysed with 6 M-HCl and the tritiated amino acids identified and counted.

Matsuo and Narita (1975) extended this tritiation procedure to the quantitative determination of *C*-terminal residues and showed that the nature of the *C*-terminal amino acid residue greatly affected the extent of tritium incorporation as shown in Table 18.2. The extent of tritiation varies with the individual amino acid but these workers established that if the tritiation is carried out in the presence of

Table 18.2. Relative ratios of tritium incorporation into various acetylamino acids relative to tritium incorporation into acetylalanine

Amino acid	Correction factor ratio
Lysine	1.52
Histidine	0.61
Arginine	1.96
Aspartic acid	0.32
Asparagine	2.30
Threonine	0.09
Glutamine	0.82
Glutamic acid	1.16
Glycine	0.68
Alanine	1.00
Valine	0.93
Methionine	2.32
Isoleucine	1.18
Leucine	2.27
Tyrosine	1.95
Phenylalanine	2.01

A weighed amount of each *N*-acetylamino acid was tritiated in the presence of a known amount of *N*-acetylalanine. After acid hydrolysis and separation of the amino acids, the ratio of the specific radioactivities of the test *C*-terminal amino acid and the alanine internal standard were measured. These data are presented with that of acetylalanine selected as 1.00 (from Matsuo and Narita, 1975).

an appropriate standard protein, then the relative ratio of the specific radioactivities of the unknown and standard *C*-terminal residues could, in conjunction with the appropriate ratio correction factor (Table 18.2), allow quantitation of the *C*-terminal residue from the protein under investigation. There are unresolved problems. Proline is not detected. Serine and threonine do not give reproducible yields. Glycine in the penultimate *C*-terminal position may depress the labelling of the *C*-terminal amino acid.

18.3.2. Hydrazinolysis

Hydrazinolysis was introduced by Akabori in 1952 to characterize *C*-terminal residues and although relatively simple in principle has often proved unsatisfactory in practice (Akabori *et al.*, 1952; Ohno, 1954; Braunitzer, 1955; Niu & Fraenkel-Conrat, 1955; Akabori *et al.*, 1956; Bradbury, 1956; Narita & Ohta, 1959; Kauffmann & Boettcher, 1959; Kawanishi *et al.*, 1964; Braun & Schroeder,

1967). Most applications have given results that are quantitatively disappointing and not reliably improved by the use of large correction factors.

The protein or peptide is heated with anhydrous hydrazine and the peptide bonds are cleaved as the carbonyl groups of the peptide bonds are converted into hydrazides. Only the *C*-terminal amino acid residue remains as the free acid (Fig. 18.5). This free amino acid is then separated from the mixture of amino acid hydrazides before determination. Modifications to the original procedure have attempted to improve either the conditons of the reaction itself, the method of separation of the hydrazides from the free *C*-terminal amino acid or the procedures used to identify and quantify the *C*-terminal amino acid.

$$\text{Peptide—NH—}\underset{\displaystyle\;}{\overset{\displaystyle R_{n-2}}{\overset{|}{\text{CH}}}}\text{—}\underset{\displaystyle O}{\underset{\|}{\text{C}}}\text{—NH—}\overset{\displaystyle R_{n-1}}{\overset{|}{\text{CH}}}\text{—}\underset{\displaystyle O}{\underset{\|}{\text{C}}}\text{—NH—}\overset{\displaystyle R_{n}}{\overset{|}{\text{CH}}}\text{—}\underset{\displaystyle O}{\underset{\|}{\text{C}}}\text{—OH}$$

$$\Big\downarrow \text{ Anhydrous } H_2N\text{—}NH_2$$

$$\text{.....} + NH_2\text{—}\overset{\displaystyle R_{n-2}}{\overset{|}{\text{CH}}}\text{—}\underset{\displaystyle O}{\underset{\|}{\text{C}}}\text{—NH—}NH_2 + NH_2\text{—}\overset{\displaystyle R_{n-1}}{\overset{|}{\text{CH}}}\text{—}\underset{\displaystyle O}{\underset{\|}{\text{C}}}\text{—NH—}NH_2 + NH_2\text{—}\overset{\displaystyle R_{n}}{\overset{|}{\text{CH}}}\text{—}\underset{\displaystyle O}{\underset{\|}{\text{C}}}\text{—OH}$$

Mixture of amino acid hydrazides — C-terminal amino acid

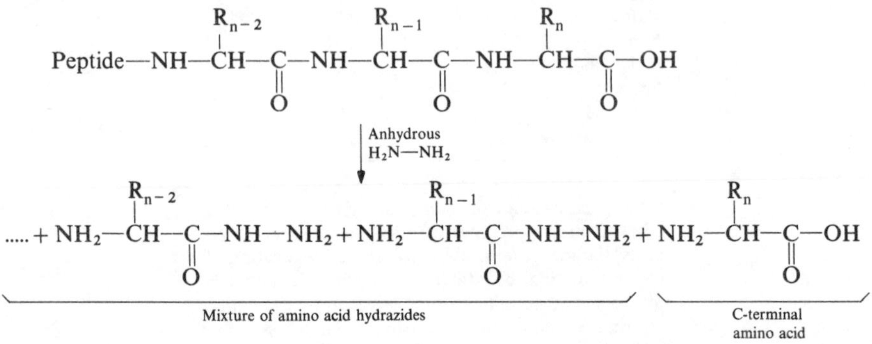

Figure 18.5. Hydrazinolysis of peptide with liberation of free *C*-terminal amino acid.

The original procedure of Akabori *et al.* (1952) involved heating the protein with anhydrous hydrazine for 10 h at 100°C. Under these conditions cystine and cysteine (but not cysteic acid, *S*-carboxymethylcysteine or *S*-aminoethylcysteine) are completely destroyed (Bradbury, 1956), arginine is decomposed and partly converted into ornithine and guanidine (Niu & Fraenkel-Conrat, 1955), and many other residues are partially destroyed, due in part to the high temperature of the reaction (Braun & Schroeder, 1967). Bradbury (1956) and Kawanishi *et al.* (1964) claimed that the addition of the acid catalyst, hydrazine sulphate, to the anhydrous hydrazine permitted the reaction to be carried out under much milder conditions (60–80°C) with a corresponding increase in the yield of *C*-terminal amino acids. Braun and Schroeder (1967) used the H^+ form of the ion-exchange resin, Amberlite CG-50 to give improved yields. They also showed that there were large variations in the rates of release of different *C*-terminal amino acids and considerable differences in the stability of various amino acids to the conditions of hydrazinolysis. Thus it is not possible to recommend a single reaction time which would be optimal in all cases. Instead a series of samples should be taken and heated for differing time periods.

The separation of the free *C*-terminal amino acid from the mixture of amino acid hydrazides must be done quickly since the hydrazides are rather unstable and are readily converted to free amino acids in the presence of water. The sensitivity depends on the method used for separation and detection of the amino acid. In the original procedure (Aabori *et al.*, 1952) freshly distilled benzaldehyde was added to an aqueous solution of the hydrazinolysate. The benzaldehyde converted the amino acid hydrazides to oily dibenzal derivatives which were readily separated from the free amino acids by centrifugation. However, the amino acid hydrazides were not completely removed by a single treatment with benzaldehyde. Modifications employing isovaleraldehyde (Akabori *et al.*, 1956), enanthoaldehyde (Boissonnas & Haselbach, 1953) and *p*-nitrobenzaldehyde (Braunitzer, 1955) did not make any appreciable improvement. Ohno (1954) and Kawanishi *et al.* (1964) treated the hydrazinolysate with 2,4-dinitro-1-fluorobenzene. The hydrazides yielded neutral di- or tri-DNP derivatives, which were readily separated from the acidic DNP-*C*-terminal amino acid by fractional extraction with ethyl acetate and ether. The acidic DNP derivatives were identified by two-dimensional paper chromatography (Levy, 1954). Niu and Fraenkel-Conrat (1955) combined the two approaches by using benzaldehyde treatment first, followed by dinitrophenylation and differential extraction. The *C*-terminal DNP-amino acid was identified by paper chromatography.

As Blackburn (1970) pointed out, many of these complicated modifications were made before modern methods of amino acid analysis became available, and are now out of date. Although the hydrazinolysate (after removal of excess hydrazine by lyophilization) can be directly applied to the columns of an amino acid analyser this is not recommended because the hydrazides are very difficult to remove from the column and may require up to five times the normal amount of NaOH to regenerate the resin. It is preferable to separate the hydrazides from the free *C*-terminal amino acid residue by rapid chromatography on a cation-exchange resin, such as phosphocellulose or Amberlite IR-120, as described below.

18.3.2.1. Procedure for hydrazinolysis

The method of Braun and Schroeder (1967) is described.

Preparation of anhydrous hydrazine

Hydrazine is distilled from sodium hydroxide under nitrogen in a conventional glass distillation apparatus. Place in a 250 ml round-bottom flask 100 ml of 'anhydrous' hydrazine (95 % +) with 40 g sodium hydroxide pellets and flush the apparatus with dry nitrogen that has been passed through conc. sulphuric acid. The hydrazine is heated under nitrogen at atmospheric pressure to its boiling point and then cooled to room temperature. The hydrazine is distilled between

16 and 18°C under oil pump vacuum with nitrogen introduced into the flask via a sturdy capillary tube and collected in a 100 ml round-bottom flask at the temperature of dry ice. The distilled hydrazine may be stored in a stoppered flask over P_2O_5 at 2°C, or better in sealed ampoules (5 ml per ampoule). Hydrazine absorbs atmospheric moisture rapidly and becomes unsatisfactory for hydrazinolysis.

Hydrazinolysis reaction

The resin (Amberlite CG-50, less than 200 mesh) is prepared by washing successively with 1 M-NaOH, water, 2 M-HCl and then with water until free of chloride. The resin is dried in an oven at 80°C and stored under vacuum in a desiccator over P_2O_5.

Dry Amberlite CG-50 (50 mg) is placed in a heavy-walled Pyrex tube (15 × 125 mm). Lyophilized protein (about 0.25 μmol) is added and the tube immediately drawn out at the upper third to a narrow neck. When the tube has cooled, anhydrous hydrazine (2.0 ml) is placed in the tube and drawn through the neck onto the sample by cooling the lower part of the tube in an alcohol-dry ice bath. It is important to avoid absorption of CO_2 by the hydrazine otherwise a large amount of ninhydrin-positive material will be produced which interferes with subsequent chromatography. The frozen sample is then sealed *in vacuo* and placed in an oven at 80°C for 10–100 h. The sample should be shaken gently 3 to 4 times a day during the heating. If the reaction mixture is not to be examined immediately after heating, it may be stored in the unopened tube at −20°C.

After hydrazinolysis, the reaction mixture is transferred to a *dry* 25 ml round-bottom flask, care being taken to avoid the absorption of moisture by the hydrazine during this transfer. Use a dry box if possible. The tube is rinsed twice with 1.0 ml of hydrazine and the entire material lyophilized. The hydrazine is completely removed from the frozen mixture in 3 h.

Chromatographic removal of hydrazides from the released *C*-terminal amino acid

In the original paper (Braun & Schroeder, 1967) three cation-exchange resins, Amberlite IR-120, Amberlite CG-50 and phosphocellulose were examined for the chromatographic separation of the free *C*-terminal amino acid from the amino acid hydrazides. Chromatography on Amberlite IR-120 required the use of two volatile buffers (pH 3.1, pH 5.2), a jacketed column held at 30°C and a run time of about 8 h. Chromatography on phosphocellulose was simpler as it required only 4 h with a single developing buffer at room temperature and allowed recovery of all amino acid residues as well as the monohydrazides of aspartic and glutamic acids. However, it had a lower capacity and a 10 × 300 mm column could not be loaded with more than 10 mg of protein. Amberlite CG-50 offered no advantages. The fractionation using phosphocellulose is described here.

Just prior to chromatographic separation, the lyophilized residue of hydrazides, free amino acid(s) and resin catalyst is suspended in 3.0 ml of water, centrifuged and the resin washed twice with 1.0 ml of water. The solution with washings (pH 8.5–9) is adjusted to pH 2 with a few drops of 2 M-HCl. Apply to a 10 × 300 mm column of phosphocellulose (Whatman cellulose phosphate powder P-70, unsifted) equilibrated with 0.4 M-pyridine formate buffer pH 3.2 (32.2 ml pyridine, 47 ml 98 % formic acid, water to 1 l) and chromatograph at room temperature with pH 3.2 buffer at a flow rate of 30 ml/h. Collect 2 ml fractions and monitor by ninhydrin reaction. Acidic and neutral amino acids elute between 20 and 37 ml of effluent volume, lysine, ammonia and the monohydrazides of aspartic and glutamic acids between 42 and 75 ml, histidine between 80 and 90 ml and arginine between 93 and 112 ml. The amino acid hydrazides are strongly fixed and the column is discarded. The appropriate fractions are pooled, taken to dryness and analyzed. The monohydrazides of aspartic and glutamic acids must first be hydrolysed before determination as aspartic and glutamic acids.

Final comments

Because the *C*-terminal amino acids vary in their rates of release and in their stability it is not possible to make any general recommendation for the length of the hydrazinolysis reaction. It is suggested that a series of samples be used, each heated for times varying from 10 to 100 h. The procedure should be applicable to the quantitative detection of all amino acid residues at the *C*-terminus except arginine, asparagine, glutamine and possibly lysine. On hydrazinolysis *C*-terminal asparagine and glutamine are released as their monohydrazides with the side chain amide group substituted. These monohydrazides are very similar in structure and chromatographic properties to the monohydrazides produced by internal aspartic and glutamic acids (Fig. 18.6). Their determination is com-

$$\begin{array}{c} COOH \\ | \\ (CH_2)_{1-2} \\ | \\ NH_2\text{—}CH\text{—}CONHNH_2 \end{array}$$

Internal acid hydrazide

$$\begin{array}{c} CONH.NH_2 \\ | \\ (CH_2)_{1-2} \\ | \\ NH_2\text{—}CH\text{—}COOH \end{array}$$

C-terminal amide hydrazide

Figure 18.6. Structures of the hydrazides produced by internal aspartic and glutamic acids and *C*-terminal asparagine and glutamine.

plicated by the observation that the two monohydrazides of aspartic acid seem to be interconvertible (Narita & Ohta, 1959).

18.3.3. Selective reduction

Chemical reduction of the *C*-terminal amino acid to the corresponding alcohol was the first chemical procedure employed to characterize the *C*-terminal group of a protein. Fromageot *et al.* (1950) reacted insulin with $LiAlH_4$ in a suspension of *N*-ethylmorpholine at 55°C for 8 h and examined the resulting amino alcohols by paper chromatography. Chibnall and Rees (1958) employed the reduction procedure after first esterifying all carboxyl groups in the protein. They used the much milder reducing agent $LiBH_4$ in tetrahydrofuran and quantitatively detected the amino alcohols derived from the *C*-terminal amino acid. They concluded that the procedure was complicated by the simultaneous reductive cleavage of 1–2 % of the total peptide bonds and reduction of peptide carbonyls to methylene groups.

Saund *et al.* (1973) used the commercially available mixed alkoxyl-halide, sodium dihydrobis-(2-methoxyethoxy) aluminate to reduce *C*-terminal amino acid residues in di- and tripeptides to their corresponding amino alcohols without the need to first convert them into esters. The amino alcohols were identified by paper chromatography. The method has not yet been applied to proteins.

Hamada and Yonemitsu (1973) reported that sodium borohydride in aqueous solution, rather than in organic solvents, selectively and almost quantitatively reduced the esterified *C*-terminal amino acid to its alcohol. When applied to lysozyme, bovine insulin and conconavalin A, the expected end group derivatives were obtained without appreciable side reactions. This may provide a practical method for *C*-terminal determinations in proteins and its future application will depend on the establishment of reliable analytical methods for all amino alcohols derived from the usual amino acids.

18.3.4. *C*-terminal analysis by aldehyde formation

Miller and Loudon (1975) reported a new approach for the identification of *C*-terminal amino acid residues. The reactions involved are summarized in Fig. 18.7. *O*-pivaloylhydroxylamine reacts quantitatively with carbodiimide-activated carboxylic acids to form an *O*-pivaloylpeptidohydroxamate. Ionization of the hydroxamate N-H bond (pK_a 6.4–7.4) initiates in peptides a Lossen rearrangement to a peptidyl isocyanate which in turn is converted to a mixture of peptidylureas. An aliquot of suitable size can then be hydrolysed in 6 M-HCl and subjected to subtractive amino acid analysis because the *C*-terminal amino acid is converted to an aldehyde. Alternatively, the aldehyde can be generated by adjusting the pH of the reaction mixture to pH 1–2 and heating at 50°C for 2 h.

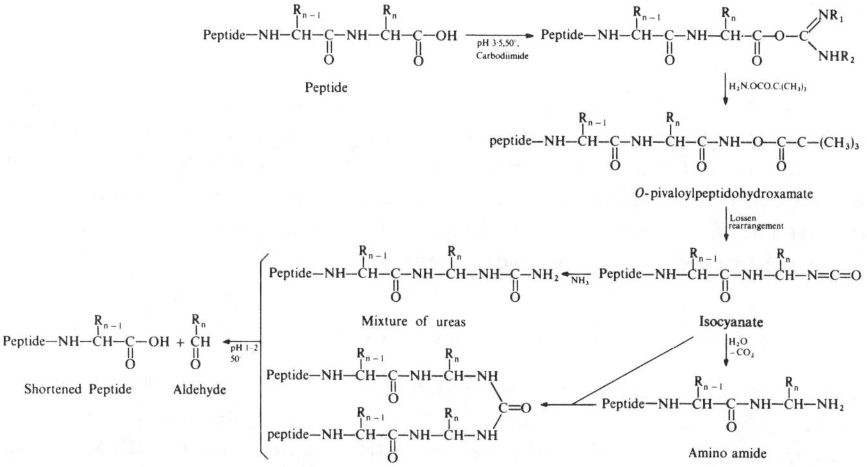

Figure 18.7. Reactions involved in *C*-terminal analysis by aldehyde formation. *O*-pivaloylhydroxylamine reacts quantitatively with the carbodiimide-activated *C*-terminal carboxylic acid to form the hydroxamate. Ionization of the hydroxamate N-H bond initiates a Lossen rearrangement and eventual conversion of the carboxyl terminal amino acid to an aldehyde (after Miller & Loudon, 1975).

The procedure was successfully applied to eighteen di-, tri-, tetra- and pentapeptides as well as the A- and B-chains of insulin, the *C*-terminal residue being identified by subtractive amino acid analysis. In some cases the aldehyde was quantitatively identified by paper chromatography as the 2,4-dinitrophenylhydrazone. The advantages of this procedure are the use of aqueous solutions (including urea if necessary), the high yields of the procedure (70–100 % for most residues) and the successful removal of *C*-terminal proline, asparagine and presumably glutamine. The limitations are the low degradation for *C*-terminal glutamic and aspartic acids, presumably because they form cyclic anhydrides on activation with carbodiimide which regenerate the original amino acid on hydrolysis, and the current restriction of the method to peptides small enough to be amenable to subtractive amino acid analysis.

18.3.4.1. O-pivaloylhydroxylamine

The method of Miller and Loudon (1975) is described.

O-pivaloylhydroxylamine HCl (154 mg, 1 mmol) (OPHA–HCl) is dissolved in 4 ml of water and the pH adjusted to pH 3.5 with 1 M-NaOH. The peptide solution (1 μmol) in 2.0 ml of H_2O is added followed immediately by 0.4 ml of a

0.5 M aqueous solution of EDC; pH 3.5 is maintained with 0.1 M-HCl in a pH stat. Three more 0.4 ml aliquots of the EDC solution are added at 15-min intervals. Excess EDC and OPHA are quenched by adding 0.5 ml of 5 M-formate buffer at pH 3.5 and stirring for 20 min. The Lossen rearrangement is initiated by adjustment of the pH to 8.5 and the temperature to 50°C. The pH is maintained with 1.0 M-NaOH and the reaction continued for approximately 20 h. The progress of the rearrangement is easily monitored by the base consumption on the pH stat. A suitable aliquot is then removed, hydrolysed with 6 M-HCl and subjected to subtractive amino acid analysis.

18.3.5. *C*-terminal analysis by oxazole alcoholysis

The conversion of the *C*-terminal peptide function into an imidoether which could then be cleaved by alcohols under acid conditions was proposed by Previero & Coletti-Previero (1977) as the basis for a new *C*-terminal sequencing procedure. The cyclic imidoethers are formed by an intramolecular dehydration between the last peptide bond and the *C*-terminal carboxyl group (oxazolinone formation) or its ester derivative (oxazole formation). Oxazole formation is preferable since it is quantitatively cleaved by acid alcoholysis to yield the *C*-terminal amino acid ester. Acid alcoholysis of oxazolinones yields the free *C*-terminal amino acid but also the uncleaved peptidyl ester as a major competitive side product. The reactions involved are shown in Fig. 18.8. The procedure should

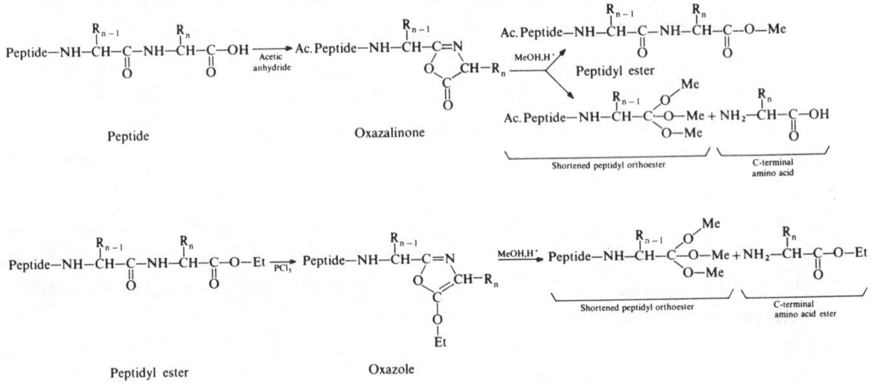

Figure 18.8. Reactions involved in *C*-terminal analysis by oxazolinone or oxazole alcoholysis. Acid-catalysed alcoholysis of oxazolinones can occur in two ways resulting in the formation of the uncleaved peptidyl ester or the desired free *C*-terminal amino acid. Acid-catalysed alcoholysis of the corresponding oxazole does not suffer from this disadvantage and results in the release of the *C*-terminal amino acid ester (after Previero & Coletti-Previero, 1977.)

be applicable to the identification of *C*-terminal residues and attempts are being made to develop it into a stepwise *C*-terminal sequencing method.

18.4. SEQUENCE DETERMINATION FROM THE *C*-TERMINUS

The methods for determining the sequence of amino acids from the *C*-terminal end of peptides or proteins are not successful. Both enzymic and chemical procedures are available but neither approach to date has been capable of extended sequence analysis beyond a few amino acid residues. The use of carboxypeptidases is still the major approach employed by most investigators, although it suffers from the problem of differential rates of cleavage for different amino acids and the difficulty of maintaining a tightly controlled stepwise degradation. The use of carboxypeptidases will be discussed in section 18.4.4.

18.4.1. Thiocyanate degradation

The thiocyanate degradation of Stark (1968) is the most thoroughly investigated chemical method. It has been modified in subsequent publications to improve the procedure in free solution (Cromwell & Stark, 1969; Kubo *et al.*, 1971; Yamashita, 1971; Stark, 1972) and extended to a solid-phase version (Williams & Kassell, 1975; Rangarajan & Darbre, 1976). Neither form of the procedure has gone successfully beyond a few steps.

In the thiocyanate degradation procedure the peptide or protein is incubated with ammonium thiocyanate in acetic acid–acetic anhydride solution to form the peptidylisothiocyanate which rearranges spontaneously to the peptidyl thiohydantoin (Fig. 18.9). The peptidyl thiohydantoin is then selectively cleaved by either acid or alkali to yield a *C*-terminal amino acid thiohydantoin and a shortened peptide or protein with a new *C*-terminal amino acid residue.

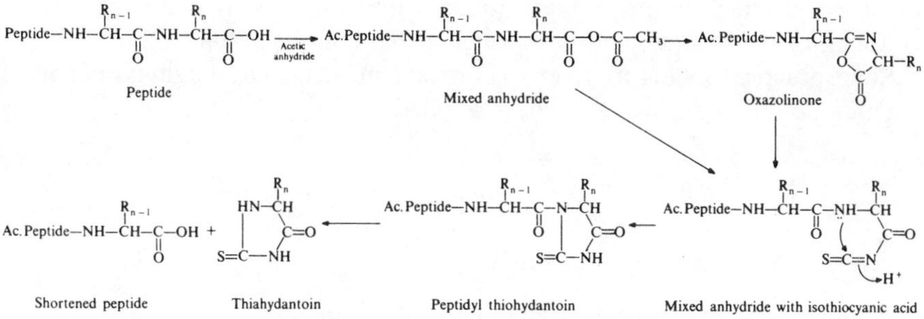

Figure 18.9. Reactions involved in the thiocyanate degradation of peptides and proteins (after Stark, 1968).

The method resulted from the work of Johnson and Nicolet (1911) on the conversion of acylamino acids to acylthiohydantoins by ammonium thiocyanate and acetic anhydride. Schlack and Kumpf (1926) included a cleavage step (1 M-NaOH, 3 h, room temperature) to allow the reaction to be applied sequentially to peptides. This procedure was subsequently applied with limited success to determine the end groups of a variety of proteins and peptides (see review Greenstein & Winitz, 1961). Stark (1968) reinvestigated the conditions required for both derivatization and cleavage and published a milder, improved procedure for peptides. This was subsequently extended by Cromwell and Stark (1969) for use with proteins. However, under the conditions used neither proline nor aspartic acid were degraded.

Aspartic acid forms a cyclic anhydride which does not react with ammonium thiocyanate. This may be overcome by modifying both side chain and terminal carboxyl groups with carbodiimide (Hoare & Koshland, 1967) and then liberating the terminal carboxyl group for sequencing by enzyme treatment as suggested by Kassell *et al.* (1977).

Other modifications include the substitution of trifluoroacetic acid and acetyl chloride for acetic anhydride (Kubo *et al.* 1971), to ensure that the reaction proceeds via the mixed anhydride rather than the oxazolinone (Yamashita & Ishikawa, 1972), and the use of thiocyanic acid rather than thiocyanate salts (Kubo *et al.*, 1971). Under these conditions successful degradations were carried out on peptides including those with *C*-terminal proline. Yamashita (1971) added trifluoroacetic anhydride to the reaction mixture and it was later reported that 14 cycles were successfully carried out on papain and 10 cycles on ribonuclease (Yamashita & Ishikawa, 1971).

The method in free solution involves usually Sephadex chromatography to separate excess reagent from the peptidyl thiohydantoin and a further step to isolate the thiohydantoin cleaved from the *C*-terminus of the peptidyl thiohydantoin. These separations and the lyophilization of the column fractions are time consuming. Solid-phase sequencing eliminates many of these problems, but insufficient work has been reported to encourage the novice in this field, which involves selection (or preliminary preparation) of an activated support and coupling of the peptide before sequencing steps can be attempted.

Some practical details are given to illustrate, in particular, the simple chemical reaction conditions which have been applied.

18.4.1.1. Thiocyanate degradation of peptides

Formation of peptidylthiohydantoin

1. Stark (1968). Dissolve the peptide in 0.5 ml or less of 50% acetic acid. A fresh solution of ammonium thiocyanate (100 mg), acetic anhydride (4.0 ml) and glacial acetic acid (1.0 ml) is added slowly with swirling to the peptide solution.

Heat the mixture at 50°C for 6 h. Additional ammonium thiocyanate (100 mg) is added with swirling to dissolve the salt and heated at 50°C for a further 18 h. At the end of the reaction add water (3.0 ml) and allow to stand a few minutes to hydrolyse excess acetic anhydride.

2. Suzuki *et al.* (1976). Treat the peptide with 1.0 ml of acetylating reagent acetic acid/acetic anhydride/pyridine (2:10:1, by vol.) for 30 min at 50°C, then add 1.0 ml of ammonium thiocyanate (20 mg/ml acetic acid) and heat with vigorous stirring for 30 min at 50°C.

3. Kubo *et al.* (1971). Dissolve the peptide in a mixture of trifluoroacetic acid (10 μl) and acetylchloride (200 μl) and incubate the reaction mixture at 30°C for 15 min. A solution (200 μl) of 3% thiocyanic acid in dioxane is added and the mixture kept at 30°C for a further 60 min. After formation of the peptidylthiohydantoin the solvent can be removed *in vacuo*.

Isolation of peptidylthiohydantoin

1. Stark (1968). The peptidylthiohydantoin is separated from excess reagents by desalting on a Sephadex G-25 column (20 × 500 mm) using 50% acetic acid. Fractions (5.0 ml) are collected, the peptidylthiohydantoin located by u.v. absorption and the peak tubes pooled.

2. Suzuki *et al.* (1976). No chromatographic separation is used. The cleavage is carried out directly following evaporation of the reaction mixture to dryness.

Cleavage of peptidylthiohydantoin

1. Stark (1968). The tubes containing the peptidylthiohydantoin are pooled, the solvent removed *in vacuo* and the residue redissolved in 0.5 ml of 0.1 M-acetohydroxamic acid dissolved in 50% pyridine. Heat the solution at 50°C for 2 h and then evaporate to dryness. The dry residue is dissolved in 3.0 ml of 50% acetic acid and the released *C*-terminal thiohydantoin separated from the residual peptide by chromatography on a Sephadex G-25 column (20 × 500 mm).

2. Suzuki *et al.* (1976). Evaporate the unfractionated thiocyanate mixture to dryness. Dissolve the residue in water (3 ml) and cleave by stirring with 2 g of cation exchanger (Dowex 50 W X 8) for 30 min at 50°C.

3. Kubo *et al.* (1971). Cleave with 0.5 M-triethylamine.

18.4.1.2. Thiocyanate procedure for proteins

Proteins are not very soluble in the acetic anhydride-acetic acid reaction mixture and to overcome this problem Cromwell and Stark (1969) introduced hexafluoroacetone trihydrate and Yamashita (1971) introduced trifluoroacetic anhydride.

Formation of proteinylthiohydantoin

1. Cromwell & Stark (1969). The protein (0.015–1.0 μmol) is dissolved in a mixture of hexafluoroacetone trihydrate (1.0 ml) and water (0.35 ml). A homogeneous solution (prepared immediately before use) of ammonium thiocyanate (100 mg) in hexafluoroacetone trihydrate (1.0 ml) and acetic anhydride (4.5 ml) is added dropwise, with swirling. The mixture is heated at 50°C for 2 h and a further 100 mg of ammonium thiocyanate is added and heated for a further 18 h. Water (3.0 ml) is added to destroy excess acetic anhydride and the proteinylthiohydantoin is separated on a column of Sephadex G-25.

Cleavage of proteinylthiohydantoin

1. Cromwell & Stark (1969). The tubes containing the proteinylthiohydantoin are pooled and the solvent is removed *in vacuo*. The residue is dissolved in 1.0 ml of 12 M-HCl and left at room temperature for 30 min. The HCl is then removed rapidly at room temperature with a vacuum pump. The mixture is redissolved in 50 % acetic acid and the *C*-terminal amino acid thiohydantoin separated from the remaining polypeptide by gel filtration as before. The solvent is removed from the thiohydantoin immediately, a small volume of methanol is added and identification by TLC commenced without delay, because thiohydantoins are not very stable in light and air.

18.4.1.3. Identification of C-terminal thiohydantoins

For simple peptides the degradation can be followed by subtractive amino acid analysis of the residual peptide after each step. For proteins, direct identification of the removed thiohydantoin must be made. This can be done by TLC on pre-coated silica gel plates impregnated with fluorescent indicator using either heptane/butan-1-ol/99 % formic acid (95:65:30, by vol.) or chloroform/95 % ethanol/glacial acetic acid (100:50:15, by vol.) (Stark, 1968; 1972; Cromwell & Stark, 1969).

A better system is two-dimensional TLC on 100 × 100 mm pre-coated polyamide sheets using acetic acid/water (7:13, v/v) in the first dimension and chloroform/95 % ethanol/glacial acetic acid (20:10: 3, by vol.) containing 0.025 % 2,5-bis (5-*t*-butyl-benzoxazol-2-yl)thiophen in the second dimension (Rangarajan & Darbre, 1975). Under u.v. light at 254 nm the amino acid thiohydantoins show up as dark spots on a pale blue fluorescent background (Fig. 18.10).

Amino acid thiohydantoins can also be identified directly by GLC after trimethylsilylation (Rangarajan *et al.*, 1973; Dwulet & Gurd, 1977), by mass spectrometry (Rangarajan *et al.*, 1973; Darbre, 1975; Suzuki *et al.*, 1972, 1976), or by analysis of the amino acids after hydrolysis using ion-exchange (Stark, 1972) or GLC (Rangarajan & Darbre, 1975).

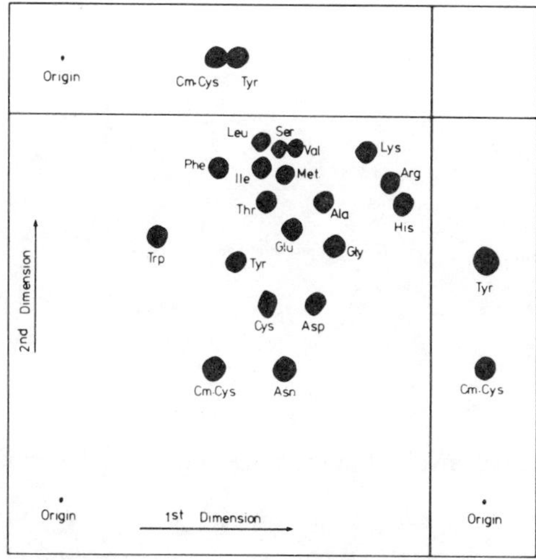

Figure 18.10. Two-dimensional thin-layer separation of amino thiohydantoins on 100 × 100 mm pre-coated polyamide sheets. Chromatography is in glacial acetic acid/water (7:13, v/v) in the first dimension and chloroform/95% ethanol/glacial acetic acid (20:10:3, by vol.) containing the fluorescent indicator 2,5-bis(5-*t*-butylbenzoxazol-2-yl)thiophen (0.025%). Under u.v. light at 254 nm the thiohydantoins show up as dark spots on a pale blue fluorescent background (after Rangarajan & Darbre, 1975).

18.4.2. Cyanamide degradation

A new procedure for sequencing proteins or peptides from the *C*-terminus was studied (Tarr, 1975). It appears to possess distinct advantages over the Stark thiocyanate degradation. The reactions involved are shown in Fig. 18.11. The peptide is first coupled to an *S*-alkylisothiuronium salt with carbodiimide, then the *C*-terminal derivatized amino acid is cleaved off in aqueous base at pH 10–11.5. The cleavage reaction occurs in two distinct stages. The alkyl portion of the acylated isothiourea is removed by base to produce the acyl cyanamide. The cyano group then undergoes nucleophilic attack by the nitrogen of the peptide bond and cyclizes with the concomitant cleavage of the iminohydantoin from the residual peptide. The iminohydantoins are fairly polar derivatives and, except for charge effects, resemble their parent amino acids in chromatographic behaviour. They can be detected either by using a nitroprusside/ferricyanide spray after TLC or by GLC of the trimethylsilylated derivatives (Tarr, 1975).

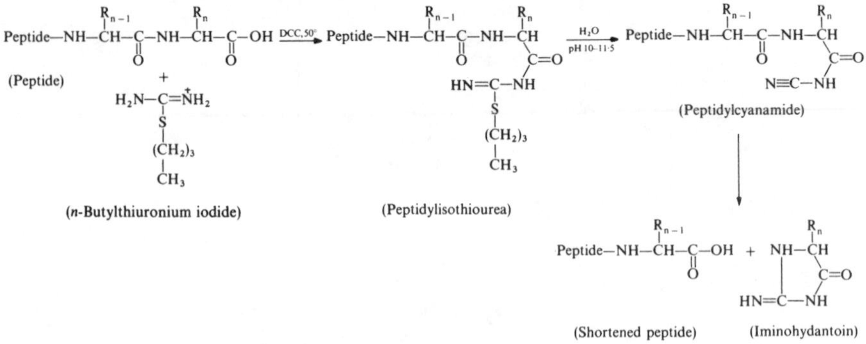

(Peptide) + (n-Butylthiuronium iodide) → DCC, 50° → (Peptidylisothiourea) → H_2O, pH 10–11·5 → (Peptidylcyanamide) → (Shortened peptide) + (Iminohydantoin)

Figure 18.11. Reactions involved in the cyanamide degradation (after Tarr, 1975).

S-(n-butyl)thiuronium iodide (BTU) was found to be the most reactive of the isothiuronium salts tested, and rapid coupling to about 99 % efficiency was achieved in pyridine (anhydrous or containing about 1 % conc. HCl) with 0.25 M-dicyclohexyl carbodiimide (DCC) and BTU in 10–15 min at 50°. A similar efficiency could be achieved in 20 min with 0.1 M reagent solutions. The presence of water slowed down the coupling reaction, particularly with DCC, thus requiring larger excesses of carbodiimide to force the reaction to completion. The cleavage reaction can be carried out in the presence or absence of excess reagent, the only requirement being an aqueous base at pH 10–11.5. Nearly optimal cleavage was obtained by pyridine/1 M-aqueous trimethylamine (1:1, v/v) for 10 min at 50°, and an overall efficiency of 96 % was obtained for one complete cycle of degradation on a model peptide. Development of the method is still in an exploratory stage. It suffers from the disadvantage (Tarr, 1975) that it cannot degrade *C*-terminal proline residues (due to steric constraints) or aspartic acid residues (due to conversion by carbodiimide to cyclic anhydrides).

18.4.3. Other chemical degradations

Several other procedures have been proposed for *C*-terminal sequencing. Khorana (1952) suggested the use of *p*-tolylcarbodiimide to form an acylurea which is then degraded at room temperature with 0.01 M-NaOH in aqueous ethanol, cleaving off the *C*-terminal amino acid as an *N*-tolylcarbonyl toluidide. However, the acylurea is also significantly hydrolysed back to the original peptide, thus making it unsuitable as a sequential procedure.

Saund *et al.* (1973) suggested that their selective reduction end-group method (see section 18.3.3) could be extended to a *C*-terminal sequencing procedure. The peptidylaminoalcohol can be cyclised by treatment with carbonyldiimidazole to produce an *N*-acyl-2-oxazolidone derivative which can then be cleaved under mild conditions (0.3 M acid, 1–2 h) to give the 2-oxazolidone and the residual peptide which now has a new carboxyl terminus. The scheme of reactions is shown in Fig. 18.12.

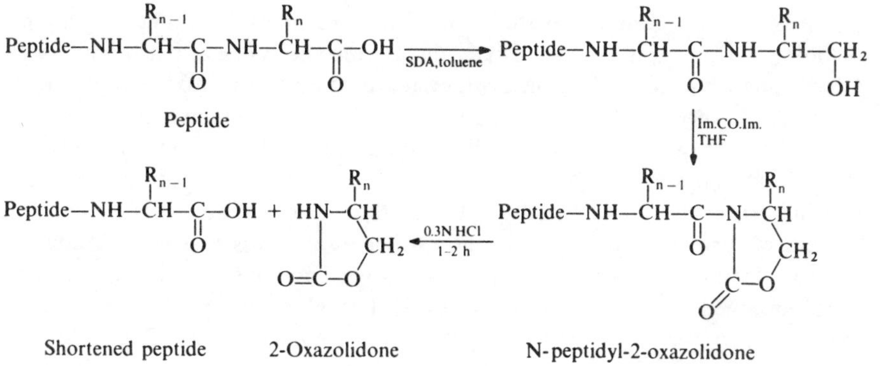

Peptide—NH—CH(R_{n-1})—C(=O)—NH—CH(R_n)—C(=O)—OH —SDA, toluene→ Peptide—NH—CH(R_{n-1})—C(=O)—NH—CH(R_n)—CH_2OH

Peptide

↓ Im.CO.Im. THF

Peptide—NH—CH(R_{n-1})—C(=O)—OH + HN—CH(R_n)—CH_2—O—C(=O) ←0.3N HCl, 1-2 h— Peptide—NH—CH(R_{n-1})—C(=O)—N—CH(R_n)—CH_2—O—C(=O)

Shortened peptide 2-Oxazolidone N-peptidyl-2-oxazolidone

Figure 18.12. Reactions involved in *C*-terminal sequencing by selective reduction with sodium dihydro-bis-(2-methoxyethoxy)aluminate (SDA), and conversion of the acyl-aminoalcohol to the 2-oxazolidone derivative with carbonyldiimidazole in tetrahydrofuran (THF). Hydrolysis under mild conditions yields the shortened peptide and the *C*-terminal 2-oxazolidone (after Saund *et al.*, 1973).

The procedure of Miller and Loudon (1975) is being explored as a possible sequential *C*-terminal degradation method (Loudon & Parham, 1978; Parham & Loudon, 1978). However, the chemistry involved is not simple. Previero and Coletti-Previero (1977) are developing their oxazole alcoholysis procedure (see section 18.3.5) as a *C*-terminal sequencing method. Problems still to be overcome include the continued release of the same *C*-terminal residue in several successive cycles of the reaction (due either to incomplete or competititve reaction at one or more steps in the procedure) and the conversion of the new carboxyl derivative (produced by the release of the *C*-terminal residue) into an ester suitable for stepwise degradation. This sequencing approach is being developed as a solid-phase procedure.

18.4.4 Carboxypeptidases

The first use of carboxypeptidase in *C*-terminal residue analysis was made by Grassmann *et al.* (1930). They used a crude preparation of enzyme to remove the *C*-terminal glycine residue from glutathione. Since then, carboxypeptidases have been used in almost every major protein sequence determination. The use of carboxypeptidase for the end-group analysis of proteins and peptides was extensively reviewed by Fraenkel-Conrat *et al.* (1955).

Three types of carboxypeptidases A, B and C are commercially available for *C*-terminal analysis and have different specificities. Carboxypeptidase A (from bovine pancreas) preferentially removes *C*-terminal residues with aromatic or large aliphatic side chains such as phenylalanine, tyrosine, tryptophan, leucine, isoleucine, methionine and valine, but it also removes threonine, glutamine, histidine, alanine, homoserine, asparagine, serine and lysine at reasonable rates. Glycine, aspartic acid, glutamic acid, cysteic acid and *S*-carboxymethylcysteine are removed very slowly while *C*-terminal proline and arginine are not removed at all. The rate of release of the acidic amino acids can be increased by lowering the pH of the digestion (Slobin & Carpenter, 1966). Carboxypeptidase B (from porcine pancreas) preferentially releases the basic amino acids lysine and arginine very much faster than any of the other amino acids. Carboxypeptidase C has been isolated from a variety of plant sources including citrus fruits (Zuber, 1968), orange leaves (Sprössler *et al.*, 1971), French beans (Wells, 1965), barley malt (Visuri *et al.*, 1969), moulds (Ichishima, 1972; Shaw, 1964). Carboxypeptidase C preparations have the advantage that they combine the specificities of carboxypeptidases A and B and in addition can release *C*-terminal proline. The rates of release of all amino acids are similar except for glycine, which is released slowly (Ambler, 1972; Tschesche, 1977). Carboxypeptidase P is obtained from *Aspergillus* (Ichishima, 1972) and *Penicillium* (Yokoyama & Ichishima, 1972) and is available from Takara-shuzo K.K., Japan. It is active at pH 2.5 and has a specificity similar to that of carboxypeptidase C. Another similar enzyme denoted as carboxypeptidase Y was isolated from baker's yeast: it sometimes fails to release aspartic acid and basic amino acids (Hayashi, 1977). Studies were reported on a method for solid-phase *C*-terminal sequencing using carboxypeptidases (Tsugita & Broek, 1980).

Basically, *C*-terminal end-group or sequence determination with carboxypeptidases involves digesting the peptide or protein sample with the appropriate enzyme or mixture of enzymes and withdrawing aliquots of the digested sample at various time intervals. The free amino acids present in these aliquots are identified and quantified and the moles of amino acid released/mole of original protein or peptide are plotted against the time of reaction (see Ambler, 1972 and Fig. 18.13).

The major problems which must be taken into account when employing carboxypeptidase digestion are as follows. Firstly, many preparations of the pancreatic carboxypeptidases A and B contain significant endopeptidase activity, mainly trypsin and chymotrypsin, which must be inactivated by pretreatment with diisopropylfluorophosphate (DFP). DFP-treated carboxypeptidases A and B are available commercially. Secondly, the differential rates of release of specific amino acids, which depend not only on the nature of the *C*-terminal residue being removed but also in part on the nature of the penultimate amino acid, can make it difficult to deduce the sequence. When a residue that is

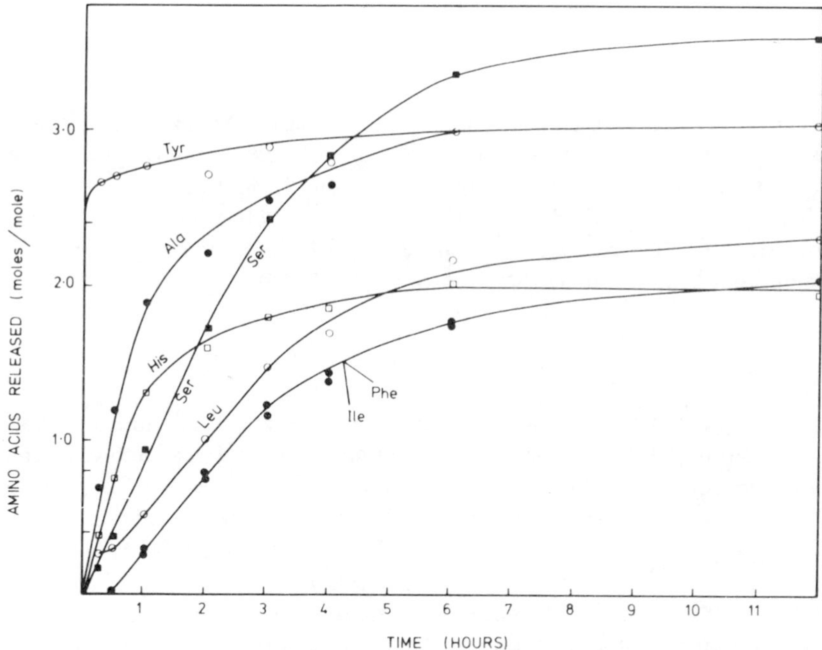

Figure 18.13. Release of amino acid residues from rabbit muscle aldolase during digestion with carboxypeptidase A (after Winstead & Wold, 1964).

hydrolysed slowly is followed by one that is removed rapidly then the rates of release of those two amino acids are almost indistinguishable. A further complication is the presence of more than one residue of a particular amino acid in a *C*-terminal sequence. For these reasons it is usually difficult to deduce with certainty more than the first three or four residues of a *C*-terminal sequence using carboxypeptidases. Finally, the peaks of asparagine, glutamine and serine overlap each other under standard conditions of amino acid analysis. When amino acids corresponding to this peak are released an aliquot of the carboxypeptidase digest must be freed of enzyme and protein and then hydrolysed before amino acid analysis. The presence of aspartic or glutamic acids indicates whether amides were present in the 'serine' peak.

18.4.4.1. Method with carboxypeptidase A or B

Enzyme pre-treatment

Both carboxypeptidase A and B normally contain free amino acids and proteolytic enzymes. The endopeptidases can be inactivated by pre-treatment of the enzyme (1 mg/ml) in ammonium bicarbonate or *N*-ethylmorpholine acetate buffer (0.1 M, pH 8.5) with a 50-fold excess of DFP (15 μl of 0.1 M-DFP in dry isopropanol per mg enzyme) for 1 h at room temperature. The reagent and free amino acids are then quickly removed by dialysis or gel filtration on Sephadex G-25.

The free amino acids can also be removed from carboxypeptidase solution by suspending the enzyme in water and centrifuging. Both carboxypeptidase A and B are insoluble at low ionic strength and readily form a pellet.

The supernatant which contains the contaminating amino acids can be discarded. The pelleted enzyme can be dissolved in a small volume of 2 M-ammonium bicarbonate (Fraenkel-Conrat *et al.*, 1955).

Preliminary experiment

Before a quantitative analysis is attempted, a preliminary experiment should be carried out to determine whether any reaction takes place at all and to establish the optimal conditions of enzyme, substrate and time of digestion. For proteins it is generally necessary that the protein be denatured and in solution. Carboxypeptidase digestion can be carried out in 6 M-urea (Halsey & Neurath, 1955) and 0.056 M-SDS (Guidotti, 1960). The amount of material taken depends on the sensitivity of the analytical procedures to be employed.

Denatured protein or peptide (5–50 nmol) is dissolved in 100–200 μl of 0.2 M *N*-ethylmorpholine acetate buffer, pH 8.5. DFP-treated carboxypeptidase A (10–50 μg) or carboxypeptidase B (5–20 μg) is added and the mixture incubated at 37°C for 4–5 h. An enzyme blank should always be run. At the end of the incubation the reaction mixture is acidified (trichloroacetic acid, acetic acid or HCl) to pH 2.0. Any precipitate formed is removed by centrifugation and the supernatant subjected to amino acid analysis.

If the preliminary experiment shows that many amino acids are released, a low enzyme to substrate ratio of 1:100 or 1:200 should be employed, or the digestion should be carried out at room temperature. If the amino acids are released only slowly then a higher enzyme to substrate ratio of 1:20 may be used. If significant amounts of amino acid are not released by the trial digestion with either carboxypeptidase A or carboxypeptidase B, a chemical end-group procedure or isolation of the *C*-terminal peptide may indicate why the protein is resisting digestion with carboxypeptidases.

Quantitative determination

When the optimum conditions have been established a quantitative experiment is carried out. The amount of sample employed depends on both the sensitivity of the analytical techniques to be used and the number of aliquots to be taken. Convenient time intervals are 0.5, 1, 2, 4, 8 and 16 h and aliquots taken should contain 1–10 nmol of sample depending on the analyser.

Interpretation of data

A typical example of the procedure and the precautions that are necessary when interpreting the results of carboxypeptidase digestions is illustrated by the data of Winstead and Wold (1964) on carboxypeptidase A acting on rabbit muscle aldolase (Fig. 18.13). No additional amino acids were released when carboxypeptidase B was included in the digestion mixture. They concluded that the *C*-terminal sequence of rabbit muscle aldolase was:

Ser-Phe-Ile-Leu-Ser-His-Ala-Tyr-OH
8 7 6 5 4 3 2 1

One puzzling feature of the data was the decreased yield of the third, fifth, sixth and seventh residues, suggesting that in some aldolase subunits the histidine at the third position was either not released, or did not occur. The higher yields for 'serine', the fourth residue, suggested that there was more than one serine residue in this *C*-terminal sequence although the accurate placement of the second serine was impossible. As Ambler (1972) pointed out it is rarely possible to deduce with certainty more than the first three or four residues of a *C*-terminal sequence when using carboxypeptidases, although there is a great temptation to over-interpret the data.

Inspection of the results in Fig. 18.13 allows the sequence of the first four *C*-terminal residues to be assigned as -Ser-His-Ala-Tyr-OH. Winstead and Wold (1964) assumed that leucine was the fifth residue from the *C*-terminus and that it was followed, in order of release, by isoleucine and phenylalanine. Phenylalanine and isoleucine were released at identical rates but can be assumed to occur in the suggested order because it is known that carboxypeptidase A releases phenylalanine some 26 times faster than isoleucine. However, the assignment of leucine at the fifth position is doubtful since, as shown in Fig. 18.13, its rate of release was identical to that of isoleucine and phenylalanine, with only the absolute values at each time period approximately 0.25 moles higher.

When the amino acid sequence of rabbit muscle aldolase was determined (Lai *et al.*, 1970; Lai, 1975), the *C*-terminal sequence was shown to be:

Ser-Leu-Phe-Ile-Ser-Asn-His-Ala-Tyr-OH
9 8 7 6 5 4 3 2 1

This agrees with the carboxypeptidase derived assignments of His-Ala-Tyr-OH, and with the deduction that the isoleucine residue is closer to the *C*-terminus than the phenylalanine. However, it differs from the other carboxypeptidase-based assignments in several respects. Firstly, the 'serine' at position four was shown to be asparagine which, as mentioned earlier, overlaps with serine in a conventional amino acid analysis chromatogram and cannot be distinguished from serine without resorting to other means. Secondly, the extra residue of 'serine' obviously present from the data in Fig. 18.13 occurs at position 5 from the *C*-

terminus, not position 8. Thirdly, the leucine which was released at the same rate as the isoleucine and phenylalanine residues occurs at position 8 from the *C*-terminus, not position 5. The decreased yields in the release of histidine and later residues was found to be due to partial deamidation of the asparagine residue at position 4, the Asp-His linkage being very resistant to the action of carboxypeptidase A (Lai *et al.*, 1970).

18.4.4.2. Method with carboxypeptidase C

Commercial preparations of carboxypeptidase C generally from orange leaves (Tschesche, 1977) or from yeast (carboxypeptidase Y) (Hayashi, 1977) are available. The commercial preparation from orange leaves may contain varying amounts of trypsin-like, chymotrypsin-like and aminopeptidase activities which can all be inhibited by pre-incubation with 0.001 M-DFP at pH 5.3–8.0. Carboxypeptidase C from citrus fruit is inhibited partially and carboxypeptidase Y from yeast completely by DFP treatment.

The strategy and procedure employed in using carboxypeptidases C or Y in *C*-terminal sequence determinations is the same as that outlined for carboxypeptidases A and B. The reaction conditions are 0.05 M sodium citrate or pyridine acetate buffer pH 5.5 with enzyme to substrate ratios of approximately 0.1–1 % (w/w). Carboxypeptidase Y sometimes fails to release aspartic acid and basic amino acids (Hayashi, 1977).

18.5. CONCLUSION

Successful, extended carboxyl terminal sequence analysis is still an unattained goal. The improvements to the thiocyanate degradation and the new approaches such as cyanamide degradation, selective reduction, aldehyde formation and oxazole alcoholysis offer the hope that a universally applicable procedure, capable of extended degradation may be forthcoming in the future. However, many problems still need to be overcome. As Richard Perham pointed out in his concluding remarks to the Second International Conference on Solid Phase Methods in Protein Sequence Analysis, those involved in the development of new approaches to the question of sequencing from the *C*-terminus should be admired for their ingenuity and courage. Many before them have tried and failed to make this a workable proposition (Perham, 1977).

18.6. REFERENCES

Akabori, S., Ohno, K. & Narita, K. (1952). *Bull. Chem. Soc., Japan*, **25,** 214–218.

Akabori, S., Ohno, K., Ikenaka, T., Okada, Y., Hanafusa, H., Haruna, I., Tsugita, A., Sugae, K. & Matsushima, T. (1956). *Bull. Chem. Soc., Japan*, **29,** 507–518.

Ambler, R. P. (1972). *Methods Enzymol.*, **25,** 143–154.

Baba, T., Sugiyama, H. & Seto, S. (1972). *J. Biochem.* (*Tokyo*), **72,** 1571–1573.
Barrell, B. G., Air, G. M. & Hutchison, C. A. (1976). *Nature* (*Lond.*), **264,** 34–41.
Berget, S. M., Moore, C. & Sharp, P. A. (1977) *Proc. Natl. Acad. Sci. U.S.A.*, **74,** 3171–3175.
Blackburn, S. (1970). *Protein Sequence Determination*, Dekker, New York, pp. 191–199.
Blombäck, B. (1967). *Methods Enzymol.*, **11,** 398–411.
Boissonnas, R. A. & Haselbach, C. H. (1953). *Helv. Chim. Acta*, **36,** 576–581.
Bornstein, P. (1970). *Biochemistry*, **9,** 2408–2421.
Bradbury, J. H. (1956). *Nature* (*Lond.*), **178,** 912–913.
Braun, V. & Schroeder, W. A. (1967). *Arch. Biochem. Biophys.*, **118,** 241–252.
Braunitzer, G. (1955). *Chem. Ber.*, **88,** 2025–2036.
Cappugi, G., Nassi, P., Treves, C. & Ramponi, G. (1971). *Experientia*, **27,** 237–239.
Carlton, B. C. & Yanofsky, C. (1963). *J. Biol. Chem.*, **238,** 636–639.
Catterall, J. F., O'Malley, B. W., Robertson, M. A., Staden, R., Tanaka, Y. & Brownlee, G. G. (1978). *Nature* (*Lond.*), **275,** 510–513.
Chibnall, A. C. & Rees, M. W. (1958). *Biochem. J.*, **68,** 105–111.
Cromwell, L. D. & Stark, G. R. (1969). *Biochemistry*, **8,** 4735–4740.
Darbre, A. (1975). *Methods Enzymol.*, **47,** 357–369.
Degani, Y. & Patchornik, A. (1974). *Biochemistry*, **13,** 1–11.
Dopheide, T. A. & Ward, C. W. (1981). *J. Gen. Virol.*, **52,** 367–370.
Duggleby, R. G. & Kaplan, H. (1975). *Anal. Biochem.*, **65,** 346–354.
Dwulet, F. E. & Gurd, F. R. N. (1977). *Anal. Biochem.*, **82,** 385–395.
Fiers, W., Contreras, R., Haegeman, G., Rogiers, R., Van de Voorde, A., Van Heuverswyn, H., Van Herreweghe, J., Volckaert, G. & Ysebaert, M. (1978). *Nature (Lond.)*, **273,** 113–120.
Fraenkel-Conrat, H., Harris, J. I. & Levy, A. L. (1955). *Methods Biochem. Anal.*, **2,** 359–425.
Fromageot, C., Jutisz, M., Meyer, D. & Penasse, L. (1950). *Biochem. Biophys. Acta*, **6,** 283–289.
Furka, A., Sebestyen, F. & Karacsonyi, T. (1970). *FEBS Lett.*, **6,** 34–36.
Grassman, W., Dyckerhoff, H. & Eibeler, H. (1930). *Hoppe Seyler's Z. Physiol. Chem.*, **189,** 112–120.
Greenstein, J. P. & Winitz, M. (1961). *Chemistry of the Amino Acids*, vol. 2, Wiley. New York, pp. 1512–1687.
Guidotti, G. (1960). *Biochem. Biophys. Acta*, **42,** 177–179.
Halsey, Y. D. & Neurath, H. (1955). *J. Biol. Chem.*, **217,** 247–252.
Hamada, T. & Yonemitsu, O. (1973). *Biochem. Biophys. Res. Commun.*, **50,** 1081–1086.
Hargrave, P. A. & Wold, F. (1973). *Int. J. Peptide Protein Res.*, **5,** 85–89.
Hayashi, R. (1977). *Methods Enzymol.*, **47,** 84–93.
Hoare, D. G. & Koshland, D. E. (1967). *J. Biol. Chem.*, **242,** 2447–2453.
Holcomb, G. N., James, S. A. & Ward, D. N. (1968). *Biochemistry*, **7,** 1291–1296.
Horn, M. J. & Laursen, R. A. (1973). *FEBS Lett.*, **36,** 285–288.
Hsieh, W. T., Gundersen, L. E. & Vestling, C. S. (1971). *Biochem. Biophys. Res. Commun.*, **43,** 69–75.
Ichishima, E. (1972). *Biochem. Biophys. Acta*, **258,** 274–288.
Jacobson, G. R., Schaffer, M. H., Stark, G. R. & Vanaman, T. C. (1973). *J. Biol. Chem.*, **248,** 6583–6591.
Johnson, T. B. & Nicolet, B. H. (1911). *J. Am. Chem. Soc.*, **33,** 1973–1978.
Jones, B. N., Pååbo, S. & Stein, S. (1981). *J. Liquid Chromatogr.*, **4,** 565–586.
Jörnvall, H. (1973). *Abstr. 9th Internat. Congr. Biochem. Colloquium D*, Abstract Db 13, p. 454.

Kassell, B., Krishnamurti, C. & Friedman, H. L. (1977). In *Solid Phase Methods in Protein Sequence Analysis*, Proc. Second Internat. Conf., Previero, A. & Coletti-Previero, M.-A. (Eds.), North Holland, Amsterdam, New York, Oxford, pp. 39–48.
Kauffmann, T. & Boettcher, F.-P. (1959). *Chem. Ber.*, **92,** 2707–2716.
Kawanishi, Y., Iwai, U. & Ando, T. (1964). *J. Biochem. (Tokyo)*, **56,** 314–324.
Khorana, H. G. (1952). *J. Chem. Soc.*, **1952,** 2081–2088.
Kubo, H., Nakajima, T. & Tamura, Z. (1971). *Chem. Pharm. Bull.*, **19,** 210–211.
Lai, C. Y. (1975). *Arch. Biochem. Biophys.*, **166,** 358–368.
Lai, C. Y., Chen, C. & Horecker, B. L. (1970). *Biochem. Biophys. Res. Commun.*, **40,** 461–468.
Laursen, R. A. (1977). *Methods. Enzymol.*, **47,** 277–288.
Leder, P., Tilghman, S. M., Tiemeier, D. C., Polsky, F. I., Seidman, J. G., Edgell, M. H., Enquist, L. W., Leder, A. & Norman, B. (1977). *Cold Spring Harbor Symp. Quant. Biol.*, **42,** 915–920.
Levy, A. L. (1954). *Nature* (*Lond.*), **174,** 126–127.
Loudon, G. M. & Parham, M. E. (1978). *Tetrahedron Lett.*, no. 5, 437–440.
Machleidt, W. & Wachter, E. (1977). *Methods Enzymol.*, **47,** 263–277.
McReynolds, L., O'Malley, B. W., Nisbet, A. D., Fothergill, J. E., Givol, D., Fields, S., Robertson, M. & Brownlee, G. G. (1978). *Nature* (*Lond.*), **273,** 723–728.
Matsuo, H. & Narita, K. (1975). In *Protein Sequence Determination*, Needleman, S. B. (Ed.), 2nd edn, Springer, Berlin, Heidelberg, New York, pp. 104–113.
Matsuo, H., Kawazoe, Y., Ohnishi, M., Sato, M. & Tatsuno, T. (1964). *Abstracts of Third Symposium on Peptide Chem.*, Osaka University, Osaka, Japan, Nov. 1964, p. 30.
Matsuo, H., Fujimoto, Y. & Tatsuno, T. (1965). *Tetrahedron Lett.*, no. 39, 3465–3472.
Matsuo, H., Fujimoto, Y. & Tatsuno, T. (1966). *Biochem. Biophys. Res. Commun.*, **22,** 69–74.
Miller, M. J. & Loudon, G. M. (1975). *J. Am. Chem. Soc.*, **97,** 5295–5297.
Narita, K. & Ohta, Y. (1959). *Bull. Chem. Soc. Japan*, **32,** 1023–1028.
Naughton, M. A. & Hagopian, H. (1962). *Anal. Biochem.*, **3,** 276–284.
Neurath, H. (1960). In *The Enzymes*, Boyer, P. D., Lardy, H. & Myrbäck, K. (Eds.), 2nd edn, vol. 4, Academic Press, New York, pp. 11–36.
Niu, C. & Fraenkel-Conrat, H. (1955). *J. Am. Chem. Soc.*, **77,** 5882–5885.
Ohno, K. (1954). *J. Biochem.* (*Tokyo*), **41,** 345–350.
Parham, M. E. & Loudon, G. M. (1978). *Biochem. Biophys. Res. Commun.*, **80,** 7–13.
Perham, R. N. (1977). In *Solid Phase Methods in Protein Sequence Analysis*, Proc. Second Internat. Conf., Previero, A. & Coletti-Previero, M.-A. (Eds.), North Holland, Amsterdam, New York, Oxford, pp. 293–296.
Previero, A. & Coletti-Previero, M.-A. (1977) *ibid*, pp. 49–56.
Previero, A., Derancourt, J., Coletti-Previero, M.-A. & Laursen, R. A. (1973). *FEBS Lett.*, **33,** 135–138.
Rangarajan, M. & Darbre, A. (1975). *Biochem. J.*, **147,** 435–438.
Rangarajan, M. & Darbre, A. (1976). *Biochem. J.*, **157,** 307–316.
Rangarajan, M., Ardrey, R. E. & Darbre, A. (1973). *J. Chromatogr.*, **87,** 499–512.
Sakano, H., Rogers, J. H., Huppi, K., Brack, C., Traunecker, A., Maki, R., Wall, R. & Tonegawa, S. (1979). *Nature* (*Lond.*), **277,** 627–633.
Saund, A. K., Prasad, B., Koul, A. K., Bachhawat, J. M. & Mathur, N. K. (1973). *Int. J. Peptide Protein Res.*, **5,** 7–10.
Schlack, P. & Kumpf, W. (1926). *Hoppe-Seyler's Z. Physiol. Chem.*, **154,** 125–170.
Shaw, R. (1964). *Biochem. Biophys. Acta*, **92,** 558–567.
Simpson, R. J., Neuberger, M. R. & Liu, T.-Y. (1976). *J. Biol. Chem.*, **251,** 1936–1940.
Slobin, L. I. & Carpenter, F. H. (1966). *Biochemistry*, **5,** 499–508.

Smyth, D. & Utsumi, S. (1967). *Nature* (*Lond.*), **216,** 332–335.
Smyth, D., Stein, W. H. & Moore, S. (1962). *J. Biol. Chem.*, **237,** 1845–1850.
Sprössler, B., Heilmann, H. D., Grampp, E. & Uhlig, H. (1971). *Hoppe-Seyler's Z. Physiol. Chem.*, **352,** 1524–1530.
Stark, G. R. (1968). *Biochemistry*, **7,** 1796–1807.
Stark, G. R. (1972). *Methods Enzymol.*, **25,** 369–384.
Stark, G. R., Stein, W. H. & Moore, S. (1960). *J. Biol. Chem.*, **235,** 3177–3181.
Suzuki, T., Matsui, S. &Tuzimura, K. (1972). *Agric. Biol. Chem.*, **36,** 1061–1063.
Suzuki, T., Song, K.-D., Itagaki, Y. & Tuzimura, K. (1976). *Org. Mass Spectrom.*, **11,** 557–568.
Tarr, G. E. (1975). In *Solid Phase Methods in Protein Sequence Analysis*, Proc. First Internat. Conf., Laursen, R. A. (Ed.), Pierce, Rockford, Illinois, pp. 139–147.
Tschesche, H. (1977), *Methods Enzymol.*, **47,** 73–84.
Tsugita, A. & Broek, R. Van den (1980). In *Methods in Peptide and Protein Sequence Analysis*, Proc. Third Internat. Conf., Birr, C. (Ed.), Elsevier/North Holland, Amsterdam, New York, Oxford, pp. 359–369.
Visuri, K., Mikola, J. & Enari, T. M. (1969). *Eur. J. Biochem.*, **7,** 193–199.
Ward, C. W. & Dopheide, T. A. (1979). *Virology*, **95,** 107–118.
Wells, J. R. E. (1965). *Biochem. J.*, **97,** 228–235.
Williams, M. J. & Kassell, B. (1975). *FEBS Lett.*, **54,** 353–357.
Winstead, J. A. & Wold, F. (1964). *J. Biol. Chem.*, **239,** 4212–4216.
Yamashita, S. (1971). *Biochem. Biophys. Acta*, **229,** 301–309.
Yamashita, S. & Ishikawa, N. (1971). *Proc. Hoshi Coll. Pharm.*, **13,** 136–138.
Yamashita, S. & Ishikawa, N. (1972). In *Chemistry and Biology of Peptides*: Proc. Third American Peptide Symp., Meinhofer, J. (Ed.), Ann Arbor Science Publishers, Ann Arbor, pp. 701–703.
Yokoyama, S. & Ichishima, E. (1972). *Agr. Biol. Chem.*, **36,** 1259–1261.
Zabin, I. & Fowler, A. V. (1972). *J. Biol. Chem.*, **247,** 5432–5435.
Zuber, H. (1968). *Hoppe-Seyler's Z. Physiol. Chem.*, **349,** 1337–1352.

Practical Protein Chemistry—A Handbook
Edited by A. Darbre
© 1986, John Wiley & Sons Ltd

ANNE DELL
Department of Biochemistry,
Imperial College of Science and Technology,
London SW7 2AZ, U.K.

19

The Application of Electron Impact Mass Spectrometry in the Structural Analysis of Peptides and Proteins

CONTENTS

19.1. INTRODUCTION

The purpose of this chapter is to summarize the methodology and expertise required for the sequencing of mixtures of peptides (as their acetyl permethyl derivatives) by low resolution electron impact mass spectrometry (MS). The strategies described have been used successfully in the sequencing of a number of proteins including ribitol dehydrogenase (Morris *et al.*, 1974b), chloramphenicol acetyl transferase (Shaw *et al.*, 1979; Dell & Morris, 1981), a *Pseudomonas* azurin (Dell & Morris, 1977) and dihydrofolate reductase from *L. casei* (fully sequenced by MS alone) (Batley & Morris, 1977; Morris, 1979), as well as numerous peptides from various sources whose structures, for a variety of reasons, e.g. blocked *N*-termini, novel amino acids, purification difficulties, could not be determined by chemical means (see, for example, Hughes *et al.*, 1975; Morris & Dell, 1975a; Morris *et al.*, 1976; Stone *et al.*, 1976; Hofmann *et al.*, 1979; Nystrom *et al.*, 1979; Fitton *et al.*, 1980; Vinson *et al.*, 1980; McDonagh *et al.*, 1981).

19.2. PROPERTIES OF PEPTIDES AMENABLE TO ANALYSIS

19.2.1. Size

In general, volatility considerations limit the length of a peptide which can be examined by electron impact MS to approximately 10 residues. However, valuable information can often be obtained from longer peptides because (1) the presence of a high proportion of small hydrophobic amino acids (e.g. Gly, Ala, Val) in the sequence renders peptides between 10 and 20 residues sufficiently volatile and (2) thermal decomposition of large peptides within the mass spectrometer source may yield fragments which ionize to give interpretable mass spectra.

19.2.2. Quantity

The minimum amount of peptide required for a successful analysis is governed by a number of factors, the most important of which are (1) the sensitivity of the instrument being used, (2) the size and polarity of the peptide and (3) the fragmentation properties of the peptide. If a mass spectrometer with the sensitivity capabilities of the KRATOS MS50 or VG Analytical ZAB is used, approximately 30–50 nmol of a 10 residue peptide is normally sufficient for a single experiment. However, the unambiguous interpretation of the complex spectra derived from peptide mixtures nearly always requires an additional deuterium-labelling experiment (see section 19.6.3), thereby doubling the quantity of sample required for a total analysis.

19.2.3. Purity

The MS sequencing strategy outlined in this chapter is based upon the direct examination of peptide mixtures. It is therefore obvious that the purity of a sample with respect to minor peptide contaminants need not be as great as required for dansyl–Edman sequencing. Indeed, it is sometimes possible to sequence the 'contaminants' as well as the peptide(s) of interest! However, non-peptide impurities, e.g. Parafilm, salt, grease, plasticizers, must be rigorously excluded from samples. Even trace amounts of silicon-containing contaminants such as grease completely suppress the electron impact ionization of peptides.

19.3. TYPE OF INSTRUMENTATION REQUIRED

To successfully sequence mixtures of peptides of the sizes and quantities given above, two instrumental requirements must be satisfied: (1) good sensitivity and (2) a mass range of at least 1000 a.m.u. at maximum sensitivity. The author has used KRATOS MS902, KRATOS MS50 and VG Analytical ZAB mass spectrometers for peptide work but other suitable instrumentation is available. Although the high capital cost of instruments capable of both high mass and high sensitivity may seem prohibitive for the average protein sequencing group it should be borne in mind that the versatility of the mass spectrometer and its potential use for other important biochemical problems renders it considerably more cost effective than instruments whose sole function is protein sequencing.

19.4. SAMPLE HANDLING PROCEDURES

Because of the wide variety of functional groups present in proteins, the methodology for the conversion of polar, water-soluble peptides of unknown composition into chloroform-soluble volatile derivatives suitable for analysis by EI-MS took more than a decade to perfect. The knowledge gained from detailed studies of a wide variety of reagents (Das *et al.*, 1967; Lederer, 1968; Vilkas & Lederer, 1968; Aplin *et al.*, 1969; Morris *et al.*, 1969) has enabled the development of chemical procedures which are applicable to all peptides, regardless of their sequence (Morris, 1972; Morris *et al.*, 1973) and which can be routinely used for peptide mixtures at the 5–100 nmol level. The chemical modifications performed are outlined in Fig. 19.1 and result in an *N*-acetyl *N*, *O*, *S* permethyl derivative of the peptide. The methods are applicable to all amino acids except arginine which is normally converted to ornithine prior to acetylation (Fig. 19.2) (Morris *et al.*, 1973). The presence of arginine in a peptide mixture is ascertained either by amino acid analysis on a fraction of the sample or by analytical electrophoresis followed by an arginine-visualizing stain (Yamada & Itano, 1966; see p. 261). The latter method is readily incorporated into the general protein sequencing strategy described later since the chromatography step is monitored by analytical electrophoresis.

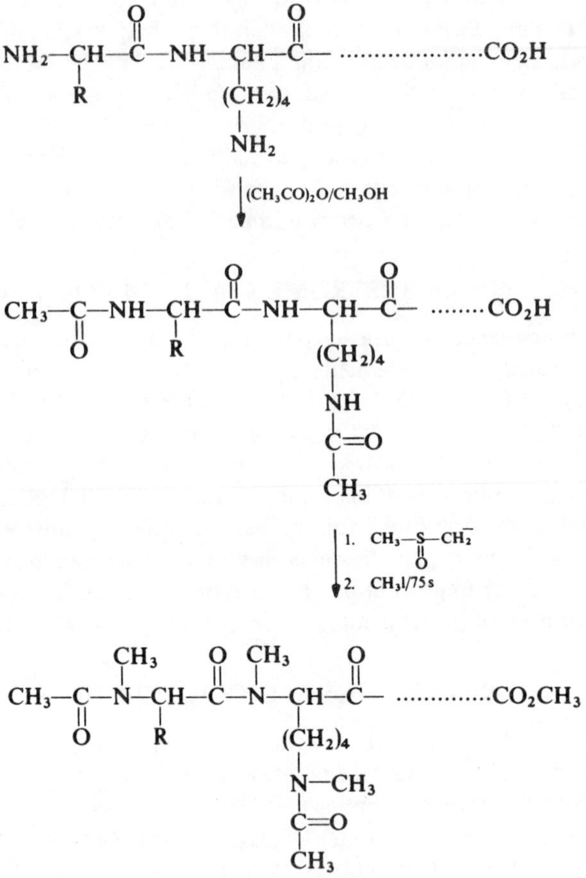

$$NH_2{-}CH(R){-}C({=}O){-}NH{-}CH[(CH_2)_4{-}NH_2]{-}C({=}O){-}\cdots\cdots CO_2H$$

$$\Big\downarrow\ (CH_3CO)_2O/CH_3OH$$

$$CH_3{-}C({=}O){-}NH{-}CH(R){-}C({=}O){-}NH{-}CH[(CH_2)_4{-}NH{-}C({=}O){-}CH_3]{-}C({=}O){-}\cdots CO_2H$$

$$\Big\downarrow\ 1.\ CH_3{-}S({=}O){-}CH_2^-\quad 2.\ CH_3I/75\,s$$

$$CH_3{-}C({=}O){-}N(CH_3){-}CH(R){-}C({=}O){-}N(CH_3){-}CH[(CH_2)_4{-}N(CH_3){-}C({=}O){-}CH_3]{-}C({=}O){-}\cdots CO_2CH_3$$

Figure 19.1. Acetylation and permethylation of peptide for mass spectrometry (Morris *et al.*, 1973).

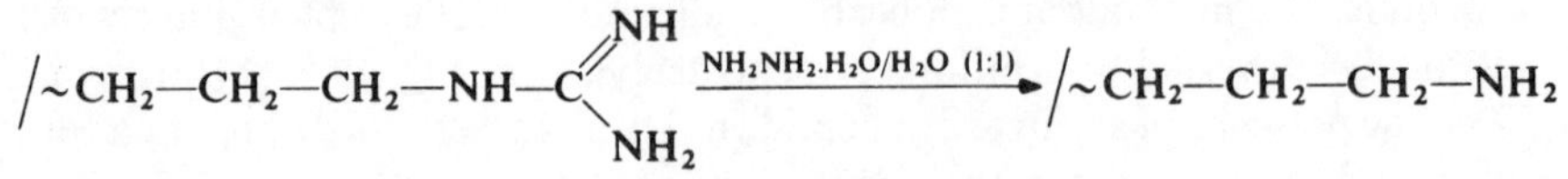

$$/{\sim}CH_2{-}CH_2{-}CH_2{-}NH{-}C({=}NH){-}NH_2 \xrightarrow{NH_2NH_2.H_2O/H_2O\ (1:1)} /{\sim}CH_2{-}CH_2{-}CH_2{-}NH_2$$

Figure 19.2. Conversion of peptidyl arginyl residue to peptidyl ornithine residue (Morris *et al.*, 1973).

19.4.1. Purification of reagents

Analar grade methanol, chloroform and acetic anhydride can be used without further purification. The quality of commercial methyl iodide is very dependent on the supplier and should be tested for impurities by derivatizing a peptide whose mass spectrum is known and checking for spurious signals. If contamination is suspected the methyl iodide should be discarded because distillation which removes added stabilizers is not recommended. Dimethylsulphoxide should be purified by distillation from calcium hydride at reduced pressure and stored over calcium hydride. Sodium hydride, which is purchased as an oil suspension, should be cleaned by repeated washing with sodium-dried ether (CARE: keep away from water) and finally dried *in vacuo*. The grey powder can be stored in an airtight container for up to a year provided care is taken to minimize exposure to moist air each time the container is opened.

19.4.2. Apparatus

All chemical manipulations should be performed without sample transfer in order to minimize losses and are conveniently carried out using small test tubes (100 × 10 mm) with ground glass stoppers. Disposable Pasteur pipettes are ideal for the addition of reagents.

19.4.3. Hydrazinolysis

Peptide samples (5–100 nmol) are hydrazinolysed by reaction with 50 μl of a mixture of hydrazine-hydrate in water (1 : 1, v/v) at 80°C for 12 min. The reaction mixture is cooled, water (50 μl) added, frozen by immersion in dry ice–ethanol or liquid nitrogen and evaporated to dryness by attachment to a high vacuum pump (a freeze-drier is suitable) whereupon the sample slowly melts and warms to room temperature. This is accompanied by the evolution of dissolved gases. Residual hydrazine is removed by extracting in water (2 × 50 μl) followed by evaporation *in vacuo*.

19.4.4. Acetylation

The α-amino group of a peptide is fully blocked within 1 min when treated with acetic anhydride in methanol while the less reactive ε-amino group requires approximately 3 h at room temperature. In the presence of a suitable base both groups are acetylated within 1 min. Hence, either of the procedures given below is suitable for peptides of unknown composition.

19.4.4.1. Long acetylation

The sample is dissolved in water (100 μl) to which is then added 500 μl of a mixture of acetic anhydride in methanol (1:3 v/v) and allowed to stand for 3 h at room temperature. Reagents are removed by evaporation *in vacuo*. Bumping is not normally a problem using this procedure.

19.4.4.2. Short acetylation

The sample is dissolved in 100 μl of sodium bicarbonate solution (1 mg/ml) and left to stand for 30 s prior to the addition of 500 μl of acetic anhydride in methanol (1:3, v/v). After approximately 1 min the sample is dried down, firstly using a water pump (CARE: bumping) followed by a rotary pump or freeze-drier.

19.4.5. Permethylation

The acetylated peptides are permethylated by the following procedure. Methyl sulphinyl carbanion base is prepared by heating sodium hydride (2 heaped micro-spatulas) in dimethyl sulphoxide (3–5 ml) at 90°C for approximately 20 min. The progress of the reaction is judged by its colour and the rate of evolution of hydrogen. Reaction is complete when the suspension becomes orange-brown (if it turns green or red the base should be discarded) and the hydrogen evolution slows down. If hydrogen release ceases prior to formation of the orange-brown colour a further micro-spatula of sodium hydride is added and heating continued until the desired colour is achieved. The mixture is then cooled and centrifuged at 3 000 rev/min for 10 min. The supernatant which contains the methyl sulphinyl carbanion should be a clear honey colour. The acetylated peptide is dissolved in one drop of dimethyl sulphoxide and 20–30 drops of the base are added. To test that excess base has been added a small sample of the solution is added to a crystal of triphenylmethane. The crystal turns deep red in the presence of excess base. After approximately 60 s methyl iodide (500 μl) is added and the reaction allowed to proceed for 70 s (Morris, 1972; Morris *et al.* 1973). Water (2 ml) is added to quench the reaction and the products are immediately extracted into chloroform (1 ml) which is washed with water (2 × 2 ml) and evaporated under a stream of nitrogen. Residual water and dimethylsulphoxide are removed by evaporation *in vacuo*.

19.4.6. Running mass spectra

The derivatized sample is dissolved in a drop of chloroform and carefully transferred to the quartz sample tip of the direct insertion probe of the mass spectrometer. The chloroform is removed in a stream of air (this can be conveniently achieved by gently blowing through a Pasteur pipette which has been drawn out to a fine tip using a Bunsen flame) and the sample inserted *via* vacuum locks into the source housing of the mass spectrometer. Best results are achieved using a probe with a retractable sample tip coupled with indirect radiative heating within the ion source for evaporation of the peptides (direct heating of the sample using a heated probe is less satisfactory for mixture analysis). Volatile impurities, e.g. dimethylsulphoxide, hydrocarbons, are evaporated by inserting the tip at a source temperature of 100°C. The tip is withdrawn when the source pressure has dropped to its original value. Mixture analysis by MS is then achieved by applying a source temperature gradient up to 350°C. At intervals of approximately 30°C the sample is inserted into the source region and the spectrum is repetitively scanned from m/e 1 000 to m/e 90.

19.5. STRATEGIES OF PEPTIDE AND PROTEIN SEQUENCING BY MASS SPECTROMETRY

The mixture analysis technique, which allows the unambiguous sequencing of a mixture of peptides, is a fundamental feature of MS sequencing strategy (Morris *et al.*, 1971, 1974a). Without this facility the mass spectrometer would not be competitive with classical procedures for routine sequence analysis because it is considerably more expensive than manual dansyl-Edman methods and it does not offer a higher sensitivity. However, as protein chemists are well aware, the rate-determining step of classical sequencing is the isolation and purification of individual peptides after enzyme digestion. Even if automated procedures are used to examine large fragments it is always necessary to complete the sequence by fairly extensive sub-digestion followed by manual sequencing of the many small peptides produced. Using MS, a single purification step is normally sufficient to effect the required partial separation into mixtures of peptides suitable for analysis. The principle of mixture analysis is the following: individual peptides are fractionally distilled from the mixture on the probe tip by taking advantage of their varying volatilities; the signals associated with any one peptide in the mass spectrum will rise and fall together and by comparing spectra at different temperatures each component can be sequenced unambiguously. The method is applicable to mixtures of up to five peptides.

To effectively exploit the speed at which large numbers of relatively small peptides can be sequenced by MS, a suitable overall strategy for sequencing a protein should be planned. Clearly, if the full sequence of an unknown protein is desired a combined classical/MS approach would be most productive. A recommended strategy is the following:

1. *N*-terminal sequences of the protein and large peptides are obtained by automated procedures;
2. peptides produced by specific enzymic cleavages (e.g. trypsin, chymotrypsin) are sequenced using manual dansyl-Edman methods;
3. mixtures of small peptides containing potential key overlaps are examined by

MS techniques after non-specific (e.g. elastase, subtilisin) enzymic digestion (Morris *et al.*, 1974a; Dell & Morris, 1977; Shaw *et al.*, 1979).

For certain sequencing problems, MS alone is the method of choice. For example, the DNA sequencing of a particular gene may be facilitated by a knowledge of a partial sequence of the protein for which it codes. This information can be obtained within a few weeks using the procedures given below. Similarly, the MS approach is ideal for rapidly assessing the degree of homology of related proteins (Dell & Morris, 1977). Finally, MS is unsurpassed for amide assignments and the sequencing of tryptophan in peptides.

A general procedure for sequencing is as follows:

1. Digestion of the protein with a non-specific protease, e.g. elastase, subtilisin. The progress of digestion can be monitored by analytical electrophoresis at pH 6.5.
2. An optional gel filtration step to separate undigested protein or large fragments from the main bulk of small peptides.
3. Separation of the peptides using a cation-exchange column, e.g. Dowex (Morris *et al.*, 1974a; Dell & Morris, 1977) or HPLC methods. Elution is effected using volatile buffers, e.g. pyridine/acetic acid for cation exchange (Morris *et al.*, 1974a; Dell & Morris, 1977) and acetic acid/propanol for HPLC (Morris *et al.*, 1980) and the eluent is monitored by analytical electrophoresis of a suitable quantity (usually 1/100) of each fraction.
4. Pooling and drying down appropriate fractions to yield samples containing up to five components.
5. Derivatization followed by mixture analysis MS.

Using this strategy at least 70% of the sequence of a 20000 dalton protein can be obtained within two months provided the research worker has been trained in the interpretation of peptide mass spectra.

19.6. INTERPRETATION OF PEPTIDE MASS SPECTRA

19.6.1. Introduction

The unambiguous assignment of peptide sequences from complex mass spectra requires considerable experience in interpretation which can only be acquired by the detailed examination of spectra from a variety of samples. This section is intended to be a 'beginner's guide' to interpretation and therefore only a brief introduction to the major fragmentation pathways of peptides is presented. For a better understanding of sequencing the reader is advised to examine the interpretations of spectra reported in references (Dell & Morris, 1974, 1981; Hughes *et al.*, 1975; Morris & Dell, 1975a, b; Morris *et al*, 1971, 1974b, 1976; Stone *et al.*, 1976; Morris, 1979).

19.6.2. Fragmentation pathways

1. The major cleavage observed in electron impact spectra of acetylated permethylated peptides is at the amide bond. This is referred to as *C–N* cleavage and is outlined in Fig. 19.3. Ions generated by cleavage at the peptide bond are called *sequence ions.* Since the mass differences between the sequence ions corresponds to the masses of the amino acid residues the sequence of the peptide is obtained by firstly identifying the *N*-terminal ion which is to be found in the mass range m/e 114 to m/e 257 and then locating sequence ions at higher masses. For example, the hyopthetical peptide A-B-C-D (where A, B, C and D represent derivatized amino acid residues) will fragment to yield A^+, AB^+, ABC^+ and $ABCD^+$. The masses of the *N*-terminal ion and in-chain residue of each of the amino acids are given in Table 19.1. It should be noted that Leu and Ile are not distinguished by the procedures described in this chapter.

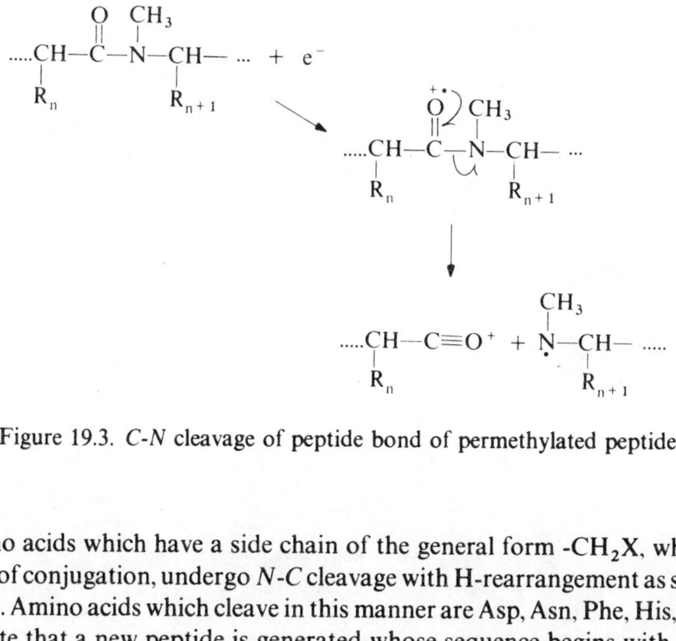

$$\ldots CH(R_n)-C(=O)-N(CH_3)-CH(R_{n+1})-\ldots + e^- \rightarrow \ldots CH(R_n)-C(=O^{+\bullet})-N(CH_3)-CH(R_{n+1})-\ldots$$

$$\downarrow$$

$$\ldots CH(R_n)-C{\equiv}O^+ + \dot{N}(CH_3)-CH(R_{n+1})-\ldots$$

Figure 19.3. *C-N* cleavage of peptide bond of permethylated peptide.

2. Amino acids which have a side chain of the general form -CH_2X, where X is capable of conjugation, undergo *N-C* cleavage with H-rearrangement as shown in Fig. 19.4. Amino acids which cleave in this manner are Asp, Asn, Phe, His, Tyr and Trp. Note that a new peptide is generated whose sequence begins with the *N-C* cleaved residue and corresponding sequence ions should be observed in the mass spectrum. The masses of the *N*-terminal ions resulting from *N-C* cleavage are given in Table 19.2.

3. Glu and Gln, regardless of their position in the sequence, usually partially cyclize to form *N*-terminal pyrrolidone carboxylic acid (PCA). This results in a characteristic ion at m/e 98 (19.I); the corresponding *N*-terminal sequence ion (19.II) at m/e 126 is usually weak or absent.

Table 19.1. Masses of the *N*-terminal ion and the in-chain residue of each amino acid after acetylation and permethylation obtained by electron impact mass spectrometry.

Amino acid	*N*-terminal mass	Residue mass
Gly	114	71
Ala	128	85
Pro	140	97
Val	156	113
Ser	158	115
Leu	170	127
Thr	172	129
Cys	174	131
Asp	186	143
Met	188	145
Asn	199	156
Glu	200	157
Phe	204	161
His	208	165
Gln	213	170
Orn	227	184
Tyr	234	191
Lys	241	198
Trp	257	214

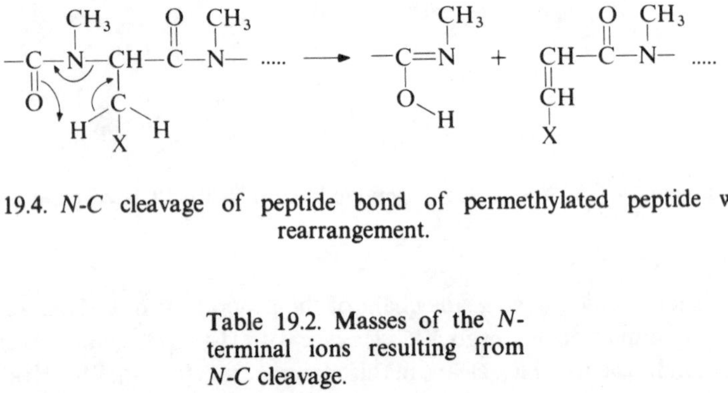

Figure 19.4. *N-C* cleavage of peptide bond of permethylated peptide with H-rearrangement.

Table 19.2. Masses of the *N*-terminal ions resulting from *N-C* cleavage.

N-C cleavage ion	Mass
. . . Asp	113
. . . Asn	126
. . . Phe	131
. . . His	135
. . . Tyr	161
. . . Trp	184

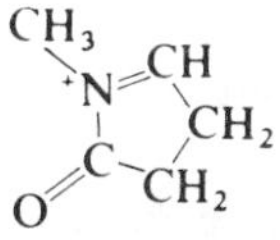

CH_3
$^+N{=}CH$
CH_2
$O{=}C$ CH_2

19. I

CH_3
$N{-}CH{-}C{\equiv}O^+$
CH_2
$O{=}C$ CH_2

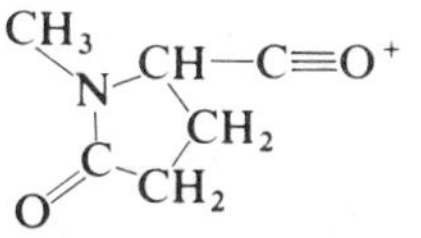

19. II

4. Side chain fragmentation of a number of amino acids is often a useful aid to assignment. The most important examples are serine and threonine which undergo elimination of methanol with the latter also exhibiting complete side chain cleavage with or without H-rearrangement. Methionine behaves analogously, losing CH_3SH (48 m.u.) and/or its complete side chain.

19.6.3. ^{2}H-Labelling

Confirmation of sequence assignments is normally sought by substituting $^2H_6$-acetic anhydride or $^2H_3$-methyl iodide for their unlabelled counterparts during acetylation and permethylation. A number of ambiguities which may be encountered during sequencing are thereby resolved. The following example illustrates this point. An ion at m/e 126 may be due to any or all of the following: (a) *N*-terminal PCA, (b) *N-C* cleavage of Asn or (c) *N*-terminal serine minus methanol. ^{2}H-labelling distinguishes these alternatives as follows: (1) $^2H_6$-acetic anhydride and unlabelled methyl iodide results in (a) and (b) remaining at m/e 126 while (c) shifts to m/e 129; (2) unlabelled acetic anhydride and $^2H_3$-methyl iodide results in (a) and (c) shifting to m/e 129 while (b) shifts to m/e 132.

19.7. SPECIAL APPLICATIONS

19.7.1. Blocked *N*-termini

Peptides which lack a free α-amino group are not amenable to classical sequencing procedures but pose no problems for mass spectrometry (Morris & Dell, 1975a; Stone *et al.*, 1976; Hofmann *et al.*, 1979; Nystrom *et al.*, 1979; Fitton *et al.*, 1980; Vinson *et al.*, 1980). If a blocked *N*-terminus is suspected, derivatization is performed under standard conditions except for the substitution of $^2H_6$-acetic anhydride in the acetylation step. Incorporation of $^2H_3$-acetyl at the *N*-terminus will take place only if a free α-NH_2 group is present. If absent, the nature of the blocking group can be ascertained from the mass of the *N*-terminal sequence ion.

19.7.2. Determination of the *N*-terminal sequence

In the majority of cases the sequenator is the instrument of choice for determining the *N*-terminus of a protein. However, for certain applications, notably if the protein is blocked or if the N-terminal sequences of several subunits are desired, the necessary information can be obtained more conveniently by MS.

The protein (50–100 nmol) is dissolved in 98% formic acid (200 μl) to which acetic anhydride (100 μl) is then added. After 30 min at room temperature the reaction is terminated by evaporation at reduced pressure. The formylated protein is dissolved in 20 mM-NH_4HCO_3, pH 8.5 and digested with chymotrypsin or elastase. The digest is freeze-dried and then permethylated. Mass spectra are then obtained as described earlier. Only peptides containing the N-terminus of the protein are observed in the mass spectra since peptides having free α-amino groups form quaternary ammonium salts during permethylation and are removed by the water washes. Clearly, the presence of an enzymic cleavage site at a suitable distance from the N-terminus is crucial to the success of the method.

19.7.3. Proteins containing γ-carboxyglutamic acid

The unambiguous assignment of γ-carboxyglutamic acid (Gla) in peptides is best achieved using MS (Morris *et al.*, 1976; Thorgersen *et al.*, 1978). Peptides are acetylated and permethylated in the standard manner (see sections 19.4.4 and 19.4.5) and sequenced by mixture analysis techniques. Gla at the N-terminus of a sequence is established by the presence of ions at m/e 112 (19.III) and 170 (19.IV) while the mass differences 171 (19.V) and 229 (19.VI) are diagnostic of the Gla residue.

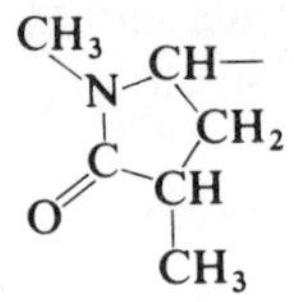

```
CH3
   \
    N—CH—
    |    \
    C     CH2
   //\   /
  O   CH
      |
      CH3
```

19. III

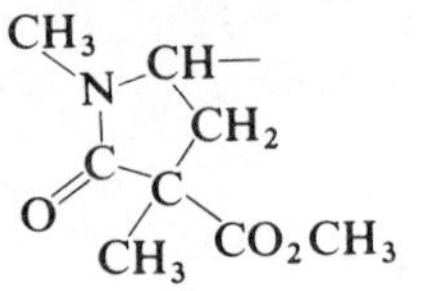

```
CH3
   \
    N—CH—
    |    \
    C     CH2
   //\   /
  O   C
     / \
  CH3   CO2CH3
```

19. IV

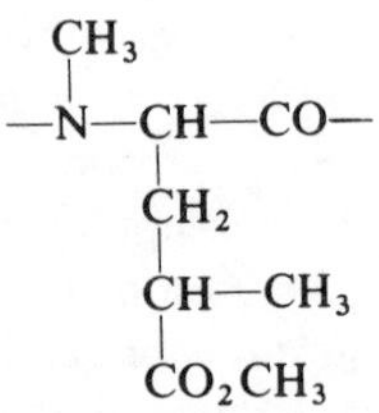

```
   CH3
   |
 —N—CH—CO—
    |
    CH2
    |
    CH—CH3
    |
    CO2CH3
```

19. V

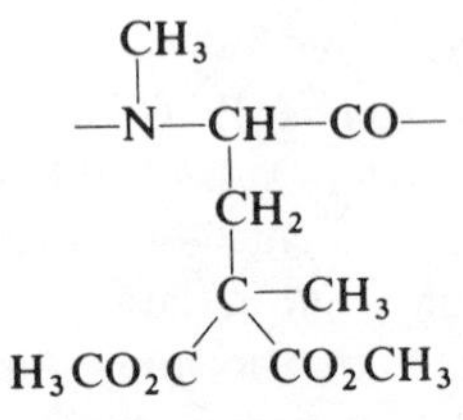

```
      CH3
      |
    —N—CH—CO—
       |
       CH2
       |
       C—CH3
      / \
H3CO2C   CO2CH3
```

19. VI

19.8. CONCLUSION

The sequencing of peptide mixtures by EI-MS is now a relatively routine procedure. Protein chemists who have access to suitable instrumentation and who are prepared to gain the necessary experience in spectral interpretation should find the mass spectrometer a useful complement to other sequencing procedures. It is especially valuable for examining mixtures of peptides which are difficult to purify, for amide assignments and for tryptophan-containing peptides. For some applications, e.g. blocked *N*-termini and novel structures, it is the most useful method of sequencing.

Recently a new ionization technique, fast atom bombardment mass sectrometry (FAB-MS) has been developed for the analysis of involatile, polar and thermally unstable molecules (Barber *et al.*, 1981). FAB-MS is now being successfully applied to peptide sequencing. Underivatized peptides can be analysed by FAB-MS and data is obtained from 15 pmol in favourable cases. See references for a good introduction to this exciting new technique (Morris *et al.*, 1981; Rinehart *et al.*, 1981; Morris *et al.*, 1982; Williams *et al.*, 1982).

19.9. REFERENCES

Aplin, R. T., Eland, I. & Jones, J. H. (1969). *Org. Mass Spectrom.*, **2,** 795–799.

Barber, M., Bordoli, R. S., Sedgwick, R. D. & Tyler, A. N. (1981). *Nature (Lond.)*, **293,** 270–273.

Batley, K. E. & Morris, H. R. (1977). *Biochem. Biophys. Res. Commun.*, **4,** 1010–1014.

Das, B. C., Gero, S. D. & Lederer, E. (1967). *Biochem. Biophys. Res. Commun.*, **29,** 211–213.

Dell, A. & Morris, H. R. (1974). *Biochem. Biophys. Res. Commun.*, **61,** 1125–1132.

Dell, A. & Morris, H. R. (1977). *Biochem. Biophys. Res. Commun.*, **78,** 874–880.

Dell, A. & Morris, H. R. (1981). *Biomed. Mass Spectrom.*, **8,** 128–136.

Fitton, J. E., Dell, A. & Shaw, W. V. (1980). *FEBS Lett.*, **115,** 209–215.

Hofmann, T., Kawakami, M., Hitchman, A. J. W., Harrison, J. E. & Dorrington, K. J. (1979). *Can. J. Biochem.*, **57,** 737–748.

Hughes, J., Smith, T. W., Kosterlitz, H. W., Fothergill, L., Morgan, B. A. & Morris, H. R. (1975). *Nature (Lond.)*, **258,** 577–599.

Lederer, E. (1968). *Pure Appl. Chem.*, **17,** 489–517.

McDonagh, R. P., McDonagh, J., Petersen, T. E., Thorgersen, H. C., Skorstengaard, K., Sottrup-Jensen, L., Magnusson, S., Dell, A. & Morris, H. R. (1981). *FEBS Lett.*, **127,** 174–178.

Morris, H. R. (1972). *FEBS Lett.*, **22,** 257–260.

Morris, H. R. (1979). *Phil. Trans. Roy. Soc. London A*, **293,** 39–51.

Morris, H. R. & Dell, A. (1975a). *Biochem. J.*, **149,** 754.

Morris, H. R. & Dell, A. (1975b). In *Instrumentation in Amino Acid Sequence Analysis*, Perham, R. N. (Ed.), Academic Press, London, New York, San Francisco, pp. 147–191.

Morris, H. R., Geddes, A. J. & Graham, G. N. (1969). *Biochem. J.*, **111,** 38 p.

Morris, H. R., Williams, D. H. & Ambler, R. P. (1971). *Biochem. J.*, **125,** 189–201.

Morris, H. R., Dickinson, R. J. & Williams, D. H. (1973). *Biochem. Biophys. Res. Commun.*, **51,** 247–255.

Morris, H. R., Batley, K. E., Harding, N. G. L., Bjur, R. A., Dann, J. G. & King, R. W. (1974a). *Biochem. J.*, **137,** 409–411.

Morris, H. R., Williams, D. H., Midwinter, G. G. & Hartley, B. S. (1974b). *Biochem. J.*, **141,** 701–713.

Morris, H. R., Dell, A., Petersen, T. E., Sottrup-Jensen, L. & Magnusson, S. (1976). *Biochem. J.*, **153,** 663–679.

Morris, H. R., Etienne, A. T., Dell, A. & Albuquerque, R. (1980). *J. Neurochem.*, **34,** 574–582.
Morris, H. R., Panico, M., Barber, M., Bordoli, R. S., Sedgwick, R. D. & Tyler, A. N. (1981). *Biochem. Biophys. Res. Commun.*, **101,** 623–631.
Morris, H. R., Dell, A., Etienne, A. T., Judkins, M., McDowell, R. A., Panico, M. & Taylor, G. W. (1982). *Pure Appl. Chem.*, **54,** 267–279.
Nystrom, L.-E., Lindberg, V., Kendrick-Jones, J. & Jakes, R. (1979). *FEBS Lett.*, **101,** 161–165.
Rinehart, K. L. Jr., Guadioso, L. A., Moore, M. L., Pandey, R. C., Cook, J. C. Jr., Barber, M., Sedgwick, R. D., Bordoli, R. S., Tyler, A. N. & Green, B. N. (1981). *J. Am. Chem. Soc.*, **103,** 6517–6520.
Shaw, W. V., Packman, L. C., Burleigh, B. D., Dell, A., Morris, H. R. & Hartley, B. S. (1979). *Nature* (*Lond.*), **282,** 870–872.
Stone, J. V., Mordue, W., Batley, K. E. & Morris, H. R. (1976). *Nature* (*Lond.*), **263,** 207–211.
Thorgersen, H. C., Petersen, T. E., Sottrup-Jensen, L., Magnusson, S. & Morris, H. R. (1978). *Biochem. J.*, **175,** 613–627.
Vilkas, E. & Lederer, E. (1968). *Tetrahedron Lett.*, no. 26, 3089–3092.
Vinson, G. P., Whitehouse, B. J., Dell, A., Etienne, A. T. & Morris, H. R. (1980). *Nature* (*Lond.*), **284,** 464–467.
Williams, D. H., Bradley, C. V., Santikarn, S. & Bojesen, G. (1982). *Biochem. J.*, **201,** 105–117.
Yamada, S. & Itano, H. A. (1966). *Biochim. Biophys. Acta*, **130,** 538–540.

Practical Protein Chemistry—A Handbook
Edited by A. Darbre
© 1986, John Wiley & Sons Ltd

DAVID DOVER
Department of Biophysics,
King's College London,
Strand,
London WC2R 2LS, U.K.

20

X-ray Crystallography and Electron Microscopy

CONTENTS

20.1. X-RAY DIFFRACTION

20.1.1. Introduction

The ultimate result desired by the molecular biochemist is the three dimensional structure of the molecule he is studying dynamically displayed, showing all the conformational changes which can occur during the functional operation of that molecule. X-ray crystallography as a method of visualizing three-dimensional structure suffers from the defect of being able, at best, to show only a few of the conformations of proteins during activity. However, it is as yet the only way of acquiring detailed information about complete three-dimenstional structures. Several other physical methods exist for indicating whether a molecule is approximately spherical or elongated and what its size and asymmetry might be, as do others methods for examining, in detail, small parts of a molecule. Electron microscopy while lacking the atomic detail of X-ray methods does provide images of molecules of considerably more overall detail than solution methods.

20.1.2. Imaging molecules—structure analysis

When a specimen is examined in a light microscope the light scattered by the object is collected by the objective lens to form a magnified image (Fig. 20.1). In forming this image the objective lens preserves the *phases* of all the scattered waves in their correct relationship to one another so that their recombination reproduces the *image* of the object (Fig. 20.2). The *resolution*, or detail, visible in the light microscope image is restricted by the wavelength of the light employed—typically about 600 nm. Finer detail would be visible if an X-ray microscope could be constructed using, for example, the common CuK_α X-ray line at 0.154 nm. Unfortunately X-ray lenses cannot be constructed, thus we have no direct imaging facility. The best that can be done is to record directly the intensity of X-rays scattered by the material, typically by allowing them to impinge directly on a photographic film. It is the loss of the phase of the scattered wave which prevents this information being utilized in its raw form to construct an image of the object.

The specimen that we have considered so far could consist of a collection of molecules of interest in different orientations. The images of these different molecules would be produced in their proper separate positions and orientations if we could image them by a lens system. Collecting, as we are, the raw scattered data we cannot divide the intensity between the different molecules or even between them and the solvent. This problem is overcome by using the regular arrangement of molecules such as are found in crystals (Fig. 20.3). Such a regular arrangement, concentrates the scattering from the molecules into intense *diffraction* peaks, not only separating the macromolecular scatter from the solvent and glass support scatter, but also improving dramatically the macromolecular

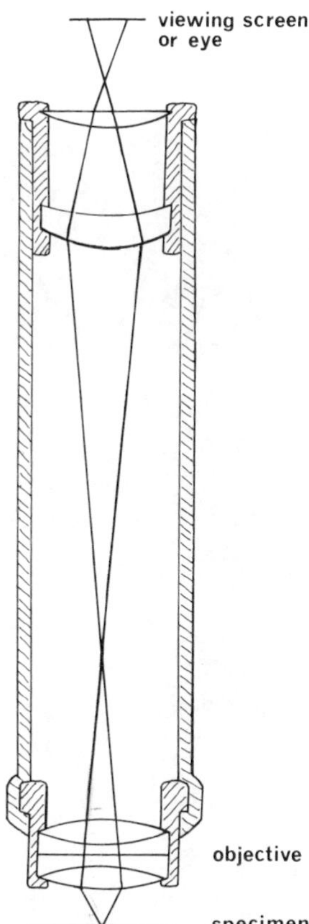

Figure 20.1. The light microscope uses glass lenses to collect the light scattered by a specimen and recombine it into an image. In doing so the lenses preserve the relative phases of the scattered waves.

signal/noise ratio. Consequently, an average molecule as packed in the crystal is seen, rather than an image of one freely floating in solution. Using a suitably designed X-ray camera the diffraction pattern from a crystal can be recorded, an example of which is shown on Fig. 20.4. The diffraction from a crystal must be thought of as three-dimentional, corresponding to the three-dimensional nature of the crystal itself. Thus a full pattern would consist of a series of photographs, such as the one shown, stacked one above the other to display a three-dimensional reciprocal lattice. The spacing of this lattice is (inversely) related to the lattice in which the molecules are packed. The intensity of the various spots contains the information about the molecular structure. If we knew the relative phase of the waves as they arrived at the photographic film we would be in a position to immediately reconstruct, by calculation or otherwise, an image of the structure. It is to the solution of this phase problem that most attention has been given.

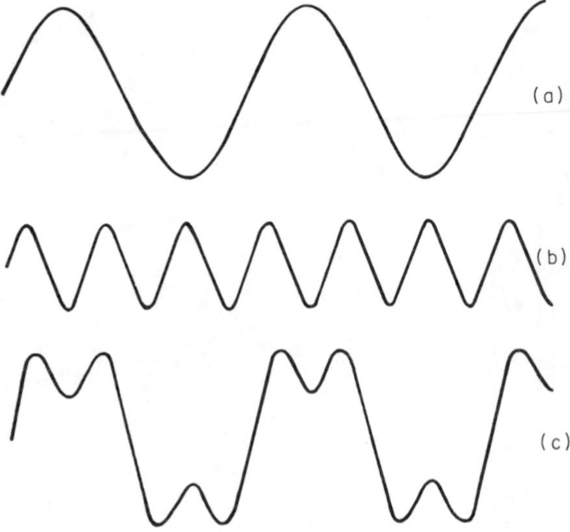

Figure 20.2. Two waves of different amplitude and phase (a, b) are combined to produce a resultant variation (c). Both the amplitude and phase are necessary to produce this particular resultant.

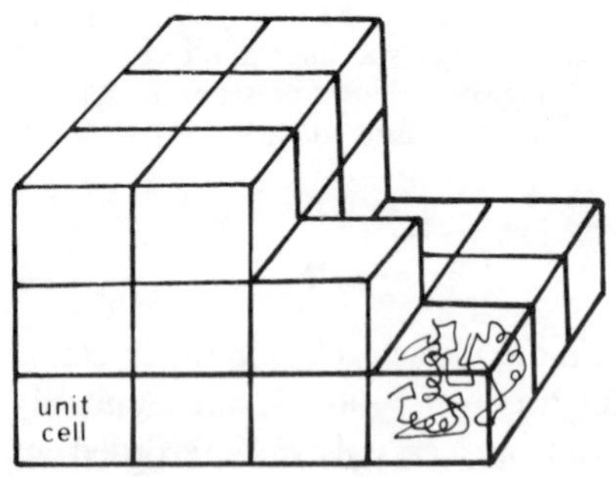

Figure 20.3. A crystal consists of unit cells regularly packed. The contents of each unit cell (one or more molecules) are identical. The effect of this regularity on a diffraction pattern is to concentrate the diffraction into intense peaks.

20.1.3. The meaning of structure

Before we start to consider the problems and procedure of X-ray crystallography let us look at the kind of information that we may hope to acquire by this approach. Structural analysis of well-formed crystals of small molecules yields values of the electron density in a unit cell of the crystal at regular, closely spaced intervals. This three-dimensional *Fourier Synthesis* can be presented as a series of two-dimensional contour maps of sections through that unit cell. The intervals at

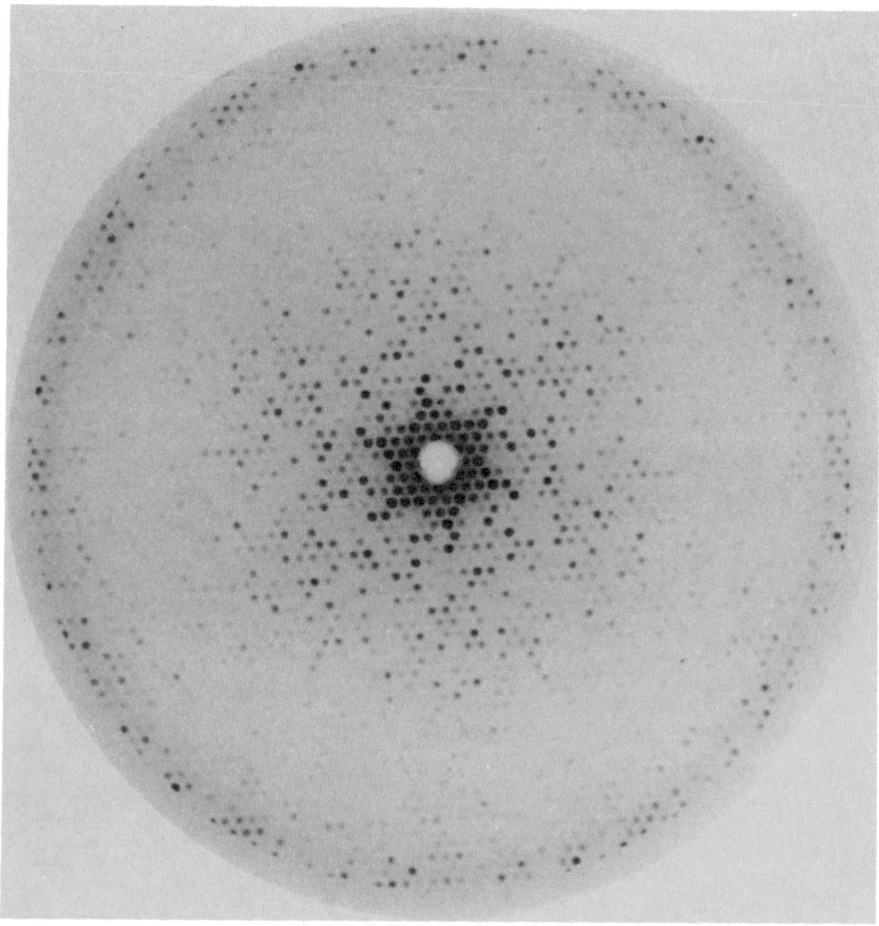

Figure 20.4. A diffraction pattern (precession camera) of the protein C-phycocyanin. The pattern records the intensity (but not the phase) of the diffraction spots.

which the densities are evaluated are sufficiently close that atoms can be seen as separated resolved peaks (Fig. 20.5). The high resolution of the maps, perhaps as high as 0.05 nm, means that a structure can be determined *ab initio* and it has become common practice in a number of organic chemistry laboratories to use this as a method of determining covalent chemical structure (see, for example, Dunitz, 1979).

With similar maps of protein structures the resolution is more limited (say 0.15 nm) and, as Fig. 20.6 shows, actual atoms cannot be resolved. Under these circumstances a structure can be solved only by introducing additonal information, viz. the covalent bonding scheme, as given by the knowledge that the chains

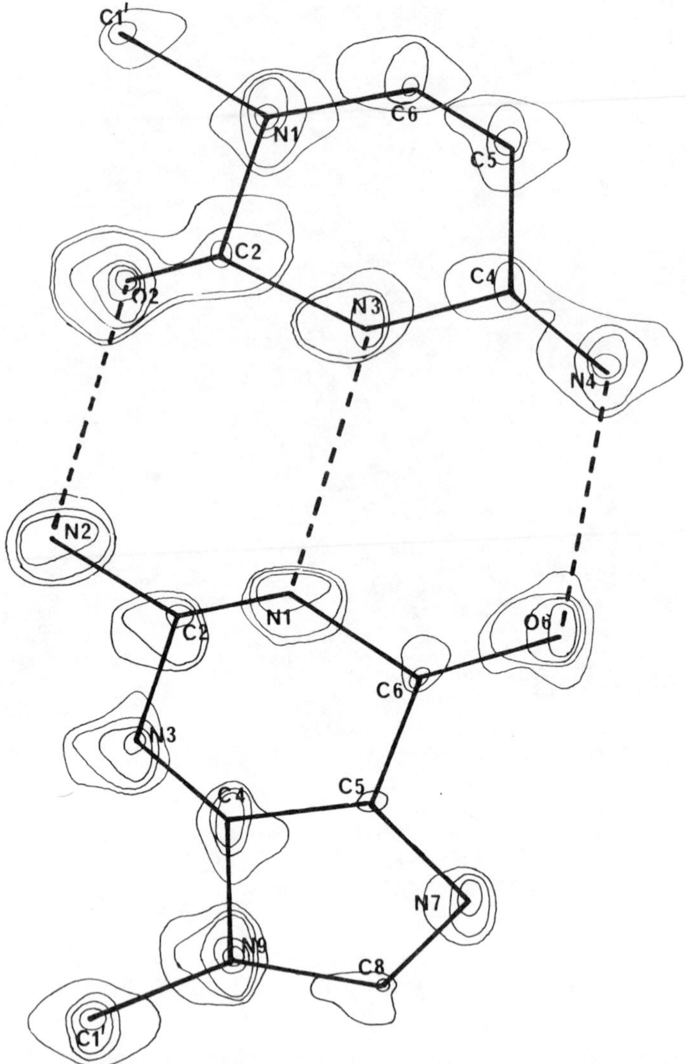

Figure 20.5. An electron density map of a small structure on which is superimposed a diagram of the structure. In the map the individual atoms can be seen resolved (photograph kindly supplied by Dr. S. Neidle.).

consist of, for example, a polypeptide backbone of defined stereochemistry with known primary sequence.

Fig. 20.7 shows how such chemical information (derived from small structure crystallography) is fed into the model and is used to build a backbone and side chains to fit the lower resolution electron density data of protein crystals.

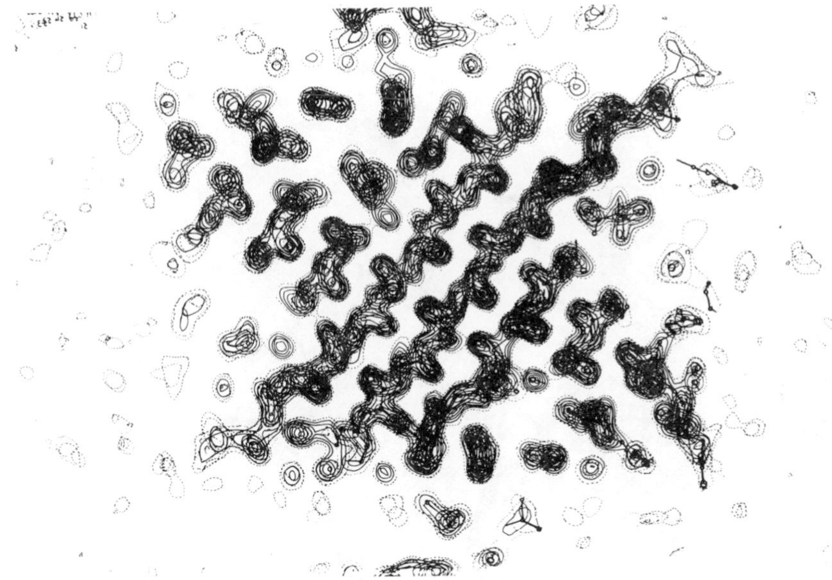

Figure 20.6. A portion of the electron density map of protein pre-albumin. The β-pleated sheet can be seen clearly but individual atoms are not usually resolved (1.8Å resolution map kindly provided by Dr. C. C. F. Blake).

20.1.4. Protein crystallography

The end-result of any crystallographic study which incorporates all the experimentally determined information is the electron density. This can be expressed as Fourier Synthesis.

$$\rho(x, y, z) = \sum_{hkl} A(hkl) \cos 2\pi(hx + ky + lz) + B(hkl) \sin 2\pi(hx + ky + lz)$$

where ρ is the electron density at the point x,y,z in the unit cell. A and B are given by

$$A(hkl) = F(hkl) \cdot \cos\phi$$

$$B(hkl) = F(hkl) \cdot \sin\phi$$

where $F^2(hkl)$ is the observed intensity of the diffraction peak indexed as hkl and ϕ is its relative phase.

It is in the method of determination of this phase that protein crystallography differs from the crystallography of smaller structures. For these smaller structures methods can be tried, usually successfully, which reveal the phase from the statistics of the intensity data of the crystal of the original compound. With proteins it is necessary to measure the intensity diffracted by a native crystal and crystals of one or more derivatives of the original protein. These derivatives are

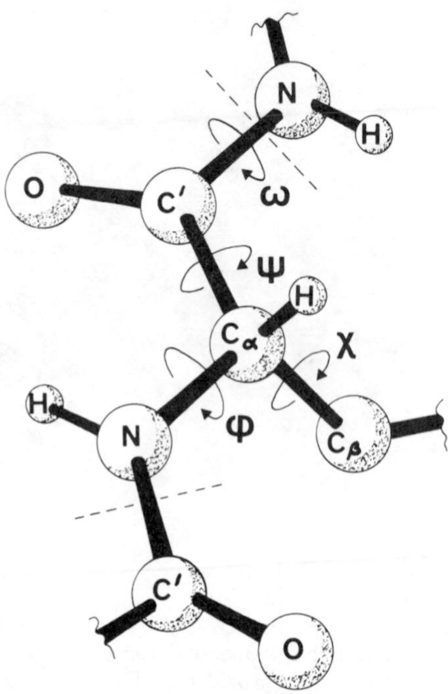

Figure 20.7. A perspective diagram showing how a portion of the polypeptide backbone can be regarded as a chain of fixed stereochemistry and whose structure is defined by rotations about bonds of fixed length.

made by adding heavy metal atoms either before or after crystallization. The derivative crystals apart from their additional metal atoms should have their proteins in the same conformation, orientation and packing as the native crystals, in which case they are said to be *isomorphous*.

Once one has been provided with a native crystal and a suitable number of isomorphous derivatives the next stage is to collect the intensities, $F^2(hkl)$, for all these different crystals. Once collected, these intensities can be combined in such a way as to reveal the positions of the added heavy metal atoms and from this information the phase of the reflexions from the native crystals can be computed (Blundell & Johnson, 1976).

20.1.4.1. Stages of analysis

A flow diagram of the procedure for solving a protein structure is shown in Fig. 20.8. There are several points where problems may be expected to occur and the

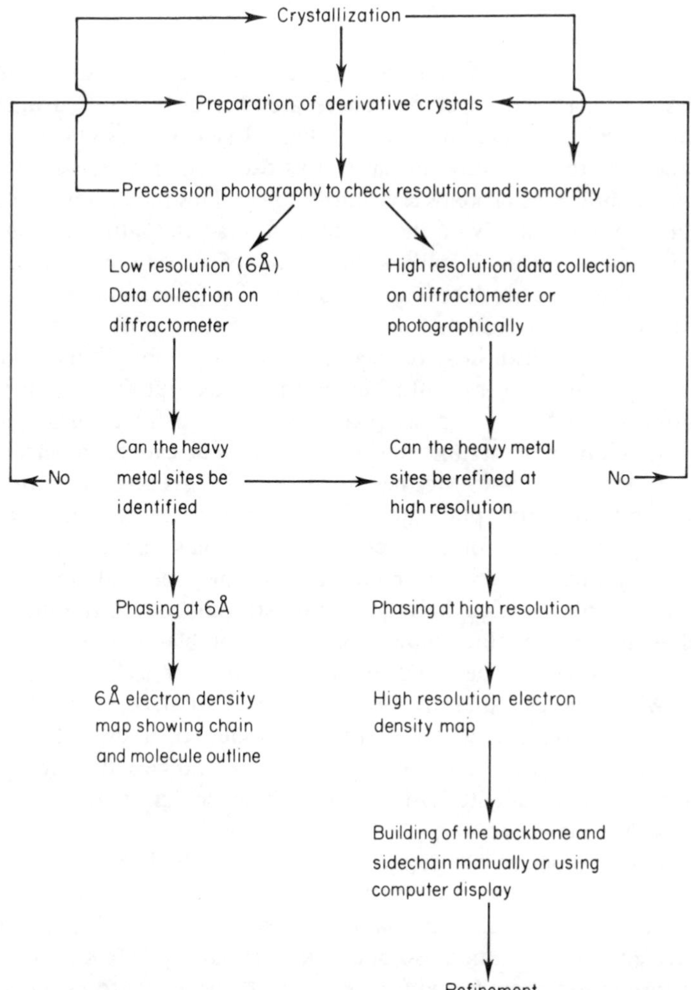

Figure 20.8. A flow diagram showing the various stages of protein crystallography and where review of results becomes necessary before proceeding.

consequent back-tracking is indicated. However, it is worth highlighting some areas where the biochemist might expect to be involved in a practical way.

20.1.4.2. Crystallization

For X-ray diffraction work, crystals of several tenths of a millimetre in size are required, substantially larger than those formed as a silky sheen in routine biochemical crystallizations. Nevertheless the knowledge that microcrystals have been prepared can be a useful incentive in the time-consuming task of trying to find a method and conditions which will produce the larger crystals necessary for a high resolution structural study. There are several methods available for attempts at crystallization but none are infallible. In fact given sufficient material it would be wise to try several in parallel as each requires the variation of a number of conditions as proteins crystallize slowly. This requires great patience. The various methods all rely on the production of a supersaturated solution of the protein and the growth within this liquid of nucleus of a crystal. The supersaturation may be produced homogeneously by, for example, simply mixing with sufficient saturated ammonium sulphate (salting out). Alternatively, a gradient of saturation can be produced by temperature, by diffusion of a salt across a dialysis tube or by diffusion of the water out from a droplet by evaporation. Similarly, the vapour phase can be used for diffusion into the crystallizing solution of organic solvents which will lead to supersaturation (Blundell & Johnson, 1976, McPherson, 1976, 1982). The quality of protein crystals was improved after purification by isoelectric focusing (Bott *et al.*, 1982).

The choice of technique for initial crystallization trials whether on a large or a small scale, by microdialysis or microdiffusion, may depend on how much material is available. As a guide to quantities, starting concentrations of about 20 mg/ml would be suitable for a moderately soluble protein. Trials using droplets of solution consume one drop of about 10 μl for each set of conditions examined. Microdiffusion and microbatch cells of a similar size can be made so that 1 ml or 200 mg of protein would suffice for about 100 trials. With care, reuse of material, and a few good crops of crystals this would be sufficient to provide enough crystals for a full study. It ought at least be sufficient to explore the possibilities of future uork.

Production of sufficiently large crystals is not, in itself, the end of this stage of the work. Firstly, these crystals must be shown, by diffraction, usually on a precession camera, to be sufficiently well ordered so that the diffraction data can be collected to high resolution. Secondly, the crystal type must be suitable, as crystallization in too complicated (or too simple) a packing arrangement would make the structure impossible to solve. Thirdly, the crystals must not be too easily damaged by the X-rays that will be used to study them. It must be possible to take several high-resolution photographs from each crystal. Fourthly, the crystals must be reproducible so that a supply of identical crystals can be made, each (before it is destroyed by irradiation) providing a portion of the X-ray diffraction data.

20.1.4.3. Derivatives

Although, theoretically, a heavy metal atom may be bonded to a protein molecule in solution and the protein derivative solution crystallized as before, the most popular method for derivative formation is the soaking of native crystal in suitable heavy metal salts. The rationale for this is that one needs a native and derivative crystal with the same molecular packing, differing only by the addition of a few metal atoms per protein molecule. It is sensible to start with the protein molecules already packed correctly. This method suffers only from the possibility that the crystalline arrangement may prevent access to an otherwise suitable binding site.

The problem in the preparation of derivatives is that in most cases it involves a form of educated trial-and-error rather than obvious logical decision. In a few favourable cases the metal ions of metallo-proteins may be replaced by other metal atoms providing the possibility of a series of heavy metal derivatives. Similarly, chemical modification of the protein itself, or the addition of a cofactor or an inhibitor may open the way for specific attachment of the metal atom. However, in most cases the binding is by a process which will be understood only after the structure has been solved and is due to the particular local three-dimensional arrangement of the amino acid side chains. As the arrangement of the surface side chains is likely to be affected by pH and salt concentration these two factors have to be considered as variables in the preparation of heavy atom derivatives. Particular attention will be paid to them in the later stage of trying to optimize conditions for a derivative. Here, attempts must be made to find a reproducible derivative with a small number of well-occupied sites. There is also the possibility that with different soaking times and/or concentration one heavy-metal salt will yield two derivatives, one with one site and the other with two sites occupied. The *occupation* of a site is a quantity that must be determined crystallographically and corresponds to the probability that that site will contain a metal atom in any one of the many copies contained in a crystal. What is important is that conditions are found for preparing derivatives with heavy metal atoms attached in a few particular sites with high and reproducible occupancy (Blundell & Johnson, 1976; McPherson, 1976).

20.1.4.4. Mounting a native crystal or derivative

Once prepared, the crystal of a native protein or a derivative must be examined by X-ray diffraction to see if it is suitable for further work. The two primary criteria are *resolution* and *isomorphism*. For any X-ray work the crystal will be mounted in a thin-walled glass capillary tube (Fig. 20.9). This is done by using a Pasteur pipette to pick up the crystal in a drop of mother liquor and transferring it to the thin-walled capillary. To avoid any change in the ionic conditions during exposure (in particular drying) a drop of mother-liquor is placed at one end of the tube which is then sealed with dental wax. Thin strips of filter paper are used to remove excess liquid from around the crystal. This reduces the scattering of X-rays by the excess liquid and ensures that the crystal is firmly held in position by the surface tension of the trace of liquid that surrounds it. The tube is finally

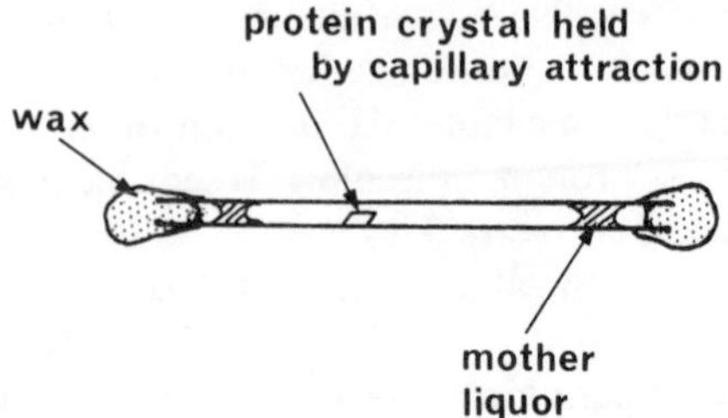

Figure 20.9. Diagram of a protein crystal mounted in a thin-walled capillary, with mother liquor ready for diffraction.

sealed with wax after another protecting drop of mother-liquor has been placed at the opposite end as shown in Fig. 20.9. It is usual to start off with a thin-walled capillary of about 30 to 40 mm and to coat the ends with dental wax before starting to insert the crystal or drops of mother liquor so that it can be sealed at each end rapidly and efficiently.

20.1.4.5. Testing a crystal

The first test of a crystal will be revealed by a precession photograph such as shown in Fig. 20.4. This will show the general quality of the crystal which should give sharp even spots on a regular lattice extending, hopefully, to the edge of the photograph. The resolution of the crystal is indicated by how far out diffraction spots can be observed. Because of the geometry of the precession method, care is necessary. For example, a crystal should not be rejected because it appears to diffract only to 0.6 nm on a photograph on which the limit would have been 0.1 nm. Using a smaller precession angle designed for say 0.3 nm limit, diffraction may be found extending to the edge of the photograph.

Stability in the X-ray beam or susceptibility to damage can also be examined by taking photographs at suitable intervals under identical conditions and looking for changes (usually showing reduced resolution) with exposure time. A crystal should be able to withstand at least one day's exposure without observable change in the precession photograph.

For derivatives, the first check for isomorphism is on the dimension of the crystal lattice as shown by the reciprocal lattice which is displayed on the precession photograph. Laying the derivative photograph over that of the native photograph, the spot positions should coincide. It is from the *intensity changes* between the native and the derivative that we are going to work out the position of the heavy atoms and thereby phase the X-ray reflections. These changes should be visible (but not too large because this would indicate conformational changes in the protein). If they are not visible another derivative should be sought.

Otherwise an attempt should be made to locate the positions of the atoms by the specialized methods of protein crystallography. Failure to solve the positions of the adducted heavy metal atoms may be due to lack of isomorphism from conformational or orientational changes in the protein, or may be due to too many heavy metal atoms having been added which lead to a confused picture.

20.1.4.6. Some technical problems

Protein crystallography proceeds *via* the determination of the heavy atom positions and the subsequent phasing of the X-ray diffraction amplitudes, to the calculation of the electron density map of the native crystal unit cell. During this process technical problems arise because of the large scale of protein crystallography compared with conventional small structure crystallography. This results in an exceptionally large number of X-ray reflections to be measured, a large map to be constructed and a large model of the protein to be built from it.

20.1.4.7. Data collection

During the development of protein crystallography automatic diffractomers were built which measured the diffracted intensity using a counter to quantify the scattered photons. The diffractometer positioned the counter so that the reflexion would enter it and orientated the crystal correctly for that reflexion to occur. The machine was also designed to correct for normal imperfections of the crystal and to allow for background scattered radiation. It would automatically measure whole sets of X-ray reflexions recording the resultant intensities on punched cards, paper tape or in some other suitable computer-readable form. Subsequently the diffractometers were connected on-line to their own computers which made their operation substantially easier, more reliable and more accurate. They continue to be widely used for low-resolution (0.6 nm) work on proteins, where the number of data points to be collected is small. Their use has tended to be restricted in this way because of their inefficiency in terms of data collection. This inefficiency arises because protein crystals with their large unit cells frequently display several X-ray reflexions at the same time in different directions. A simple counter instrument can observe only one of these, thus wasting potentially useful information. Not only does this extend the time taken to measure a set of data from a crystal, but because of the finite lifetime of a crystal in the X-ray beam many crystals are required. This results in the necessity for combining correctly data measured from different crystals.

At low resolution and for the smaller proteins the finite lifetime of the crystal presents little problem. As the protein mol. wt. rises and the desired resolution increases, the need for efficiency in data collection grows. What is required is a two-dimensional detector, corresponding to an array of counters, capable of observing every photon scattered by the crystal. Such devices on-line to computers are under development and represent the ultimate efficiency in collection.

As an intermediate solution photographic data-collection has been developed. If precession photographs are used the *indexing* (assignment of *h,k* & *l* values to a reflexion) is straightforward as is the actual measurement of intensity. However, precession photographs are made using a camera containing a metal screen which moves in such a way as to block out the unwanted reflexions from layers other than the one being recorded. They are thus not as efficient as totally screenless cameras. The photographs from such cameras must be analysed using a computer to index the reflexions and add up the relevant density areas. Such an analysis involves digitizing the optical density of the photograph as a two-dimensional map using an automatic densitometer.

20.1.5. Model building

The experimental results lead after analysis to an *electron density map* stored within the computer. As this is not at the level of atomic resolution, a model incorporating the known stereochemistry of the polypeptide backbone and its side chains must be built to fit it. This is usually done by first contouring transparent plastic sheets according to the electron density map printed out by the computer to a suitable format and scale. A stack of all these sheets together would show the whole unit cell but it would be difficult to see through. Therefore selected levels around the region currently being constructed or altered in the model are placed in a vertical frame and illuminated from the side and back. Using a half-silvered mirror an image of the wire model being constructed is reflected in such a way that the observer can view the reflexion of the model superimposed on the sections of the map as if the model were being built within the map (Fig. 20.10) (Richards, 1968).

20.1.6. Computer-aided model building

While this system of model-building has the distinct advantage of 'feel' and of producing a completely visible model it suffers from certain disadvantages. Computer-aided construction is an area of development at the moment where visual display units can be used to show models of different types superimposed on contour maps. An orthogonal view or stereo-pairs can be displayed simultaneously and three-dimensional contours can be drawn corresponding to a net thrown over the electron density at a certain level (Fig. 20.11). Typical advantages of such a system would be:

1. the possibility of introducing automatic refinement of a residue once it had been initially positioned in the map;

Figure 20.10. A device for building a wire model of a protein seen by reflection to be 'within' the protein electron density map. The mirror has been removed so that the maps may be seen. It normally lies between the model frame and the map frame (Richards, 1968; photograph kindly supplied by Dr. C. C. F. Blake).

2. incorporation of Van der Waal's repulsion to avoid the possibility of over-close contacts;
3. the addition of symbols such as dotted lines, showing possible H-bonds.

These advantages must be weighed against the cost of the hardware and programming and the limitations of speed and size which would currently reduce the display so that it shows only a very few residues of the chain at any one time.

20.2. ELECTRON MICROSCOPY

20.2.1. Introduction

Electron microscopy can be contrasted with X-ray diffraction most obviously in its accessibility and actual imaging. Microscopes themselves are widely available together with assistance in their use and with the techniques of specimen preparation. The results of microscopy are immediate direct images which can, with due caution, be interpreted to yield useful information as to the structure of the molecules being examined. In normal use a minimum of theoretical understanding of the microscope is necessary although as with any such

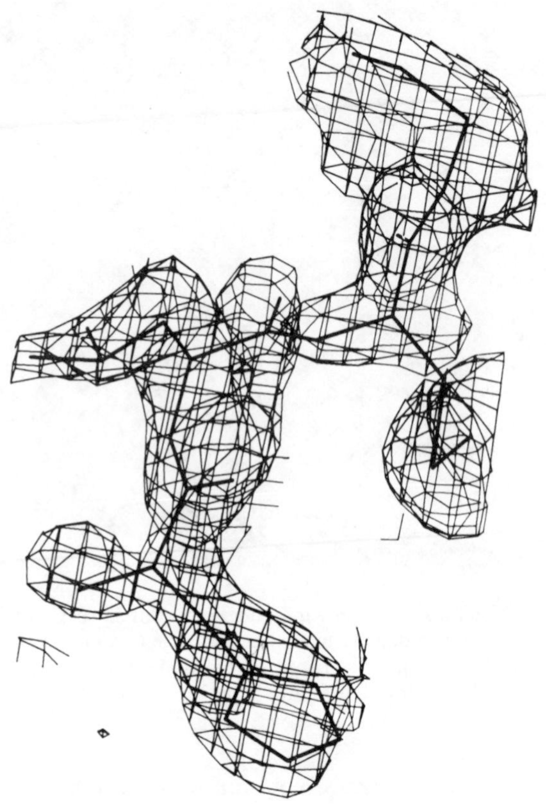

Figure 20.11. A computer-aided fitting of residues B5-B8 of the hormone insulin. The computer driven display shows a three-dimensionally contoured electron density map and model simultaneously (photograph kindly supplied by Dr. A. C. Evans & Prof. A. C. T. North).

instruments, the more extensive the understanding (and experience) the quicker the results will be obtained and the better and more reliable the micrographs produced.

The difference between X-ray diffraction and electron microscopy arises because of the existence of magnetic electron lenses which enable a direct image to be formed. This ability has made the microscope so obviously convenient that a large enough market existed for manufacturers to develop their instruments with a view to ease of usage, reliability, quality and convenience. The net result is that a biochemist can expect to have a good quality microscope in his immediate neighbourhood. He may also expect to have an experienced operator to examine his specimens for him in the first instance and, should they appear suitable, to request a few lessons before using the microscope himself.

20.2.2. Resolution and limitations

The *resolution* of a microscope is determined by the wavelength of the radiation used. With light, wavelength 500 nm, this could be around 250 nm. With X-rays using diffraction and a computer to form the image this could be improved to 0.10 nm with the wavelength 0.15 nm. An electron microscope uses electrons accelerated to about 100 keV when they behave as waves of about 0.003 nm wavelength. For technical reasons the resolution of microscopes is only at best about 0.3 nm, but this would be ample to enable one to distinguish many details of molecular structure. Unfortunately, with biological materials the specimens themselves cannot easily be prepared so that they preserve their structure at this resolution. It is not the intrinsic resolution of the microscope which limits its usefulness, but the difficulties of specimen preparation and preservation. The latter is necessary because of the hostile environment of the vacuum within the microscope and the damage produced by the electron bombardment during the microscopy.

The traditional method of obviating these problems has been to stain the specimen with heavy metals. One then observes an image of the stain and infers the molecular structure from it. Using this technique a limit of 1.5 nm resolution is the most one can expect. New techniques are being developed for molecular microscopy at fairly high resolution (0.6 nm) but these require (as does X-ray diffraction) the amplification of the signal/noise ratio produced, by using an ordered array of molecules. Let us first of all consider the conventional approach before examining in more detail these recent improvements.

20.2.3. The electron microscope

An electron microscope (Fig. 20.12) is analogous to an optical microscope in that it consists of components (lenses) which bend the rays to form an image. The rays in this case are a stream of electrons emitted by a hot filament in the gun and accelerated by the high voltage between the gun and the rest of the microscope. This voltage is specially stabilized so that speed of the electrons is constant. The electrons behave as if they are waves of identical wavelength (monochromatic). This wave treatment enables all the theory of light behaviour and of the light microscope to be applied to the electron microscope.

The main part of the microscope begins with the specimen stage which is designed to hold a copper grid which itself supports a carbon film (Fig. 20.13) on which is the specimen itself. The specimen stage is designed to provide an easy means of positioning the specimen within the microscope and to eliminate drifting or vibration during direct observation or photography.

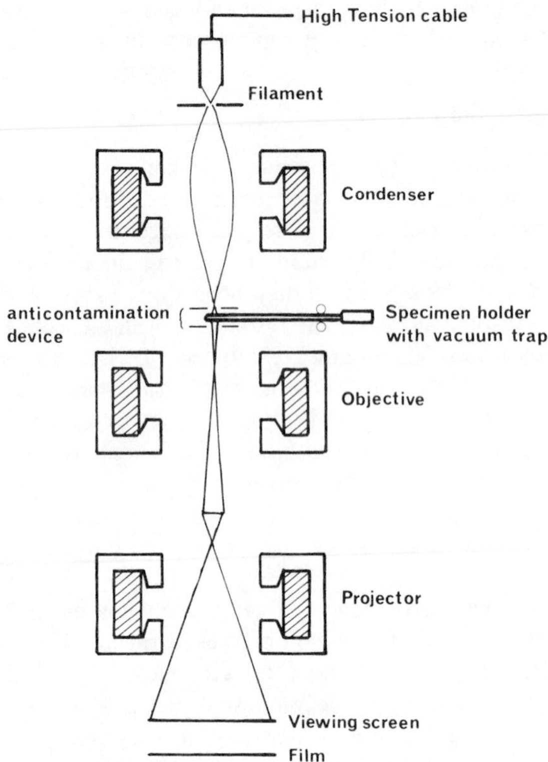

Figure 20.12. A schematic diagram of an electron microscope showing the main (magnetic) lenses. Unlike the optical microscope the electron microscope usually has its illumination at the top and the final, magnified image at the bottom.

The stage rests within an anticontamination device which is kept at the temperature of liquid nitrogen by being attached to copper wires or rods leading to a liquid nitrogen reservoir. The device condenses on to itself any organic or other molecules which may have evaporated from the specimen either naturally or under the electron bombardment during observation, or other contaminants within the microscope. These molecules would otherwise be susceptible to conversion into very reactive free radicals in the electron beam. These radicals would react with the specimen or support film, thus building up a layer of contamination, reducing resolution and producing artifacts.

Below the specimen lies the first lens of the electron optical system, the objective. This is followed by a series of subsequent lenses which finally produce an image either on a fluorescent screen for direct observation or on a piece of photographic film for the recording of an electron micrograph.

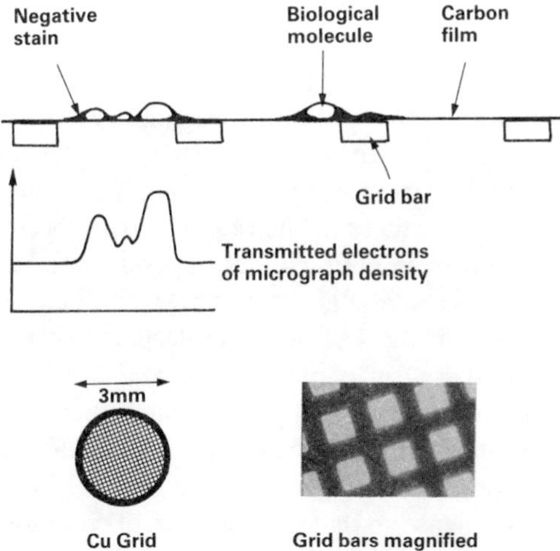

Figure 20.13. The specimen in an electron microscope is supported on a carbon film which is itself supported on a copper grid (for electrical and thermal conductivity). The upper part of the diagram shows the effect of negative stain.

The magnification of the microscope, which can be up to × 250 000 during use (and the resolution may be good enough to justify a further three-fold magnification of the micrograph during printing), is altered by alteration of the focal length of the lenses below the objective. This is done electronically because the focal length (and hence magnifying power) of an electron lens is determined by the electric current through the coil which surrounds the magnetic pole pieces. This is very convenient for the microscope user but it does require electronics which provide him with the facility to alter the current and then to maintain it at the new value with a very high degree of stability (otherwise the focal length would be continually changing).

Focusing the microscope is achieved by separately altering the current through the objective lens and this must generally be done after every change of magnification. The overall procedure is to alter the lower lenses indirectly by setting a value for the magnification (and allowing the manufacturer's electronics to decide precisely how the individual lenses are set). Focusing is then done using the objective lens current.

20.2.4. The microscope environment

As we are using the electrons in the microscope to visualize molecules we would expect that any extraneous atoms interfering with the electron beam would reduce the quality of the resultant image. Thus the inside of an electron microscope is evacuated and the specimen will become dehydrated. We shall discuss later methods of avoiding the effect of dehydration on specimen structure.

The second unwanted consequence is due to the effect of the electrons themselves on the specimen. Not only do these high energy electrons produce direct damage but indirectly, by converting other molecules into highly reactive species, they cause contamination of the specimen. This contamination can be reduced by having copper surfaces in the region of the specimen cooled to the temperature of liquid nitrogen thus condensing out most volatile molecules. Direct damage by electrons can be eliminated only by reducing the specimen dose and this will be discussed in the section on molecular microscopy.

These problems of dehydration, contamination and damage can be avoided by examining not the true specimen but one *stained* with a heavy metal. The specimen is stained before drying and the stained molecules are considerably less susceptible to dehydration and damage than the original light atoms. They are also much more electron dense so that the problem of contamination is reduced. However, it is still necessary to take all possible steps to minimize contamination by keeping the microscope and specimen support grids as clean as possible and to adopt any convenient techniques to reduce the electron exposure of the specimen.

20.2.5. Specimen preparation and staining

Although large biological structures are often sectioned by embedding them in a suitably hard support and slicing with a knife made of glass or diamond, examination of individual molecules is usually made by starting with a solution of the molecules in an appropriate buffer. Two factors govern the choice of buffer. The first is volatility during the drying process and the second is interference or chemical reaction with the stain. Perhaps the most frequently used buffer is ammonium acetate. Although this is volatile it must be remembered that during the drying process the concentration of buffer increases dramatically. This can sometimes affect molecular structure and aggregation to such an extent that the stain pattern, which is to be observed, is distorted or destroyed.

The specimen is supported in the microscope on a thin layer (about 15 nm) of carbon. This layer is made by evaporating carbon on to a freshly cleaved piece of mica. The carbon is obtained in the form of rods of about 5 mm diameter. A length of about 10 mm at the end of one rod is ground to 1.5 mm diameter and given a pointed tip. This thin portion of the rod is then evaporated by passing a current of about 30 A through it while it is pressed against the flat end of another rod. Evaporation occurs at the pointed tip and constant pressure is maintained by a spring (Fig. 20.14).

The carbon is deposited on the mica which is placed about 200 mm from the source. It is convenient to place next to the mica a piece of clean white paper on which rests a small object the darkness of whose 'shadow' gives a good impression of the thickness of the carbon layer.

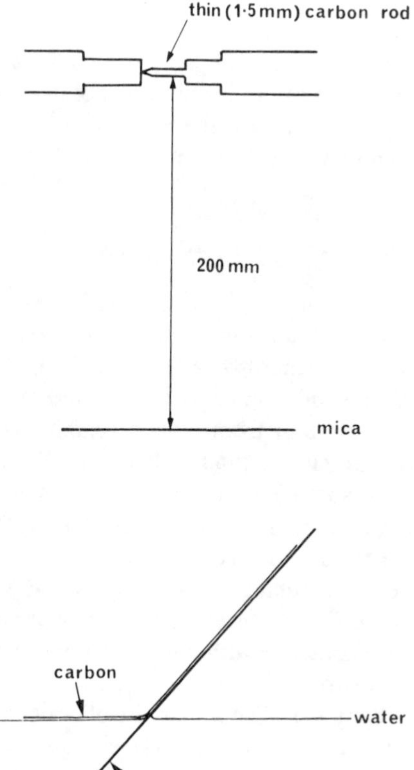

Figure 20.14. Carbon is evaporated on the mica in an evacuated bell jar and the thin film 'stripped' onto a clean water surface.

The carbon layer is easily removed from the mica sheet because of the extremely hydrophilic nature of the mica surface. The coated sheet is lowered into a large dish of clean water with the mica held at about 45° to the water surface. The carbon layer will float off intact and can be moved on the surface by touching it with a hair. Cleanliness is most important at all stages in the preparation of the carbon film. The mica should be trimmed with a fine pair of scissors, cleaved and handled with forceps. The dish used for the water should be washed in detergent, thoroughly rinsed and heated with a flame before being filled with glass distilled water. These precautions will avoid any grease on the mica or water surface which will interfere with the stripping of the carbon film.

The thin carbon film must itself be supported by a fine mesh (200 or 400) copper grid of 3 mm diameter. It is convenient to place a dozen or so grids on a tray or on a piece of wire gauze beneath the water surface before stripping the carbon from the mica. This tray can then be raised or the water surface lowered, while holding the carbon film in position. As the grids pass through the water surface they pick up pieces of the carbon film. If carefully done most of the grids will have an intact carbon layer over their whole area. Again the gauze or tray and the grids must be kept grease-free by cleaning with solvent or heating. The grids are dried and kept in a desiccator.

20.2.5.1. Staining

There are two types of staining—positive and negative. In the case of positive staining a heavy metal compound reacts with part of the molecule being studied and remains attached to it (Richardson & Davies, 1980). With negative staining the molecule is viewed embedded in a 'glass' of stain. Fig. 20.13 shows how this results in the exclusion of stain from certain regions which are thereby less electron dense. (Huxley & Zubay, 1960; Fabergé & Oliver, 1974).

In positive staining the specimen is reacted with the stain before being loaded on to the grid. With negative staining a droplet containing the specimen is placed on the grid and about a minute allowed for molecules to 'settle' on to the carbon. The grid is almost dried by touching the edge with a filter paper. Just before it is dry a droplet of negative stain is placed on it. After about 15 s this droplet is removed as completely as possible with filter paper and the grid allowed to dry.

It is important that the carbon film is wettable. This can be seen when removing the excess liquid with filter paper. The wettability of grids varies with atmospheric conditions, history and age. A variety of techniques exist to improve the wettability should the specimen and stain droplets not wet the surface. First one should simply try leaving the specimen droplet for longer, then one can try leaving the grids in a damp petri dish for a couple of hours before use. Finally one can try exposure to u.v. radiation or ion bombardment. These last two, however, may make the surface so hydrophilic that the droplets wet both sides!

20.2.5.2. Shadowing

Another method of revealing structure is to coat the surface of the specimen in such a way that heavy metal atoms are deposited by evaporation from a hot source. The source is placed so that the stream of atoms strikes the surface at an oblique angle casting 'shadows' of raised parts of the specimen. Another way of looking at this is to consider the build-up of stain rather than the shadows. Quantitative results can then be obtained (Smith & Kistler, 1977) which relate the stain density to the specimen contours. Any problems created by the direction of shadow can be obviated by rotating the specimen during the process but such rotary-shadowed grids are not susceptible to quantitative analysis.

The resolution of any shadowed specimen must be regarded as very low. The process succeeds because the surface is uneven and all water has been removed. The distortions from native structure will be very great although gross features may be retained.

20.2.6. Image processing

The final result of an electron microscope examination is the micrograph formed when the electrons are allowed to strike a photographic plate and the resultant latent image is developed. Frequently molecules which are being examined can be induced to form two-dimensional regular arrays which provide the opportunity for image enhancement. The technique is most readily applied using an optical diffractometer (Fig. 20.15). In this the micrograph acts as a two-dimensional diffraction grating giving a series of diffraction spots (Fig. 20.16). These spots can be thought of as arising from the *average* of the repeating structure in the specimen and any deviations from average (or noise) give diffraction between the spots. If we introduce another lens into our optical diffractometer we image or 'reconstruct' the original micrograph. If, in addition, we remove the diffraction from the noise by introducing a filter, the reconstruction is then of an array composed of the average repeating unit with the noise removed. Extension of this technique enables us to separate an image which originally consisted of two overlapping arrays by filtering out the diffraction spots due to one, while still allowing the other set to continue to form the reconstructed image. In both these cases we have substantially improved the information yielded by the original micrograph.

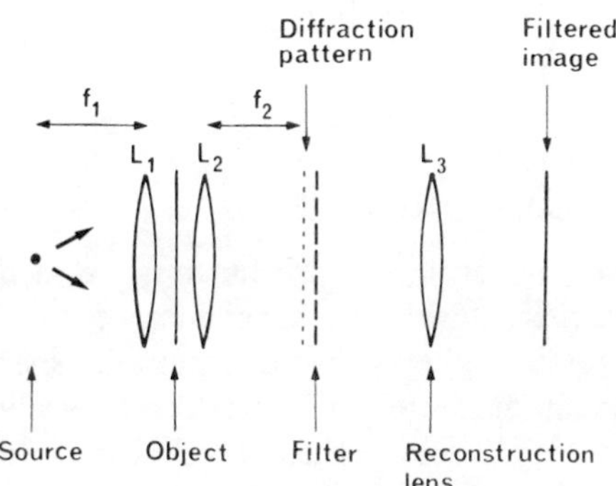

Figure 20.15. An optical diffractometer in which an electron micrograph is illuminated by a parallel beam of light and its diffraction pattern observed. After filtering a reconstructed image is produced.

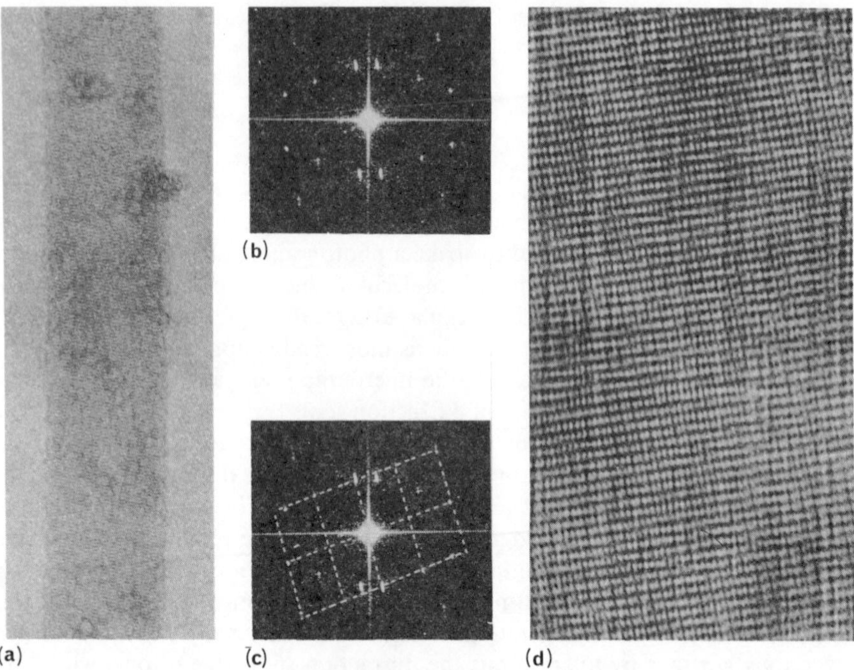

Figure 20.16. Optical processing as shown by a two-sided specimen. (a) Micrograph of an actin tube. (b) The original diffraction pattern which has spots corresponding to diffraction from both sides of the tube. The peaks arise from the ordered structure and the intensity between the peaks from disorder or noise. (c) The lattice of spots selected by the filter with the spots shown joined by dotted lines. (d) The filtered (reconstructed) image. (From M. J. Dickens, Actin tubes, *Proc. Royal Microscopical Soc.*, **13,** 80–81, 1978. These micrographs were awarded the Royal Microscopical Society Glauert Medal.)

20.2.7. Three-dimensional microscopy

The details that we see in micrographs are due to two-dimensional projections of the three-dimensional structure of the original specimen. Using a microscope with a goniometer (or tilting) stage we can view the specimen from different angles. These individual images might be extremely revealing in terms of detail. More interesting, however, would be their combination into a three-dimensional reconstruction. This can be done for some specimens, although it is a lengthy process. It uses mathematical techniques akin to those of X-ray crystallography. These are applied to the digitized images of the specimen obtained using an optical densitometer to feed the information to a computer either directly or via some intermediate such as magnetic tape. The resultant three-dimensional image can be presented in a variety of ways but it is usually viewed initially as a series of sections analogous to a section that a microtome would cut from an embedded specimen (Dover & Elliott, 1979; Dover *et al.*, 1981).

With a highly ordered stained specimen a good three-dimensional electron density map is obtained. Fig. 20.17 shows such a map from ribosome crystals (Unwin, 1977). The map is analogous to a crystallographic electron density map at low (5 nm) resolution.

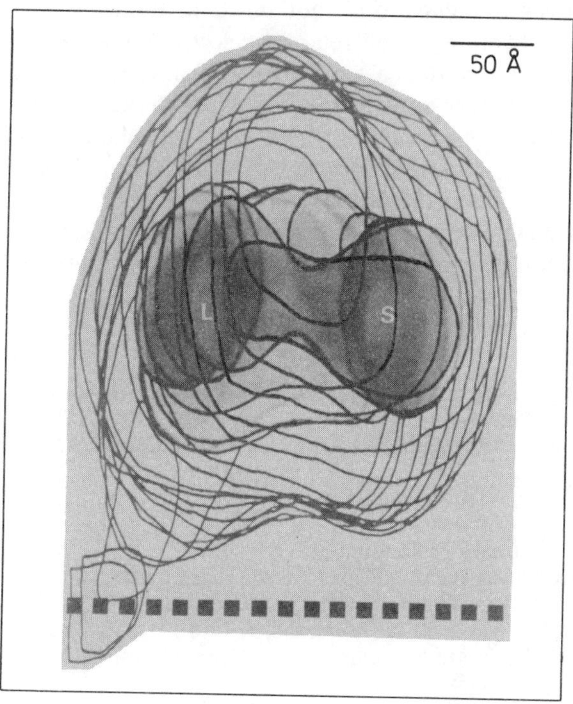

Figure 20.17. The three-dimensional reconstruction from micrographs of tilted specimens, of the ribosome as found in crystalline arrays in the oocytes of the lizard *Lacerta sicula* (Unwin, 1977). L: Large subunit; S: small subnit.

20.2.8. Molecular microscopy

The possibility of three-dimensional microscopy of *stained* specimens leads on to the idea of looking at unstained specimens particularly at high resolution. This implies directly imaging the atoms of molecules themselves. This has been done (Henderson & Unwin, 1975; Unwin & Henderson, 1975), but it requires special methods in using the microscope and even more complicated treatment of the images obtained.

First the specimen must be protected against dehydration in the microscope by surrounding it with a 'glass' of sucrose. As we are looking at the 'molecule' atoms rather than 'stain' atoms extremely short exposures must be taken so that the electron damage of the specimen is negligible. These exposures are so short that the averaging obtainable from a two-dimensional array is absolutely essential for obtaining any information from the very weak image.

The special methods of image processing are necessitated by the fact that the specimen consists of light atoms, not of stain. Rather than absorbing any electrons these light atoms simply retard the electrons, thus altering their 'phase'. An in-focus image of such a phase specimen is featureless. To visualize the atoms we must take a number of out-of-focus images which can be computer processed to produce a *phase-contrast* image whose peaks will show atomic positions.

This process of combining information from a number of images also requires a high signal/noise ratio in the data and this is produced only by a specimen which is a two-dimensional array. Only such specimens are at present susceptible to the possibility of molecular microscopy.

20.3. REFERENCES

Blundell, T. L. & Johnson, L. N. (1976). *Protein Crystallography*, Academic Press, London, New York, San Francisco.

Bott, R. R., Navia, M. A. & Smith, J. L. (1982). *J. Biol. Chem.*, **257**, 9883–9886.

Dover, S. D. & Elliott, A. (1979). *J. Mol. Biol.*, **132**, 323–341.

Dover, S. D., Elliott, A. & Kernaghan, A. K. (1981). *J. Microsc.*, **122**, 23–33.

Dunitz, J. D. (1979). *X-ray Analysis and Structure of Organic Molecules*, Cornell University Press, Ithaca & London.

Fabergé, A. C. & Oliver, R. M. (1974). *J. de Microscopie*, **20**, 241–246.

Henderson, R. & Unwin, P. N. T. (1975). *Nature (Lond.)*, **257**, 28–32.

Huxley, H. & Zubay, G. (1960). *J. Mol. Biol.*, **2**, 10–18.

McPherson, A. (1976). In *Methods of Biochemical Analysis*, vol. **23**, Glick, D. (Ed.), Wiley, New York, pp. 249–345.

McPherson, A. (1982). *Preparation and Analysis of Protein Crystals*, Wiley, New York.

Richards, F. M. (1968). *J. Mol. Biol.*, **37**, 225–230.

Richardson, W. D. & Davies, H. G. (1980). *J. Cell Science*, **46**, 253–278.

Smith, P. R. & Kistler, J. (1977). *J. Ultrastr. Res.*, **61**, 124–133.

Unwin, P. N. T. (1977). *Nature (Lond.)*, **269**, 118–122.

Unwin, P. N. T. & Henderson, R. (1975). *J. Mol. Biol.*, **94**, 425–440.

Practical Protein Chemistry—A Handbook
Edited by A. Darbre
© 1986, John Wiley & Sons Ltd

B. ROBSON
*Department of Biochemistry, The Medical School,
University of Manchester, Oxford Road,
Manchester M13 9PL, U.K.*

21

The Prediction of Peptide and Protein Structure

CONTENTS

21.1. INTRODUCTION

The art of predicting biopolymer conformation is now more than 30 years old. How has it fared in this time? It is probably true to say that no prediction on theoretical grounds alone has ever been accepted with *absolute* confidence. This is because many approximations must be made in the calculations in order that a conclusion can be reached in reasonable time, even with the aid of a computer. The consequent uncertainty makes any conclusions ambiguous: structure A may look a slightly better bet than structure B, but unless the evidence against B is strong, it must still be considered a plausible candidate. This ambiguity also applies, by varying extents, to experimental data. Fortunately, however, the ambiguities are not generally of the same type. Historically, the true power of the theoretical approach was that, combined with an experimental approach, it could often lead to an unambiguous conclusion. This combination has led to those major breakthroughs in molecular biology, the structures of protein α-helices and pleated sheets, and the structure and function of DNA. More recently, a variety of problems have been tackled of which polysaccharide conformation and rational drug design are but two examples. In certain areas, such as the attempted refinement of protein X-ray data by energy calculations, the theoretical approach has become almost routine. Thus, while it would be misleading to imply that the art is in anything more than its adolescence, it would be equally misleading to suggest that it is still in its infancy.

Workers interested in predicting molecular conformation prefer to speak of calculating the conformation, and of practising the discipline of *conformational analysis.* (More recently, of 'molecular mechanics'.) The discipline allows both interpretation and prediction. The general aim is to calculate molecular properties (of which the favoured conformation is but one example). Many molecules do not have a specific conformation (and strictly speaking no molecule does except at absolute zero).

Many oligopeptides such as the neuropeptides may be open and relatively flexible, to an extent that it may not be useful to calculate a unique structure at all. One can, nonetheless, calculate average properties such as the average vicinal proton–proton coupling constant that would be obtained by nuclear magnetic resonance spectroscopy. Comparison with experimental values can then serve to interpret the experimental observation in terms of conformational behaviour.

Prediction of protein structure remains, however, a topic of considerable interest and success in this area has long been expected to be of socio-economic importance (Robson, 1976, 1980). The Commission of the Economic Communities highlighted this in document EC COM(84)230. Predictive and design tools for peptide and protein engineering are now generally accepted as important for realizing a market of 10–100 billion US$ by 2000 AD, with important pharmaceutical, agricultural and social benefits. This has stimulated advances in predictive technology of a highly mathematical nature, which it would be inappropriate to discuss here. Instead, emphasis will be placed on the basic physical principles and the kind of calculations involved. A more technical treatment has been presented elsewhere (Platt & Robson, 1983; Robson & Garnier, 1985; Sternberg, 1983).

21.2. TRADITIONAL APPROACHES

As progress continues to be made, there is a 'steady state' of tension between the experimentalist critical of new developments, and the theoretician who sees established approaches as approximate and limited. This section considers in a very general way some predictive tools with which the experimentalists seem fairly happy, to the extent that they sometimes appear within otherwise experimental papers. Later, some theoretical objections and improved methodologies will be discussed.

The *hard sphere* model assumes that atoms can be represented as spheres with a characteristic van der Waals' radius. Here, all conformations are permissible in which no atoms overlap and all conformations in which at least one pair of atoms overlap is excluded. The method is limited because it allows too many conformations and provides no way of assigning them in any kind of order so that one conformation can be chosen as the 'best bet'. It implies that all forces between atoms are neglected, except for the repulsive van der Waals' force which is taken to be of infinite energy if atoms overlap, and zero energy otherwise. However, combined with a qualitative treatment of other, fairly strong, forces, such as hydrogen bonding, it is nevertheless a powerful technique for stereoregular molecules. Thus, Pauling and Corey arrived at the α-helical and β-pleated sheet structure of polypeptide chains, and Watson and Crick arrived at a structure for DNA.

This approach is implied in the use of bench top manipulation of 'space filling' molecular models. It can also be programmed into a computer, following the mathematical formalism below:

$$E(X) = \sum_{i=2}^{n} \sum_{j=1}^{i-1} e_{ij}, \tag{21.1}$$

In equation 21.1 the energy of conformation X is taken as the sum of the pairwise interactions of all atoms in a molecule. Indices i and j relate to a numbering of the atoms sequentially from an arbitrary first to last and e_{ij} depends on the types of atoms i and j. The interaction energy e_{ij} between atoms i and j can be defined in various ways, one choice of which implies the hard sphere model:

$$e_{ij} = \begin{cases} \infty, & \text{if} \quad r_{ij} \leqslant R_{ij} \\ 0, & \text{if} \quad r_{ij} > R_{ij} \end{cases} \tag{21.2}$$

Here e_{ij} is a function of r_{ij}, the distance between the atom centres, and discontinuous, the discontinuity occurring at the distance $r_{ij} = R_{ij}$ where R_{ij} is the sum of the van der Waals' radii of the two atoms i and j. Such radii are held to be dependent only on the types of atoms (aliphatic carbon, ketone oxygen, etc.) and are traditionally derived by the inspection of contact distances in X-ray data from crystals of simple molecules. Generally, e_{ij} is described as a potential function and in this case the set of the R_{ij} constitute the only potential parameters for that function.

It is simple to extend equation 21.2 to handle hydrogen bonding, and then only conformations which (a) contain no steric overlaps and (b) make maximum use of hydrogen bonding, are considered as fully allowed.

The problem with the hard sphere model is that atoms are neither hard nor spherical. Moreover, forces other than those arising from steric over-lap are neglected (indeed, as noted above, it is so advantageous to include hydrogen bonding that the hard sphere model is rarely used in its purest form). Electrostatic forces are also strong, arising between atoms carrying a unit or partial charge (due to the differing electronegativities of atoms in the molecule). Such forces, unfortunately, cannot be represented as simple discontinuous functions of r_{ij}. It is pertinent to note that the hard sphere model works only because the densest region of distribution of electrons around isolated atoms is not radically different to that in a molecular context. However, departures from an approximate spherical shape, hydrogen bonding possibilities and partial charges, all arise from significant changes which occur in placing each atom in the molecular context.

The *soft sphere* model exploits equation 21.1 but redefines e_{ij} as a continuous function of r_{ij}. In principle, the continuous potential functions are little more than analytical forms which approximate the required way in which the energy changes with r_{ij}, and there is no obligation to identify their components as having separate physical significance. However, it is natural, convenient and, to some extent, justified to make the continuous potential function a sum of terms, each of which describes the van der Waals' repulsive energy, or an electrostatic component, or a hydrogen bonding component, and so on. The potential functions used have varied widely but the following is the most general and simplest description compatable with modern usage and notions.

$$e_{ij} = \frac{A_{ij}}{r_{ij}^{a}} - \frac{B_{ij}}{r_{ij}^{b}} + \frac{C_{ij}}{r_{ij}^{c}}, \quad a > b > c \tag{21.3}$$

A_{ij}, B_{ij}, C_{ij} are the potential parameters which are dependent on the two types of interacting atom (aliphatic carbon, ketone oxygen, etc.). The first (A) term is that which is closest to the hard sphere description equation 21.2 and indeed it approaches it if the value of a is increased, raising r_{ij} in this term to a very high power. However, to make it fall rapidly enough towards zero as r_{ij} is increased, the second (B) term is required. Usually, the second (B) term is interpreted as expressing the weak van der Waals' attractive forces due to the mutual induction of transient dipoles on the interacting atoms (London dispersion). This contribution is, however, very small compared to the overall error in the calculation, and the major role of the second term is to introduce greater flexibility to the description of the repulsive force. The third (C) term is interpreted as representing the electrostatic (coulombic) interaction due to the partial charges when localized for convenience at the atom centres. A further hydrogen bonding term has not been included here on the grounds that it is possible to interpret hydrogen bonding as a predominantly electrostatic effect in biological molecules (this cannot be true elsewhere, e.g. in liquid hydrogen fluoride). Thus, both the energy and directionality of a hydrogen bond are held by some to fall naturally out of the use of equation 21.3.

The first two terms (A and B) are generally treated together, which must obviously be the case when the second term is envisaged as simply introducing flexibility to the description of the repulsive force. Let e_{ij}^{vw} be the description of the van der Waals' interaction between atoms (equation 21.3 with the electrostatic (C) term omitted). Methods of assigning values to A_{ij} and B_{ij} for the different types of atom pair are better appreciated if these parameters are envisaged as functions of e_{ij}^0, the minimum value of e_{ij}^{vw}, and r_{ij}^0, the distance at which this minimum occurs. The existence of a minimum is necessitated by the relationship $a > b$, and is supposed to represent the equilibrium situation for a pair of atoms i and j when strong van der Waals' repulsive forces and weak van der Waals' attractive forces balance each other. Note that such a situation will only be by chance an equilibrium one in a molecule, because of all the other interactions between the remaining pairs of atoms. Nonetheless, we may write:

$$A_{ij} = \frac{b}{a-b} e_{ij}^0 (r_{ij}^0)^a \tag{21.4}$$

$$B_{ij} = \frac{a}{a-b} e_{ij}^0 (r_{ij}^0)^b \tag{21.5}$$

Here r_{ij}^0 is the important parameter in the sense that e_{ij}^0 does not vary greatly from -0.4 kJ/mole and qualitative calculations are not markedly sensitive to any reasonable choice. Parameter r_{ij}^0 obviously has something to do with the size of the atoms, and indeed used to be taken as R_{ij}, the sum of the van der Waals' radii. However, the radii observed in crystals are unlikely to represent this equilibrium value, precisely because all the other interactions, including the electrostatic ones, have a say in determining the crystal structure. Generally, taking $r_{ij}^0 = R_{ij}$ would make r_{ij}^0 too small and more sophisticated ways of assigning the parameters are required (see, for example, Hagler *et al.*, 1974). The point is, however, that some interpretation of atomic sizes is the essential and critical feature of the A and B terms of equation 21.3, as it was in the hard sphere model.

The remaining C term is usually parameterized by setting

$$C_{ij} = \frac{q_i \cdot q_j}{\varepsilon} \tag{21.6}$$

where q_i and q_j are the partial charges on atoms i and j and ε is a parameter held to represent the effective dielectric constant. While taking the product of q_i and q_j as the best theoretically founded aspect of the whole function e_{ij}, the use of ε_{ij} is problematic. When determined empirically, as a residual 'fudge factor' to bring calculation and experiment into agreement, it rarely corresponds to the experimental macroscopic dielectric constant of the system. When assigning values of q_i and q_j by quantum mechanical calculation or from experimental studies in the vapour phase, there is little doubt that $\varepsilon = 1$ should be used. Previously, the best that could be done with any very simple calculation is to try all values of ε from one to infinity and to ascertain what effect this has on the result and the conclusions to be drawn from it. Now, various models for the solvent effect replace arbitrary choices of ε.

The parameters a, b and c, are not usually described as potential parameters because they are usually held constant within the calculation. Their choice varies from author to author, and the values of a-b-c are used to define the class of potential function used, e.g. 12-6-1, 9-6-1, 9-6-2. There are good theoretical reasons for choosing $b = 6$ and $c = 1$, although the value of c is sometimes increased in the context of hydrogen bonds or when the calculation is time consuming (since r_{ij}^2 is faster to compute than r_{ij} from atomic coordinates). The value of a is more problematic; in reality the A term is better justified theoretically as an exponential function $A_{ij}\exp(-\mu_{ij}\cdot r_{ij})$. However, for rough calculation the choice is not too critical as long as a is taken as large. An exponential form has been used by several authors, but requires an extra parameter (μ_{ij}) and is slower to compute. Note that the values of parameters A_{ij}, B_{ij}, C_{ij} depend on a, b, c; it is not possible to alter the values of the latter without being prepared to calculate new values for the former.

Finally, it should be noted that both hard and soft sphere models were in the

Table 21.1. Hagler-Huler-Lifson parameters for use with the 9-6-1 potential functions

	A_{ii} Kcal/mole A^9	B_{ii} Kcal/mole A^6	C_i electron units (charge on electron = −1 e.u.)	
H	445	15	+0.11	aliphatic
C	38 900	1230	(chose to neutralize CH_n)	
C′	12 500	335	+0.46	carbonyl
O′	45 800	1410	−0.46	
N	36 900	2020	−0.26	amide
H	0	0	+0.26	

* These parameters are to be used in the equation

$$e_{ij} = \frac{A_{ij}}{r^9} - \frac{B_{ij}}{r^6} + \frac{C_i C_j}{r}$$

where $A_{ij} = (A_{ii})^{0.5}(A_{jj})^{0.5}$ and similarly for B_{ij}. They were derived by Hagler *et al.* (1974) and Hagler and Lifson (1974) from crystals of peptide-like molecules, and shown to be consistent both with experimental data on conformational behaviour (Robson *et al.*, 1979) and *ab initio* quantum mechanical calculations (Hillier & Robson, 1979). They should be used in conjunction with a valence force field (see text). Although they were arguably amongst the best known parameter sets, they are being further refined in several laboratories, including those of Dr. Hagler and the author.

past almost always used in conjunction with the rigid geometry approximation. This is to say that in varying the conformation in order to assess the different conformational energies, only rotation angles (dihedral angles) round single bonds are varied. Bond lengths, bond valence angles and torsion angles round double, triple (or partially double or triple) bonds were usually held constant, to standard values generally obtained by inspection of X-ray data for crystal structures. It is assumed that the standard values thus assigned are equilibrium values and that a great deal of energy is required to distort them. There are two practical reasons for this neglect: (1) it vastly simplifies the problem, since it constrains the number of conformations to be explored and (2) the above calculations would be insufficient for these further interactions which represent energies dependent on covalent effects. However, the most important reason for this approach seems to have been the feeling that variation of the geometry of bond lengths and valence angles from standard values is simply not an important factor.

21.3. DEFICIENCIES IN THE TRADITIONAL APPROACHES

Criticisms to the above traditional approaches are legion, and some aspects are treated later in this review. However, Robson *et al.* (1979) have listed some particularly serious deficiencies, the overcoming of which seems to be possible at the present time, if not on a routine basis.

1. Better potential functions must be obtained, which means that they should be parameterized more objectively and tested more critically.
2. Flexible geometry, e.g. variation of bond lengths and particularly bond angles may need to be considered.
3. Since biological peptides exert their effect in water, and since most experiments involve solvents of some kind, some way of properly treating the solvent should be developed.
4. The above traditional methods predict that the observed conformation should be that of least energy $E(X)$ whereas, in fact, the observed conformation is well known to be that of least free energy $F(X)$. It seems essential as a general predictive method to calculate the free energy contribution, and this means adding in the vibrational entropy contribution.
5. The above traditional methods largely relied on locating the least energy conformation by inspection of the results for many conformations. However, this is not feasible for a complex molecule like a protein. Indeed, even automatic minimization procedures which ought to be efficient for finding the least value of a function like $E(X)$ or $F(X)$ are unsuitable in practice because the conformational energy surface contains many local minima in which the procedure becomes trapped. Thus, there is for complex peptides a multiple minimum problem which must be overcome.

Finally, observations on non-peptide molecules have recently obliged us to extend this list of 'problems to be tackled'. The potential functions usually used are based on interactions between atom centres, which essentially means the positions of the nuclei. However, atoms also have electronic orbitals, such as lone pair orbitals, which place an electronic charge at some distance from the nucleus. In other words, some atoms are far from spherical. This may emerge as a problem which is also important for a proper treatment of peptide-like molecules. This leads to the Orbital Force Field (OFF) method of Platt and Robson (1983).

21.4. THE ARMOURY OF MODERN THEORETICAL TOOLS

The above deficiencies in the traditional approach are overcome by the powerful battery of techniques which theoreticians have provided. The interrelationships between these techniques are shown in Fig. 21.1. Further, their relationships to the underlying principles are indicated. These are described in terms of successive approximations to the fundamental ideas as indicated in the vertical columns on the left of the diagram. The purest theoreticians, like Dirac, seem to have considered that the problems have been essentially solved even when the solutions cannot be applied to answer questions of practical importance in reasonable time. In practice, no single unified procedure exists but one or more must be selected at an appropriate level of approximation. However, the theoreticians continue to make a limited number of studies of a rather exact nature, and the aim of these studies should be to provide better approximations for more routine application. It is always important to distinguish between such 'exact' studies from those suitable as routine tools, though of course, procedures which are very expensive today are likely to become routine tools of tomorrow. Much depends on future progress in computing machinery.

21.4.1. Quantum mechanical methods

The use of a quantum mechanical approach implies exploitation of very exact and fundamental ideas. Rightly or wrongly, quantum mechanical methods of outstanding quality have occasionally been considered as delivering results which have the same status as experimental data, though unfortunately only for simple situations (e.g. *in vitro*) of minimal biological interest. Whether or not this optimism is justified, they are certainly the only alternative to the development or testing of the more approximate approaches when experimental data is lacking or poor.

The topmost level of Fig. 21.1 suggests that all properties of a system can be calculated from the Hamiltonian equation.

$$H\Psi = E\Psi \tag{21.7}$$

Where Ψ is the wave function, E the total energy of the system, and H is the

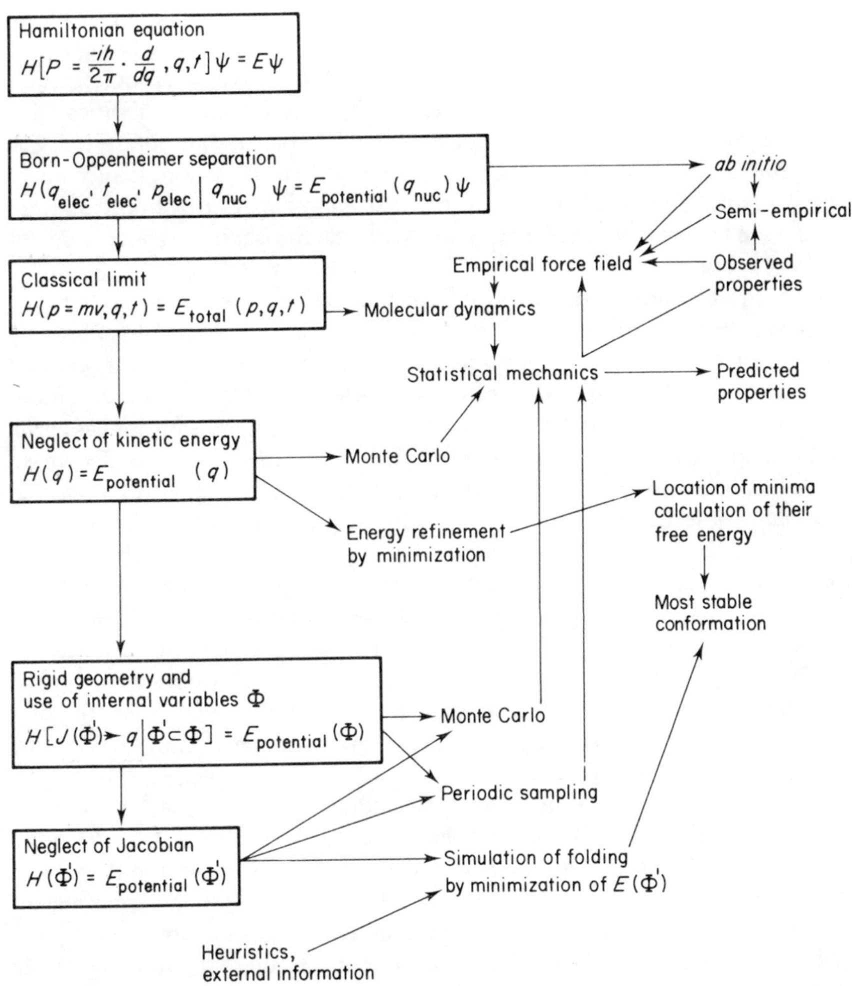

Figure 21.1. Some relationships between concepts discussed in the text.

Hamiltonian operator. The wave function Ψ gives, on solution of the equation, as Ψ^2 the probabilities of finding particles at positions q at moments t. E on solution gives the observable energy of the system. It is the Hamiltonian H which is the input to the problem, and the principles of its use are fairly easy to grasp. Essentially it is an operator (a recipe, like the instruction to differentiate, for converting functions to other functions). To formulate it, one 'simply' writes the classical expression for the energy of the system in terms of the positions q and momenta p of the constituent particles at time t, and then substitutes momenta p

by the operator $(-ih/2\pi)(d/dq)$. Thus, the Hamiltonian is an operator precisely because it is an expression which contains these operators. As may be expected, there are many solutions to the Hamiltonian equation. However, only certain eigen values satisfy physical realism. Those that do are all possibilities with calculable probabilities. The big problem is that any particle i in the system not only affects the behaviour of all the others, but in turn its behaviour is affected by them. To handle this mathematically, the first approximation is to introduce a *perturbation theory* as was earlier applied to handle the mutual purturbations of planetary orbits.

However, a big step of practical significance to conformational work is the next, in which the 'Born-Oppenheimer Separation Principle' is invoked. This effectively means that it becomes meaningful to speak of the energy of a system associated with a conceptually fixed configuration of nuclei. The positions of the nuclei represent the conformation of the molecule, with which we can now assign a specific conformational energy. The calculation of this energy is shown in Fig. 21.1 as being possible via *ab initio* calculation. In practice, however, true *ab initio* calculations can be carried out only for very simple chemical systems. To approximate the energy of any conformation quantum mechanically one may proceed in a variety of ways, each themselves implying different levels of approximation and various degrees of cheapness. We can consider these as still being at the same major level of approximation, because all of these are at least in some sense quantum mechanical or analytical approximations to quantum mechanical results. Further, they all concentrate on evaluating the energy of a single conformation as permitted by the Born–Oppenheimer separation. Thus, we may digress sideways in Fig. 21.1.

In order to treat more complex cases (including simple peptides) in the most exact way possible, two further principles are invoked. One is known as the principle of the *Linear Combination of Atomic Orbitals.* It implies that the atomic orbitals are one possible, but unlikely, solution to the situation in a molecule, so that a more realistic if more complex molecular solution can be found by adding together the atomic orbital contributions, each contributing a different weighting to the sum via a coefficient. The other principle, the *variation principle*, rests on the premise that all situations are possible, but that the most probable is the situation of least energy. Thus the coefficients and possibly the form of the atomic orbitals can be varied in a search for the situation of least energy. The set of atomic orbitals are input in the form of parameters for three-dimensional Gaussian functions or the rather similar Slater-type functions. Typically, several added functions are used to represent each atomic orbital. The more that are used, the better, since the computer has more flexibility allowed it in order to find the true situation of least energy. The set of functions for the orbital description constitute the *basis set*, and basis sets are generally classed as being of two types, *minimal* and *extended.* Extension of a basis set implies introducing more orbital functions, giving greater flexibility with which the computer can vary the description of the system and lower final energies. Any calculation giving a lower final energy than another will usually imply a more realistic result. Given enough flexibility, the calculation would be completely independent of the starting data, and thus '*ab initio*' in the proper sense. In practice, this would be prohibitively time consuming, and lead to mathematical difficulties. As it is, a good quality calculation may take about one hour for one conformation of a 20–30 atom molecule. Nonetheless, the choice of basis set merely implies a certain level of approximation, and in this sense the method is considered parameter-free, and still worthy of the name '*ab initio*'.

The time-consuming step is the calculation of certain integrals. A semi-empirical calculation, though essentially quantum mechanical, saves time by introducing *ad hoc* adjustable parameters. Such calculations are cheaper but inevitably less credible. The various semi-impirical approaches are given various names, usually in the form of initials (EHT 'Extended Huckle Theory', CNDO 'Complete Neglect of Differential Overlap', PCILO 'Perturbative Configuration by Interaction of Localized Orbitals'). PCILO is ingenious by retaining information about excited states, normally an unused spin-off from the calculation, in order to calculate the contribution from the dispersion (van der Waals' attractive) contribution. Admittedly, this would be too time consuming at an *ab initio* level, and seems useful. However, it may be recalled that this contribution to the conformational energy is very weak (say, 0.4 KJ/mole pair of touching atoms), and is at least almost one order of magnitude smaller than the overall errors anyway.

Programmes for these calculations are available as standard packages. Typical *ab initio* packages are GAUSSIAN 70 and ATMOL/3, while the semi-empirical packages may go under the name (e.g. PCILO) for the technique used. Typically, one provides the Cartesian coordinates of the nuclei, though some packages allow the coordinates to be calculated in a prior step from a bond geometry description, from bond lengths, valence angles and torsion angles. To seek the least conformational energy one must therefore carry out several calculations, one for each slightly different conformation. The result is a representation of the potential surface in tabular or graphic form, i.e. of the way the energy changes as a function of conformation. Some packages, however, have facilities for automatically carrying out several conformational calculations in a way most pertinent to finding the point of least energy in the potential surface. Normally, however, this adjustment of conformation is restricted to bond lengths and valence angles, and is described as *optimizing the geometry*. There may still be many conformations as defined by the torsion angles.

A further approximation in the area of simplifying the calculation of the potential surface is to approximate quantum mechanical results by simple analytical functions, i.e. to use potential functions, as outlined in section 21.2. Traditionally the parameters for these have come, however, from experimental data.

21.4.2. Molecular dynamics

The Born–Oppenheimer separation principle allowed calculation of the quantum mechanical energies of single conformations. The real significance of this principle is that it makes possible separate considerations of electronic and nuclear motion, and hence of the kinetic energy contributions of electrons and nuclei. However, this means that carrying out calculations on individual conformations, we have as yet neglected the kinetic energy contribution due to nuclear motion. How can this be added back in?

In principle, quantum mechanical calculations can consider changes in a molecular system with time, but such an all-out quantum mechanical approach is intractable in practice. The practical replacement of the kinetic energy depends on a further approximation in which the system is taken to the classical limit, pretending that the nuclei behave as macroscopic bodies. This implies that the momenta of the nuclei are not written as $p = (-ih/2\pi)(d/dq)$, but as $p = mv$, the product of mass and velocity. Then, the Hamiltonian operator is no longer operating on the wave function, but is itself a function of which the value *is* the energy of the system. It becomes the *Classical Hamiltonian*. Energies are no longer quantized, there are no quantum mechanical uncertainties, and the nuclei behave according to Newton's laws of motion. In reality, nuclear motion, like electronic motion, is quantized, though hopefully neglect of this may only be serious for vibrations of the chemical bonds. Again, in principle, it would seem possible to calculate the potential energy of each conformation quantum mechanically even if the resulting nuclear motion is now treated classically, but this is still far too time consuming. In practice, quantum mechanics makes no appearance whatsoever, and the potential energies of each conformation are evaluated from empirical potential functions. Nonetheless, the quantum mechanical approach is often used to develop the analytical functions.

The essential feature of a molecular dynamics calculation is that the potential energy calculation, using analytical functions, immediately allows the force on each particle to be calculated as the derivative of the energy as a function of position. Knowing the masses of the nuclei, Newton's law (force = mass × acceleration) allows the acceleration to be calculated, and hence the position, momentum and kinetic energy can be ascertained after a short specified time interval. Unlike the traditional method of attempting to find the most favourable conformation, i.e. that of least energy, the user is no longer free to sample conformations and assess their energies at will; rather, an algorithm dependent on Newton's laws determines which conformations must be sampled next. The elegant simplicity of Newton's laws is reflected in the simplicity of the algorithms used, though practical considerations of speed of computation and accuracy have resulted in the development of several different algorithms. One, due to Beeman (1976), is particulary popular because of its speed, robustness and simplicity. Apart from specifying an initial conformation and the potential functions to be used, the algorithm demands that a short time interval for the numerical solutions to the equations, and initial velocities of the particles, be specified. Too large a time interval will save computing time, but leads to non-conservation of total (potential plus kinetic) energy. Velocities must be specified initially because the average kinetic energy of the system determines the temperature. It makes no great sense to start the calculation at absolute zero. On the other hand, the typical high energy of an arbitrary starting conformation will naturally be partly converted to kinetic energy as the system 'relaxes', or 'anneals'. Thus, rather than end up doing a calculation on a biological system at thousands of K°, the velocities are rescaled at intervals to maintain a biological temperature. This implies non-conservation of energy, but is justified in that no system of interest will really be isolated, but in equilibrium with its environment and capable of losing excess thermal energy to it.

Despite neglect of quantum mechanical effects, molecular dynamics is still enormously time consuming. Computing time of many minutes (depending on hardware!) may be required to calculate the time course of events over real time periods of less than 10^{-10} seconds. Since a protein folds up in roughly the order of one second, simulation of protein folding would require hundreds of years of computer time, which is hardly the basis for a cheap predictive method. However, the technique has been applied to study the vibrational process of the native state of a small protein (McCammon *et al.*, 1977) and the behaviour of a simple peptide in aqueous solution (Rossky & Rahman, 1976).

It could be argued that since the favoured structure of a molecule is the equilibrium one, then a description of how it achieved that state, kinetically speaking, is not essential for a prediction. However, molecular dynamics does provide one rigorous method of calculating the free energy contribution, which depends on molecular motion in the vicinity of a minimum in the potential energy surface (the potential energy as a function of conformation). Further, it is not absolutely clear that the native structure corresponds to the global, or deepest energy minimum, but merely to that which it can reach in reasonable (biological) time. If this is so, then following the kinetic history of the molecule may be precisely what is needed in order to predict what the native structure should be.

21.4.3. Monte Carlo techniques

Providing that the interest is in equilibrium situations only, and not in the kinetic aspects of attaining that equilibrium, it is the custom to neglect the kinetic energy component except to the extent that this is inherent in kT.

The advantage of neglecting the kinetic aspects is that the nuclei cannot be considered as possessing inertia, and thus we are free to sample the conformational energies in any way we like. In some sense, the technique which otherwise most resembles molecular dynamics is *Monte Carlo*. Certainly, the form of the input data, and the problems of storing and analyzing the results, are similar, and the same problems tend to be treated. While by neglecting the kinetic energy the Monte Carlo approach is more approximate, it is often cheaper. The Monte Carlo approach is characterized by sampling in a particular way, and this is essentially at random. The basic principle is that of statistics: one samples at random, being careful not to introduce any bias, and keeps sampling till the results converge within a sufficient degree of accuracy. In this case, the results are anything we wish to calculate, such as the free energy.

Procedures are available for generating random numbers which are used to assign random conformations, usually as a small random perturbation of the previous conformation. The random numbers are, in fact, only 'pseudorandom'. The random number algorithm requires an input number, or 'seed', and generates a new number which becomes the input for the next call to the algorithm. Thus, unless the algorithm is deliberately 'jostled' by linking its behaviour to the computer clock or junk numbers in the computer memory, the same seed will always produce the same results. This may be desirable, since the study is then reproducible. This reproducibility does not necessarily mean that the numbers generated have any more significant relationship to each other than truly random ones. However, this is a matter to be tested, and one obvious danger is that if the random number algorithm produces the original seed as output (and subsequent input) at any stage, the whole sequence of numbers will be repeated. The algorithm and seed must be chosen so that the cycle period is very long. Otherwise, even if the programme is so constructed as to minimize the more obvious consequences of this, there may still be a subtle bias in the results.

Having stated that a bias must be avoided at all costs, it is confusing to find that almost all Monte Carlo procedures rely on some kind of biasing in order to be efficient. However, the point, and elegance, of these algorithms is that they do not bias the results: the bias is cancelled or later removed in some way. The reason for including a judicious bias is that without it too many high energy conformations of no significance would be generated: it is necessary to enrich the population of low energy conformations generated. These biasing techniques are the real meat of the Monte Carlo method; without them the approach would not only be trivially obvious, but impractical. Several different techniques give different names to the Monte Carlo approach, e.g. *polymer unit biasing*, *Metropolis*, and *Force Bias*. The Metropolis technique is so popular that it is often (erroneously) taken as synonomous with Monte Carlo, and will be considered here. However, the polymer unit biasing technique should be mentioned as being particularly suited for studying random coil behaviour of polypeptides (Permilat & Hermans, 1973; Premilat & Maigret, 1976). This is because it enriches those conformations most prevalent in a fully denatured random coil homopolypeptide.

The essence of the Metropolis algorithm (Metropolis *et al.*, 1953) is as follows. Certain coordinates of the system are displaced over a preset range from their starting position. The potential energy of the new conformation is calculated, and its difference with respect to the potential energy of the previous conformation is recorded a ΔE. Note that 'conformation' here need not imply the conformation of one molecule, but the configuration of, for example, a system including peptides and water molecules. If ΔE is less than zero, the new conformation is accepted. If ΔE is greater than zero, then a random number is generated on the interval 0 to 1, and $e^{-\Delta E/kT}$ is compared with this random number. If it exceeds it, the new conformation is accepted. Otherwise, it is rejected and the previous position is retained as the current position.

An important feature is the way in which the Metropolis procedure allows calculation of average properties (averaged, that is, over all conformers). To do this, one simply adds together the values of the property associated with each conformer generated, and divides this sum by the total number of generated conformers. Here, by 'generated' is meant 'successfully generated'. Trial conformers which were rejected are not counted, though the previous conformer, to which the algorithm returns when a trial is rejected, is counted again. This method of averaging is not the usual one of statistical mechanics (see below), and is valid only in the context of the Metropolis procedure of which it is an intrinsic part. The distinction is that one usually calculates explicitly a *weighted* average, the weighting factor being $e^{-E/kT}$. In the Metropolis procedure, however, the probability of successfully generating conformers is proportional to this weighting factor and thus the correctly weighted average is cleverly obtained implicitly, and efficiently. The major contribution of Metropolis was to demonstrate this and concomitantly the *ergodicity* of the algorithm (which essentially means that, in principle, all significant conformational possibilities can be explored).

This procedure is particularly suited to the study of aqueous solutions of peptides or protein. The number of conformational variables is the time-consuming factor, not the size of the system *per se*. Thus, as long as the protein molecule is not allowed to change conformation, its effect on the surrounding solvent is a moderately cheap computer calculation.

An interesting test system is a hydrated crystal, since the position of at least some of the more rigidly held water molecules can be compared with the X-ray result. Thus, the studies by Hagler and Moult (1978) and Hagler *et al.* (1980a) of the lysozyme crystal and the hydrated crystal of a cyclic dipeptide is of considerable importance. These studies also illustrate how to avoid the problem of surface effects. One can happily treat 350–400 water molecules, but even this is a small number and thus implies a small droplet of solution with a vacuum at its surface. This surface is actually avoided by calculating the behaviour in a unit cell of the hydrated crystal, which is considered to have the other unit cell interfacing with it and thus completely filling space out to infinity. The behaviour of the water in one unit cell is copied by the behaviour of an equivalent water in all the others, and a water molecule leaving the bottom left of every cell will re-enter at the top right of every cell. In this way, one imposes 'periodic boundary conditions'.

Although the periodic boundary conditions in that study exploited the natural unit cell of the crystal, they can also be employed in the study of solutions. It is merely necessary to chose the size of the unit cell and the distances up to which energies are calculated (the 'cut off' distance) such that no long-range order effects are manifest. Hagler *et al.* (1980b) carried out a study of *N*-acetyl alanyl *N*'-methylamide in aqueous solution, on this basis. The molecule is of interest as the classic choice of a peptide containing chemical features encountered in a polypeptide. This solute molecule has sufficiently few rotatable bonds (5) that one can try different conformations, and thus consider not just the effect of the solute on water, but the effect of the water on the preferred conformation of the solute. Thus it was possible to arrive at a number of assumptions which can be exploited in cheaper calculations of conformational behaviour in solution. These are that:

1. as may be expected, water molecules are strongly bound to peptide hydrogen bonding groups, and are likely to have some steric effect on conformational possibilities;
2. the effect on the water behaviour is largely confined to the innermost hydration shell of the solute;
3. polar solvents tend to stabilize the conformations of the solute which have the highest dipole moments.

In this case, however, an internally hydrogen bonded conformation with a low dipole moment was still preferred because of steric factors within the solute molecule. Incidentally, these conclusions seem to confirm the validity of three well-known models for treating solvent effects approximately, namely the *supermolecule, solvation shell* and *reaction field* models, respectively, each of which presumed the corresponding finding above, though individually treating only one aspect of the solvation problem. This well illustrates the value of more 'exact' calculation in exploring the possibility of approximate methods. Conversely, such an 'exact' method cannot be used routinely, as a direct attack on the problem. For example, to obtain the free energy of a simple peptide-solvent system can take up to 40 hours of computing time.

21.4.4. Statistical mechanics

The purpose of this discipline (the theoretical counterpart of thermodynamics) is to calculate average properties of a system which can adopt many configurations, as all molecules do except at absolute zero. Molecular dynamics, Monte Carlo, and a simple tabulation of energies for different conformations, are all essentially sampling procedures, and the application of statistical mechanics can be regarded as a second stage in which the data obtained from this sampling is analysed. Note, however, that in some cases, e.g. the Metropolis procedure, the methods of sampling and statistical analysis are inseparable.

In essence, the method of obtaining statistical mechanical averages is identical to any statistical method. If each conformation X is associated with a value of property $V(X)$, and has a probability $P(X)$, then the observed property is the 'expected' or average $\bar{V}$:

$$\bar{V} = \int P(X).V(X).dX \tag{21.8}$$

The special feature of statistical mechanics, however, is that the probability $P(X)$ is proportional to the Boltzmann weighting factor $e^{-H/kT}$

$$P(X) = Q^{-1}.e^{-H/kT} \tag{21.9}$$

Here H is the total energy of the system as calculated through the classical Hamiltonian and should, strictly, contain the kinetic energy contribution

$$H = H(p,q) = KE(p) + E(q) \tag{21.10}$$

where KE is the kinetic energy dependent on momenta p, and E is the potential energy dependent on nuclear positions q and hence on conformation X. Q is the *partition function*, from which in principle, all thermodynamic properties can be calculated. It relates to the free energy:

$$F = -kT\ln \alpha\iint e^{-H/kT}dpdq = -kT\ln \alpha Q \tag{21.11}$$

Factor α depends on the numbers of indistinguishable and therefore interchangeable nuclei (see Landau & Lifshitz, 1958, pp. 90–91) and is a common cause of misunderstanding. Fortunately, it can be neglected whenever it is only relative free energies of the same molecule which are of interest.

A fundamental difference between molecular dynamics and the Monte Carlo approach is, of course, that the latter neglects $KE(p)$, the kinetic energy contribution. Neglect of kinetic energy leaves us with the *configurational partition* function Z. This is valid at equilibrium.

The value of the property $V(X)$ can be anything which is conformation dependent, such as the NMR coupling constant associated with a conformation X. It may also be a vector, such as the circular dichroism spectrum at specified sample values of wavelength, or a matrix, as in Flory's theory for random coil behaviour of polymeric chains in which $\bar{V}$ would represent the average of the matrix which determines the position of each polymeric unit relative to its predecessor. Quantities of thermodynamic interest are the enthalpy, which is obtained by choosing $V = E$ (though here pressure volume changes are neglected), and the entropy, which is obtained by setting $V = k\ln P(X)$, the negative information content of the event of adopting conformation X. Further, specific heat is related to the standard deviation of the energy, while temperature is related to the average kinetic energy.

21.4.5. Minimization

From one point of view, minimization is a sampling procedure as are Molecular dynamics and the Monte Carlo approach. However, it does not by itself make any reference to statistical mechanics precisely because it concentrates on the calculation not of average properties, but of most probable conformation. This it does by minimizing the energy as a function of conformation, and discovering the most probable conformation as that of least energy.

Strictly, of course, it should be the free energy which is minimized as a function of conformation. What is done in practice is to make the reasonable assumption that a minimum in the potential surface must correspond to a minimum in the free energy surface. Then, one can make the further, and much less reasonable, assumption that the minima are of similar shape and width, so that the minimum of least energy also corresponds to that of least free energy. When there is no justification for this (and there rarely is), the latter assumption should not be made and the free energies at the local minima should be calculated and compared. The several methods of calculating free energy all necessarily make the prediction that a broad minimum of high potential energy can still have a lower free energy than a narrow minimum of low potential energy, providing the potential energy difference between them is not too great. The methods include the obvious one of integrating in the vicinity of each minimum to evaluate the local partition function Z (see above), calculation of free energy as a function of second derivatives and calculation of free energy based on vibrational frequencies (see for example, Hagler *et al.*, 1979b). Thus, statistical mechanical reasoning can and should be brought in at a second stage, to analyse the minima located by minimization of the potential surface alone. However, this further step is a fairly recent and rare innovation in peptide conformational analysis. It should also be said that it is theoretically possible to evaluate the free energy from the potential surface at all points, not just in the vicinity of the minima, and that a judiciously programmed procedure could provide a cheap method of minimizing free energy from the outset.

Minimization of the potential energy as a function of conformation is widely used. It is the basis of the computer simulations of protein folding, and of the energy refinement of protein structures deduced by X-ray analysis. A variety of programme packages are available for doing the minimization part of the calculation; most often these are general packages for which the user must write his own function for calculating energy from conformation. The variety of packages reflects the variety of mathematical approaches to determination of the minimum value of a function. However, these basically fall into two types.

The first type is represented by *gradient* methods. Although the general principle is to locate that point for which the first derivatives with respect to conformation are zero, and the second derivatives positive, some procedures do not require derivatives to be provided. The methods can be envisaged as members of a family tree of minimizers, one characterizing feature of each member being the way in which $E_{i+1} - E_i$ (where i and $i+1$ are successive points in conformational space) is expanded as a series of derivative terms, and used to select conformation X_{i+1} from X_i and the derivatives at point i. The *steepest descent* method depends on one of the most approximate expansions, but has still proved popular in conformational work because of its convergence properties. Convergence is, however, slow. The Newton–Raphson procedure is a more sophisticated expansion requiring the evaluation of the second derivatives; there is a recent renewed interest in second derivative minimizers. The Fletcher–Reeves methods avoid explicit evaluation of the second derivatives, gaining information about curvature of the energy surface by assuming a quadratic form. Procedures due to Davidon (1968) and Fletcher and Powell (1963) take advantage of features in both Newton–Raphson and Fletcher–Reeves methods. These and subsequent techniques have a considerable degree of efficiency and sophistication, but all suffer from the fact that the procedures stop at the first minimum located. Indeed, to exit from a minimum would be contrary to their basic philosophy.

The second type, represented by *search* methods, can also be fairly sophisticated as exemplified by the SIMPLEX procedure. It is well suited to minimizing energy as a function of all band rotations rather than, x, y, z coordinates. A 'simplex' ($n+1$ points in n-dimensional space) samples the function value at its points and on this basis decides where to move its highest value point in an attempt to find a point of even lower energy. This involves operations on the SIMPLEX such as reflexion, then expansion or contraction of the probe point through the centroid of the remaining points. It would not be too misleading to speak of the alternative gradient methods as resembling a ball rolling down hill, while the SIMPLEX method resembles a blind octopus groping with its tentacles for its hole in a coral reef. While this understates the sophistication of the SIMPLEX method, it conveys something of its robustness. It is not confounded by discontinuities in the derivatives or in the energy function itself, and is indeed ideally suited to handling artificially imposed constraints, either on the conformational variables or on the energy or some other function of the conformational variables. Further, it has a chance of escaping from shallow minima if a probe point encounters a region of even lower energy. The chances of this happening depend on the depth and shape of the minimum encountered and the width of the energy wall, and also on control parameters for the SIMPLEX procedure. By allowing a new point to be generated at a considerable distance from the centroid of the SIMPLEX, the chances of encountering new minima are obviously increased. In practice, modifications of the original SIMPLEX procedure may be required to treat particular problems more efficiently (e.g. COMPLEX). The price of the considerable robustness of the SIMPLEX approach is its considerable slowness.

On the one hand, therefore, gradient methods locate local minima rapidly, but are less robust and run into trouble if the function has certain properties (such as the possession of discontinuities). On the other hand, the SIMPLEX approach can attempt to handle most functions and may permit escape from local minima thus increasing the chances of locating the global minimum. There is much to be said for combining the two for peptide-protein work. Thus, Robson and Osguthorpe (1979) employed a double minimization in which the gradient method would switch to a SIMPLEX mode if it became trapped in minima or ran into trouble, and conversely the SIMPLEX would switch back to a gradient method when the vicinity of a new minimum was located. The switching was entirely automatic, being governed by diagnostic variables generated by each minimizer and by the current degree of convergence. This avoided the need to implement special procedures for escape from local minima (see, for example, the 'thermalization' method of Levitt & Warshel, 1975).

21.4.6. Rigid geometry and neglect of the Jacobian

Although Monte Carlo calculation and minimization can be carried out using the Cartesian coordinates of the atoms, this is usually done only in special cases (e.g. in energy refinement). Usually, rigid geometry is assumed, so that only rotations round single bonds are permitted to vary, and other variables (bond lengths, valence angles, etc.) are fixed at standard values. Obviously this saves much time, since the search of the potential surface is constrained to the standard geometry values. Further, reducing the number of variables to be handled has considerable practical advantage.

The problem with this approach is that the new system of internal coordinates used is no longer conjugate to the real Cartesian coordinates of the molecule. This simply means that sampling bond rotation angles (at random or at constant intervals) does not produce an even sampling of the real conformational space defined by the Cartesian variables of all the atoms. Mathematically, the partial derivatives of the Cartesian variables with respect to the internal variables define the elements of a (Jacobian) determinant which can be regarded as the correction factor to be introduced into the calculation. In practice, the Jacobian is almost always neglected, except when its formulation is particularly simple (as in the case of rotating a water molecule in a Monte Carlo simulation of solution behaviour). Probably, its neglect is not too serious for a molecule with a deep potential energy minimum and particularly when we are interested only in the minimum values (as located by a minimization procedure).

Rigid geometry is not usually employed in a molecular dynamics calculation, where the variables are the Cartesian coordinates of the atoms. To include it at all would seem inconsistent with the basis of the method, although a constraint on bond lengths, etc. would obviously save computer time. In fact, it is possible to include constraints, although this is by no means trivial. Such constraints would, nonetheless, detract from a real description of the dynamic behaviour of the system, though it could be argued that these need not be much of a loss in relation to bond vibrations since quantization is neglected anyway.

21.4.7. Periodic sampling and mapping

Periodic sampling is a sampling procedure which almost always implies rigid geometry, the sampling being made at regular intervals of the bond rotation angles. The energy at each point is calculated and stored in an array. The array produced is often two dimensional because one usually wishes to obtain certain information by inspection. To facilitate this, the tabular representation is usually transformed into a map, i.e. a contoured representation of the energy points in the array. When the molecule of interest contains more than two rotatable angles, those considered as having less importance are usually 'minimized'. More precisely, the energy of the molecule at each point in the space defined by the two mapped angles is minimized as a function of the remaining, unmapped angles. However, it is also reasonable to calculate their statistical mechanical average energy, as opposed to the minimum energy.

Since the map implies a fairly representative account of the potential surface (subject to the resolution implied by the interval between sample points), it can be used for a variety of purposes. One may rapidly identify the conformation of least and of most energy, and the forms of the minima. The array of energies prior to contouring can also be used to calculate statistical mechanical averages such as the configurational entropy, average CD spectrum, and so on. The contouring itself implies, however, interpolation between the energy points which may be useful for obtaining average properties when the sample points are sparse. The averaging carried out should, but usually does not, include the Jacobian (see above). When one is purely interested in calculation, not inspection, it is not of course, necessary that the array be one or two dimensional. The array can also be used to locate the minimum energy value automatically, in which case this is often described as minimization by an 'exhaustive periodic scan' (since the 'whole' of the energy surface is estimated). When the array corresponds to the potential surface of one unit in a random coil polymer, hydrodynamic and light scattering properties of the whole polymer molecule can be estimated according to assumptions made in the satistical theory of polymer behaviour.

Periodic sampling is not the only method of producing a contoured map. Minimization procedures can be adapted to interpolate and extrapolate the potential surface in the region through which the minimization procedure passes.

21.4.8. Heuristic methods

Generally, a 'heuristic' is any trick which facilitates the search for the solution to a complex problem, usually by reducing the amount of time which would otherwise be required to reach that solution. Time saving tricks have been widely exploited in attempts to predict protein conformation. In order that the definition of a heuristic is not so broad as to include any approximation, or the Metropolis or minimization techniques, attention is here directed towards procedures which (1) do not demand modification of the potential surface, and (2) imply constraints, or bias to the search of the conformational space according to parameters which are provided *a priori* and not according to the form of the potential surface as discovered at run time.

The principal difficulty in predicting protein structure is the number of local minima and the application of heuristic techniques is thus primarily directed towards tackling the multiple minimum problem. Constraint methods are intended to eliminate the sampling of regions of conformational space in which the global minimum is unlikely to occur *a priori*, or to constrain the sampling to regions of conformational space which are believed to contain an acceptable approximation to the true solution.

In the pioneering attempt of Levitt and Warshel (1975) to simulate pancreatic trypsin inhibitor folding, a principal point was the use of a simplified protein representation. This was in order to smooth out the details of the potential surface, including the local minima. However, this aspect could be considered as representing a particular choice of potential functions, and may certainly be considered as implying a further (intentionally crude) level of approximation. Thus, this is not a heuristic in the specific sense meant here, though to opt for a level of such drastic approximation does imply a specific philosophy and approach to the folding problem which is a 'technique' in its own right. The present author feels, however, that gross simplification of the problem should only be a last resort, when more exact representations are not possible. The danger in modifying the potential surface needlessly is of course that the result may have little relationship to the energy surface of the real molecule.

Robson and Osguthorpe (1979) used an alternative philosophy which is in principle applicable to *exact* representation of a protein. It is well known that the true variables of a protein are not independent (at the level of single residues). The value of one variable (typically a bond rotation angle) predetermines the possible values allowed to certain others. Thus, there is no need to waste time trying all combinations from the outset, since many combinations are known to be highly unlikely. The trick is to chose effective variables which keep the true variables away from these unlikely combinations.

This procedure may be formalized as follows. First, consider the minimization, in the usual manner, of the conformational energy as a function of bond rotation variables $v_1, v_2, \ldots$

$$E = E(v_1, v_2, \ldots, v_m) \tag{21.12}$$

Next, consider that subsets of these variables are chosen, each subset grouping those variables to which it is required to assign interdependent behaviour. For example, variables v_{i-m} through to v_{i+m} might be one of the subsets selected. A coupling function is then written for the variables in each such subset, the value of the function for the (j)th subset being w_j:

$$w_j = f(v_{i-m}, \ldots v_i, \ldots v_{i+m}) \tag{21.13}$$

This coupling function is chosen to represent a specific curve (trajectory) in the portion of conformational space defined by real variables $v_{i-m}, \ldots v_i, \ldots v_{i+m}$. The function is also chosen such that w_j represents the distance round that trajectory. Thus, specifying a value of w_j defines a point in the conformational space $v_{i-m}, \ldots v_i, \ldots v_{i+m}$, but necessarily only a point which lies on the trajectory. The precise vector $P = (v_{i-m}, \ldots v_i, \ldots v_{i+m})$ which represents the point defined by the value of w_j is calculated through the inverse of f

$$(v_{i-m}, \ldots v_i, \ldots v_{i+m}) = f^{-1}(w_j) \tag{21.14}$$

Carrying out this operation for each and every subset of variables allows all variables v to be evaluated and, hence, the energy to be calculated via equation 21.12. The net effect, albeit indirect, is that the energy can be minimized as a function of the smaller number of w.

$$E = E(w_1, w_2, \ldots w_N) \tag{21.15}$$

Generally, f^{-1} is any algorithm which will interpret an effective variable w in terms of several real variables v. It is useful, though not essential, that the function f of which it is the inverse has a unique value. A conformation defined in terms of variables w can always be made as similar as possible to one defined in terms of variables v by a fitting procedure, in which the conformation is varied as a function of variables w in order to minimize some measure of difference between the conformations.

How are functions f^{-1} chosen in practice? At present, studies are being carried out either with very simple functions defining very simple trajectories, or by approximate interpolation of more complicated trajectories. Robson and Osguthorpe (1979) chose f^{-1} to be the equation for a tilted ellipse in the conformational space of the two backbone dihedral angles (Φ and Ψ) of each residue. The 'distance' D round the ellipse (corresponding to the value of w) was actually expressed as the angle between two vectors with origins at the ellipse centre, one vector indicating the current point on the ellipse circumference and one representing a fixed reference vector.

The consequence of this was that it was possible, by judicious choice of parameters, to interpret $D = 0°$, $D = 360°$ as a roughly extended, pleated sheet type backbone conformation, $D = +90°$ as approximately classical right hand α-helix, $D = -90°$ as approximately left-hand α-helix, and so on. The time saving due to the implementation of this particular heuristic, was such that the folding of pancreatic trypsin inhibitor protein could be simulated while retaining a more realistic and much less controversial backbone representation than that of Levitt and Warshel (1975).

It is apparent that such methods are constraint methods because all conformations explored must lie along such trajectories. In the conformational space of the backbone of each residue, care must be therefore taken that the trajectory passes through all important conformations (e.g. α-helical conformations) and close to all possible ones. In this sense, choosing an elliptical form seems a rather severe constraint, even though its tilt can be selected so that pleated sheet, helical and reverse turn conformations are allowed (Robson & Osguthorpe, 1979). The danger is, of course, that the conformation of a residue in a protein for which a tertiary structure prediction is to be made may lie off the elliptical path and thus cannot, in any circumstances, be predicted. In the current state of the art, however, slight errors in residue conformation would be permissible provided that the general course of the polypeptide backbone is duplicated correctly. In practice, it appears to be the case that a serious discrepancy in the conformation of one residue can be cancelled by compensating movements in adjacent residues, so that the general course of the backbone can be fitted. This is shown by imposing the elliptical constraint on the conformation of every residue, and minimizing not the conformational energy of the protein but the root mean squared (rms) deviation of certain crucial atoms (in this case, C_α atoms) from the experimental values. A remarkably low rms deviation of 1.0–1.1 Å has been obtained for several proteins, which is very reasonable when compared with the rms deviation of 4–6 Å typically obtained in a prediction of protein structures by energy minimization.

Even better fits can be obtained by more complex trajectories implied by certain choices of f^{-1}. Interestingly, however, the procedure implied by f^{-1} need not be particularly complicated as far as programming is concerned. By way of a simple example, consider a trajectory as approximated by a series of short straight lines, the $(I+1)$th line round the trajectory being defined by its terminal points Q_I and Q_{I+1}. Let the distance round such a trajectory be again D, such that if $D = 6.3$ then the conformation is three-tenths along the 7th straight line connecting Q_6 and Q_7. More generally, for $D = I + X$ (where I is the whole, X the fractional part of D), so that conformation (X)th of the way along the $(I+1)$th straight line connecting Q_I and Q_{I+1}, the conformation $P_D = (v_{i-m}, \ldots v_i, \ldots v_{i+m}) = f^{-1}(w_j)$ for $w_j = D$ is

$$P_D = Q_I + X \,.\, (Q_{I+1} - Q_I) \tag{21.16}$$

Note that P_D and Q_N, Q_{N+1} are vectors, with as many elements as there are variables v being coupled, and that vectors Q will be prespecified to the computer as a look-up stack of such vectors representing the points to be linked in defining the trajectory. In contrast, P_D is the continuously changing vector representing the trial conformations $P = (v_{i-m}, \ldots v_i \ldots v_{i+m})$ corresponding to distance D along the trajectory. In practice, the trajectory will generally be chosen to be cyclic, so that Q_J generally replaces Q_{I+1} and is selected via a modulus function of the current value of I and the number of straight lines by which the trajectory is approximated. Unlike the ellipse as used by Robson and Osguthorpe (1979), this approach can be extended to couple any number of variables, in any particular way. It suffers from the disadvantage of leading to discontinuous derivatives of energy as a function of the set of distances D, but the resultant difficulties can be overcome by using a SIMPLEX method of minimization. As discussed by Robson and Osguthorpe (1979) this is already a desirable minimization procedure for predicting protein structure, so the existence of discontinuous derivatives need not necessarily be seen as a serious deficit. On the other hand, there are alternative general ways of coding trajectories so that the derivatives of the energy are continuous; these are more complicated and will be discussed elsewhere.

21.5. PREDICTION OF SECONDARY STRUCTURE—A SPECIAL TYPE OF CALCULATION

Predictions considered here are a special case in that they do not, in themselves, usually involve energy calculations. However, they are also arguably heuristic methods (see section 21.4.8), in that they are useful devices for selecting suitable starting conformations for a globular protein prior to energy minimization. Because of the multiple minimum problem this choice of starting conformation is critical. Conversely, a secondary structure prediction cannot hope to give the tertiary structure directly and always requires the subsequent energy calculation. There are two principal reasons for this. The first is that secondary structure prediction algorithms neglect those interactions responsible for maintaining the compact, tertiary structure, and consider only those interactions most relevant to secondary structure formation. Because the secondary structure is still partly determined by the neglected interactions, a perfect prediction is impossible. Second, even a perfect prediction of secondary structure implies, in practice, that each backbone dihedral angle is correctly assigned to a range of angles. A small uncertainty in a dihedral angle can mean, for a long-chain molecule, an enormous uncertainty in the spatial coordinates of many of the atoms and hence in the tertiary structure. Some uses have been found for secondary structure predictions without a subsequent energy calculation and indeed such applications or investigations have been so widespread that the technique deserves special emphasis. Nonetheless, too much must not be expected of a secondary structure prediction method used in isolation, and its limitations must be borne in mind.

If secondary structure prediction methods are not in themselves energy calculations, what are they? With the exception of some which use helix-coil transition data from experiments on artificial polypeptides, the vast majority are statistical in nature. They depend on counting frequencies of events in tables of sequence-conformation relationships for protein of known structure. A crude example of a prediction approach would be one which starts with the observation that proline never occurs within a helical region in that data base (except occasionally at the helix *N*-terminus). Thus, it would not be assigned as being within a helical region in the initial protein conformation from which minimization is to be performed. In practice, one develops algorithms and parameters, which account for the observed sequence-secondary structure relation and which have proven value in predicting the conformations of proteins not in the data base.

Early prediction methods tended to classify each of the twenty amino acid residues as helix formers or helix breakers. The observation that no type of residue consistantly adopted the same conformation was then overcome by simple rules (e.g. that at least four helix forming residues had to occur in a run before helix would be formed). However, there is no reason to suppose that binary assignments of residues as helix formers or breakers reflect the real situation, as demonstrated by helix-coil transition studies of artificial polypeptides. Therefore, Pain and Robson (1970) assigned the twenty amino acids to a scale of helix-forming propensity. In an attempt to develop a predictive procedure, which made minimal assumptions, and rested only on the most objective possible interpretation of the data base, a more rigorous information theory approach was introduced by Robson and Pain (1971). This was to avoid attaching any prejudicial physical significance to the problem and instead to treat the sequence and conformation of proteins in the data base as two messages related by an unknown code. In order to reach objective conclusions, a Bayesian approach was combined with the information theory method. This took into account the fact that all interpretations of finite data have a subjective component, which, having been identified, can be minimized. The formal basis of the procedure, and its applications, are discussed by Robson (1974).

The demonstrable power of the secondary structure predictive methods, and their computational cheapness, have inspired a variety of approaches by many authors. However, emphasis is given here to the information approach. This is justified because the approach does not constitute a specific prediction algorithm, but a general formalism for seeking optimal predictive algorithms. The popular predictive method of Chou and Fasman (1974a, b), for example, can be shown (Robson & Suzuki, 1976) to correspond closely to the information theory approach taken by Robson and Pain (1971) when terms representing interactions between residues are neglected. The aspect of the information theory description which is related to the basis of the Chou and Fasman approach can be used to exemplify the method. This aspect ay be formulated as follows:

$$I(S;R) = \#[f(X,R)] - \#[f(\bar{X},R)] - \#[e(X,R)] + \#[e(\bar{X},R)] \tag{21.17}$$

With the function # [ ] defined as below (equation 21.18), this represents the information which a residue type R (say alanine) carries about its own conformational state S (say α-helical). Here $f(X,R)$ is the observed frequency (number of times) with which R adopts the state $S = X$ in the data base, and $f(\bar{X},R)$ is the frequency with which R adopts some other conformation $S = \bar{X}$. The frequencies $e(X,R)$ and $e(\bar{X},R)$ are the corresponding 'expected frequencies' as defined in the chi-squared test. For example, $e(X,R) = f(X).f(R)/f_{\text{total}}$. Function # [ ] is the 'information content' of these frequencies, defined as:

$$\#[0] = 0$$

$$\#[i] = \sum_i \frac{1}{i} \tag{21.18}$$

Function values for non-integral arguments (e.g. expected frequencies) can be obtained by interpolation.

Robson (1974) expressed i for $i > 1$ in equation 21.18 as the observed frequency minus one, but this depends on a theoretical point which, though widely accepted, is arguably somewhat arbitrary. As the above formulation proposes, the simple value of a frequency can be taken directly as an argument of the function. There are no significant changes to the results. Any addition or subtraction of small values implies differences in the form of our degree of belief about the outcome *before* seeing the data, and the effect of this should be small. Addition or subtraction of larger values would provide a means of introducing, in a formal manner, strong if justifiable prior prejudices.

Any particular application of the information approach depends on which terms are neglected in the expansion of $I(S_j; R_1, \ldots R_n)$. Here j is the residue whose state S is to be predicted, and $R_1, \ldots R_n$ is the whole sequence of the protein whose secondary structure is to be predicted by ultimately considering all values of j. The simplest technique which produces results at least as good as those of other methods when a large number of predictions is made (Garnier *et al.*, 1978), limits this expansion to

$$I(S_j; R_1, \ldots R_n) = \sum_{M=-8}^{M=+8} I(S_j; R_{j+m}) \tag{21.19}$$

Here j is the residue whose conformation is to be predicted, S_j a conformation and R_{j+m} the type of residue (alanine, etc.) m residues away down the sequence. A positive value of m implies a residue in the C-terminal direction of j, and a negative value a residue in the N-terminal direction.

The terms are negligible beyond $M \pm 8$, though this cut-off is somewhat arbitrary. The case $m = 0$ implies the type of residue j itself. This prediction is done for each residue in the sequence from $j = 1$ to $j = n$. This is then repeated for every possible conformational state S which one wishes to consider. For each residue, the conformation S with which the highest information content is associated is the one predicted. In practice, it is useful to first subtract a 'decision constant', dependent only on the conformation S, for the value of the information for each residue in each of its conformational states. This can be envisaged as extra information from circular dichroism studies, and it is not necessary to know the secondary structure contents exactly but only to classify the proteins a helix rich, pleated sheet rich, etc. There is physical justification for this. For example, a protein rich in potential sheet regions will further stabilize pleated sheet formation by a cooperative hydrogen bonding between the strands. More importantly, long helices are more stable than short ones.

The study by Garnier *et al.* (1978) is still fairly typical in that four conformational states (right-hand α-helix, extended chain as potential pleated sheet, reverse turns and periodic structure) are considered. In considering assignment of four types of conformational state, roughly 60% of residues will be correctly predicted as being in, or not being in, that state. This is, however, over many proteins, and a few proteins show marked deviations from this. The procedure is readily coded in about one page of a high level computer language, and for short sequences can even be carried out manually. The procedure was used to assign starting conformation in the folding simulation by minimization carried out by Robson and Osguthorpe (1979).

21.6. GROUNDWORK

So far, the techniques available have been considered but little has been said about the extent to which they are successful in practice. The critical factor limiting success is the degree of realism of the potential surface. All aspects of behaviour (for example, conformation of least energy, kinetic energy, vibrational entropy, solvent effects) are only as good as the calculation of the potential surface from which this behaviour is deduced.

Having a good potential surface means having good potential functions. Each point in the conformational space could alternatively be evaluated by a quantum mechanical technique. A good (*ab initio*) calculation, however, would require of the order of one hour's computer calculation for each conformation of a roughly twenty atom system. Whatever the method by which the energy is calculated, the Monte Carlo approach, for example, may demand the evaluation of the energy of some six or seven million conformations of a several thousand atom system before a reliable (well-converged) free energy is obtained. Obviously, a quantum mechanical approach is out of the question here. Everything depends, then, on obtaining good potential functions.

The question of how one can know that potential functions are good poses a dilemma. In essence, a set of potential functions can be deemed valid if they predict a large variety of experimental properties, which is to say that the calculated values fall in the experimental range. However, using the potential functions obviously requires the use of a particular technique, like the Monte Carlo technique for example, with many implicit approximations. If agreement with experiment is bad, how can we know whether it is the potential functions, rather than the technique with which they are applied, which are at fault?

This dilemma has been emphasized in the laboratories of Hagler and of Robson, and steps taken towards resolving it. Here we review in outline some of the main conclusions of this work, which are described in more detail elsewhere (Hagler & Lifson, 1974; Hagler *et al.*, 1974; Robson *et al.*, 1978; Hillier & Robson, 1979; Robson *et al.*, 1979; Hagler *et al.*, 1980a, b; Robson, Stern & Hagler, 1985).

Other groups have of course, for a long time made major contributions in this area, (see Table 21.2 and, for example, the review by Nemethy and Scheraga, 1977) but it is probably true to say that a concerted attack on assessing the adequacies of the potential functions and procedures employing them has not previously been a major feature. Whereas studying potential functions *and* procedures seems

Table 21.2. Some practical applications of calculations on peptides and proteins other than protein folding simulations.

System studied	Authors
Understanding of peptide group and the *cis-trans* conversion	Hagler *et al.* (1976) Ramachandran & Mitra (1976) Zimmerman & Scheraga (1976)
Understanding of conformational preferences of residues in polypeptides	Lewis *et al.* (1973a, b) Nishikawa *et al.* (1974) Zimmerman & Scheraga (1977) Robson *et al.* (1978) Hillier & Robson (1979) Robson *et al.* (1979)
Understanding of β-bends	Nishikawa *et al.* (1974) Howard *et al.*, (1975) Hurwitz & Hopfinger (1976) Melnikov *et al.* (1976)
Melanotropin H-Pro-L-Leu-Gly-NH_2	Ralston *et al.* (1974)
Thyrotropin Pyroglu-His-Pro-NH_2	Burgess *et al.* (1975)
Enkephalins Tyr-Gly-Gly-Phe-Met and Tyr-Gly-Gly-Phe-Leu	Beddell *et al.* (1977) Isogai *et al.* (1977) Momany (1977) Gorin *et al.* (1978) Smith & Griffith (1978)
Luteinizing-hormone releasing factor (decapeptide)	Momany (1976a, b)
Cyclic peptides	Dygert *et al.* (1975) White & Morrow (1979)
Collagen	Miller & Scheraga (1976) Okuyama *et al.* (1976)
X-ray refinement of globular proteins	Levitt & Lifson (1969) Warme & Scheraga (1974) Swenson *et al.* (1977)
Protein vibration	McCammon *et al.* (1977)
Prediction of protein structure from that of homologous proteins	Warme *et al.* (1974) Swenson *et al.* (1977) Robson & Timms (1980)
Calculation of protein structure by packing known secondary structure features	Ptitsyn & Rashin (1975) Cohen *et al.* (1979) Sternberg & Cohen (1981)

ambitious, the essence of the dilemma is, of course, that the two aspects are difficult to study independently. However, Hagler and Lifson (1974) and Hagler *et al.* (1974) made the important first step by parameterizing potential functions against data for a variety of crystals of peptide-like molecules (e.g. structural data, heats of sublimation). The functions obtained, by automatically varying parameter assignments to optimize the least squares fit of calculation to experiment, bore a particularly simple relationship to the experimental properties. Vibrational effects are probably minimal, and the position of every atom is to a good approximation definable. This is not true for any solution system, where the positions and orientations of water molecules are not experimentally defined. Thus, here is one instance where the technique relating the potential functions to the properties of the system is almost certainly reliable, and the potential functions thus obtained are intuitively more reasonable. Nevertheless, they still have to be tested. As they stand, they relate to intermolecular interactions in a crystalline environment. Are they transferable to the case of intramolecular interactions in a more biological environment?

As a first step the traditional methods of calculating properties from potential functions were assumed adequate (obviously, it makes sense to try the simplest approaches first). These traditional approaches (see section 21.2), imply neglect of kinetic energy and the use of rigid geometry. The solvent is neglected, except for the effect of varying dielectric constants. Properties were then calculated according to the dictates of statistical mechanics. If no choice of dielectric constant, or any adjustment of potential functions still retaining consistency with the crystal data was capable of producing agreement with the conformational data, then the adequacy of the traditional approach was re-examined. Each approximation was dropped in turn and the consequence noted.

21.6.1. Tests of crystal-derived potential functions by traditional techniques

As described by Robson *et al.* (1979) the potential functions obtained by Hagler and colleagues lead to satisfactory predictions of a variety of conformation-dependent properties. These include the vicinal NH---CH coupling constants as obtained by NMR spectroscopy for a variety of *N*-acetyl amino acyl *N*′-methylamides, and dipole moments of a variety of oligopeptides. The former NMR data is of particular interest because of all conformation-dependent properties, the theory and technique for calculation seems best founded here. Thus, general agreement with many *N*-methylamides and *N*-dimethylamides was particularly promising. For example, the calculated value of the coupling constant for the classic test molecule *N*-acetyl alanyl *N*′-methylamide was 7.8 Hz, in the middle of the experimental range of 7.8 ± 0.5 Hz.

Nonetheless, computer experiments soon demonstrated that the calculation of coupling constants was not a particularly good test. Since choices of potential function are to some extent degenerate with respect to the values of properties calculated from them, there seems to a diversity of potential functions which can produce reasonable agreement with experimental coupling constants. Thus, even the good general agreement here used to be treated with caution. On the other hand it did seem possible to disqualify several of the earlier sets of potential functions obtained by other workers and to choose between two alternative sets of potential functions proposed by Hagler *et al.* (1974). These workers had shown that both 9-6-1 and 12-6-1 potential functions (independently parameterized) were consistent with the crystal data: although there is a considerable difference between the functionalities of the repulsive terms, the crystal data could not at that time distinguish between them. The calculation of NMR coupling constants, and indeed of all properties, supported only the softer 9-6-1 function. This choice has subsequently been further supported by a study on carboxylic acid crystals (Hagler *et al.*, 1979a) and quantum mechanical studies of the functionality of the steric overlap term (Hillier & Robson, 1979). It must be remembered that this functionality is really variable with atom type, although so far, accumulative evidence suggests that the 9-6-1 function is generally far more appropriate to biological molecules than the more usual 12-6-1 type of function.

It is most informative when disagreement is found between theory and experiment. However, one has to ensure that it is genuine negative evidence against the validity of the calculations. Many so-called 'experimental properties' are in fact summary parameters deduced from raw data by some theoretical reasoning. The uncertainty so implied must also be taken into account, and then it becomes rather more difficult to find cases in which calculation and experiment are unequivocably disparate. The only serious disagreement was, in fact, between the calculated and observed characteristic ratios of random coil polypeptides. These characteristic ratios are expressions of the end-to-end distance of the polymeric chain, and hence of its radius of gyration and hydrodynamic and light scattering properties, (they are, however, normalized to be independent of chain length for long chains). The calculated value for a homopolymer unbranched at the side chain C_β atom was 19, clearly outside the experimental range which, even in the most cautious interpretation of experimental errors, is 5–12 (Brant & Flory, 1965a, b; Miller *et al.*, 1967; Hawkins & Holtzer, 1972).

However, even if the experimental interpretation of random coil behaviour is accepted there are difficulties even here, as pointed out by Hawkins and Holtzer (1972). The standard method of calculating characteristic ratios from the potential functions (Brant & Flory, 1965a, b) implies many approximations. Thus, we attempted a number of more exact methods of calculation, including Monte Carlo methods and a detailed study of the effect of ionic shielding on the behaviour of polyglutamate. While these studies were very informative in other ways, they failed to resolve the disparity and thus suggest that the functions are not transferable to the intramolecular case.

Computer experiments were carried out to show what features of the potential surface, and indirectly of the potential functions, would have to be changed in order to bring all types of value into the experimental range. In a preliminary study, Hagler and Robson (see Robson *et al.*, 1978) found that the reduction of the heights of barriers in the potential surface could readily achieve this. Further, it would also better account for the population of these barriers by residues in globular proteins. Reduction of the implied sizes of certain atoms would be consistent with smaller potential barriers. Unfortunately, two areas of study suggested that a reduction in atomic dimensions, or in atomic 'hardness', was not justified. First, Hagler's laboratory reported to us that any such 'refinement' was then inconsistent with the crystal properties against which the potential functions were originally parameterized, (recalculation of crystal properties were carried out not only with modified parameters, but also with new (softer) exponential functions). Second, Robson *et al.* (1978) and Hillier and Robson (1979) had shown that the potential surface as calculated by the original 9-6-1 potential functions were in excellent agreement with extended basis set *ab initio* calculations. Indeed, there was agreement up to 25 kcal/mole or more above the minimum in the potential surface. There was no evidence that atoms should be 'shrunk' or 'softened'.

Thus, a new dilemma was encountered. How could one retain the same potential functions, but bring calculated properties into the experimental range?

21.6.2. Tests of crystal-derived potential functions by less approximate techniques

While the above dilemma may be brushed aside by simply saying that the methods of calculating properties from the given potential functions are inadequate, one seems obliged to consider what kind of approximation could cause the overestimation of barrier heights, even though the potential functions might be correct. In fact, there is only one simple answer: the use of rigid geometry. In reality, the bond lengths and valence angles of the molecule are not invariant. The valence angles in particular can change to relax atomic clashes. Especially interesting in the case of relaxing the A . . . D interactions in a four-atom bonded configuration A-B-C-D by opening the angles A-B-C and B-C-D. However, the relaxation of other interactions, as between A and E in A-B-C-D-E, is also likely to be important.

Proper treatment of these relaxation phenomena demands a valence force field, i.e. a set of potential functions which describes the variation of energy with bond length and valence angle, dependent on the natures of the chemical bonds. A preliminary valence force field obtained by Hagler and Stern (unpublished work) by fitting the vibrational frequencies and structure of *N*-methyl acetamide was introduced and tested (Robson *et al.*, 1979). Recalculation of properties then showed that valence angles were flexible enough to produce the required reduction of barrier heights. Indeed, this particular valence force field seemed to confer too much flexibility on peptide molecules. For example, the characteristic ratio was brought down from 19 to 5, at the very bottom of the experimental range. While undoubtedly needing further refinement, flexible geometry is an extremely important factor and cannot be neglected. In attempts to test and possibly refine the preliminary valence force field, studies have also been carried out on a valence force field parameterized against the extended basis set *ab initio* calculations of valence angle bending in peptides (Hillier & Robson, 1979). This valence force field seems to imply a slightly less flexible peptide geometry than did the empirical valence force field, but valence angles could still readily change by $\pm 5^\circ$ and the general conclusions remain the same.

When flexible geometry must be considered, it is little extra effort to calculate the free energy contributed by all degrees of internal freedom. Then, one may assess whether this also is likely to be an important factor. Hagler *et al.* (1979b) carried out such a calculation on hexapeptides of alanine, methionine and glycine (for which experimental data is available). It turns out that this entropic contribution to the free energy can be of the same magnitude as the potential energy, and for accurate calculation of behaviour cannot be neglected.

If the vibrational freedom of a biological molecule can make such a significant contribution to the conformational free energy, it also seems likely that the solvent makes a very significant contribution. Indeed, it is, of course, well known that the stability of a globular protein depends on the hydrophobic effect, which is to say on the entropic contribution of the surrounding solvent. In addition, there are likely to be important solvation enthalpy contributions. These include the effects of solute–solvent hydrogen bonding, and the accumulative effect of bulk polar solvent on the electrostatic interactions within the molecule.

Hagler and Moult (1978) have calculated, by the Monte Carlo technique, positions of water molecules in hydrated crystals of lysozyme and a cyclic peptide. The solute conformation was constrained to that observed in the crystal. Although several features of interest emerged, it was the tolerable agreement between calculated and observed water positions which are of particular interest. Thus, when the same technique was later applied to simulate behaviour of a peptide solution (Hagler *et al.*, 1980b), one could begin to have some measure of confidence in the results, even though equivalent experimental data is not available in this case. In any event, some of the more important general findings of the latter study seem unlikely to depend on the potential functions or the approximations used in a Monte Carlo approach. For example, it was shown that the solvent contribution to the conformational energy of a small peptide molecule can be of the same order as the potential energy. All these less approximate studies appear to reveal a rather depressing picture in which all previous 'quantitative' calculations neglecting such aspects appear to be of little value. The potential energy contribution as typically calculated is just one contribution amongst others of comparable magnitude. On the other hand, this rather depends on the nature of the system studied. Perhaps it is a much less drastic approximation to neglect the solvent when the solvent is non-polar. Further, it can be seen that more exact calculations *are* possible, and although they are currently expensive, future development in computing may yet make them routine. Finally, the rigorous calculations have also served to identify valid approximations. For example, (1) the observation that a major effect of the solvent on the conformation of the solute is due to the reaction field is of considerable importance, since this is readily represented analytically, and (2) the fact that effects of the solute on the solvent were mainly confined to the first shell supports 'hydrophobic bond' models as used in folding simulations (Levitt & Warshel, 1975; Robson & Osguthorpe, 1979).

21.6.3. Attempts to predict the tertiary structure of a globular protein

The idea of predicting the three-dimensional structure of a globular protein from the amino acid sequence has long excited workers in the field. Nonetheless, it is only recently that direct attacks on the problem have been made. Earlier studies represented a preparative 'clearing of the ground'. However, many would argue that the ground work is far from complete, and that current, more ambitious studies are premature.

It is not the size of the protein molecule, but the number of minima in the potential surface, which creates the difficulties. The first attempts were directed towards smoothing out the potential surface, so that only major trends in slope, leading down to major minima, remained. This smoothing was achieved by correspondingly smoothing out the details in the structural representation of the protein. Although as a methodology this was first emphasized in the historically important study by Levitt and Warshel (1975), it was nonetheless an implicit part of a pioneering attempt by Ptitsyn and Rashin (1974), to predict myoglobin structure from the known helical regions. Since this was not a computer simulation, the helical regions were literally represented by cylinders. These were assigned hydrophobic surface patches, and packed together to optimize hydrophobic contacts. The study achieved some success in that the observed native structure corresponded to one of those packing arrangements considered as plausible candidates for the native one.

The Levitt-Warshel approach was more general, and allowed any backbone conformation, not just a helical one, to be modelled in an approximate way. Further, the method was implemented on a computer, though this did not protect it from criticism. The CONH peptide group atoms of the backbone were neglected, C_α carbon atoms were joined directly by a single virtual bond and side chains were replaced by 'united atoms' (large dummy atoms replacing whole groups). This model was first applied to the small protein pancreatic trypsin inhibitor (PTI) by Levitt and Warshel (1975). Some measure of success was claimed, though this attempt is now largely seen to be of historical importance. Although it had many ingenious aspects, there were nonetheless deficiencies both in concept and in procedure which have been much discussed (see, for example Némethy & Scheraga, 1977; Hagler & Honig, 1978). That debate probably played a major part in establishing rigorous standards for predicting protein structure. In particular, it is now clear that the procedure should (1) be wholly automatic, (2) have a parameter input intended to apply to all proteins, (3) be quantitatively reproducible and (4) provide no possibility of altering procedures in any way to suit each protein. Levitt has subsequently been a major force in developing such methods. Ideally, the only input peculiar to each protein molecule should be the amino acid sequence. Modern criteria seem to run against the earlier much celebrated proposal (Levinthal, 1968) of folding a protein molecule by human interaction with a computer visual display system. In particular, in the hands of Professor Blundell's team at Birkbeck College, London, interactive graphics is emerging as a powerful tool in protein engineering. In the view of this author it is still better that it should be treated cautiously. It is probable that many decisions made by a human observer can be made by a non-interactive computer programme, and if the computer is, for certain tasks, much slower than the human brain, this is a small price to pay for objectivity and reproducibility. To introduce subjectivity as a time-saving feature may nontheless be very desirable provided that objective solutions are shown to be possible.

How is the success of a tertiary structure prediction judged? For secondary structure, the number of residues in the correct secondary structure state is an obvious (though still not trouble-free) measure of success. For tertiary structure, one type of measure is obviously the extent to which residues in the calculated structure preserve the same relative positions as in the observed structure. Thus, Levitt and Warshel employed the rms deviations between the inter-residue distances in the calculated structure and the corresponding inter-residue distances in the observed structure. This is not the same as the rms deviation between observed and calculated positions as used by crystallographers to compare structures and must not be confused with it. Difficulties arise from the fact that no one residue can be considered as unambiguously correctly or wrongly placed, and that the performance of the prediction averaged over the protein obscures the distinction between a fluctuating performance or a uniformly mediocre one throughout the molecule. For this reason, as well as because of the insensitivity of the measure (see below), such rms estimates should be combined with the use of tables of all residue–residue distances, often expressed as 'Ooi-Plots' (Phillips, 1967; Ooi & Nishikawa, 1973) and stereoscopic pictures.

One of the simulated PTI folds obtained by Levitt and Warshel had a 6 Å rms to the native, which at that time was judged promising. As indicated by Hagler and Honig (1978) a 6 Å result does not necessarily indicate a native-like structure. In principle, there are other criteria of native-like character which could indicate a significant success even in conjunction with a 6 Å result, but Hagler and Honig (1978) were clearly of the opinion that these criteria had not been met in this case. More recently, many authors including Sternberg have pointed out that the rms measure used alone is almost worthless when of this magnitude, and several, including Levitt, have subsequently demonstrated that circa 6 Å can be attained by a random, compact structure. Nonetheless, it must be emphasized that the measure is very useful for making internal comparisons of the quality of different

Direct determination of the rms distance between calculated and observed residue positions is possible, but requires that the calculated and observed structures be superimposed by translating and rotating one structure with respect to the other so as to minimize the rms deviation.

results for the same molecule, and that when the measure approaches zero it becomes progressively better as an unambiguous indicator of a good fold. Based on studies in our laboratory of comparisons between homologous proteins, a result of less than 3 Å must be judged significant. Unfortunately, only very slightly larger values (circa 4 Å) can be obtained between rather different structures. Combining this with the lack of alternative objective measures of success, the fact that no simulation taking no account of the known result has penetrated the 3 Å range makes it rather difficult to estimate whether real success is being achieved. When Levitt and Warshel later applied their simplified representation to parvalbumin, the results were poor and thus the early procedure as applied cannot be generally useful even if the PTI result were deemed significant.

An alternative approach by Kuntz *et al.* (1976) achieved in several runs an rms of 4.7–6.5 for PTI and 4.0–6.0 Å for rubredoxin. This, however, was not an energy calculation but an optimization under a large number of penalties based on known protein structures. A penalty is a function which rises continually as the calculated value of any property departs from the expected one. It is not obvious which penalties are reasonably deducible in the absence of the experimental PTI data and which are not. Also, the result obtained is not very detailed since whole residues are represented as 'giant atoms'. On the other hand, the results include some of the best achieved by 1980, and this procedure could still be useful for setting up the starting conformations for a detailed energy minimization.

A description of the earlier approaches of the Scheraga school is given in the Némethy–Scheraga review (1977). See also Tanaka and Scheraga (1975). Briefly, these are predominantly Monte Carlo techniques of the Metropolis type (see 21.4.3), with emphasis on the good prediction of secondary structures as a first step. It is rather difficult to judge the quality of the results obtained but they certainly fare no worse than the Levitt-Warshel simulation.

21.7. THE CURRENT STATE OF THE ART

There have been important developments in the last five years and raised optimism that the protein folding problem will be tractable in the near future. A detailed analysis will be presented elsewhere, but several conclusions may be noted. We should first realise that in practical terms the word 'tractable' no longer has the same unique significance. At one end of the scale, the laboratories of Levitt and of Scheraga recognise that prior knowledge of conformational features can be usefully used to direct protein folding simulations because NMR spectroscopy is providing increasingly detailed descriptions of more and more interatomic distances in proteins. The question is rather one of how little experimental data is required to achieve a successful result, and with progress in theory less and less data is required. At the other end of the scale, the author's laboratory has begun to emphasise the applicability of folding simulation techniques to 'miniprotein' biological peptides in the decapeptide range, while for proteins in the 100 to 200 residue range it is emphasised that the required quality of the result depends on the purpose to which the result is to be put. For example, it has been shown that fast, crude, and ambiguous calculations are adequate to aid considerably the design of artificial peptide vaccines which, when rendered immunogenic by attachment to a carrier molecule, raise antibodies well against the protein studied. It is likely that *de novo* simulations will make important contributions to biotechnology long before X-ray crystallographers are satisfied with the quality of the predictions. There are therefore indications that much effort is being put into predictions of proteins of unknown conformation, which although much less satisfying academically than for proteins of known conformation (allowing verification), are in fact of much wider interest. A pharmacologist, neurochemist, immunologist, or genetic engineer has little direct interest in a calculation on a protein when the conformation is well known experimentally, but a considerable interest in any insight which a calculation may provide concerning the protein of interest to him.

The speed of the simulation software is an important factor in determining the degree of quality which can be obtained and the size of molecule which can be treated. Though speed enhancement is of technical rather than biological interest, many laboratories are doubtless putting considerable effort into speeding software. In our laboratory, the speed has been enhanced at least one thousand-fold since 1979. However, a contributing factor is the availability of high speed vector machines such as the CRAY ONE and CYBER205. New algorithms can be developed to exploit the special features of vectorisation, beyond simply 'vectorising' the program. A basic 30-fold speed enhancement of the CYBER205 over the CDC 7600, has been achieved. These speed enhancements have allowed a dramatic increase in the number of folding simulations which can be carried out, and it is interesting to note what these have taught us. For one thing, they provide greater insight into the form of the potential energy surface, as a function of conformation. Putting together our results with those of other workers, it seems clear that this surface as calculated consists of several large valleys, each nonetheless populated by tens, hundreds, and in some cases thousands of local minima. It turns out, an analysis by Levitt of the form of the potential surface is in fact a description of the form of the valley containing the native state. Our laboratory and that of Levitt and Scheraga seem to have adequate technology for approaching the native state *once* the valley containing that native state is located. The philosophy of Blundell, Sternberg, Thornton and Taylor at Birkbeck College, London can be interpreted as one of compiling dictionaries of conformational motifs, and of predicting which motif is appropriate to the primary sequence of interest in order to start off as closely as possible to the valley containing the native state. Our approach is to develop criteria by which the 'native-likeness' of the conformations both in and in the vicinity of each valley can be rapidly assessed. Energy and free energy alone are not useful criteria. The arbitrary minimization starting point within a valley can be of very high energy. Moreover, several workers are now obtaining results which show that, in the light of our interpretation in terms of valleys, the floors of different valleys can have comparable energies to that containing the native state. At least this is as calculated, and Karpluse's laboratory has indicated that the factor distinguishing the native valley is the arrangement and/or exposure of nonpolar (hydrophobic) groups which reasonably suggests a deficiency in the way that the solvent effect is represented. Building on a solvent model suggested to us by Sternberg for a better treatment of internal hydrogen bonding, we have developed a technique which appears to guarantee that the native state is always more stable than that of more open conformers (properly relaxed): this effectively means that many of the non-native valleys are now considerably raised in free energy with respect to that containing the native state. The method is, however, still too slow to direct a folding simulation to 'native-like' regions of the potential surface, and is in any case only applicable close to the floor of the valley. What is required is to bias the simulation against even beginning a minimization attempt in an unlikely valley. In fact, this problem appears to have been resolved in our laboratory, but now reveals a new mountain range of theoretical problems to be overcome. These appear to be tractable, but demand techniques as bizarre as folding proteins not in three dimensions of Euclidean space, but in four, the concomitant increase in simulation time being offset by taking account of recent advances in topology.

This in turn opens the possibility that the limiting factors in development of technology for predicting protein conformation from sequence alone will not be the further development of potential functions, treatment of solvent, or speed of software and of hardware, but continuing advances in basic mathematics. We have already encountered one area in which a solution to an apparently resolvable but not yet resolved problem in topology would have been helpful. Probably this view is needlessly pessimistic, however. In essense, we are still talking about techniques which save time, albeit by more efficiently searching the conformational space, and extensive computer time may be a reasonable price to pay for a successful fold while awaiting progress in mathematics. Because using a lot of computer time takes in practice a lot of real time, we can as yet only report on the degree of success in proteins which are still folding on the computer. For example, the observation that domains of protein structure have very promising character in agreement with the native, but that the domains have not yet associated (in some instances because not all the domains have yet formed) seems to be increasingly common.

It must be emphasised that much needs still to be done, and optimism in the author's laboratory is rather due to the fact that insight into the problem is mounting rapidly with the road to progress ahead becoming clearer. At the time of writing we still cannot report an accurate fold based on sequence alone for a reasonably sized protein. What must be remembered, however, is the fact that many applications of the same kind of software, in peptide drug design, energy refinement of crystallographic structures, modelling of proteins using the known conformations of homologous proteins, and approximate models of protein structure derived *de novo* for applications in artificial vaccine design, are becoming increasingly routine.

21.8. REFERENCES

Beddell, C. R., Clark, R. B., Lowe, L. A., Wilkinson, S., Chang, K. J., Cuatrecasas, P. & Miller, R. (1977). *Br. J. Pharmacol.*, **61,** 351–356.

Beeman, D. (1976). *J. Computational Phys.*, **20,** 130–139.

Brant, D. A. & Flory, P. J. (1965a). *J. Am. Chem. Soc.*, **87,** 2788–2791.

Brant, D. A. & Flory, P. J. (1965b). *J. Am. Chem. Soc.*, **87,** 2791–2800.

Burgess, A. W., Momany, F. A. & Scheraga, H. A. (1975). *Biopolymers*, **14,** 2645–2647.

Chou, P. Y. & Fasman, G. D. (1974a). *Biochemistry*, **13,** 211–222.

Chou, P. Y. & Fasman, G. D. (1974b). *Biochemistry*, **13,** 222–245.

Cohen, F. E., Richmond, T. J. & Richards, F. M. (1979). *J. Mol. Biol.*, **132,** 275–288.

Davidon, W. C. (1968). *Comput. J.*, **10,** 406–410.

Dygert, M., Gō, N. & Scheraga, H. A. (1975). *Macromolecules*, **8,** 750–761.

Fletcher, R. & Powell, M. J. D. (1963). *Comput. J.*, **6,** 163–168.

Garnier, J., Osguthorpe, D. J. & Robson, B. (1978). *J. Mol. Biol.*, **120,** 97–120.

Gorin, F. A., Balasubramanian, T. M., Barry, D. C. & Marshall, G. R. (1978). *J. Supramol. Struct.*, **9,** 27–39.

Hagler, A. T. & Honig, B. (1978). *Proc. Natl. Acad. Sci. U.S.A.*, **75,** 554–558.

Hagler, A. T. & Lifson, S. (1974). *J. Am. Chem. Soc.*, **96,** 5327–5335.

Hagler, A. T. & Moult, J. (1978). *Nature* (*Lond.*), **272,** 222–224.

Hagler, A. T., Huler, E. & Lifson, S. (1974). *J. Am. Chem. Soc.*, **96,** 5319–5327.

Hagler, A. T., Leiserowitz, L. & Tuval, M. (1976). *J. Am. Chem. Soc.*, **98,** 4600–4612.

Hagler, A. T., Lifson, S. & Dauber, P. (1979a). *J. Am. Chem. Soc.*, **101,** 5122–5130.

Hagler, A. T., Stern, P. S., Sharon, R., Becker, J. M. & Naider, F. (1979b). *J. Am. Chem. Soc.*, **101,** 6842–6852.

Hagler, A. T., Moult, J. & Osguthorpe, D. (1980a). *Biopolymers*, **19,** 395–418.

Hagler, A. T., Osguthorpe, D. J. & Robson, B. (1980b). *Science*, **208,** 599–601.

Hawkins, R. B. & Holtzer, A. (1972). *Macromolecules*, **5,** 294–301.

Hillier, I. H. & Robson, B. (1979). *J. Theor. Biol.*, **76,** 83–98.

Howard, J. C., Ali, A., Scheraga, H. A. & Momany, F. A. (1975). *Macromolecules*, **8,** 607–622.

Hurwitz, F. I. & Hopfinger, A. J. (1976). *Int. J. Peptide Protein Res.*, **8,** 543–550.

Isogai, Y., Némethy, G. & Scheraga, H. A. (1977). *Proc. Natl. Acad. Sci. U.S.A.*, **74,** 414–418.

Kuntz, I. D., Crippen, G. M., Kollman, P. A. & Kimelman, D. (1976). *J. Mol. Biol.*, **106,** 983–994.

Landau, L. D. & Lifshitz, E. M. (1958). *Statistical Physics*, Pergamon, London, Paris.

Levinthal, C. (1968). *J. Chimie Phys.*, **65,** 44–45.

Levitt, M. (1976). *J. Mol. Biol.*, **104,** 59–107.

Levitt, M. & Lifson, S. (1969). *J. Mol. Biol.*, **46,** 269–279.

Levitt, M. & Warshel, A. (1975). *Nature* (*Lond.*), **253,** 694–698.

Lewis, P. N., Momany, F. A. & Scheraga, H. A. (1973a). *Biochim. Biophys. Acta*, **303,** 211–229.

Lewis, P. N., Momany, F. A. & Scheraga, H. A. (1973b). *Isr. J. Chem.*, **11,** 121–152.

McCammon, J. A., Gelin, B. R. & Karplus, M. (1977). *Nature* (*Lond.*), **267,** 585–590.

Melnikov, P. N., Akhmedou, N. A., Kipkind, G. M. & Popov, E. M. (1976). *Bioorg. Khim.*, **2,** 28–42.
Metropolis, N. A., Rosenbluth, A. W., Rosenbluth, M. N., Teller, A. H. & Teller, E. (1953). *J. Chem. Phys.*, **21,** 1087–1092.
Miller, M. H. & Scheraga, H. A. (1976). *J. Polym. Sci.* (*Polymer Symposia*), **54,** 171–200.
Miller, W. G., Brant, D. A. & Flory, P. J. (1967). *J. Mol. Biol.*, **23,** 67–80.
Momany, F. A. (1976a). *J. Am. Chem. Soc.*, **98,** 2990–2996.
Momany, F. A. (1976b). *J. Am. Chem. Soc.*, **98,** 2996–3000.
Momany, F. A. (1977). *Biochem. Biophys. Res. Common*, **75,** 1098–1103.
Némethy, G. & Scheraga, H. A. (1977). *Quart. Rev. Biophys.*, **10,** 239–352.
Nishikawa, K., Momany, F. S. & Scheraga, H. A. (1974). *Macromolecules*, **7,** 797–806.
Okuyama, K., Tanaka, N., Ashida, T. & Kakudo, M. (1976). *Bull. Chem. Soc. Japan*, **49,** 1805–1810.
Ooi, T. & Nishikawa, K. (1973). *The Jerusalem Symposia on Quantum Chemistry and Biochemistry*, **5,** 173–187.
Pain, R. H. & Robson, B. (1970). *Nature* (*Lond.*), **227,** 62–63.
Phillips, D. C. (1967). *Proc. Natl. Acad. Sci. U.S.A.*, **57,** 484–495.
Pincus, M. R., Zimmerman, S. S. & Scheraga, H. A. (1976). *Proc. Natl. Acad. Sci. U.S.A.*, **73,** 4261–4265.
Platt, E. & Robson, B. (1983). In *Computing in Biological Science*, Geisow, M. J. & Barrett, A. N. (Eds.), Elsevier Biomedical Press, Amsterdam, New York, Oxford, pp. 91–131.
Premilat, S. & Hermans, J. (1973). *J. Chem. Phys.*, **59,** 2602–2610.
Premilat, S. & Maigret, B. (1976). *C. R. Acad. Sci. Paris, C*, **282,** 225–228.
Ptitsyn, O. B. & Rashin, S. (1974). Report of the Academy of Science of the U.S.S.R.
Ptitsyn, O. B. & Rashin, A. A. (1975). *Biophys. Chem.*, **3,** 1–20.
Ralston, E., De Coen, J.-L. & Walter, R. (1974). *Proc. Natl. Acad. Sci. U.S.A.*, **71,** 1142–1144.
Ramachandran, G. N. & Mitra, A. K. (1976). *J. Mol. Biol.*, **107,** 85–92.
Robson, B. (1974). *Biochem. J.*, **141,** 853–867.
Robson, B. (1976). *Trends Biochem. Sci.*, **3,** 49–51.
Robson, B. (1980). *Trends Biochem. Sci.*, **7,** 240–244.
Robson, B. & Garnier, J. (1986). *Introduction to Proteins and Protein Engineering*, Elsevier Biomedical Press. In press.
Robson, B. & Osguthorpe, D. J. (1979). *J. Mol. Biol.*, **132,** 19–51.
Robson, B. & Pain, R. H. (1971). *J. Mol. Biol.*, **58,** 237–259.
Robson, B. & Suzuki, E. (1976). *J. Mol. Biol.*, **107,** 327–356.
Robson, B. & Timms, D. (1980). *Trends Biochem. Sci.*, **7,** 240–244.
Robson, B., Hillier, I. H. & Guest, M. (1978). *Chem. Soc. Faraday Trans. II*, **74,** 1311–1317.
Robson, B., Stern, P. S., Hillier, I. H., Osguthorpe, D. J. & Hagler, A. T. (1979). *J. Chimie Phys.*, **76,** 831–834.
Robson, B., Stern, P. & Hagler, A. T. (1985). To be submitted.
Rossky, P. J. & Rahman, A. (1976). Cecam Workshop Models for Protein Dynamics (Cecam Report, Orsay, Paris), pp. 107–152.
Smith, G. D. & Griffith, J. F. (1978). *Science*, **199,** 1214–1216.
Sternberg, M. J. E. (1983). In *Computing in Biological Science*, Geisow, M. J. & Barrett, A. N. (Eds.), Elsevier Biomedical Press, pp. 143–177.
Sternberg, M. J. E. & Cohen, F. E. (1981). *Philos. Trans. R. Soc. London, Ser. B*, **293** (1063), 177–189.
Swenson, M. K., Burgess, A. W. & Scheraga, H. A. (1977). Reported at Symposium on Frontiers in Physico-chemical Biology, Paris.
Tanaka, S. & Scheraga, H. A. (1975). *Proc. Natl. Acad. Sci. U.S.A.*, **72,** 3802–3806.

Warme, P. K. & Scheraga, H. A. (1974). *Biochemistry*, **13,** 757–767.
Warme, P. K., Momany, F. A., Rumball, S. V., Tuttle, R. W. & Scheraga, H. A. (1974). *Biochemistry*, **13,** 768–782.
White, N. J. & Morrow, C. (1979). *Computers and Chemistry*, **3,** 33–48.
Zimmerman, S. S. & Scheraga, H. A. (1976). *Macromolecules*, **9,** 408–416.
Zimmerman, S. S. & Scheraga, H. A. (1977). *Biopolymers*, **16,** 811–843.

Index

Numbers in bold print have been used to direct the reader to more important references. A list of abbreviations is given at the beginning of the book.

purif. = purification of

QP 551 .P65 1987

Practical protein chem

GEMCO